D0850150

Minding the Modern

MINDING *the* MODERN

Human Agency, Intellectual Traditions, and Responsible Knowledge

THOMAS PFAU

UNIVERSITY OF NOTRE DAME PRESS
NOTRE DAME, INDIANA

Copyright © 2013 by University of Notre Dame
Notre Dame, Indiana 46556
www.undpress.nd.edu
All Rights Reserved

Manufactured in the United States of America

Library of Congress Cataloging-in-Publication Data

Pfau, Thomas, 1960–
Minding the modern : human agency, intellectual traditions, and responsible knowledge / Thomas Pfau.
pages cm
Includes bibliographical references and index.
ISBN 978-0-268-03840-3 (cloth : alk. paper) — ISBN 0-268-03840-6 (cloth : alk. paper)
1. Humanism. 2. Agent (Philosophy) 3. Philosophical anthropology. 4. Free will and determinism. 5. Humanities. I. Title.
B821.P45 2013
190—dc23
2013022543

∞ *The paper in this book meets the guidelines for permanence and durability of the Committee on Production Guidelines for Book Longevity of the Council on Library Resources.*

CONTENTS

PART III. PROGRESSIVE AMNESIA: WILL AND THE CRISIS OF REASON

PART IV. RETRIEVING THE HUMAN: COLERIDGE ON WILL, PERSON, AND CONSCIENCE

ABBREVIATIONS

ADT St. Augustine of Hippo, *The Trinity* [*De Trinitate*]

AR Samuel Taylor Coleridge, *Aids to Reflection*

AV Alasdair MacIntyre, *After Virtue*

BL Samuel Taylor Coleridge, *Biographia Literaria*

BT Martin Heidegger, *Being and Time*

BTr Boethius, *Tractates, De Consolatione Philosophiae*

CCS Samuel Taylor Coleridge, *On the Constitution of Church and State*

CD St. Augustine of Hippo, *The City of God against the Pagans* [*De Civitate Dei*]

CF Samuel Taylor Coleridge, *The Friend*

CL Samuel Taylor Coleridge, *The Collected Letters of Samuel Taylor Coleridge*

CLS Samuel Taylor Coleridge, *Lay Sermons*

CM Samuel Taylor Coleridge, *Marginalia*

CN Samuel Taylor Coleridge, *Notebooks,* 5

CPP William Blake, *The Complete Poetry and Prose*

DCD John Henry Newman, *An Essay on the Development of Christian Doctrine*

EPA Francis Hutcheson, *An Essay on the Nature and Conduct of the Passions and Affections, with Illustrations on the Moral Sense*

GHA Johann Wolfgang von Goethe, *Werke*

HC Hannah Arendt, *The Human Condition*

HI Francis Hutcheson, *Inquiry into the Original of Our Ideas of Beauty and Virtue*

HT David Hume, *A Treatise of Human Nature*

Lev. Thomas Hobbes, *Leviathan*

LHP Samuel Taylor Coleridge, *Lectures on the History of Philosophy, 1818–1819*

LMA Hans Blumenberg, *Legitimacy of the Modern Age*

MFB Bernard Mandeville, *The Fable of the Bees: or Private Vices, Publick Benefits*

OM Samuel Taylor Coleridge, *Opus Maximum*

PG G. W. F. Hegel, *Phänomenologie des Geistes*

PS G. W. F. Hegel, *Phenomenology of Spirit*

Quodl. William of Ockham, *Quodlibetal Questions*

SC Anthony Ashley Cooper, Third Earl of Shaftesbury, *Characteristicks of Men, Manners, Opinions, Times*

ST St. Thomas Aquinas, *Summa Theologiae*

SW&F Samuel Taylor Coleridge, *Shorter Works & Fragments*

SZ Martin Heidegger, *Sein und Zeit*

TI Emmanuel Levinas, *Totality and Infinity: An Essay on Exteriority*

TMS Adam Smith, *The Theory of Moral Sentiments*

TT Samuel Taylor Coleridge, *Table Talk*

WMA Johann Wolfgang von Goethe, *Wilhelm Meister's Apprenticeship*

WWR Arthur Schopenhauer, *The World as Will and Representation*

Portrait of a Gentleman in his Study, 1528–30, Lorenzo Lotto (c.1480–1556) / Galleria dell' Accademia, Venice, Italy / The Bridgeman Art Library

EXORDIUM

Modernity's Gaze

The young man's forlorn, abstracted, and blank gaze suggests disorientation and incipient melancholy: we cannot meet his eyes, and they will not meet ours. Indeed, the beholder of Lorenzo Lotto's canvas may feel somewhat flustered, as though he or she had accidentally intruded on a scene of intensely personal, albeit ineffable anguish. For Lotto's young man, whose identity remains unknown, seems utterly alone in the world—the quintessentially modern, solitary individual confined to his study in ways familiar from the candle-lit interior of Descartes's *Meditations* all the way to the cork-lined refuge where Proust would labor on his magnum opus. Yet Lotto's youth also appears bereft of the dynamism, confidence, and sense of purpose usually claimed for the modern, autonomous self—be it Descartes's *cogito,* John Locke's "consciousness," or Johann Gottlieb Fichte's "founding act" (*Tathandlung*). The cold-blooded lizard and discarded ring on the table hint at the loss of *ēros* as a source of motivation, an impression compounded by the fact that lute and hunting horn, emblems of conviviality and worldly pleasure, are now hung up on the wall in the background.[1] Instead, the glimpse of the outside world that the painting affords us shows dusk encroaching. The pendulum swings; time moves on. We have

1. Dating of Lotto's canvas varies, with some (Berenson) dating it as early as 1524, and others suggesting dates of 1526 (Brown et al.) or even 1530 (Humfrey). For discussions, see Humfrey, *Lorenzo Lotto,* D. A. Brown, *Lorenzo Lotto,* and Berenson, *Lorenzo Lotto.*

happened upon a scene of palpable melancholy. Thus, even as a massive folio dominates the picture, the young man's irresolute posture intimates that books no longer hold answers, perhaps because the right questions elude him. On one widely accepted interpretation, the tome is a business ledger. Other, earlier accounts view the massive folio as emblematic of a life of study to which the man now means to dedicate himself. Either way, the relation of the young man's body to the book suggests a state of incapacitation and inertia, rather than gathering resolve. Moreover, the enigmatic knowledge contained in the folio may well account for the young man's distracted and withdrawn expression. Hence it is that the book's ponderous mass supports the young man only physically. For his body, leaning on it, strikes a twisted, faintly artificial pose, and his left hand betrays his distracted and indifferent attitude toward the book. Moreover, the absence of a chair, of paper and quill in this study, as well as the miscellaneous array of a half-opened letter and a ruffled blue silk cloth casually bunched up beneath the folio all suggest a psychological state of abstractive loitering rather than focused and purposive study.

Meanwhile, the unwieldy folio appears more as dead mass than as a repository of learning. We suspect that the unspecified past wisdom contained in it has but the most tenuous hold on the young man whose consciousness, to judge by his withdrawn gaze, appears altogether adrift. If the book seems incapable of answering questions, it is so because for Lotto's youth to articulate those questions would require contact with an outside world of experience from which he has quite obviously withdrawn. Sequestered into gathering darkness, the young man appears wholly bereft of sense experience, interpersonal relations, and commitments such as define the world outside his study. That world has been reduced to a narrow slice of landscape faintly illumined from the horizon and soon to be expunged from sight by the nocturnal clouds gathering overhead. Yet, to return to the heart of the painting, the book: does the massive tome with its worn leather binding constitute a bona fide repository of learning, or is it but an emblem of the futility or sheer elusiveness of knowledge? Do the fading rose petals, conventional emblems of transience and of time lost, stand in metonymic relationship to the book's vellum leaves so distractedly fingered by the man? Is it truly a book, or are we to take it as an emblem of a lost plenitude, an allegory of the premodern cosmos that has been displaced by numberless theoretical perplexities liable to induce the terror of Blaise Pascal's silent, "infinite spaces"? Indeed, if the book no longer stands for the plenitude of (past and future) meanings but, instead, allegorizes the terminal loss of certainty in matters of both speculative and practical reason, can art (including the art of this painting) be said to fare any better? Aside from confronting us with modernity's pervasive loss of intellectual orientation and practical purpose, might Lotto's canvas also suggest that art itself can only tabulate, yet never remedy that very predicament?

To be sure, Lotto's painting, part of an oeuvre sometimes credited with having inaugurated the modern psychological portrait, should not be freighted with excessive significance for the arguments to follow. Still, its eloquent tonal composition furnishes a poignant and compact illustration of this book's principal concerns. First, there is the increasingly embattled, seemingly untenable status of *action,* practical reason, and a coherent model of human agency as both *self-aware* and *responsible.* Irresolute and metaphysically perplexed, Lotto's young man suggests that the nexus between human flourishing and action, and indeed the very legitimacy of these basic concepts, has become acutely problematic. Does melancholy (*acedia*) still belong to the realm of choice and will? Does it still name a condition of "sin"? Or has it been reconstituted as an irreversible existential "condition," thereby destabilizing the very underpinnings of what it means to be human—viz., notions of judgment, will, choice, intentionality, action, responsibility, and relationality? While some connection between action and ultimate ends may yet exist at the periphery of Lotto's portrait, the notion of the human appears more than ever an enigma. It appears to elude the theoretical (syllogistic) type of explanation that in the modern era (certainly by the beginning of the seventeenth century) has largely established itself as the *only* model of reason. As a result, the premodern, Aristotelian view that had posited action as the consummation of practical reason and its commitment to a communal and normative set of ends now appears strangely illegitimate and almost incomprehensible. Indeed, Lotto's portrait gives little hope that whatever thought process may be unfolding behind those mournful eyes could ever be translated back into the realm of action that proceeds on the strength of habits, judgments, and traditions whose meaning is inseparable from our acknowledgment of their authority and their dialectically reasoned transmission to the future.

The physiognomy of Lotto's modern melancholic individual vividly captures one of my principal claims: viz., that beginning with the advent of nominalism and voluntarism in the fourteenth century, theoretical inquiry and practical reason have terminally parted ways. In Part I, I explore how that parting of ways came about, the premise being that any coherent and meaningful understanding of action—in contrast to strictly naturalist conceptions of "process" or sociological accounts of the "behavior" of individuals and groups—presupposes a profound alignment of will and intellect. Central to that narrative is the story of how that integral relation between cognition and commitment, intellect and will, gradually unravels in the aftermath of Aquinas's synthesis of Aristotelian realism with the Augustinian conception of the human will—at once incontrovertibly self-aware, eminently fallible, and yet responsible for its elections. Both the theological origins of modernity's disaggregation of practical and theoretical reason and the innumerable speculative problems and perplexities to which this development gives rise are vividly captured in the withdrawn countenance and hesitant posture of Lotto's melancholic young man. He seems above all *irresolute*, that is, bereft of the capacity to be, or even imagine himself as, a creative, committed, and responsible agent in the world.

A second objective of this book is to clarify the increasingly confused understanding of what role concepts play in humanistic inquiry, and what constitutes the ground or source of their authority. As we shall find, the principal issue here concerns the wholesale and often unreflected migration of modern scientific methods into a domain of thought that is essentially interpretive, and where acts of inquiry aim at clarifying and realizing a notion of the good, rather than at sifting quantifiable and ostensibly value-neutral "information." If we accept the older view of that massive folio in Lotto's painting as an emblem of humanistic learning, then the young man's distracted and perplexed countenance truly embodies a distinctly modern type of individuality, at once bewildered by the seeming illegibility of inherited traditions of moral inquiry and, thus, unable to grasp the very nature and significance of tradition per se. Instead of a dialogic principle that allows a given generation to orient itself by engaging, extending, and transforming the reach of inherited conceptions, tradition now appears but dead weight. Even if it were to be shuffled off, what could possibly take its place? The young man's conspicuous loitering over the folio in utter isolation suggests a profound bewilderment as to just how knowledge is to be achieved now that the Scholastic model of *disputatio* has collapsed, a model premised on the productive dialectical encounter with past attempts at grasping questions of the good, human flourishing, responsibility, and ultimate ends. No longer understood as an "open transcendental" (C. Gunton), specific intellectual traditions (as indeed the very notion of *traditio* itself) appear to concern modernity only as an object of perplexity and indignation, or impending oblivion. At once distraught and distracted by the chimera of the *new* and riveted onto a future with a mix of manic anticipation and growing dread, the modern subject only knows *that* it has forgotten something but appears unable to recall just what it was. This book is an attempt to retrieve this twofold enigma: the unique nature of humanistic, interpretive concepts and frameworks enabling our quest for articulate and responsible knowledge in the realm of practical reason, and the distinctive dialectical process whereby such concepts (e.g., will, person, judgment, action, and the Platonic triad of the good, the true, and the beautiful) are received, rethought, and transmitted to future generations.

Though much of this book took shape in a solitary study not unlike the one depicted by Lotto, it is also a palimpsest of many intellectual debts and was made possible by countless acts of personal friendship and collegial support. Of the many voices that enriched this book and contributed greatly to whatever merits it may have, some belong to people whom I have never met, yet whose intellectual personae and integrity have spoken powerfully to me through their published work. Thus what follows was often inspired and is gratefully indebted to the work of Hans-Georg

Gadamer, Alasdair MacIntyre, Louis Dupré, Charles Taylor, and Robert Sokolowski, to name some of the most compelling writers to have traversed similar ground. I hope to have emulated not just the scope of their ambition but, however imperfectly, the exemplary sense of intellectual purpose and responsibility that speaks from their published work. Now in my twenty-first year of teaching and writing at Duke University, I am acutely aware of the enormous debt that this book, and indeed my overall intellectual flourishing at this most mercurial institution, have owed to the generosity, wisdom, and unflagging support of some of my closest and most trusted colleagues and friends. David Aers read the entire manuscript and commented on it with the degree of care and detail that only someone possessed of his unfathomable learning and intellectual passion could have summoned; his late-night emails, alerting me to invariably crucial primary and secondary readings, have done much to deepen and consolidate the arguments I sought to develop, and they greatly helped sharpen my sense of responsibility and humility as I strayed farther and farther out of my main area of expertise (European Romanticism) and into the refreshingly coherent traditions of inquiry of which philosophical theology and moral philosophy are composed. Stanley Hauerwas and Paul Griffiths at the Duke Divinity School both read the book in its later stages and commented on it with characteristic generosity and a sharp eye for the nuances and complex etiology of theological and philosophical argument. I was also a deeply grateful auditor of Reinhard Hütter's seminars on Aquinas, which led to some inspiring and enriching conversations about the challenging and topographically complex borderlands connecting theology, philosophy, and aesthetics. A unique debt of gratitude I owe to Vivasvan Soni, who has followed this book's evolution with truly unparalleled care and attentiveness. Viv's preternatural ability to grasp an argument's basic intent, while sympathetically and constructively drawing out its larger potential and significance, never ceases to astonish. His detailed and trenchant written responses to individual sections have left a lasting imprint on whatever is of merit in this book.

A book of such immodest (though, I hope, not altogether irresponsible) scope and ambition obviously takes time to gestate, and much of that fermentation takes place in discussions and conversations such as follow the presentation of some selection from it. Truly invaluable in this regard was my stay at the National Humanities Center—generously funded by the NHC (via the Duke Endowment) and by an ACLS fellowship—during the 2010–2011 academic year. By "genial coincidence," as Coleridge might have called it, James Engell was also a fellow there that year, and I am ever so grateful for the keen interest that he took in the book project as a whole and in the sections on Coleridge in particular. My conversations with James Engell, Miguel Tamen, Bernie Levinson, and Geoff Harpham on the serene, sun-dappled terraces and meeting spaces of the National Humanities Center are among my fondest memories of a year spent in what, surely, has to be every academic's "paradise." While the book was taking shape, I also was fortunate enough to be invited to present portions of it at Johns Hopkins, Stanford, Oxford, Cambridge, Harvard, Yale, Brown, Rice, Michigan, SUNY Buffalo, Indiana, the Catholic University of Louvain, and the University of Oregon, occasions when treasured friendships were forged and countless intellectual debts were accumulated. Though no doubt I now fail to recall all of them, many colleagues and graduate students at these and other institutions provided me with often invaluable suggestions and probed the book's arguments with a degree of attentiveness and dialectical rigor that reassures me that the Scholastic ethos of *disputatio* has not (at least not yet) been completely vanquished by a self-regarding and superficial professionalism. In this regard, I wish to acknowledge my particular gratitude to David Collings, David Clark, Richard Macksey, Noel Jackson, Denise Gigante, Nicholas Halmi, David Wellbery, Paul Fry, Jacques Khalip, Fritz Breithaupt, Joshua Kates, Bill Rasch, Eyal Peretz, Tres Pyle, Nancy Yousef and, here at Duke, Rob Mitchell, Jakob Norberg, Frank Lentricchia, Tom Ferraro, Natalya Chuchinsky, Rachel Stern, and William Revere.

Finally, throughout the writing of this book I have enjoyed the collegial and personal support of some great colleagues and friends here at Duke: Len Tennenhouse and Bill Donahue have been most generous colleagues and, in working heroically to reestablish sound standards for constructive and responsible chairmanship in the two departments to which I belong, they also did much to create the supportive work environment that has allowed this project to flourish. Finally, my wife and partner in life, Sandra—gifted artist and pedagogue in her own right and a font of common sense in all matters artistic and pedagogical—was at once shrewd and gentle in helping me maintain the right balance between genuine intellectual passion and outright self-absorption. In this she had the unflagging support of our young daughter, Naomi, who along with my fully grown daughters, Natalie and Elisa, has given my life more meaning and love than I could have ever imagined. Together, they have been a steady prompt for me to balance life and work, and to be alert to the myriad and often unpredictable ways in which those two spheres show themselves to be entwined.

Durham, July 2012

Part I

PROLEGOMENA

The present is a text, and the past its interpretation.

—John Henry Newman

1

FRAMEWORKS OR TOOLS?

On the Status of Concepts in Humanistic Inquiry

This is a study of two closely related concepts—"will" and "person"—which have proven indispensable to Western humanistic inquiry and its ongoing, albeit enormously diverse, attempts to develop a satisfactory account of human agency. More implicitly, what follows is also a study of our changed relationship to concepts and, hence, to the nature, purpose, and responsibility of thinking and knowledge. The argument to be advanced hinges on a number of interlocking claims and objectives that should be sketched right away, if only in preliminary fashion. A first claim is *historical* in kind, albeit just as emphatically not *historicist*. Its purport is that, for reasons to be considered shortly, both will and person—as well as a number of other key concepts of humanistic inquiry entwined with these notions—undergo momentous and, I argue, deeply problematic change in European modernity. First, the scope of their relevancy to humanistic inquiry, as indeed that very project itself, *contracts*. Second, for a variety of reasons having to do with transformations internal to philosophical theology and the rise of naturalist and reductionist approaches sponsored by the emergence of a scientific culture in the sixteenth and seventeenth centuries, the internal coherence of these key concepts and their centrality within humanistic (interpretive) inquiry *erodes* over time. Finally, given modernity's accelerating commitment to an ostensibly value-neutral ideal of knowledge anchored in efficient causation alone, conceptions of a responsible will and a person defined by its relation to others are progressively relegated to the margins of philosophical inquiry.

Along with a host of contiguous notions (e.g., judgment, responsibility, self-awareness, teleology, etc.), they ultimately succumb to a process of pervasive *forgetting*. As remains to be seen, such forgetting was inevitable considering the extent to which post-Hobbesian thought had lost sight of, or had rejected outright, the ancient view that both the meaning and the significance of humanistic concepts are inseparable from their complex and often agonistic history of transmission.

To approach modernity as a condition of progressive conceptual amnesia, which in turn results in an increasingly stunted outlook on human agency, undoubtedly will ruffle some feathers in what (often at its own peril) for the past thirty-five years or so has been reconstituted as the "profession" of the humanities. A first way of arguing the point would be to establish a causal connection between modernity's diminishing grasp of concepts as dialectically evolving, hermeneutic frameworks and the professionalization of humanistic knowledge that, in David Simpson's pointed formulation, has all but become "divorced from content" and is vaguely presumed to be "useful in itself."[1] For however one may feel about it, there can be no question that for the past four decades or so, the humanities (especially in North America) have undergone enormous change as regards their institutional cache, their methodological orientation, and, ultimately, their perceived object of inquiry. Notably, as the preoccupation with finding a "definitive" method of inquiry intensified, the identity of the object or core questions to be engaged by humanistic study seemed to grow more obscure. Post-structuralism (in its various psychoanalytic, philosophical, anthropological, or aesthetic guises), deconstruction, new historicism, cultural materialism, queer studies, post-colonialism, and the more recent incursion of neuroscientific methodologies into the humanities are just some of the more conspicuous instances of this shift. Cumulatively these approaches reveal how a proliferation of methodologies tends to shift the object of inquiry and inflate the number of subspecializations, while simultaneously shrinking their intellectual scope; one is left with the impression of a rather dubious mathematical procedure, something we might call multiplication-by-division. To be sure, the quest for a sharply defined method, reliable in its application and guaranteed to produce marketable results, hardly amounts to a new development; it had crucially shaped European modernity in the era of Bacon, Boyle, Gassendi, Newton, and Leibniz, and if anything its much belated arrival and euphoric reception in (American) humanistic inquiry in 1966 at the newly inaugurated Johns Hopkins Humanities Center seemed to betoken a new, heightened legitimacy for the humanities as a bona fide science.

Not considered, however, was the question, previously raised by Hans-Georg Gadamer's *Truth and Method,* as to whether a commitment to some determinate

1. *Academic Postmodern,* 7.

method within the humanities might not entail unwarranted and unsustainable assumptions about the kind of knowledge to be thus produced. Indeed, the proliferation of increasingly short-lived, and often adversarial methodological prescriptions since the 1960s suggests that while the emergence of "theory" had undeniably taken control of departments and schools of inquiry throughout North America, its outlook on the long *durée* and complex genealogies of inquiry implicitly at stake was not so much as a body of work to be diacritically engaged but as so much fossilized intellectual substance to be historicized, syllogistically disproven, or in some other fashion overcome. Simply put, the ethos underlying the practice of "theory" since the 1960s in North America has been typically one of emancipation, and as a result its approaches have been axiomatically conceived as so many methods or techniques, to be applied to various fields and objects of inquiry. Compounded by a pragmatist and anti-metaphysical stance whose long history in Anglo-American and British culture David Simpson has traced some time ago, the history of "theory" that has shaped North American academia for several decades now has largely devolved into a quest—rather in the tradition of Bacon—for an inductive and universally applicable method of reducing contingent phenomena to infinitely repeatable certitudes.[2]

With barely concealed irony, Paul de Man's 1982 essay "The Resistance to Theory" thus concludes with the faintly dispiriting observation that "technically correct rhetorical readings may be boring, monotonous, predictable and unpleasant, but they are irrefutable." Modernity's most ardent wish—the wish not to be deceived or, in de Man's parlance, to engage written works in a way "that would stay clear of any undue phenomenalization or of any undue grammatical or performative codification of the text"—can ultimately never be granted.[3] Given modernity's conception of knowledge as a series of deductions inexorably following from our embrace of an all-encompassing methodological template, there can be no conclusive triumph of theory (or, in the present instance, a method known as "rhetorical reading") but only an endless sequence of performative misadventures. Perhaps de Man's sardonic reflexivity was meant as a gibe at his many followers, so doggedly intent on proselytizing his interpretive approach as a definitive method and, as Nietzsche had put it, repaying their teacher poorly by remaining forever disciples. And yet, to suppose that the professional theorist inhabits a "state of constant suspension" or "undecidability" is to vacillate between a narcissistic indifference concerning basic human questions and an incipient despair over the entire project of theoretical inquiry; for Terry Eagleton, it is the professional narcissism and blatant disregard for historical specificity that have defined postmodern theorizing ("those who are privileged

2. Simpson, *Romanticism, Nationalism,* esp. 19–63 and 126–148.

3. De Man, "Resistance to Theory," 20.

enough not to need to know, for whom there is nothing politically at stake in reasonably accurate cognition, have little to lose by proclaiming the virtues of undecidability"), whereas for David Simpson "the sheer emotional and rhetorical difficulty of remaining in a state of constant suspension ... seems to have made a place for a headlong retreat from theory and from the dissatisfactions it seems to prescribe."[4]

While these criticisms are not without merit, de Man's argument nevertheless goes to the very heart of method—viz., its speculative and seemingly deluded confidence in the eventual attainment of total certainty and impregnable authority. Even a casual reading of Bacon or Newton shows modernity's quest for objective method to be thoroughly steeped in the spirit of utopia, its heart stirred by that quintessentially modern fantasy: the *libido dominandi*'s conclusive possession of all phenomena rather than letting them speak and conceiving knowledge as our *adequatio* to and participation in them. More recently, the proliferation of methodologies (mislabeled as "theory") and the concurrent multiplication and division of their professed objects of inquiry have only accelerated, at least in part because of the humanities' increasingly frantic quest for greater institutional prestige and also in consequence of their rather naïve attempt to incorporate themselves as a modern *profession*. While highly effective for information-based sciences, professionalism turns out to be inapposite to interpretive disciplines that require our sustained immersion in a many-layered past composed of intellectual genealogies and their often conflicting lines of transmission. Not surprisingly, the price of "professionalization" (to use a word of which college accrediting organizations, graduate school deans, and funding agencies seeking to maximize their returns are equally enamored) has been steep. Far too often, individuals working in the humanities are tempted to tailor their research projects to minor grants made available by (non-researching) career administrators keen to promote research on topics whose importance they have mimetically deduced from other administrators. The projects in question tend to be labeled (often well before their completion) as "cutting-edge," interdisciplinary, or multidisciplinary while raising doubts as to whether those pursuing them any longer enjoy a clear grasp of what constitutes *disciplinarity*. A particularly farcical aspect of the humanities' "professionalization" involves the haphazard and naïve uses of instructional technology urged upon faculty by university administrators, themselves gullible captives of corporations sensibly minding their own business interests. Conceivably, a Power-Point presentation may be a sensible tool for conveying information to a panel of experts in oncology or marketing; yet one need not be a Luddite to recognize it as a wholly inapposite medium for developing and presenting a nuanced and sustained interpretive effort.

4. Eagleton, *Illusions*, 5; Simpson, *Academic Postmodern*, 26.

Many of these and other symptoms of what Raymond Tallis has provocatively termed "the suicide of the humanities" strongly correlate with the humanities' prolonged bout of "science envy."[5] What Coleridge had already indicted as his contemporaries' "asthmatic" style of thinking and writing—riveted by new information yet ill at ease with sustained reflection—is particularly evident in the current preoccupation, unparalleled in the history of humanistic inquiry, with devising forever new techniques, concepts, and methods.[6] The missing link between the recent phenomenon of a fully professionalized humanities and the latter's pervasive misapprehension of method as "theory" is modernity's quintessentially utopian nature—its nervous or, in Coleridge's combative phrase "finger-active, brain-lazy" (*CM,* 2:648), quest for anticipating and seizing the new. Concurrently, a humanistic inquiry legitimated primarily by its professional organization and methodological sophistication will naturally reenact modernity's iconoclastic, not to say allergic reaction against the mere suggestion that to *know* might depend on the cultivation of moral and intellectual virtues such as patience, good sense, moderation, and studiousness (rather than blind and fleeting "curiosity")—most of which modernity had so unwisely anathemized. Overall, then, the way that the humanities have constituted themselves as an aggregate of disciplines obeying, by and large, a historicist framework and procedural ethos since the mid-nineteenth century shows them to be a specific epiphenomenon of modernity and, thus, inauspiciously positioned as regards a critical and comprehensive assessment of the modern project's limitations and antagonisms.

Still, the objective of what follows is not to indulge in a jeremiad but, rather, to show by example that if there is to be a future for humanistic and interpretive knowledge, it will hinge less on the contrivance of yet another theory or slate of technical terms than on the sustained retrieval and critical engagement of some key concepts that (so my argument) have proven indispensable for meaningful humanistic inquiry since its beginnings in Plato and Aristotle. Inasmuch as such a retrieval is successful, it will also restore a clearer understanding of the distinctive nature and function of concepts within those disciplines committed to the cultivation of interpretive knowledge. To make that case in responsible and hopefully convincing fashion it is imperative to recognize the central and indispensable role of *agency* and *action* to any interpretive discipline. More than anything, it is the naturalist and, especially, the reductionist legacy of Hobbes, Locke, Mandeville, Hume, and others that has estranged us from the abiding and unique phenomenon of human intelligence as it is

5. Tallis, "Suicide of the Humanities."

6. "I can never ... affect a style which an ancient critic would have deemed purposely invented for persons troubled with the asthma to read, and for those to comprehend who labour under the more pitiable asthma of a short-witted intellect" (*CF,* 1:20).

realized in *action*—in contradistinction to a mechanistic and literally mindless notion of *process* or *behavior*. To that end, what follows will seek to recover the history of two conceptions that are always in play when questions of action and agency are being considered: those of will and person. The first sections of Part II thus trace the idea of the will (to specify it as the *human* will would be to commit a pleonasm) with a strong focus on its relation to the emotions, the intellect, and their respective involvement with the Platonic *logos*. In time, the notion of the will crystallized by absorbing and recalibrating a number of other concepts (desire, self-possession, judgment, teleology, etc.) into a complex and progressively self-aware hermeneutic tradition that dates back to ancient Greek thought, its subsequent cultivation in Stoic and neo-Platonic philosophy, and that first culminates in Augustine's supple and profound synthesis of these traditions with the relatively new field of Christian theology. Likewise, an intellectual archeology of the idea of person in both Christian and (to a lesser extent) Jewish philosophical theology, will be undertaken in Part III, which traces Coleridge's profound investment (unique among his contemporaries) in that tradition.

Over the course of some 1,800 years spanning from fifth-century Athens to the Dominican synthesis of Aristotelian and Augustinian thought in thirteenth-century Paris, Western philosophy and theology had gradually evolved a coherent and supple conception of human agency as embodied, capable of intellectual self-awareness, constitutively related to other rational agents, and hence incontrovertibly *capable* of making (and being responsible for) choices. It should go without saying that the *capacity* of choosing implies both a reflexive awareness of the agent invested with it, as well as the perennially looming possibility of his or her failing to exercise that capacity in timely and responsible fashion, which is not to say that what follows means to deploy the concept of awareness as synonymous with some version of Cartesian self-presence or certainty. On the contrary, in both its genesis and eventual awareness the self is essentially bound up with its relatedness to other persons—a relation that is only consummated when the other person becomes a "thou" rather than an impersonal he or she. As remains to be seen, that point explains why the phenomenology of conscience was to assume such pivotal importance in the later Coleridge's philosophical theology.

The premise for an inquiry into human agency and action—from Plato, Pythagoras, and Aristotle onward to the Stoics, Augustine, and all the way to Aquinas—had been that, far from being antagonists, will and intellect were essentially and productively entwined. Hence a coherent account of responsible action had to resist the temptation, sometimes observable in the late Augustine and especially conspicuous in his modern descendants (Martin Luther, René Descartes, et al.) to make one or the other aspect wholly dominant and to construe the lesser one as being merely epiphenomenal. What supported this classical view of human agency was, ultimately, the "onto-theological" axiom that contingency and doubt were but natural entailments of our manifestly imperfect modes of apprehension and cognition. Thus one divine form of reason was taken to have created and continued to pervade the cosmos—which, after all, signifies not a mere inventory of objects but the permanent and rational "order" of things; and the *telos* of a meaningful and justifiable life could only be to apprehend and participate in that *logos* as fully as possible, be it in the kind of rational contemplation (*theoria*) that Aristotle unfolds at the end of the *Nicomachean Ethics* or the mystical *visio beatifica* that St. Augustine and his mother, Monica, share in Book 9 of the *Confessions,* and which continued to be the *terminus ad quem* organizing most narratives of human flourishing well into the early modern era. The great objective of human existence did not involve the epistemological conquest and material domination of the world but the sustained engagement and approximation of the *logos* of which that world was a fluctuating and inscrutable manifestation.

The contrasting vision, and the principal antagonist of Platonic, Christian, and other more "secular" forms of humanism, is found in the reductionist, naturalist, and quasi-legalistic accounts of mind and reason pioneered by William of Ockham and his voluntarist successors. It is in the work of Hobbes, Gassendi, Locke, Mandeville, Hartley, La Mettrie, Helvetius, Hume, Priestley, and Godwin, among others, that we encounter its methodological and, in time, emphatically secular legacy in fully developed form. What Coleridge would somewhat polemically label the "corpuscular school" of inquiry, of whose distant sources in Democritus, Leucippus, Protagoras, Epicurus, and Lucretius he is well aware, eventually culminates in the strident anti-psychologism of Gottlob Frege, Ludwig Wittgenstein, and Gilbert Ryle, and in concurrent behaviorist attempts at tethering human action to mono-causal input/output ratios. In our own time, this project has been refashioned into a neuroscientific utopia of a wholly deterministic account of human consciousness and action—one that, to the extent that it expects to succeed, must logically abandon the notion of the human as a distinctive intellectual and ethical agent in favor of a strictly quantitative conception of our biological, carbon-churning species. Even this thumbnail sketch already suggests that humanistic inquiry faces enormous challenges that are further compounded by a recrudescent utilitarianism of state legislators, funding agencies, and university administrators unabashedly and single-mindedly committed to the "bottom line," fixated on the grant- and publicity-getting potential of some disciplines and, as a result, prone to confuse means with ends.

To be sure, the idea of humanistic inquiry—to say nothing of its recent and troubling corporate incarnation as "the profession of the humanities"—had never really enjoyed a "golden age." Far more plausibly, its history can be read as a series of focused and often intensely adversarial exchanges with a variety of competing intellectual projects—among them Ockham's divine-command ethic; Luther's dystopic, quasi-Manichean theory of the will; Hobbes's aggressively voluntarist and artificial concept of personhood; Locke's psychological hedonism and its underlying nominalist epistemology; Mandeville's and Hume's non-cognitive model of the passions; and so on. With varying degrees of success (and by no means always taking the same view), Thomas More, Erasmus, the Cambridge Platonists, Anthony Ashley Cooper, Earl of Shaftesbury, Francis Hutcheson, Coleridge, John Henry Newman, and others thus rise to defend some version of a Platonic cum Christian model of human agency where consciousness is inseparable from self-awareness, and where the integrity and uniqueness of the human person arises both from a productive alignment of will and intellect and from the person's prima facie ethical being—viz., as an agent constitutively related *and* obligated to other persons. Agency here is not conceived epistemologically—that is, as involving (or lacking) some technical skill for solving situation-specific and ostensibly value-neutral puzzles of what to *do*. Rather, it pivots on the far more complex and value-saturated mystery of what kind of person one seeks to *be*.[7]

In truth, while humanistic inquiry has always been a dialectical, not to say agonistic endeavor, born of contradiction and indeed thriving on it, such a pronouncement is easy to make yet hard to sustain once its implications begin to reveal themselves. The postmodern response to that challenge has all too often been to radicalize the impulse to historicize to the point "where continuities simply dissolve [and] history becomes no more than a galaxy of current conjectures, a cluster of eternal presents, which is to say hardly history at all."[8] Still, even as the myriad positions and antagonisms comprising the flow of intellectual history is liable to be experienced as bewildering and seemingly pointless, it would be a mistake to think of contradiction as simply a gratuitous obstacle to be removed or, better yet, circumnavigated on some imagined royal road toward clear and definitive insight. In fact, there is no such road, quite simply because contradiction "lies at the heart of movement, whether that movement takes place in things, in ideas, or in language. Contradiction generates movement. Contradiction does not threaten ideas, but it suggests their unreal-

7. On this distinction, see C. Taylor, who notes that contrary to modern, "single-term moralities" that "offer us a homogenous, calculable domain of moral considerations" and in their "caculability fit with the dominant models of disengaged reason," a truly capacious ethic "involves more than what we are obligated to do. It also involves what it is good to be" (*Dilemmas and Connections*, 6–9); see also Murdoch's insistence on "goodness" as holding axiological priority over rational choice, and as furnishing "a permanent background to human activity." Her pivotal question—"are there any techniques for the purification and reorientation of an energy which is naturally selfish, in such a way that when moments of choice arrive we shall be sure of acting rightly?" (*Sovereignty*, 52–53)—had arguably been answered, albeit by a tradition of thinkers whom Murdoch instinctively avoids; for the question goes to the heart of why and for what end human beings ought to cultivate habits and virtues; on the issue of habituation in Aquinas, and contrasted to modern behaviorism, see below 360–369.

8. Eagleton, *Illusions*, 46.

ized potentiality, the inadequacy of a present formulation, or the becoming which is their actual form of being."[9]

Michael Buckley's eminently Hegelian formulation does not consider an alternative scenario—one altogether central to my own argument: viz., there is no guarantee that the recurrent tension between two distinct conceptions of human agency and, implicitly, between two modes of knowing—a Christian-Platonic framework and an ancient atomist/modern naturalist one—will necessarily (nor, indeed, inadvertently) advance knowledge. To speak of "contradictory" views is to prejudge the conflict of views as dialectically generative. Another way of approaching the conflict between humanist-interpretive and strictly naturalist (or deterministic and reductionist) models of human agency would be to view the two paradigms as outright "incommensurable." On that account, what Plato attempts vis-à-vis the Sophists, what Augustine seeks vis-à-vis the Pelagians, what Shaftesbury and Hutcheson seek to accomplish in their running battle with Hobbesian and Lockean naturalism; and what Coleridge pursues vis-à-vis Hobbes, Locke, Hume, and countless other thinkers, is a concerted attempt at securing basic humanistic concepts (judgment, responsibility, teleology, transcendence, personhood) against reductionist attempts to quarantine and ultimately reject these concepts as merely subjective and, in time, altogether irrational and unintelligible. Recoiling from what Charles Taylor calls "the illusion of the rational 'obviousness' of the closed perspective," those thinkers to whom the present study is unabashedly sympathetic (Plato, Augustine, Aquinas, Coleridge, and Newman; and in our time, Hans-Georg Gadamer, Alasdair MacIntyre, Robert Sokolowski, and Louis Dupré) diversely seek to defend, preserve, and further elucidate a nuanced and differentiated conception of human agency against persistent attempts by modern rationalism, mechanism, and determinism to reject it as indefensible on epistemological and methodological grounds.[10] The principal tension, then, is not between two opposing conceptions of the human but, more fundamentally, between two strictly incommensurable views of how even to approach human phenomena to begin with.

A first and decisive question to be taken up thus concerns the status and operative logic of concepts in humanistic inquiry: viz., whether we recognize them to have a history, to signify *for us* only on the condition of our gradually internalizing that history—understood not as a lifeless catalogue of earlier usages but as a dynamic tradition indispensable for orienting ourselves in the present by honing our basic intuitions and interpretive capacities. As the following case studies in philosophy, theology, and (occasionally) literature seek to illustrate, modernity's gradual flattening out and ultimate forgetting of the hermeneutic and normative dimension of those

9. Buckley, *Origins*, 336.
10. C. Taylor, *Secular Age*, 556.

key concepts here under consideration was inevitable given how "modernity" understood itself. To be sure, there is no "singular modernity," as Fredric Jameson has rightly cautioned; rather, there are "the many narratives," several of which Charles Taylor has recently sought to disentangle as so many epiphenomena of the secular, and they certainly do not all date from the same period. Following Louis Dupré, Stephen Gaukroger, Charles Taylor, and Alasdair MacIntyre the present study locates the breakdown of the onto-theological conception of the *logos* in the Franciscan critique of Aquinas launched by Bonaventure, Duns Scotus, and, especially, in Ockham's startling proposition that reason is a function, indeed a projection of power, rather than the criterion for its responsible exercise. Others might prefer to locate that break later, say, in the debate waged by Luther and Erasmus regarding the freedom of the will (traced in Michael Gillespie's recent work), in the seventeenth century's preoccupation with putting natural rights on a strictly secular footing (an argument first advanced by Leo Strauss and more recently inflected by Knud Haakonssen), or in modernity's evolving preoccupation with models of self-possession or autonomy (a story unfolded in great detail by Jerome Schneewind). As regards the startling disintegration of classical models of teleology and the underlying, axiomatic view of nature as *entelecheia* that Aristotle develops in his *Physics* and elsewhere, one may certainly come to different conclusions as to whether the contestation of this model begins with Bacon, Boyle, Descartes, or Newton, or perhaps as late as Hume's *Dialogues*.

Yet the present study is not concerned with pinpointing an *origin* or even multiple *origins* of modernity, which indeed "is not a concept, philosophical or otherwise, but a narrative category."[11] Rather, the objective of what follows is to illustrate a fundamental change in the *habitus* and self-understanding of humanistic inquiry and with tabulating the costs of that shift, which has transformed the very idea of reason itself. To consider the changing understanding of concepts in the modern era (a transformation that prima facie defines modernity as a distinctive *epochē*) also means to apprehend the costs of several other, closely related shifts: viz., from a contemplative to an active stance; and from a mode of knowing that takes itself to be *participating in* reason to one that takes itself to be *producing* rational order by applying concepts to what is now posited as a universe replete with puzzling and ostensibly isolated objects and phenomena—a world seemingly devoid of rational order except such as we can authoritatively ascribe to it. This takes us to the second major claim advanced by this study, a claim that is *philosophical* or *meta-conceptual* in nature. For the main characteristic of modernity's changed intellectual *habitus* is an acutely self-conscious quest for an accumulative, inter-subjectively demonstrable,

11. Jameson, *Singular Modernity*, 40.

and systemic model of knowledge qua "information."[12] By inaugurating itself as an *epochē,* a break with the past, by repudiating cosmological and metaphysical frameworks (i.e., substantial forms, entelechies, and the divine source of reason [*logos*] itself), and by supplanting ontological *truth* with a quest for contingent *certainties* sought in the methodical, accumulative, and increasingly compartmentalized study of nature (including human nature), modernity developed a fundamentally changed and far more restrictive understanding of the quintessentially human act of conceptualization and articulacy. While modern scientific inquiry "retains the formal meaning of the one all-encompassing science, the science of the totality of what is [*Totalität des Seienden*]," it no longer understands that objective as ontologically given but as a *project,* an edifice to be predicatively realized: "in a bold elevation of the meaning of universality, begun by Descartes, this new philosophy seeks nothing less than to encompass, in the unity of a theoretical system, all meaningful questions in a rigorous scientific manner, with an apodictically intelligible methodology, in an unending but rationally ordered progress of inquiry."[13]

What Hans-Georg Gadamer has traced as the widening gap between truth and method stirs to life in Ockham's strident repudiation of Aristotelian elements in Dominican theology, and in the nominalist pathos with which he shifts the locus of truth from intrinsically rational and timeless universals to singularities whose meaning and authority pivot on their being ordained and licensed by a divine will that consequently appears not just inscrutable but potentially discontinuous. While the secular implications of Ockham's theological arguments would not reveal themselves for some time, a fundamental shift had taken place. Thus Ockham restricts human cognition to what can be demonstrably and verifiably conceptualized—that is, to the isolated, unrepeatable, and non-generalizable singular object. In time, Bacon would treat each of these singularities as a building block for a systematic edifice of abstract, lawful, and mathematical "idealities." Thus Ockham clears the ground for modernity's reductionist idea of knowledge as "information" that can be expressed as a mathematical constant. As Hans Jonas puts it, "for the modern idea of understanding nature, the least intelligent has become the most intelligible, the least reasonable the most rational. At the bottom of all rationality or 'mathematics' in nature's order lies the mere fact of their being quantitative constants in the behavior of matter, or

12. Husserl, *Crisis,* esp. §9 on Galileo's and Descartes's mathematization and quantitative transformation of nature; Gaukroger, *Emergence,* 400–451; Dupré, *Passage,* 42–64; Buckley, *Origins,* 68–85; on the correlated migration of the attribute of infinity from a "fulfilling dignity" to a mere "predicate of indefiniteness," see Blumenberg, *LMA,* 77–87.

13. Husserl, *Crisis,* 8–9; see also E. Cassirer, *Erkenntnisproblem,* 1:442–482; Dupré, *Passage,* 65–91; Gaukroger, *Emergence,* 159–195; Pippin, *Modernism,* 22–25; and M. Polanyi, *Personal Knowledge,* 3–65.

'the principle of uniformity' as such, which found its first statement in the law of inertia—surely no mark of immanent reason."[14] As we shall see in Hobbes, Locke, Mandeville, and above all in Hume, naturalist and, especially, reductionist accounts tend to beg the question of action and agency on a large scale, quite simply because from the outset they only accept as "proof" something that must be non-human, a-semantic, a-rational, and ultimately unintelligible; reductionism begins by positing (without arguing the point) that all causation is mechanical, rather than something imagined, reasoned, chosen, and enacted. Inasmuch as there are to be only efficient (never final) causes, causation itself is pared down to the strictly unintelligible instant and, in effect, consumes itself in its mechanical occurrence; for modernity to recognize a cause as efficient, the latter must be denuded of all memory or awareness.

Yet to take that view, for which Ockham's preoccupation with God's *potentia absoluta* had crucially prepared the ground, means to quarantine what can be known—including the isolated, gratuitous, and inexplicable acts of will exercised by an omnipotent (if enigmatic) God—as strictly *singular* occurrences that bear no discernible relationship to any other act, event, or phenomenon. As a result, the relation of human inquiry to divine reason is fundamentally thrown into doubt and, indeed, has been suspended indefinitely. For the ancient conception of the *logos* had implied the objective, indeed ontological, continuity and hierarchy of the cosmos, even if Aristotle already had to defend that premise against Democritus's and Leucippus's atomism and its irrational, radically skeptical implications. For Edmund Husserl, the rise of positivism "in a manner of speaking decapitates philosophy." For it dramatically narrows the very concept of reason, such that questions of human flourishing, and of interpersonal obligation and responsibility are effectively quarantined and, as I shall argue, gradually forgotten: "The positivistic concept of science in our time is, historically speaking, a *residual concept* [*Restbegriff*]. It has dropped all the questions which had been considered under the now narrower, now broader concepts of metaphysics, including all questions vaguely termed 'ultimate and highest.' Examined closely, all the excluded questions derive their inseparable unity from the fact that they contain ... the *problems of reason* in all its particular forms. For reason is the explicit theme in the disciplines concerning knowledge (i.e., of true and genuine, rational knowledge), of true and genuine valuation (genuine values as values of reason), of ethical action (truly good acting, acting from practical reason)."[15]

Written against the backdrop of fast-rising irrationalism and political violence, Husserl's 1935 Prague lectures on *The Crisis of European Sciences* raise questions that

---

14. Quoted in Dupré, *Passage*, 68.
15. Husserl, *Crisis*, 9; trans. modified.

bear pondering no less today, and which the following study means to keep in play throughout. Is a strictly procedural (methodological) outlook on reason even conceivable, or might the notion of the *logos* ultimately prove intrinsically *normative?* Conversely, what could possibly legitimate modernity's conception of rationality as a "historically specific" *consensus* (social, scientific, moral)—in short, as literally nothing more than a "convention" (Lat. *convenire*) and hence as endlessly negotiable, reversible, and liable to fragmentation into a plurality of rationalities? As Brad S. Gregory has recently shown in impressive detail, that development constitutes the Reformation's lasting and powerful legacy. In particular, the "transformation from a substantive morality of the good to a formal morality of rights" was dramatically accelerated by magisterial Protestantism's inability to contain the fragmentation of radical Protestant communities of belief; and it was just this "constitution of exclusive moral communities [that] would eventually suggest to some people that morality itself is contingent and constructed, or at least that its basis and precepts are separable from religion."[16] It is this disintegration of a coherent, normative, and supra-personal framework (one not based on claim rights) that was eventually ratified as the supposedly self-evident truth of Hume's fact/value distinction, one widely, if unthinkingly embraced by a great many individuals working in the humanities today. What Leo Strauss has analyzed as the "noble nihilism" of Weberian sociology—arguably the most salient instance of an entire discipline premised on the fact/value divide—will at various turns be of concern in this book. For now, the question is simply whether there can truly be multiple (and supposedly competing) rationalities, and whether, as Max Weber had argued, a conflict between "values cannot be resolved by human reason."[17]

Echoing some of the most salient points made by his one-time teacher, Martin Heidegger in his 1938 essay "The Age of the World Picture" indexes several traits distinctive of modern knowledge: increasing specialization; precisely quantifiable findings; the transposition of natural phenomena into hypothesized idealities; an insistence on their repeatable, experimental verification, etc. He then asks whether "every epoch has its distinctive world picture … or whether it is a distinctly modern form of conceptualization that raises the question concerning the world picture." As it turns

16. *Unintended Reformation*, 184, 205.

17. Strauss, *Natural Right and History*, 48, 64. Woven into Weber's pluralism is a profoundly agonistic view of human life as "essentially an inescapable conflict" (ibid., 65). In tracing the implicit theology of Weber—a bowdlerized Calvinism in which the drive toward peace and salvation is firmly planted in the human individual, even as the means for its attainment have been withheld—Strauss draws attention to the Machiavellian and Nietzschean assumptions *from* which Weber's sociology proceeds, yet which themselves never came under scrutiny: "Weber, who wrote thousands of pages, devoted hardly more than thirty of them to a thematic discussion of the basis of his whole position. Why was that basis so little in need of proof? Why was it self-evident to him?" (ibid., 64).

out, "world picture" for Heidegger constitutes not merely, indeed, not even primarily, some second-hand depiction (*Abklatsch*) of the world as it is ostensibly at hand. Rather, it furnishes us with a distinctively modern kind of orientation. That is what is meant by the colloquial phrase of "we get the picture" (*wir sind über etwas im Bilde*). Not only does such a picture "represent" the world *for* us, but it denotes "all that belongs to it and all that stands together in it—as a system" (*daß es in all dem, was zu ihm gehört und in ihm zusammensteht, als System vor uns steht*).[18] For Heidegger, who in this regard sounds far more sanguine about the modern project than the late Husserl, what resonates in the German idiom of "getting the picture" (*im Bilde sein*) is this "being prepared and adjusted to the world" (*Gerüstetsein und sich darauf Einrichten*). This attitude of a resourceful, autonomous self fully equipped for its encounter with and mastery of the world—words of rather ominous import in 1938—Heidegger intends as a kind of allegory of the modern era:

> World picture ... does not mean a picture of the world but the world conceived and grasped as picture. What is, in its entirety, is now taken in such a way that it first is in being and only is in being to the extent that it is set up by man, who represents and sets forth. Wherever we have the world picture, an essential decision takes place regarding what is, in its entirety. The Being of whatever is, is sought and found in the latter's representational character. However, wherever being [*das Seiende*] is *not* interpreted in this manner, the world also cannot enter into a picture; there can be no world picture. The fact that whatever is comes into being in and in the modality of representation transforms the age in which this occurs into a new age in contrast with the preceding one. The expressions "world picture of the modern age" and "modern world picture" have the same tautological meaning, for they assume something that never could have been before, viz., a medieval and an ancient world picture. The world picture does not change from an earlier medieval one into a modern one, but rather the fact that the world becomes picture at all is what distinguishes the essence of the modern age.[19]

18. Heidegger, "Die Zeit des Weltbildes," 86; Eng. *Question Concerning Technology*, 129.

19. Heidegger, *Question Concerning Technology*, 129–130 (trans. modified). Ger. "*Weltbild, wesentlich verstanden, meint daher nicht ein Bild von der Welt, sondern die Welt als Bild begriffen. Das Seiende im Ganzen wird jetzt so genommen, daß es erst und nur seiend ist, sofern es durch den vorstellend-herstellenden Menschen gestellt ist. Wo es zum Weltbild kommt, vollzieht sich eine wesentliche Entscheidung über das Seiende im Ganzen. Das Sein des Seienden wird in der Vorgestelltheit des Seienden gesucht und gefunden. Überall dort aber, wo das Seiende nicht in diesem Sinne ausgelegt wird, kann auch die Welt nicht ins Bild rücken, kann es kein Weltbild geben. Daß das Seiende in der Vorgestelltheit seiend wird, macht das Zeitalter, in dem es dahin kommt, zu einem neuen gegenüber dem vorigen. Die Redewendung 'Weltbild der Neuzeit' und 'neuzeitliches Weltbild' sagen zweimal das-*

In what follows, Heidegger's essay strikes a rather sinister note by suggesting that modern individualism and humanism are anachronistic, quasi-nostalgic reactions against the essentially "corporate" reality that has already established itself; his parsing of "subject" and "individual" is particularly instructive here. In ideologically less troubling language, Louis Dupré has described the "fateful separation" of nature from grace and the resulting conceptualization of nature in increasingly anthropomorphic categories of efficient causality, objective "representation"—a development that begins with Duns Scotus (the subject of Heidegger's doctoral thesis) and continues well into the late phase of natural theology in the eighteenth century.[20]

Its ideological encumbrances notwithstanding, Heidegger's "Age of the World Picture" highlights a number of key points. First, the arrival of the *Weltbild* entails the displacement of two founding concepts of human thought—grace and narrative. Leaving aside the second of these for the moment, we can see how a model of immanent, as it were homespun rationality comes to supplant a notion of divine *ratio,* rendering it all but unfathomable, incoherent, and ultimately obsolete. Particularly Ockham's work, to be taken up later on, widens the gap between divine reason and finite, human intellection to the point that the former is no longer an ontological datum on the order of Aquinas's (in provenance Augustinian) conception of grace. Rather, eternal, benevolent, and providential divine reason disclosed qua grace has mutated from a premise to an inference, and an increasingly tenuous one at that. Consequently, the late medieval and early modern era finds itself far more dependent on elaborating a systematic model, or *Weltbild,* independent of any transcendent guarantees or presuppositions. For Descartes, truly the prototypical modern thinker, it is thus "mind, not the universe, [which] bears the evidence for the divine existence. Just as the divine truth guarantees the external physical world, so the divine infinity removes from this universe any discernible final order and purpose."[21] That is, following Blaise Pascal's proto-existentialist reflections, infinity is no longer *plērōma* but emptiness, a metaphysical void that all but invalidates any talk of final causes and thus denudes the material world of the *logos* previously taken to organize all things and manifest them as phenomena susceptible of progressively deepening experience.

What distinguishes the modern *Weltbild* from the Christian-Platonic *logos* is not just its strictly immanent character and its ongoing legitimation by an exclusively

---

selbe und unterstellen etwas, was es nie zuvor geben konnte, nämlich ein mittelalterliches und ein antikes Weltbild. Das Weltbild wird nicht von einem vormals mittelalterlichen zu einem neuzeitlichen, sondern dies, daß überhaupt die Welt zum Bild wird, zeichnet das Wesen der Neuzeit aus" ("Die Zeit des Weltbildes," 87–88).

20. Dupré, *Passage*, 167–189.

21. Buckley, *Origins*, 97.

human quest for "clear and distinct" representations. Of equal (if more embarrassing) import is modernity's acute bewilderment when confronted with ancient frameworks that seem increasingly unintelligible and illegible to its naturalist conception of knowledge. Put differently, "ancient" and "modern" do not so much identify competing frameworks as they are the flags flown by the proverbial two ships passing each other in the night. Thus it makes little sense to construe the ancient/modern divide as a momentous rupture within a single vector of historical progress extending confidently toward some utopian future; for such an explanation can only ever issue from within a modern perspective to begin with and thus begs the central question. What Heidegger's portrait of modernity as the "age of the world picture" hints at but, given his ideological entanglements, fails to say outright is that modernity's quest for capturing all present and future phenomena in causally determinative "representations" and aggregating them in a single, unifying *Weltbild* shows it to be absolutely committed to a strictly mimetic construction of the experiential world. The world may be captured *as* a totalizing image—that is, a comprehensive "system" of scientifically warranted and putatively "self-evident" propositions. Yet as a result, modernity's dependency *on* the image risks deteriorating into an utter entrapment by some strictly immanent or naturalist frame. "Representation" (*Vorstellung*) becomes modernity's version of the golden calf, and whatever cannot be assimilated and rendered legible within the specific terms of our modern *Weltbild*—and that includes above all earlier, so-called premodern frameworks—can also no longer be dialectically engaged.

In this regard, at least, modern and premodern constructions of the world truly *are* incommensurable; for prior to G. W. F. Hegel's retrieval of dialectical thinking, modernity can only *anathemize* the foreign, unassimilable, and wholly other phenomenon, whereas ancient Platonic and early Christian frameworks seek to engage it dialectically. Hegel's powerful critique of the Enlightenment's struggle with superstition as de facto ensnared by the otherness that it takes itself to oppose had shown modernity's major liability to be precisely this lack of dialectical thinking or genuine reflection. What derails the Enlightenment project is its unreflective, undialectical "struggle with otherness [*als ihr Anderes*]" and the categorical supposition that "what is not rational has not *truth*, or, what is not grasped conceptually, is not."[22] The quintessential age of the world picture, Enlightenment thinking is incapable of grasping

22. The Enlightenment's "notion [*Begriff*] is all essentiality and there is nothing outside of it ... As insight, therefore, it becomes the negative of pure insight, becomes untruth and unreason [*Unwahrheit und Unvernunft*] ... It entangles itself in this contradiction through engaging in dispute, and imagines that what it is attacking is something other than itself [*etwas Anderes zu bekämpfen meint*]. It only imagines this, for its essence as absolute negativity implies that it contains that otherness within itself" (*PS*, 332–333/*PG*, 388–389); on this momentous chapter in the *Phenomenology*, and on the Enlightenment's implicit evolution of "utility" as a new gold standard of truth, allegedly supplanting emotivist conceptions of faith, see Pinkard, *Hegel's Phenomenology*, 165–180.

truth as a movement—by which we mean not its appropriation by an isolated self, but the dialectical movement of an idea progressively clarified by the inadvertent miscarriage of that very attempt. Blind to any possible mediation of the intelligible with the foreign, the Enlightenment thus rejects and pathologizes *per definitionem* (i.e., as sheer superstition or as illegitimate, threatening otherness) all those phenomena that resist integration into value-neutral, conceptual idealities. Notably, that includes those *qualia* (feelings, beliefs, commitments, moral obligations, virtues, aspirations toward transcendence, etc.) whereby the individual is alerted to its a priori relatedness to others and to the world of phenomena at large. Not until Hegel's generous tribute to Aristotle's concept of entelechy in the "Preface" to the *Phenomenology* do we have a genuine attempt to overcome the exclusionary logic of modernity's strictly propositional take on the world. The age of the "world picture" captures the world of phenomena by liquidating their specificity, their distinctive and incontrovertible valence and resonance as *qualia* within the human agent. Yet as a result, the *Weltbild* also confines the knower; as Wittgenstein was to put it, "a *picture* held us captive. And we couldn't get outside it, for it lay in our language, and language seemed only to repeat it to us inexorably."[23]

A second implication of Heidegger's thesis concerns modernity's changed outlook on concepts and the uniquely human act of conceptual thinking. For Aquinas, whose oeuvre can justly be taken as the most comprehensive and lucid articulation of a premodern framework, "our experience of things is not a confrontation with something utterly alien, but a way of absorbing, and being absorbed by, the world to which we naturally belong. The mind does not primarily depict, reflect or mirror the world; rather, it assimilates the world as it is assimilated to the world."[24] Progressively estranged from this integrative and unified framework, to which we shall return in due course, post-Thomist thought appears increasingly preoccupied with explaining the discontinuity or seeming randomness of natural phenomena and human action. It thus begins to accord a far more prominent and, in time, near-exclusive role to efficient causation and in so doing recasts human cognition as a fundamentally pragmatic, instrumental endeavor.[25] No longer do concepts function as vehicles for articulating the manifest structures of the *logos* and the character of our participation

---

23. *Ein* Bild *hielt uns gefangen. Und heraus konnten wir nicht, denn es lag in unserer Sprache, und sie schien es uns nur unerbittlich zu wiederholen*" (*Philosophical Investigations*, 53 [§115]); see also C. Taylor's discussion of the "immanent frame" (*Secular Age*, 539–593).

24. Kerr, *After Aquinas*, 31; see also Hyman, *Short History*, 47–66, and Blumenberg, *LMA*, 325–337.

25. Quoting David Braine, Kerr notes that "whereas our ordinary workaday pre-philosophical concept of causing is occluded by the model of the interaction of impersonal forces, ... the much older and richer premodern conception of irreducibly distinctive modes of agency 'has been lost sight of or repudiated in an attempt to reduce all agency to the material or mechanical model, or to mysterious mentalistic variants of this'" (*After Aquinas*, 47).

in it; instead, concepts are deployed, in contingent and occasional fashion, as mere *tools* for representing or "depicting" (very much in the sense of Heidegger's *Weltbild*) isolated and fleeting phenomena or substantially alien "objects." From here on, "one 'knows' only what one has built up from within. In [Robert] Lenoble's pithy expression: '*Connaître c'est fabriquer*.'"[26]

To the extent that the world's coherence as "cosmos" is no longer guaranteed but, on the contrary, is hypostatized as a system incessantly demanding further elaboration and verification, the function of modern concepts is no longer integrative but disjunctive. A particularly apt instance involves the shift from Pythagorean tuning, which "harmonizes the octave," to Vincenzo Galilei's rationalization of tuning, which partitions the scale into equal intervals. Where "the Pythagorean ratios of 2:1, 3:2, 4:3 and 9:8 ... [had] enabled the inaudible sounds of the heavens to vibrate within the early soul, and, conversely, for the audible tones of human music to reflect the celestial spheres," the modern, rationalized conception of equal temperament "collapsed music into 'reality' as an audible *fact* divorced from celestial *values*."[27] Descartes's reasoning from God to the world reinforces what we shall find lurking in Ockham's *Quodlibetal Questions:* meaning is no longer deemed intrinsic to experience, and knowledge is won only at the expense of its terminal divorce from any type of sensation. Given nominalism's assertion of the utter incommensurability of God and creation, and given its insistence that "whatever is asserted must be asserted hypothetically with the theological recognition that it may be totally otherwise," it cannot surprise that Descartes's project of a *mathesis universalis* should eventually have stripped the senses of any evidentiary role.[28] As early as in the writings of Ockham and in Nicholas of Autrecourt's subsequent revival of atomism, concepts—rather than enabling us to articulate our participation *in* phenomena invested with unconditional reality and rationality—instead come to function referentially and predicatively; they serve to juxtapose discrete empirical objects "out there" to a hermetically enclosed observing consciousness, or *cogito*. Central to the modern epistemological stance is the axiom of a *cogito* permanently estranged from the phenomena with which it is engaged. Indeed, because it can engage them only on the premise of their radical heterogeneity, the "representational character" (*Vorgestelltheit*) of objective

26. Dupré, *Passage*, 66.

27. Chua, *Absolute Music*, 15, 18. As he sums up his case: "ancient rationality unifies; modernity divides" (20).

28. Confronting the mutation of rationalism "from a comprehensive natural theology to a comprehensive skepticism" and "the progressive temptation of the intellect to destroy itself first by overweening pretensions and then by ineluctable disappointment," Descartes's preoccupation with certitude—and his emphatic dissociation of such certitude from the testimony of the senses—seems inevitable (Buckley, *Origins*, 74, 70, 72).

phenomena implies a strictly referential model wherein cognition and abstraction have become fully convertible. On a modern, post-Copernican understanding, to know is to render something visible *as such,* albeit in a medium (universal mathematics) essentially different from the phenomenon at hand and without making normative claims about either the phenomenon or its relation to the epistemological agent. Not only does such a model of cognition require the methodical cultivation of distance and detachment, but it also implies the neutrality, the indifference (perhaps even the outright incommensurability) of the knower and the known.

Often remarked upon, the Enlightenment's preoccupation with visibility, with bringing "to light," or making "plain" and "evident" knowledge in the here and now, also points to the changed function of narrative—whose authority now pivots on its emancipation *from,* not relation *to,* the past. Johann Gottfried Herder's shrewd remark that "in our century we have, alas, so much light" points toward what Michael Polanyi has called the "separation of reason and experience" and the "attempt rigorously to eliminate our human perspective from our picture of the world."[29] The beginnings of that shift may indeed date back as early as the Ionian school of Democritus for whom, contrary to Pythagoras, "numbers and geometrical forms were no longer assumed to be inherent as such in Nature."[30] And yet, in embracing a counterintuitive theory such as the one ventured by Copernicus, modern scientific inquiry abides in "the expectation of an indefinite range of possible future confirmations of the theory." Moreover, it can only defer—yet never obviate or supplant—our return to what Husserl calls the "natural attitude" (*natürliche Einstellung*). That is, confirmation of a new theory cannot be strictly immanent to its own mathematical design but must eventually become intuitable; for "any critical verification of a scientific statement requires the same powers for recognizing rationality in nature as does the process of scientific discovery, even though it exercises these at a lower level."[31] In his 1910–1911 lectures, Husserl had drawn attention to this often obscured fact that "every natural science, insofar as it presupposes the theses of the natural world-perspective [*natürliche Weltansicht*] *is a priori bound up with the ontology of the real* [*reale Ontologie*]." Thus it "presupposes as valid what is prescribed for it in terms of the general sense of nature as a datum of experience."[32]

Even as he approaches questions of method from a strictly scientific perspective, Michael Polanyi leaves no doubt that no method can ever be entirely self-certifying, but that it presupposes what Newman, speaking in a different context, had called

29. Herder, quoted in Dupré, *Enlightenment,* 219; M. Polanyi, *Personal Knowledge,* 3.
30. M. Polanyi, *Personal Knowledge,* 8–9.
31. Ibid., 5, 13.
32. Husserl, *Basic Problems,* 24.

"antecedent probability." At stake here is the axiological priority of subsidiary over focal awareness, of "fore-meaning" (*Vorhabe, Vorbedeutung*) over intention. Any specific act of inquiry presupposes a subject's teleological orientation vis-à-vis a particular "life-world" (Husserl's *Lebenswelt*), a world whose sheer givenness alone enables us to conceive and articulate specific epistemic objectives. In Heidegger's nomenclature, all *Dasein* involves a subsidiary "attunement" (*Stimmung*), an antecedent grasp of what is in light of what ought to be; Heidegger calls it "care" (*Sorge*). For a particular scientific method to have been conceived at all, let alone to have undergone purposive "application," there has to be this subsidiary orientation—a non-transcendable "horizon" that can neither be unilaterally suspended nor objectively dissected by some particular methodology.[33] Approaching the issue from the perspective of philosophical theology, rather than the hard sciences, the same point emerges no less forcefully in the recent work of Jean-Luc Marion. "Method," Marion insists, "should not ... secure indubitability in the mode of a possession of objects that are certain because produced according to the a priori conditions for knowledge. It should provoke the indubitability of the apparition of things, without producing the certainty of objects ... The method does not run ahead of the phenomenon, by *fore*-seeing it, *pre*-dicting it, and *pro*-ducing it, in order to await it from the outset at the end of the path (*meta-hodos*) onto which it has just barely set forth."[34] Not only, then, is it "of the essence of the scientific method to select for verification hypotheses having a *high* chance of being true," but the application of theoretical maxims or "rules of art" presupposes "a good deal of practical knowledge of the art. They derive their interest from our appreciation of the art and cannot themselves either replace or establish that appreciation."[35] Discussing the case of highly complex symmetries in crystals, Polanyi thus notes that our ability to identify an object of inquiry as apposite cannot itself be licensed by some theory but, instead, depends on an antecedent "aesthetic ideal, closely akin to that deeper and never rigidly definable sensibility by which the domains of art and art-criticism are governed." The point emerges most clearly from the counterfactual scenario of utter randomness. The truly random is by definition unintelligible; it "can never produce a significant pattern" quite simply because its sole criterion involves "the absence of such a pattern."[36] Once order and

33. Nietzsche's famous depiction of a secular modernity that has "wiped away the entire horizon" (*Wer gab uns den Schwamm, um den ganzen Horizont wegzuwischen?*) foreshadows Husserl's and Gadamer's subsequent use of the same trope (*Gay Science*, §125; p. 120).

34. *Being Given*, 9.

35. M. Polanyi, *Personal Knowledge*, 30–31.

36. Ibid., 48, 37. "Any numerical assessment of the probability that a certain event has occurred by chance can be made only with a view to the alternative possibility of its being governed by a particular pattern of orderliness" (ibid., 33).

knowledge, *logos* and cognition, have been understood as essentially convertible, it also becomes apparent that reason and structure can never simply be predicated *of* objects but must truly be found *in* them.

Put differently, reason is not some attribute of autonomously conceived, higher-level propositions about specific phenomena; rather, it informs how those phenomena themselves are apprehended to begin with. This is even (indeed especially) true where the initial stance is one of principled and thoroughgoing skepticism. As Husserl notes with regard to the Cartesian *cogito,* whenever "in our phenomenological attitude we are focused on a perception, we apprehend it as a completely immediate This!" There is no second-guessing of the phenomenon as such. To be sure, we may certainly suspect that "something only appears to have being" and consequently doubt "whether it really exists ... Yet precisely thereupon, this appearing, this perceiving, remembering, judging, and so forth, are presupposed as given, just as they are indeed given [*aber eben damit ist dieses Erscheinen, dieses Wahrnehmen, Erinnern, Urteile usf. als gegeben vorausgesetzt, wie es in der Tat gegeben ist*]." As Husserl sums up the case (herein anticipating Marion's recent reintegration of phenomenology with theology), "doubt presupposes the givenness, the indibutable givenness of the meaning that is posited in the doubt [*Jedenfalls setzt also der Zweifel Gegebenheit voraus, die zweifellose Gegebenheit der Meinung, die in Zweifel gesetzt ist*]. Consequently, this perception, this phenomenon of an abiding empirical givenness ... is given absolutely."[37]

If this is true of the hard sciences, it is eminently more true yet of the interpretive sciences which—to the extent that they have recently sought to emulate a strictly procedural concept of inquiry—have not only misconstrued their own mission and object but, as it turns out, also distorted the idea of scientific method. "To the extent to which our intelligence falls short of the idea of precise formalization," Polanyi remarks, "we act and see by the light of unspecifiable knowledge and must acknowledge that we accept the verdict of our personal appraisal." In other words, the authority of a specific method of knowing inevitably rests on, and is circumscribed by, the art of judgment—a term that, not coincidentally, we shall also find to be uniquely enmeshed with concepts of will and person. While obviously a crucial and indispensable tool for cognition and its communication, no method can ever be entirely self-authorizing. It rests on a "view" or judgment that, however provisionally and tenuously, charts the course for a given method's progressive application.

It is in this, by definition pre-theoretical domain that the primacy of practical over theoretical (or speculative) reason reveals itself, and along with it the indispensable role of tradition. For at the moment of "application" we encounter "the principle

37. Husserl, *Basic Problems,* 54–55 (§24).

of all traditionalism that practical wisdom is more truly embodied in action than expressed in rules of action."[38] If our engagement with concepts is to be responsible and capacious, it cannot simply unfold in quasi-nominalist, over-focused fashion on their pragmatic use as seemingly neutral tools that fortuitously happen to be at hand. Michael Buckley's distinction between four conceptions of method, while helpful, needs to be amended here. Only two of the methodologies that he identifies, the operational and the logistic, are truly methods in the modern sense of being "applied" to objects or phenomena held to be distinct from (and unrelated to) the agent of knowledge. The other two, the dialectical and the problematic methods, are not properly concerned with objects but with entire "conceptions" of knowledge; their concern lies not with some local object or phenomenon but with a historically conditioned discursive formation. This is true of the "problematic" method of Aristotle and Aquinas, particularly the latter's method of *disputatio,* which progresses toward knowledge by staging a rigorous contest between the strongest versions of competing arguments. Likewise, Platonic and Hegelian dialectics constitute a second-level, as it were meta-discursive operation, and the strength of both—indeed, their inherent superiority over the other two—pivots on their showing knowledge to be a movement, a teleological progression. That crucial implication can already be located in the etymology of method (from Greek *μέθοδος* = a way, road, journey), which carries with it a strong narrative dimension, and as such is not focused on the application of a specific procedure but on the transformation of the agent of knowledge. Contrary to its fleeting and misleading association with "hunting" and sexual conquest—which, tellingly, is only suggested by the stranger in Plato's *Sophist* (218[d])—the dominant meaning is that of "a pilgrimage to the presence of a goddess."[39]

In taking up the question "What Is a Concept, and How Do We Focus on It?" Robert Sokolowski emphasizes that concepts do not "represent" or "depict" *objects* but enable us "to focus on the thing in its intelligibility." As he adds, it is better "to say 'the thing *in* its intelligibility' than 'the thing *and* its intelligibility,' because the latter suggests that the intelligibility and the thing are two different 'entities,'" when in fact "the thing subsists only by being intelligible . . . It wouldn't be *what* it is without it, and it wouldn't *be* without it."[40] The last point will prove crucial to the historical exploration of the concepts of will and person, for it underscores that the object of

38. Ibid., 54. Polanyi's subsequent distinction between subsidiary and focal awareness, wholes and meanings, tools and frameworks, and his emphasis on the indelible role of "commitment" in focused inquiry (55–65) reveals his intellectual proximity to modern phenomenology, arguably the most capable philosophical stance from which to rethink modernity as a problem without *eo ipso* being ensnared in its conceptual and ideological premises.

39. Robinson, *Plato's Earlier Dialectic,* quoted in Buckley, *Origins,* 22.

40. *Phenomenology of the Human Person,* 177.

inquiry—and indeed the idea of human agency adumbrated by these concepts—is inseparable from the sustained interpretive effort by which it is gradually distilled and articulated. In short, object and concept (Hegel's *Gegenstand* and *Begriff*) are not related referentially, as word and object, but instead are mutually constitutive. The structure of concepts thus mirrors what Gadamer characterizes as "the ontological structure of understanding [*Verstehen*]." That is, a philosophically reflective and responsible engagement with concepts aims "not to develop a procedure of understanding, but to clarify the conditions in which understanding takes place." When approached as historically grown frameworks at once complex and dynamic, our conceptions never serve to *produce* knowledge *ex nihilo* but, instead, facilitate our encounter with what Husserl had called the world's radical and indisputable anteriority, or its "absolute givenness."[41] Within the domain of humanistic inquiry at least, to work with concepts thus means to enter into an ethical—as opposed to a straightforward pragmatic—relation to the reality that these concepts prima facie allow us to apprehend and, in so doing, to acknowledge the rich and often agonistic history of uses to which they have been put in the past. Our relationship to concepts thus should mirror that to other persons; that is, it ought to rest "not on the subjection and abdication of reason but on an act of acknowledgment and knowledge." Hence, whatever intellectual authority concepts possess "cannot actually be bestowed but is earned, and must be earned if someone is to lay claim to it. It rests on acknowledgment and hence on an act of reason itself."[42] For that to happen, and for us to inhabit concepts as living frameworks with a deep history, rather than occasionally wielding them as tools (such as resonates in the sadly common phrase of "applying a theory") also means to conceive rationality not as a correlate of self-possession but of what, echoing Hegel, Gadamer calls "recognition" (*Anerkennung*). At issue here is a sustained, deliberative, and potentially creative reflection on the "antecedent probability" of a concept's truth value along the lines explored by John Henry Newman in his *Development of Christian Doctrine* (1845) and further scrutinized in his *Grammar of Assent* (1870).

To take that view also means to recognize that the intelligibility of our conceptions is never simply achieved by us as individual agents of knowledge, but that it pivots on our dialectical engagement of intellectual traditions—viz., the complex record of others' articulations of those very concepts. Only so does their intentional correlate—what Hegel calls *die Sache selbst* (a notably more apposite term than "object")—disclose itself as the focal point of a jointly cultivated awareness. Not coincidentally, the antagonism between knowledge as a shared and participatory

41. Gadamer, *Truth and Method*, 295; Husserl, *Basic Problems*, 54.
42. Gadamer, *Truth and Method*, 281.

process, and knowledge as commodity and capital—a conflict long in the making—has of late erupted into full view, such as in the current legal contestation of the fair use clause in international copyright law, particularly as it applies to academic instruction.[43] Such legal disputes over the economic disposition of knowledge are but an inevitable entailment of modernity's gradual redefinition of knowledge as a state of hermetic, "inner" certitude and, thus, as an object of possession rather than a phenomenon of disclosure. Humanistic inquiry, if it is to remain a meaning-generating (*sinnstiftend*) undertaking, would be especially ill advised to borrow reductionist models from the sciences, no matter how vexed its current practitioners may be by the humanities' supposedly inferior (because less "rigorous") public image.

In fact, humanistic inquiry not only cannot succeed but will positively vitiate its raison d'être if it deploys concepts on a purely occasional basis, viz., as tools to pry open the resistant casing of some putatively alien object or text.[44] Within humanistic inquiry, concepts are received and inflected as we attempt to respond to questions we have inherited (not conceived *ab novo*); and to these questions we can only respond as a community of ethical beings whose *responsibility* extends both synchronically to our fellow beings and diachronically to the history of earlier respondents and to future generations who will inherit a world shaped by the values and commitments of our practical reason in the here and now. Far from the totalitarian specter or metaphysical menace as which it is commonly portrayed at present, normativity is simply the ethical framework (Newman calls it "implicit reason") absent which intellectual work would be nothing more than a type of professionalized curiosity—that is, mere transaction rather than bona fide action. Both will and person, the terms most central to this study, can only signify if we recognize them as intrinsically normative. Throughout their complex and often conflicted hermeneutic history, they are deployed as imperfect articulations of a *good,* of a value or ethical ideal to the realization of which we take ourselves to be committed (notwithstanding our inevitable lapses in honoring that commitment); and their true province is that of practical reason, not theoretical speculation. For value concepts, as Robert Spaemann has pointed out, cannot be understood independently of their historical evolution and transmis-

---

43. A striking example is the case of Georgia State University, sued by Cambridge University Press, Oxford University Press, and Sage Publishers, over its facilitation of electronic course readings. On the initial suit, see "Publishers Sue Georgia State on Digital Reading Matter" (*NY Times*, 16 April 2008 www.nytimes.com/2008/04/16/technology/16school.html); the case was decided in favor of the defendant, Georgia State University, in August 2012. On the perils of treating knowledge as private property and asserting exclusive ownership over it—and thus perverting *intelligibilia* into *sensibilia*, verities into consumables—see Griffiths, *Intellectual Appetite*, esp. 154–159.

44. "The human sciences cannot be adequately described in terms of this conception of research and progress, ... [because] what the[y] share with the natural is only a subordinate element of the work done in the human sciences" (Gadamer, *Truth and Method*, 284).

sion; rather "to make their meaning understood we must again tell a story; but this time it is not the story of the referent, but of the term itself."[45] They do not have a referent vis-à-vis which their truth-content or "correctness" could be objectively verified. Rather, they are linchpins of our hermeneutic situation within a "process of tradition" (*Überlieferungsgeschehen*) of the kind that Gadamer, MacIntyre, et al. have affirmed, quite self-consciously, *against* modernity.

Following similar arguments by Gertrude Elizabeth Anscombe, Iris Murdoch, and Alasdair MacIntyre, the present study's principal concern lies with modernity's apparent inability to grasp this trans-generational, hermeneutic dimension intrinsic to conceptual activity within the humanities. Within the interpretive disciplines, I argue, concepts ought to be engaged as hermeneutic frameworks, not as tools but as prima facie objects of inquiry; and their elusive perfection and authority cannot be separated from the trajectory of their previous applications or "effective history" (*Wirkungsgeschichte*), which in turn circumscribes, focuses, and indeed motivates our engagement with these concepts. As Gadamer had worked out with much care, "interpretation [*Auslegung*] is not an occasional, post facto supplement to understanding [*Verstehen*]; rather, understanding is always interpretation, and hence interpretation is the explicit form of understanding." Concepts thus do not serve to "decode" a text or set of phenomena ostensibly unrelated to us and only of objective or, as the case may be, historical interest. Rather, our reliance on concepts in humanistic, interpretive practice reflects the bilateral nature of all understanding as a process that "always involves something like applying [*Anwendung*] the text to be understood to the interpreter's present situation."[46] Concepts thus *disclose* or *unveil* something; as Sokolowski argues, they draw out the intelligibility of the thing, its immanent essence and perfection, and they do so not merely to satisfy a questioner's professional curiosity but, crucially, to articulate the knowledge so produced *for another.* Humanistic concepts, that is, acquire reality only within a shared hermeneutic space—a domain that, as remains to be seen, cannot be thought of in isolation from the evolving history of its guiding conceptions.

In his late work, Husserl took up what he had come to regard as a perilously limiting and abstract understanding of concepts, at once prone to estrange modern man from the reality of his "prescientific" experience and as modernity's dangerous illusion of their supposedly neutral and objective "application." While there certainly is an indispensable element of "correctness" to the way concepts function, he insists

45. *Persons,* 17.

46. Sokolowski, *Phenomenology of the Human Person,* 306–307. Gadamer's point is (surprisingly) echoed by Jameson, who notes that "what passes for modernity ... is itself little more than the projection of its own rhetorical structure onto the themes and content in question: the theory of modernity is little more than a projection of the trope itself" (*Singular Modernity,* 34; see also 94).

that their truth value is by no means exhausted in it.[47] For there is also what Husserl calls the "truth of disclosure," which is no longer concerned with the utility-function of concepts in a proposition but with their reflective evaluation as modes of "disclosing" the intrinsic logic of a thing *for others*. Hence, to *know* or *understand* something necessarily involves more than a strictly factual, detached, and value-neutral pronouncement about the object at hand. In fact, the very supposition of a thing's "intelligibility" is intimately entwined with what Sokolowski calls "the goodness or perfection of those things. We never work with things simply as they are; we always see and understand them against the background of what they *can* be and what they *should* be." There is an important temporal dimension to knowledge, in that it would be redundant to content ourselves with conceptualizing objects merely as they happen to be in the here and now. In fact, "the thing is not just what it is at the given moment in which we come to name it," but it is a correlate of our intentional and conceptual activity precisely because it is susceptible of transformation, either from within or from without: "Only ends bring out the full intelligibility of things."[48]

---

47. Husserl, *Formal and Transcendental Logic*, 120–127 (§§44–45).
48. Sokolowski, *Phenomenology of the Human Person*, 186–188.

2

FORGETTING BY REMEMBERING

Historicism and the Limits of Modern Knowledge

To return once more to Heidegger's notion of the modern *Weltbild,* it appears that yet another change wrought by the age of the "world picture" concerns a thoroughgoing shift in the form, function, and scope of narrative. The structure of narrative mutates from the mnemonic to the emancipatory, from the genre of epic to that of utopia, and from an evolving, deepening, and transformative engagement with received concepts and meanings to the methodical cultivation of a detached and critically objectifying stance whose principal concern lies with overcoming the past. Developing their critique of modernity from diametrically opposed points of view, both Schopenhauer and Coleridge recognize that what impels and legitimates modernity's changed concept of narrative is a deep-seated fear of error, be it as a result of the constant possibility of deception perpetrated by Descartes's specter of a *dieu trompeur* or because of our supposed propensity to become mired in the past, a habit that for Descartes spells mere stasis and mindless repetition; hence modernity's preoccupation with both remembering and overcoming the past, which accounts for the modern era's simultaneous cultivation of vigilance and forgetfulness. To fend off this perceived threat of the past as sheer *recurrence,* modern narrative unfolds as a utopian quest for a radically autonomous and entrepreneurial model of agency—one that *produces* and *consumes* both its own conceptual inventory and those social, moral, economic, and political meanings to whose construction that inventory is deemed uniquely conducive. Defining of modern "progress," Hans Blumenberg notes, is "the continuous self-justification of the present, by means of the future that it gives itself, before the past, with which it compares itself" (*LMA*, 32).

While some of these issues will be taken up more fully at the beginning of Part IV, it is necessary to identify more precisely the kind of narrative of modernity that is being presented in what follows and, in particular, how it differs from a by now fairly established model of intellectual history or some such historicist survey that aspires to (or presumes outright) the essential "pastness" of the past and its merely archival interest for the present. Neither the historical evolution of the concept of the will nor that of the person admits of being treated as some kind of prehistory, be it in the spirit of our having overcome its alleged inadequacies or finding ourselves as the putative *telos* of the trajectory of either idea as it migrates from Greek philosophy into the modern era. If, then, modern narrative conceives (by default, as it were) the past in essentially *historicist* form—viz., as something concluded, alien, and incommensurable with present and future exigencies—it is also true that modernity has proven a fertile ground for the production of a very different kind of narrative. The basic impetus and objective pursued in modern narrative is a notion of the event as essentially unprecedented and singular, that is, a *novum* or, indeed, a "novel." Not only must modernity find forever new ways to impress on us the sublimity of its very occurrence, but it must simultaneously cut from whole cloth the intellectual template whereby this event is to become intelligible for us. One of the first "discoveries" of the modern era—and structurally cognate with the emergence of Heidegger's *Weltbild*—thus involves the proposition that the past is a historical object, deemed intelligible because (and only insofar as) it has definitively expired and thus no longer constrains our self-awareness. If, as Heidegger concludes, "the fundamental event of the modern age is the conquest of the world as picture" (*Der Grundvorgang der Neuzeit ist die Eroberung der Welt als Bild*), its Achilles heel will be a one-sided conception of narrative as a strictly archeological endeavor concerned with preserving, and thus containing, what is peremptorily construed as other.[1] Under conditions of modernity, all history is merely prehistory.

Staging a curious version of Sigmund Freud's *fort/da* game, modernity thus compensates for its original dilemma by simultaneously engaging with and disengaging from the past. It invents the notion of a "past" as strictly *passé*, as archival, fossilized ("sedimented"), and inert stuff. Already the etymology of *modernus* (first attested around A.D. 500) shows the word denoting less a particular span of time than a fundamentally changed perspective on temporality itself. Derived from *modo* (Lat., only, merely, just), a word that also means "lately" and carries a strong associ-

1. *Question Concerning Technology*, 134/*Holzwege*, 92.

ation with the present, *modernus* is "one of the last legacies of vulgar Latin" among related temporal terms, the only one to perform "the exclusive function of designating the historical now of the present."[2] What defines the modern is less the idea of novelty and the "new" than a present viewed in sharp contrast with what was formerly held to be of timeless validity. Gradually establishing itself in contradistinction to *antiquitas,* the "modern" rejects the notion of the distant past as a reservoir of exemplary meanings: "The twelfth century *moderni*'s experience of time is ... typological, not cyclical. Typology takes moments separated in time and relates them to one another as the intensification of the old in the new. The new preserves the old; the old lives on in the new. The old is redeemed in the new, and the new is built on the foundation of the old."[3]

An analogous history characterizes the increasingly prominent role of the words "secular" and "epoch" in the early modern era, though there is no space to trace it here. In each case, a concept seemingly designating a particular span of recent time introduces the postclassical notion of time as a linear progression or sequence that no longer sees the past as having an enduring and indispensable "presence" within our ongoing quest for rational orientation. Instead, modernity's dominant conception of time is one of chronometric and value-neutral accountancy, a series of discrete and fungible epochs occasionally punctuated by threshold moments or "hot chronologies" (1648, 1707, 1789, 1815, 1848, etc.).[4] Both this computational model of time and the partitioned conception of epochs that it helped spawn rest on one crucial, albeit unexamined assumption: that neither time nor history is to be credited with *meaning,* that both are categorically devoid of "plenitude" in the strong neo-Platonic and eschatological sense of *parousia* (fulfillment, presence). In their linear and monochrome progression, concepts such as "modern," "secular," and "epoch" thus institute estrangement and loss as the affective signature of human experience since the late Middle Ages. What has vanished is what Husserl's "phenomenological reduction" so painstakingly seeks to recover: viz., the *persistence of time in consciousness.*[5] As remains to be seen, the dismantling of time into heterogeneous, incessantly

---

2. E. R. Curtius, quoted in Le Goff, *History and Memory*, 27; Jauss, "Modernity and Literary Tradition," 333. See also Gillespie, *Theological Origins*, 1–18; Jameson, *Singular Modernity*, 17–41; and Buckley, *Origins*, 25–26.

3. Friedrich Ohly, "Synagoga and Ecclesia: Typologisches in Mittelalterlicher Dichtung" (1966), quoted in Jauss, "Modernity and Literary Tradition," 336.

4. On this concept, see Chandler, *England in 1819*, 67–84; on the modern conception of the "epoch," see Blumenberg, *LMA*, 27–51; Koselleck, *Futures Past*, 93–104.

5. Recalling more comprehensive arguments to the same effect from his earlier *On the Phenomenology of the Consciousness of Internal Time* (1905–1906), Husserl in his 1910–1911 lectures on *The Basic Problems of Phenomenology* points out how any instance of an "intentional relation" yields a "phenomenological datum" whose identity "in diverse acts of consciousness ... is not an

"lapsing" units of measurement correlates with the (in origin nominalist) dissolution of the person into a series of states whose connectivity Locke is only prepared to accept as a hypothesis in urgent need of the kind of "demonstration" that Hume with good reason eventually declared to be impossible. At the same time, ever watchful that the past might not be sufficiently dead but might inopportunely rise again—not as a truly living presence, to be sure, but as the "undead" of the modern Gothic imagination—modernity spawns an entirely new discipline aimed to ensure that this will not happen. It is called historicism.

To characterize modernity's outlook on intellectual genealogies and traditions as one of amnesia is to suppose, minimally, that what has taken place is not a radical, terminal "forgetting" but, rather, a prolonged failure to remember—with the proviso that "remembering" here means engaging the history of an idea or conception in such a way as to recognize ourselves to be implicated in it. In a post-historicist account of the kind here attempted, remembering thus entails less the past's possession *by* than its dialectical transformation *of* the subject. While this failure to recognize history as a genuinely interpretive process ought to be seen as self-inflicted, it should not be construed as a case of "repression" of the standard Freudian variety. For it is not that the content of a given idea or conception—its "topicality" (Freud's *Besetzung*)—is being repressed. Rather, we will find that *how* we apprehend and relate to conceptions and ideas has been decisively altered and, in part, become deeply confused. To be sure, the content and thematic scope of conceptions, particularly those inherited from the premodern era, undergoes much scrutiny as the modern project of critique develops an ambitious, explicit, and often iconoclastic outlook on that past. That much is readily apparent when considering William of Ockham's rejection of Aquinas's ontology of a timeless and uncreated divine *logos* or Hobbes's assault on free agency, self-awareness, and Aristotelian, teleological models of human flourishing.

Yet what is being elided in modernity's methodical elaboration of a *critical* perspective on past frameworks and ideas is a fuller understanding of how ideas and conceptions actually develop over time—viz., as a long, if uneven dialectical progression. Indeed, it is only by tracing their evolution over time, rather than by seizing on their specific meaning at any given historical moment, that we are able to grasp the reality, significance, and truth value of ideas. What Michael Buckley has shown to characterize modern conceptions of atheism, viz., that its central terms "function

extra-phenomenological fact, but itself something phenomenologically given ... Not only do we now have an expectation of the datum, then a perception of it, then a memory as retention, then a recollection, then a repeated recollection, but these series of acts also stand as series before our consciousness in the recollecting reflection [*stehen als Reihen in der wiedererinnernden Reflexion vor unserem Bewußtsein*]" (*Basic Problems*, 68).

more like variables than like constants in intellectual history," also holds true for the conceptions of "will," "judgment," and "person" throughout this study.[6] Seneca's caveat—"If ever you want to find out what a thing really is, entrust it to time"—thus stands in stark contrast to modernity's impatience with contemplative forms of knowing and, as Hobbes so supremely exemplifies, its axiomatic view of knowledge as a type of *property* supposedly freed from the interpretive contingency said to vitiate human expression, belief, and inner certitude.[7]

A major impediment to achieving a comprehensive grasp of our historical situation (inasmuch as such attempts are undertaken from within the humanities at all anymore) has to do with the fact that in describing historical processes and interpreting specific aesthetic forms we tend to rely almost without thinking on a vocabulary of breaks, ruptures, and caesurae, and a nomenclature of "epochs." Yet to understand modernity simply by looking for discontinuities of the kind so loudly asserted by its intellectual progenitors surely amounts to a case of the "imitative fallacy" and as such begs the question of modernity on a grand scale. For it fails to consider the alternative possibility, viz., that concepts—far from being mere "tools" or heuristic devices—acquire legitimacy and meaning within the human sciences only by virtue of their complex history of transmission and their gradual elucidation of an underlying idea. To be sure, the claim here is not that there *are* no breaks or that the very idea of modernity as somehow constituting (or instituting) a break with the very idea of "tradition" is false. Rather, we must learn to disentangle the performative character of modernity's self-descriptions—which tend to *create* the intellectual discontinuities that they purport to have uncovered in the form of past "error"—from their truth value. Among the more powerful arguments to that effect, Gadamer's view of understanding as the "immersion in a process of tradition" and Blumenberg's reading of modernity as the unwitting "reoccupation" of the ancient and intractable legacy of Gnosticism stand out, and both will at various turns inform the arguments that follow. Yet the principal emphasis of this book is rather different. To read against the grain of those self-certifying, "epoch-making" accounts that modernity has periodically proffered (William of Ockham, Luther, Bacon, Descartes, Locke, Kant) is to become aware of a pervasive, if often nearly imperceptible weakening of basic concepts that had been central to humanistic inquiry since Plato and Pythagoras. Modernity's fading awareness of the deep histories circumscribing these concepts stems from a changed idea of the very act of "conceptualization" itself. Beginning with Machiavelli, Bacon, and Hobbes, the focus now is on the sheer *efficacy* of political, social, and economic reasoning. Rather than being understood as outgrowths of histories

6. *Origins*, 7.
7. *De Irā*, 3.12.4.

and traditions that ought to be reflexively engaged, concepts come to be appraised *instrumentally*. The focus now is on their methodological tidiness and their demonstrable fitness for achieving a specific quantifiable "objective" to whose pursuit we are committed beforehand.

In focusing on "will" and "person," this study seeks to tabulate the costs of modernity's principled and "progressive" forgetting of what I take to be an elemental aspect of all ideas and conceptions related to the human: viz., that they achieve meaning only within the long *durée* of historical time, and that their value and import is not secured by a singular, interventionist act of *definition* but by our steadily deepening interpretive engagement with their historical transmission and development. Yet precisely this outlook was short-circuited by a modernity whose self-image as a decisive break and "unprecedented" epoch implied a fundamentally altered notion of progressive, secular time conceived in chronometric and equivalent, rather than rhythmic and epiphanic, terms. This story, to be considered here only briefly, has been told from a variety of disciplinary viewpoints, albeit with sharply divergent emphasis. We know it as Schiller's "wound upon modern humanity" inflicted by "culture," Hegel's unhappy consciousness propelled into self-awareness by the "self-movement of the concept" (*Selbstbewegung des Begriffs*), Max Weber's "disenchantment" (*Entzauberung der Welt*), Karl Polanyi's "Great Transformation," Heidegger's "loss of the gods" (*Entgötterung*), Michael Buckley's "self-alienation of religion," Hannah Arendt's displacement of "action" by "behavior," and Michel Foucault's emergent regime of systemic disciplinary and discursive formations. Alternatively, the shift has also been conceptualized, by Hans Blumenberg, as modernity's "second overcoming of Gnosticism" and, more recently, by Anthony Giddens, Louis Dupré, Marcel Gauchet, and Charles Taylor as variously inflected narratives of secularization or the "great disembedding" paradoxically ushered in by post-Scholastic Christianity and, eventually, by Protestantism's insistence on a human-engineered, individualistic salvation.[8]

Given modernity's self-description as an "epoch" unlike any other, any engagement with its intellectual legacy must be on guard against merely reenacting its avowed discontinuities. In arguing that humanistic inquiry depends on a sustained and reflective grasp of conceptual histories, the following exploration of will and person (and their shifting affiliation with notions of judgment, responsibility, and self-awareness) suspends the distinctively modern antithesis between a sublime, apocalyptic model of historical time and a blandly chronometric model, as they have

8. Schiller, *Aesthetic Education*, 39; Hegel, *PS*, 44/*PG*, 57; Weber, "Science as a Vocation," 15; K. Polanyi, *Great Transformation*, esp. 35–70; Heidegger, *Question Concerning Technology*, 116; Buckley, *Origins*, 348; Arendt, *HC*, 41; Blumenberg, *LMA*, 126; C. Taylor, *Secular Age*, 146; see also Gunton, *The One, the Three, and the Many*, 11–21.

variously been realized in "decline-and-fall" and "rise-and-progress" narratives of modernity. Instead, the following critical readings of some pivotal texts and voices seeking either to give fuller articulation to or to dismantle basic concepts of humanistic inquiry locates the meaning of historical time in these articulations. In so doing, the present argument stages a dialectical conversation between a Christian-Platonist tradition, broadly conceived, and a naturalist-reductionist tradition that spans from the Greek atomists to David Hume and his contemporary, neuro-scientific descendants. Instead of a narrative of progress or decline homogenizing the intellectual contents wherein the ebb and flow of historical time become prima facie legible, what follows is an attempt to trace two basic ways of inhabiting historical time, one hermeneutic and the other methodical in kind. If the latter tends to draw on a utopian or dystopic conception of time (rise-and-progress/decline-and-fall), the hermeneutic model conceives of ideas and concepts as continuously evolving realizations of a *truth*—as opposed to a mere aggregate of propositions—taken to have informed the concepts in question from their very beginning. Its purest form, and one to which this study is openly committed, is to be found in Platonic *anamnēsis,* itself subject to intricate modern re-articulations in. Hegel's phenomenology, Newman's theory of development, and Hans-Georg Gadamer's philosophical hermeneutics.

Where concepts are grasped as conduits for the successive distillation of a truth, rather than as propositions contingently advanced by (putatively) autonomous selves, time is liable to be experienced as epiphanic rather than linear in nature. Its phenomenology is one of sudden disclosure, as opposed to the flat-line temporality that characterizes modernity's procedural and methodological self-portrayal as an age of "progress," one that with inexorable logic gave rise to and, in turn, was sanctioned as "necessary" by the modern discipline of sociology from Auguste Comte to Max Weber. It bears recalling here that, as conceived in the "enchanted" world of ancient myth (Egyptian, Greek, and early Roman), time was experienced and conceived as cyclical, recurrent, and inherently rhythmic—something memorably captured in the elegiac, if also unabashedly belletristic opening of Johan Huizinga's *Waning of the Middle Ages* (1919). In a less mournful idiom, Charles Taylor maps the differentiated, premodern conception of time involving ordinary, quotidian time, the "higher time" realized in sacred ritual and sacramental practice, and two models of eternity—the *nunc stans* to "which we aspire by rising out of time; and God's eternity, which doesn't abolish time, but gathers it into an instant."[9] With greater emphasis on

9. C. Taylor, *Secular Age,* 57; on modern chronological or wholly distended conception of time, see ibid., 322–351. On the transformed conception of time, see E. Cassirer, *Symbolic Forms,* 2:104–118; Gehlen, *Urmensch und Spätkultur,* 251–275; Koselleck, *Futures Past,* 93–104. Huizinga's famous opening meditation is worth recalling: "To the world when it was half a thousand years

critical method, Erich Auerbach extends Huizinga's vivid portrayal of late medieval time as sharply accented and internally differentiated, and of quotidian life punctured by moments of heightened spiritual significance. Thus he notes how the established, typological reading of "an occurrence like the sacrifice of Isaac" (viz., as prefiguring the sacrifice of Christ) conceives a relation "between two events ... linked neither temporally nor causally—a connection which it is impossible to establish by reason in the horizontal dimension." Their simultaneity is not defined temporally but, in fact, "can be established only if both occurrences are vertically linked to Divine Providence ... [Thus] the here and now is no longer a mere link in an earthly chain of events, it is *simultaneously* something which has always been, and will be fulfilled in the future." Benedict Anderson reaffirms Auerbach's sense that such a conception of time cradled by eternity, divine providence, and hence a simultaneity that has nothing to do with mere "coincidence" or "chance" is deeply alien to us today.[10]

Beginning in the seventeenth century and culminating in Hobbes, the modern "idea of a sociological organism moving calendrically through homogeneous, empty time" gradually displaces the older model. For Anthony Giddens, this shift coincides with "the separation of time from space" and the consequent emergence of a "radical historicity" that "depends upon modes of 'insert' into time and space unavailable to previous civilizations."[11] Speaking of the "complex" experience of time that prevailed for the first thousand years of Christianity, Charles Taylor notes that aside from the "secular time of ordinary 'temporal' existence, in which things happen one after another in an even rhythm, there was ... Platonic eternity," as well as the "eternity of God, where he stands contemporary with the whole flow of history." Finally, there was "a higher time of original founding events, which we can periodically reapproach at certain high moments," that is, in religious ritual. As J. G. A. Pocock has persuasively argued, it is in Hobbes that this model of time as a strictly transcendent and self-sufficient framework begins to break down; modern thought initially faces the theoretical challenge of defining the apparent coexistence of two models of time, a monotheistic concept of time intelligible only in relation to "divine actions and

younger, the outlines of all things seemed more clearly marked than to us. The contrast between suffering and joy, between adversity and happiness, appeared more striking. All experience had yet to the minds of men the directness and absoluteness of the pleasure and pain of child-life. Every event, every action, was still embodied in expressive and solemn forms, which raised them to the dignity of a ritual. For it was not merely the great facts of birth, marriage and death which, by the sacredness of the sacrament, were raised to the rank of mysteries; incidents of less importance, like a journey, a task, a visit, were equally attended by a thousand formalities: benedictions, ceremonies, formulæ" (*Waning of the Middle Ages*, 1).

10. Auerbach, *Mimesis*, 64; Anderson, *Imagined Communities*, 25.

11. On Hobbes's conception of time, see Pocock, *Politics*, 148–201; Giddens, *Consequences*, 20.

utterances" and "a rich texture of the acts, words and thoughts of personal and social beings" for which empirically "observable continuities, recurrences, and occurrences" could no longer be axiomatically thought "vertically" but, instead, had to be "recast ... in terms of process, change and discontinuity."[12]

Here, then, lie the origins of Walter Benjamin's much-quoted characterization of modern time as "homogeneous [and] empty." Benjamin faults nineteenth-century historicism for ignoring the distinction between historical and messianic time and contenting itself with a flat-line notion of history as nothing more than "a causal connection between various moments." Yet to string up "a sequence of events like the beads of a rosary" fails to recognize that what converts a mere fact into an explanatory cause cannot itself be historical; facts only become "historical posthumously." For Benjamin, historicist knowledge is de facto impossible unless supplemented by a complex, speculative, and eschatological (as opposed to a strictly chronometric) conception of time, which alone (given the right "constellation" of inquiry) may reveal how the void of our present "is shot through with chips of Messianic time."[13] In their gnomic rejection of historicism, however, Benjamin's *Theses on the Concept of History* implicitly concede modernity's dominant conception of time as chronometric, homogeneous, and inexorably forward-moving. Indeed, nineteenth-century historicism is merely the most conspicuous instance of modernity's acquiescence in the downward transposition of time from a dynamic trajectory punctuated by epiphanic intensities into a mere unit of measurement. Thus modernity's methodically constructed "world picture" (to recall Heidegger's apt phrase) as it emerges from the canonical writings of Bacon, Descartes, and Leibniz for the most part understands time as merely "lapsing" and incessantly receding into a "past" now conceived as history—a vast inventory of essentially equivalent or, rather, indifferent and disaggregated, nominal "facts" awaiting their opportunistic retrieval as evidence in some explanatory scheme shaped by present exigencies.

Inasmuch as modernity—at least prior to the crucial revaluations of Hegel, Coleridge, and Newman—understands time only ever as lapsing and expiring—and hence as incapable of realizing or fulfilling antecedent meanings—its figuration of time as inherently *historical* also carries with it a strong, if often unacknowledged implication of loss. The price paid for embracing a model of time that with uniform dullness extends forward into an endlessly hypostatized future is the psychopathology of what Anthony Ashley Cooper, Earl of Shaftesbury, calls a "distracted universe"—a continually nagging, albeit inarticulate expectancy, and an alternately passive (consumerist) or hyperactive (mindless) hunger for the "nothing new."

---

12. C. Taylor, *Secular Age*, 96; Pocock, *Politics*, 151–152.
13. Benjamin, *Illuminations*, 262–264.

Meanwhile, what *does* happen, and what this book seeks to chart in some detail, is a persistent forgetting of the past that had once saturated our conceptual frameworks and the intricate translation and reinterpretation of the ideas comprised by them across various cultural and linguistic boundaries.

Unlike mythical time, modern temporality and history not only can never *recur* but can only ever be experienced as "passing" into oblivion or as the anxious projection of an uncertain future. As Gadamer saw so clearly, modern historicism not only does not remedy this situation but, since its beginnings in the early nineteenth century has only reinforced and perpetuated it. For its "paradoxical tendency toward restoration—i.e., the tendency to reconstruct the old because it is old"—only magnified the Enlightenment's prejudice toward tradition as nonsensical. For the "historical consciousness that emerges in Romanticism," which had been the exception in the Enlightenment (viz., the idea of tradition as an obstacle to progress), now "has become the general rule." Gadamer's opposition to the strenuous iconoclasm of Enlightenment critique is rooted in the supposition that "reason exists for us only in concrete historical terms."[14] Yet even as it is being magnified by Romantic historicism, this conception of the past as lapsed and irretrievably "lost" time—to be embalmed by modern philological and archival methods devoted to a value-neutral "reconstruction" of the past—is being contested. For one thing, the procedural ethos of modern historicism comes to be challenged by the work of historical drama and fiction (e.g., Friedrich Schiller, Sir Walter Scott, Victor Hugo, C. F. Meyer, Theodor Fontane, et al.). It also emerges as a central and deeply vexing premise for an aesthetic and philosophical response to modernity as a traumatic, indeed persistently re-traumatizing, development. Newman's gnomic remark that "the present is a text, and the past its interpretation" is unwittingly echoed in Gadamer's observation that "our historical consciousness is always filled with a variety of voices in which the echo of the past is heard. Only in the multifariousness of such voices does it exist: this constitutes the nature of the tradition in which we want to share and have a part." His much-quoted remark that "understanding is to be thought of less as a subjective act than as an immersion into a process of tradition (*Einrücken in ein Überlieferungsgeschehen*) furnishes a cue for much literary and philosophical writing beginning in

14. *Truth and Method*, 275, 277. Gadamer continues: "In fact history does not belong to us; we belong to it. Long before we understand ourselves through the process of self-examination, we understand ourselves in a self-evident way in the family, society, and state in which we live. The focus of subjectivity is a distorting mirror. The self-awareness of the individual is only a flickering in the closed circuits of historical life. *That is why the prejudices of the individual, far more than his judgments, constitute the historical reality of his being*" (278); for a somewhat critical account of Gadamer's argument, see Auerochs, "Gadamer über Tradition."

early Romanticism.[15] It drives the archeological ethos of Wordsworth's "Spots of Time" no less than that of Blake's early prophetic books, agitating with iconoclastic fervor and prophetic urgency for the restoration of spiritual time to a nation (Albion) whose rabid commercialism and imperial ambition have trapped it in what Hegel was to call the "bad infinity" (*schlechte Unendlichkeit*) of undifferentiated, secular "progress." Though profoundly complicated by the very different, post-human(istic) models of temporality set forth by Darwin and Nietzsche, the Romantics' project of reconstructing a more complex, dynamic, and potentially eschatological model of time can be found to culminate in the great novelistic and philosophical projects of European modernism, such as in Thomas Mann's rich and persistent meditations on time in *The Magic Mountain* (1924), Husserl's 1905–1906 *Lectures on Inner Time Consciousness* (publ. 1928), and Proust's eponymous magnum opus on lost time (1913–1927). The latter's concept of "involuntary memory" constitutes a specifically modernist revision of the prevailing concept of time as a monochrome vector rendering equivalent and so threatening to denature all human experience. Thus the facts "recalled by voluntary memory, the memory of the intellect . . . preserve nothing of the past itself." Instead, "the past is hidden somewhere outside the realm, beyond the reach of intellect, in some material object (in the *sensation* which that material object will give us) of which we have no inkling. And it depends on chance whether or not we come upon this object before we ourselves must die."[16] However obliquely, Proustian "sensation" echoes a central tenet of Christian theology; being radically contingent, it recalls the notion of "grace," just as its unsought-for plenitude appears to mark the manifestation of messianic time or "revelation" within an otherwise undifferentiated model of time as the sheer succession of equivalent units of (secular) experience.

Still, by its very serendipity, Proust's *mémoire involontaire* reveals how under conditions of modernity time is rarely experienced as "unfolding" or "revealing" itself *in* and *as* the present. If Proust and other modernists still cling to the possibility of an aesthetic epiphany belatedly rupturing a flat-line model of time as pure *durée*, the latter model is positively embraced by the existentialist stance of Heidegger's *Being and Time* (1927). For Heidegger, modernity's prevailing notion that life

15. J. H. Newman, in Ker, *John Henry Newman*, 206; Gadamer, *Truth and Method*, 284, 291. To render the German *Überlieferung* as "tradition" risks activating inapposite connotations of stasis and primordial determinacy, especially in an Anglo-American cultural context; and yet, Gadamer himself emphasizes that *Überlieferung* involves a living, fluid, and dialectical process of "transmission" that most definitely "does not persist because of the inertia of what once existed. It needs to be affirmed, embraced, cultivated, . . . and it is active in all historical change."

16. Proust, *In Search of Lost Time*, vol. 1 (*Swann's Way*), 59–60; italics mine. On recollection, time, and the ennui of modern bourgeois psychology in Thomas Mann, see Pfau, "From Mediation to Medium."

"*consists* of a succession of experiences 'in time'" effectively forecloses on any methodological analysis of *Dasein*. Still, even the "vulgar interpretation of the 'connectedness of life' does not think of a framework spanned 'outside' of Da-sein ... but *correctly* looks for it in Da-sein itself" (*BT*, 343; italics mine). As the telling qualification ("correctly") makes clear, Heidegger's analyses of *Dasein* are meant to be carried out free of any transcendent presuppositions or expectations. Instead, his argument is firmly anchored in an existentialist stance embodied by the "God is dead" pronouncement of Nietzsche's madman, Max Weber's "disenchantment of the world" (*Entzauberung*), and Georg Simmel's 1910 essay on "The Metaphysics of Death." Thus *Dasein* "does not first fill up an objectively present path or stretch 'of life' through the phases of its momentary realities, but stretches *itself* along in such a way that its own being is constituted beforehand as this *stretching along* [*Erstreckung*]." When conceived as mere extension or Bergsonian *durée*, human secular time is by definition circumscribed by the contingent endpoints of birth and death: "Factical Da-sein exists as born, and, born, it is already dying in the sense of being-toward-death [*Sein zum Tode*]" (*BT*, 343).

Heidegger's bleak framing of human existence within a temporality utterly emptied of all dynamism and meaning also shapes our conception of history (*Geschichte*) and of historical knowledge (*Historie*): "How history can become a possible *object* for historiography can be gathered only from the kind of being of what is historical, from historicity [*Geschichtlichkeit*] and its rootedness in temporality." Decisive for our purposes is Heidegger's contention that the being of *Dasein* "is not 'temporal,' because it 'is in history,' but because, on the contrary, it exists and can exist historically only because it is temporal in the ground of its being" (*BT*, 344–345). However cogent, such an outlook is flawed in that it posits (without taking into account countervailing arguments or indeed the possibility of its own falsification) a model of time devoid of all transcendent points of reference or forms of expectancy; for Heidegger, to understand *Dasein* as sheer temporality means *eo ipso* to be committed to a strictly formal, linear model of time as incessant "vanishing" (as Hegel's *Phenomenology* and *Logic* define it), a quintessentially modern position that, as we shall see, is also elegiac to its very core.

To suppose, as Descartes does (still rather covertly) vis-à-vis Aristotle, or as Hume, Pierre Louis Maupertius, Claude Adrien Helvetius, and Baron d'Holbach, among the philosophes do much more flamboyantly, that past conceptions and ideas are fundamentally inert, calcified, and, as so many prejudices obstructing "progress" is questionable at best. The true casualty of modernity's amnesia, then, is not this or that idea or concept *taken as a proposition*—a term that in any event fails to grasp the nature of ideas. Propositions, though crucial and indispensable to rational conversation, are by their very nature subject to what John Henry Newman calls *notional* assent.[17] Yet formal assent presupposes an antecedent view or framework of commitments that prompts us in a given situation to *bestow* our assent—not only to a proposition's formal correctness but to the reality of its constituent terms and its potential truth value as such. Logically, then, this antecedent "view" or framework belongs to a categorically different realm, that of an idea in which fact and value are inextricably woven together, a domain a fortiori beyond the reach of propositional ratiocination. We here encounter the (originally Platonic) insight that key conceptions of humanistic inquiry—such as will, judgment, teleology, person, action, and a normative idea of the good—constitute the starting premise for a process of dialectical clarification and development whose inclusive and open-ended nature ensures the vitality of intellectual life qua tradition. To sharpen the point, it will help to juxtapose my account of an amnesiac modernity to the by now classical view of modernity as a story of progressive loss, depletion, and intellectual impoverishment, a narrative typically associated with a wide array of psychological ailments (melancholy, anomie, depression, ressentiment, dissociated sensibilities, etc.). In cautiously distancing itself from such "subtraction stories," Charles Taylor's recent account of secularization identifies its overarching concern to be the origination of a "disenchanted world, a secular society, and a post-cosmic universe." While no longer conceiving "fullness" as a condition of lived experience such as "point[s] us inescapably to God," modernity in Taylor's telling amounts to an "evolutionary history," an "*Entstehungsgeschichte* of exclusive humanism."[18]

Right away, a qualification is in order inasmuch as neither will nor person can be approached straightforwardly as a concept in the prevailing, modern sense as a sortal term or a predicate of generic traits. Rather, each term constitutes an idea whose historical scope, effectiveness, and reality go well beyond the pragmatic and definitional spirit of ordinary concepts. Hence, if the term "concept" is to be applied to person and will at all, its meaning lies closer to what Hegel calls a "conception" (*Begriff*), that is, a comprehensive and dialectically mutating framework affording human beings some basic orientation about their distinctive and indisputable self-awareness as *ethical* agents. Yet the Hegelian attempt to capture the idea as a sequence of distinct

17. For Newman's discussion of notional assent—which "seems like inference" and which he parses into profession, credence, and opinion, see *Grammar of Assent*, 49–65; on Newman's theory of assent and its broader objectives and quasi-phenomenological orientation, see Jay Newman, *Mental Philosophy*, 14–29; and Richardson, *Newman's Approach*, 67–92. On the *Grammar*'s inconsistent underpinnings, partially informed by a radical empiricism (in the spirit of Reid rather than Locke or Hume) ingeniously mobilized on behalf of a critique of Protestant fideism and, on the other hand, an "essentially modern" ("all too English, all too Anglican") probabilistic conception of faith, see Milbank, "What is Living and What is Dead." For an earlier critique, see Price, *Belief*, 315–348.

18. *Secular Age*, 26.

historical modes of appearance risks becoming a strictly neutral *method*, a parade of successive, embodied conceptions reviewed by a modern, impersonal, and disengaged philosophical "we" that takes itself to be in possession of history as an *inventory* of the varied appearances of consciousness but no longer takes itself to be implicated in the underlying idea *of which* these appearances are the manifestation. To be sure, Hegel's historicism is obviously nothing like the positivist enterprise of Leopold von Ranke, Heinrich von Treitschke, and Jules Michelet. Yet his mode of argument, particularly in his Berlin lectures on the philosophy of history, religion, and art, undeniably laid the groundwork for a historicism that, as Gadamer has pointed out, was erroneously premised on the idea of an observer standing aloof from the dialectical "self-movement" (*Selbstbewegung*) of his or her subject matter and thus remaining unaware of his or her hermeneutic entanglement and contingent self-understanding. The historically circumscribed and conditioned knowledge of what Hegel calls "natural consciousness" (*natürliches Bewußtsein*) is categorically distinct from the reflexive awareness of Hegel's philosophical "we," just as the local, pragmatic meanings and their significance belatedly captured by dialectical thinking no longer stand in any substantive relation. Simply put, there always remains something adventitious and incalculable about what Hegel calls "determinate negation."[19] Where the hermeneutic structure of understanding goes unrecognized, intention soon will be supplanted by the adventitious movement of the idea; action morphs into process, and the uniqueness of the human person is sublated into the impersonal authority of what Hegel calls *System*.

Inevitably, the question—answered powerfully in the affirmative by Terry Pinkard and Robert Pippin, among others—becomes whether Hegel might not be the one thinker to have charted for us how post-Cartesian modernity at last succeeded in overcoming its debilitating, undialectical (in origin nominalist) model of knowledge as sheer "sense-certainty," thereby supplanting an adversarial outlook on the past with a containment strategy that throughout the nineteenth century held nearly unimpeded sway under the name of historicism. Put as a question, is the specifically modern kind of forgetting (of intellectual traditions) traced in this book merely an instance of what Hegel calls *Aufhebung?* To begin answering that question, it helps to stay with Taylor's markedly Hegelian account. Echoing some key passages from the "Preface" to Hegel's *Phenomenology,* Taylor observes that "our sense of where we are is crucially defined in part by the story of how we got there" and that "our past is sedimented in our present."[20] And yet, unlike Hegel's organic trope according to which

---

19. On this key problem of "transition" in Hegel, see Pippin, "You Can't Get There from Here" and Pinkard on "Philosophy as Communal Self-Reflection" in *Hegel's Phenomenology*, 260–268.

20. "Just as little as a building is finished when its foundation has been laid, so little is the achieved Notion [*Begriff*] of the whole the whole itself. When we wish to see an oak with its massive

the beginning lives on to the extent that it has been thoroughly "conceptualized" (*be-griffen*) by what ensues, Taylor's geological metaphor of sedimentation underscores a point that the argument that I shall unfold below emphatically contests. The past is not a residue; it does not live on in fossilized trace amounts, nor indeed in the virtual, abrasion-free domain of the "concept." Nor indeed does the past unilaterally "define [our] sense ... of how we got there." Indeed, in Faulkner's and T. S. Eliot's poignant formulation, it is not even past. Eliot, in particular, had famously stressed how tradition "cannot be inherited" but, instead, must be "obtain[ed] by great labour. It involves, in the first place, the historical sense" and that, in turn, "involves a perception, not only of the pastness of the past, but of its presence." Moreover, while this sense "makes a writer most acutely conscious of his place in time, of his own contemporaneity," it also enables him to rearticulate the past's deeper significance in light of present dynamics; thus one "will not find it preposterous that the past should be altered by the present."[21] When allowed to operate within a hermeneutic, rather than informational, model of knowledge, inherited conceptions and ideas furnish us—less in the spirit of a "definition" than that of motivation and opportunity—with the intellectual and spiritual meanings that stand to be continuously husbanded and cultivated further if we are ever to achieve any orientation in our own present.

Unlike Gadamer, MacIntyre, and, well before them Newman—whose strongly related conceptions of tradition we will consider momentarily—Taylor seems notably vague about this crucial question: what is the relation that responsible knowledge—viz., knowledge not merely sought and appraised with regard to its causal efficacy and contingent utility but integrated into an articulated framework of human ends—bears to the past? Consider the following passage:

> It is a crucial fact of our present spiritual predicament that it is historical; that is, our understanding of ourselves and where we stand is partly defined by our sense of having come to where we are, of having overcome a previous condition. Thus we are widely aware of living in a "disenchanted" universe; and our use of this word bespeaks our sense that it was once enchanted. More, we are not only aware that it used to be so, but that it was also a struggle and an achievement to

---

trunk and spreading branches and foliage, we are not content to be shown an acorn instead. So, too, Science, the crown of a world of Spirit, is not complete in its beginnings ... Everything turns on grasping and expressing the True, not only as *Substance*, but as *Subject* ... The True is the whole. But the whole is nothing other than the essence consummating itself through its development [*das durch seine Entwicklung sich vollendende Wesen*]" (*PS*, 7, 10–11).

21. As Eliot continues, "Someone said: 'The dead writers are remote from us because we *know* so much more than they did.' Precisely, and they are that which we know." "Tradition and the Individual Talent," in *Selected Prose*, 37–45.

> get where we are; and that in some respects this achievement is fragile. We know this because each one of us as we grew up has had to take on the disciplines of disenchantment.[22]

What is left unclear is whether modernity's self-authorizing claims to "having overcome a previous condition" are at all commensurable with our "sense" (Taylor's notably vague word) of "where we are." No doubt, beginning with Luther, Galileo, and Bacon, and continuing through Bunyan's *Pilgrim's Progress,* Defoe's *Moll Flanders,* Wordsworth's *Prelude,* and the nineteenth-century *Bildungsroman,* modernity has been deeply enmeshed with the genres of auto-narration and a familiar language of struggle and emancipation that Northrop Frye has identified as generic features of the quest romance. Yet it is a mistake to conclude, as Taylor does, that the fragility of "this achievement" is merely a consequence of modernity's apparent incompletion. Instead, modernity's precariousness—so vividly attested by its epiphenomenal psychopathologies of *Angst,* paranoia, and melancholy, and by its ubiquitous rhetoric of "crisis"—stems from how that quest itself was being conceived and pursued. For by its self-legitimation as the overcoming of "a previous condition," and as the ongoing repudiation of so-called premodern frameworks, conceptions, and ideas no longer engaged dialectically but unilaterally declared irrelevant or inimical to the endeavor now at stake, modernity made disorientation its founding premise and enduring condition.

In taking up these questions, Leo Strauss's *Natural Right and History* (1953) proves to be a good point of departure. Among the aspects of modern historicism repeatedly flagged by Strauss is its emphatic "this-worldliness," its principled, positivist rejection of transcendence as an authentic form of experience. Characteristic of historicism—as of Heidegger's notion of modernity defined by a "world picture"—is the totalizing claim that "history was thought to supply the only empirical, and hence the only solid, knowledge of what is truly human." What is glossed over is the question as to what exactly it is that should allow a historicist mode of explanation to compel the "assent" of those individuals whose intellectual, spiritual, economic, and cultural coordinates it purports to draw in exhaustive and authoritative detail. The historicist conception of knowledge as *technique* and *method,* in other words, takes as a given our assent to the specific narratives thus produced. Here Strauss demurs,

22. Ibid., 28, 29. Taylor rightly notes that modernity for the past several centuries has certainly not yet entered upon a phase where "there could be unbelief without any sense of some religious view which is being negated." Thus far, at least, "unbelief . . . is understood as an achievement of rationality. It cannot have this without a continuing historical awareness. It is a condition which can't only be described in the present tense, but which also needs the perfect tense: a condition of 'having overcome' the irrationality of belief" (269).

for to take that view is to have "obscured the fact that particular or historical standards can become authoritative only on the basis of a universal principle which imposes an obligation on the individual to accept, or to bow to, the standards suggested by the tradition or the situation which has molded him."[23] Having thus evacuated any transcendent, sacred dimension from history and construing it as a strictly factitious and sequential occurrence, the authority of modern knowledge is precariously entwined with the methodological and conceptual protocol that governs the telling of its results.

As Strauss argues, we are left with an unbridled historicism forever struggling to legitimate its account of human experience and bereft of any transcendent notions (the good, the beautiful, the just, reason) *for which* one will inevitably be searching when faced with the choice of accepting this or that account of human experience. Historicism is the very embodiment of Heidegger's "business" (*Betrieb*) and sheer "talk" (*Gerede*): that is, a transactional, impersonal, and open-ended accumulation of "facts" aimed at furnishing the answer to a question that has never been properly asked. It is the quintessentially post-charismatic discourse of modernity, in the sense that it no longer conceives narrative as capable of breaking from chronological time but as merely accumulating knowledge within the matrix of empirical, equivalent, and as such a-semantic units of measurement.[24] In engaging early twentieth-century theories of agency (in particular, Stuart Hampshire's *Thought and Action*), Iris Murdoch thus characterizes the rhetorical and conceptual stance of modernity as one of overwhelming "dryness." Though not quite sharing Murdoch's unvarnished Platonism, Leo Strauss's critique of historicism reaches fundamentally similar conclusions: "the historical standards, the standards thrown up by this meaningless process, could no longer claim to be hallowed by sacred powers behind that process. The only standards that remained were of a purely subjective character, standards that had no other support than the free choice of the individual ... Historicism culminated in nihilism. The attempt to make man absolutely at home in the world ended in man's becoming absolutely homeless," quite simply because any instance of "thought that

---

23. *Natural Right and History*, 15, 17.

24. To understand how post-charismatic historicist narrative has supplanted the premodern possibility of "revelation" with a monochrome notion of "information," it helps to recall the etymology of "charisma"—viz., "a free gift or favour specially vouchsafed by God; a grace, a talent" (*OED*); while in modern English the term first surfaces in R. Montagu (1642), its etymology goes back to the Greek *kharisma* ("favor, divine gift") and the verbal form, *kharizesthai* ("to show favor to"), which in turn derives from *kharis* ("grace, beauty, kindness"). Also pertinent is the relation to the Greek *kairos,* which signifies the "right or opportune moment" and thus stands in express antithesis to the other word for time: *chronos*. *Chronos* denotes merely chronological or sequential (i.e., quantitative) time, whereas *kairos* refers to the puncturing of merely chronological time by an interlude, a moment where the sacred, unanticipated, and revelatory may occur.

recognizes the relativity of all comprehensive views has a different character from thought which is under the spell of, or which adopts, a comprehensive view. The former is absolute and neutral; the latter is relative and committed."[25] Bearing out this dichotomy to the fullest extent, Hegel's philosophy shows how human flourishing and philosophical cognition have terminally parted company. Yet the question remains whether this impersonal, detached, and supposedly value-neutral mode of cognition is a viable strategy for humanistic inquiry. The point on which Hegel seems to be hedging concerns the question of whether humanistic inquiry could ever admit of anything like the fact/value distinction that has played such a crucial role in the modern era. For even as the "theoretical analysis of life is noncommittal and fatal to commitment," life can never be of the same kind as a theory of it: "life means commitment."[26] If that much can still be agreed on, it is hard to see how historicism can ever furnish a viable framework for humanistic inquiry. Again, it ought to be stressed that the choice here is not between historicism and *a-historical* knowledge, as is often suggested by the former's stalwart defenders. Rather, the question is whether a historicist mode of inquiry can possibly do justice to our irreducibly *ethical* involvement with such elemental and indispensable conceptions as will, person, teleology, judgment, self-awareness, responsibility, introspection—ideas without which humanistic inquiry is not even conceivable.

---

25. *Natural Right and History*, 18, 25; somewhat polemically, Strauss urges the central point: "The epoch which regarded Aristotle's fundamental questions as obsolete completely lacked clarity about what the fundamental issues are" (ibid., 23).

26. Ibid., 27.

3

"A LARGE MENTAL FIELD"

Intellectual Traditions and Responsible Knowledge after Newman

Leo Strauss's critique leaves us with the impression of modern historicism as above all a *distancing technique,* driven by modernity's visceral fear of the unknown and its consequent resistance to any transcendent or otherwise heteronymous authority. Echoing and elaborating Strauss's view, Hans-Georg Gadamer was to argue that "our usual relationship to the past is not characterized by distancing and freeing ourselves from tradition [*Überlieferung*] ... We do not conceive of what tradition says as something other, something alien. It is always part of us, a model or exemplar, a kind of cognizance." The first manifestation of reason, and the basis for all subsequent acts of understanding, thus involves our intuitive awareness as being *related to,* rather than estranged from, the specific phenomena under investigation. By its very nature, human inquiry is never a purely random product of gratuitous spontaneity but, instead, belongs to the realm of *action.* It constitutes a response to a calling, that is, to phenomena soliciting our attention and engaging our intelligence. This they do because, in a strictly pre-discursive (indeed ontological) sense, we achieve self-awareness and purposive orientation in our life world only because we are already embedded in and committed to it in what Heidegger calls an attitude of "care" (*Sorge*).

For Gadamer, "the anticipation of meaning that governs our understanding of a text is not an act of subjectivity, but proceeds from the commonality that binds us to the tradition. But this commonality is constantly being formed in our relation to tradition. Tradition is not simply a permanent precondition; rather, we produce it ourselves inasmuch as we understand, participate in the evolution of tradition."[1] What (with an oblique nod to Husserl) Gadamer calls "the ontological structure of understanding" we shall find to be at the very heart of Coleridge's phenomenology of the human person, conscience, and the responsible will. Against the autistic models of human agency proffered by Descartes and Hobbes, Coleridge's focus on personhood is prima facie ethical rather than epistemological. His *Aids to Reflection* and *Opus Maximum* thus conceive of personhood as essentially relational. The person originates in, and is subsequently sustained by, her or his relation with another being—a metaphysical truth (as Coleridge and, following him, John Henry Newman were to argue) first made apparent by the phenomenology of human "conscience"—that is, by an incontrovertible awareness that the sense of relatedness and obligation to the other is sanctioned by the vertical rapport (however latent, tenuous, and/or susceptible to misconstrual and neglect) that all persons have with the divine *logos*.

The same metaphysical truth thus revealed in the person's relation with the other—apprehended as a "thou" rather than an impersonal he or she—also relates to our continuous appraisal of ambient phenomena to a supra-personal, normative *logos*. Colin Gunton thus emphasizes how tradition "involves a personal relatedness to others in both past and future time," as well as our "recognition of the uniqueness and value of that which is given ... To deny the salutary character of tradition is to say that we can only be ourselves by freeing ourselves *from* others."[2] Though Gunton himself does not make the connection, what he later elaborates under the heading of "open transcendentals" characterizes rather precisely the notion of tradition that this book means to reconstitute, specifically with reference to the concept of the will and the idea of the person. They, too, qualify as an open transcendental,

> a notion in some way basic to the human thinking process, which empowers a continuing and in principle unfinished exploration of the universal marks of being. The quest is indeed a universal one, to find concepts which do succeed in some way or other in representing or echoing the universal marks of being. But it is also to find concepts whose value will be found not primarily in their clarity and certainty, but in their suggestiveness and potentiality for being deepened and enriched, during the continuing process of thought, from a wide range of sources in human life and culture.[3]

1. *Truth and Method*, 283, 293–294.
2. Gunton, *The One, the Three, and the Many*, 95.
3. Ibid., 142–143.

Intellectual traditions, and the concepts of which they are variously composed, thus attest both to the transcendent and universalizing *telos* that impels human thought and to the necessary incompleteness and boundless variety that tradition-bound understanding will display over time. Far from a merely impersonal method or abstract procedure, all "understanding" constitutes a response to the calling of a specific phenomenon, a person or thing whose apparent rationality solicits our attention and sustained engagement. As Gadamer has argued, such a calling marks the beginning of a sustained, quasi-dialogic progression wherein acts of judgment or, rather, prejudgment undergo continual development and revision as a result of our openness to, and reflection on, emergent evidence and competing interpretations. The objection, famously advanced by Jürgen Habermas, that such a view reifies tradition as a single, monolithic, and oppressive superego of sorts altogether misconstrues Gadamer's argument. Tradition (*Überlieferungsgeschehen*) is not some metaphysical notion gratuitously constraining the ebb and flow of rational thought. Rather, in the manner of Kant's regulative principles, it posits the continuity of ideas over significant stretches of historical time, absent which rational conversation on any variety of issues could not be effectively pursued, indeed could not even be conceived as a project. Rationality is not a property either intrinsic to or (under certain conditions) ascribed to the mere temporal *punctum* of the present. Instead, reason only ever crystallizes to the extent that present objectives and exigencies are interpretively framed and reexamined as variations, and thus as more or less apparent manifestations of an idea dialectically transmitted from the past.[4]

In his seminal 1845 work on *The Development of Christian Doctrine*, Newman makes an especially eloquent and compelling case for a conception of knowledge as interpretive and evaluative, rather than impersonal and computational in kind—an argument that by a series of intermediate steps leads him to regard humanistic (and specifically religious) knowledge as inextricably entwined with our deepening grasp of historical continuity. As Newman puts it, "it is the characteristic of our minds to be ever engaged in passing judgment on the things which come before us. No sooner do we apprehend than we judge: we allow nothing to stand by itself: we compare, contrast, abstract, generalize, connect, adjust, classify: and we view all our knowledge in the associations with which these processes have invested it."[5] What

---

4. For Kant's notion of "regulative" or "heuristic" principles," see *Critique of Pure Reason*, esp. A509ff. and A616f. On the Gadamer-Habermas debate, see bibliographical entries for both authors below. On that exchange, see Mendelson, "Habermas-Gadamer Debate"; Rauch, *Hieroglyph of Tradition*, 151–178; and Scheibler, *Gadamer*, 9–70.

5. Newman, *Development*, 33 (henceforth cited parenthetically as *DCD*); for discussions of Newman's *Essay*, see Carr, *Newman & Gadamer*, 111–131; Lash, *Newman on Development*; and, arguably the most compelling account of Newman's anti-liberalism and intellectual persona, Pattison, *Great Dissent*.

distinguishes judgment from mere opinion, and so underwrites its greater probity and significance, is its continuity over time; whereas opinions "come and go," judgments are gradually recognized to be "firmly fixed in our minds, with or without good reason, and have a hold upon us, whether they relate to matters of fact, or to principles of conduct, or are views of life and the world, or are prejudices, imaginations, or convictions."

Crucially, the apparent durability of judgments is not a sign of rigidity or inertia; on the contrary, their phenomenology within the mind is altogether dynamic. Whereas opinion is merely reactive and tends to expire along with the transient impression that had solicited it, judgment is the catalyst of an ongoing dialectic transformation of both the judging subject and the *idea* or phenomenon with which it is engaged: "The idea which represents an object or supposed object is commensurate with the sum total of its possible aspects, however they may vary in the separate consciousness of individuals; and in proportion to the variety of aspects under which it presents itself to various minds is its force and depth, and the argument for its reality. Ordinarily an idea is not brought home to the intellect as objective except through this variety" (*DCD*, 34). Notwithstanding its momentous doctrinal implications, even the "idea" of Christianity's substantive identity over 1,800 years remains subject to the same hermeneutic circle that governs all human cognition: "in all matters of human life, presumption verified by instances, is our ordinary instrument of proof," though Newman hedges ever so slightly by adding that "if the antecedent probability is great, it *almost* supersedes instances" (*DCD*, 113–114; italics mine).

Yet wherever the tension between a metaphysical and a historical conception of truth and evidence threatens to unravel his argument, Newman neither eschews the antagonism nor simply commits to one position. Rather, he mines the conflict itself for further insight. Time and again he insists that history is not some antimetaphysical, nuts-and-bolts sphere, but that its material development is suffused with hermeneutic, interpretive commitments on the part of the human agents involved in it: "the event which is the development is also the interpretation of the prediction ... [and] provides a fulfillment by imposing a meaning" (*DCD*, 102). Newman cautiously navigates between a view of history as the incremental revelation of a transcendent truth and an existentialist (Gadamerian) framework that conceives hermeneutic activity as a case of Aristotelian *phronēsis*—that is, of practical reason at once generative of and continually tested and revised by its dialectical "process of transmission" (*Überlieferungsgeschehen*).[6] Any suggestion of a conflict

6. Both Lash and Carr identify "the tension between this transformative pressure history exerts upon Christian ideas and the Catholic belief in the immutability of dogma" as the *basso continuo* of Newman's entire career (Carr, *Newman & Gadamer*, 115); for Lash, the Platonism behind Newman's "idea" is that of the Alexandrian fathers, and that the *Essay*'s "'progressive' view of the his-

between tradition and development, memory and innovation, merely exposes the ignorance of either term on the part of those venturing such a claim. As Michael Buckley has forcefully argued,

> there is only an apparent contradiction between discovery and tradition. The disclosure of what is new only superficially excludes the transmission of what is old. Actually, discovery can only light upon what is hidden within the given, while a tradition can possess significance, can perdure, only if that which is past is continually made present, changed, reinterpreted, and transposed—if only to be understood by succeeding generations. Discovery is the grasp of new meaning; tradition is its mediation ... A tradition in the history of ideas, then, presents theological discovery with its own prior and repeated discoveries and verifications ... Vital traditions are the situations of the present. Tradition is the contemporary presence of the past.[7]

Inasmuch as humanistic inquiry necessarily unfolds *within* history, it can never be reduced to an aggregate of interconnected, logical propositions, let alone to a definitive knowledge *of* and utopian emancipation *from* the past. A specific idea will crystallize only by means of "fore-judgments" (*Vorurteile*) destined to undergo continuous revision. In the course of such a process, "aspects of an idea are brought into consistency and form," and what Newman's eponymous work means by "development" is a dialectical working out of sorts, "the germination and maturation of some truth or apparent truth on a large mental field" (*DCD,* 38). Such a process is never simply carried forward by individual acts of occasional introspection or private, ritualized meditation; rather, the movement of an idea is necessarily trans-generational, inter-subjective, and materially concrete. Against the Cartesian or (more pertinent to Newman) Lockean idea of a "punctual self," Newman's theory of development rests on "a phenomenological account of what *actually* happens when a person comes to know what he or she knows," and to trace the development of an idea within such a framework is to understand knowing as "the activity not of a mind in isolation but of the whole, living person."[8]

Ideas and the intellectual traditions to which they incrementally give rise are the very catalysts and source of any ethical community seeking to articulate and realize

---

tory of Christianity ... is untypical; much depends on whether Newman can demonstrate the 'comparability of terms' in the development of the idea of Christianity—something the *Essay* tackles in the first of its Notes on Development, on the 'Preservation of Type' [*DCD,* 171–178]" (*Newman on Development,* 59).

7. *Origins,* 35–36.

8. Carr, *Newman & Gadamer,* 91, 96.

supra-personal meanings and ends. At the other end of the spectrum we find social formations associated with classical (Lockean) liberalism and contemporary libertarian ideologies, that is, a contract-based, adventitious, and transient "enterprise association" (to borrow Michael Oakeshott's term) of competitive individuals pursuing their economic interests in grudging fulfillment of certain enumerated legal obligations. As the impoverished character of the public sphere and political discourse in the United States amply demonstrates, any purely interest-based social formation is prone to the hyper-pluralism whose Protestant origins Brad Gregory has recently traced in such compelling detail; and being so preoccupied with the conflict and apparent incommensurability of individual values, beliefs, and rights it necessarily fails to articulate a trans-generational and supra-personal vision for itself as a *community*. Simply put, a political community no longer capable of distinguishing between engaging an idea and holding an opinion—and hence bereft of a culture of reflection, imagination, and "negative capability" (as John Keats had called it)—is almost certainly in a phase of advanced decline. Pondering utilitarianism's rapid emergence as Britain's dominant political and economic framework, Coleridge (as we shall see in Part IV) was among the first to realize that a society defined solely by private interests and personal claim rights has effectively lost sight of reason, the faculty concerned not with contingent propositions but with ideas. For ideas are necessarily concerned with ends, not means, and unlike propositions they have themselves agency. Their force and significance stems less from their logical conclusiveness than from their charismatic presence within a social imaginary. Echoing Hegel's thesis about social process as the progressive "working out" or "realization" (*Verwirklichung*) of an underlying conception—which Gadamer would later analyze under the heading of "effective history" (*Wirkungsgeschichte*)—Newman sets out his notion of "development" in a particularly eloquent passage that warrants quoting in full:

> when some great enunciation, whether true or false, about human nature, or present good, or government, or duty, or religion, is carried forward into the public throng of men and draws attention, then it is not merely received passively in this or that form into many minds, but it becomes an active principle within them, leading them to an ever-new contemplation of itself, to an application of it in various directions, and a propagation of it on every side … At first men will not fully realise what it is that moves them, and will express and explain themselves inadequately. There will be a general agitation of thought, and an action of mind upon mind. There will be a time of confusion, when conceptions and misconceptions are in conflict, and it is uncertain whether anything is to come of the idea at all, or which view of it is to get the start of the others. New lights will be brought to bear upon the original statements of the doctrine put forward; judgments and aspects will accumulate. After a while some definite

> teaching emerges; and, as time proceeds, one view will be modified or expanded by another, and then combined with a third; till the idea to which these various aspects belong, will be to each mind separately what at first it was only to all together ... The multitude of opinions formed concerning it in these respects and many others will be collected, compared, sorted, sifted, selected, rejected, gradually attached to it, separated from it, in the minds of individuals and of the community. It will, in proportion to its native vigour and subtlety, introduce itself into the framework and details of social life, changing public opinion, and strengthening or undermining the foundations of established order. Thus in time it will have grown into an ethical code, or into a system of government, or into a theology, or into a ritual, according to its capabilities: and this body of thought, thus laboriously gained, will after all be little more than the proper representative of one idea, being in substance what that idea meant from the first, its complete image as seen in a combination of diversified aspects, with the suggestions and corrections of many minds, and the illustration of many experiences. (*DCD,* 36–38)

Among the many things that are striking and instructive in this passage is Newman's insistence on the catholicity—the breadth and universality—of an idea; there is no hint of sectarianism here, nor of the self-regarding relativism of, say, Stanley Fish's professional or "interpretive communities"—where disagreement is peremptorily taken as evidence of incommensurability, and where the Platonic, Thomist, and Hegelian models of *disputatio* and dialectics have been all but supplanted by the narcissism and "delirious nonstop monologue of ... so many in-group narratives."[9] For an idea to become effective as "an active principle" within a community, it must be apprehended as a conception of apparent significance and potential, supra-personal authority. Ideas and conceptions are not so much "thought out" or conjured up in a hermeneutic vacuum; nor, for that matter, are they something possessed in the manner of a commodity, or "held" qua "opinion." Rather, they are received on trust and, as such, stand to be "compared, sorted, sifted, selected"—in short, to be engaged dialogically, interpersonally, and in ways bound to transform both the knower and the known.

For rather obvious reasons, Newman's principal exhibit in support of his thesis concerns the "antecedent probability ... that the Christianity of the second, fourth, seventh, twelfth, sixteenth, and intermediate centuries is in its substance the very religion which Christ and His Apostles taught in the first" (*DCD,* 5). Yet his main

9. Jameson, *Postmodernism,* 368; see also Eagleton's review of Fish, *Professional Correctness* ("Death of Self-Criticism").

contention extends well beyond the religious controversies of the day and his own prolonged theological and personal struggles with conversion. It is a point that, in a different idiom, proves just as critical for modern phenomenology: viz., that if there is to be such a thing as responsible and responsive human understanding, its initial presumption has to be in favor of the substantive identity and temporal continuity of those phenomena with which the intellect takes itself to be engaged. For Newman, that basic premise constitutes "not a violent assumption ..., but rather mere abstinence from the wanton admission of a principle which would necessarily lead to the most vexatious and preposterous scepticism."[10] The first step in any legitimate intellectual progression must be one of *assent,* not to a proposition or opinion, but to the truth of the phenomenon's sheer givenness: "to be just able to doubt is no warrant for disbelieving." Yet that ultimately means to accept, indeed positively embrace and continuously sift the rich and intricate historical filiations of all our conceptions. Having long struggled with Anglo-Protestantism's self-image as a caesura akin to the revolutions that Bacon, Descartes, and Hobbes had wrought in science, philosophy, and politics, Newman insists on the continuity of Christianity and implicitly rejects the sectarian, fideist, and denominationalist character of modern religious culture: "To be deep in history is to cease to be a Protestant" (*DCD,* 5–8). We can detect here an essential analogy between how Husserl or Michael Polanyi understand human knowledge, viz., as constituted through our sustained participation *in* (rather than unilateral domination *of*) specific phenomena, and what Newman means by the development of an idea.

At the same time, Newman readily concedes that the development of an idea may in many cases come to nothing, that ideas may eventually be "rejected" either because of their false premises or their tendency to license mistaken, even absurd conclusions. Consequently, a "process will not be a development, unless the assemblage of aspects, which constitute its ultimate shape, really belongs to the idea from which they start."[11] What makes the distinction between opinion and judgment so pivotal is the antagonism—beyond remediation for Newman by the time he writes his 1845 book—between a Protestant and a Catholic conception of Christianity; in

10. Curiously, Newman himself would late in his career have to defend himself against the charge of skepticism, as leveled against him by Andrew M. Fairbairn in an article entitled "Catholicism and Modern Thought." Newman's reply in the subsequent issue of the journal emphasized that in limiting reason to strictly probabilistic conclusions he had availed himself of a colloquial understanding of reason. On this debate, see Richardson, *Newman's Approach,* 99.

11. *DCD,* 38; echoing Newman almost verbatim, Buckley views history as a "demonstration" of ideas, their varying plausibility, the soundness (or lack thereof) of their premises and entailments, and their adversarial relation with other conceptions; and he concludes that precisely "for these reasons, the history of theological ideas is not external to theology, but an essential moment within it" (*Origins,* 334–335).

Newman's strident formulation, the former rests on the "hypothesis ... that Christianity does not fall within the province of history,—that it is to each man what each man thinks it to be, and nothing else," that it is but a set of Wittgensteinian "family resemblances" or, as the *Essay* puts it, "a mere name for a cluster or family of rival religions all together [or] ... at variance one with another" (*DCD*, 4). Consistent with Newman's lifelong opposition to any conjunction of belief with "private judgment" (*DCD*, 6) and "opinion," his *Essay* above all attempts to work out a dialectic that fuses together judgment, knowledge, and history. The development of an idea, and the immanent law of intellectual traditions and the responsible knowledge to be achieved in them, is not attained by contingent claims or definitions. Rather, such knowledge must begin by immersing oneself in the history of usages, adaptations, distortions, and abuses that cumulatively circumscribe the valence and significance of a specific conception.

Yet in a rather more controversial move that has occasionally been likened to Darwin's theory of natural selection, Newman also insists that the development of an idea (specifically that of Christian doctrine) is not to be misconstrued as some inexorable deduction of consequences from a single premise.[12] Instead, there is something markedly adventitious to Newman's theory of development, similar to the vacillating and contingent ways in which, beginning with Plato and Pythagoras, the concepts of will and person gradually took shape under often highly adversarial conditions. More than Hegel, Newman seems prepared to recognize the vicarious and uneven nature of historical processes, including the discursive and meta-discursive operations wherein the manifold entailments of an idea, and indeed its "antecedent probability" itself, are incrementally realized:

> its action being in the busy scene of human life, [an idea] cannot progress at all without cutting across, and thereby destroying or modifying and incorporating with itself existing modes of thinking and operating. The development then of an idea is not like an investigation worked out on paper, in which each successive advance is a pure evolution from a foregoing, but it is carried on through and by means of communities of men and their leaders and guides; and it employs their minds as its instruments, and depends upon them, while it uses them. (*DCD*, 38)

---

12. Regarding affinities between Newman's theory of development and Darwin's theory of evolution, see Pattison, *Great Dissent*, 194–196; Ker, *Newman*, 300; on Newman's theory as such, see Lash, *Newman on Development*, 46–79; "Literature and Theory," and Chadwick, *From Bossuet to Newman*, 96–119. For background on the writing of Newman's *Development of Christian Doctrine*, see Ker, *Newman*, 257–315; on Newman's theory of development in relation to Gadamer's hermeneutics, see Carr, *Newman & Gadamer*, esp. 89–150.

Against the hardening line of the Catholic magisterium, particularly during the later years of Pius IX's papacy (as embodied in his 1864 *Syllabus*), the "ecclesiastical vagabond" Newman took considerable intellectual risks by opposing both any metaphysical claims regarding the inerrancy of dogma and all contemporary, liberal doctrines of historical "progress."[13] His view of knowledge as sustained "investigation" (as opposed to gratuitous, and skeptical "inquiry") eschews a doctrinaire and over-reaching foundationalism while at the same time avoiding the anti-foundationalism that fuels contemporary historicism, and which in time would meet its foreordained end in the radical perspectivalism of the later Nietzsche. The idea of tradition, then, allows Newman to tether human knowledge to empirical particulars without therefore losing all capacity for understanding how it is implicated in a notion of transcendent truth. To frame knowledge as an immersion in a process of transmission is to embrace "a rationality appropriate to created knowers in a world with which they are continuous."[14]

To be sure, Newman's idea of an "investigation" still amounts to an "advancement" of insight into the specific subject at hand, as well as a deepening sense of responsibility for the knowledge thus achieved. Yet the dialectical movement known as "tradition" that binds knower and known in a reciprocal dialogue in which what we receive is not prized as a possession (*dominium*) but as something to be cultivated and gifted (*donum*) to those who come after us, can no longer be construed as "progress" in the ordinary Enlightenment sense. Nicholas Lash's claim that Newman "believed in religious progress as little as he believed in secular progress" rings fundamentally true, though Robert Pattison's observation that "Newman was the creature of the liberalism he despised" seems no less to the point.[15] To resolve this tension, one

---

13. Pattison, *Great Dissent,* 53; for Pattison, it is not in spite but because of his "comprehensive failure" that Newman was able to emerge as "a lone voice standing outside the first principles of the whole age ... whose counterpart in intellectual history is not Carlyle or Arnold so much as Nietzsche."

14. Gunton, *The One, the Three, and the Many,* 135; Gunton quotes M. Polanyi, who characterizes his great work (*Personal Knowledge*) as an attempt "to achieve a frame of mind in which I may hold firmly to what I believe to be true, even though I know that it might conceivably be false."

15. *From Bossuet to Newman,* 98; Pattison, *Great Dissent,* 5; see also Lash, *Newman on Development,* 61–62; in both instances, Lash concurs with Chadwick. Prior to F. L. Cross's 1933 essay ("Newman and the Development of Doctrine") vituperative accounts of Newman as a closet ultramontane ideologue masquerading as a German historicist (or vice versa) tended to be widespread; notably, Newman's adversaries seemed divided on whether his position amounted to "German infidelity communicated in the music and perfume of St. Peters" (as James Mozley had put it), or whether "the ultra-liberal theory of Christianity" was to conceal his "join[ing] the Church of Rome" (quoted in Carr, *Newman & Gadamer,* 112). For his part, Lash views Newman's theory of development as genuinely "influenced by the continental school of history"—albeit less by Hegel and Comte than by the Catholic Johann Sebastian Drey—leader of the Tübingen theological seminary, whose 1819 *Introduction to the Study of Theology* and a newly founded journal (*Theologische Quartalschrift*)

must first desynonymize progress from development. Doing so, at least briefly, is called for inasmuch as a non-teleological yet dynamic conception of intellectual traditions is to provide a framework for the inquiry into the concepts of will and person that is to follow. That is, concepts are hermeneutic frames that evolve and are transformed by the "effective history" of their application and by the contested and shifting interpretations put on them. In Newman's words, "power of development is a proof of life, not only in its essay, but especially in its success"; and, in a remark likely to have disquieted many of those in Rome later tasked with reviewing the merits of his proposed elevation to cardinal, Newman baldly states that "the idea never was that throve and lasted, yet, like mathematical truth, incorporated nothing from external sources" (*DCD*, 186). Not only, then, do the criteria Newman adduces as prima facie evidence of an idea's intrinsic truth ("Preservation of Type," "Continuity of Principles," "Power of Assimilation," "Logical Sequence," etc.) attest to his profoundly dynamic view of how conceptions and ideas evolve over time; they also affirm the priority of practical reason over theoretical argumentation and description. Like Hegel, Newman thus approaches history as a succession of material changes and intellectual developments whose underlying focal point—the idea of freedom for Hegel; the "antecedent probability" of Christian truth for Newman—and operative logic remain perforce elusive to the individual and communal agents instrumental to its advancement. Inasmuch as an idea "employs their minds as its instruments, and depends upon them, while it uses them" (*DCD*, 38), the implicit rationality of conceptions taken up "on faith" and worked through by successive generations will gradually divulge itself, provided the idea in question had sufficient weight and significance not to expire in its struggle with competing notions or succumb to inner contradictions or corruptions. As Newman puts it, "logic is brought in to arrange and inculcate what no science was employed in gaining ... [For] intellectual processes are carried on silently and spontaneously in the mind of a party or school, of necessity come to light at a later date and are recognized, and their issues are scientifically arranged" (*DCD*, 190).

Precisely because Newman's theory is so pointedly dialectic, indeed agonistic, it is impossible to distill the idea in question from any one of the discrete stages through which it successively passes. Thus it would be incorrect to say that an idea *undergoes* development. Rather, a process of development gradually fleshes out, fills in, and so "realizes" the meaning and significance of a specific, and at first cryptic idea or motif. Indeed, it is the sheer persistence of a conception throughout its numerous adversarial encounters with competing ideas that incrementally corroborates its truth value—

of the same year "were marked by the spirit of Roman Catholic liberalism" and aimed at "the restatement of Catholicism with the aid of the new historical and critical and philosophical instruments" (*Newman on Development*, 108).

though even after 1,800 years the latter is not to be taken as a metaphysical truth but merely as a state of heightened probability:[16]

> An idea not only modifies, but is modified, or at least influenced, by the state of things in which it is carried out, and is dependent in various ways on the circumstances which surround it. Its development proceeds quickly or slowly, as it may be; the order of succession in its separate stages is variable; it shows differently in a small sphere of action and in an extended; it may be interrupted, retarded, mutilated, distorted, by external violence; it may be enfeebled by the effort of ridding itself of domestic foes; it may be impeded and swayed or even absorbed by counter energetic ideas; it may be coloured by the received tone of thought into which it comes, or depraved by the intrusion of foreign principles, or at length shattered by the development of some original fault within it. But whatever be the risk of corruption from intercourse with the world around, such a risk must be encountered if a great idea is duly to be understood, and much more if it is to be fully exhibited. It is elicited and expanded by trial, and battles into perfection and supremacy. (*DCD*, 39–40)

Truth and "survival" appear nearly convertible terms in this passage, whose vivid depiction of an idea's abrasive and transformative "intercourse with the world" not only highlights Newman's empiricist sympathies and his very English "preoccupation with the concrete" but also explains why some readers have read the *Essay* as anticipating Darwin's theory of natural selection.[17]

---

16. Arguably, Newman's conception of probability vacillates between a premodern and a modern one, an ambiguity liable to complicate his theory of development. For Milbank, "Newman seems un-alert to the radical distinction between a premodern sense of the probable as involving a kind of ineffable intuition which approximates to an unreachable truth, and a modern sense of the probable as concerning a *calculable* approximation to certainty" ("What is Living and What is Dead," 51); alternatively, one might read Newman here equivocating on an intractable theoretical issue, something he frequently does.

17. Lash remarks that Newman's preoccupation with the problem of continuity, though "more easily handled in a 'linear' perspective") nonetheless has him approach the issue "episodically." Like Darwin's "punctuated equilibrium," Newman's vision of development involves a fundamentally irregular series of uneventful periods interrupted by spikes of intense and momentous controversy and change; see Lash, *Newman on Development*, 57–60. It ought to be pointed out that "natural selection," not "evolution," is where Newman's and Darwin's account of development converge; in fact, Étienne Gilson notes that "the word ['evolution'] is to be met with nowhere, either in the first edition (1859) nor in any of the subsequent editions until the sixth" of Darwin's *Origin of the Species*, and that the term's eventual attribution to Darwin is largely the result of misidentifying Herbert Spencer's views with those of Darwin, a fusion that was made as it were "official" in the famous ninth edition of the *Encyclopedia Britannica* in the entry on "Evolution" (*From Aristotle to Darwin*, 58, 86).

However that may be, the logic of development set out here furnishes an apt matrix for our hermeneutic engagement with intellectual traditions and the possibility of an expanded "self-understanding" (Gadamer's *Selbstverstehen*) opened up by that encounter.[18] Whereas a number of impressive accounts of modernity (e.g., those of Charles Taylor, Hans Blumenberg, John Milbank, Louis Dupré, Hannah Arendt, Michael Gillespie) have unfolded as high-altitude surveys of intellectual shifts and diverse, often competing strands of inquiry, the following argument seeks to capture the intrinsic idea of will and person through a series of forensic readings of representative arguments. For any account of competing or intersecting intellectual traditions has to rest on the kind of close, textual analysis that, at its best, has always been the bread and butter of literary studies. To render intellectual history vivid and engaging, and so become alert to the profound stakes of its contested ideas and genealogies of inquiry, one must pay scrupulous attention to the rhetorical maneuvers, metaphoric shifts, ellipses, competing translations, and countless stylistic quirks and symptoms of its preeminent voices. Like Newman, whose pellucid style rarely fails to make us feel the heat and stakes of a specific argument, the following readings (though undoubtedly falling woefully short of his rhetorical gifts) proceed from a view of intellectual culture, and of the life of the mind, as a profoundly dialectical, indeed agonistic process; and like Newman, arguably one of the great controversialists of his age, I believe that controversy, even polemic, can at times help restore clarity, especially where (as in our contemporary, self-consciously "professionalized" academic landscape) substantive and informed argument often appears on the verge of being supplanted by what Freud called the narcissism of minor differences.

Still, notwithstanding his gifts as a polemicist and writer of so many tracts, sermons, essays, and lectures seemingly tailored to transient occasions, Newman never lost sight of his underlying purpose, viz., to demonstrate how the flourishing of the human individuals pivots on their commitment in thought and action to the reality of a transcendent idea. What in the *Essay* secures the integrity of development as a complex, agonistic, and trans-generational progression is a single and continuously discernible motif. Already in the tenth of his *Oxford University Sermons,* Newman had emphasized how all inquiry hinges on anticipations of meaning—Gadamer's "pre-understanding" (*Vorverständnis*)—since in the absence of such *praejudicata*

18. In marked contrast with classical or quadratic notions of form-as-architecture, Newman's model of development is essentially one of transformation, metamorphosis, or the type of postclassical, "open" variational form that we encounter in Beethoven's late quartets. Lash thus characterizes Newman's intellectual style as a type of "fugal writing," that is, as deploying "literary (or 'real') as distinct from 'theoretical' (or 'notional') patterns of argument." For a fuller discussion of the retrieval of Aristotelian *entelecheia* in Romantic notions of organic form, and of the persistence of such models in contemporary aesthetics and biology, see Pfau, "All is Leaf."

opinioni it would be logically impossible to correlate the evidence that is to either confirm or disprove them at all. Far from originating in some incidental and passive apprehension of brute facts, all understanding begins with a "view," a commitment to a hypothesis or idea whose hold on the intellect is as palpable as it is destined to undergo continual revaluation and revision. Newman's "antecedent probability" thus amounts to a hermeneutic projection "that gives meaning to those arguments from facts which are commonly called the Evidences of Revelation"; and he adds that if "mere probability proves nothing, mere facts persuade no one; that probability is to fact as the soul to the body; that mere presumptions may have no force, but that mere facts have no warmth."[19]

Newman's conception of "development" thus cannot be dismissed as a *petitio principi,* that is, as an illusory and gratuitous imposition of "consistency and form" on a supposedly random concatenation of interpretations, usages, and adaptations.[20] In a rather daring inversion of a common trope, Newman thus portrays a living tradition as an open-ended process of clarification, a successive deepening and consolidating of an initially cryptic meaning first introduced in his *Oxford University Sermons* under the heading of "implicit reason." Filled with intellectual struggle, conflict, and trial, the passage of historical time thus exposes—in ways that no syllogistic method ever can—the deeper semantic strata of an idea, and in so doing attests to the "antecedent probability" of that idea's truth by acknowledging its undiminished capacity for engaging and transforming individuals and communities in continued and focused hermeneutic activity:

> It is indeed sometimes said that the stream is clearest near the spring. Whatever use may fairly be made of this image, it does not apply to the history of a philosophy or belief, which on the contrary is more equable, and purer, and stronger, when its bed has become deep, and broad, and full. It necessarily rises out of an existing state of things, and for a time savours of the soil. Its vital element

19. *Fifteen Sermons,* 200.

20. That, of course, was to be Nietzsche's famously anti-teleological polemic in the *Genealogy of Morals:* "there is a world of difference between the reason for something coming into existence in the first place and the ultimate use to which it is put, its actual application and integration into a system of goals ... The entire history of a 'thing,' an organ, a custom may take the form of an extended chain of signs, of ever-new interpretations and manipulations, whose causes do not themselves necessarily stand in relation to one another, but merely follow and replace one another arbitrarily and according to circumstance. The 'development' of a thing, a custom, an organ does not in the least resemble a *progressus* towards a goal ... Rather, this development assumes the form of the succession of the more or less far-reaching, more or less independent processes of overpowering which affect it—including also in each case the resistance marshaled against these processes, the changes of form attempted with a view to defence and reaction, and the results of these successful counteractions. The form is fluid, but the 'meaning' even more so" (*Genealogy of Morals,* 57–58 [Pt. II, §12]).

> needs disengaging from what is foreign and temporary, and is employed in efforts after freedom which become more vigorous and hopeful as its years increase. Its beginnings are no measure of its capabilities, nor of its scope. At first no one knows what it is, or what it is worth. It remains perhaps for a time quiescent; it tries, as it were, its limbs, and proves the ground under it, and feels its way. From time to time it makes essays which fail, and are in consequence abandoned. It seems in suspense which way to go; it wavers, and at length strikes out in one definite direction. In time it enters upon strange territory; points of controversy alter their bearing; parties rise and fall around it; dangers and hopes appear in new relations; and old principles reappear under new forms. It changes with them in order to remain the same. In a higher world it is otherwise, but here below to live is to change, and to be perfect is to have changed often. (*DCD*, 39–40)

Echoes of Newman's idea of "development," some casual and others far more systematic, abound. One may recall John Ruskin, fellow Oxonian, who in his preface to the final volume of *Modern Painters* remarks on his own evolving aesthetic conceptions: "all true opinions are living, and show their life by being capable of nourishment; therefore of change. But their change is that of a tree—not of a cloud." The same ratio of malleability and continuity informs Gadamer's characterization of understanding as vicariously inserting the subject into "a process of transmission" (*Überlieferungsgeschehen*); likewise, Michael Buckley argues for a movement of intellectual history "towards control, not in the sense of technical use, but in the sense that wonder or puzzlement advance toward an adequate grasp of a state of affairs, as the internal coherence of its material elements and their formal relationships is determined."[21]

The present argument is above all an attempt at retrieving the deep history of two concepts whose centrality to a nuanced understanding of human agency had gradually crystallized in the complex philosophical and theological landscape of the Hellenistic era and its amalgamation with earlier, Aristotelian thought by high Scholasticism. By the time we reach Hobbes and Locke, yet also in the work of those eighteenth-century moralists (Anthony Ashley Cooper, Earl of Shaftesbury, Francis

21. Ruskin, *Modern Painters,* 5:xi. Buckley, *Origins,* 14; striking intellectual affinities between Buckley's project and that of Gadamer (both of which bear strong affinities to Newman's theory of development) emerge when, late in his book, Buckley describes his project as "tracing the historical logic of [theological] concepts." He goes on to insist that "attention to the historical experience of theological ideas, the consciousness of their intellectual roots, growth, and full flower, constitutes an indispensable prerequisite for their assessment ... Tracing out the entailments of ideas and charting the influence of the forms in which they are proposed exhibits their fullness of meaning and the capacities conferred upon them by the modes in which they are specified" (ibid., 334–335).

Hutcheson, Adam Smith) opposed to naturalist and reductionist critiques of human agency, and uneasy with the overweening claims of a (science-based) epistemology as the sole legitimate intellectual framework for its description, it becomes apparent that the concepts themselves have grown opaque to their detractors and defenders alike. What Cora Diamond has identified as a "conceptual amnesia" of sorts is not simply a contingent predicament haunting eighteenth-century moral philosophy but in effect furnished the conditions under which that curiously de-contextualized and ultimately incoherent enterprise came to take shape. Much of what follows can thus be described as a comprehensive attempt to answer Diamond's key questions: viz., "what kind of good a concept is, [and] what kind of loss it is to lose concepts?" Following the ground-breaking arguments by Gertrude Elizabeth Anscombe and Alasdair MacIntyre, Diamond suggests that, significant differences notwithstanding, both agree "that certain concepts require for their content or intelligibility background conditions which are no longer fulfilled." Yet she also factors in Stanley Cavell's objection that such a loss, were it to have taken place as definitively as MacIntyre's *After Virtue* suggests, could logically no longer be experienced *as* "loss" at all. Rather, we "should have lost the very notion of morality itself."

Diamond then raises the key question: if indeed we inhabit "a world in which the concept of morality is missing" and in which we appear bereft of "the capacity to recognize that it is missing, is this, as MacIntyre would have it, a true portrayal of our world, or is it, as Cavell suggests, a reflection of [our] blindness to what we still have?" Precisely here we find that a philosophy solely committed to a formal-syllogistic model of knowing—to what Husserl had called the truth of "correctness" in contradistinction to the truth of "disclosure"—will neither be in a position to register the *fact* of conceptual amnesia nor be able to articulate its *significance*. For "how you see the good of having particular concepts or kinds of concepts ... depends on at least two things: first, your *view* of the relation between experience (taking that in a very broad sense) and thought ... Second, how you see the good of having these or those concepts or kinds of word[s] depends on the significance you attach to thinking well about certain things." Diamond's insight that "the way you consider the values of modes of thought itself depends on your *view* of what thought is" seems fundamentally correct, even though she does not expressly connect it with the hermeneutic tradition to which it rather obviously pertains.[22]

This may be the point to address a question that may well have arisen in light of what has been said thus far: viz., whether the argument to be unfolded in this book is driven by an underlying sense of nostalgia. To address that question—likely to be raised about any account critical of the modern project—one should probably begin

---

22. "Losing Your Concepts," 255, 256, 260, 269–270 (italics mine).

by clarifying what nostalgia is ordinarily taken to mean, and what its conceptual premises are. The longing for a past plenitude, as indeed the supposition that it had once existed, rests on two closely related assumptions: first, that historical time is linear rather than cyclical, monochrome in its forward motion rather than recursive and imbued with various kinds of "higher time" or spikes of semantic intensity. For it is this premise that sanctions the axiom of "loss" without which there could not be any nostalgic affect. Second, nostalgia implies that our relationship to the past is one of disaffection, even terminal estrangement, a premise borne out by the self-certifying affect of "longing" at the heart of nostalgia. Yet precisely these premises also show nostalgia to be a distinctively modern phenomenon inasmuch as it acquiesces in the modern (historicist) view of time as a monochrome vector pointing toward the future, which renders the past as strictly passé, that is, as sheer *inventory* to be, perhaps, objectively known but most definitely incapable of signifying *for* (let alone transforming) us. Yet as I argued in the previous chapter ("Forgetting by Remembering"), this underlying conception of history as a repository of expired meanings and outmoded practices, so strenuously opposed by the young Nietzsche's *On the Use and Abuse of History* (1874), I take to be categorically inapposite, indeed positively inimical to genuine hermeneutic engagement of any kind.

In fact, in our engagement of intellectual traditions we should not let ourselves be forced into the false choice between a nostalgic and an agnostically "objective" stance. A more productive approach to humanistic inquiry and its discrete intellectual traditions, and the road followed in this book, is dialectical and agonistic in nature. It holds that the intellectual traditions of philosophical theology at the center of my argument only took shape in a struggle with radically materialist and reductionist accounts, such as in Plato's ongoing disputes with the atomists and Sophists, Augustine's various controversies with Manichean, Pelagian, and Donatian views, or Aquinas's hard-won synthesis of Platonist, Aristotelian, and Augustinian positions. The fact that humanistic inquiry and the broadly speaking Platonic tradition from which it springs had only ever constituted itself in a prolonged and richly inflected struggle with the competing projects of naturalism and reductionism means that the retrieval attempted in the following pages is focused on the internal logic and underlying stakes of a prolonged *debate*, rather than of some self-contained, homogeneous, and monolithic tradition. It is for this reason, too, that the present argument accords a strong presence to voices strenuously opposed to the Platonic-Christian-humanist line of reasoning (e.g., Hobbes, Locke, Mandeville, Hume, and various other representatives of what Charles Taylor dubs the "school of natural indifference").

At the same time, however, what follows is not (I hope) driven by a mere slogan such as "teach the conflict" by means of which literary studies of the early 1990s had sought to shore up its professional credit against growing evidence of that field's conceptual incoherence. Like virtually every one of the writers engaged in this book (including those with whose premises and conclusions I find myself in sharp disagreement), I believe that reasoned inquiry not only does not preclude an inner commitment but, in fact, positively demands it. Hence, in tracing the conflict between a humanist and an emergent, hyper-naturalist and reductionist account of will, person, and closely associated concepts, I do not interpret their agonistic encounters in the course of Western intellectual history as prima facie evidence of some underlying "symmetry" or "equivalence." In fact, in the case of Hobbes, I not only take his account of human agency to be extremely restrictive and limiting (which to me seems rather obvious) but also as vitiated by an internal, performative contradiction and therefore *untrue*. Yet to be committed to a particular view of things as having far greater truth value (or "antecedent probability") is not *eo ipso* to indulge in a nostalgic outlook on the past any more than it betokens sheer subjective opinion or some milieu-specific prejudice. For one thing, in embracing a particular intellectual tradition, which in any event is a complex and shifting phenomenon, we do not thereby adopt some triumphalist view of intellectual history. In fact, it may well turn out that the substance of the Platonic-Christian-humanist model of will, person, action, judgment, and responsibility has been irretrievably misconstrued or by now lost outright. The resulting stance, then, would be not one of nostalgia but of lucid and articulate mourning, a perspective on history variously cultivated by writers of the Baroque (Andreas Gryphius, Pedro Calderon), Romanticism (Novalis, Joseph Freiherr von Eichendorff) all the way forward to Walter Benjamin and Theodor Adorno in their alternately tragic, lyric, and critical idioms.

The bigger point at issue is that (pace Hegel) not all dialectical tension will issue in a productive outcome. Not every narrative can be deemed inherently "progressive" merely on account of its underlying dialectical organization. Moreover, the (often hidden) costs and the presumptive yield implied in *Aufhebung* cannot be authoritatively balanced from a perspective that is itself generated and circumscribed by Hegel's dialectical narrative. We simply do not have at our disposal an independent point of view from which objectively to judge whether the recurrent confrontation between naturalist and Platonic legacies has truly advanced our thinking or, perhaps, left it impoverished—possibly to such an extent that (as Diamond argues) we are no longer consciously experiencing the loss in question and, hence, incapable of articulating its implications. Here, again, what accounts for the dilemma is the fact that merely being locked in a prolonged *agon* does not per se render the two frameworks equivalent or symmetrical. The dialectical movement or debate concerning the nature of the human that is being re-engaged in the following pages cannot be construed a priori as axiomatically progressive, generative, and beneficial except insofar as each of the competing views—crudely put, the Platonic-Christian and the naturalist/reductionist one—gains in sharpness, internal consistency, and force over time. Newman's epigrammatic conclusion that, in the development of ideas, "to be perfect is to have changed often" (*DCD,* 40) only asserts that a qualitative improvement in our ability to apprehend the truth of the idea and, hence, in the degree of its "antecedent probability" has taken place. Newman does not, however, mean to reason us into a view of progress as an impersonal, systemic, and necessary "occurrence." Rather than arising by means of syllogistic demonstration, the validity and realization of an *idea* is bound up with the Aristotelian category of "action" (*praxis*). It pivots on the degree of commitment and clarity with which we assent to its meaning and subsequently inhabit this "idea" *in practice.*

Hegelian dialectics fails to acknowledge the degree to which the truth of an idea can never be simply the product of a *systemic* process but, for its ultimate "realization" (*Verwirklichung*), will always depend on the "real assent" of the individual person. Hence it is that a strictly impersonal and procedural idea of reason will always struggle to defend itself against the heckling "it's too soon to tell" with which Zhou Enlai in 1971 had famously responded to Henry Kissinger's polite conversational opener regarding the "meaning of the French Revolution." What the Chinese premier meant to suggest is not so much that we must patiently wait for the meaning of the past to disclose itself but, rather, that it is continually being determined by our active (revolutionary) engagement with its legacy. Hence, too, the logical possibility always remains that what Hegel envisions as so many instances of generative, "determinate negation" might be punctuated by cases in which distinct systems of thought will prove positively incommensurable. They might lack even the most elemental shared premises: for example, that will, judgment, choice, action, and responsibility are not merely *notional* or discursive but incontrovertibly *real.* Augustine's contention that self-knowledge is not some epistemological hypothesis but an ontological *datum* may already take as a given what modern, hyper-naturalist accounts (e.g., Hobbes, Hume) just as emphatically contest.[23] If so, dialectics furnishes us not so much with a linear progress narrative as with a complex history of misconstruals and misunderstandings.

Like Augustine or, much later, Newman and Gadamer, I do indeed believe that the proper point of departure for hermeneutic inquiry is not some instance of objective certainty or "first principle" to be syllogistically proven and conveyed in propositional form. The motional gesture of "understanding," as Gadamer frequently stresses, is not one of coercion but play, not an inexorable sequence of logical steps but a series of recognitions (or, as the case may be, misprisions) such as characterize all genuine "conversation" (*Gespräch*). Humanistic inquiry thus begins with a

23. "Let the mind then not go looking for a look at itself as if it were absent, but rather take pains to tell itself apart as present. Let it not try to learn itself as if it did not know itself, but rather to discern itself from what it knows to be other" (*ADT,* 10.12).

moment of certitude, a "view" (Newman's term again) to the elucidation of which we take ourselves to be committed, it being understood that the quest for clarifying that view is one of unceasing dialogue and learning and, as such, destined to transform and, hopefully, deepen the view that first prompted it. In the case of this study, its underlying "view" might bear reformulating thus: philosophies that peremptorily exclude all questions of value, commitment, and final causes—for example, modern science-derived epistemologies, reductionist accounts of mind, or the logical minimalism of much analytic philosophy—are by and large incapable of correlating thought and existence, life and action, for they only attend to the propositional structure of our locutions insofar as these seem to lead (with seemingly efficient and inexorable causality) to some kind of "outcome." And yet, from Plato to Augustine, Aquinas, Shaftesbury, Coleridge, and Newman, it is not the seamless conjunction of our locutions and actions but our insight into their persistent asymmetry and frequent collision that has yielded the most significant and capacious descriptions of human agency: "The words in which one thinks about one's life and actions do not themselves go to make the moral character of what one does. The philosophy of mind leads to a separation between what a person is *like,* where that is tied to his style of thought, and his capacities as a moral agent." The dominance of analytic models of philosophy in the twentieth century thus has at least partially deprived us of grasping the primacy of practical reason and the inescapability of judgment and value as integral components of human cognition: "Disagreement about the significance of the loss of the earlier notion depends on seeing the possibility of the loss, but the possibility itself may be invisible to us."[24]

For humanistic inquiry to grasp and articulate that dissonance, it must recognize both the primacy of practical reason over theoretical inquiry and the myriad ways in which its key concepts are saturated with an often conflicting array of historical usages and valuations. Like Iris Murdoch and Charles Taylor, I believe that what has occurred is a de facto loss of a differentiated and historically informed moral conception, and that the result has been a growing inarticulacy within humanistic inquiry and indeed the public sphere broadly speaking, which has rendered modern theories of agency and practical reason increasingly marginal and often incoherent. At the same time, a long historical perspective such as this book seeks to develop shows challenges to practical reason, a responsible will, and the uniqueness and incommunicability of the human person to have always been with us. From Protagoras through Leucippus, Democritus, Epicurus, Lucretius, next resurfacing in the extreme voluntarism and irrationalism of the late nominalists (Gabriel Biel, Nicholas of Autrecourt), and taken to their logical conclusion in the mechanistic and deter-

24. Diamond, "Losing Your Concepts," 271–272.

minist theories of mind spawned by Hobbes, Locke, Hume, Hartley, Priestley, and Schopenhauer—there is ample evidence of a competing, naturalist, and reductionist outlook. Contesting the human as a unique phenomenon distinguished from all other forms of life by practical rationality and an indelible awareness of moral obligation toward other persons, these thinkers not only prove radically at odds with Platonic, Augustinian, and Thomist thought as regards its premises and conclusions, but their very methods and culture of argument prove incommensurable with humanistic frameworks of any kind. In the early twentieth century, building on Gottlob Frege's pioneering critique of introspection, the anti-humanist strain diversifies considerably, such as in ordinary language philosophy, philosophy of mind (Ludwig Wittgenstein, Stuart Hampshire, A. J. Ayer, Gilbert Ryle, et al.), modern behaviorism and, most recently, contemporary neuro-scientific accounts of "mind."

Indeed, it seems clear that, very much along the lines suggested by Newman and Gadamer, the conception of human agency evolved dialectically, taking shape and acquiring depth precisely through its ongoing engagement with competing accounts. For that reason, Murdoch's subtly hopeful characterization of conceptual amnesia seems fundamentally right: "the conceptual losses we have indeed suffered have not [at least not yet] actually changed us into human beings limited to the interests and experiences and moral possibilities we can express in our depleted vocabulary."[25] If anything, then, it is my hope that the present argument will help clarify why a deep historical awareness of key concepts is a *desideratum*, and why its sweeping displacement by (or naïve assimilation to) neuro-scientific and informational methodologies is bound to render humanistic inquiry stunted, inarticulate, and ultimately obsolete.

Finally, let me offer a brief word on the selection of writers whose arguments will be sifted in what follows. While many of the authors and texts explored here are canonical, it was not their widely acknowledged prominence but the particular force and exemplary (or symptomatic) nature of their arguments that made them compelling choices. This is particularly true as regards the evolving conception of the will, from its initial, oblique emergence in Aristotle, the Stoics, and Plotinus all the way through to its sudden displacement or elision in eighteenth-century moral philosophy, which—reacting to Hobbes's momentous, not to say disastrous premises and deductions—eclipses the cognitive and ethical dimensions of *voluntas* by

25. Ibid., 263; Diamond subsequently claims to disagree with Murdoch insofar as for the latter "the failure of contemporary moral philosophy ... rests on inadequacies in philosophy of mind," whereas for Diamond it is an "underlying inadequacy in a philosophical view of language that ties description to classification." Yet that disagreement seems somewhat contrived, since Murdoch herself also registers strong criticisms regarding Frege's view of language, in particular, its tendency to "make invisible the character of the difference between the concepts *member of the species Homo sapiens* and *human being*" (ibid., 266).

abruptly recasting questions of will in terms of seemingly non-cognitive and mechanical "passions." Arguably, there is less discretion as regards selecting representative figures when exploring the concept of person. For as soon as the legal and dramatic connotations of *prosōpon* and *persona* have been taken up into early Christian theology (by the Cappadocian fathers, Augustine, and Boethius), a highly self-conscious, indeed canonical tradition of voices is established that subsequent writers such as Richard of St. Victor or Aquinas recognize as the inevitable point of departure and guiding framework for further reflections on personhood and agency.

The great exception in all of this is Coleridge. Long suspicious of modernity's indifference to the complexity, inner dynamism, and historical depth of basic humanistic conceptions, Coleridge around 1808 embarks on a quest of singular, indeed impossible ambition. Significantly inspired by Ralph Cudworth, Henry More, and other Cambridge Platonists, he attempts to oppose the hegemony of modern naturalist and reductionist methods of argument (in philosophy, theology, social theory, and political theory) by rebuilding a comprehensive Platonic-Christian archive of the history of key humanistic conceptions: viz., will, person, action, responsibility, obligation, conscience, and judgment. Through patient, albeit forcefully urged close readings, Coleridge means to impress on his readers that—far from having emancipated itself from premodern conceptions of human agency and responsibility—modernity had substantially failed to comprehend what it dismissed as "premodern" notions. For Coleridge, it is this principled and unilateral refusal to *engage* intellectual traditions that accounts for the conceptual flatness and, ultimately, ethical lapse of modern thought since Hobbes and Locke. Needless to say, Coleridge's extraordinary intellectual ambitions were profoundly at cross-purposes with his mercurial and irresolute personality. As a result, the often enigmatic presentation of his critique, to say nothing of his oeuvre's vexing incompleteness, limited its impact and largely prevented most of his alternately puzzled or bemused Victorian readers from apprehending the depth, scope, and intellectual rigor of his late philosophical theology.

Yet his unparalleled range of reading in premodern writers was not the only deterrent; for what perplexed his intellectual heirs was Coleridge's distinctive critique of modern instrumental reason's insidious and corrosive effect on what he regarded as an unconditional truth: viz., that all human agency originates in and remains circumscribed by an ethic of interpersonal relations and obligations. Unlike the more familiar voices of political reaction or religious nostalgia (Joseph de Maistre, François-René de Chateaubriand, Adam Müller, the late Friedrich Schelling, et al.), Coleridge's ethical and religious philosophy overleaps his entire century, with the partial exception of Newman. Thus his ontology of human conscience and the primacy of the I-Thou relation in constituting personhood not only harkens back to early Christian theology but also bears (up to a point) striking affinities to the anti-systematic strand of twentieth-century Jewish philosophical theology as we encounter it in Martin Buber and Emmanuel Levinas. In the end, Coleridge's great insight—cryptically anticipated in the so-called "Conversation Poems" of his early years—was that modernity's strictly epistemological approach to questions of human agency and responsibility was doomed to fail simply because it had no theory of a "Thou" but only ever juxtaposed the *ego* or *cogito* to some impersonal he, she, or it.

Inevitably, any attempt at mapping complex and extensive intellectual genealogies, no matter how ambitious, will remain an incomplete and often unsatisfactory undertaking. Other figures and seminal texts—Luther's "The Bondage of the Will"; Kant's moral philosophy; or Hegel's parsing of volition, intention, purpose, etc. in Part II ("Morality") of his *Philosophy of Right,* for example—would certainly have warranted inclusion. Specifically Kant's conception of will and moral agency may strike readers as a culpable omission; yet the complexity of Kant's arguments (briefly alluded to at the opening of Part III, below), his uniquely ambivalent role within modern political liberalism, as well the vast number of competing interpretations that have accrued around his practical philosophy just in the last couple of decades would altogether have exceeded the limits of an argument that is committed to close textual analysis as its principal method. To be sure, other accounts of modernity have settled for different strategies of presentation, being either more allusive in their treatment of large patterns (Charles Taylor, Hans Blumenberg, Louis Dupré) or, alternatively, settling for a largely non-argumentative, quasi-encyclopedic presentation of its major players (Jerome Schneewind on the *Invention of Autonomy*). Yet such approaches, too, have their risks, seeing as they present us either with a strong argument but potentially insufficient evidence to clinch it, or with an abundance of proof for an account whose relevance to our own historical moment risks never quite coming into focus. Whatever its shortcomings (and they may be many), what follows seeks to argue a single, perhaps polemical thesis (rather than offering a detached scholarly survey): viz., that absent a sustained, comprehensive, and evolving critical engagement with the history of its key concepts of human agency (will, person, judgment, teleology), humanistic inquiry will not only find itself increasingly marginalized in the modern university, but will eventually discover itself to have been the principal agent of its own undoing.

Part II

RATIONAL APPETITE

AN EMERGENT CONCEPTUAL TRADITION

We have suffered a general loss of concepts, the loss of a moral and political vocabulary. We no longer use a spread-out substantial picture of the manifold virtues of man and society. We no longer see man against a background of values, of realities, which transcend him. We picture man as a brave naked will surrounded by an easily comprehended empirical world. For the hard idea of truth we have substituted the facile idea of sincerity. What we have never had, of course, is a satisfactory Liberal theory of personality, a theory of man as free and separate and related to a rich and complicated world from which, as a moral being, he has much to learn ... We have never solved the problems about human personality posed by the Enlightenment.

—Iris Murdoch

4

BEGINNINGS

Desire, Judgment, and Action in Aristotle and the Stoics

If there is a single aspect of modernity that sets it apart from classical and Scholastic thought, it is the supposition that the spheres of human knowledge and human action, theoretical and practical rationality, are fundamentally distinct and possibly altogether unrelated. Such a partitioning of the order of fact from that of value and of cognition from willing, which eventually finds its consummate expression in Hume's *Treatise,* is also remarkable because it strips the emotions—that is, those states wherein the will is said to manifest itself—of any cognitive dimension. Beginning with Hobbes and continuing in the work of his empiricist and pessimist heirs, the sources of action are considered purely appetitive, emotive, and (so it is premised) of fundamentally irrational, somatic provenance. How, then, are we to assess modernity's disjunctive view of will and intellect without finding ourselves constrained by its intellectual legacies—for example, voluntarism, empiricism, radical skepticism, associationism, scientific determinism, behaviorism? Quite possibly the only available safeguard here is to reconstruct the genesis of the modern will by tracing various conceptual shifts, transpositions, and translations as these occur both within a single philosophical tradition and, more typically, between different social and intellectual cultures. Central to the project of *critically retrieving,* rather than merely inventorying, an intellectual tradition is thus charting its genesis before it understood itself *as* a tradition.

In this regard, a first question has to be how it came to pass that the will would eventually come to be appraised as the inscrutable and non-cognitive causality that Hobbes bequeathed his successors. Related to that question is the further peculiarity that seventeenth-century rationalism (whose origins we shall find to reach back into the early fourteenth century) locates the source of will and action in an equally non-cognitive and discontinuous emotion, Hobbes's "last appetite," Locke's "uneasiness," Hume's "passion"—which is to say, in some mental state allegedly incapable of self-awareness and thus impervious to philosophical conceptualization. Any alternative conception of human agency and personhood—viz., as endowed with the potential for self-awareness and with the ontological fact of its ethical responsibility—will thus have to examine how a highly differentiated vocabulary covering desire, emotion, and those *qualia* whereby such states attain phenomenological distinctness *for* consciousness prepared the ground for the formation of the early Christian conception of the will as "free choice." If, as has often been argued, pre-Christian thought did not have a concept of the will, the reason for that state of affairs has to be sought in the ancient Greeks' starkly different model of the emotions and their subtle interplay with human cognition. Even so, Aristotle in particular expended much thought and energy on articulating a mental faculty concerned with deliberate choosing (*prohairesis*) that significantly anticipates the modern idea of the will.[1] Still, it was only by attempting to translate and appropriate a uniquely differentiated psychological vocabulary that Roman and early Christian thought was able to articulate a coherent and enduring conception of human agency. So as to understand how the notion of the will, understood as a form of rational commitment, arose out of the confluence of several Aristotelian and Stoic concepts, an archeology of the relationship between emotion, desire, and cognition in Aristotle is in order.

It is in the *Rhetoric* that Aristotle explores the status of the emotions (*ta pathē*) most directly, primarily because he understands rhetoric to be principally concerned with "deliberate choice" or "judgment" (*prohairesis*) and, concurrently, because in targeting the emotions rhetoric shows them to have a direct bearing on judgment. If one leaves aside the question of factual proof, Aristotle notes, "there are three things which inspire confidence in the orator's own character ... good sense [*phronēsis*], excellence [*aretē*], and goodwill [*eunoia*]" (*Rhetoric*, 1378ᵃ8–9). Specifically the last of these, "goodwill," prompts Aristotle to ponder further the relation of the passions to judgment as evinced by both the orator and the audience seeking to appraise his character: "Emotions are all those feelings that so change men as to affect their judgments, and that are also *attended* [or "followed" = ἕπεται] by pain [λύπη] or pleasure

1. See Arendt, who argues for *prohairesis* as "the precursor of the Will" (*Life of the Mind*, 2:62).

[ἡδονή]. Such are anger, pity, fear, and the like, with their opposites."[2] Of strategic importance for our discussion here is the relation of the emotions to judgment and, specifically, whether pleasure and pain are to be regarded as distinct from the emotions. For by arguing that pain and pleasure "accompany" the emotions themselves, Aristotle implicitly tells us that these affective charges are fundamentally distinct from the emotions (*pathē*) themselves. Understood as strictly non-cognitive *qualia*, pleasure and pain instead reveal how a particular emotion is phenomenologically registered *in* the mind, even as it retains its distinctive character and its underlying, specific propositional content.[3]

A first and momentous question thus arises: do emotions *condition* judgment in a determinative sense, or do they merely "color" or influence it in some secondary way? Do they qualitatively alter a particular judgment, or are they themselves properly constitutive of it? Does, say, the emotion of envy determine the propositional content of our judgment, or does it merely (albeit significantly) color a judgment that already has independent standing? In analyzing the contrast between "i) change of judgment as a consequence of emotion, [and] ii) change of judgment as a constituent of emotion," Stephen Leighton turns to a passage in the *Nicomachean Ethics*, where Aristotle remarks how "anger [θυμός] seems to listen to reason [τοῦ λόγου] to some extent, but to mishear it, as do hasty servants who run out before they have heard the whole of what one says, and then muddle the order ... so anger by reason of the warmth and hastiness of its nature, though it hears, does not hear an order, and springs to take revenge."[4] It appears, then, that emotions (here that of anger) compromise judgment by not allowing it to run its course but, instead, drawing precipitous and partial conclusions as to the propositional content presented to the judgment itself. While such distortion of judgment is naturally undesirable and prone to create confusion, Aristotle's wording here suggests that the emotions are not, strictly speaking, incommensurable with judgment; rather, they account for the latter's imperfection.[5] Were Aristotle to take the opposing view—viz., of emotion as constituting and

2. *Rhetoric*, 1378ᵃ20–23 (emphasis mine).

3. There is no time to pursue the question as to whether Aristotle means to distinguish between different *kinds* of pleasure or pain; Leighton certainly does think so, and Aristotle's *Rhetoric* (1175ᵃ22–28) would appear to prove his point. For our purposes, what matters is that only the "emotions" (*pathē*) but not the *qualia* of pleasure or pain (which announce the presence of the emotions *in* the mind) are propositional and cognitive in nature.

4. Leighton, "Aristotle and the Emotions," 154; Leighton contends that emotions in Aristotle are "that on account of which judgments change, [but are] not ... themselves changes of judgments" (ibid., 148). *Nicomachean Ethics*, 1149ᵃ24–31 in Aristotle, *Complete Works*, hereafter cited parenthetically as *NE*.

5. See also Aristotle's discussion of how emotions may distort perception in *De Somnis* (460ᵇ1–16).

determining judgment outright—he would hazard a reductionist account of human cognition and, thus, jeopardize the (originally Platonic) conception of human knowledge as a form of *participation* in divine reason (*logos*).

At first blush, such a pessimistic view appears to inform Homer's conception of a person's "will" (*thūmos*) as sheer non-cognitive energy.[6] In his discussion of Book 1 of the *Iliad,* Alasdair MacIntyre thus notes how "someone's *thūmos* is what carries him forward: it is his self as a kind of energy." Setting side by side George Chapman's, Alexander Pope's, and Robert Fitzgerald's translation of *Iliad,* 1.189–192, MacIntyre notes how each version reflects "some contemporary well-articulated account of the determinants of action." Yet in Homer, he insists, there is no sense of a contest between passion and reason. There can be no antinomy between the somatic and the mental because "all psychology in Homer is physiology." Even so, Homer's *thūmos* is not identical with but, rather, is fueled by contingent passion. Taken as such, *thūmos* points to an enduring disposition or substratum that, however palpable its physiological scaffolding, anticipates ever so faintly Augustine's notion of the will in that it is nourished by (but not identical with) the passions. It is a type of internal causation, albeit one overwhelmingly unintelligible to the person in the grip of it. Yet what is unintelligible is, strictly speaking, only the *object* (lust for Briseis? jealousy of Agamemnon?) toward which the subject experiencing the emotion is oriented. As such, the emotion must be reflected since a "feeling of agitation all by itself will not reveal to me whether what I am feeling is fear or grief or pity. Only an inspection of the thoughts discriminates."[7] That this should be so, MacIntyre argues, shows that the relation of agent to action is always already in place by the time a particular passion intervenes. Action and, consequently, a sense of self as capable agent does not arise from autonomous reflection or discursive practice; even less is action contingent on some fluctuating emotional state. Rather, "the agent already has envisaged the action that he or she is to perform; what he or she reasons to is *either* a reminder that he must curb his *thūmos* if he is to perform it or else must suffer baneful consequences." Thus all conclusions about "what to do next" can be drawn "only because [agents] already know independently of their reasoning what act it is that they are required to perform."[8]

6. To render *thūmos* as "will" is, of course, question-begging and certainly anachronistic. Yet it is a reasonable heuristic device given that the objective for now is to show how Greek thought about human agency relies on several basic concepts—*thūmos, boulēsis, prohairesis, hekousion*—all of which came to play a significant role in the eventual conception of the will as developed by Augustine.

7. MacIntyre, *Whose Justice?* 29.

8. Ibid., 15–19.

If we fast-forward to Aristotle, we find that for him emotions are quasi-cognitive frameworks that color the way in which concrete or "incidental objects of sense" (*συμβεβηκὸς αἰσθητόν*, or objects *per accidens*) are apprehended, known, and judged. Yet as such they already rely on non-contingent perceptions or objects per se—say, a "white thing" subsequently identified as the special case of "the son of Diares."[9] Thus an object of emotion such as fear of violence is not gratuitously superimposed on a contingent perception, such as tanks rumbling into a public square, but instead conditions, albeit in a notably precipitate way, how that particular sensation is apprehended. Emotions are ambient grids of evaluation or, to use a more modern idiom, hermeneutic frames. They constitute a kind of basic interpretive "disposition" or "mood" (a type of *Stimmung* or *Befindlichkeit* as Heidegger was to call it) and as such are inseparable from, though never identical with, judgments. As Martha Nussbaum remarks, "emotions are not *about* their objects merely in the sense of being pointed at them and then let go, the way an arrow is released toward its target. Their aboutness is more internal, and embodies a way of seeing," as well as "beliefs—often very complex—about the object."[10] As such, emotions reveal that perceptions are rarely, if ever, judged with complete neutrality but, instead, are focal points toward which the mind is oriented in an intrinsically engaged, evaluative, and interested manner. Certainly, any philosophical school of antiquity would have been rather mystified by the post-Cartesian conception of the world as a neutral inventory of medium-sized dry goods awaiting impersonal perception, analysis, and use. On the contrary, the world has to be understood as a dynamic and profoundly interconnected grid of phenomena toward which we relate in prima facie evaluative form, viz., as focal points of interpretive curiosity and, potentially, as sources or means for our continued flourishing.

Before proceeding, some clarification is in order regarding the idea of "flourishing," which some strands of modern moral philosophy have perhaps employed with the same excessive ease with which other (non-cognitivist or naturalist) thinkers have anathemized it. One may begin by recalling Gertrude Elizabeth Anscombe's observation "that getting one another to do things without the application of physical force is a necessity for human life, and that far beyond what could be secured by … other means." For it reminds us of the strong link between altruism and an "Aristotelian necessity"—viz., something "that is necessary because and insofar as a good hangs on it."[11] Put differently, "flourishing" does not denote a type of mystical good

9. Aristotle, *De Anima,* 418^{a}10–22.

10. *Upheavals,* 27–28. For a discussion of how the Romantic era rethinks "mood" and "emotion" in an effort to overcome their strictly non-cognitive status in the materialist and reductionist epistemologies of the mid-eighteenth century, see my *Romantic Moods,* esp. 27–74.

11. Anscombe, "On Promising," 18.

or some New Age version of shiny, happy existence gratuitously superinduced on (human) animals otherwise red in tooth and claw. Rather, it signifies the presence of conditions, and our practical quest for their attainment, such as will "determine what it is for members of a particular species to be as they should be, and to do that which they should do."[12] In other words, flourishing hinges on an understanding of organic beings—plants and animals no less than humans—as entelechies, beings whose form *implies* and *embodies* a rational *development* rather than some seemingly invariant carbon-churning process. Aristotle's conception of living things as entelechies—while not exhaustive as a descriptor of human beings—thus has to be presupposed if there is to be any meaningful understanding of organic life. Philippa Foot thus argues for the substantive continuity between natural goodness and moral goodness, that is, for the possibility "that the concept of a good human life plays the same part in determining goodness of human characteristics and operations that the concept of flourishing plays in the determination of goodness in plants and animals."[13] Only a teleological being whose form constitutes an intrinsically purposive (rational) trajectory of *becoming* could ever be credited with the capacity for seeking out optimal conditions for its existence and recognizing them when they are present; to which it needs to be added, as Foot (quoting Aquinas, *ST,* Ia IIae Q 1 A 2) well recognizes, that "in doing something for an end animals cannot apprehend it *as an end*."[14]

By contrast, it is only in the modern era, where the organic and physiological processes come to be assessed as effects wrought by the mechanical interaction of discrete parts, and where the idea of life as substantial form has effectively been lost, that reason is no longer derived from but unilaterally imposed on embodied existence. As Edmund Husserl was to point out, philosophy thus finds itself stranded between its "naïve faith in reason and the skepticism which negates or repudiates it in empiricist fashion." Thus, even as it insists "on the validity of the factually experienced world [*die tatsächlich erlebte Welt*]," modernity's axiomatically disjunctive and skeptical stance finds in that world "nothing of reason or its ideas. Reason itself and its [object], and 'that which is,' become more and more enigmatic." Consequently, any outright rejection of, or skeptical challenge to, the idea of flourishing takes for granted a quintessentially modern, skeptical point of view that is only ever prepared to ascribe or impose, yet never *find,* reason in a world said to be composed of strictly

12. Foot, *Natural Goodness,* 15; "the way an individual *should* be is determined by what is needed for development, self-maintenance, and reproduction: in most species involving defence and some rearing of the young" (ibid., 33); the point is also made by Sokolowski (*Phenomenology of the Human Person,* 186–188).

13. *Natural Goodness,* 44.

14. Ibid., 54.

equivalent and quantifiable objects bereft of all agency. And yet, reversing course to an astonishing degree late in life, Husserl himself points out how utterly modernity's axiomatic skepticism begs the crucial question: "Are reason and that-which-is to be disaggregated, given that reason, as knowing, determines what is?"[15]

The classical conception of reason as embodied vision and purposive action—anathemized in modernity until its conditional recovery by Hegel—takes us to the next concept, that of "desire," which modern, post-Cartesian thought quite unhelpfully conflates with emotion. "Desire" (*orexis*) is a word Aristotle may well have invented so as "to indicate the common feature shared by all cases of goal-directed animal movement."[16] Ordinarily, three types of desire are distinguished in Aristotle: "appetite" (*epithūmia*); "spiritedness" or "non-rational desire" (*thūmos*); and a rational desire for the good (*boulēsis*).[17] Of fundamental importance here is to understand that for Aristotle all desire (rational and non-rational alike) is fundamentally distinct from emotion. For defining of desire is its intentionality, its relation to an object or objective to be attained. At the same time, desire lacks the epiphenomenal *qualia* of pleasure and pain that Aristotle regards as integral to emotion—viz., as indices allowing us to understand that we *are* experiencing a specific emotion when we do. However irrational its aim, all desire is accompanied by a basic consciousness of its presence. Even in the case of a purely appetitive desire (*epithūmion*) such as hunger, being hungry fundamentally equates with being conscious *of* one's hunger. Consequently, desire "does not satisfy the pleasure/pain test as emotion does, but, at most, as perception or thought does."[18] This crucial insight, which in the wake of early fourteenth-century voluntarism begins to slip away, Coleridge would much later recover as the "essential inherence of an intelligential Principle (*φῶς νοερόν*) in the Will (*ἀρχὴ θελητική*) or rather the Will itself thus considered, [which] the Greeks expressed by an appropriate word (*βουλή*)" (*AR*, 260). The contrast with the emotions is particularly marked in the case of a "rational desire for the good" (*boulēsis*)—a term in Aristotle's oeuvre that has an especially significant bearing on

15. *Crisis,* 13, 11. The translation of the last passage has been modified: "*Ist Vernunft und Seiendes zu trennen, wo erkennende Vernunft bestimmt, was Seiendes ist?*" (*Krisis,* 11).

16. Nussbaum, *Fragility,* 273. According to Nussbaum, there is only one mention of *orexis* to be found prior to Aristotle, viz., in some Democritean fragments whose philological status and reading, moreover, is in dispute.

17. For passages where Aristotle distinguishes between rational desire (*boulēsis*) and the non-rational desires of *thūmos* and *epithūmia,* respectively, see *De Anima,* 414^{b}1–5, 433^{a}23ff.; *Eudemian Ethics,* 1223^{a}27; and *Magna Moralia,* 1187^{a}36–37. On *boulēsis,* see *Rhetoric,* 1369^{a}1–7: "All actions that *are* due to a man himself and caused by himself are due either to habit or to desire; and of the latter, some are due to rational desire, the others irrational. Rational desire is wishing, and wishing is a desire for good [*βούλησις ἀγαθοῦ ὄρεξις*]." See Frede and Striker, eds., *Rationality,* 8; Kahn, "Discovering the Will," esp. 239ff.; and Leighton, "Aristotle and the Emotions," 160f.

18. Leighton, "Aristotle and the Emotions," 160.

the Augustinian notion of the will.[19] For here the object of desire has to be cognitively apprehended *as* good for it to be desired: "it is a desire for the good which one does not have unless one takes something to be good."[20] It is a characteristic of the human being to develop a conscious perspective on his or her hunger, sexual longing, etc. *Epithūmia,* in other words, is not intelligible as a type of desire exercising compulsive dominion over consciousness but, on the contrary, is reflectively apprehended as a representation.

It might be said that, in contrast to the "emotions" (*pathē*), desire is at once closer to and farther removed from the *qualia* of pleasure and pain. It is farther removed in that, say, in a case of extreme hunger, desire is inseparable from a consciousness of the particular "need" or "lack" (*δεήσεις*) of those goods that promise to remedy the situation. For in being thus aware, consciousness is not so much in the grip of desire (*orexis*) as it is already reflexively preoccupied with its remediation. At the same time, desire is also closer to the *qualia* of pleasure or pain in that—again, taking a case of extreme hunger—the "desire" in question is not so much "accompanied by" but positively indistinguishable from the *qualia* of pain or "uneasiness." Hunger is not "accompanied by" pain; it is itself painful. Crucially, then, the appetites are cordoned off from the emotions and vice versa, a point on which modern voluntarist thought from Hobbes onward takes a different and, it must be said, notably confused view. With good reason, then, Nussbaum observes that close study of "*orexis* reveals its intentionality and selectivity; we can also say that the practices of education and exhortation in which we engage would be unintelligible if *orexis* were, as Plato (and [Terence] Irwin) say, purely mindless."[21] It is furthermore apparent that for Aristotle the emotions *do* have an intrinsic relation to reason precisely because they belong to the realm of propositions. By contrast, an appetite (rational or otherwise) cannot be a proposition, though its presence certainly can be articulated in speech or be dealt with in rational, remedial action. Yet in that case, the formal coherence and purpo-

---

19. Kenny, *Aristotle's Theory of the Will,* frequently translates *boulēsis* as "will."

20. Frede and Striker, eds., *Rationality,* 8, here paraphrasing Aristotle's *Rhetoric:* "Nobody wishes for anything unless he supposes [*οἰηθῇ*] it to be good" (*Rhetoric,* 1369^{a}4). The distinction between contingent desires and an underlying *telos*—"the pursuit of the good" (*τὸ ἀγαθὸν ἄρα διώκοντες*)—had first been worked out in Plato, *Gorgias,* 467–468. Following his long discussion of diseases of the body, Plato in the *Timaeus* sharply distinguishes volition (*boulē*) from the physiological realm. Its corruption, always a possibility, is the result of intellectual misapprehension of the good, "for no one is voluntarily wicked" (*Timaeus,* 86^{d}).

21. *Fragility,* 286; admittedly, there are slippages in Aristotle's writings on this topic. As Leighton ("Aristotle and the Emotions," 166ff.) points out, Aristotle at times will include the non-rational appetites (*epithumia*) among the emotions; see, for example, *Eudemian Ethics,* 1220^{b}12; and *Nicomachean Ethics,* 1105^{b}21.

sive character of a statement identifying oneself as hungry or asking for food confirms the latently propositional nature of the desire that had led up to it.

Aristotle's distinction between "voluntary" action (*hekousion*) and deliberate action involving conscious choice (*prohairesis*) is of particular interest here because it responds to a logical dilemma broached in both the *Eudemian Ethics* and *Magna Moralia*. It concerns the apparent asymmetry between wishing and voluntary action, both of which seem at first glance instances of what William James calls "ideo-motor action"—that is, "a movement [that] *unhesitatingly and immediately* follows upon the idea of it ... brought about by the pure flux of thought."[22] In true dialectical fashion, Aristotle unfolds a paradox that, as it turns out, will require the introduction of a new mental faculty for its resolution. In the *Eudemian Ethics,* Aristotle takes up the distinction between "what is the voluntary and the involuntary [*ἑκούσιον καὶ τί τὸ ἀκούσιον*]" so as to get at how "goodness and badness are defined." As it happens, a paradox looms here in that wishing (*βούλομαι*) and voluntary action (*ἑκούσιον*) appear to be convertible terms: "what a man does voluntarily he wishes, and what he wishes to do he does voluntarily." And yet, there is the phenomenon of a man "acting incontinently" (*ἀκρατευόμενος*),

> through appetite contrary to what [he] thinks best; whence it results that the same man acts at the same time both voluntarily and involuntarily; but this is impossible ... But if it is impossible for a man voluntarily and involuntarily to do the same thing at the same time in regard to the same part of the act, then what is done from wish [*boulēsis*] is done more voluntarily than that which is done from appetite or anger; and a proof of this is that we do many things voluntarily without anger or desire.[23]

The "impossible" scenario here concerns the split of the person into playing simultaneously an active and a passive part, a clear violation of the founding principle of all thought (e.g., the law of contradiction). It is a paradox with a rich history eventually restated by Jean-Paul Sartre, who notes that in what is ordinarily called "self-deception" the self would simultaneously have to know of the deception (as an agent) and not know of it (as a patient).[24] This logical impasse prompts us to desynonymize

---

22. "Psychology: Briefer Course," in *Writings, 1878–1899*, 394–395.

23. *Eudemian Ethics*, 1223^{a}22, 1223^{b}5–10 and 25–30. The programmatic relevance of this passage has long been recognized; see Arendt, *Life of the Mind*, 61f.; and Kenny, *Aristotle's Theory of the Will*, 13–26.

24. Sartre's position is elaborated in Bk. 1, Ch. 2 of *Being and Nothingness*. For a fuller account of the history of this paradox, see the entry on "Self-Deception" in the *Stanford Encyclopedia of Philosophy*, at www.plato.stanford.edu/entries/self-deception/ (accessed 29 March 2011).

"willing" (as Anthony Kenny renders *boulēsis*) from the merely voluntary. Some five hundred years later, Plotinus registers the same contradiction when pointing out that a strictly mechanical account of the soul (viz., as enslaved by its own nature) leads to a logical contradiction: for "to speak of being enslaved to one's own nature is making two things, one which is enslaved and one to which it is enslaved. But how is a simple nature and single active actuality not free, when it does not have one part potential and one actual?"[25] Recognizing "that voluntariness cannot be defined in terms of will," Aristotle confirms that a different type of mental faculty has to be recognized, one capable of deliberation and choice without being passively cued by contingent desire.[26] Thus he introduces the concept of *prohairesis,* which as we shall see proved of great significance for late antiquity's development of the concept of the will. Just how to translate that term has long puzzled classical scholars, who thus have proposed terms as disparate as choice, purposive choice, judgment, decision, ethical intent, or commitment.[27]

More about that momentarily; for now, what matters is Aristotle's insistence on distinguishing between merely voluntary and positively deliberate action, a distinction that hinges on the degree to which the emotions are involved in action. In the case of strictly voluntary action, they are likely to be a significant factor inasmuch as such action is more on the order of an impulse (*hormē*) and thus prone to lead the individual to act in a "rushed" and non-deliberative fashion. By contrast, Aristotle leaves no doubt that genuinely deliberate action such as terminates in a "choice" is aimed at identifying the means toward a goal that is already in existence:

> The origin of action—its efficient, not its final cause—is choice [*προαίρεσις*], and that of choice is desire and reasoning with a view to an end. This is why choice cannot exist either without thought and intellect or without a moral state; for good action and its opposite cannot exist without a combination of intellect and character. Intellect itself, however, moves nothing, but only the intellect which aims at an end and is practical; for this rules the productive intellect as well, since every one who makes makes for an end, and that which is made is not an end in the unqualified sense (but only relative to something, i.e. of something)—only that which is *done* is that; for good action is an end, and desire aims at this. Hence choice is either desiderative thought or intellectual desire [*ὀρεκτικὸς νοῦς ἡ προαίρεσις ἢ ὄρεξις διανοητική*]. (*Nicomachean Ethics,* 1139^{a}32–^{b}5)

25. *Ennead,* 6.8.4.

26. Kenny, *Aristotle's Theory of the Will,* 22.

27. See Chamberlain ("Meaning of *Prohairesis*"), whose suggested rendering of *prohairesis* as "commitment" has much to recommend it.

In a few sentences, this programmatic passage establishes some of the most salient points that were to define for centuries to come the conception of judgment (*prudentia*) and choice (*electio*) in relation to the intellect, as well as that of intellect to desire. As Hannah Arendt notes, "the starting-point of Aristotle's reflection on the subject [of willing] is the anti-Platonic insight that reason by itself does not move anything." Indeed, Aristotle's repeated characterization of *prohairesis* as "the deliberative desire of things in our power" (*ἡ προαίρεσις ἂν εἴη βουλευτικὴ ὄρεξις τῶν ἐφ' ἡμῖν*] (*Nicomachean Ethics,* 1113^{a}10; see also *Eudemian Ethics,* 1226^{b}17) confirms that for him "desire retains a priority in originating movement, which comes about through a playing together of reason and desire."[28]

Central here is the contention that the goal or "end" (*telos*) constitutes the ambient and indispensable framework for all rational deliberation since in its absence we would not know *wherefore* we are deliberating. Our commitment to it, which is to say, our fundamental awareness of what constitutes the nature of our flourishing or the good is never some merely notional product such as can be generated by discursive or material action. It is important, then, not to overstate the distinction between means and ends, and to read this or similar programmatic passages (e.g., *Nicomachean Ethics,* 1111^{b}1–6; 1113^{b}4–5) as some straightforward division of mental labor to the effect that "*boulēsis* sets the end and *prohairesis* determines the means to this end." In fact, it is by virtue of *prohairesis* that "our objects of deliberation become also the objects of our desires, and so, finally, the substance of our actions."[29] With reference to *Nicomachean Ethics,* 1113^{b}4–5, Danielle Allen notes that the phrase "our end is a thing desired, while the means to that end [*βουλευτῶν δὲ καὶ προάιρετῶν τῶν πρὸς τὸ τέλος*] are the subjects of deliberation and *prohairesis*" should not be interpreted as narrowly instrumental; for "the phrase [Aristotle] uses to describe the means we seek is grammatically complicated: *tōn pro telos*. It is a prepositional phrase made into a substantive by means of an article. We choose 'things' that are for the sake of something else. That is, when we choose our means, we necessarily appeal to narratives about our ends ... *Prohairesis* is thus the process by which we moralize our actions: when we choose our means, we find ourselves obliged to acknowledge the ends we already desire ... and also our orientation toward those ends."[30]

---

28. *Life of the Mind*, 2:57–58; the standard discussion of Aristotelian *prohairesis* is by Kenny, *Aristotle's Theory of the Will*; see esp. 69–107.

29. Kahn, "Discovering the Will," 239–240; D. Allen, "Talking about Revolution," 200.

30. "Talking about Revolution," 200–201; Kahn ("Discovering the Will") likewise notes how *prohairesis* "marks the point of confluence between our desire for a goal and two rational judgments: first, our judgment that the goal is a good one, and, second, our judgment that this action is the best way to pursue it" (241). Arendt (*Life of the Mind*, 62) fails to notice the extent to which the choice of means also clarifies our awareness of, and deepens our commitment to, ends. O. O'Donovan's working definition of judgment—"an act of moral discrimination that pronounces upon a preceding act

Aristotle thus takes great care to embed "judgment" within a rational framework of habit, itself honed by a complex intergenerational process of role-specific and *praxis*-oriented socialization. Thus, while judgment in Aristotle's ethics involves a volitional element that is in turn consummated in "action," the commitment in question pivots on the person's capacity to bear in mind when it most matters that "the principles of the things that are done consist in that *for the sake of which* they are to be done." Judgment and choice thus are rational only because they unfold in an ontological framework of things and purposes hierarchically and teleologically ordered. In the *Nicomachean Ethics* we thus find Aristotle taking pains to desynonymize "judgment" from mere "conjecture" or "opinion" (1142^{b}5), the latter denoting mere guesswork or capricious self-realization. Aristotelian "judgment" and "choice" thus stand in sharp contrast with Hobbes's or Carl Schmitt's "decision" (*Entscheidung*), which by definition pivots on the absence of any good providing guidance for the work of deliberation, thus rendering judgment little more than the *imposition of an irrational will*—an emphatic toss of the coin designed above all to put an end to any further deliberation. Understood as "the right discrimination of the equitable" (*Nicomachean Ethics,* 1143^{a}20) and as that which "implies ... the right reason," Aristotelian *prohairesis* (what Aquinas would translate as *electio*) is bound up with the rational "deliberation"—Aquinas will call it "counsel" (*consilium*)—of how a proposed action relates to a broader network of practices; and to make that determination means scrutinizing how an action fits into the hierarchy of goods aimed at realizing the highest political and spiritual goods—viz., the flourishing of the *polis* and wisdom achieved in contemplation (*theoria*), respectively.[31]

Put differently, judgment is rarely a discrete, isolated pronouncement but, typically, unfolds as a series of related insights as these eventuate in conversation or in quasi-conversational, internal deliberation. It involves a sustained engagement with a specific problem. By contrast, "if we focus on a single judgment, we lose all the flexibility and nuance that are in play when things are made to show up through speech."[32] Thus *prohairesis* in the *Nicomachean Ethics* cannot be located outside of, or construed as indifferent to, the common. Rather, it signals an agent's heightened

---

or existing state of affairs to establish a new public context"—recognizes the temporal dimension and, consequently, hermeneutic rather than syllogistic character of judgment. It is simultaneously concerned with a scenario already given *and* with "a prospective object of action" and, thus, "both *pronounces retrospectively on,* and *clears space prospectively for,* actions that are performed within a community." By nature, then, judgment is "both subject to criteria of *truth* ... and to criteria of *effectiveness*" (*Ways of Judgment,* 7–9).

31. By contrast, inasmuch as Hobbes's thinking "is rooted in an individualistic account of the will, oblivious to questions of its providential purpose in the hands of God, it has difficulty in understanding any 'collective making', or genuinely social process" (Milbank, *Theology,* 14).

32. Sokolowski, *Phenomenology of the Human Person,* 118.

responsibility to achieve rational articulacy in those situations where the course of action does not imply itself and, precisely by virtue of that intellectual effort, to strengthen the rational and normative framework of community itself. In ways that eluded all but a handful of nineteenth-century thinkers (Coleridge, John Henry Newman, and George Eliot being notable exceptions), Aristotelian judgment and the Thomistic concept of the will as rational appetite must be understood, not as a formal-epistemological dilemma confronted by the solitary "self" but as a more elemental, hermeneutic realization: viz., that rational personhood can be achieved only within an existing, normative, albeit imperfect moral community.

Here it is essential to remember that for classical thought—Platonic, Aristotelian, and Stoic—knowledge is inseparable from our commitment to its practical realization. Ancient philosophy has no concept equivalent to the modern idea of "information" as neutral and instrumental knowledge; in fact, the proposition that there might be a type of knowledge that can be agnostically or indifferently appraised by means of some "view from nowhere" would have struck thinkers of that era as bizarre. Instead, for knowledge to have a claim on us *as* knowledge presupposes that it is perceived to have a bearing on the overall order of life—be it of an individual or a community—which is to say, that it is framed by an ambient, albeit mostly implicit conception of the good and of human flourishing. Only so can knowledge, including its deceptively neutral manifestation as technical competence (*epistēmē*), be appraised as relevant and meaningful. All knowledge presupposes our implicit awareness of, and practical commitment to, specific goals, as well as a sense (not always fully developed) that in so pursuing these goals or ends the person participates in the larger rational order (cosmos). As Aristotle acknowledges in the *Eudemian Ethics,* "choice [*prohairesis*] is neither opinion nor wish singly nor yet both (for no one chooses suddenly, though he thinks he ought to act, and wishes, suddenly)." As he notes on this occasion, "the very name is an indication. For choice is not simply a picking but picking one thing before another; and this is impossible without consideration and deliberation" (1226^{b}4–7).

As has been pointed out, Aristotle's conception of *prohairesis* is etymologically linked to *hairēsis* and the verb *hairēo* ("to take with the hand"), which suggests that to deliberate is never just to decide on means but to reflect on their conduciveness to and commensurability with an overarching end. It thus stands in vivid complementarity to "desire" (*orexis*), whose root and primary meaning (*orēgo*) "indicates a stretching out of one's hand to reach for something nearby."[33] Consequently, there is

33. Arendt, *Life of the Mind*, 58. The proximity of *hairēo* to *orēgo*—the root-verb behind Aristotelian *orexis* ("extending one's hand" and in the medio-passive voice, "to reach out for, stretch oneself towards, grasp at, aim at, or hit at")—emerges in Nussbaum's discussion of *orexis* (*Fragility*, 274–275).

a narrative dimension at work, "a notion of trajectory," as Allen puts it, which markedly differs from the instantaneity of "decision" and the purely efficient causation or instrumental rationality of "choice" as a mere seizing of specific "means." Consequently, "*prohairēseis* do not involve sudden choices" but, instead, "entail consistency over time."[34] In what follows, I shall render *prohairesis* as "judgment" (though Charles Chamberlain's proposal of "commitment" is just as viable), in part so as to accentuate the difference between Aristotle's conception and modern utilitarian or pragmatist notions of "rational choice" and "decision." For Aristotle, that is, all practical reason is appraised, ratified, and integrated into a normative teleological order by means of what he calls "judgment." Of pivotal importance to Stoic thought, judgment in due course emerges as a complementary term for Aquinas's theory of the will (*voluntas*) and "free choice" (*electio*). Yet before taking up the rather convoluted migration of Greek concepts into classical and, eventually, Scholastic Latin, a closer examination of judgment in Aristotelian and Stoic thought is in order. For the reflective operation of judgment is yet another casualty of modern accounts of the will and also of the person.

By definition, judgment constitutes a hermeneutic and evaluative act. In it, we draw on our powers of discernment regarding what matters and what is peripheral or positively distracting. That is, we seek to respond to and resolve a situation that involves some palpable conflict of goods and thus is initially experienced as a cause of perplexity. Resolution here involves our choosing an *action,* which implicitly also has us affirm an order of values and ends. All judgment thus is prompted by a tension between what Alasdair MacIntyre terms the "goods of efficiency" and those "of excellence," between *causa efficiens* and *causa finalis*. Beginning in the post-Homeric era, that antagonism leads to a more acute discrimination between actions undertaken "in order to" and those pursued "for the sake of."[35] Since their proper calibration is nothing less than essential to the flourishing of the *polis*—itself the highest good "for the sake of" which all action is ultimately to be undertaken—the sustained, measured, and cooperative pursuit of communal flourishing can only get underway qua judgment. Confronting the *polis*'s dramatically weakened cosmological and political framework following Athens's defeat in the Peloponnesian War, political reasoning undergoes a fundamental shift as regards the sources of authority. Danielle Allen has persuasively argued that with the generation of leaders (Themistocles, Pericles, Solon, Alcibiades, et al.) whose genealogies had extended "back to the mythological period" now giving way to leaders "who were simply wealthy," and with generals

---

34. "Talking about Revolution," 199, 201.

35. The distinction first surfaces in MacIntyre, *Whose Justice?* 32; see also Arendt, *HC*, esp. 22–78.

(*stratēgoi*) being rapidly replaced by professional orators (*dēmagōgoi*), "much of the political discourse of the first half of the fourth century ... was an effort to figure out how to justify political leadership without reference to military expertise."

By the time Aristotle addresses that question head-on in his *Rhetoric* (lectures first presented around 356 B.C.), the concept of "ethical intent" or "judgment" serves to consolidate this "revolutionary conceptual change" from a command ethic to a culture of deliberative argumentation that seeks to counterbalance the evident risk of outright deception and dissimulation—always present in rhetoric—with a cultivation of judgment.[36] Singling out the benefit of "maxims" (*gnōmai*) for persuading a less intelligent audience, Aristotle is careful to offset their rhetorical effectiveness with an account of "choice" (*prohairesis*) by remarking on "another advantage [of using maxims] which is more important—[viz., that] it invests a speech with character [*ἦθος*]. There is character in every speech in which the choice is conspicuous" (*Rhetoric,* 1395^{b}10–14). To translate *prohairesis* as "choice" is to exaggerate the term's pragmatic or instrumental connotations—in the sense of "mere rhetoric"—while obscuring the teleological framework within which alone such choice unfolds as a rational act. For unlike the demagogue or Sophist, Aristotle's political orator understands rhetoric to be aimed at persuading an audience about more than the choice of specific means. Instead, public oratory unfolds in explicit acknowledgment of (and aims to secure broad-based assent to) a rational end or good. The true orator, unlike the Sophist, not only seeks to persuade 51 percent of his audience but all of them. More than aiming at a favorable decision by a majority, however slim, the Aristotelian orator seeks to achieve recognition of his reasons. Thus it makes sense to translate *prohairesis* as "ethical intent," "moral purpose," or "commitment." In due course, it is just this teleological, narrative dimension of judgment that shows it to be a pivotal source for the Augustinian notion of the will. As Allen notes, the growing appeal of Aristotle's conception of "a way of life" (*prohairesis tou biou*) to fourth-century orators further underscores "this notion of *prohairesis* as entailing consistency over time" and of choice as "a *series of actions all oriented toward the same goal or project.*"[37]

Consistent with his stress on the indelible nexus between "choice" and "character," Aristotle thus discriminates more sharply than his precursors between "theoretical knowledge" (*epistēmē*) and "practical wisdom" (*phronēsis*), and between rational and irrational forms of desire; and it is this heightened concern with

---

36. Allen, "Talking about Revolution," 197–198, 192. Partially consistent with Allen's thesis, though not concerned with the important political and social changes that she highlights, is Dihle, *Theory of Will,* 20–47.

37. Allen ("Talking about Revolution"), 201–202; for a prominent instance of the phrase, see *Politics* (1280^{a}30): "a state exists for the sake of a good life [*ἀλλὰ μᾶλλον τοῦ εὖ ζῆν*], and not for the sake of life only."

understanding the *sources* of action as the foci of our ethical commitments as persons that culminates in Aristotle's introducing *prohairesis* as a central philosophical concept. Whereas the term only surfaces a handful of times before Aristotle—once in Plato (*Parmenides,* 143^{c}) and a few times in speeches by Isocrates and Demosthenes—the word appears 156 times in Aristotle's corpus. The *Nicomachean Ethics* defines knowledge to be "what we suppose … is not capable of being otherwise," the product of deductive processes that "start from what is already known" (1139^{b}20). Precisely because of its strictly deductive etiology, however, knowledge does not constitute a wholly self-sustaining edifice. Rather, each instance of deductive thought necessarily rests on principles and thus "proceeds *from* universals" that in turn can only be furnished by induction. Noting how "the reasoned capacity to act is different from the reasoned capacity to make" (1140^{a}1), Aristotle discriminates between practical and theoretical reason, albeit without therefore declaring these realms incommensurable or framing them as competitors.

An important challenge, then, is to understand how these two modes of knowledge might be linked. What Aristotle explores under the heading of "practical wisdom" and "deliberation" (*prohairesis*) in Book 6 of the *Nicomachean Ethics* is, in effect, the first fully articulated account of "judgment" or ethical intent in Western thought, and it already hints at the precariousness of any strictly post-cosmological notion of rationality:

> The man who is capable of deliberating has practical wisdom. Now no one deliberates about things that cannot be otherwise nor about things that it is impossible for him to do. Therefore since knowledge involves demonstration, but there is no demonstration of things whose first principles can be otherwise (for all such things might actually be otherwise), and since it is impossible to deliberate about things that are of necessity, practical wisdom cannot be knowledge nor art; not knowledge because that which can be done is capable of being otherwise, not art because action and making are different kinds of thing. It remains, then, that it is a true and reasoned state of capacity to act with regard to the things that are good or bad for man. (1140^{a}30)

Deliberation or judgment of the kind here adumbrated by Aristotle involves a distinction that, as Hannah Arendt notes, has since withered: viz., that between the "grounds" that may render a specific action instrumental ("in order to") or give it teleological legitimation ("for the sake of"). Where that distinction is preserved, as in the case of Pericles, practical wisdom involves "temperance" and only so is able to discriminate between individual interestedness and "man in general" as a proper end. Ultimately, judgment, deliberation, temperance, and practical wisdom all but converge in one's capacity to bear in mind when it most matters that "the principles of the things that are done consist in that *for the sake of which* they are to be done."[38]

What remains, of course, is the basic dilemma of wisdom's apparent indemonstrability, which is to say, the fact that "wisdom" by its very nature remains "opposed to comprehension," a point also acknowledged by Aristotle (*Nicomachean Ethics,* 1142ᵃ25). It is thus to be achieved only by a sustained inductive process (*epagōgē*) whereby deliberation moves "from a set of particulars to a universal, to the concept of the form that those particulars to different degrees exemplify ... This dialectic method, in which a particular thesis or theory justifies itself over against its rivals through its superior ability in withstanding the most cogent objections from different points of view" was in due course to become the focal point of Stoic logic.[39] What is at stake is the status of an "object, not of knowledge but of perception" or what Aristotle calls the "ultimate particular"—and to it alone belongs the name of "judgment" (*Nicomachean Ethics,* 1143ᵃ20). Implicit in this inductive, as well as social and discursive, elaboration of Aristotelian *phronēsis* is another feature of strategic importance: viz., all "judgment" is intricately tied to the idea of *praxis,* which in turn is at all times circumscribed by the hierarchically ordered sociality of the *polis.* Rationality, justice, and the discrete acts of judgment required for their continual elaboration—whose goal Aristotle captures under the heading of "excellence" (*aretē*)—are only achievable *within and for the sake of* a community. Hence the *Nicomachean Ethics* takes care to desynonymize "judgment" from mere "conjecture" or "opinion" (1142ᵇ5), notions that, like our contemporary expression of "making a judgment call," strongly suggest the lack of any good guiding the work of deliberation, thus rendering judgment little more than a toss of the coin. By contrast, "the right discrimination of the equitable" (1143ᵃ20), which in turn "implies ... right reason" as the very foundation of Aristotelian judgment, is constitutively tied to a particular practice and to an assessment of how that practice fits into the hierarchy of goods ultimately meant to ensure the greatest good itself—that is, the flourishing of the *polis.* In its verbal form, *prohairēomai,* "choosing" comes "very close to our concept of will" and, like so "many words for cognition or thought, inevitably impl[ies] the semantic element of decision or intention which results from intellectual activity."[40]

38. 1140ᵇ15 (italics mine); this convergence is expressly noted by Aristotle in *Nicomachean Ethics*, 1143ᵃ25, and it suggests why a theory of judgment forever risks succumbing to a tautological method, and why modernity's strong commitment to neutral "method" as "salvation" (Adorno and Horkheimer) could not succeed given its principled refusal to specify normative "ends" in the strong Aristotelian sense.

39. MacIntyre, *Whose Justice?* 91.

40. Dihle, *Theory of Will*, 21; as Dihle later notes, this pervasive conjunction of cognition with intention is reflected in the lexical roots of several key concepts; thus "the root 'gno-' (as in *γνώσις* = knowledge and *γιγνωσκω* = to discern; and γνωμη = maxim; formulated wisdom) always

There can be no judgment without an ambient, normative frame relative to which it is to be exercised; which is to say, that there can be no judgment without thought. As Simone Weil was to put it in her famous essay on the *Iliad,* "where there is no room for thought, there is no room either for justice or prudence."[41]

Particularly crucial is a qualification that Aristotle makes just as his discussion of "practical wisdom" and "judgment" in Book 6 of the *Nicomachean Ethics* winds down. As he notes, Socrates was on the right track when defining choosing and doing "in accordance with the right reason" as a particular kind of "excellence," one that in turn already presupposes an intuitive sense of the "right reason." Still, Aristotle insists, it is not enough to "divine" the "right reason" or "end" and, having done so, to act "in accordance with it." Rather, what ultimately defines the ethical status of human *praxis* is that this "state of excellence ... impl[y] the *presence* of the right reason [μετὰ τοῦ ὀρθοῦ λόγου]."[42] Aristotle's stress on the "*presence* of the right reason" hints that the ultimate criterion of rationality involves not merely its formal-syllogistic correctness but, rather, the *explicitness* with which that reason is embraced and "realized" (in the strong Hegelian sense of *verwirklichen*) in practice. What Aristotle terms "right reason" thus has to be understood as a "concrete universal," rather than an abstraction arrived at by deductive argument and syllogistic predication, and his thinking here rests on a notion of rational qua good, that is, of reason as a perfection and value term rather than an ordinary predicate.[43]

The most crucial features of Aristotelian *prohairesis* to be kept in mind are the following: (1) it is oriented not toward "knowledge" (*epistēmē*) as such but toward its configuration with "practical wisdom" (*phronēsis* or *nous praktikos,* as it is called in *De Anima*); (2) judgment's rational potential is achievable only within and on behalf of a community; (3) it unfolds as an inductive and dialectical process (*epagōgē*) whereby knowledge arises from a public confrontation of competing arguments;

---

implies an element of intention. Specifically *gnomē* "includes both intellectual activity and volition" (ibid., 29).

41. "The Iliad, Poem of Might," in *Simone Weil Reader*, 163.

42. *Nicomachean Ethics*, 1144^{b}26. At first glance, Kant's insistence in his (strikingly neo-Stoic) *Grounding for the Metaphysics of Morals* (1785) that an action is to be undertaken not merely in "conformity" with, but positively "for the sake of the moral law" (3) would appear to echo Aristotle's stress on the "explicitness" of "right reason." Yet Kant's moral philosophy no longer sees an essential connection between cognition and praxis; rather, it locates reason exclusively in an inner attitude or intention that has effectively been sealed off from the contingent empirical realm of practical life.

43. On the Stoic theory of "assent," see Brennan, "Stoic Moral Psychology." On the question of normativity in relation to the Stoics' formal-linguistic criterion of explicitness, see Brandom, *Making it Explicit*, 3–66; following Peter Geach, Foot insists that to speak of something as "good" is to employ it as an "attributive adjective," which is "logically different" from a "predicative adjective"—as in "the car is *red*" (Foot, *Natural Goodness*, 2–3).

(4) far more than a merely conceptual (or mentalist) attribute, Aristotle's pivotal criterion of "explicitness" means that judgment and rationality are never simply epistemic or notional but political and social in their very essence. Underlying all of these criteria is the Aristotelian notion of knowledge as a gradual process; for "wisdom is concerned not only with universals but with particulars, which become familiar from experience, but a young man has no experience, for it is the length of time that gives experience" (*Nicomachean Ethics,* 1142a10). To judge well we thus begin by reining in our propensity to extend judgment beyond its bounds of competency. MacIntyre's description of the seemingly circular logic of judgment points us in the right direction here:

> We cannot judge and act rightly unless we aim at what is in fact good; we cannot aim at what is good except on the basis of experience of right judgment and action. But the appearance of paradox and circularity are deceptive. In developing both our conception of the good and the habit of right judgment and action—and neither can be adequately developed without the other—we gradually learn to correct each in light of the other, moving dialectically between them.[44]

Echoing Plato's model of education into the virtues, Aristotle understands the dialectic of judgment very much as one of progressive socialization and its basic logic as one of continual revision. By contrast, Stoic logic was to approach judgment through a formal analysis of moral and cognitive predicates to be (ideally) pursued and completed by the individual independent of (or prior to) its eventual socialization. Judgment thus is increasingly equated with the prolonged "suspension of judgment" or with what the Stoics call the withholding of subjective "assent" to appearances whose relation to bona fide ends remains as yet elusive. What drops out is the social character of "dialectic" (*epagōgē*), which in Stoicism has been supplanted by a far more solitary and nominalist conception of individual experience and, consequently, of judgment as a process of skeptical self-scrutiny.

Before exploring how Aristotelian judgment is both appropriated and reoriented by the Stoics, it will be necessary to head off a terminological confusion liable to be wrought by modernity's misapprehension and mistranslation of "judgment" as "choice" or "decision." For Aristotle, and indeed for the Stoics, "judgment" (*prohairesis*) at no point involves any uncertainty about the *telos* of action but only about the commensurability of means with ends that are at all times hierarchically ordered and accepted as normative. Distinguishing between "(rational) wish" (*boulēsis*) and judgment (*prohairesis*)—the latter being rendered as "choice" by the standard Oxford

44. *Whose Justice?* 118.

edition of the *Complete Works*—Aristotle repeatedly notes that "wish relates rather to the end, choice to what contributes to the end" (*Nicomachean Ethics,* 1120^{b}25). Hence to choose without relating, in an act of explicit judgment, that which is chosen to a specific end is to fail both as a rational and just being: "The origin of action—its efficient, not its final cause—is choice, and that of choice is desire and reasoning with a view to an end" (*Nicomachean Ethics,* 1139^{a}30). For a variety of (initially theological) reasons, this notion of "choice" constrained by the intrinsic rationality of forms and the normative authority of a supra-individual tradition came under increasing pressure in late Scholasticism and was rejected outright by the emerging discourse of "rights" in the seventeenth century.[45] Whereas "the conception of a single, albeit perhaps complex, supreme good is central to Aristotle's account of practical rationality, . . . it is just this conception which most, if not all, recent moral philosophers find quite implausible." For them, "there can be no uniquely rational way of ordering goods within a scheme of life, but rather there are numerous alternative modes of ordering, in the choice between which there are no sufficient good reasons to guide us." The argumentative fallacy of such objections, as Alasdair MacIntyre and Charles Taylor have variously argued, lies in the fact that they rest on an unexamined, quasi-naturalistic set of commitments that would have seemed utterly alien, indeed irrational, to citizens of the Aristotelian *polis.* Above all, they hold that the "individual human being confronts an alternative set of ways of life from a standpoint external to them all. Such an individual has *ex hypothesi* no commitments."[46]

Yet to premise theoretical (and moral) arguments on the axiom of a "punctual self" supposed to be the bearer of inherent rights is methodologically spurious for at least two reasons. First, it *presupposes* (though evidently cannot demonstrate) the ability of any individual at any given point in time making any variety of choices or new beginnings, and so taking itself to be unconstrained by the genealogy of those concepts through which alone it could ever hope to make the rationale of its eventual choice articulate and meaningful *for others.* Second, the notion of a completely unconstrained choice also implies that commitments implicit in an individual's *eventual* choice cannot be rationally justified by appeal to any normative *ends.* Rather, in

45. This is not the place to take up the relationship between judgment, will, and the emergence of a theory of natural rights. For a discussion of the latter, and for very different accounts of its historical emergence and textual (scriptural) sources, see Tierney, *Idea*, 43–77, who locates the beginning of a modern rights conception in twelfth-century legal practice; Strauss, *Natural Right and History*, 165–251, focuses on Hobbes and Locke; whereas Wolterstorff, *Justice*, 35–131, implicitly rejects MacIntyre's "right order" or "right reason" conception by dating back the emergence of claim rights ("rights to specific goods") to certain "inherent rights" in Hebrew and Christian scripture; see also Milbank, "Against Human Rights" and below, 380–392.

46. MacIntyre, *Whose Justice?* 133; for a strong, more detailed version of this argument, see C. Taylor, *Sources*, 25–52.

what turns out to be a significant misappropriation of Augustine's conception of "free choice" (*liberum arbitrium*), the modern idea of a "free will" remains incoherent, indeed irrational on account of its strictly occasional and preferential nature. Lacking all objective justification, such choice can only be *inferred* to have been taken "in order to" realize a contingent objective—that is, gratify a subjective impulse or desire whose meaning remains as enigmatic to the one (putatively) choosing as to those witnessing the choice and potentially victimized by its unconditional pursuit as a private "right."[47] Aristotle's painstaking analysis of judgment in the *Nicomachean Ethics* already hints at the extent to which competing models of rationality, justice, and excellence had come to divide the self-understanding of the Athenian *polis*—with older military and aristocratic models of virtue being challenged by the professional *dēmagōgoi* (the Sophists), and Platonic and Aristotelian models of "character formation" (*paideia*) soon yielding to the rise of Zeno's Stoic school. Yet beginning with the shift toward a strictly nominalist concept of reason as it is advanced by the Franciscans of early fourteenth-century Oxford (often in polemical opposition to Aquinas's alleged Aristotelianism), modernity views judgment increasingly as a dilemma and source of perpetual discomfiture.[48]

Given the hierarchical order of goods and the overall clarity about ends, "judgment" (*prohairesis*) in Aristotle does not concern itself with ontological uncertainties but primarily with calibrating the ratio of what MacIntyre calls the goods of excellence and those of efficiency. For Aristotle as for the Stoics, who in significant ways absorb and develop his ethical arguments, "judgment" is nothing like our modern concept of "decision." If, in the spirit of Coleridge, one were to "desynonymize" the two, Carl Schmitt's theory of "decision" (*Entscheidung*) would furnish a particularly apt contrast. For Schmitt, any "decision" is inherently arbitrary because of an unbridgeable gap between the particular and the universal, between the "concrete fact" and the "standard of judgment [*Maßstab der Beurteilung*]."[49] As the sheer "intercession of authority (*auctoritatis interpositio*)," any "decision" at once recognizes and

---

47. Perhaps no other writer grasped potentially disastrous implications of a notion of choice reduced to the assertion of absolute *dominium* over a claim right *at the expense of any communal norms* better than Heinrich von Kleist; see esp. his *Michael Kohlhaas* (1808).

48. On this threshold, see Blumenberg, *LMA*, 125–226; for readings of nominalism as a major catalyst of modernity, see C. Taylor, *Secular Age*, 90–99; Dupré, *Passage*, 88–89 and 121–125; Gauchet, *Disenchantment*, 47–62; Aers, *Salvation and Sin*, 25–54; and Gillespie, *Theological Origins*, 19–43.

49. Schmitt, *Politische Theologie*, 37; all translations are my own. For an alternate translation, see Schmitt, *Political Theology*, trans. George Schwab. See also Heidegger's observation that "every decision ... bases itself on something not mastered, something concealed, confusing, else it would never be a decision [*Entscheidung*]" ("Of the Origin of the Work of Art," in Hofstadter, *Philosophies*, 681).

capitalizes on that very disjunction of fact and value. By projecting arbitrary force into a complex realm of human affairs deemed inherently resistant to the pragmatics of political necessity, Schmitt's "decision" is also "instantaneously emancipated from any argumentative reason-giving." Evidently, then, what has disappeared in such a model of "decision" is the (for Aristotle crucial) notion of a gradual inductive transitioning from particular to universal (*epagōgē*), a process that for Aristotle centers on a maximally explicit giving of reasons *to*—and their inter-subjective, argumentative testing *by*—members of the *polis*. Instead, Schmitt sees all lines of communication between "grounding norms [*zugrundeliegende Normen*]" and a given "decision" as having been severed, which in turn gives rise to his startling conclusion that "decision, normatively considered, arises *ex nihilo* [*aus dem Nichts geboren*]." Clearly, then, there is no longer a shared, normative conception of the good that guides deliberation and legitimates a specific "decision" as rational. Rather, only a practical objective to be realized qua decision "determines what a norm is and what has normative authority." Schmitt thus sees all "decision" characterized by an "indifference of content [*inhaltliche Indifferenz*]." Inasmuch as it is the authority that "makes the decision" that is performatively consolidated by that very act, all decision is at least "relatively and, under certain circumstances, even absolutely independent of its proper content [*unabhängig von der Richtigkeit ihres Inhaltes*]."[50] A far cry from Aristotelian *prohairesis*, Schmitt's "decision" at once presupposes and responds to "disorientation" as an ontological and thus irreversible condition of modern existence; indeed, far from seeking to remedy that predicament, it positively relishes, much in the spirit of Hobbes, the irrational conclusions and opportunities seemingly licensed by it.

When contrasted with Aristotelian *prohairesis*, the Stoic view of mental life, and of judgment in particular, proves less affirmative than skeptical. Much of that shift has to do with the fact that both the source and the *telos* of judgment have at once contracted and expanded in extreme ways; for the source is now the deliberative individual, one who is no longer guided by a normative social framework but, instead, seeks virtue against the backdrop of a political and social reality that appears substantially irrational. The causes that had rendered the Hellenistic world so disorienting bear striking resemblance to the "great disembedding" (as Anthony Giddens and Charles Taylor have called it) of the later seventeenth and early eighteenth centuries when previously static and autochthonous communities found themselves dislodged by scientific innovation, speculative finance, social mobility, and the rise of professionalism. Stoicism—which unsurprisingly experiences a major resurgence during the same period—had formerly arisen as a concerted response to the geopolitical

50. Schmitt, *Politsche Theologie*, 36–38.

transformation of the Mediterranean world once Alexander the Great "stripped the individual of the insulated shelter of his little city-state and forced him to come to terms with and find a place in an enormously expanded polity."[51] As a result, the Aristotelian model of ethical rationality had to be fundamentally rethought so as to account for the simultaneous jurisdictional "contraction" and "expansion" of judgment already mentioned. For when "all human beings are fundamentally not members of families or cities, but *kosmopolitai,* members of the 'city-state of the universe,'" the *ends* "for the sake of" which judgment had defined a course of action now involve the cosmos as a whole.[52] For the first time, that is, ethics presents itself as universal in both its conceptual architecture and its social intent. At the same time, however, Stoicism's moral scrutiny now scrutinizes the self's inner disposition prior to its potential or actual socialization, a process deemed inherently unstable and full of distractions (so-called indifferents).

Notwithstanding its dramatically altered circumstances, Stoic thought retains and even intensifies some features of Aristotelian *prohairesis.* First, Stoicism develops Aristotle's condition regarding the "explicitness" of all judgment into a complex theory of "predication" that anticipates significant features of modern analytic philosophy. In so rendering ethics and logic all but inseparable, Stoicism effectively treats rationality and accountability, the cognitive and the discursive, as convertible notions. "The wise man," Diogenes Laertius notes, "is always a dialectician."[53] In contrast with the situational dynamic of Aristotelian *epagōgē,* Stoicism's grounding of judgment in an epistemological *method* introduces a significant ascetic and asocial element. Judgment now demands above all the *suspension* or (potentially indefinite) postponement of what would likely be an individual's premature "assent" to sensory impressions. A central objective of Stoic mental life thus involves to be perpetually on guard against the deceptive and perilous impact of all kinds of "impulses," "desires," and "opinions" on our perceptions. In drawing out a skeptical dimension intrinsic to Stoic thought (and in effect turning it against the Stoics), Sextus Empiricus thus remarks how our externally grounded (cataleptic) sense impressions may at all

---

51. Moses Hadas, in Seneca, *Stoic Philosophy,* 20. As regards the reemergence of Stoic motifs in modernity, particularly salient instances would include Justus Lipsius, beginning with his highly influential Stoic dialogue *De Constantia* (1569) and culminating in his late *Manuductio ad Stoicam philosophiam* ("Guide to Stoic Philosophy") and *Physiologia Stoicorum* ("Physical Theory of the Stoics") of 1602; an early instance of neo-Stoic thought can be found in John Calvin's early commentary on Seneca's *De Clementia* (1532); while less consistently observable in Montaigne's *Essais,* Stoic motifs are altogether central to the work of Descartes, the work of Shaftesbury, and also the aesthetics of Nicholas Poussin.

52. Nussbaum, *Upheavals,* 359.

53. Quoted in Long and Sedley, *Hellenistic Philosophers,* 184; all references to this edition are to vol. 1 and will henceforth be made parenthetically as *HP.*

times yet deceive us "like incompetent messengers."[54] Hence "to withhold assent is no different from suspending judgment. Therefore the wise man will suspend judgment about everything" (*HP,* 255).

Second, the resulting methodological *askēsis* and the envisioned state of *ataraxia* with which Stoicism came to be principally (and often erroneously) identified in later periods could thus be described as an "introjection" of the dialogical principle that had guided the cultivation of rational sociality in the Aristotelian *polis.*[55] Commenting on a diametric reversal of Aristotle's association of freedom with the public realm and of necessity with the *oikos,* Louis Dupré remarks on the Stoics' "distinction between the internal realm of freedom and the external world of compulsion ... In the huge Hellenistic and Roman empires ... the Stoa presented a new ideal of freedom that, while not avoiding political or social duties, consisted in an inner attitude, independent of external circumstances," and which fostered "attitudes of withdrawal rather than of dominance."[56] It does not surprise, then, that the Stoics' unrelenting stress on identifying and expunging all traces of "opinion" from genuine cognition effectively alters the notion of truth itself. No longer conceived in terms of its social significance but, rather, as a quest for a dispassionate, formal correctness, truth—and by extension the Stoic idea of mental life overall—moves away from Aristotelian "action" as the supreme human achievement and toward a proto-modern idea of "critique." The goal, of course, is for the individual to extirpate any admixture of opinion or impulsive judgment from the myriad "impressions" that constitute its world. As A. A. Long and D. N. Sedley note, "unlike most previous philosophers ... no Stoics officially recognize the existence of *true* opinions" (*HP,* 258). For the Stoics, mental life achieves legitimacy not by supporting a socially elaborated framework of rationality but by continually scrutinizing the public realm in open-ended, critical contemplation. Theoretical and practical rationality now begin to diverge to the extent that the Aristotelian idea of *praxis* is being deemed inherently premature on account of its supposed lack of conclusive theoretical legitimation.

Far more explicitly than either Plato or Aristotle, the Stoics' conception of judgment is anchored in a philosophy of language. As the fifth-century compiler Joannes Stobaeus notes, "all impulses are acts of assent ... But acts of assent and impulses actually differ in their objects: propositions are the objects of acts of assent, but impulses are directed toward predicates" (*HP,* 197). As the foundational criterion for rationality, "explicitness" is now being conceived in linguistic form. Thus legitimate

---

54. *HP*, 461; on the debate between the Stoics and the Academics concerning the authenticity and reliability of sense impressions, see Hankinson, "Stoic Epistemology."

55. For a succinct account of popular and persistent misconstruals of Stoic commonplaces, see Rorty, "Two Faces of Stoicism."

56. Dupré, *Passage*, 121.

"assent" can only be given to a proposition that makes fully transparent the subject's relation to both the concrete object of which it has received an "impression" and its place within the world as a hierarchy of "ends." An "impulse," by contrast, shows our relation toward an impression to be unreflected and inarticulate. As Long and Sedley comment, the Stoics moved beyond Aristotle's identification of meanings with thoughts "by distinguishing rational impressions from sayables" and so demonstrating "that the meaning of a thought is something which is transferable, through language, across minds. I cannot pass on to you the physical modification of my mind, but I can tell you what I am thinking about" (*HP,* 201). Stoicism's philosophical program thus pivots on the gradual disaggregation of mere auto-affections of the mind—so-called non-cataleptic impressions or "figments" (*phantasmata*) and similar acts of "imagination" (*phantastikon*)—from genuine, "cataleptic sense impressions" (*phantasiai*) that are anchored in an external source.[57] The latter stand to be converted into articulate "propositions" or "sayables" (*lecta*), for only they legitimately warrant "assent" and so can bring the individual closer to the Stoic ideal of intellectual autonomy (*autarkeia*).

Third, and perhaps most decisively, in taking a far more skeptical view of universals, particularly as conceived in Plato's doctrine of ideas, Stoicism foreshadows fourteenth-century nominalism's logical prioritizing of the isolated particular and the exclusive causal role that the isolated material phenomenon is assigned by Hobbesian and Lockean empiricism. As Long and Sedley note, Stoicism views universals strictly as "concepts" (*ennoēmata*) which, lacking any corresponding sense impression, are thus regarded as "figments" (*HP,* 181–182). A central task in Stoic epistemology thus involves accurately discriminating between a unique sensory impression and a "conception" (*ennoēma*) that is to be gradually and explicitly distilled from it. Stoicism's sharply increased stress on epistemological technique at once constrains "judgment" and alters its principal aim. As the expression of a fundamentally *critical,* rather than (Aristotelian) *practical,* notion of rationality, judgment for the Stoics has in effect become a technique or method designed to assist the individual in its quest for cognitive *autonomy* and moral *self-legitimation.* As Diogenes Laertius notes, "without the study of dialectic the wise man will not be infallible in argument, since dialectic distinguishes the true from the false, and clarifies plausibilities and ambiguous statements" (*HP,* 184). To the eclectic Cicero, it is just this kind of contraction of the scope of judgment, and its reappraisal as a kind of sustained prevarication, that ultimately exposes the limitations of Stoic thought: "Every thorough account of argument has two parts, one concerned with invention and the other with judgment . . . The Stoics, however, have exerted themselves only in one of these. With

57. See Hankinson, "Stoic Epistemology," 60f., and Sellars, *Stoicism,* 64–74.

that science that they call dialectic they have thoroughly pursued the methods of judgment, but they have completely neglected the art of invention called topics" (*HP*, 185). Cicero's critique, in Book 4 of *De Finibus*, of the younger Cato's version of Stoicism raises legitimate concerns about Stoicism's hostility to rhetoric. In flagging the imaginative and creative force of rhetoric, however, Cicero also reveals that to resist this "positional power of language" (as Paul de Man was to put it in our time) is to embrace a strictly defensive and reactive conception of ethics. As I have argued elsewhere, Jean-Jacques Rousseau's imaginative amalgamation of Stoicism's dialectical concept of judgment with the rhetorical complexity of epistolary form forges a unique way beyond the threatening calcification of judgment into a strictly formalist parsing of propositional language.[58]

Two related problems opened up (though arguably never solved) by Stoicism turn out to be especially salient for eighteenth-century, neo-Stoic thought. One concerns the apparent role of extreme idealization in Stoic moral psychology, a field ordinarily understood to explore human action descriptively rather than normatively, and hence to adopt a realist rather than ideal mode of representation. The other issue has to do with what the later Stoics regard as cases of false (or precipitous) judgment: strong emotion, or passion (*pathē*).[59] What prompts the Stoic critique of the emotions as de facto failures or lapses of judgment is their strongly evaluative character. Emotion misconstrues as intrinsically good or bad what ought to be regarded as "indifferents" or, at most, as "preferred indifferents" (wealth, fame, health, etc.). It thus constitutes an instance of precipitous "assent" (*sunkatathēsis*) to an impression, which it validates merely on the strength of its relation to the subject. At the same time, it is hard to see how we could even wish to scrutinize an impression carefully unless we had already judged it to be a priori meaningful for us in some elemental way. As Tad Brennan observes, one may be tempted to regard such "assent" to an impression as "a quasi-deliberative or discursive process, like the investigation of a witness's bona fides before their testimony is admitted as evidence." Yet such scrutiny would seem possible only if "the agent has already suspended judgment, at least temporarily. There is no more elevated standpoint from which one can decide whether to

58. Pfau, "Letter of Judgment."

59. "The Stoics postulate pandemic error when it comes to matters of evaluation: all of the individuals around us, as well as our cultures, laws, and institutions, are wildly misguided in our assessments of what is good and bad. . . . These false beliefs . . . are known to the Stoics as *pathê*, or emotions" (Brennan, "Stoic Moral Psychology," 264). There is tension between Zeno's (early Stoic) position that emotions are the *products* of judgments and Chrysippus's and the later Stoics' view of emotion as being ipso facto a judgment; on this issue, see Sellars, *Stoicism*, 114–120; for further discussion of the place of emotion within Stoic thought, see Rorty, "Two Faces of Stoicism," 343–344; Nussbaum, *Upheavals*, 358–372 (esp. on "compassion"), and Nussbaum's earlier study of Stoicism (*Therapy*, 359–401).

assent or suspend; the very fact of scrutiny entails that one has suspended judgment."[60] Inasmuch as for the Stoics judgments occur without the individual being fully aware of his or her assent to an impression or the rejection of it as a mere "figment" (*phantasma*), their epistemology tends to critique judgment as nearly identical with the "impulse" (*hormê*) that makes us "assent" to a given impression and that ultimately discharges itself in an action.

There are, however, rare instances when what we judge to be a cataleptic impression undergoes genuine scrutiny, such as when deliberation does not "involve actual intensified scrutiny of the *same* impression, but rather the deliberate acquisition of a *distinct* impression (i.e., taking another look)."[61] Triggered by the affective and psychosomatic surfeit of emotion, such cases of concerted revaluation throw into relief the basic programmatic intent of Stoic moral psychology. As Brennan puts it, "the most important moment of our ethical progress comes in the replacement of emotions by selections (i.e., the correction of our false beliefs about values)."[62] It is precisely this second, counterintuitive reappraisal of emotion as a "product of mistaken judgments, namely assents to impressions that include unwarranted ascriptions of value,"[63] that fuels the Stoic quest for neutralizing the emotions altogether. To succeed in this quest is a prerequisite to our achieving virtue, for only in this manner can the individual affirm and consolidate its teleological constitution as a rational being. Central to the objective of *apatheia*—which in turn serves the ultimate end of rational and tranquil existence—is *an explicit and necessarily didactic staging of judgment*. In other words, judgment unfolds as a dialectic, narrative drive toward expunging the emotive underpinnings of our epistemological and moral commitments. It constructs truth by methodically disaggregating fact and value, the formal proposition from the affective hold it may have on the individual entertaining it. With regard to this proto-nominalist quality of Stoic virtue, Lawrence Becker has pointed out that "motivated norms are to be found *only* within the psychological structures of the actual endeavors of individual agents" and that "every norm (as a fact about the world) is internal to *some* agent's project." Yet his immediately following paraphrase also reveals the extent to which such "internalism" is, in fact, already fully enmeshed

60. Brennan, "Stoic Moral Psychology," 262. It is just this insight that prompts Newman, in his *Grammar of Assent* (1870), to distinguish sharply between "notional" and "real" assent. The latter involves "the unconditional acceptance of a proposition" which, as Newman stresses, hinges on our unconditional embrace of "the images in which it lives." By contrast, notional assent embraces a proposition only "on the condition of an acceptance of its premisses" and thus cannot offer positive "certitude" but only the "certainty" that derives from inferential reasoning (*Grammar of Assent*, 76, 86).

61. Ibid., 263.

62. Ibid., 272.

63. Sellars, *Stoicism*, 117.

with the sociality of language: "We simply cannot *find* any norms—*as opposed to sentences about them in writing or speech*—that are external to agents."[64]

It is this socially constructed aspect of Stoic reason that was to be developed (ultimately against Stoicism's core axioms) in Augustine's moral psychology. In his repeated engagement with Stoic thought, which culminates in the critique of Stoic *apatheia* in Book 19 of *De Civitate Dei,* Augustine specifically targets the Stoic ideal of self-mastery and the extirpation of all heteronomous affect as both misguided in its intention and unrealistic in its ambition. To begin with, there is the question concerning the sources that generate and sustain Stoicism's quest for a strictly neutral and self-contained mode of being (*apatheia*). Such a model of virtue strikes Augustine as incoherent, first and foremost because it is intrinsically *reactive,* whereas true virtue must unconditionally assent to a vision of the good: "true virtues can exist only in those in whom there is true godliness" (*CD,* 19.4). Stoic virtue is furthermore contradictory in that it asserts the nullity of external attachments and ills but simultaneously allows for suicide in the event of extreme adversity. The contradiction is not merely logical but, as such, reveals an underlying metaphysical confusion as to whether the ultimate good is to be realized in this world or not. Stoic virtue thus misjudges the nature of its antagonist by perennially mistaking for "external vices" what, in fact, are "internal ones, not the vices of others, but clearly ours and only ours" (*CD,* 19.4). Once the nature of vice, suffering, and disorder is appropriately identified as internal to the self—indeed, properly constitutive of it—the Stoic program of complete self-possession falls to the ground unless it is situated within a supervening economy of divine grace.[65] In brief, Augustine here identifies the central dilemma bedeviling the Stoic idea of "reason as deflecting passion or emotion in the manner of a fortress wall deflecting an outside enemy." For to the extent that "the passion is itself an expression of some judgment of value, ... reason is implicated in the experience and the enemy is within the gates. Even should the sage act against the dictates of passion, that very dissent would indicate an opposition within reason itself, not a conflict between rational and blindly irrational sources of motivation."[66]

At the heart of Augustine's moral psychology, then, lies the need for a fully articulated conception of the will—inevitably divided and self-alienated—as the *tertium*

---

64. Becker, *New Stoicism*, 76–77.

65. See Augustine's claim that "we cannot live rightly unless, while we believe and pray, we are helped by Him Who has given us the faith to believe that we must be helped by Him. The philosophers, however, have supposed that the Final Good and Evil are to be found in this life. They hold that the Supreme Good lies in the body, or in the soul, or in both, ... in rest, or in virtue, or in both" (*CD*, 19.4). On the seeming circularity that the very faith and strength of will on which the recognition of divine grace depends is itself a gift bestowed by it, see Stump, "Augustine on Free Will."

66. Wetzel, *Augustine and the Limits of Virtue*, 53.

comparationis. For without it we cannot grasp, positively and explicitly, what Stoicism had only ever been able to think disjunctively: sense and intellect, passion and reason. Yet this agency of the will is not simply some psychological term gratuitously *added* to the existing inventory of desire, intention, impulse, passion, etc. so as to remedy certain impasses within Stoic and Platonist philosophy. Rather, Augustine in *De Trinitate* and *De Civitate Dei* takes care to show that the will subtends and coordinates the operation of all other psychological concepts. More crucially yet, the argument will be that the will (a synecdoche for the human person as a spiritual being) is axially linked to the Trinitarian God whose gift it is. Augustine's model of the human person fundamentally differs from the modern (in origin Stoic or Pelagian) idea of a self committed to the twin ideals of autonomy (*autarkeia*) and indifference (*apatheia*) for which method, rather than grace, is to provide the basic resource, and whose realization hinges on the projection of unfettered individual power (*libido dominandi*) rather than on a participation in, and love for, a supra-personal, transcendent good. Against the latent voluntarism of his Stoic precursors and Pelagian contemporaries, Augustine maintains that willing never constitutes an outright and successful (let alone legitimate) mode of self-possession and self-assertion. The will here is not some "unqualified *arbitrium*" presaging modern inwardness and its vaunted claims to Cartesian self-scrutiny or Kantian self-legislation. For in that case, Augustine's self would have to "exercise *arbitrium* as a power or choice unqualified by other essential divine predicates and without motive beyond power's exercise."[67] We must now examine why such a reading is implausible, not only as an account of Augustine's model of human flourishing but, potentially, as a sustainable philosophical argument even for those who, beginning in the early seventeenth century, identified autonomy as both the very foundation and ultimate *telos* of purposive (human) agency.

---

67. Hanby, *Augustine and Modernity*, 93. Arguing that Augustine's differences with the Pelagians not only involve the interpretation of grace but, by implication, whether human nature—including the capacity for faith and righteous willing—is itself a divine gift (*donum*), Hanby convincingly views the Augustinian self as constitutively situated vis-à-vis the Trinity and, hence, as incommensurable with "the Pelagian introduction of another kind of self, alien to this economy, into Christian thought and practice" (91).

5

CONSOLIDATION

St. Augustine on Choice, Sin, and the Divided Will

To understand the adaptation of ancient philosophical concepts to changed social and intellectual purposes, what Hans Blumenberg calls their "reoccupation," one has to be mindful of how intricately that history is enmeshed with issues of translation. In the case of the will, translation holds particular significance because it is only by transposing and reconfiguring *hekousion, boulēsis, eph'hemin,* and *prohairesis* into classical Latin and early Christian culture that the notion of the will (*voluntas*) came into existence. Once established, it challenged thinkers from Augustine onward to clarify the will's relationship to the intellect and reason. In significant measure, the lexical and cultural migration of concepts is itself a catalyst in philosophy's ongoing appraisal of the human capacity for conceptualization. In tracing the transposition (and partial fusion) of the above Aristotelian concepts, Charles Kahn notes that, beginning with Cicero, *voluntas* translates *boulēsis.* Merging the Aristotelian notion of *hekousion*—viz., an action done voluntarily but not deliberately—with *voluntarium,* post-Ciceronian Latin construes an agent's simple awareness of an act as a form of intentionality. The mere fact that a deed is accompanied by the agent's consciousness *of* the action is taken as evidence of its standing in instrumental relation to an underlying aim. Summarizing Stoic doctrine in his pastoral retreat from politics, Cicero thus notes how "as soon as the semblance of any apparent good presents itself, nature of itself prompts [people] to secure it. Where this takes place in an equable and wise way the Stoics employ the term *βούλησις* for this sort of longing; we should employ the term wish [*nos appellemus voluntatem*]."[1] The result is a somewhat elliptic conception of the will in that *voluntas* partially obscures the distinction between a voluntary and a deliberative action. Only in the second case does it make sense to impute to the agent an awareness of the distinction between means and ends, efficient and final causality. Though ostensibly justified by the exact parallel between *velle* and *boulomai,* the concurrent fusion of *hekousion* with *voluntarium* nonetheless gives rise to a conceptual asymmetry inasmuch as a link is established between willing and the strictly voluntary, "whereas nothing in Greek connects *hekousion* with *boulēsis*."[2] Willing, after all, does not signify a merely voluntary action such as reaching for an apple to still one's hunger; rather, it denotes a choice deliberatively arrived at in consideration of an overarching good or end.

A second fusion involves the translation of Aristotle's *eph'hemin,* which Aquinas will later render as "in our power" (*in nostra potestate* [*ST,* Ia Q 83 A 3]), yet which early Christian theology partially assimilates to the notion of "free choice" (*liberum arbitrium*). Beginning with the early Augustine, this strikingly new conception is closely entwined with the notion of "will" (*voluntas*). In contrasting Aristotle and Aquinas, Charles Kahn observes that where Aristotle had "analyzed the process of decision-making on the basis of three or four concepts that were only loosely related to one another: the voluntary, what is in our power or up to us, *boulēsis,* and *prohairesis* ... in Aquinas all four concepts are defined by reference to *voluntas,* the will."[3] The result is the emergence of an entirely novel conception central to accounts of human flourishing—viz., freedom. This framework markedly differs from the Aristotelian model, in which "rational desire" (*boulēsis*) sets the end and judgment only affirms, as it were *ex post facto,* that end by means of explicit "assent" and then chooses means conducive to and commensurable with it. As remains to be seen, the agent's relationship to the end is profoundly changed in Augustinian thought. Kahn also observes how Aristotle's *prohairesis* is eventually absorbed into Aquinas's notion of *liberum arbitrium voluntatis,* which consists "not in 'freedom of the will,' but in the exercise of 'free choice' *by* the will." Aquinas's term for the act proper wherein that free choice is realized—and the moral nature of the agent determined—is *electio.*[4]

Concurrently, the Stoic concept of "assent" (*sunkatathesis*) and its eventual appropriation by neo-Platonism's postulate of an immaterial soul suggests that, in building on Aristotelian psychology, Hellenistic thought adds a crucial component to what would soon be articulated by Augustine as the idea of a responsible and divided will. For the concept of "assent" implies not merely an intellectual *orientation*

---

1. *Tusculan Disputations,* 4.6.12.
2. Kahn, "Discovering the Will," 241.
3. Ibid., 242.
4. Ibid., 241, 250.

toward a particular object or objective, such as is implied by Aristotelian *boulēsis*. Rather, in conceiving the mind-world relation as a function of "assent," we are made to see that to aspire to something also implies one's *committing oneself to* the value of the object or idea in question. To will is not simply to want but to "consent to" what we take to be the value of that thing or idea at which we aim. Stoic assent thus highlights, in ways eventually writ large in Augustine's theory of the will, how the instant of "free choice" wherein the will becomes manifest both realizes an agent's spiritual condition and discloses that condition as the focal point of her or his self-awareness. To will *eo ipso* means "revealing" the agent's distinctive nature, an insight first given due weight by Epictetus, whose "use of *prohairesis* serves to expand the notion of consent into the broader notion of moral character and personal 'commitment,'" and who thus moves decisively beyond the Platonic *nous* as a "principle of reason most fully expressed in theoretical knowledge." Thus Kahn sees the Stoic notion of "assent" as part of a broader current moving toward the Augustinian model of introspective consciousness and, indeed, anticipating Aquinas, "who says [that] ... 'Consent belongs to the will.'"[5] With good reason, Aquinas takes much pains to desynonymize assent and consent, primarily (one has to surmise) because he is drawing on Augustine's conception far more than on that of his Hellenistic precursors. As he points out, to assent (*assentire*) "implies a certain distance from that to which assent is given." It simply means "to feel toward something" (*ad aliud sentire*). By contrast, "consent (*consentire*) is 'to feel with,' and this implies a certain union to the object of consent. Hence the will, to which it belongs to tend to the thing itself, is more properly said to consent: whereas the intellect, whose act does not consist in a movement toward the thing, but rather the reverse ... is more properly said to assent" (*ST,* Ia IIae Q 15 A 1). To the question of whether such consent belongs itself to the "higher faculty" or (as the objection would have it) whether it is solicited merely by a prospect of "delight" (*delectatio*), Aquinas responds by stressing that consent is by nature propositional, not affective. Its realization involves a "final decision" (*finalis sententia*). That which the will urges "to be done" (*agendum*) is conceived and ratified only by consent, and "consent to the act belongs to the higher reason (*pertinet ad rationem superiorem*)" (*ST,* Ia IIae Q 15 A 4).

Contrary to the modern notion of willing as sheer appetition, Stoic, Augustinian, and Thomist models of the will not only affirm its (subsequently obliterated) connections with Aristotelian "judgment," but they also show—in ways that Aristotle himself had not yet worked out—how all *prohairesis* constitutes an act of assent. Beyond that, Aquinas (herein following Augustine) also emphasizes that to assent to a proposition goes well beyond a purely syllogistic, intellectual operation. It also im-

5. Ibid., 253, 247.

plies "consent" or (to recall an alternative translation of *prohairesis*) a person's "commitment" *to* the moral value that stands behind the proposition. Indeed, the linguistic "act" (*sententia*) enunciating such consent is the very substance of the will, which can only be grasped *in actu*, and never properly as a faculty among others. All assent thus draws not only on a logical matrix that decides on the formal correctness of what is asserted, but it also discloses what Robert Sokolowski calls the agent's "veracity"—"the *eros* involved with rationality." Aquinas here extends what Stoic grammar and logic were unable or unwilling to acknowledge: viz., that "when we enter into the space of reason, we do not float up into a kind of distilled detachment that places us beyond human involvement. There is an ethics to disclosure; we have to *want* to be logical for others and for ourselves, and this wanting can be cultivated in either a virtuous or vicious way."[6]

It is as a result of the Stoic and neo-Platonic clarification of willing as propositional, as assent, and, ultimately, as consent to a supra-personal good that "*voluntas* and its cognates play a role in Latin thought and literature for which there is no parallel term in Classical or Hellenistic Greek."[7] The result of these closely related transpositions is the emergence of an entirely novel conception central to accounts of human flourishing: freedom. We recall how in Aristotle's ethics it is "rational desire" (*boulēsis*) that sets the end, whereas judgment only affirms, as it were *ex post facto*, that end by means of explicit "assent" and then chooses means conducive to and commensurable with it. It is precisely the agent's relationship to the end that is profoundly changed in Augustinian thought. Commenting on a remark by Cicero about the recently murdered Caesar ("what he wants is no great matter, but what he wants, he wants with a will [*quidquid volet valde volet*]"), subsequently rendered somewhat misleadingly as *sphoda bouletai* by Plutarch, Albrecht Dihle remarks on a palpable shift of terminology; whereas the Greek *boulēmai* entails an element of deliberation and planning, there is "a lack of psychological refinement in the Latin vocabulary" that may account for the "indiscriminate use of *velle* and *voluntas* for various kinds of impulse and intention [which] undeniably contributed to the voluntaristic potential in Roman thought."[8]

Questions of will and free choice span the entirety of St. Augustine's career. From his early *Of Free Choice of Will* (*De Libero Arbitrio*) to the *Confessions*, especially

---

6. Sokolowski, *Phenomenology of the Human Person*, 21, 66.

7. Kahn, "Discovering the Will," 248. The role of Plotinus in this emergent tradition is rather more extensive than can be addressed within the confines of this argument. Undoubtedly central to discussions of the will is *Ennead*, 6.8. Kahn notes how "from the Neoplatonists Augustine gratefully accepted the notion of a purely intelligible, noncorporeal domain of reality, to which the human will belonged together with the intellect" (255); see also Dihle, *Theory of Will*, 123–130.

8. Dihle, *Theory of Will*, 133.

Book 8, to the middle books of *De Trinitate* and *De Civitate Dei,* to his late anti-Pelagian writings, in particular, *Of Grace and Free Will,* Augustine continues to ponder the theological, epistemological, and psychological dimensions of a concept whose centrality to philosophy he was among the first to argue. Yet unlike the neo-Platonists and Stoics before him, Augustine does not so much see himself developing his account of the will by engaging already extensive traditions of philosophical inquiry. To be sure, Roman law had made extensive use of the will (*voluntas*) as a legal concept; yet in so doing it had treated the will solely as a heuristic device "invented to grasp the intention which underlies words or formalized actions," while remaining uncurious about its intrinsic ethical, anthropological, and psychological nature.[9] Likewise, the Aristotelian psychological vocabulary of *boulēsis, prohairesis,* and *hekousion* only reaches Augustine in its substantially altered, Stoic inflection. Thus the principal sources for Augustine's arguments and insights into the psychology of the human will and its perplexing entanglement with divine grace and (most vexingly) with predestination tend to be personal experience and scripture, above all the letters of St. Paul. Hannah Arendt's somewhat breezy observation that Augustine was "the first man of thought to draw his deepest inspiration from Latin sources and experiences," while true as far as it goes, warrants some clarification. For Augustine's account of human agency pivots on his careful *dis*engagement from Stoic and neo-Platonist philosophy.[10]

Yet instances of this rhetorically scrupulous swerve away from pagan thought, while abundant throughout his corpus, are not our principal concern here, and we shall limit ourselves to just one example: in closing Book 1 of the *Confessions,* Augustine recalls how he guarded "my wholeness, the imprint of that most hidden Oneness from which I took my being [*vestigium secretissimae unitatis ex qua eram*]." This remark (1.20.31) clearly echoes a late passage in Plotinus's Sixth *Ennead* (6.9.11), though Augustine is quick to limit that debt by emphasizing, in the very next sentence, his guardianship over the quasi-Trinitarian architecture of his senses: "and by an inner sense I watched over the integrity of my senses [*custodiebam interiore sensu integritatem sensuum meorum*]." Though familiar with Porphyry's and Marius Victorinus's Plotinian accounts of *esse/vivere/intellegere* and with Platonism's overall insistence on the primacy of thought over will, Augustine notably does "not use the Porphyrian triad, however modified, to explain the Trinitarian creed of his church." In fact, he decisively breaks with the quintessential Platonic axiom that all intention

9. Ibid., 143.

10. Arendt, *Life of the Mind,* 2:85; likewise, Dihle notes that even as Seneca appears to intuit "that will should be grasped independently of both cognition and irrational impulse," it was not until Augustine that philosophy was able to "develop the distinct notion of will" (*Theory of Will,* 134–135).

arises from cognition and, as Dihle has shown, decisively "separate[s] will from both potential and achieved cognition."[11] In a similar vein, James Wetzel notes how for Augustine "the power of knowledge to motivate is not enough to determine action. We must *choose* to be motivated by rational rather than habitual desires, and if our choice fails to carry into action, we have not the triumph of appetite over reason, but a failure of will." In sharp divergence from the Platonic tradition, then, Augustine's way out of the dilemma of sin chosen and grace refused "is not through knowledge but through love."[12]

More about that shortly; first, though, a basic matter of translation stands to be addressed. With regard to Augustine's notion of the *liberum arbitrium*, it is best to translate it as "free choice." The still widespread alternative rendering of that concept as "free will" (a notion Eleonore Stump retains for her discussion) is bound to create misunderstanding since the will itself is only disclosed in the choice *actually made*. As such, however, the will cannot be free in the modern, libertarian sense of a subject indifferently sampling some buffet of possible identities. Indeed, as the later Augustine was to argue with much force and to rich, if disconcerting, effect, the will—being wholly dependent for its realization on a notion of inscrutable divine grace—is inseparable from the idea of predestination which, beginning with *Ad Simplicianum* (A.D. 397), he developed with reference to Romans 9:11. Summing up the implications of Augustine's anti-Pelagian writings, Stump thus observes that "a person can be morally responsible for a sinful act of will even when it was not possible for her not to will to sin." Consequently, in Augustine's view "free choice" and moral responsibility do not require "that an agent have the ability to do otherwise ... A person who is unaided by grace cannot do otherwise than sin, and yet she is morally responsible for the sin she does."[13] From a theological perspective, the status of the Augustinian will coincides with the relation of the human individual toward grace and the "theological dynamite" of Augustine's doctrine of predestination.[14]

Yet our present objective is not to take up the rich theological implications of the Augustinian will and its deeply vexing entanglement with grace. This is not to deny that the rather arbitrarily compartmentalized fashioning of "an 'Augustine of philosophical interest'" preoccupied with "establish[ing] the mind's self-relatedness" risks

11. Dihle, *Theory of Will*, 124–125. The Platonic axiom is stated very clearly in the *Hippias Minor*, which arrives at this position by having examined two types of deficiency (in practical art and moral reasoning alike), viz., those arising from "involuntary" (ἀκούσίως) and "voluntary" (ἡκούσίως), respectively. Unsurprisingly, a vast body of critical literature has accumulated on the role of the will in Augustine's thought; for introductory accounts, see Stump, "Augustine on Free Will," Wetzel, "Snares of Truth," and Dihle, *Theory of Will*, 123–144.

12. Wetzel, *Augustine and the Limits of Virtue*, 3–4.

13. Stump, "Augustine on Free Will," 131.

14. See Wetzel's superb discussion of predestination and will in Augustine ("Snares of Truth"); quote from 123.

distorting the overall design and purposes of Augustine's thought and occluding its ontological sources.[15] Still, to remain true to this study's overall intent the present discussion will have to limit itself to Augustine's quasi-phenomenological account of how the will registers *in* the self, which is developed most fully in the *Confessions* and *De Trinitate*.[16] Staying within this more confined anthropological matrix, we may begin by drawing out an internal contradiction at the heart of the modern, libertarian conception of the will "as the power of choice—the power to choose one's own motives" and hence "conceptually distinct from desiring." For to take that view is to argue simultaneously that we act "under some representation of the good" and yet frequently and "willfully refuse to act in accordance with what we judge to be best." Inevitably, such a multiple-choice conception of the will renders us "unintelligible to ourselves, at least in so far as our refusal outruns our available motives." On both theological and psychological grounds, Augustine's rejection of the will qua multiple choice targets the Pelagians' "reification of choice, not only because it gave them rather than God the last word on redemption, but because it amounted in essence to a denial of God's power to transform human agents." Moreover, what (following Étienne Gilson) James Wetzel calls a "Pelagian fiction" also proves epistemologically incoherent: "The theory of will as the power of choice, informed by but independent of desire, makes every action to some degree unintelligible, for if the theory were true, no action could ever be sufficiently explained by its motives."[17]

Against this view, Augustine uncompromisingly insists on a "corrupted will" (*mala voluntas*) that is strictly the result of Adam and Eve's Fall and cannot be causally explained as a product of contingent circumstances and wayward desires. While a fuller discussion of the will does not begin until Book 8 of the *Confessions*, a single, early reference to the adolescent Augustine's emergent sexuality correlates the will with "the restlessness of youth" (*inquieta indutum adulescentia*) and a consequent fascination with the created body at the expense of the creator. He characterizes this

15. Hanby (*Augustine and Modernity*, 7) goes on to argue how "both the soul and the city 'answer' to each other by obeying the same dynamic of sin, dissolution and conversion, just one of many macrocosmic/microcosmic isomorphisms that complicate the meaning of Augustinian interiority" (10).

16. Stump, "Augustine on Free Will," 131; Stump offers a concise account of how, "in one treatise after another, Augustine grapples with the problem of making God the sole source of all goodness in the post-Fall human will without taking away from human beings control over the wills ... In the end, Augustine makes it clear that he cannot solve this problem and that he knows it" (ibid., 139).

17. Wetzel, *Augustine and the Limits of Virtue*, 7–8. Hanby concurs that "the Pelagian failure is ... attributable to those stoic debts and the ontological baggage they bring. In employing a conception of self-hood and agency whose original ontological register was the immanentist and monist cosmology of the stoics, the Pelagian self will carry into Christian thought and practice the tensions intrinsic to this cosmology" (*Augustine and Modernity*, 92).

state as being "drunk on the invisible wine of its own will, perverse as it is [*de vino invisibili perversae atque inclinatae in ima voluntatis suae*] and bent on lower things" (2.3.6). Crucially, this passage and its metonymic link with the young Augustine's notorious theft of pears has been set up by an acknowledgment that "I had abandoned you, and was drifting wherever the tide of my own desire took me" (2.2.4), the main point being that a "perverted will" precedes the drama of contingent desires and temptations rather than being caused by them. Thus "sin is its own motive," indeed insatiably so: it is "desire without end."[18] According to Augustine, who never wavers on this point, the disorder and division of the will must not be misconstrued as something externally obtruded but, rather, as the inescapable manifestation of inherited sin. In ways that Blaise Pascal and Coleridge would echo with similar intensity, Augustine's will operates not only "prior to and independent of the act of intellectual cognition" but is also "fundamentally different from sensual and irrational emotion."[19] The point emerges forcefully in Augustine's account of his theft of pears (*Confessions,* 2.4.9ff.), which he is careful not to attribute to any need or want but, instead, explains as manifesting "a love of rebellion itself" (*sed defectum meum ipsum amavi*). More than two millennia later, Coleridge echoes the central point: "a Sin is an Evil which has its ground or origin in the Agent, ... Sin is Evil having an *Origin*. But inasmuch as it is *evil,* in God it cannot originate: and yet in some *Spirit* (i.e. in some *supernatural* power) it *must*. For in *Nature* there is no origin. Sin is therefore spiritual Evil: but the spiritual in Man is the Will ... the corruption must have been self-originated" (*AR,* 266, 273). As Augustine was to put it in *De Libero Arbitrio,* "we sin by the will, not by necessity," since any other position effectively vitiates the very idea of responsible choice and, ultimately, the very idea of human agency itself: "our will would not be a will if it were not in our power ... just as no one sins unwillingly by his own thought, [nor] yields to the evil prompting of another unless his own will consents" (3.3, 10).

It is this startling and unprecedented view—of a corrupted and internally divided will deeply enmeshed with human cognition—that shows Augustine's decisive break with both Stoic and Platonist thought, even as these models continue to inform his argument and conceptual inventory in important ways. As Augustine's critique of Stoic moral theory in *De Civitate Dei* (esp. Book 19) makes clear, the Stoics' ideal of *apatheia* presupposes a strictly external understanding of the main antagonist, viz., as the threatening "intrusion of *phantasiai* into the citadel of consciousness." Stoicism thus restricts "the realm of the voluntary to what we can *control,* not what we *want.*" By contrast, Augustine views that particular fixation as a sign of

18. Wetzel, "Snares of Truth," 132.

19. Dihle, *Theory of Will,* 127; see also Arendt, *Life of the Mind,* 2:95.

"enduring unfreedom," to which he opposes the notion of freedom as "the single-minded love of God." Related to this is the fact, briefly hinted at above, that Stoicism's defensive regimen concerning the senses and external contingency fails to identify its motivating source. The Stoics cannot explain "how we are moved to sin involuntarily, but [also] how we might even be moved to desire beatitude."[20] It is a problem that in due course would also vex Immanuel Kant, who, in good neo-Stoic fashion, is able to furnish a concise definition of the moral "ought" but—having peremptorily committed himself to an agnostic, value-neutral conception of *Kritik*—finds himself unable to determine why anyone *ought* to adopt the "ought" of his moral law. Clearly, Augustine no longer regards the senses or the emotions as the true antagonist of the intellect, which they had been for the Stoics; nor does he see *curiositas* resulting from a distorted appraisal of reality, as the followers of Plotinus had argued. Instead, as illustrated by the theft-of-pears episode in Book 2 of the *Confessions,* Augustine's account of the will originates in and is guided by two metaphysical problems: the origin of evil and the nature of grace.

Recalling his transition from Manichean to Platonic thought early in Book 7, Augustine confronts the startling proposition "that free will was the reason why we commit evil [*liberum voluntatis arbitrium causam esse ut male faceremus*]" (7.3.5). It is here that we find Augustine making one of his most salient points about the will, viz., that willing is inseparable from knowing. Indeed, it is precisely by willing that self-awareness is raised from a latent apperceptive state to a fully explicit one:

> What raised me up toward your light was the fact that I knew that I had a will just as much as I knew that I was alive. Thus, when I willed or did not will something, I was wholly certain that it was I and no one else who was willing it or not willing it; and I was now on the point of perceiving that therein lay the reason for my own sin. As for the evil I did against my will, I saw that I was suffering it rather than committing it, and adjudged it not so much my guilt as my punishment. (7.3.5)

In a striking reversal of Platonic doctrine, Augustine insists that what accounts for the sinful nature of an act is precisely the self's total awareness of the act *as something willed by oneself.* Knowledge of the act and knowledge of the self prove inseparable in ways unprecedented in the moral literature of the Stoics and neo-Platonists. To illustrate the shift, one might turn to Plotinus's late treatise (*Ennead,* 1.4) "On Well-Being," where he deploys the mirror analogy to suggest that reflexive awareness of an activity does not *produce* but at most clarifies the object in question: logically, "there must be an activity prior to awareness," just as in the event that the mirror "is not in

20. Hanby, *Augustine and Modernity,* 98.

the right state the object of which the [mirror] image would have been is all the same actually there." Anticipating the exact same argument that Novalis would advance against Johann Gottlieb Fichte's performative theory of self-creation in his 1794 *Wissenschaftslehre,* Plotinus regards consciousness not so much as thought but as an "after-thought" (*nach/Denken*): in the absence of conditions facilitating a mirror-like reflection of the self *in actu,* "intellectual activity takes place without a mind-picture." Indeed, there are "a great many valuable activities, theoretical and practical, which we carry on both in our contemplative and active life, even when we are fully conscious, which do not make us aware of them." Moreover, there is reason to suppose that act and awareness are ultimately mutually exclusive: "Conscious awareness, in fact, is likely to enfeeble the very activities of which there is consciousness."[21] If for Plotinus, "consciousness is thus more of a memory than a presence," that view hints at a deep chasm separating act from awareness, such that "the more intense an activity is, the less it is conscious."[22]

The contrast with Augustine's moral psychology could hardly be greater. Particularly in the *Confessions,* self-awareness never involves the acquisition—timely or belated—of some unified object or focal point; rather, awareness of self for Augustine almost always amounts to awareness of an indelible conflict. Indeed, it consists of the vivid, present, and all-encompassing recognition of the self *as* constitutively divided rather than subject to occasional "conflicted" feelings. Stump's distinction between first- and second-order desires helps us understand Augustine's divided will and its oblique debt to the earlier, Aristotelian and Stoic antagonism between desire or impulse on the one hand, and deliberate choice or acts of assent on the other hand: "it is a commonplace of medieval philosophy that the higher faculties of human beings are characterized by reflexivity. The intellect can understand itself; the memory can remember itself and its acts; and the will can command itself, as well as other parts of the willer."[23] The principle of unconditional and unconditioned reflexivity, first introduced in *De Libero Arbitrio* (2.4), also constitutes a core premise of Plotinus, and its neo-Platonic version appears again in Ralph Cudworth's discussion of "Freewill" and, very prominently, in the later Coleridge. Long before, Homer's *Odyssey* had furnished a vivid example of the moral agent's intrinsic self-awareness in the Sirens episode (12.197–257) where Odysseus, anticipating the overpowering force of his first-order desires (pleasure), takes measures to ensure that his second-order desires (survival) will prevail.

---

21. Plotinus, *Ennead,* 1.4.10. For a discussion of Novalis's critique of the paradox of reflexivity in his *Fichte-Studien,* see Pfau, *Romantic Moods,* 45–63.

22. Hadot, *Plotinus,* 32–33. Speaking of "the secret of Plotinian gentleness," Hadot observes how "there is no struggle against the self, no spiritual 'combat' in Plotinian asceticism" (95).

23. Stump, "Augustine on Free Will," 126.

Displaying their author's unique "mastery of the psychology of self-contradiction," Augustine's *Confessions* derive their narrative coherence from a metonymic progression toward the idea of a self whose inner phenomenology is shaped by a conflict between "my two wills, the old, carnal will, and the new, spiritual will [whose] discord rent my soul in pieces" (*ita duae voluntates meae, una vetus, alia nova, illa carnalis, illa spiritalis, confligebant inter se atque discordando dissipabant animam meam* [8.5.10]).[24] To sharpen the main point, one might link Augustine's metaphor for the divided will—viz., of domestic strife ("I stirred up a great quarrel in the house of my inner being" [8.8.19])—with a no less powerful and horrific moment conceived by one of modern literature's most Augustinian temperaments. The pivotal chapter (27) of Leo Tolstoy's *Kreutzer Sonata* finds Pozdnyshev recalling how he murdered his wife, an act toward which both the narrative and his own will have been inexorably spiraling:

> With my left hand I seized her hands. She disengaged herself. Then, without dropping my dagger, I seized her by the throat, forced her to the floor, and began to strangle her. With her two hands she clutched mine, tearing them from her throat, stifling. Then I struck her a blow with the dagger, in the left side, between the lower ribs. —When people say that they do not remember what they do in a fit of fury, they talk nonsense. It is false. I remember everything. —I did not lose my consciousness for a single moment. The more I lashed myself to fury, the clearer my mind became, and I could not help seeing what I did. I cannot say that I knew in advance what I would do, but at the moment when I acted, and it seems to me even a little before, I knew what I was doing, as if to make it possible to repent, and to be able to say later that I could have stopped. —I knew that I struck the blow between the ribs, and that the dagger entered. —At the second when I did it, I knew that I was performing a horrible act, such as I had never performed, an act that would have frightful consequences. My thought was as quick as lightning, and the deed followed immediately. The act, to my inner sense, had an extraordinary clearness. I perceived the resistance of the corset and then something else, and then the sinking of the knife into a soft substance. She clutched at the dagger with her hands, and cut herself with it, but could not restrain the blow.[25]

Re-enacting mankind's inaugural trauma of original sin and, more specifically, Cain's version of it, Tolstoy's protagonist acknowledges what most modern philosophy—

---

24. Wetzel, "Snares of Truth," 130. The proof-text for Augustine's divided will is Romans 7:15: "For that which I work, I understand not. For I do not that good which I will; but the evil which I hate, that I do" (*quod enim operor non intellego non enim quod volo hoc ago sed quod odi illud facio*).

25. Tolstoy, *Kreutzer Sonata*, 166–167.

from Hobbes to Locke, Hume, Godwin, Bentham, and on to modern behaviorist and neuro-science accounts of the will as a type of radical and wholly enigmatic, external or internal *compulsion*—so strenuously seeks to deny: viz., that willing is profoundly enmeshed with self-awareness. As he recalls how, prior to his conversion, "my will was perverted, and became a lust; I obeyed my lust as a slave, and it became a habit [*consuetudo*]; I failed to resist my habit, and it became a need" (*Confessions,* 8.5.10), Augustine most definitely does not consider himself a victim of some extraneous or unconscious causality. Each of the steps that cumulatively yield the dystopic narrative of sinful habituation involves a micro-judgment undertaken with acute awareness that the object or action assented to constitute a step in the wrong direction. Consequently, the "chain" (*catena*) of interconnected links both defines the substance of his moral persona and distorts the ways in which it is rendered phenomenologically distinct *by and for the self.*[26] Inner disorder is not obtruded on the will from without; it is its very essence.

Against the Manichean view that "these two minds belong to two Principles, one good, the other evil" (*Confessions,* 8.10.22), Augustine insists that such attempts at tracing the mind's inner division back to some external and metaphysical cause is tantamount to an evasion of responsible agency and, thus, an instance of sin ("they themselves are truly evil, as long as they hold these evil opinions"); for "sin commits our identity and destiny to what must inevitably pass out of existence."[27]

> The mind orders the mind to will [*imperat animus ut velit animus*]; it is only one mind, but it does not do as ordered. Whence is this strange situation? And why is it so? I repeat: the mind that gives the order to will could not give the order if it did not will to do so; but it does not do what it orders. It does not will with its whole being, therefore it does not order with its whole being. The mind orders in so far as it wills, and its orders are not obeyed in so far as it does not will them; for it is the will and nothing else that gives the order that the will should exist [*quoniam voluntas imperat ut sit voluntas, nec alia*]. (8.9.21)

For quite some time, it has been a commonplace to credit Augustine with the "discovery" of inwardness. Hannah Arendt speaks of his "discovery of an *inward* life,"

---

26. As remains to be seen, in discussions of both Aquinas and Adam Smith below, "habit" in Augustine functions in markedly different ways. Overwhelmingly, he views habit as sinful, indeed as "the law of sin ... by which mind is held fast and dragged along as punishment for slipping willingly into it" (*Confessions,* 8.5.12). Because for Augustine any instance in which the soul is acted upon by the body is inherently sinful, he cannot envision—as Aquinas in his account of habit (*habitus*) eventually does—the possibility of mind and spirit being strengthened by repeated, purposive acts of the body. On this topic, see Prendeville, "Idea of Habit."

27. Wetzel, *Augustine and the Limits of Virtue,* 67.

and Albrecht Dihle insists that Augustine's conception of grace and the Trinity are only intelligible if "seen in the wider context of the change from the ontological to a psychological approach to religion and ethics which he initiated."[28] More recently Charles Taylor has argued how "Augustine shifts the focus from the field of objects known to the activity itself of knowing"; and he sees Augustine's "fateful" conceptual innovation of a "first-person standpoint" characterized by a type of "radical reflexivity" such as "brings to the fore a kind of presence to oneself which is inseparable from one's being the agent of one's experience." Of particular concern to Taylor is the hermetic, seemingly asocial, and epistemologically unverifiable model of the Augustinian self, which proves "asymmetrical" inasmuch as no outside observer can ever reconstitute this inner experience.[29] While Augustine does indeed regard the division of the will as "a conflict, and not a dialogue," there is no tension between mind and body, for only the will is inherently "minded." In fact, the conflict and "exchange is entirely mental" and essential to the will since "a will that would be 'entire' without a counter-will, could no longer be a will properly speaking."[30]

What defines the Augustinian will, then, is not the nature of its objects, let alone some hypostatized, external cause. Rather, it is the will's agonistic operation—an inner theater of constant division, strife, and perplexity—which shows it to be inseparable from self-awareness. Alluding to the parable of the lost sheep (Luke 15:3–7) and the prodigal son (Luke 15:11–32), Augustine muses why it is that the soul "delights in finding or having returned to it the things that it loves more than if it had always held on to them" (*Confessions*, 8.3.7). What accounts for the peculiar appeal of the "wax and wane, repulsion and attraction" (8.3.8) that Goethe would later develop into a theory of life as polarity, that Freud would rediscover in little Hans's *fort/da* game, and that Arnold Toynbee would identify as a pattern of "withdrawal and return" permeating much of Western narrative? Augustine's answer, here intimated by a set of examples but not fully conceptualized, appears to be that by our very nature we prefer emergent or restored gifts and capacities over undifferentiated continuity.

28. Arendt, *Life of the Mind*, 2:85; Dihle, *Theory of Will*, 132.

29. *Sources*, 130–131; extending Hauerwas's and Matzko's critique of Taylor's account as one-sided and inattentive to the ways in which the self's inner constitution becomes intelligible and justified only as part of a supra-personal narrative movement toward the City of God, Aers emphasizes how Augustine "reflects, characteristically, on the limits of our self-awareness." Indeed, even the post-conversion Augustine is portrayed, in the *Confessions* (10.32.48), as an "enigma" (*quaestio*) to himself, and his self-reflexivity yields neither clarity nor certainty but lamentable darkness in which his obscure potentials are hidden" (*Salvation and Sin*, 8, 10). For a similar critique that reads Taylor as oblivious of "Augustine's politics [as] a counter-interpretation of his [Taylor's] account of the modern self," see Hanby, *Augustine and Modernity*, 6–12 (quote from p. 10). Taylor has a point, nonetheless, in that the history of Lutheran and Cartesian modernity is inseparable from its often one-sided appropriation of Augustine's understanding of selfhood as a divided inwardness.

30. Arendt, *Life of the Mind*, 2:95.

Phenomenologically speaking, both the reality of the will and human self-awareness hinge on the experience of difference, tension, alternation, or rupture—that is, on a "happening" of some kind.[31]

It is just this view of unrest and conflict as an indelible substratum of human personhood (and of inexpungable sin) that reveals the full extent of Augustine's break with Greek thought, both Stoic and neo-Platonic. That break manifests itself in two fundamental ways. First, Augustine places unprecedented emphasis on self-awareness, albeit not in the sense of Stoic "self-possession" (*autarkeia*) but as the self's indelible awareness of its defective moral vision and consequently, of its utter dependency on divine grace. Thus it is rather ironic that Augustine should so often be credited with having originated a modern, proto-Cartesian conception of "the self" in which agency and autonomy are construed as wholly interchangeable and as the *terminus ad quem* of a methodical quest for a metaphysically uncurious, "buffered self" (as Charles Taylor has recently dubbed it). The second break with pagan philosophy involves Augustine's contention that sense perception and intellectual activity are not anchored in an impersonal order of being (cosmos). Instead, his psychology "seems to be self-sustaining, at least with regard to man's intellectual activity ... Both the raw material of cognition and the drive towards understanding can be found in the soul without an indispensable point of reference in the outside world."[32] Central to Augustine's theory of the will, and decisively setting it apart from Stoicism, is the concept of "grace" unaccountably and freely bestowed by a personal God who, unlike Stoicism's pantheist deity, constitutes an internally differentiated Trinity entwined with the human individual whom Augustine views as the imperfect *imago* of the creator God: "that supreme and most high being of which the human mind is the unequal image [*impar imago*], but the image nonetheless" (*ADT*, 10.4.19).[33]

31. Hegel, herein also concurring with Augustine, connects this anthropological link between self-awareness and discontinuity to the phenomenology of inner-time consciousness, and he aptly defines time as "absolute unrest" (*die absolute Unruhe*) and "absolute self-division" (*absolute Selbstzerissenheit* [*PS*, 487/*PG*, 558]).

32. Dihle, *Theory of Will*, 125–126. C. Taylor's initial suggestion that "Augustine gives us a Platonic understanding of the universe as an external realization of a rational order" (*Sources*, 128) seems rather at odds with his overall reading of Augustine's "calling us within" (*in interiore homine*).

33. On Augustine's "Stoic appropriation of Plato," see Wetzel, who notes that "Stoic rather than Neoplatonic influence informed [Augustine's] early views of virtue, autonomy, and the good life and disposed him to think Stoically about ethics throughout his career as a philosopher and theologian" (*Augustine and the Limits of Virtue*, 10–11); for fuller accounts of Augustine's inconsistent uses of and evolving relation to Stoicism, see Colish, *Stoic Tradition*, 142–238, and Verbeke, "Augustin et le Stoïcisme." Hanby is right to question Taylor's reading here, "which makes the trinitarian context" for Augustine's psychology "appear incidental" (*Augustine and Modernity*, 9).

While this is not the place to engage the apparent tension between free choice of will and grace—a topic whose exploration spans almost all of Augustine's writings—a couple of basic aspects of that issue need to be kept in mind. First, and arising out of Augustine's protracted disputes with Pelagius and his followers, there is the basic distinction between "enabling" and "cooperative" grace. Far from posing a challenge to freedom of will, enabling grace is generally accepted as its very condition of possibility. It is a "gift of power," of the capacity to form judgments and to resolve and embark upon a specific course of action. Rather more problematic is Augustine's insistence (against the Pelagians) that the gap between an intentional action and its successful execution cannot be closed by human means alone. Following at times violent confrontations with the Pelagians in the wake of Pope Zosimus's decision to uphold their view that men have no need for cooperative grace, the Council of Carthage at the end of April A.D. 418 had anathemized that view.[34] Comparing God to the eye without which we cannot see,[35] Augustine insists that absent divine, cooperative grace the success of our endeavors—and indeed the integrity of the goal at which they are aimed—is not conceivable. The very conception or act of judgment whereby the will orients us as practical agents in a particular situation is for Augustine (and also for Jerome) impossible without God's cooperative grace. Referencing Romans 8:28, the argument here is that not just the successful execution of a particular action but even the clarity of the intellect that had previously resolved upon it hinges on "God's cooperation"—the Latin *cooperantur* here rendering the Greek *συνεργεῖ ὁ θεὸς*. Yet precisely this indispensable role of divine "synergy" threatens to undermine the basic notion of human freedom and self-determination.

The matter comes to a head eight years later in Augustine's essay "On Grace and Free Choice." Writing around Easter of A.D. 426, Augustine here defends himself against a certain construction put on his account of divine grace that he had set forth in a letter to the then bishop of Rome (and future Pope Sixtus III) in A.D. 418 when the Pelagian controversy had been at its peak. Against the misreading of that (subsequently disseminated) letter by the "rustic and less educated" monks at Hadrumetum, Augustine seeks to counter the impression that the enigmatic and absolute character of divine grace effectively corrupts or denies the possibility of free choice. As so often, Augustine turns to St. Paul as he tackles the question, in this case to 2 Corinthians 6:1 ("Working together with him, then, we appeal to you not to receive the grace of God in vain"). "Why," he asks, "does he beg them if they received grace in such a way that they lost their own will? [*Utquid enim eos rogat, si gratiam sic suscep-*

34. For more detailed accounts of Augustine's conception of free will in relation to grace, see Kirwan, *Augustine*, 82–128; on the historical background of the controversy, see P. Brown, *Augustine of Hippo*, 340–366.

35. *De Peccatorum Meritis*, 2.4.4 (quoted in Kirwan, *Augustine*,108).

erunt, ut propriam perderent voluntatem?]." Notably, in broaching the issue as a rhetorical question that implies the correct and only possible answer, Augustine minimizes his own intellectual agency—a point also evident when at the close of his essay he insists "that it is not so much I as it is the divine scripture itself which has spoken with you by the clearest testimonies of the truth."[36] Such a claim is no mere rhetorical conceit; rather it embodies what Augustine never tires to affirm—viz., that the human intellect's engagement with the "gift" of God's revelation in the medium of scripture constitutes prima facie evidence of the ways in which the human will and divine grace coexist.

Meanwhile, his reading of Paul carefully balances two points warranting affirmation: that the human will has its own, discrete psychological reality, and the soteriological fact of its unconditional dependence on divine grace. On the one hand, Paul's insistence that "his grace has not been without effect in me" (*gratia eius in me vacua non fuit* [1 Cor. 15:10]) affirms that grace does not *determine* the nature of finite, human consciousness in the manner of a causal mechanism. Rather, grace is *for* consciousness, an intentionality at once distinct from consciousness and yet uniquely capable of focusing and elevating it. Hence, too, Paul affirms that "by the grace of God I am what I am [*gratia autem Dei sum id quod sum*]," thereby reinforcing the distinct reality and free agency of his empirical persona while remaining fully aware of its metaphysical source. Grace, as Augustine construes the passage in Paul's letter, is itself the precondition for disclosing what is empirically and psychologically experienced as the sheer plenitude of human life, its distinctive and enduring character (*ēthos*) or self-identity as manifested by a narrative of past failings. Thus Paul had just acknowledged (1 Cor. 15:9) his previous role as "persecutor" of the church. Not only, then, are human agency and identity not incommensurable with divine grace, but in Augustine's view they are positively unintelligible without it. Indeed, Augustine's entire psychology hinges on this productive tension between grace and free choice. In his exchange with St. Jerome about this seemingly insoluble question, he acknowledges that the soul "is not a part of God. For if it were this, it would be utterly immutable and incorruptible … it would not become worse and make progress for the better."[37] Likewise, in *De Civitate Dei* Augustine also affirms that the image of God that we recognize in ourselves "is not equal to God" but, in fact, "is very far removed from Him, … it is not of the same substance as God." Even so, it is "nearer to God in nature than anything else made by him [*tamen qua Deo nihil sit in rebus ab eo factis natura propinquius, imaginem Dei*]" (*CD*, 11.26).

36. "Of Grace and Free Choice" (12; 20.41), in *Answer to the Pelagians*, 4.99.

37. *Epistel* 166.2.3 (as quoted in Teske, "Augustine's Theory of Soul," 118).

In an idiom less emotionally and rhetorically charged than either his *Confessions* or his late writings against the Pelagians, Augustine's *De Trinitate* integrates an understanding of the will into a comprehensive psychological theory whose logic bears retracing so as to guard against misunderstandings and internal tensions that befall modern faculty-psychology (e.g., in Kant). Augustine's arguments here arise out of a critical discussion of the Delphic injunction to "know thyself" (*gnothi seauton*) which, Augustine maintains, ought not to be misconstrued as enjoining the mind to furnish some syllogistic self-definition. In fact, mind (*mens*) is intrinsically self-aware: "when it seeks to know itself, it already knows itself seeking" (*ADT,* 10.2.5). By definition, mind lives, for "it cannot be mind and not be alive … But it knows that it lives; therefore it knows its whole self. Finally, when the mind seeks to know itself it already knows that it is mind; otherwise, it would not know that it is seeking itself" (*ADT,* 10.2.6). Among the most original and compelling arguments in *De Trinitate,* Augustine's critical engagement with the Delphic principle of self-knowledge takes that exhortation to mean that mind "should think about itself and live according to its nature" (*secundum naturam suam vivat*). In what amounts to a reorientation of the Stoic doctrine of assent, Augustine thus notes how the mind, so richly embedded in and engaged with the external world, has become perilously attached to things by converting them into images whose status, for Augustine no less than Plotinus, remains deeply ambivalent. "Image" (as *eidos/imago*) may certainly be appraised as the representation of an already secondary, ectypal, and imperfect world and, hence, as something deficient and potentially misleading—an "illusion" (*eidolon*).

Yet at the same time, Augustine is close enough to the Platonic tradition to recognize that the image is never merely determined by its putative referent but also points back to the source from which that referent—the ectypal world of created things-of-sense—derives its existence. In this latter sense, each image constitutes a trace of its archetype, a point strongly affirmed by the evident fact that the essence and seat of the image is in the human mind rather than in the three-dimensional world to which it ostensibly refers. "A 'slide' in the estimation of the image is therefore possible. It may be thought as having very little share in reality," or even as dialectically revealing the ultimate enigma of the material world, a point forcefully made by Plotinus's characterization of the ghostly nature of "matter" (*ὑλη*) and still prominent in early patristic thought (Origen, Eusebius).[38] Both dimensions of the image—

38. A. H. Armstrong, "Neoplatonic Valuations," 42. See Plotinus, who puzzles over matter, concluding that it is "not soul or intellect or life or form or rational formative principle or limit—for it is unlimitedness—or power—for what does it make?—but, falling outside all these, it could not properly receive the title of being but would appropriately be called non-being … [or] truly not-being [*μη όν*]; it is a ghostly image [*είδωλον και φάντασμα*] of bulk … a phantom which does not remain and cannot go away either" (*Ennead*, 3.6.7); see also Eusebius's letter to Constantia (of dis-

as a form of reference and as trace of the divine, respectively—operate in Augustine's work on the Trinity. Inasmuch as the mind's commerce with the world involves an elaborate web of imagistic representations, it is forever at risk of becoming enslaved by its own projections. Having given "something of its own substance to their formation," the mind risks becoming attached to its images "with the glue of care, [and] it drags them along with itself even when it returns after a fashion to thinking about itself [*eisque curae glutino inhaeserit, attrahat secum etiam cum ad se cogitandam quodam modo redit*]" (*ADT,* 10.2.7). As James Wetzel notes, "Augustine tends to redescribe problems of knowledge ... as problems of will or agency, thereby making the appropriation of knowledge and not knowledge per se his explanandum."[39] By its very nature, mind is acquisitive and proprietary as it converts objects into images and,

> in its destitution and distress ... becomes excessively intent on its own actions, and the disturbing pleasures it culls from them; being greedy to acquire knowledge of all sorts from things outside itself, which it loves as known in a general way and feels can easily be lost unless it takes great care to hold onto them, it loses its carefree sense of security, and thinks of itself all the less the more secure it is in the sense that it cannot lose itself.[40]

This is a fine instance of how for Augustine epistemological and moral questions are essentially entwined; representation of the world by means of the "image" (*imago*) is prone to induce an acquisitive and proprietary state of mind, and the security of these virtual possessions tends to render mind oblivious of its strict metaphysical dependency on divine grace—a point nicely captured by the chiasmic construction that concludes the above passage.

Augustine's basic view that the things with which we are engaged surreptitiously come to define our very being ("when the mind thinks of itself like that, it thinks like a body" [*ADT,* 10.3.8]) thus traces the epistemological sources of the divided will whose spiritual burden he had already explored in the *Confessions.* Rather than zeroing in on itself as the object of some eventual, clarifying proposition, the remedy

---

puted authenticity), which echoes "the Origenian idea of an instrumental body added onto the soul as a result of sin" (Besançon, *Forbidden Image,* 120).

39. *Augustine and the Limits of Virtue,* 14; see also Wetzel's subsequent discussion of "Sin and Entropy" (ibid., 37–44).

40. *ADT,* 10.2.7: *ideoque per egestatem ac difficultatem fit nimis intenta in actiones suas et inquietas delectationes quas per eas colligit; atque ita cupiditate acquirendi notitias ex iis quae foris sunt, quorum cognitum genus amat et sentit amitti posse, nisi impensa cura teneatur, perdit securitatem, tantoque se ipsam minus cogitat, quanto magis secura est quod se non possit amittere.*

involves instructing the mind in the (originally Stoic) art of developing distance vis-à-vis its own projections: "when it is bidden to know itself, it should not start looking for itself as though it had drawn off from itself, but should draw off what it has added to itself . . . Let the mind then not go looking for a look at itself as if it were absent, but rather take pains to tell itself apart as present. Let it not try to learn itself as if it did not know itself, but rather to discern itself from what it knows to be other [*sed ab eo quod alterum novit dignoscat*]" (*ADT*, 10.3.11–12). To do so is, at last, to "turn on to itself the interest of its will" (*sed intentionem voluntatis . . . statuat in semetipsam*).[41] Thus, simply by grasping the nature of self-knowledge the mind has already fulfilled the injunction: "it is being commanded to do something which it automatically does the moment it understands the command." The basic psychological schema—composed of memory, intelligence, and will—breaks down into different types of awareness, achieved by *memoria* and *intellegere*, and the agency proper, the will, which "is there for us to enjoy them or use them" (*ADT*, 10.3.12).

In Books 10–12 Augustine works out a homology between the internal relation of the senses, the faculty of the intellect, and the Trinity. Parsing memory, understanding, and will, Augustine construes the mind as structurally cognate with the relation of the divine persons. While "not three lives but one life, not three minds but one mind, . . . each of them is life and mind and being with reference to itself." Augustine's careful insistence on the co-presence of *memoria, intelligentia,* and *voluntas* (*ADT*, 10.4.18) constitutes a shrewd rhetorical move. Thus the more challenging aspects of Trinitarian theology are illustrated and affirmed by seemingly self-evident psychological realities, while at the same time suggesting that the structure of the mind *in actu* reflects a divine and providential arrangement. At least in part, Augustine's reciprocal account, in which Trinitarian and psychological argumentation and evidence mutually sustain each other, has to be seen against a contentious history in which the volatile political divisions of late imperial Rome had become entangled with complex theological and social issues, such as the ongoing dispute over the understanding of the Trinity between the Councils of Nicaea (A.D. 325) and Constantinople (A.D. 381), the political and social status of pagan religions, and the widening gap between the Eastern and Western churches.[42] In taking up the "relation" of the mind to itself, Augustine's *De Trinitate* enters a debate that had barely quieted down following the defeat of Arianism at the Council of Constantinople whose resolutions he felt still required a great deal of analytic work. Crucially, *De Trinitate* seeks to clar-

41. The inward focus is conveyed by the compound *semetipsum* (derived from the Greek *ekenôsen*).

42. For a concise overview, see MacCulloch, *Christianity*, 211–222; for detailed accounts, see Ayres, *Nicea and its Legacy*, 41–130, and Cassiday and Norris, eds., *Cambridge History of Christianity*, esp. the essays by Khaled Anatolios ("Discourse on the Trinity"), Alan Brown ("The Intellectual Debate between Christians and Pagans"), and Mark Edwards ("Synods and Councils").

ify the strictly organic—not hierarchical—interaction of the mind's psychological faculties which, consistent with Trinitarian thought, Augustine views as distinct manifestations of a single essence. Mind is defined by their inter-relation and, as such, "perceives its whole self and nothing else." As he puts it, in the soul (*anima*) the different functions "are, so to say, all rolled up and have to be unrolled in order to be perceived and enumerated—substantially or being-wise, ... and not as in a subject [*essentialiter, non tanquam in subjecto*], like color or shape in a body, or any other quality or quantity" (*ADT,* 9.1.5).

The relation of will to intellect and memory first established in the *Confessions* (10.11.18) is integrative by nature. Mind is not an aggregate of three discrete faculties, but "these three are one thing, and when they are complete they are equal" (*ADT,* 9.1.4). In characterizing the relation between mind, love, and knowledge (as well as the triad of memory, intelligence, and will), Augustine frequently stresses that the *difference* of these faculties is not substantive but functional: "thus mind is of course in itself, since it is called mind with reference to itself, though it is called knowing or known or knowable relative to its knowledge; also as loving and loved or lovable it is referred to the love it loves itself with" (*ADT,* 9.1.8). Moreover, the discrete functions always interpenetrate one another, such that "the mind loving is in love, and love is in the knowledge of the lover, and knowledge is in the mind knowing" (*ADT,* 9.1.8). This at once differential and integrative (Trinitarian) conception of mental functions—for which Augustine, wishing to avoid the undifferentiated collective *omnes,* fashions the somewhat idiosyncratic Latin *tota vero in totis quemadmodum sint* ("how they are all in all")—is crucial to Augustine's anthropology and psychology. For if each faculty were to be credited with operating independently of the others, not just as regards its function but doing so *incommunicado,* as it were, one might very well conclude that an act might be undertaken without full responsibility, that an act of will might occur without consciousness, etc. Against this disjunctive (skeptical) view, Augustine insists on the outright convertibility of mind and self-awareness:

> nobody surely doubts, however, that he lives and remembers and understands and wills and thinks and knows and judges. At least, even if he doubts, he lives, if he doubts, he remembers why he is doubting, if he doubts, he understands he is doubting, if he doubts, he has a will to be certain, if he doubts, he thinks, if he doubts, he knows he does not know; if he doubts, he judges he ought not to give a hasty assent.[43]

43. *ADT*, 10.3.14; having been misled by Cicero into thinking of Aristotle's *De Anima* as a materialist account of mind, Augustine resists the notion, later adopted by Aquinas, that the soul is the form of the body (*ADT*, 10.3.15).

Thus the most elemental operation of sight (e.g., of some thing outside ourselves) already reveals the presence of a "conscious intention" (*animi intentio*) inasmuch as we are not merely gazing at something but positively sustain our focus on the object in question. Even in a blind person "the desire to see remains intact" (*ADT,* 11.1.2). Behind every event or act stands the will to focus "attention" on the thing seen, a point that emerges clearly when, in the absence of the object in question, memory retrieves it. This it can only do because the initial vision had involved an act of "conscious attention" (*acies animi*). Indeed, already the very act of sight—by virtue of the agency of the will—transposes raw sensory data into an image or internal vision, and it is only the latter that can ever be retrieved by memory later on.

Though seemingly neutral in its engagement with an ambient world of sensory objects, every mental operation presupposes a will if it is to have any coherent and sustained orientation. Hinting how the psychology of perception is enmeshed with the hermeneutic of the human person and, ultimately, with the ontology of original sin, Augustine thus notes that "the will was already there before sight occurred [*prius enim quam visio fieret, jam erat voluntas*], and it applied the sense to the body to be formed from it by observing it."[44] Mind is dependent on the senses, and the quality of their interaction points to the agency of the will as the indispensable *tertium quid.* To be sure, this connection has to be established in somewhat oblique fashion since Augustine must take care not to indict the created, material world as intrinsically defective. Rather, the divided will is said to correlate to human life's intrinsic split between the sensory and the intelligible:

> By the very logic of our condition, according to which we have become mortal and carnal, it is easier and almost more familiar to deal with visible than with intelligible things, even though the former are outside and the latter inside us, the former sensed with the senses of the body and the latter understood with the mind, while we conscious selves are not perceptible by the senses, not bodies, that is, but only intelligible, because we are life [*nosque ipsi animi non sensibiles simus, id est, corpora, sed intelligibiles, quoniam vita sumus*]. (*ADT,* 11.1.1)

At first glance, we seem to have been returned to the world of Gnostic speculation or, perhaps, to its belated resurgence in William Blake's *Book of Urizen* (1793), which

44. *ADT,* 11.3.9. A famous instance of the sense of sight being enslaved by a sinful will and, as such, continuing to enslave a person by sheer force of habit would be that of Augustine's friend Alypius, mentioned in *Confessions*, caught up in voyeuristic thralldom to violent gladiatorial shows and eventually wounded in the soul by the act of sight (*Confessions*, 6.8.13). See also Augustine's emphatic indictment of "carnal sight," "bodily illusions," "unreal bodies," "empty shadows," and other "illusions of mine" in *Confessions*, 3.6.10.

depicts man's creation as a kind of fall ("rent from Eternity") into an embodied and confined "state of dismal woe" vividly depicted by the restriction of eternity to what can be apprehended by the senses, including the eyes—"two little orbs ... fixed in two little caves" (*CPP*, 76). Traces of the Gnostics' division between a demiurge responsible for the abject state of material creation and a redeemer God capable of restoring the fallen beings to a state of divine fullness (*plērōma*) seem to linger in Augustine's argument.

Yet such a reading of Augustine and of his neo-Platonic precursors needs to be revised.[45] To be sure, Augustine does regard the mind's very dependence on the senses as prima facie evidence of inherited sin; yet this is not to say that sensory mediation is per se the *cause* of sin. Rather, our dependence on the senses is for Augustine the source of profound and abiding confusion, and only in an entirely mediate sense can such confusion be viewed as the proximate cause of sin. The argument in question begins with Augustine noting how, by dint of its perceptual habits, the self is liable to lose track of the distinction between an inner sense of "sight" (*visio*) furnishing the condition of possibility for perception and the contingent activation of that sensory apparatus by external appearances. His analysis here differs markedly from Aristotelian and Stoic epistemology (subsequently echoed by Aquinas), according to which the human being's reliance on the senses constitutes a neutral anthropological fact and indeed a natural consequence of how finite and embodied beings have been created. Mounting a far more evaluative argument, Augustine not only demurs our habitual obliviousness of the distinction between the power of sight and its externally triggered activity, but he further distinguishes between the faculty *in actu* and the will that directs its attention: "the conscious intention [*animi intentio*] which holds our sense on the thing we are seeing and joins the two together not only differs in nature from that visible thing, since it is consciousness while that is body, but it also differs from the sense itself and the sight, since this intention belongs only to the consciousness [*quoniam solius animi est haec intentio*]" (*ADT*, 11.1.2). Since perception is not some merely reactive and occasional event but a form of *intentionality*, it cannot be understood as a value-neutral, "innocent" act but reveals the basic inclination of the will.

Augustine's position was to be recovered by a strand of modern phenomenology. Extending arguments by Edmund Husserl, Robert Sokolowski thus argues that sensory perception is premised on what he calls "veracity ..., an inclination that needs to be wanted." Contrary to some "indifferent impulse," the process of sensory

45. Often misattributed to Plotinus, the view of the natural (created) world as fallen, ectypal, and reprehensible is anything but neo-Platonic. For a nuanced discussion of Platonist thought and its tempering influence on Augustine, see Armstrong, "Neoplatonic Valuations."

perception reveals our pre-conceptual stance vis-à-vis the world. To be sure, Heidegger, and long before him Kant, had elaborated how rationality invariably originates in a pre-conceptual ground—some type of aesthetic "mood" or existential "attunement" (*Stimmung*). Yet for Augustine, such a position would seem precariously impersonal and adventitious—a function of shamanistic contingency rather than an elective commitment. Thus, he attributes our responsiveness to sensory perception to the inner disposition of our will, which either allows or denies phenomena to disclose themselves to us in their fullness. Likewise, "it is not the case that our reason gathers in various truths in an impersonal manner, and that our responsibility starts up only after such acquisitions have taken place." Rather, "the very disclosure of things, the initial manifestation that makes the options available, depends on our responsiveness to the way things are."[46] Far from being some impersonal and universal mode of uptake or default of our being "thrust into the world" (Heidegger's *Geworfenheit*), a person's "responsiveness" even to seemingly elemental, sensory phenomena invariably presupposes, and potentially reveals, her or his spiritual condition or will; all perception always implies (and potentially reveals) something qualitative about both the percept *and* the perceiver.

Even at a basic epistemological level, the operation of the will is the necessary condition for the reality of the soul (*anima*) and the specific operation (*actus*) of the tripartite mind (*mens*): "the will, then, turns the attention here and there and back again to be formed, and once formed keeps it joined to the image in the memory [*voluntas vero illa quae hac atque illac fert et refert aciem formandam, conjungitque formatam, si ad interiorem phantasiam tota confluxerit*]" (*ADT,* 11.2.7). Imagination (*phantasia*), a key term for the Stoics and Plotinus, furnishes evidence that "the image of the visible thing is formed in our sense when we see it" (*formari in sensu nostro imaginem rei visibilis* [*ADT,* 11.1.3). Indeed, even in the absence of such a thing, we may reconstitute or contrive objects of possible experience (to use a Kantian phrase) "by increasing, diminishing, altering, and putting ... together" memory images in entirely novel ways.[47] There are, then, three building blocks of sensory experience: "that form of the body which is seen, and its image imprinted on the sense which is sight or formed sense [*impressa ejus imago sensui quod est visio sensusve formatus*], and the conscious will which applies the sense to the sensible thing and holds the sight on it"; and all are "compounded into a kind of unity" because "the will exerts such force in coupling the [other] two together" (*ADT,* 11.1.5). Echoing less the relaxed vision of Plotinus than the otherworldly stress of Porphyry, Augustine does in-

46. Sokolowski, *Phenomenology of the Human Person*, 93–94.

47. See *ADT*, 11.2.8. Notably, the terminology here shifts from the morally neutral *imago* to a concern with potential (self-)deception at the hands of "imaginative fancies" (*imaginata phantasmata*).

deed view the body as an antagonist of the spirit, albeit only insofar as the world of finite creation tends to distract the soul from its ideal upward trajectory.[48] Yet even if Augustine may have missed or ignored some of its more flamboyant elements, Plotinus's version of Platonism nonetheless appears to have furnished an important counterweight to Augustine's more stridently dualist, Pauline views; by helping "to overcome the tendency to a sometimes positively frenzied dislike of this world and of the body, which seems to have curiously deep roots in the religion of the Incarnation."[49] As he was to observe later in *De Civitate Dei,*

> we are pressed down by the corruptible body, ... yet we know that the cause of our being pressed down is not the nature and substance of the body, but its corruption; and, knowing this, we do not wish to be divested of the body, but to be clothed with its immortality ... Those who suppose that the ills of the soul derive from the body are in error, ... for though this corruption of the flesh results in some incitements to sin and in sinful desires themselves, we still must not attribute to the flesh all the vices of a wicked life ... It is not by having flesh ... that man has become like the devil. Rather, it is by living according to his own self; that is, according to man [*sed vivendo secundum se ipsum, hoc est secundum hominem, factus est homo similis diabolo*]. (14.3)

An analogous, threefold relation of "memory and internal sight [*internae visionis*] and the will" (*ADT,* 11.2.6) obtains even when objects of sense are not present but only remembered. Mining the etymology of *cogo* (a contraction of *coago* = to coerce or bring together), Augustine thus sets forth a synthetic model of knowledge whereby discrete mental functions "are *coagitated* into a unity the result [of which] is called cogitation, or thought [*quae tria cum in unum coguntur, ab ipso coactu cogitatio dicitur*]" (*ADT,* 11.2.6). Augustine's principal challenge now becomes to show that the will is neither the antecedent and superior source ("quasi-parent") of cognition nor its belated "quasi-offspring." For in the first case, the result would be a quintessentially modern theory of knowledge as sheer projection of a will denuded of all self-awareness—one whose variously mechanist, necessitarian, associationist, or pessimist form we find developed by Hobbes, Locke, Gassendi, Mandeville, Hartley, Godwin, Priestley, and Schopenhauer, among others. The obverse scenario involves

---

48. What Augustine read of Plotinus, and to what extent he received Plotinus's teachings from Porphyry or Marius Victorinus also remains a matter of much debate; following O'Connell's earlier work, Hilary Armstrong suggests that "Augustine missed just those elements in the thought of Plotinus which ... are capable of stimulating the artistic imagination and inducing a positive attitude to the body and the sense world" ("Neoplatonic Valuations," 39).

49. Ibid., 38.

an account that demotes the will to a secondary, epiphenomenal agency exclusively tasked with implementing what cognition has furnished, thereby returning us to the Socratic maxim that moral action naturally arises from cognition. In fact, Augustine insists, memory and intelligence only ever operate in relation with the will, such that "in every one of them you find these three; the thing stowed away in the memory even before it is thought about, and the thing that is produced in thought when it is looked at, and the will joining the two together" (*ADT,* 11.3.12). As the dynamic principle of the mind, the will is the agency of choice and commitment in that it alone directs and sustains our focus of attention and so creates a quasi-intentional state (*acies animi*). Consequently, the specific orientation of mind (*mens*) involves an evaluative commitment that defines the person's "soul" (*anima*): "just as it is the will which fastens [*conjungit*] sense to body, so it is the will which fastens memory to sense and the thinking attention to memory" (*ADT,* 11.3.15). Likewise, it is the will that "leads the thinking attention where it pleases through the stores of memory in order to be formed, and prompts it to take something from here out of the things we remember, something else from there, in order to think things we do not remember" (*ADT,* 11.3.17).

Augustine's psychology of willing thus stands in sharp contrast to modernity's overwhelmingly non-cognitive and irrational concept of volition as some externally induced compulsion. Far from being one of several faculties seemingly enjoying free-agent status, Augustine's will constitutes the enduring and all-pervading substratum of personhood. It defines the essence of the person, which is to say, discloses her or his unique (if invariably divided and conflicted) identity. As such, it must not be mistaken for a mere attribute or faculty of the mind or the "self"—that is, some generic capacity supporting or, as the case may be, obstructing the intellect's conceptual labors. In fact, Augustine's will is neither subsidiary nor opposed to the intellect; strictly speaking, it is not even a "concept" at all, certainly not in the way that "intellect" (*mens*) or "reason" (*ratio*) may be said to constitute such. Neither can *voluntas* be predicated of multiple selves but, being incommunicable, it only discloses itself in the evolving narrative of a life realized by a singular and necessarily unique person. Still, the disclosure of the will is inseparable from the person's habits of deliberation, choice, and action as these take place within the intellect, a point frequently overlooked by modern hyper-Augustinian writers such as Luther who, in rather Manichean fashion, construe the will as an antagonist of reason forever chafing at the constraints that any enduring commitment to a rational order would allegedly impose on it. It is just this development that now stands to be traced.

6

RATIONAL APPETITE AND GOOD SENSE

Will and Intellect in Aquinas

The intellectual dimension so prevalent in Aristotle's account of "choice"—yet crucially fused with Augustine's metaphysics of grace—was to find its consummate articulation in Thomas Aquinas's *Summa Theologiae,* particularly in his discussion of the will in the so-called "Treatise on Man" and at the beginning of the *Prima Secundae.* Unlike in Aristotle's ethics, the will now presents itself in the two distinct forms of the divine (uncreated) and the human, finite person. Both are, for Aquinas, ontologically related, specifically because the very idea of the "person" as a rational being is intelligible only on the basis of its participation in "sanctifying grace" (*donum gratiae gratiam faciens*). For Aquinas, that ontological framework must be accepted by all those who wish to engage in a rational and sustained exchange about pretty much anything at all. Even heretics and pagans can be engaged, within limits, provided they accept the premise of some non-contingent relation between the human and the divine. Minimally, such a relation becomes legible in the inner teleological structure or purposiveness of being, a point famously set forth in the "five ways" of Quaestio 2 of the *Summa.*[1] Thus "each and every part exists for the sake of

1. On the *quinque viae,* see Rudi te Velde's scrupulous account of the first "way" in *Aquinas on God,* 37–63. Te Velde effectively dismantles the longstanding misreading of the "five ways" as

its proper act, as the eye for the act of seeing; secondly, that the less honorable parts exist for the more honorable, as the senses for the intellect ... thirdly, that all parts are for the perfection of the whole, as the matter for the form [*sicut materia propter formam*] ... Furthermore, the whole man is on account of an extrinsic end, that being the fruition of God [*ut fruatur Deo*]" (*ST,* Ia Q 65 A 2). However distant it may seem to us now, Aquinas's position constitutes not simply a statement of belief but, consistent with Anselm's notion of "faith seeking understanding" (*fides quaerens intellectum*), concurrently enacts a commitment to reason itself. For Aquinas, the very intelligibility of any particular thing—that is, of matter concretized qua form—hinges on its teleological orientation toward a superior end. Yet if the rationality of things inheres in their *relations,* it does so not merely by being of instrumental use to higher creatures but also by affirming the teleological ordination of the whole: "Every creature exists for the perfection of the entire universe."[2] It bears recalling that in its root *universum* means "turned toward unity," and that for Aquinas a singular being cannot be known except as something "active, self-manifesting, and self-communicating through action."[3] The entire "Treatise on the Work of the Six Days" (*ST,* Ia 65–74), as well as the "Treatise on Man" (*ST,* Ia QQ 75–102) that follows it, pivot on this relational model of rationality. In the case of the human being, however, some special circumstances apply and now need to be briefly recalled.

First and foremost, Aquinas insists that unlike all other created beings, "rational creatures ... can attain to [God] by their own operations" [*quem attingere possunt sua operatione* (*ST,* Ia Q 65 A 2)]. To understand Thomas's conception of will and intellect is to encounter a model of personhood radically different from modern notions of individuality or subjectivity as they are consolidated under the heading of self-possession, self-discipline, and autonomy by the alternately neo-Stoic, anti-Aristotelian, and anti-clerical modern projects of Luther, Justus Lipsius, Cornelius Jansen, Descartes, Hobbes, Kant, and many others besides. Unlike the Platonic model, in which *eidos* and *physis* remain antagonists, all existence—including, especially, that of the human person—constitutes for Aquinas a vivid instantiation of the ontology of reason. Following Aristotle, he thus conceives being as the realization of a substantial form. In a daring synthesis of Aristotelian metaphysics with Augustinian spirituality, Aquinas also posits all existence as a divinely created gift

---

supposed evidence of Thomas's commitment to some version of natural theology. See also Kerr, *After Aquinas,* 52–72.

2. Discussion of teleology in Aquinas, and on the possibility of a non-teleological and post-theistic modeling of reason continues unabated. For the varied reception of Aquinas in twentieth-century existentialist thought from Étienne Gilson to Heidegger to Hans Urs von Balthasar, see Kerr, *After Aquinas,* 80–93; for an instructive, contrasting account of Aquinas's and Hegel's conceptions of being and teleology, see Lakebrink, *Perfectio Omnium Perfectionum,* esp. 38–74.

3. Clarke, *Explorations,* 215.

(*donum*)—something that can never be occasionally sought out as an object of interest or excluded from further consideration, simply because it is always already "given." As such, existence enjoins the created human individual prima facie to participate in the "concreated" forms that confer on each particular thing its reality as a distinctive specimen. Time and again Aquinas thus emphasizes the dynamic, operative character of all being. To be is an act: "From the very fact that something exists in act, it is active" (*Summa Contra Gentiles,* I, Chapter 43). In ways that will make a surprising reappearance in Goethe's formalist account of plant development, Aquinas sees all being as, literally, "actual." To be actual means for a thing to be an agent, to act by actively manifesting its being *for others:* "It is the nature of every actuality to communicate itself insofar as it is possible."[4] What Jacques Maritain calls "the basic generosity of existence" thus implies that the identity of a thing (*res*) is guaranteed not by its inert and self-contained otherness vis-à-vis a subject but by its active relatedness to other beings.[5]

Following Augustine, Aquinas accords "relation" a unique place, above and beyond the other nine Aristotelian categories (*praedicamenta*). Thus he insists that "relation" is not simply an intrinsic quality or "accident": "The true idea of relation is not taken from its respect to that in which it is, but from its respect to something outside ... [Relations] signify a respect which affects a thing related and tends from that thing to something else." In other words, relatedness is not an accidental, occasional event befalling otherwise self-contained singularities; rather, it is an ontological characteristic of being: "In so far as relation has an accidental existence in creatures, relation really existing in God has the existence of the divine essence in no way distinct therefrom" (*Sic igitur ex ea parte qua relatio in rebus creates habe esse accidentale, relatio realiter existens in Deo habet esse essentiae divinae* [*ST,* Ia Q 28 A 2]). The relational character of all created, finite beings thus offers an imperfect reflection of the consummate relation, viz., of the persons within the triune God—a conception that, as remains to be seen, the late Coleridge will recover with great urgency for a reluctant audience. To sharpen the point, we might say that persons in Aquinas cannot be objects because the latter term is not even properly applicable to nonhuman things. *Res* is not *objectum,* is not something merely "in itself" that *opposes* us in the sense of the German *Gegenstand.* We do not "confront" being now and then at our choosing but, instead, always "participate" in it. Perhaps unwittingly, Hegel's eventual parsing of "in itself" (*an sich*) and "for itself" (*für sich*) reoccupies this Scholastic understanding of being as relational and participatory—that is, as an embedding of beings,

---

4. *Natura cuiuslibet actus est, quod seipsum communicet quantum possibile est* (*De Potentia,* Q 2 A 1).

5. Quoted in Clarke, *Person & Being,* 9.

a network, system, or "community" (*communio*) always already in place when we seek to break down the world into discrete singularities and object-representations.

Now, because Aquinas does not conceive of relations in terms of subject and object, the perspective that we have on the world is for him never as a static aggregate of isolated entities. Rather, it is participatory in the sense that it is the very nature of a thing to communicate its existence to others: "Action, 'passion' (being acted upon), and relations are inseparably linked up together."[6] To confront being in merely appetitive, acquisitive, or otherwise utilitarian fashion is to ignore this participatory and relational framework—and, hence, to fail to acknowledge it as a gift. A world in which subjects fashion representations with the sole intent of getting "a purchase on" objects would be devoid of grace or, at least, one in which grace is no longer a framing and acknowledged reality. Notably, for Aquinas such a world would *eo ipso* also be wholly irrational. To be sure, it seems quite evident that the abandonment of notions like grace and gift in relation to the world has been an accomplished fact for some time now; yet the question remains whether modernity's voluntarist and nominalist epistemologies can succeed on their own terms or whether they might yet succumb to the implicit charge of irrationalism that Thomism has repeatedly leveled against them. While it makes sense to read the rise of nihilism in the nineteenth century as de facto conceding the internal contradictions and ultimate incoherence of the Enlightenment project, the dilemma is ultimately not to be solved by "a hermeneutic strategy that is entirely parasitic on the interpretations it challenges," as Daniel Conway has argued about Nietzsche's *Genealogy*.[7] A key question that will stay with us, then, might be formulated thus: is a modernity that has substantially given up on the primacy and reality of "person"—and that has consequently abandoned (or simply "forgotten") the twin concepts of relation and participation—capable of producing a coherent account of lived existence?

Perhaps sensing the precariousness of the Franciscan model—which stresses the radical singularity of Christ at the expense of a fully integrative theology—Aquinas takes great care to distinguish the will from the blind and impulsive craving of "the irascible and concupisciple appetites." Instead, the human being is understood as constitutively deliberative and enjoined to make judgments and choices. Lacking the

---

6. Clarke, *Person & Being*, 14. Echoing Clarke's account (in *Explorations*, 45–64) of Aquinas's ontology of action, which conceives all "being as actualization," Kerr quotes the *Summa*—"things exist for the sake of what they do [*omnes res [sunt] propter suam operationem*]" (Ia Q 105 A 5); as Kerr goes on to argue, in sharp contrast to modernity's "substantialist ontology of self-enclosed monadic objects ... Thomas's cosmological picture is, rather, of a constantly reassembling network of transactions, beings becoming themselves in their doings" (*After Aquinas*, 48); the "postmodern" overtones of Aquinas's framework seem almost too obvious to point out.

7. "Genealogy and Critical Method," 318; for a fuller version of that argument, see MacIntyre, *Three Rival Versions*, 32–57.

instinctual guidance and unequivocal directedness that characterize animal life, the human person "awaits the command of the will, which is the superior [or intellectual] appetite" (*ST,* Ia Q 81 A 3). The relevant sections in the *Summa* offer a nuanced and precise account of the will, at once acknowledging the intrinsic "necessity" with which it operates while showing how such necessity not only does not conflict with but positively supports "free choice" (*electio*). In so "insisting on the transcendence of the ultimate end, ... Aquinas can press his analysis to offer complete freedom to a person, without generating the paradoxes that accompany the notion of freedom as autonomy or absolute indeterminacy."[8] Maintaining the viability of free choice proves crucial to sustaining the *Summa*'s core argument regarding the intellectual nature and central function of the virtues. Thus, even as the will obeys necessity, what impels it is itself a good both apprehended and assented to by our intellect. In taking up, later in the *Summa,* the crucial question of "Whether the Act of Reason is Commanded," Aquinas draws a crucial distinction. On the one hand, there is the intellect that "apprehends the truth about something. This act is not in our power: because it happens in virtue of a natural or supernatural light. Consequently the act of reason is not in our power, and cannot be commanded." Yet on the other hand, there is "the act of the reason ... whereby it assents to what it apprehends" (*ST,* Ia 2ae Q 17 A 6), and it is here that the will operates.

In Aquinas's "non-subject-centred approach to human experience" the phenomenology of the will's inner determinacy does not register as an experience of "coercion." Rather, it involves the person's growing awareness of an "end" voluntarily embraced.[9] Indeed, not only is "necessity of the end ... not repugnant [*Necessitas autem finis non repugnat voluntati*]" (*ST,* Ia Q 82 A 1) to the will; it is indispensable since in the absence of a normative criterion regulating its choices (even when we fail to choose well) there would be no ground for having chosen or willed anything to begin with. If, then, "the will must of necessity inhere in the last end [*voluntas ex necessitate inhaereat ultimo fini*], which is happiness" (*ST,* Ia Q 82 A 1), we can also

8. Burrell, *Aquinas,* 142. See also Kahn ("Discovering the Will"), who traces the continuities between Aristotelian *prohairesis* ("choice") and Aquinas's *electio* but also remarks on the crucial difference that sets them apart—viz., that Aquinas's concept of a human will is premised on the archetype of the divine will as wholly aligned with divine reason.

9. Kerr, *After Aquinas,* 27. See *ST,* Ia Q 83 A 1, where Aquinas specifies that "Man has free-will [*homo est liberi arbitrii*]" because he "acts from judgment, because by his apprehensive power [i.e., the intellect] he judges that something should be avoided or sought. But because this judgment, in the case of some particular act, is not from a natural instinct, but from some act of comparison in the reason, therefore he acts from free judgment and retains the power to be inclined to various things." For discussions of Aquinas's conception of the will, see McCabe, *On Aquinas,* 79–99; Arendt, *Life of the Mind,* 2:113–125; Burrell, *Aquinas,* 141–146; and McInerny, "Ethics," esp. 196–202.

understand why there are at all times two dimensions entwined within it. The first, intellectual one concerns our awareness of the end, however particular or finite, which we seek. "Awareness" for Aquinas not only means to focus on it as a particular goal or objective but, in a reflexive sense, to deliberate and arrive at a judgment as to how a specific objective fits into an overarching hierarchy of goods or ends. The second, as it were kinetic quality, which concerns the actual commitment of the will as *actus,* "regards not the end, but *the means to the end*" (*ST,* Ia Q 82 A 1). Aquinas here is clearly following (indeed quoting) Aristotle, as he will again at the end of his discussion of free will (*ST,* Ia 82 A 3).

Still, our ability to determine whether a particular objective conforms to the supreme end of "happiness" relies not, as it does for Aristotle, on rational pedagogy (*epagōgē*) alone but requires "the certitude of the Divine Vision" (*ST,* Ia Q 82 A2) and the twofold grace as something habitual manifested in human action and as divine "assistance" (*auxilium*).[10] "Free will is not sufficient ... unless it be moved and helped by God" (*ST,* Ia 82 A 2); hence our ability to choose rationally and to select appropriate means for particular ends depends on a framing vision of the "ultimate end" or "hyper-good" (Charles Taylor's phrase) that *eo ipso* transcends the realm of finite, empirical praxis and cannot itself be chosen.[11] As David Burrell observes, "ends are consented to, not chosen." Indeed, they are not even quite "decided upon. Rather, they grow on us. Or is it that we grow into them?"[12] Willing for Aquinas thus stands in sharp contrast to the two assumptions altogether central to modern secular agency: its autonomy and its indeterminacy. By contrast, Aquinas takes all acts of will to be intimately entwined with an intellectual practice that involves far more than conceptualization. Thus, even as the "will can tend to nothing except under the aspect of the good [*sub ratione boni*]," an element of discriminating choice is always involved "because good is of many kinds" (*ST,* Ia Q 82 A 2). With characteristic both/and logic, Aquinas's next article (Q 82 A 3) proceeds to parse whether the will or the intellect holds priority in human affairs. As he will eventually conclude, "the intellect understands that the will wills, and the will wills the intellect to understand [*intellectus intelligit voluntatem velle, et voluntas vult intellectum intelligere*]."[13] In short, only be-

10. On Aquinas's account of grace in the *Summa* (*ST,* Ia IIae QQ), see Wawrykow, "Grace," and te Velde, *Aquinas on God,* 147–169.

11. *Sources,* 62–75.

12. Burrell, *Aquinas,* 144; see also McInerny, who notes how "what Aquinas sometimes calls the object of an action—cutting cheese, chopping wood, binding wounds, running in place—is the proximate end of the action, what individuates it." Only by observing how a variety of discrete acts is coordinated by the same end do we recognize how "any individual act is an act of a given type and its type is taken from its end or objective" ("Ethics," 199).

13. *ST,* Ia Q 82 A 4; later in the *Summa,* when investigating "That which moves the will," Aquinas reiterates this crucial point (*ST,* Ia IIae Q 9 A 1).

cause and to the extent that "the good itself is apprehended under a special aspect as contained in the universal true" (*ST,* Ia IIae Q 9 A 1) can the will manifest itself as a particular and purposive act. Hence, *actus* for Aquinas "is a single complex operation involving both will and intellect," less an instant discharge of causal power than a hermeneutic process whereby human beings achieve a coherent perspective on and commitment to the specific goods in which they take themselves to participate.[14] Rather than single-handedly *originating* or *implementing* some impersonal, rational calculus, action is integrally related to the hermeneutic activities of deliberation (*consilium*) and choice (*electio*). In sharp contrast to the outward, instantaneous, and gestural model that post-Hobbesian modernity devises, action in Aquinas unfolds as a complex and sustained interpretive and evaluative process.

To act is to reveal both our apprehension of and assent to an evaluative framework which, however partial or inadequate, furnishes the source and motive for any specific act of will. Action thus is neither a matter of sheer "compliance" with the framework of norms, goods, and ends within which it is unfolding, nor does it involve an outright quasi-iconoclastic assault on it. Rather, there is always an element of transcendence at work in action inasmuch as it deepens our grasp of those values and meanings that make up the order of things. As Maurice Blondel puts it, "we do not act, if we do not draw from ourselves the principle of our action, if this principle does not surpass past experiences, if we do not sense in it something else, if we do not make of it a kind of transcendent reality. One is never interested in one's own acts unless they are mixed in with some passionate ideology ... We die, as we live, only for a belief."[15] Only on the premise of that transcendent, imaginative, or counterfactual aspect is action intelligible at all; and only so can it manifest or actualize a person that is at once unique, rational, and free. Freedom for Aquinas is thus never gratuitously aspirational, such that a self decides to be or become someone else and transform its identity by selecting from some buffet of imagined identities. Commenting on Aquinas's subtle economy of choice, will, and action, Alasdair MacIntyre has remarked on the *Summa*'s crucial departure from a core aspect of Aristotelian moral rationality. Noting how "it is the presence of *intentio* which distinguishes a genuine act of will from a mere wish," MacIntyre observes that Aquinas's translation of *prohairesis* as *electio*—in turn rendered as "choice" in modern English—masks an important conceptual shift. Whereas for Aristotle "it is only desire as disciplined and directed by right moral habit which accords with reason, ... Aquinas takes the component of action which expresses *prohairesis,* rational desire, to be an act of the will. And the will is always free." The result is that *electio* "does not so much render

14. McCabe, *On Aquinas,* 81; as Aquinas puts it: "Men's acts and choices are in reference to singulars" [*actus et electiones hominum sunt circa singularia* (*ST,* Ia IIae Q 9 A 2)].

15. Blondel, *Action,* 114.

'*prohairesis*' into Latin as offer instead an alternative concept." Following Augustine, that is, the *Summa* refuses to partition the human person into a pre-rational and inherently *akratic* bundle of juvenile impulses on the one hand and a rationally self-governing adult on the other. Instead, "Aquinas sees every human being as held responsible from a relatively early age for his or her choices."[16] As an Augustinian, Aquinas naturally takes the *inadequacy* of these choices—viz., selecting means for ends as the person is able to apprehend them at a given stage in his or her life—to be beyond dispute.

The remedy for this predicament, and the indispensable source for whatever rational orientation the individual may conceivably achieve, thus has to be located in an economy of operative and cooperative grace. The reality of grace, Thomas argues, is brought home to us in the ascending order of speculative thought: from *prima philosophia* (metaphysics) to *sacra doctrina* (theological wisdom) to a mystical *visio beatifica*. In its various dimensions and manifestations, then, the Scholastic doctrine of grace implies (and reaffirms) the intrinsic and permanent heteronomy of reason (*ratio*). As Andrew Moore puts it in a fine essay, "[Anselm and Aquinas] used reason with exemplary rigour and precision, but they did so in a way that was dedicated to understanding and communicating the substance of Christianity to which the assent of faith had already been given, or which, as in the case of Anselm's *Cur Deus Homo,* set the terms for debate ... Reason was seen as a human faculty used in the orderly exposition of what was given in scripture and tradition; it was not an independent source or norm of Christian belief. Reason and revelation were not distinct and potentially competing sources of knowledge in the way that foundationalism has taught us to think of them." Central to Anselm's *fides quaerens intellectum* is thus the proposition that reason seeks to grasp in progressively fuller and more adequate ways a reality to which the mind has already given real assent. This is eminently true of both speculative and practical reason. In the former context it means that "the church locates the doctrine of creation *within* the structure of the creed: it is ingredient within that faith. By contrast, under the pressure of the search for rational justification for Christianity, Christian natural theologians have sought to argue from the world to its having been brought into being by a creator. Belief that the world has been created by God migrates from being the substance of faith to being a condition of faith ... But in doing so ... inferential reason becomes an independent source of belief."[17]

At the same time, the intrinsic dependency of reason on a complex notion of grace—which thus deems *ratio* incapable of either self-origination or self-perfection—also applies to practical, moral knowledge. Thus practical reason embraces, but does

16. MacIntyre, *Whose Justice?* 189.
17. Moore, "Reason," 395–396, 398.

not per se establish the ends for whose realization it seeks to devise means at once effective and responsible. While offering a more differentiated account of these degrees of speculative knowledge, Aquinas still follows Aristotle's account of *theoria* in Book 10 of the *Nicomachean Ethics*. Like Aristotle, Aquinas holds that, instead of some occasional and contingent judgment, only contemplation allows rational created beings to conceive the ultimate end of happiness; and while such knowledge is neither discursive (being incommunicable) nor "certain" (not belonging to the order of propositions), it is indispensable for all other cognitive acts. For "we reason theoretically *to* and *about* that ultimate end which is the *archē* [Lat. *principium*] of practical enquiry and reasoning, but *from* that *archē* it is by practical reasoning that we are led to particular conclusions as to how to act."[18] However indispensable, practical reason with its realization in the development of habits, as well as the intellectual and moral virtues, cannot on its own be foundational. To complete the formation of rational personhood, the Aristotelian model of *praxis* and *phronēsis* also requires a framing vision of ends *to* which all individual action is oriented. Not to be confused with the contingent purpose of individual action, the end in question is that "highest good" absent which all deliberation and choice of means would prove impossible. Indeed, without a conception of ends that no individual can ever *autonomously achieve* but, in David Burrell's words, can only "grow into," all choice and action would prove unmotivated and unintelligible—mere instances of caprice or "decision" of the sort so curiously extolled by Carl Schmitt.

Crucially, the *ultimus finis* for Aquinas cannot be something voluntarily chosen, let alone performatively generated by some individual act of will. In fact, that latter (deeply incoherent) view of the modern individual as self-originating and self-determining could only take hold once crucial features of Aquinas's thought had been misinterpreted and/or rejected for reasons to be taken up later.[19] For his part, Thomas understands freedom as a transition from potentiality to actuality and, as such, conceives it *narratively*—viz., as involving an agent's practical cum intellectual movement toward a fuller realization of his or her essence in relation to the absolute fact of sanctifying grace. The different degrees of knowledge more recently scrutinized in Jacques Maritain's work are one particular version of that narrative; Aquinas's account of the "perseverance" of habitual grace, itself enabled by divine

18. MacIntyre, *Whose Justice?* 193.

19. Among the most strident critics of the modern, post-Thomist understanding of freedom as radical autonomy and indeterminacy is Schopenhauer; see his *Prize Essay*, which all but demolishes a naïve, and intellectually bankrupt, self-certifying account of "free will" that, nonetheless, has gained quasi-axiomatic authority in modern liberalism and, most aggressively, in contemporary libertarian thought. Yet Schopenhauer's argument, far from making common cause with Scholastic theology, only clinches its main point by stipulating the radically non-cognitive and un-intellectual nature of the will; in that sense, Schopenhauer is an extreme descendant of Ockham and Hobbes.

auxilium, foreshadows the process-character of human flourishing as it is developed in the *Prima Secundae.* Inasmuch as "we are free not because we act at random, but because we act for reasons, and there are many possible available reasons," those reasons that ultimately prevail and thus nudge the agent from potentiality to actuality will reveal something essential about that agent's personhood or character.[20] For "the movement of the will is from within [*ab intrinseco*]" and "that alone, which is in some way the cause of a thing's nature, can cause a natural movement in that thing" (*ST,* Ia IIae Q 9 A 6).

At issue here is the primacy of human intentionality which, as Aquinas stresses well before the term entered our modern philosophical vocabulary, cannot be subjected to reductionist accounts. For if acts of thinking, knowing, judging, and willing are "dismissed as contentless epiphenomena, the objective world disappears with them, a world that is only there *for us* in the first place thanks to such acts." Robert Spaemann, whom I have been quoting here, continues:

> If we are to be clear what it means to pursue something, we must speak of conscious willing and acting. This has led some people to conclude that the only form that end-directedness can have is the conscious choosing and willing of ends. All other use of teleological language is taken to be improper or, at best, metaphorical. But this will not do, for we can only bring our will to a resolution in the first place as we experience within ourselves a primordial orientation [*ein ursprüngliches Aussein-auf*] that is already there. Without such an interest the world would be a matter of indifference to us; we would have no reason to will one thing rather than another.[21]

In defining the relation of will and intellect, Aquinas insists on the priority of the intellect since it alone presents the will with "the very idea of appetitible good" (*ST,* Ia Q 82 A 2), which is to say, with an "object" or "aspect." The "event-character" of the will is twofold, viz., "as [regards] the exercise of its act; secondly, as to the specification of its act, derived from the object" (*ST,* Ia IIae Q 10 A 2). Were we to consider the will only in the second, circumstantial sense as responding to a given object, it would

---

20. McCabe, *On Aquinas,* 68.

21. *Persons,* 53, 56 (trans. modified); Spaemann's chapter on "Intentionality" offers an especially lucid refutation of neuro-scientific reductionism, itself but a latter-day variant of the mechanistic turn that voluntarism takes in the seventeenth century. For a superb exploration of teleological thinking, see Spaemann and Löw, *Natürliche Ziele,* esp. 11–20 and the discussion of action, causality, and teleology in Aquinas (68–79). Regarding Aquinas's understanding of the human being—which presupposes, but must not be conflated with, the unity of body and soul—see Eberl, "Aquinas on the Nature of Human Beings," as well as related scholarship by Eleonore Stump, Robert Pasnau, and Joseph Bobik engaged in that article.

become a merely reactive faculty and thus cease to be properly a will enacting a "choice" (*electio*).

To be sure, occasional causes do indeed operate where will is concretized as *actus*. Yet what matters most is how the performance of an act of will—viz., a *choice* of suitable *means*—pivots on an antecedent "view" (to borrow a key term from John Henry Newman); and only the intellect, as a contemplative (though by no means passive) agency, can furnish such a view. Hence, the intellect "excels and precedes the will, … whereas the appetite moves and is moved" (*ST,* Ia Q 82 A 3). Central to this argument is Aristotle's distinction between final and efficient causes, which notably recedes as theological voluntarism begins to construe the will solely in terms of efficient (quasi-mechanistic) causation. Yet by showing how "a thing is said to move in two ways," viz., as "end" or "as an agent" (*ST,* Ia Q 82 A 5), the *Summa* emphasizes that what moves or solicits the will is not the particularity of a thing qua object but its value as an "essence" (*quidditas*) or substantial form embedded within a framework of normatively ordered goods or ends. Indeed, the modern nomenclature of "objects" confronted and experienced by a "subject" as possible catalysts for willing or desiring is misleading. As Michael Buckley notes, the word *objectum* prior to the fourteenth century "did not denote a thing, … [for] things were subjects of their own actualization in being, in attributes, in processes, and in the realizations of their potentialities. Metaphysically, things were not *objecta;* they were *res* as they were indeed beings."[22] The notion of value-neutral objects as a kind of "raw material" or "stuff" only arose once nature itself had been de-potentiated, stripped of its mediating role as the substantial form teleologically orienting the finite intellect toward the divine will that is both its source and *telos*. This "autonomization of nature was the first timid step towards the negation of all super-nature."[23] Whereas subsequent theology and philosophy seek to dismantle the intellect's *dominio sui* as merely an epiphenomenon of fluctuating psychological states associated with the will—itself little more than an oblique mental convulsion on the order of Hobbes's "last appetite"—Aquinas situates rational personhood within an ontology of "being and truth" under which "is contained both the will itself, and its act, and its object" (*ST,* Ia Q 82 A 4). Hence the extent to which the finite will is capable of aiming at the right object depends on the acuity with which we grasp the key terms of that ontology; willing thus is "more like an inclination than a push."[24] Later in the *Summa,* Aquinas reinforces this notion of the will as an "intellectual appetite" (*appetitus intellectivus* [*ST,* Ia IIae

22. Buckley, *Denying and Disclosing God,* 94; see also Sokolowski, *Phenomenology of the Human Person,* 112–116; for a striking "reoccupation" of this Scholastic insight in Goethe's botanical writings, see Pfau, "All is Leaf."

23. C. Taylor, *Secular Age,* 91.

24. Burrell, *Aquinas,* 141.

Q 9 A 2]) by stressing how "the will can tend to the universal good, which reason apprehends; whereas the sensitive appetite tends only to the particular good, apprehended by the sensitive power" (*ST*, Ia IIae Q 19 A 3).

Two key implications need to be identified here. First, in sharp contrast to the modern conception of "society" as an imagined community of autonomous "individuals" or "selves"— terms whose conflation of the particular with the generic must give us pause—Aquinas's argument is that we can "apprehend" a good only because of our indissoluble commonality. As MacIntyre puts it, "to achieve an understanding of good ... we shall have to engage with other members of the community in such a way as to be teachable learners."[25] *Communitas* thus is not some distant, aspirational utopia to be realized *eventually* by so many individuals. Rather, it is the premise for rational personhood, a point succinctly captured in John Macmurray's 1953–1954 Gifford Lectures. Entirely in the spirit of Aquinas, Macmurray insists that "existence cannot be proved; it is not a predicate." Instead, "we know existence by participating in existence," viz., in *action*. It is only when this understanding has been displaced by solipsistic and endlessly prevaricating (Cartesian) introspection that the reality and sheer givenness of existence appear in need of philosophical legitimation. The result is a notion of the disengaged and "buffered self" (Charles Taylor's phrase) that amounts to "a *reductio ad absurdum* of the theoretical standpoint." Insofar as the "self exists only in dynamic relation with the Other," the very conception of a "self" turns out to produce many of the theoretical dilemmas that steer modern philosophy toward some version of skepticism, naturalism, or existentialism. Part of that incoherence is already coded into the odd nomenclature of a self, that peculiar "combination of singularity—as the 'I,' with generality—as 'all thinking beings,' [and that] is possible only if we postulate the *identity* of all the particulars denoted by that term." For Macmurray, the leveling and pre-emptive erasure of the person by the modern nomenclature of a self, or *cogito*, amounts to a "logical sleight of hand [bound] to conceal the essential differences between individual people; and particularly, the formal distinction between 'I' and 'You.'"[26]

The second implication is that the will per se cannot, for Aquinas, simply be cathected onto a particular object as such. Were this to be the case, it would effectively cease to be will (i.e., choice) and prove itself to be a strictly re/active and mind/less appetition. "Choice" (*electio*), however, always rests on an active and intellectual conception or view that demands our implicit or explicit assent; and to assent here means to interpret how a particular goal and the specific means contemplated for its attainment fits into an overall hierarchy of ends. Consequently, a goal is never simply

25. *Three Rival Versions*, 136.

26. J. Macmurray, *Persons in Relation*, 17, 19.

seized by the will or vice versa. Rather, having come into focus by virtue of its inherent essence or form, a goal will acquire motive-force only because our intellectual apprehension of it justifies our *assent* to it. Much later, William James will restate the seminal point, observing how "a great part of every deliberation consists in the turning over of all the possible modes of *conceiving* the doing or not doing of the act in point ... *In action, as in reasoning, then, the great thing is the quest of the right conception*."[27] The deliberation on possible action thus is less concerned with the contingent object (or objective) *at* which we aim than with whether willing it, and pursuing it in this or that specific manner, is commensurable with our broader conception of personal flourishing in a social and moral space. Opening Part II of the *Summa* with what is commonly known as the "Treatise on the Last End," Aquinas puts the matter thus:

> Man must, of necessity, desire whatsoever he desires for the last end. This is evident for two reasons. First, because whatever man desires, he desires it under the aspect of good [*sub ratione boni*]. And if he desire it, not as his perfect good, which is the last end, he must, of necessity, desire it as tending to the perfect good, because the beginning of anything is always ordained to its completion [*ut appetatur ut tendens in bonum perfectum, quia semper inchoatio alicuius ordinatur ad consummationem ipsius*] ... Now it is clear that secondary moving causes do not move save inasmuch as they are moved by the first mover. Therefore secondary objects of the appetite do not move the appetite, except as ordained to the first object of the appetite, which is the last end.[28]

Corresponding to this basic distinction between an *ultimus finis* and "secondary objects" is that between the "substantial forms" that constitute the source of all particular existents and the concrete and fleeting materiality of the latter. It is just this concept of form mediating the finite, created with the divine will that became the target of a critique directed against the Aristotelianism (or Averroism) by the early nominalists.[29] For Aquinas, "the substantial form is not produced by the operation of

27. *Writings, 1878–1899*, 400.

28. *ST*, Ia IIae Q 1 A 6; as McInerny notes, the multifarious nature of human actions does not imply a corresponding diversity of ends. Rather, whatever kind of action it may be in which a person is engaged, "action is undertaken on the implicit assumption that to act in that way is perfective of the agent ... That is Aquinas's basis for saying that all human agents actually pursue the same ultimate end" ("Ethics," 201).

29. On the 1269–1270 debate regarding the unicity of substantial form—a position strongly affirmed by Aquinas and just as vehemently denounced by Robert Kilwardby, as well as various other Dominicans and virtually all major Franciscan scholars of the time—see Torrell, *Saint Thomas Aquinas*, 1:187–196.

nature" but, instead, has to be understood as something "produced by creation" (*ST,* Ia Q 45 A 8). It is itself the condition of possibility (to use the anachronistic, Kantian phrase) of existence, that by virtue of which a concrete entity *has* its being: "the form of the natural body is not subsisting, but is that by which a thing is ... It does not belong to forms to be made or to be created, but to be concreted [*formarum non est fieri neque creari, sed concreata esse*]" (*ST,* Ia Q 45 A 8). Guided by the intellect, then, the will responds to the form, rather than the material particularity of the thing; for the latter is but the vessel for the meaning or value that the intellect apprehends and that the will (as *electio* and *actus*) realizes.

For Aquinas, an act of will never involves the mindless tropism of contingently embodied desire fixating on (or recoiling from) disconnected and transient objects according to the laws of efficient causality. For any such model would have struck Aquinas as deeply incoherent in that it pre-emptively disaggregates mind and world. Moreover, to construe being as somehow separate from the intellect's ontological embeddedness and participation in the world is to reduce human thought to an apperceptive correlate of seemingly random appearances and, hence, permanently estranged from reason. Yet even for the most radical skeptic, Hume being a fine case in point, reason is necessarily being presupposed by any model of intelligent life and activity, including one bent on questioning the objectivity and interaction of various "impressions," as well as their capacity to ground personal identity.[30] As we shall see, Hume's own dilemma is a distinctly modern one: viz., his reductionist construal of the mind-world relation defies all solutions and peremptorily rejects any premises regarding the continuity of consciousness and the identity of the person; it thus can succeed only if one has already assumed that nothing—neither mind nor world—shall ever be accepted as *given.* That injunction, a key premise of the Pyrrhonist and skeptical tradition going back to Sextus Empiricus, gratuitously contracts being to what can be propositionally demonstrated—which, in Hume's case anyway, turns out to be very little indeed. Yet to confine the real to the tenuous methodological precinct of "warranted assertibility" is to ignore how the skeptic's diffidence regarding what is to count as "real" presupposes a deep reservoir of motivation—a *passion* for doubt that ironically reveals the skeptic's rich, if unexamined embeddedness in the world.

30. The locus classicus here is Hume's critique of Locke in his chapter "Of Personal Identity" in the *Treatise* (1.4.6). Characteristically, Hume rejects "the notion of a *soul,* and *self,* and *substance*" as a mere "disguise" or "fiction" that seeks "to feign the continue'd existence of the perceptions of our senses" and the manifest "interruption or variation" observable of all impressions made on the mind. Hume further radicalizes his position in the "Appendix" to the *Treatise,* observing that "no connexions among distinct existences are ever discoverable by human understanding ... All my hopes vanish, when I come to explain the principles, that unite our successive perceptions in our thought or consciousness." *Treatise,* 166, 400; for a lucid critique of Hume's position, see Spaemann, *Persons,* 143–147.

Hegel was to capture this dilemma rather well when noting how "what calls itself fear of error reveals itself rather as fear of the truth" (*das, was sich Furcht vor dem Irrtume nennt, sich eher als Furcht vor der Wahrheit zu erkennen gibt* [*PS,* 47/*PG,* 65]).

It was Augustine who had bequeathed philosophical theology the obverse of this self-certifying skepticism, viz., the idea of being as the primordial "gift" (*donum*).[31] Likewise, Aristotle's realism enables his Scholastic readers to approach being not as an aggregation of inert and meaningless material particulars but as a rational coordination of so many things (*res*) or entelechies. Thus understood, the individual thing, no less than the individual mind relating to it, is by definition an active presence and, thus, was never excluded from the *logos* in the way that objects appear bereft of reason to Descartes, Locke, or Hume. The conjunction between Aquinas's thinking about the will as rational appetite and his acceptance of Aristotle's theory of substantial forms is most apparent in Question 6 of *De Malo,* a work that Jean-Pierre Torrell dates sometime after Easter 1269, and thus prior to Aquinas's renewed exploration of the human will in the *Prima Secundae* (QQ 9–10).[32] Aquinas here responds to some twenty-four objections, several of which seem uncannily prescient of later, reductionist arguments against free choice as we find them in Luther, Hobbes, and Schopenhauer. Throughout, Thomas insists that to regard the will as externally coerced is not only "heretical" but will inevitably lead to the downfall of moral philosophy and rational personhood altogether. To him, the entire reductionist enterprise is, quite simply, "odd" (*extranea*). For "what is coerced is as contrary to what is natural as to what is voluntary since the source of both the natural and the voluntary is internal [*violentum enim repugnat naturali sicut et voluntario, quia utriusque principium est intra*] … It is not only contrary to faith but also subverts all the principles of moral philosophy [*quia non solum contrariatur fidei, sed subvertit omnia principia philosophiae moralis*]."

Central here is the distinction between a natural and a rational appetite. The former would be an animal's inclination to an object. Being by definition sensuously cathected onto a singular entity, the animal "incline[s] to act in only one way." By contrast, to human consciousness "things of nature have forms, which are the source of action, and inclinations resulting from the forms, which we call natural appetites, and actions result from these appetites." As emphasized by the argument's paratactic form, willing is itself the result of a series of transpositions that are cognitive in nature. Contrary to some mindless appetition inexorably fixated on and mechanically

31. On the meanings of *donum,* especially the juxtaposition of the "cosmos" as a gift the giving of which in no way takes away from the giver and the "already damaged" ways that gift functions within the human sphere, see Griffiths, *Intellectual Appetite,* 50–74.

32. Torrell, *Saint Thomas Aquinas,* 1:201–207; quotes from *De Malo* follow the English translation, *On Evil.*

compelled by the particular sensory object in view, the will *assents* (or refuses to assent) *to the substantial form* of it—viz., to the perceived significance rather than to the brute facticity of the given object:

> Human beings have an intellectual form and inclinations of the will resulting from *understood forms,* and external acts result from these inclinations [*ita in homine invenitur forma intellectiva, et inclinatio voluntatis consequens formam apprehensam*]. But there is this difference, that the form of a thing of nature is a form individuated by matter, and so also the inclinations resulting from the form are determined to one thing, but *the understood form is universal* and includes many individual things.[33]

Staying close to Aristotle, especially *De Anima* and the *Nicomachean Ethics,* Aquinas here again affirms the intertwined operation of intellect and will. He does so by establishing an analogy between the intellect, capable of inferentially moving "from things actually known to unknown things that were only potentially known," and a will that "by actually willing something" moves itself "to will something else." The example he offers is that of a person "willing health" subsequently consenting to take medicine. Integral to the operation of the will is thus the act of deliberation (*consilium*) whereby the contingent fact of an appetite or desire is situated within an intellectual, reflective economy. Likely recalling Aristotle's account of *prohairesis* in Book 6 of the *Nicomachean Ethics,* Aquinas thus stresses the essential self-transcendence of human willing inasmuch as it rests on a mental image to be realized by a human person. Consequently, mind cannot be reduced to a sheer, instantaneous act of will since the phenomenology of that act shows that willing is inherently *for* a consciousness. As Aquinas puts it, "because the will moves by deliberation, and deliberation is an inquiry that does not yield only one conclusion but leads to contrary conclusions, the will does not move itself necessarily" (*Cum ergo voluntas se consilio moveat, consilium autem est inquisitio quaedam non demonstrativa, sed ad opposita viam habens, non ex necessitate voluntas seipsam movet*).[34]

To be sure, deliberation proper will "necessarily" incline the will to act in pursuit of the good "since human beings cannot will the contrary." Thomas's well-known view of evil as a privation or misapprehension of good does not allow for acts planned and undertaken in pursuit of evil per se. Still, where a situation suggests such to be

---

33. *On Evil,* 257–258.

34. Ibid., 259; see also Hütter, who remarks on a noticeable shift from the more intellectualist strain running through the first part of the *Summa* toward "a later, more voluntarist leaning that comes to the fore with the inception of ST I-II, and especially *De malo* 6" ("Directedness of Reasoning," 174).

the case, two possible explanations remain. Thus an act of palpable evil deliberately planned and undertaken for the sake of evil may either be an instance of extreme misapprehension of the good or the work of a deranged individual. The latter scenario, moreover, would necessarily raise questions as to whether it makes any sense at all to speak of "act" and "action" without the presupposition of rational personhood. Be that as it may, the second, far more common scenario would be that of an individual who "at a particular time [may] not will to think about happiness." The suspension of deliberation, most likely as a result of transient emotional disorder or an enduring distemper (e.g., anger or despair) is certainly factored in by Aquinas. Notably, though, he takes care not to premise his account of rational personhood and intellectual appetite on contingent and anomalous states or dispositions. For their role in a potential misapprehension of (or indifference to) the good could not even be explained if the person had not already grasped the substantial form of the good at stake, which in turn presupposes an ontologically "given" (*donum*) awareness of the hyper-good of happiness.

In Aquinas, things acquire reality *for* human consciousness, not as material singularities but as concretions of rational form. Cognition and willing alike constitute forms of *assent* and, hence, can never be reduced to an immediate gravitation of mind toward body. Conversely, things acquire their reality *for* us by virtue of their forms, and as such they are always already part of a rational economy. Being, we might say, is for Thomas by definition a gerund form rather than an indifferent material substrate accidentally holding in balance various qualities or *praedicabilia:* "The first act is the form and integrity of a thing; the second act is its operation [*Actus quidem primus est forma et integritas rei, actus autem secundus est operatio*]" (*ST,* Ia Q 48 A 5). As the conjunction of "form" and "integrity" qua "act" suggests, "substantial form" not only guides the created will as it draws on an intellectual "view" (*apprehensio*) and thus chooses *for a particular reason.* For these very forms that constitute the ontological premise of concrete (ontic) being also reveal the continuous presence of divine reason and mediate it with our finite intellect: "God is absolute form, or rather absolute being [*Deus sit ipsa forma, vel potius ipsum esse*]."[35]

Arising from God's will and attesting to its perfection, forms in Aquinas's account do, however, also "bind" their creator. Given their perfection—so emphatically stated in Genesis—"And God saw all the things that he had made, and they were very good [*viditque Deus cuncta quae fecit et erant valde bona*]" (1:31)—it is not to be supposed that God could ever wish to remake creation. Moreover, Aquinas's divine will

35. *ST,* Ia Q 3 A 7. For Aquinas's discussion of the being of forms, see also Ia Q 45 A 5. For a thorough analysis of Aquinas's conception of God, see te Velde, *Aquinas on God,* esp. 65–93.

is no mere "prime mover," let alone some deist watchmaker but, instead, continuously and lovingly affirms all of creation:

> Since God is very being by His own essence, created being must be His proper effect; as to ignite is the proper effect of fire. Now God causes this effect of things not only when they first begin to be but as long as they are preserved in being ... As long as a thing has being, God must be present to it, according to its mode of being [*Quandiu igitur res habet esse, tandiu oportet quod Deus adsit ei, secundum modum quo esse habet*]. (*ST,* Ia Q 8 A 1)

The creative will, in other words, is an expression not only of perfection but also of love, and the forms arising from it show how for Aquinas nature is necessarily "integrated within the context of grace."[36] It is only in virtue of non-contingent, "substantial forms" that the concrete instances of process and existence summarily referred to as "nature" are possible, actual, and sustainable. Form thus crystallizes the coincidence of logical analysis and ethical evaluation in the *Summa;* for a will to act means for it to have been guided by a logically (not temporally) antecedent and superior apprehension of form, just as conversely "evil has no formal cause [but] is a privation of form [*Causam autem formalem malum non habet, sed est magis privatio formae*]" (*ST,* Ia Q 49 A 1).

It is this abiding commitment of the divine creative will to the perfection of being realized in substantial forms that Aquinas's critics would reject as an unacceptable constraint. To be sure, the introduction of an omnipotent God potentially defaulting on his commitment to the rationality of creation does not immediately entail the irrational appraisal of the will eventually found in Hobbes and Schopenhauer. Still, beginning with William of Ockham, there is a marked shift toward interpreting divine power (*potentia absoluta*) as wholly self-certifying, rather than as committed to sustaining the forms it has created. For the first time, the will is being conceptualized in a way that hints at a possible antagonism, perhaps even incommensurability, between it and the intellect. In all brevity, it bears pointing out here that what most vexed Aquinas's heirs (Duns Scotus, Thomas Bradwardine, and, above all, William of Ockham) was his account of theological language as only ever bearing an analogical relation to the divine. Thomas's notion of analogical predication is meant to compensate for the "fundamental mismatch between the tool [language] and its [divine] object." Aquinas thus seeks to navigate between the hubris of univocal predication, which assumes "that our language refers to God in the same way that it refers to

36. Dupré, *Passage,* 171. On Aquinas's conception of grace, see Torrell, *Aquinas's Summa,* 33–36; Wawrykow, "Grace," 192–221; Hütter, "Thomas on Grace and Free Will," and Kerr, *After Aquinas,* 134–148.

things in the world," and the road to despair opened once it is stipulated "that all language used of God is used *equivocally,* that is, . . . in a way completely unrelated in the way [words] are used in ordinary language." In the latter case the act of predication seems pointless, and the implicit ethos would appear to be one of, literally, bad faith inasmuch as the equivocal theory "would cut God off from any human knowledge whatsoever."[37]

Even as he stipulates that "no name belongs to God in the same sense that it belongs to creatures" (*ST,* Ia Q 13 A 5), Aquinas stresses that this fact by no means renders God a complete enigma, let alone a fiction. Inasmuch as there are created beings attesting to a creator (though not *proving* him), Aquinas specifies that perfection terms

> are said of God and creatures in an analogous sense. Now names are thus used in two ways; either according as many things are proportionate to one, thus for example "healthy" predicated of medicine and urine in relation and in proportion to health of a body, of which the former is a sign and the latter the cause: or according as one thing is proportionate to another, thus "healthy" is said of medicine and animal, since medicine is the cause of health in the animal body. And in this way some things are said of God and creatures analogically, and not in a purely equivocal nor in a purely univocal sense. For we can name God only from creatures. Thus whatever is said of God and creatures is said according to the relation of God as its principle and cause . . . Now this mode of community of idea is a mean between pure equivocation and simple univocation. For in analogies the idea is not, as it is in univocals, one and the same, yet it is not totally diverse as in equivocal. (*ST,* Ia Q 13 A 5)

Closely entwined with his negative theology, Aquinas's theory of analogy insists on the unconditional, if permanently incomplete fact of a *relation* between finite human agents and God: "since everything is knowable according as it is actual, God, Who is pure act . . . is supremely knowable." That is, inasmuch as "the ultimate perfection of the rational creature is to be found in that which is the principle of its being," and considering furthermore that "the ultimate beatitude of man consists in the use of his highest function, which is the operation of his intellect" (*ST,* Ia Q 12 A 1), God must be within reach of our intellectual faculties, albeit in necessarily partial and incomplete ways. Few propositions seem more anathema to modernity than this

37. "If an equivocal understanding of theological language is to be avoided for betraying any human communion with God, so too a univocal understanding is to be avoided for betraying the transcendence of God. This twin conviction lies at the heart of Aquinas's teaching on analogy" (Hyman, *Short History,* 50–51).

acquiescence in the heteronomy and intrinsic limitations of human cognition, to say nothing of Aquinas's suggestion that this state of affairs is actually to be welcomed. And yet, it must be so, for to suppose "that the created intellect could never see God [*nunquam essentiam Dei videre potest*]" is to imply that the intellect "would never attain to beatitude" (*ST,* Ia Q 12 A 1) and, thus, to call into question the ontological fact of grace itself. What is more, such a view would, paradoxically, constitute a claim of (radically negative) insight into God after all.

Still, it is just as important to guard against a hubris of intellectual self-sufficiency and self-assertion that would render the notion of grace superfluous and, in so doing, would essentially premise the reality of God on the cogency and formal correctness of those propositions ventured about him by finite individuals. One such misconception involves what Jacques Maritain has called Descartes's "sin of angelism," that is, the notion of "thought ... as intuitive, and thus freed from the burden of discursive reasoning; innate, as to its origin, and thus independent of material things." By contrast, Aquinas maintains that there is "no gap between mind and world, thought and things, that needs to be bridged," for which reason "his view of how our minds are related to the world is interwoven with his doctrine of God: no epistemology without theology."[38] From the outset, the *Summa* thus emphasizes that the ground or source of knowledge can never be construed as just another object of inquiry since "it is impossible for any created intellect to see the essence of God by its own natural power [*per sua naturalia essentiam*]" unless "the power of understanding should be added by divine grace [*oportet quod ex divina gratia superaccrescat ei virtus intelligendi*]" (*ST,* Ia Q 12 A 4–5). Inasmuch as all knowledge involves a progressive adequation of knower and known, knowledge of God presupposes "a certain deiformity [*intellectum in quadam deiformitate constituit*]." Yet even then, as the qualifying *quadam* is meant to convey, to know is not to merge substantially with the thing known but to cultivate within the intellect a formal "likeness [*similitudo*] which resembles the object." The knowledge in question thus is "not of the thing in itself but of the thing in its likeness [*non dicitur res cognosci in seipsa, sed in suo simili*]" (*ST,* Ia Q 12 A 6; A 9). Nonetheless, while holding direct cognition (*visio intellectualis*) of God to be an impossibility, at least "in this mortal life" (A 11), Aquinas accords the intellect ample powers for participating in God's essence *analogically*—and it is here that his focus shifts to the ways in which human cognition is enmeshed with linguistic structures. Thus, aside from enabling and sustaining what MacIntyre calls dependent rational animals (possessed of both theoretical and practical reason), grace also proves incompatible with the notion that God and finite creation, however disparate in all other ways, share the self-same ontological predicate of being.

38. Kerr, *After Aquinas,* 24, 30. Kerr is referencing Maritain's *Three Reformers: Luther-Descartes-Rousseau* (1928).

Central to Duns Scotus's epistemology, the supposition of a strict univocity of being holds that every ontic entity—quite apart from its incidental states, contingent appearance, and unique attributes—partakes of the same ontological predicate of being. While being per se does not appear, each thing—or, rather, the mental image that we have formed of it—points back to being as that *of* which it is the appearance, that which licenses the phenomenon's "self-showing" or sheer "givenness." For Gavin Hyman, it is in Duns Scotus that, ever so tentatively, the vertical (Augustinian/Thomist) axis linking—yet also keeping categorically distinct—the transcendent and uncreated God of Christianity from finite, fallible, and sinful man is being replaced by a horizontal vector. Scotus's contention that "'being is univocal to the created and the uncreated' ... aims to conceive of being as a concept quite independently of any revelatory knowledge and independently of th[e] divine-human distinction." Henceforth, the divine is "articulated in quantitative terms on a single ontological plane. God transcends humanity only in 'intensity of being,'" thereby precluding the kind of analogical reconciliation of God and man that Aquinas had worked out with unprecedented depth. Yet now, "although there was an ontological continuity between God and humanity (the 'domesticating' move), this also installed an infinite metaphysical gap between them (the 'distancing' move). The 'distancing' move was intended to compensate for the domesticating move, but the combined effect of both was to turn God into an unknowable unfathomable abyss."[39]

It is no accident that Aquinas's firm rejection of this hypothesis—viz., that "what is said of God and of creatures is univocally predicated of them" (*ST,* Ia Q 13 A 5)—should be found in the midst of his theory of language. Characteristically, the opening *disputatio* of Article 5 makes a strong case for the univocal hypothesis. Objection 2

39. Hyman, *Short History,* 70–71; similarly, Dupré reads Scotus's critique of Aquinas's theory of the Incarnation as implicitly undermining the doctrine consolidated at Ephesus (431) and Chalcedon (451). Thus, by investing humanity "with the potential of being assumed by a divine person Scotus's artificial construction, intended to protect the concept of human nature from breaking under the weight of a theological exception ... [and] specifically devised for joining Christ more intrinsically to human nature, results in a quasi-independent abstraction of a pure nature." To the conceptualist, the ontological fact of grace thus seems enigmatic, begging epistemological relief, which Scotus provides by postulating human nature's "infinite receptivity" to the divine (*Passage,* 175–176); see also C. Taylor, *Secular Age,* 93–99. A particularly strident version of this argument has been advanced by Catherine Pickstock, for whom Scotus is pivotal in preparing the turn toward a conceptualist, epistemological, and anthropomorphic "modernity." This view has been passionately argued by some and, just as fiercely, contested by others. "Strict univocity permits a proof of God's existence without reference to a higher cause beyond our grasp," thus setting into motion a slow but inexorable "move from ontology to epistemology" (Pickstock, "Modernity and Scholasticism," 12, 6). Sharply critical responses to Radical Orthodoxy's reading variously center on its exegesis of Scotus and on how to evaluate its findings; for a particularly thorough and judicious critique, see R. Cross, "Where Angels Fear to Tread" (Pickstock engages Cross in her essay, pp. 3–8, notes 2–4). For an introduction to Scotus, see the entry in *Stanford Encyclopedia of Philosophy* (www.plato.stanford.edu/entries/duns-scotus/), accessed 3 July 2012.

thus states how, unless God and created being operate on one and the same ontological plateau, there is nothing to guarantee that they stand in any relation to one another whatsoever; after all, "there is no similitude among equivocal things." Remarking on the nature of language and the act of predication, Aquinas insists that this scenario presents us with a false choice. To begin with, perfection terms such as "wise," "good," or "just" function in categorically different ways, depending on whether they are applied to God or to finite creation. In the latter case, they are used in a strictly attributive sense, such that the man we call wise is not so by virtue of his essence (i.e., being a man) but only contingently. By contrast, "when we apply it to God, we do not mean to signify anything distinct from His essence, or power, or existence." What causes the univocal hypothesis to fail—indeed to beg the central question altogether—is its obliviousness to the fact that the fundamental act of human intelligence (viz., the cultivation and articulation of knowledge) unfolds in essentially different ways when it engages divine or finite matters, respectively.

To be cognizant of that difference also opens insight into the connection between Aquinas's conception of grace, analogical predication, and the human will. Thus the unconditional and non-negotiable priority of grace (*gratia*)—that is, of the phenomenal world as freely ("gratis") given, a gift (*donum*)—negates from the outset modernity's axiomatic idea of knowledge as a strictly *autonomous* act, and of human rational agents as epistemologically self-determining and self-legitimating. For Aquinas, the ontological *datum* of each phenomenon and, indeed, of creation as a whole means that knowledge unfolds as a progressive adequation of the finite intellect—not so much to *what* has thus been given but to the incontrovertible reality of its *givenness*. Hence our predicative acts never signify or define the reality of some object or phenomenon "out there" but, instead, only make explicit the finite individual's contingent, fallible, and evolving relation to the given phenomenon in question. Knowledge does not commence as the encounter between a distinct and hermetic "subject" with some likewise self-contained "object." Rather, it begins as we find ourselves in the presence of something unconditionally "given" (*datum*), indeed "gifted" (*donum*); as Plato had long before insisted when tracing knowledge to "wonder" (*thaumazein*), to know is, first and foremost, to witness the disclosure of the phenomenon's sheer "givenness." Aquinas's basic framework is echoed, at least implicitly, by Jean-Luc Marion's profound analysis of the event (*Ereignis/évenement*) of "givenness" itself, "Being withdraws from beings because it gives them; all givenness implies that the giving disappear (withdraw) exactly to the degree that the gift appears (advances) precisely because giving demands leaving (it behind)."[40]

40. *Being Given,* 34. Notably, Marion also qualifies Heidegger's designation of "givenness" as "event" since the latter term risks eclipsing the transcendent dimension that the world qua

What matters above all here is not to construe this apparent withdrawal or absence, as well as the (divine) giver's unfathomable nature, as a case of epistemological *privation*. Even less are we asked to reconstruct the divine *plēroma* inferentially in the manner of natural theology. In fact, for Aquinas the reality of God neither calls for some type of inferential demonstration, nor could it ever be the focal point of some self-authorizing epistemological skepticism such as might arise from the recognition of natural theology as a hopelessly circular method (which, indeed, it is). What makes Aquinas's question concerning God's existence—"whether [he] exists [*an sit*]"—such a challenge to a modern reader is the latter's casual assumption that the reality of anything, indeed everything, is contingent on our ability to get a purchase on it—predicatively. Yet in the *Summa*, "the question as to whether God exists is first and foremost a matter of finding an access (*via*) to the intelligibility of God." Rudi te Velde rightly notes that "the real issue for Thomas is not whether god exists as a matter of fact, or even whether we may consider ourselves to be rationally justified in believing that god exists. His focus is in a certain sense not epistemological at all ... What Thomas is looking for is not so much rational certainty as intelligibility."

At the same time, Aquinas does not presuppose God's existence in any intuitionist manner; indeed, he readily concedes "that there is no *immediate* evidence by which God's existence forces itself upon us ... That God exists is, in itself, not mediated by something else, since in God essence and existence are one and the same. This is what it means to be God. But all the same *our knowledge* of this truth, thus our access to it, is mediated."[41] Simply put, the true object of every epistemological effort is the intellect of the subject engaged in it, rather than some putatively extraneous referent; and to the extent that the quest for knowledge crystallizes in specific predicative acts, such representation merely mediates the relation of finite consciousness to God—a point substantially recovered six centuries later by Hegel as the dialectical movement between "natural consciousness" (*natürliches Bewußtsein*) and the "Absolute."

---

phenomenon—and indeed our potential knowledge of it—discloses in the radical givenness of the saturated phenomenon. Givenness is neither an undifferentiated "intuition" (Husserl's *Anschauung*), nor can it be resolved into the impersonality of an "event" (Heidegger's *Ereignis*): "I think that the irruption of the *Ereignis* tends—without completely succeeding—to hide the fact that givenness, which Heidegger constantly uses to unveil Being, finds itself deserted by it. . . . this denial frees him from having to think givenness as such" (ibid., 38).

41. Te Velde, *Aquinas on God*, 37–38, 42; Burrell likewise notes how the use of *esse* in statements about God "functions in the first as a predicate nominative, and in that role can 'signify [God's] act of existing.' Odd as it may seem, however, this assertion does not succeed in telling us whether God exists. For its form is not that of an existential assertion, but of a definition giving the nature of the thing in question. If we accept grammatical form as the decisive clue to meaning, we will not confuse this assertion with an existential one. In fact, ... it is not a proposition at all since it links two unknowns" (*Aquinas*, 8).

At the level of verbal representation, the (predicative) act wherein alone knowledge is truly achieved and consummated, the acceptance of grace necessarily entails being committed to analogical predication: "For as we can apprehend and signify simple subsistences only by way of compound things, so we can express simple eternity only by way of temporal things ... [Thus] demonstrative pronouns are applied to God as describing what is understood, not what is sensed [*ad id quod intelligitur, non ad id quod sentitur*]" (*ST,* Ia Q 13 A 1). Recalling Aristotle's distinction between the "is" of predication and the "is" of identity, Aquinas insists that language's capacity for evoking sensory qualities here operates in a strictly analogical sense. His broader aim is "to *show* what we cannot use our language to *say* ... [and] to make us aware of how we might use those features [of our discourse] to show what something which transcended that discourse would be like."[42] What Aquinas calls "negative names" are needed in the absence of any other mode of access available to the finite intellect; yet it must be remembered that such acts of analogical predication "do not at all signify His substance, but rather express the distance of the creature from Him [*remotionem alicuius ab eipso*] ... So when we say 'God is good,' the meaning is not, 'God is the cause of goodness,' or 'God is not evil'; but the meaning is, 'Whatever good we attribute to creatures, pre-exists in God,' and in a more excellent and higher way. Hence it does not follow that God is good, because He causes goodness; but rather, on the contrary, He causes goodness in things because He is good" (*ST,* Ia Q 13 A 2).

Here we see Aquinas guarding, not only against the divine-command ethic that will soon take center stage in Ockham, but also against the supposition that we could ever predicate anything of God in the manner of an ordinary syllogism. For in all such acts the qualities in question do not exist in God by virtue of being so ascribed to him: "As to the names applied to God—viz. the perfections which they signify ... as regards what is signified by these names, they belong properly to God, and more properly than they belong to creatures ... But as regards their mode of signification [*modum significandi*], they do not properly and strictly apply to God" (*ST,* Ia Q 13 A 3). Just as the ontology of grace precludes the modern dream of radical autonomy, and thus places a firm check on a finite will otherwise prone to emancipate itself from (perhaps even oppose) the human intellect, so it also proves incommensurable with mimetic conceptions of language. In the *Summa,* language does not operate *referentially,* which is to say, by taking possession of an ostensibly "other" and inanimate entity or "object." Instead, it establishes and (potentially) refines our relation to, engagement with, and participation in the reality of the thing so named. Inasmuch as it serves "to achieve a purpose—to facilitate action or increase understanding," language in Aquinas's *Summa* bears significant affinities to modern linguistic pragma-

---

42. Burrell, *Aquinas,* 6–7.

tism.[43] Rather than laying claim to some concrete or noumenal entity, the true "referent" in any predicative utterance thus concerns the degree to which the agent of knowledge participates in a given thing's reality and activity (*operatio*): "We can give a name to anything in as far as we can understand it." Moreover, because God "is above being named [*esse supra nominationem*], ... He can be named by us from creatures, yet not so that the name which signifies Him expresses the divine essence in itself."[44] To know means to study the essence of the phenomenon, rather than to dissect it for purposes of appropriation and domination. Within the framework of enabling and cooperative grace, which Aquinas takes over from Augustine, the only legitimate epistemological stance is one of "study" (*studiositas*), rather than a gratuitous and impersonal inquisitiveness (*curiositas*) that reflects an "appetite for the ownership of new knowledge" and whose "principal method is enclosure by sequestration of particular creatures or ensembles as such."[45]

The crucial point here concerns the productive alignment of intellect and will, an unending challenge according to Aquinas because the very possibility of human knowledge hinges on the incomprehensible and indemonstrable condition of divine grace. As the case of the eminent mathematician Kurt Gödel suggests, this scenario has a way of insinuating itself even into the hardest of modernity's "hard" sciences: number theory. As Gödel recalls, it was in the summer of 1930 that, trying "to prove directly the consistency of [classical] analysis by finitary methods, I saw two distinguishable problems: to prove the consistency of number theory by finitary number theory and to prove the consistency of analysis by number theory." Tackling the second of these issues, Gödel realizes "that I had to use the concept of truth (for number theory) to verify the axioms of analysis ... [and] that the concept of arithmetic truth cannot be defined in arithmetic. If it were possible to define truth in the system itself, we would have something like the liar paradox, showing the system to be inconsistent." Arguably, this dilemma affects the formation of any theoretical framework whatsoever. Simply put, any rational system of thought requires the presence of at

43. James, *Pragmatism*, 33.

44. *ST*, Ia Q 13 A 1. Echoed in A 4: "Our intellect, since it knows God from creatures, in order to understand God, forms conceptions proportional to the perfections flowing from God to creatures, which perfections pre-exist in God unitedly and simply, whereas in creatures they are received and divided and multiplied." On the unique challenges of predication in the context of God's "being" (*esse*), see Burrell, *Aquinas*, esp. 13–62; elsewhere ("Analogy, Creation, and Theological Language"), Burrell argues that in having "to square the syllogistic requirement of univocity with the demand internal to a 'knowledge of God' (*theologia*) that human discourse about the One ... can only be analogous" (83), Aquinas significantly parted ways with Aristotle. Hence, "verbs never *refer* for Aquinas, but state the manner in which something is what it is" (ibid., 8).

45. Griffiths, *Intellectual Appetite*, 20; see also Griffiths's account of "participation," ibid., 75–91.

least one proposition that fits into it but does not follow from its axioms, that can neither be proved nor disproved. For "if there were no undecidable propositions, all (and only) true propositions would be provable within the system." The dilemma would be that such a system of thought would only admit such problems of analysis as would be a priori soluble on its terms, thus generating wholly predictable results but, alas, never enabling us to discover anything new. Realizing that arithmetic truth and arithmetic provability are not coextensive, Gödel notes that the idea of theory is incommensurable with the quintessentially modern ideal of absolute epistemological self-sufficiency (autonomy), and that truth and provability are, in fact, not isomorphous ("We would have a contradiction").[46]

Gödel's discovery in many ways runs parallel to Aquinas's quarrel with the emergent (nascently modern) idea of univocal predication—which no longer concedes that a system of rational predicates necessarily rests on, and indeed owes its cogency to, a ground that can never be predicatively secured and, hence, cannot be taken to operate on the same ontological plateau as the theory or "system" in question. Identifying the stakes of Aquinas's conception of knowledge, as well as its evident opacity to a modern sensibility, David Burrell puts the matter rather well:

> Perhaps it is the lot of our generation, chastened by the horrors of two centuries intoxicated by "autonomy," to be reminded of the need to find a way to give constructive articulation to this elusive yet constitutive *distinction* of all-that-is from its transcendent source. And when we try to do so, to explicate not merely what it is to be something of a kind (Aristotle), but what it is for things of such a kind to *be,* as creatures, we find that nothing can be unless it is internally related to what causes it to be. Yet that relation cannot be a feature of the thing, since without it there would be no such thing; rather, the being of each thing is such that it reflects its sources, from which it cannot be separated under pain of annihilation. So in itself it is nothing![47]

Every problem (*explanandum*) soliciting our intelligent, reasoning engagement presents itself as such *to* us as something unconditionally, if enigmatically "given." It cannot itself be a problem that we have "made" or conjured up in terms wholly controlled by the theory now devised for its solution. This givenness is what Aquinas calls "grace." If the term has a metaphysical ring to it, this is the case not because the *Summa* fails to grasp the nature of rational analysis and inquiry but, on the contrary, because it has also grasped the limits that are an intrinsic feature of any theoretical

---

46. "Kurt Gödel," in *Stanford Encyclopedia of Philosophy* (plato.stanford.edu/entries/goedel/#FirIncThe).

47. Burrell, "Analogy, Creation, and Theological Language," 89.

framework capable of generating insight, rather than merely reconfirming its own conceptual suppositions. To recognize rational thought's dependency on the sheer givenness or, rather, "giftedness" of the underlying phenomenon also sets a strict limit to the role of the will in human cognition. In ways that would gradually elude his successors, Aquinas thus saw with great clarity how in every act of cognition the human agent's moral and rational dimension, will and intellect, are fully entwined. For inasmuch as intellectual (theory-forming) activity rests on an "undecidable proposition" (Gödel), it can never be cordoned off from moral responsibility—viz., the province of the will and judgment (*prohairesis*) of which Aristotle had already observed that its proper concern lies with "what can be otherwise" (*Nicomachean Ethics,* $1140^{a}30$). At the beginning of all intellectual activity, then, lies the commitment to a specific outlook on the nature, scope, and ambition of knowledge itself, beginning with the choice of how (or whether) to acknowledge that all rational inquiry, in addressing itself to what is inexplicably and unconditionally given, rests necessarily on an "undecidable" proposition.

7

RATIONAL CLAIMS, IRRATIONAL CONSEQUENCES

Ockham Disaggregates Will and Reason

If analogical predication is perceived to be an unacceptable constraint on human cognition by Aquinas's successors, this is because they operate with a fundamentally weakened sense of *obligation and responsibility* that the knowledge bears to its objects of inquiry. Already in Duns Scotus's mystical speculations about the "univocity of Being," it is palpable how "talk of analogy ... became marginal rather than central" because the Thomist model appeared to fall short of the criteria of explicitness, transparency, certainty, and verifiability that had come to define knowledge as a human (autonomous) product. Because of its unwavering emphasis on the humility of the knower and on enabling and cooperative grace as the preconditions for knowledge itself, the concept of analogical predication came "to be regarded with suspicion. For it emphasized the necessary centrality of uncertainty, imprecision and the theological propriety of linguistic imprecision ... If language can now be predicated of God in the same unequivocal way that it is predicated of things in the world, the implication of this is that God is, in some sense, closer to things in the world; indeed, to such an extent, that he becomes a 'thing' himself. In other words, there is a qualitative change in what God is conceived to be."[1] He becomes a First Cause, the

1. Hyman, *Short History*, 53, 55; Hyman appears to echo Edward Craig's earlier thesis that "beginning with Galileo and Descartes, [there] is a tendency to suppose 'quantitative difference but

supreme "substance" or "power" essentially continuous with a world now conceived less as complex and infinitely variegated divine order than as an inventory of discrete things to be tabulated and appropriated at will.

Yet precisely this conceptual flattening-out of what in Aquinas are two ontologically distinct realities also has the (seemingly paradoxical) effect of rendering God more enigmatic and potentially irrelevant to the pursuits of the human intellect. Thus, "there is another strand in modern theism in which the 'quantitative' difference between the human and divine is such that God becomes far removed from human knowing, without there being any analogical mediation between them."[2] It is here that we turn to William of Ockham (ca. 1287–1347), and specifically to his project of a "positive moral science" which, in strikingly legalistic and formalist ways, explores the "human and divine laws that obligate one to pursue or to avoid what is neither good nor evil except because it is commanded or prohibited by a superior whose role it is to establish moral laws."[3] While Ockham holds fast to the Scholastic premise that the divine will is *eo ipso* rational, such reason now seems remote, if not wholly unintelligible to the human realm. Perhaps rational conceptions are still intrinsic to the divine will; yet they now are so merely by virtue of *ascription*. Beginning with Ockham, moral meanings are increasingly viewed as contingent on the divine will having ordered them so: "What God has ordained (Ordinate Power) does not exhaust what God could do (Absolute Power)" and "value terms such as 'just' and 'meritorious' do not indicate a natural quality of an act. Rather, they reflect the absolute freedom of God to constitute any possible act as morally valuable."[4] Examples of this startling shift—somewhat reminiscent of Richard Nixon's famous 1977 remark to David Frost ("Well, when the president does it that means that it is not illegal")—abound in Ockham's writings, particularly in his early *Commentary on the Sentences*. If we ask whether God can order evil to be done, "the resolution of this query is that whatever God commands is *de facto* good."[5] To be sure, the question (albeit not Ockham's solution) is of ancient provenance, such as when in the *Euthyphro* Socrates bewilders

---

qualitative identity'" (quoted in Gunton, *The One, the Three, and the Many,* 108). As Brad S. Gregory remarks, "it is self-evident that a God who by definition is radically distinct from the natural world could never be shown to be unreal via empirical inquiry." Yet if, "having absorbed and taken for granted metaphysical univocity, one imagined that God belonged to the same conceptual *and causal* reality *as* his creation, and if natural regularities could be explained through natural causes without reference or recourse to God, then clearly the more science explained, the less would God be necessary as a causal or explanatory principle" (*Unintended Reformation,* 32, 54–55). On natural theology's confinement of God to a hypostatized First Cause and the consequent displacement of that "cause" by sustained empirical inquiry into the operation of so-called secondary causes, see Thomson, *Before Darwin,* esp. 21–58.

2. Thomson, *Before Darwin,* 59.
3. Ockham, *Quodlibetal Questions,* II:14; henceforth cited parenthetically as *Quodl.*
4. D. W. Clark, "Voluntarism and Rationalism," 78, 80.
5. Ibid., 73–74.

his interlocutor with the basic question: "Is that which is holy loved by the gods because it is holy, or is it holy because it is loved by the gods? —[Euthyphro:] "I don't know what you mean, Socrates."[6]

While Ockham's answer to this dilemma might not have been as brazen as Nixon's command ethic, it certainly reflects an "extremely thin account of divine acceptance, modeled on statutory law."[7] Given how the meaning of acts by the finite, created will depends on what has been divinely ordained, "no act is necessarily virtuous" (*Quodl.*, II:14). Not only could God will the suspension of such basic precepts as the Ten Commandments, but even the one act that comes closest to being "necessarily virtuous," viz., to love God above all things, could be reversed by divine fiat.[8] As Ockham puts it so bluntly, "God cannot be bound to any act and, precisely for this reason, what God wills is the same as what is just [*Deus autem ad nullum actum potest obligari, et ideo eo ipso quod deus vult, hoc est justum fieri*]."[9] There is a hyper-Augustinian element at work in Ockham's voluntarism, seemingly supported by Augustine's claim that "when God orders something against the existing morality or social contract of any body of men, even if it has never been done in that place, it must be done" (*Confessions*, 3.8.15). Yet Augustine's true concern in this passage lies with what is *unprecedented* rather than with what has already been divinely ordered in some other way, and as such it is consistent with the Platonic notion of the *logos* being of divine provenance but, qua cosmos, also enjoying eternal divine support. Reason is not a body of concepts or propositions but, for Augustine no less than for Plato, it is *logos*—the manifestation of the abiding framework within which alone meanings of any kind are to be prima facie achieved. A more thorough account than can here be offered would thus have to evaluate the competing visions of Augustinian theology, as well as its potential commensurability (or incommensurability) with Aristotelian eudaimonism; which, in turn, leads to the thorny question of whether the Franciscan critique of Thomism hinges on underplaying or failing to recognize the deep Augustinian moorings of his philosophical theology. Among the most strident accounts in this regard is that by Bonnie Kent, who rejects Alasdair MacIntyre's reading of Aquinas as having achieved a synthesis of Aristotelian and Augustinian thought and—particularly in *Whose Justice?* and *Three Rival Versions*—relying ex-

6. Plato, "Euthyphro," 10[a].

7. Aers, *Salvation and Sin*, 32.

8. "God can command that he himself should not be loved for some stretch of time, since he can command that the intellect, and likewise the will, be so devoted to studying or to some other act that during the time it is able to have no thought at all about God" (*Quodl.*, II:14); see also Ockham's commentary on Petrus Lombardus's *Sententiae* (II/1:19) and, especially, on the contingency of virtuous action relative to the Commandments, see *Sententiae*, III Q. 12:aaa.

9. *Sententiae*, IV Q 9, e–f; quoted in D. W. Clark, "Voluntarism and Rationalism," 78 (trans. mine).

cessively on Étienne Gilson's master-narrative, which unfolds the *History of Christian Philosophy in the Middle Ages* from an unabashedly Thomist perspective.[10] In scrutinizing the "philosophical interregnum" between the condemnation of 1277 and Duns Scotus's inception as master at Oxford in 1305, Kent not only seeks to defend the Franciscan critique of Thomism but, perhaps inadvertently, reproduces some of the nominalist implications that that legacy would only divulge later in the fourteenth century: viz., the tendency to understand problems and issues by preemptively conceiving them as *sui generis* and thus being suspicious of any meaningful and continuous narrative movement. Thus Kent rejects "the tendency to describe thirteenth- and fourteenth-century thought in anticipation of developments centuries later. Knowing what will come, the modern writer easily slips into foreshadowing, dividing those masters and doctrines that were 'properly' medieval from those that anticipated, even helped to produce, the ultimate divorce of philosophy from theology."[11] Needless to say, the present argument reaffirms—albeit also seeks to demonstrate at the level of close textual interpretation—the viability, indeed the necessity, of narrative continuities in the domain of intellectual history and philosophical theology.

For his part, Ockham emphatically contests that view and in so doing inaugurates a distinctively modern conception of rationality. Reason for him is always bound up with a specific perspective, thesis or, indeed, hypothesis on an equally particular problem. As such, *ratio* in Ockham is less *logos* than *epistēmē*, a special case of inquiry forever liable to revision and diametrical reversals, and hence incommensurable with other such cases. The story of this momentous shift from the Thomistic synthesis of Aristotelianism and Augustinianism toward a Franciscan (voluntarist) theology—one wherein agents, situations, and meanings are no longer connected to an underlying rational order or substantial form but, instead, prove inherently discontinuous—has been told from a variety of disciplinary viewpoints. Typically, those who have tendered a version of this momentous shift toward a (modern) pluralist and voluntarist conception of meaning and agency either emphasize nominalism's role in shaping the scientific and liberal self-image of European modernity or voice fundamental misgivings about the soundness of its conceptualist outlook on the sacred, while also stressing the apparent disparity between its stated goals and its actual effects.[12] Arguably, the watershed moment in that narrative involves Bishop Étienne

10. See Kent, *Virtues*, 19–34.

11. Ibid., 2, 9.

12. On the momentous shift within Scholastic theology following the death of Aquinas, and on the emergence of modern, voluntarist notions of the self and the growing role of science as theological "proof," see Kent, *Virtues*, 94–149; Dupré, *Passage*, 65–90; Blumenberg, *LMA*, 145–179; MacIntyre, *Three Rival Versions*, 149–169; Gillespie, *Theological Origins*, 19–43 and 170–206; Buckley,

Tempier's 1277 condemnation of 219 philosophical and theological theses allegedly pagan (i.e., derived from Aristotle and Averroes) then being debated by the faculty of arts under his jurisdiction at the University of Paris. Capitalizing on this shift in the balance of institutional power from a Dominican to a Franciscan outlook, and reinforcing the sense that the newly formed philosophical faculty and the theological faculty at the University of Paris were not parallel but competing enterprises, Tempier's condemnation revived and dramatically expanded earlier strictures of 1210 against the study of pagans and philosophers (e.g., Aristotle) that had actually found their way into the constitution of the Dominican Order in 1228. The proscription of Aristotle's corpus faded after 1230, and study of his writings was finally permitted, even declared compulsory, in 1255. Still, resistance to an enlarged role of philosophy persisted, with the most vehement objections being raised against the employment of strictly ratiocinative methods in the area of metaphysics—the province of inquiry that the Fourth Lateran Council (1215) had decreed to be the exclusive province of the Church and its theological establishment. Looming large among the contested questions was whether moral or intellectual virtues were to be considered superior, and whether Augustinian humility or Aristotelian magnanimity was to be considered as the standard for the moral virtues.[13]

While the 1277 condemnation and Scholasticism's emergent nominalist turn arguably mark a pivotal shift in European intellectual history, our more restricted purposes here only allow us to take up a couple of its more salient implications.[14] Much of nominalism's impact on subsequent theology, philosophy, science, and political thought flows from Bishop Tempier's commitment in his 1277 condemnation to accord God unconditional priority over every other value or "divine name" (reason, justice, grace, charity, wisdom, virtue, etc.). As developed by a number of nominalist theologians over the coming century, privileging God meant that he could no longer be thought of as unconditionally committed to, or bound by, the rational order (cosmos) that he had created. The divine names now came to be understood as contin-

---

Denying and Disclosing God, 1–24; and Gaukroger, *Emergence*, 59–86; on the study of Aristotle, and the history of his preservation and transmission by Islamic scholars, see MacIntyre, *God, Philosophy, Universities*, 43–60; on Aristotle's corpus in Latin translation, see Dod, "Aristoteles Latinus."

13. See Gaukroger (*Emergence*, 62–86) for an excellent account of this contentious phase.

14. Though officially (and unusually) directed merely against "some scholars of arts at Paris" (*nonnulli Parisius studentes in artibus*) said to have "exceeded the limits of their own faculty" (*proprie facultatis limites excedentes*), Tempier's declaration clearly targeted Aquinas's teachings above all; see Wippel, "Condemnations of 1270 and 1277 at Paris," Thijssen, *Censure and Heresy*, 40–56; and Kent, *Virtues of the Will*, 68–93; on the intellectual conflicts surrounding the reception of Aristotle at Paris, see MacIntyre, *Three Rival Versions*, 105–126 and Torrell, *Saint Thomas Aquinas*, 1:296–316; of the 219 sentences proscribed, nos. 157–169 specifically concern the will; they are reproduced and discussed in Kent (*Virtues*, 76–79), who rightly comments on the frequent obscurity of the arguments marshaled against the theses in question.

gent *predicates* rather than essential *qualities* of God—which is to say, as attributes that could be ascribed to God only inasmuch and for as long as his revealed will supported doing so. The shift is anticipated by Abelard's controversial semantic theory, according to which "nouns and verbs have a univocal signification, irrespective of their case endings, to theological questions." To take that view was to argue that "verbs in propositions stating the goodness of God are univocal with those in propositions stating his creative activity."[15] However recondite at first glance, at issue here is nothing less than the nature of concepts and the source and scope of their authority. Already in the late eleventh century, a dispute between Anselm of Canterbury and Roscelin had suggested that official Church doctrine on the status of universals was vulnerable to attack from an Aristotelian position (developed in the *Categories*) according to which concepts were but derivatives of our individual and unique encounter with particulars. While the matter seemed to come to a head around the question of the Trinity (Roscelin being charged with the heresy of Tritheism at the Council of Soisson in 1092), it was Abelard, opposing his former teacher, who identified the broader stakes of the conflict. Distinguishing between the logical status of concepts and the "degree of reality or being of those things designated by universal concepts," Abelard insisted that things might well exist "out there" without anyone having a proper concept of them; conversely, to employ concepts was to be engaging a logical, not a real, object. In so moving "from knowledge of individual things, to knowledge of their concepts, to knowledge of abstract concepts," Abelard ultimately threw open the doors to a nominalist position that, rather more crudely, his former teacher Roscelin had sought to articulate.[16]

From here on the authority of a concept no longer derives from substantial forms or ideas but from the efficacy and plausibility with which it *refers* to particulars. By the time that we reach the extreme nominalism of Ockham's *Quodlibetal Questions* and Nicholas of Autrecourt's avowed atomism in his *Exigit Ordo Executionis* (ca. 1340), the status of theological concepts is limited strictly to their predicative probability. No longer, that is, are concepts understood as ontological frameworks underlying and indeed conditioning the being and reality of all things *and* our capacity for participating in the *logos* of which they are a created manifestation. It follows, too, that God is no longer understood to owe anything to his creation, including human life, but, instead, is thought as forever free to renounce or remake the created world *and* the laws defining it. Henceforth "God is not ... the executor of a world plan that

15. Gaukroger, *Emergence*, 64.

16. "Abelard's attempt to move from his understanding of the nature of universals and how we come by knowledge of them to theological questions was, it has to be said, not a success if the defence and vindication of orthodoxy was the aim, and his doctrines were condemned at the Councils of Soisson (1121) and Sens (c. 1140)" (ibid.).

is consistent in itself and makes its own uniqueness manifest, and whose ideal status means precisely that any rational being must recognize in it (and accordingly put into effect) the necessary characteristics of a world as such." This state of affairs reflects a Gnostic dilemma that, according to Hans Blumenberg, had never been successfully overcome and thus is once again being "reoccupied" by the tension between *potentia absoluta* and *potentia ordinata* encountered in late thirteenth-century theology. Thus the God of early nominalism operates within "the widest horizon of non-contradictory possibilities, within which He chooses and rejects without enabling the result to exhibit in any way the criteria governing His volition."[17]

Henceforth the integrity of creation as an ordered cosmos whose rationality at once embodies and guarantees God's continuity and conditional intelligibility appears increasingly atrophied. Cosmos becomes nature. As a result, the principle of teleology as it had functioned from Aristotle all the way through Avicenna and Aquinas is now subordinated to a strictly volitional notion of the deity. Nominalist thought rejects Aquinas's carefully elaborated arguments regarding the essential coordination of will and reason: "for supposing that He wills a thing, then He is unable not to will it, as His will cannot change" (*ST,* Q 19 A 3). Instead, by insisting that all other attributes of God remain subservient to that of his omnipotence, all creation must be the manifestation of a will that cannot even be constrained by the intrinsic rationality or goodness of what has been created.[18] Henceforth entities may only be thought as coming "into existence from nothing, ... [for] only in this way can the possibility be excluded that God might restrict his own power by creating a particular entity, because any aspect of other concrete creations that happened to be identical in species with the first could only be imitation and repetition, not creation" (*LMA,* 153). Because the nominalist God owes his creation nothing, he also cannot be swayed by any efforts that finite human beings may undertake on behalf of their salvation. Not without justification, some have perceived an "oversimplification of metaphysics" in Ockham's oeuvre, a flattening out of the universe in which singular

17. Blumenberg, *LMA*, 152–153; along with every other notion of rational order, the concept of salvation is also bracketed by what came to be known as the *facientibus*-principle, which "seemed to imply that there were standards for salvation, but that the standards were completely idiosyncratic to each individual ... The determination of sanctity and sinfulness was thus taken out of the hands of the church. No habit of charity was necess[ary] for salvation, for God in his absolute power could recognize any meretricious act as sufficient, and more importantly could recognize any act *as* meretricious" (Gillespie, *Theological Origins*, 28); see also Aers, *Salvation and Sin*, 25–54.

18. The nominalist conception of the divine qua "will" appears convertible with an anthropomorphic model of volition that Aquinas had already rejected as inapposite to understanding the divine will: "the knowledge of God ... would be variable if He knew enunciable things by way of enunciation, by composition and division, as occurs in our intellect," which is precisely *not* the case (*ST*, Q 14 A 15 R. 3). On shifts in the understanding of the will after Aquinas, see MacIntyre, *God, Philosophy, Universities*, 97–112.

entities (*res absolutae*) are all of the same degree and, except for God, everything "has the same fundamental ontological status."[19]

Capable of changing his mind at any moment, this God is no longer bound by any *lex insita* that Scholasticism, following classical sources, had retained as a key presupposition for a rational notion of the cosmos as the totality of laws governing the natural and human realms alike.[20] Consequently, the individual is challenged to assert and legitimate itself *as an individual* (Blumenberg's *Selbstbehauptung*) and to chart its course based on a law that, as representatives from Renaissance humanism to Gottfried Wilhelm Leibniz and the Romantics were to argue, has to be understood as isomorphous with individuality itself. What has dropped out from such a view is the Aristotelian insistence on a fully articulated relation between the project of self-realization and that of communal flourishing. As a consequence, we also find the distinction between "rational longing" (Grk. *boulēsis;* Lat. *voluntas*) and a merely appetitive drive (Lat. *libido, cupiditas*) so integral to Aristotle, Cicero, the Stoics, and St. Augustine to have faded from view.[21] Instead, the nominalist concept of *voluntas* clearly anticipates its Romantic apotheosis as sheer groundless volition of the kind described by F. W. J. Schelling in his 1809 essay *Of the Essence of Human Freedom:* "Will is primordial Being [*Urseyn*], and all predicates apply to it alone—groundlessness, eternity, independence of time, self-affirmation." It is this supposition that "the Good is whatever God wills" rather than "God must will whatever is (determined by nature) as good" that radically alters the human stance vis-à-vis the world; as Charles Taylor observes, the

> framework, the meaning of being, is relative not just to a vision of the world, but also to an understanding of the stance of the agent in the world. [Scholastic] Realism about essences bespeaks the predicament of an agent who sees rightful action as following patterns (essences) which must first be described in things. As against this, in nominalism, the super-agent who is God relates to things as freely to be disposed of according to his autonomous purposes. But if this is right, then we, the dependent, created agents, have also to relate to these things not in terms of the normative patterns they reveal, but in terms of the autonomous super-purposes of our creator. The purposes things serve are extrinsic to them. The stance is fundamentally one of instrumental reason ... the shift will not be long in coming to a new understanding of being, according to which, all intrinsic

19. Milton, "John Locke and the Nominalist Tradition," 131.

20. "Lex est ratio summa insita in natura, quae iubet ea, quae facienda. sunt, prohibetque contraria. Eadem ratio, cum est in hominis mente confirmata et <per>fecta, lex est" (Cicero, *De Legibus*, 1.6.18).

21. See Cicero, *Tusculan Disputations*, 4.12.

> purposes have been expelled, final causation drops out, and efficient causation alone remains.[22]

Taylor here flags for us two closely related shifts; first, it appears that nominalist thought is no longer prepared to credit material existence as the embodiment of substantial forms and, hence, as an entelechy. Instead, existence is being conceived in minimalist fashion as a static and isolated entity accessible only by means of sensory perception and hence denuded of meanings except those attributed to it by the human *sensorium* and intellect.

What Hans Blumenberg calls the "disappearance of inherent purposes" (*Telosschwund*) (*LMA*, 147) takes us to the second shift in question. Once stripped of any intrinsic *telos* or dynamism, the ancient and holistic notion of a cosmos is supplanted by the scientific and particularist study of nature as a complex mechanism of efficient and material causes. A quintessential instance of Weberian "disenchantment," the nominalist fragmentation of the cosmos into a mere aggregate of so many discrete entities lays bare a radical singularity as another entailment of the divine omnipotence that Ockham, Nicholas of Autrecourt, and Gabriel Biel, among others, had set out to defend. The nominalist attribution of the strictest omnipotence (and consequent unaccountability) to a wholly volitional God reveals another characteristic of their theology that was to prove vital to the construction of modern subjectivity over subsequent centuries. For not only is the created world being understood as an agglomeration of so many individual things that can only be signified qua proper name now that universal and normatively rational concepts have lost their authority; it also follows that a God conceived as strictly omnipotent and volitional must likewise be radically singular. At the same time, the emergence of a modern scientific method and its guiding hypothesis of a universe whose lawful, quantifiable, and predetermined organization Leibniz sought to work out in his *mathesis universalis* shows how the nominalist legacy came to furnish a prompt for sustained conceptual innovation. Indeed, the post-Galilean construction of the universe as a profoundly interconnected web of "mathematical idealities" (as Edmund Husserl was to call them)

22. C. Taylor, *Secular Age*, 97; likewise, Dupré remarks on the wide gap between mind and reality in Ockham's writings and the consequent restriction of knowledge to the interpretation of data delivered by the senses: "to know by means of a contact with physical reality, however, is essentially a process of efficient causality, wherein no form is transferred from that reality to the mind" (*Passage*, 39). See also Brad S. Gregory, who notes how, "aside from some late medieval Dominican preference for Aquinas and the persistence of Scotism, the Nominalist *via moderna* became and remained the principal intellectual framework for natural and moral philosophy . . . [As] metaphysical univocity and Nominalism spread, . . . the dominant scholastic view of God was not *esse* but an *ens*—not the incomprehensible act of to-be, but a highest being among other beings" (*Unintended Reformation*, 38; see also 318–319).

ultimately shows modern objective science seeking to recover an *apodictic* (rather than derivative) conception of the world not unlike the Aristotelian substantial forms whose repudiation in the late thirteenth century had paved the way for the modern scientific worldview.

Now, within Ockham's early nominalist framework singularity manifests itself on at least two distinct plateaus. First, it characterizes a creator-God whose unconditional power (*potentia absoluta*) nominalist theology locates at an infinite remove from the created world (*potentia ordinata*) wherein that power has found contingent expression. Noting how the 1277 condemnation, far from settling the questions it sought to resolve, "compounded the problems in many ways," Stephen Gaukroger singles out the question of divine omnipotence and, in particular, the Aristotelian thesis (anathemized by Bishop Tempier) that "the first cause cannot make many worlds." It quickly became apparent that "if it were possible for God to have created such a [non-geocentric] world, then there is a legitimate natural-philosophical question about what its physical characteristics would have been."[23] Second, in establishing such contingency as an attribute of all creation, nominalism disaggregates all finite particular being within the world thus created by divine fiat. Moreover, given that nominalism understands God as utterly separate from his creation, the singularity of created things threatens to denude them of any inner purposiveness or dynamism; no longer intelligible by dint of their inner structure as entelechies, things are on the verge of being demoted to inert objects and, as such, appear increasingly bereft of any essential relation to anything else. In the absence of a clear and abiding teleological matrix, the concept of "world" shifts from that of a rational order to a mere inventory. By contrast, Aquinas had understood all things existent to be continuously sustained in their very being (*essentia*) by their creator, and to reflect that bond through their joint participation in substantial forms. In this regard he remains on firmly Aristotelian ground by holding that "mind and object [being] informed by the same *eidos*, ... mind participates in the being of the known object, rather than simply depicting it."[24] Though Aquinas entertains the possibility of a fact/value distinction, his affirmation that "in idea being is prior to goodness [*ens secumdum rationem est prius quam bonum*]" (*ST,* Ia Q 5 A 2), he is quick to reintegrate the two domains, noting that our knowledge of things in existence hinges on our ability to grasp and articulate them as entelechies:

> Goodness, since it has the aspect of desirable [*rationem appetibilis*], implies the idea of a final cause, the causality of which is the first among causes since an

---

23. *Emergence*, 74.
24. C. Taylor, "Overcoming Epistemology," in *Philosophical Arguments*, 3.

> agent does not act except for some end; and by an agent matter is moved to its form. Hence the end is called the cause of causes [*causa causarum*]. Thus goodness, as a cause, is prior to being, as is the end to the form ... For goodness has the aspect of the end, in which not only actual things find their completion, but also toward which tend even those things which are not actual, but merely potential.[25]

This dynamic, teleological conception of being is more or less overtly rejected by Aquinas's nominalist critics. In their view, a thing achieves its identity strictly by virtue of its dissimilarity vis-à-vis all other things, from which in turn it follows that an entity can never actually enter into a substantive relation with any other entity. This latter view entails some momentous consequences, many of which nominalism (at least in its early phase) arguably neither intended nor understood. For one thing, positing God as but "an extrinsic cause" from his creation not only abandons the Scholastic view of God's constant and sustaining presence in his creation ("Therefore as long as a thing has being, God must be present to it, according to its mode of being" [*ST,* Ia Q 8 A 1]) but, in taking that view, opens a profoundly ambiguous space for human agency.[26] For not only does the nominalist God not owe anything to his creation, but that very fact—once fully realized by human agents now effectively bereft of any reasonable expectation of divine grace and salvation—invariably erodes man's sense of obligation to God, too. As a result, the Christological and soteriological dimensions of the spiritual life wither away, as does the rich and intrinsically universal category of sin. As Michael Gillespie puts it, "the nominalist doctrine of divine omnipotence ... thus undermine[s] the authority of religion in secular affairs. Therefore it is not the rejection of religion that produces modern natural and political science but the theological *demonstration* of religion's irrelevance for life in this world."[27] In ways symptomatic of modern apologists of secular, liberal, Enlightenment society (Mark Lilla, Jonathan Israel), Gillespie's tendentious phrasing ("demonstrates") glosses over the illogic with which he extrapolates his generalized claim concerning "religion's irrelevance for life in this world" from the very discourse (theology) whose authority he means to call into question.

---

25. *ST,* Ia Q 5 A 2; Aquinas subsequently (Q 13 A 5 Rp 1) reinforces his key premise that universals are indispensable to cognition by discriminating between the particularity ("univocal predication") at which all propositional language is aimed and the universality intrinsic to the principle of causation itself and without which we could not even formulate a meaningful proposition about anything.

26. Aers, *Salvation and Sin,* 29.

27. *Theological Origins,* 210 (italics mine).

Paradoxically, it was St. Francis's emphasis on Christ's individuality, singularity, and exemplarity that prompted nominalist theologians to widen the gap between the spiritual and temporal realms. Doing so also meant, however inadvertently, consolidating the relative autonomy of the secular as a realm of human experience and what Blumenberg calls "self-assertion" (*Selbstbehauptung*). For it had been above all Francis of Assisi's rededication to Christ's exemplary life of poverty, as well as the stress placed by his spiritual heir, St. Bonaventure, on the radical individuality of Jesus that initiated the disenchantment of Aristotelian and Thomistic cosmology. That is, it was the Franciscans' quest for *spiritual* regeneration based on the life of Jesus that first challenged "the primacy of the universal ... If the Image of all images [viz., the Incarnation] is an individual, then the primary significance of individual form no longer consists in disclosing a universal reality beyond itself. Indeed, the universal itself ultimately refers to the singular."[28] Arising from what the mostly uneducated early followers of St. Francis undoubtedly meant as an intensification of the spiritual life, the anti-ecclesiastic stress on Christ's "singularity" as the true locus of spiritual authority has a number of important consequences. Not only does the term signify the supremacy or uniqueness of the divine will as embodied in the individuality of Christ; but Christ's very uniqueness carries along with it an implicit challenge to the authority of the Church as the institution invested with the administration of all spiritual meaning, a matter that came to a head with Pope John XXII's 1320 repudiation of the Franciscans' doctrine of Christ-inspired poverty.

Likewise, with regard to God as creator, the nominalist preoccupation with his omnipotence implied that he would never enter into (and thus bind himself by) any structured and reciprocal interaction with his own creation. Beyond the original act of creation itself, the nominalist God cannot make a commitment to what he has called into being, a commitment that for Scholasticism had flowed from the essential goodness, entelechy, and rationality of creation understood as not only "being" but "essence"—that is, a thing imbued with intrinsic purposiveness and, thus, irreversibly entwined with, rather than merely emanating (perhaps even estranged) from the divine *logos*. By contrast, the God of fourteenth-century nominalism is increasingly taken to regard the world with an attitude of quasi-legalistic indemnity and thus as quintessentially not-him but as an otherness that, in Blumenberg's account, explains modernity's unwitting "reoccupation" of the Gnostic

28. Dupré, *Passage*, 38; the argument is extended in Charles Taylor's recent study of secularization: "The Franciscan stress on the exemplarity of Christ that we see in Bonaventure, Duns Scotus, and Ockham gradually undermined the ontological priority of *rationes aeternae* by prompting a stress on the particular 'individual form, the *haecceitas*' among whose many unintended consequences we find an altered interest in nature, now understood as strictly *potential* aggregate of particular data rather than as an *a priori* coherent, rational order or *kosmos*" (*Secular Age*, 93–98).

dilemma. Whereas the Thomist paradigm had always regarded the individual as a strictly contingent "quantification of a universal species" or form, the nominalist stress on the primacy of *forma individualis* promotes "the singularity of the individual" over and against the now waning authority of "universal forms of genus and species. Individuality, then, far from being a mere sign of contingency, constitutes the supreme form."[29] While there is no space here to address Duns Scotus's influential critique of Aquinas's metaphysical alignment of natural philosophy and theology, it was here that Ockham found his prompt: "Scotus drew the contrast between God and his creation in terms of a distinction between finite and infinite being, and what is created and finite cannot in any way determine what is uncreated and infinite on Scotus' account. God's relation to anything else must always be absolutely free, contingent, and unconditioned." Yet in so opposing Aquinas's mediation of the finite and the infinite by means of the sacraments, analogical predication, and the institution of the Church, Scotus's argument "had the drawback of separating God and his creation so radically that ... infinite being was hardly accessible at all."[30]

In passing, we note that nominalism's fragmentation of the ancient Greek notion of a cosmos and its reconstitution as a "nature" composed of wholly disaggregated entities first highlights a dilemma that was to prove particularly vexing to modern narrative, from spiritual and confessional narratives of seventeenth-century Puritanism and early eighteenth-century Pietism to the increasingly ambitious *and* dystopic conception of the modern individual encountered in the *Bildungsroman* of the nineteenth century (in Goethe, Stendhal, Gottfried Keller, Adalbert Stifter, Gustave Flaubert, George Eliot, et al.). Can narrative still aspire to exemplarity, let alone universality? Can it still be "representative" of anything beyond its own, singular occurrence? And can there be anything like a meaningful "plot" in a world composed of strictly singular and thus ostensibly a-rational entities and subjects? Would not all development seem strictly adventitious because confined to the isolated agency in which it is empirically observable? John Henry Newman was to flag that very paradox when remarking "that Knowledge, in proportion as it tends more and more to the particular, ceases to be Knowledge."[31] It is the incipient recognition that singularity marks not the culmination but, logically, the vanishing point of knowledge that now stands to be traced, however briefly, in Ockham's writings.

As evidenced by his conspicuous interest in hypothetical logical scenarios, some of them rather far-fetched, Ockham "seems more interested in exploring what God's freedom allegedly could have performed than in exploring what God has actually re-

29. Ibid., 39.
30. Gaukroger, *Emergence*, 81.
31. *Idea of a University*, 84.

vealed and done in Christ with humanity."[32] In a modern, quasi-performative sense, Ockham conceives the divine will as *instituting* rather than *affirming* meanings, viz., by projecting itself into the world as a spontaneous and inscrutable force "free from all obligation."[33] Centuries later, Coleridge is one of very few thinkers still to grasp the magnitude of the shift when, in *Aids to Reflection* (1825), he remarks on the "direful" doctrine that "swallow[s] up all the attributes of the supreme Being in the one Attribute of infinite Power, and thence deducing that Things are good and wise because they were created, and not created through Wisdom and Goodness" (*AR*, 140). Ockham's abstract *quod/hoc* formula (*quod deus vult, hoc est justum fieri*) indicates that acts of divine will no longer involve substantial forms in which finite human beings may intellectually participate. Instead, the divine will is imagined as a wholly alien, inscrutable power whose dominion over the moral life human, fallible beings must embrace as rational, if only a fortiori. Not only does the human being's ability to participate in divine or right reason appear profoundly impaired; Ockham's account also gives rise to a rather impoverished conception of what it means to be human. What seems to have evaporated is the profound confidence of Aristotelian and Thomistic teleological thinking, whereby created beings do not simply exist *under* God but are ordered *toward* him, and where knowledge *sub ratione boni*, means not merely apprehending but inclining *toward* the highest good (or "end"). Aquinas's stress on the continuity of person and being has been replaced by a theological rhetoric of hypotheses. The idea of life as a process, an ordered progression *sub ratione boni* has effectively been suspended by a syllogistic and, at times, legalistic gaming of equipossible metaphysical scenarios. No longer does order appear as something ontologically given; rather, it pivots on the logical stringency and rhetorical effectiveness of its (human) conceptualization. One of the not-so-distant entailments of that shift will be the weakening of teleology, simply because "Ockham refuses to pay attention to the processes, the forms in which divine grace and human agency work to transform the human subject towards a life lived in reconciliation, obedience, and love to God." He can only imagine God as an "extrinsic cause" whose statutes are "simultaneously rational *and* quite arbitrary in relation to the good intrinsic to their subjects."[34] Rather than being drawn toward a greater sharing in divine reason, the created will is simply expected to submit to it. What, one must ask,

32. Aers, *Salvation and Sin*, 40; Aers quotes *Quodl.*, VI:4, which offers some of the more extreme illustrations of what, in contravention of the world actually created (*potentia ordinata*) God actually *could* do, for example, such incongruous acts as "accepting a sinner for eternal life without grace."

33. Ibid., 30.

34. Ibid., 29–30.

has changed in the underlying philosophical structure for an ethics of intellection to be supplanted, at least in portions of Ockham's writings, by one of compliance?

The answer to that question, I believe, has to be that Ockham no longer works with a notion of substantial forms, which is to say, he "cuts the will off from nature," and it was this crucial part of the Aristotelian legacy that, in Aquinas's account, mediated the divine with the human realm.[35] In Aquinas, the identity and dignity of a thing, its *essentia,* is anchored in the substantial form: "The essence of anything is completed by the form ... [and] is commonly called nature [*per formam completur essentia uniuscuiusque rei quam ... vocatur natura*]" (*ST,* Ia Q 29 A 1). Aquinas's conception of a thing is striking in its richness, if also difficult to sustain as regards its implementation. At one level, substance is bound up with the "quiddity" of the thing in question (*quidditas rei*). On the other hand, a substance is that which endures through contingent change, a substrate (*suppositum*) or "hypostasis." Finally, "as it exists in itself and not in another, it is called 'subsistence' [*subsistentia*]." Crucially, Aquinas emphasizes that all three aspects are simultaneously in play when we speak of a person: "What these three names signify in common to the whole genus of substances, this name 'person' signifies in the genus of rational substances." If this claim may at first glance seem mere Scholastic subtlety, it proves crucial for several reasons. First, if the singular being was to be thought merely as *suppositum* enduring changes, it would appear to lack all identity other than exhibiting a certain reactive persistence in the face of contingent change. Yet inertia and indestructibility alone are merely negative qualities and, as such, incapable of conferring identity on the substance in question. Were it merely to be an embodiment of its species characteristics (*quidditas*), the thing would once again lack identity; its "specificity" would be absorbed, indeed negated by its membership in a given "species" or kind. Hence Aquinas insists that the reality and meaning of a being or thing hinges on the interrelation of all three features, which in the realm of "rational substances" is captured by the word "person" (*persona*). Crucially, "form" thus does not supervene to the things subsisting, "but gives actual existence to the matter and makes it subsist as an individual [*dat esse actuale materiae, ut sic individuum subsistere posit*]" (*ST,* Ia Q 29 A 2).

By contrast, for Ockham, "it is the singular ... that is known first by means of a cognition that is proper to it and simple." As regards its genesis, then, knowledge is

---

35. Adams, "Ockham on Will, Nature, and Morality," 245; likewise, Aers remarks how in Ockham "the theology of the reconciliation between God and humanity in Christ gets swallowed up in a language of power and control." Thus "we are drawn into a discourse in which the decisive matter is acknowledging divine freedom and power" while sidelining aspects of charity, and leaving "the human transformations that such divine agency elicits utterly *extrinsic* to the sinning agent" (*Salvation and Sin,* 27).

always focused on singularities and construes them disjunctively as that which "is naturally caused by the one thing and not by the other, and is not able to be caused by the other. Hence it is not because of a likeness that an intuitive cognition, rather than an abstractive cognition, is called a proper cognition of a singular thing. Rather, it is only because of causality" (*Quodl.,* I:13). Most revealing here is that the singular entity is no longer related to an Aristotelian notion of "form" but to a process of "abstraction." It thus constitutes a derivative concept rather than a real existent. Nature has become something alien, not something in which we always already participate but an enigmatic other to be acquired and remade by the kind of human conceptual labor that, for Ockham and his nominalist successors, defines all rational activity. Consequently, too, the distinction between potentiality and actuality fades since "form does not exist in a more real way in the potency of the matter than in the potency of the agent." Indeed, "before its production a form is neither existence nor essence ... it is a pure nothing" (*Quodl.*, II:8). In the absence of forms as the very linchpin of conceptualization—and of furnishing the will with a rational view (*apprehensio*) *in virtue and for the sake of which* it may act—the will comes to look peremptory, aloof, and potentially unhinged. It does so because its separation from a rational order leaves the entire project of moral deliberation, judgment, and action adrift; as Reinhard Hütter puts it,

> If the logos that elicits faith and legitimates theology is a contingent word spoken, a willful positing, such that it can in no way be related to the way things are and vice versa, and, more importantly, such that the way things are cannot be disclosed by this logos, such a "logos" only intensifies the specter of the will by placing one willful positing over against others, so that the last ground of reality is nothing but the agonism of warring wills and their contingent positings.[36]

Inasmuch as the nominalists had "desymbolized the universe" and postulated that "nature lacks cognition" it is reasonable to view Ockham as preparing, however unwittingly, the advent of the "modern notion of agency as constructing orders, rather than conforming to those already in 'nature.'"[37] Here are the origins for the long process of "theoretical self-determination" (*theoretische Selbstbehauptung*) that Hans Blumenberg and others view as defining of the modern era. For the urgency of reading the "book of nature" on our strictly human terms increases even as the prospect of rendering the divine will intelligible through such readings continues to diminish. For "if the natural world is an arbitrary expression of the otherwise inscrutable

36. Hütter, "Directedness of Reasoning," 164.

37. Funkenstein, *Theology*, 58; Adams, "Ockham on Will, Nature, and Morality," 250.

attributes of God, then it seems impossible that one should be able to ascend to knowledge of God through its study."[38] This apparent impasse can only be resolved by gradually transferring the divine capacity to make order to the realm of human, conceptual activity. Thus what had been an indispensable teleological framework is gradually supplanted by the interventionist, volitional nature of human inquiry. Though ostensibly still concerned with the "discovery" of divinely created order, human knowledge in fact imposes its own paradigms of cognition in lieu of God's, which the nominalists' preoccupation with divine omnipotence and unaccountability had effectively rendered unknowable. In due course, too, the once guiding metaphysical superstructure of final causes is being superseded by a reliance on occasional and efficient model of causality.[39]

Not surprisingly, numerous commentators have drawn attention to this "modern" or "voluntarist" streak in Ockham, albeit in the process often overstating their case and, at times, making Ockham sound rather too much like Hobbes or Schopenhauer.[40] More nuanced and responsive to the complexities of Ockham's oeuvre, readers from Frederick Copleston to Marilyn McCord Adams have noticed the apparent emergence of two value theories within his oeuvre, one "authoritarian" and the other reflecting Ockham's repeated insistence on "right reason, which would seem to imply that reason can discern what is right or wrong."[41] In some form or other, that distinction tends to persist in most political thought, and can be encountered again in Kant's juxtaposition of *Willkür* and *Wille* in his late *Metaphysics of Morals.* Still, as is the case with his (often unwitting) intellectual heirs, Ockham's attempts to reconcile

38. Brice, *Coleridge and Scepticism*, 30.

39. C. Taylor, *Secular Age*, 127; s. a. 112. Ockham's *Quodl.* illustrates this shift in the characteristically dry idiom of formal logic: "It is not always the case that an effect has a final cause distinct from its efficient cause" and "the causality of an end is being loved efficaciously" (see *Quodl.*, IV:1; 246, 248). On the expanded portfolio of efficient causes in Ockham, see *Quodl.*, VI:12.

40. See Frederick Copleston, who speaks of Ockham's "authoritarian ethic"; Gordon Leff, who remarks how "with God and His will synonymous there could be no way of judging right or wrong other than by the decrees of His will"; Armand Maurer, who remarks on Ockham's having "sever[ed] the bond between metaphysics and ethics and base[d] morality not on the perfection of human nature, nor upon the teleological relationship between man and God, but upon the obligation to follow laws freely laid down for him by God" (quoted in D. W. Clark, "Voluntarism and Rationalism," 72–73n). In the same vein, Adams quotes Maurice de Wulf, for whom Ockham's God is defined by "absolute autonomy of volition" and Paul Helm, who sees "Ockhamist Divine Command Theory" as holding "that morality is founded upon a free divine choice" ("Structure of Ockham's Moral Theory," 1); Adams's own reading struggles to counter this strictly voluntarist and seemingly arbitrary model of divine will in Ockham.

41. Copleston, quoted in Adams ("Structure of Ockham's Moral Theology," 4); in her more recent work, Adams maintains that "however distinctive, Ockham's theories of the will and morality are developed within the broad outlines of an Aristotelian theory of rational self-government" ("Ockham on Will, Nature, and Morality," 246).

these two models of the will and of moral agency tend to yield little more than carefully elaborated distinctions that gloss over the rift in question. Seeing Ockham "weaving a complex tapestry of continuity and innovation," Adams, for example, suggests how, "when suitably informed by speculation and revelation, right reason, the internal regulator of the agent's willing, finds another rule in divine commands."[42] "There is a double criterion of a morally virtuous act—the dictates of right reason, on the one hand, and divine precepts on the other." Yet to insist that Ockham's "cultured detractors" distort his moral theory because they fail to honor his distinction between "positive" and "non-positive" morality does not resolve the principal dilemma.[43] For it is precisely the *warrant* for drawing that very distinction that has to be questioned in light of Ockham's own view of the divine will as infinitely reversible, ineluctable, and beyond all mediation with human agency. The latter thus appears bereft of all guidance, such as had been furnished by Aristotelian natural teleology, Augustinian Christology, and their synthesis in Aquinas. Put differently, Adams's both/and logic cannot be legitimately applied if it should turn out that Ockham's two sources of moral action (i.e., divine precept and rational deliberation) and of reflection (i.e., positive and non-positive) either lack all coordination or, worse yet, prove altogether incommensurable.

Indeed, there is reason to conclude that both scenarios apply. Even Adams acknowledges how Ockham's "God is a debtor to no one ... He has no obligation to continue creatures in existence," and "He is under no obligation to accept morally virtuous acts or to reject morally vicious ones."[44] Undoubtedly, Ockham still expects the finite, created will to deliberate on a morally sound course of action, yet in that pursuit the self can no longer take itself to be guided by a normative framework such as it is unconditionally set forth in Aristotle's substantial forms or in the Decalogue. Instead, for Ockham, all forms hold their intrinsic rationality only for as long as God extends his support to them. That is arguably Aquinas's position, too. Yet unlike Aquinas, Ockham also holds that forms may never be construed as binding the divine will in any way whatsoever; and, in taking that view, he sacrifices the intelligibility of creation to an overriding, abstract, and seemingly legalistic preoccupation with divine omnipotence. No doubt, Ockham does not wish to imply that his vastly expanded notion of God's *potentia absoluta* be regarded as irrational and capricious, nor indeed as outright suspending Aristotelian right reason as the principal framework for the finite agent's moral deliberation. Still, that God might "change his mind," as it were, remains for Ockham a constant possibility, simply because to suppose

42. Adams, "Ockham on Will, Nature, and Morality," 246.
43. Adams, "Structure of Ockham's Moral Theology," 24, 33.
44. Ibid., 20–21.

otherwise is to set unacceptable limits to his omnipotence, even if those limits would be of his own creation.

There is no doubt that Ockham's conceptual shift had enormous implications. His contention that the rationality of moral and natural forms or concepts exists only by virtue of the fact that God has willed or ordained things to be as they are implicitly disrupts the individual's ability to progress toward, let alone participate in, the *logos*. The sheer unintelligibility of divine reason inevitably atrophies any orientation and progression toward it, such as Aquinas had still conceived of it as a narrative trajectory leading from an awareness of a basic inclination to its cultivation as an array of habits to the formation of moral and intellectual virtues to the ultimate goal of a *visio dei*. Having rendered the *logos* contingent on a terminally unintelligible and inscrutable act of divine ordination, Ockham also deepens the finite self's sense of his or her finitude and, however unwittingly, prompts human thought to devise compensatory strategies of self-legitimation premised on the coherence of our propositions and concepts rather than on a metaphysics of grace. An apt illustration of the long-term consequences of Ockham's voluntarism can be found in Hume's *Dialogues* (1779) when Cleanthes, the advocate of natural theology, remarks on the contingency of all creation: "'Any particle of matter,' it is said, 'may be *conceived* to be annihilated; and any form may be *conceived* to be altered. Such an annihilation or alteration, therefore, is not impossible.' But it seems a great partiality not to perceive, that the same argument extends equally to the deity, so far as we have any conception of him; and that the mind can at least imagine him to be non-existent, or his attributes to be altered."[45]

In so opening a chasm between the rationality of the divine will and the kind of reason-giving and deliberation that defines human agency, Ockham offers glimpses at a "modern notion of agency as constructing orders, rather than conforming to those already in 'nature.'"[46] Well before Hobbes, this shift also reveals another (clearly unintended) consequence, viz., that "the secular as a domain had to be instituted or *imagined*, both in theory and in practice" so as to compensate for the apparent lack of mediation between human and divine reason.[47] To be sure, neither in Ockham's writings nor indeed for some time thereafter does the shift in question entail the Manichean irrationalism that at times confronts us in Luther, nor the inferential theism and anthropomorphism of natural theology that we encounter in Robert Boyle, John Ray, William Paley, and so many other writers. Still, in taking exception with what he regards as the Dominicans' excessive reliance on pagan (Ar-

45. Hume, *Dialogues*, 65.

46. C. Taylor, *Secular Age*, 127.

47. Milbank, *Theology*, 9.

istotelian) thought—and particularly on substantial forms allegedly constraining divine omnipotence but, in fact, lacking any conceptual warrant—Ockham injects an element of radical inarticulacy into conceptions of the divine will. We can see the consequences in the anti-institutional, fideist hyper-Augustinianism of seventeenth-century Puritanism, Jansenism, early eighteenth century German Pietism, and in the evangelical and Pentecostal denominationalism of the early nineteenth century. More than Luther's Protestantism, it is the radical and increasingly fragmented Pietism of his late seventeenth-century heirs (Philipp Jakob Spener, Nicolaus Zinzendorf) for whom Luther's ambivalent view of the church and institutional mediation was a source of frustration and, increasingly, a cause for rebellion. If in Luther and German Pietism will and intellect appear caught in a constant, quasi-Manichean struggle, others (John Calvin, Justus Lipsius, Immanuel Kant) try to recover from voluntarism's more unsettling implications by devising alternately neo-Stoic or hyper-Augustinian projects of self-discipline and self-abnegation subsequently revived by the evangelical revival movements of the early nineteenth century.

Eventually, these strains would be subjected to an incisive critique by a wide and diverse spectrum of cultural critics—Heinrich Heine, Ralph Waldo Emerson, John Henry Newman, Matthew Arnold, Friedrich Nietzsche, and others—charging Ockham's remote descendants with having embraced outright irrationalism almost as a matter of principle. Highly influential in this regard is Matthew Arnold's *St. Paul and Protestantism* (1870), which applies and extends what in *Culture and Anarchy* (1869) Arnold had identified as evangelical Protestantism's disturbing antithesis between "thinking and doing," or between the Hellenist and the Hebraic models—"rivals not by the necessity of their own nature but as exhibited in man and his history."[48] From a different perspective, Newman found Arnold's own Anglo-Protestantism greatly wanting, in part because it lacked a coherent position from which to tackle mid-nineteenth-century evangelicalism's deep-seated anti-rational and seemingly contentless fideism. Offering a mordant characterization of "the Anglican communion as the golden mean between men who believe too much and men who believe too little," Newman left no doubt that an effective response to contemporary voluntarism and hyper-Augustinian fideism required more than Anglicanism's "national form" and "gentleman's knowledge," which is "never more than the furniture of the mind ... [and] never thoroughly assimilated with it." For Newman, to counter modern denominationalism's emotivist, anti-cognitivist, and anti-institutional pathos what is needed is not *less of the same* religion, but a different *kind*. Merely to settle, in Anglican fashion, for "having the Bible read in Church, in the family, and in private," and to regurgitate doctrines that "are not so much facts, as stereotyped aspects of facts"

48. *Culture and Anarchy*, 126.

was to blur the line between the real assent of religious practice and the notional assent to a moral philosophy of the kind that Henry Sidgwick was concurrently trying to formulate on a wholly secular basis.[49]

However unwittingly, Ockham's voluntarist account of divine omnipotence and his consequent estrangement of both God and man from an ontology of reason qua teleology furnishes the archetype of a conception that Matthew Arnold was to find so disturbing more than five centuries later. Having by then woven its way deep into the fabric of modern Anglo-Protestantism and its embrace of the modern *vita activa,* Ockham's implicitly Pelagian conception rests on a hyper-ventilating life of making and doing that is at once the cause of an apparent lack of introspection and a type of compensation for it. For Arnold (well before Hannah Arendt), *homo faber* names the inevitable endpoint of a theology that, beginning in the early fourteenth century, found itself unable to sustain the Thomistic cooperation of intellect and will: "Our preference of doing to thinking," Arnold notes, is but "another version of the old story that energy is our strong point and favourable characteristic, rather than intelligence." The modern, hyper-Augustinian "sense of the obligation of duty, self-control, and work" has all but vanquished the Aristotelian and Thomistic tradition of centering the individual not on compliant behavior but on the attainment of right reason and a concept of *action* (not behavior) as genuinely constructive and imaginative rather than defensive and conformist.[50] Certainly Ockham's heirs (Luther being an obvious, if also extreme case in point) came to regard the will as no longer resting on an underlying intellectual conception or "view" but as generating (seemingly *ex nihilo*) those meanings that henceforth shall count as authoritative.

A shift begins to take place, slowly and unwittingly for some time, yet nonetheless inexorable in its logic and outcome. As theoretical and practical rationality increasingly diverge, and as philosophy and theology begin to acquire distinct conceptual and institutional identities, the project of rational, virtue-based self-governance goes into decline because there no longer appears to be any ontological guarantee that what counts as right reason must remain so for all time. With the supposition that right reason is but a function of enigmatic, divine command, human agency and flourishing are profoundly destabilized. Rather than constituting the *source* of an *actus,* the rationality of moral and natural forms comes into being only by virtue of a particular act of (divine) will or, ultimately, by dint of our self-fashioned interpretations and ascriptions of reason to an opaque deity that, in time, is reduced to the mere formalism of a "First Cause." For inasmuch as reason itself is contingent on God's ordination,

---

49. Newman, *Grammar of Assent,* 204, 62–63; see also his essay on "Private Judgment" (1841), in *Essays Critical and Historical,* vol. 2, 336–374.

50. Arnold, *Culture and Anarchy,* 126.

"the concept of an unrestricted divine power ... weakened the intelligibility of the relation between creator and creature" and, in so doing, precipitated what Brad Gregory calls the "self-marginalization of theology" in the early modern era.[51] Moral meanings thus appear on the verge of losing their *essential* quality and thus are taken to exist solely by *ascription* of what the finite, created will takes God to have ordained.

It follows that the practical realization of such meanings through a lifelong process of habituation in a differentiated spectrum of the virtues is also being attenuated, if for no other reason than that the divine source sanctioning such meanings and enjoining their practical realization only furnishes us with decrees whose rationality remains at all times ineffable and potentially reversible. An ethic of knowledge thus is gradually supplanted by one of obligation, just as rational deliberation appears increasingly trumped, not to say defeated, by compliance with an increasingly enigmatic divine "law." Having thus extended the scope of divine power (*potentia absoluta*) "beyond its previously assumed moral and rational limits," Ockham in particular "stresses that order depends at each moment on God's resolution to abide by it. God's sovereign power is not intrinsically bound by any necessity other than that of its absolute freedom, nor is that freedom subject to what we consider ultimate rationality."[52] Only in this precarious metaphysical constellation could there have arisen a need for the type of containment strategy offered by Hobbes's *scientia civilis,* for the neo-Stoic and hyper-Augustinian projects of moral self-discipline and self-improvement, and for epistemological methods seeking to regain control over self and nature, both of which now appear in conflict, opaque, and in urgent need of theoretical (re-)legitimation.

Perhaps most symptomatic of these related developments is the sudden rise of a new type of explanatory scheme known as theodicy. "When men could no longer *praise,*" Hannah Arendt remarks, "they turned their greatest conceptual efforts to *justifying* God and his creation in theodicies." Similarly, Odo Marquardt insists that "theodicy's question ... was always blunted in earlier, premodern times, by an intact religion ... Where there is theodicy, there is modernity ... [For the] modern age is the age of distance," and in it "the *malum metaphysicum,* finitude, the *malum morale,* evil, the *malum physicum,* sufferings ... are not the taken-for-granted and normal

51. Dupré, *Passage,* 174; Gregory, *Unintended Reformation,* 93.

52. Dupré, *Passage,* 176, 123. Among others, Dupré also draws attention to the enormous consequences of this shift, particularly for the scientific investigation of nature, so clearly on the ascendant by the fifteenth century (see 37–41); see also Milbank, *Theology,* 14–18. Surprisingly, in her detailed account of the gradual emergence of psychological, ethical, and metaphysical voluntarism after 1270, Bonnie Kent expressly bypasses the third and arguably most consequential form of voluntarism, which "signifies a strong emphasis on God's freedom (or absolute power) to will anything not involving a contradiction" (*Virtues,* 95).

state of affairs for human beings." With a nod to Blumenberg's reading of modernity as an unwitting "reoccupation" of Marcion's Gnostic dilemma, Marquardt thus notes that the God of post-Scholastic and post-Aristotelian modernity "evades this burden [of Gnosticism] in the role of the alien and hidden redeemer God who at the same time no longer orders anything intelligibly in the world." The result is either war (where "human beings have to dispute—ultimately in a bloody manner—about questions of salvation") or science, whereby "the urgency of redemption must be removed by a demonstration that this world is endurable."[53] Particularly intriguing about this picture is how the epistemological drive toward comprehensive theoretical explanation and justification implicitly *creates* the anthropological phenomena that it claims to have found.

The modern idea of a disengaged and hermetic self—the hedonistic composite of so many inarticulate and fluctuating desires—finds its consummate expression in Hobbes, Gassendi, and Locke. Progressively naturalized as an authoritative account of "human nature," this premised self in time also furnishes the point of departure for the Whig apologists of commercial society (Mandeville, Defoe, Adam Ferguson, James Steuart, Adam Smith, et al.) who, against Hobbes, seek to demonstrate how such a self may yet be capable of moral goodness. And, as remains to be seen, where the conversion of irrational passions into rational interests should fail, the remedy is once again not to be looked for in some inner advance of the self toward rational personhood. Rather, correction is to be achieved by internalizing impersonal, quasi-syllogistic precepts on the order of Kant's moral law, which promise to *engineer* moral agency—not in the modality of knowledge but of compliance. Along with the project of practical reason and virtue ethics and the Thomistic model of participatory and relational *personhood,* what continues to fade away after Hobbes is the conceptual space in which the will might not be axiomatically viewed as the antagonist of the intellect (or, indeed, as the source of the latter's outright demystification). Correction, not transformation, becomes the central objective of modern moral epistemologies. Long before Michel Foucault, John Henry Newman thus remarks in 1841, it "is a chief error of the day, in very distinct schools of opinion,—that our true excellence comes not from within, but from without; not wrought out through personal struggles and sufferings, but following upon a passive exposure to influences over which we have no control."[54]

---

53. Arendt, *Life of the Mind*, 97; Marquardt, "Unburdenings," 11–13.
54. *Discussions and Arguments*, 266.

Part III

PROGRESSIVE AMNESIA

WILL AND THE CRISIS OF REASON

8

IMPOVERISHED MODERNITY

Will, Action, and Person in Hobbes's Leviathan

At times a terror, Leviathan has always been an enigma on account of an innate tendency of instrumental reason to turn into its other, rather in the spirit of William Blake's dictum that "Opposition is true friendship." Embodying those very terrors of irrational strife that it had been designed to keep at bay, the Hobbesian state thus peremptorily seizes all possible venues from which it might be materially or intellectually challenged. Most obviously, that means securing a monopoly on power (*potentia*), which now is conceived strictly in terms of efficient, instrumental causality. Our prevailing idea of the modern state has been profoundly shaped by Hobbes's notion of power as mechanical "force," that is, as the state's unconditional, legal, and material prerogative to effect a "decision" on any range of issues—including prima facie the decision of what issues stand to be decided. Hobbes's political voluntarism thus effects a downward transposition of the classical meaning of "power" (Grk. *dynamis;* Lat. *potentia*) to a strictly efficient "force." It abandons the generative meaning of *potentia*—which, as Hannah Arendt was to remark, can be "actualized only where word and deed have not parted company, where words are not empty and deeds not brutal ... Power is always ... a power potential and not an unchangeable, measurable, and reliable entity like force or strength" (*HC,* 200). Yet beginning with Machiavelli and Hobbes at the latest, political power comes to be understood as a non-cognitive and mechanistic *means.* Thus arguments for the legitimacy of power, while not abandoning an appeal to a transcendent, divine source, tend to emphasize its pragmatic *efficacy* and sustained *enforceability*. Designed to constrain the brute and inarticulate wills of its subjects—and, invariably, coming to mirror their supposedly non-cognitive nature—Hobbesian sovereignty is not, however, an entirely novel phenomenon in modern political thought. Rather, it secularizes and radicalizes some central tenets of voluntarist theology.[1] However startling it would have been to Aquinas's fourteenth-century critics, what the apologists of absolute state power from Hobbes to Carl Schmitt propose is nothing more (or less) than to draw out the irrational implications so unwittingly prepared for by William of Ockham's conceptualist approach to God as the agent whose absolute power (*potentia absoluta*) must never be constrained, not even by the reality of his own creation (*potentia ordinata*).

We can now begin to trace the evolution of modernity's dominant conception of power as *efficient force,* that is, as the outward manifestation of a non-transparent and non-cognitive will that can only be known or unmasked *after* it has projected itself into social and political spaces. In so doing, we become aware of the omnipotence, unaccountability, and consequent opacity of voluntarism's God, on the one hand, and the emergent ideal of modern "autonomy" or self-possession, on the other.[2] It also helps us understand how, by the middle of the nineteenth century, a rather flat, voluntarist notion of power could have migrated from the self-possessed and autocratic persona of the Hobbesian monarch to the abstract proceduralism of the modern, liberal-bureaucratic nation-state. Somewhere in the volatile transition from the late Enlightenment to the early nineteenth century, this transformation of state power is finally completed; and by the 1850s, the notion has effectively metastasized to a complex institutional and bureaucratic landscape that, on the face of it, has little in common with Hobbes's notion of a polity governed by an autocratic will. Yet to understand the historical emergence (and intrinsic paradoxes) of the *systemic* and *institutional* model of politics long associated with the nineteenth-century liberal, secular, and institutionally embedded state, a bit of intellectual history (albeit in highly compressed form) is in order. For only by linking the historical genesis of theological voluntarism to the modern state's view of the individual subject as begging institutional containment is it even possible for us to assess the viability of the project of Enlightenment and post-Enlightenment political culture. Absent such a counter-narrative, all thinking about modernity—and the modern state's institutional, economic, and constitutional frameworks—remains premised on an under-

1. For influential accounts of modernity as the emergence of the secular, bureaucratic, and institutionally embedded state—paralleled by the decline of ancient and medieval virtue ethics—see MacIntyre, esp. *After Virtue* and *Whose Justice?;* Pocock, *Machiavellian Moment;* Blumenberg, *LMA;* Giddens, *Consequences;* C. Taylor, *Sources* and *Secular Age;* and Arendt, *Origins of Totalitarianism*.

2. On the emergence of modern autonomous agency, see Schneewind, *Invention,* esp. 17–36.

lying (and, I would argue, deeply flawed) assumption that these frameworks are the only conceivable embodiment, indeed the very apotheosis of rationality.

A sensible place to begin this narrative—though certainly not its point of origin—is with Hobbes. For more than any other conception of sovereign power, it is the Hobbesian model that has thrown a long shadow over subsequent political thought. As Hannah Arendt observed some time ago, "there is hardly a single bourgeois moral standard which has not been anticipated by the unequaled magnificence of Hobbes's logic."[3] To which one should add that this is the case because virtually all subsequent models (from Locke and Montesquieu forward via Adam Smith and Kant to John Stuart Mill)—regardless of how their various progenitors felt about Hobbes—are dialectically conditioned by his thinking. While they may seek to *contain* the more disconcerting implications of Hobbesian voluntarism, they remain (with very few exceptions) unable to *escape* Hobbes's model of human agency. Setting aside the question of Hobbes's intellectual forebears for the moment, we merely note that his concept of power in the *Leviathan* constitutes an extension of his earlier reflections on physics and anthropology, which show Hobbes to understand matter as intrinsically "minded, or at least willed" and of humans "as bodies driven by passions."[4]

The conception of political power advanced in the *Leviathan* thus reflects its author's underlying view of nature as prima facie irrational and categorically incapable of furnishing the human individual with any purposive and coherent framework. As Hobbes insists, "notions of good, evil, and contemptible are ever used in relation to the person that useth them, there being nothing simply and absolutely so, nor any common rule of good and evil to be taken from the nature of the objects themselves" (*Lev.*, 6:7). Devoid of the substantive forms that Aristotelian and Scholastic thought had postulated, nature is instead conceived as an aggregate of inherently value-neutral forces. For Hobbes, order is never *found* in nature but has reality only as something *constructed, ascribed, and imposed*—even as the *appeal* to natural law as a putative source of authoritative meanings may well help legitimate the sovereign's political will. In partial compensation for nature's apparent lack of rational order,

3. *Origins of Totalitarianism,* 186. On Hobbes's *Leviathan* as the culmination of the political consequences wrought by post-Reformation religious strife and its specter of religious hyper-pluralism—acknowledged by magisterial Protestantism no less than by Catholic Counter-Reformers—see Gregory, *Unintended Reformation,* 148–163.

4. Gillespie, *Theological Origins,* 225, 236; Schneewind also notes how Hobbes's "psychology is intimately tied to his physics" (*Invention,* 84, 88f.); Strauss sees Hobbes fusing Epicurean materialism and Platonic idealism and conceiving of "a universe that is nothing but bodies and their aimless motions" (*Natural Right and History,* 172); for detailed accounts of Hobbes's radical mechanism and his dispute with Boyle regarding the foundations of experimental science, see Shapin and Schaffer, *Leviathan and the Air-Pump,* and Gaukroger, *Emergence,* esp. 281–289 and 368–379.

Hobbes imagines sovereignty as a *countervailing force* composed of the law, an (Erastian) church, and a monopoly on military might.[5] With latent menace, the sovereign's will casts its long shadow over the totality of all the embodied wills of which the body politic is composed—wills axiomatically viewed as a-rational and (at best) indifferent to the interests of the state. Unlike the rational coherence of Aristotelian *energeia* or the omnipresent love that Aquinas posits as the supreme quality of the creative will (*potentia absoluta*), power for Hobbes is intrinsically agonistic and a-rational. To the question first broached in Plato's *Republic* and here reformulated by Oliver O'Donovan—"Is there in the nuclear core of human judgment a shortfall of reason, which generates an exertion of force to compensate for its lack?"—Hobbes answers with an emphatic "yes."[6] The Leviathan's sovereignty thus is (dialectically) legitimated by its adversarial relation to the kinetic force of so much unruly, minded matter, the containment of which the Leviathan takes to be its principal, perhaps its only, task.

Once power is no longer legitimated by its rationality or exemplarity, its peremptory and compulsory nature rests on a view of the individual as categorically incapable of rational self-possession. Echoing C. B. Macpherson's account of the "possessive quality" surfacing in the seventeenth-century view of the individual ("essentially the proprietor of his own person or capacities, owing nothing to society for them"), John Milbank views Hobbes as the pivotal figure within a broader historical shift: "*Dominium* over oneself, 'self-government,' was traditionally a matter of the rational mastery of the passions and ... also the basis for one's legitimate control and possession of external objects ... Yet at the margins of this classical and medieval theme there persists the trace of a more brutal and original *dominium,* the unrestricted lordship over what lies within one's power" and in the "seventeenth century this original Roman sense not only returns, but for the first time advances from the margins into the center."[7] Undoubtedly, Hobbes would be Exhibit A in a more expansive version of that story, which would also tell of the virtues' gradual retreat from political and theological thought, a development certainly not initiated, though greatly accelerated by the religious, political, and scientific upheavals of the sixteenth

5. On the complicated legacy of that axis for post-Hobbesian theories of the commonwealth, see Pocock, *Machiavellian Moment,* 406–422; see also Eisenach, "Hobbes on Church and State and Religion"; Koselleck, *Critique and Crisis,* 23–40; and Oakeshott, *Hobbes on Civil Association,* 50–58; Strauss remarks on the "essential ambiguity" of Hobbesian "power" as both "*potentia,* on the one hand, and ... *potestas* (or *jus dominium*) on the other. It means both 'physical' and 'legal' power" (*Natural Right and History,* 194).

6. *Ways of Judgment,* 15.

7. Milbank, *Theology,* 3, 13.

century.[8] Of crucial importance here is Hobbes's rejection of "right reason" in *De Cive* (Chapter 2, 1n):

> By Right Reason in the naturall state of men, I understand not, as many doe, an infallible faculty, but the act of reasoning, that is, the peculiar and true ratiocination of every man concerning those actions of his which may either redound to the dammage, or benefit of his neighbours ... Although in a Civill Government the reason of the Supreme (i.e. the Civill Law) is to be received by each single subject for the right; yet being without this Civill Government, (in which state no man can know right reason from false, but by comparing it with His owne).

The scope of rationality here has contracted to whatever a given individual feels *compelled* to assent to simply because the sovereign demands and enforces such assent. Yet absent a strong state power, Hobbes insists, "no man can know right reason from false." Long before Frege, Wittgenstein, and Ryle, Hobbes is already calling into question the possibility of the inner life as a rational and purposive (narrative) progression. Indeed, by appraising human agency strictly in terms of desire and volition—to be contingently thwarted or accommodated—Hobbes effectively denies that individual life may ever coalesce into a meaningful and continuous narrative. Pared down to an agglomeration of disjointed volitional states (themselves the outward projection of so many random desires), agency appears denuded of all the formal, historical, and hermeneutic coherence implied by the idea of a "person." As Iris Murdoch (thinking of Wittgenstein and Stuart Hampshire rather than Hobbes) puts it, on this view "reasons are public reasons, rules are public rules. Reason and rule

---

8. On this point, see Schneewind's discussion of Suarez and Grotius (*Invention*, 58–81) and their uneasy vacillation between a Thomistic, virtue-based, relational, and participatory account of created life in divine being, and the Franciscans' (nominalist and proto-legalistic) understanding of right as moral self-governance of individuals seeking to compensate for their de facto abjection from God. See also Milbank's distinction between a "theological natural rights tradition [that] discovered a self-sustaining world of pure power without virtue" and a "non-Christian Machiavellian tradition derived from Polybius [which] insisted that human power was a form of virtue" (*Theology*, 25). Undoubtedly accelerated by the Reformation, the diminishing role of the virtues in the first of Milbank's senses can be measured by the collapsing belief in their inculcation by way of Aristotelian and Augustinian *habitus*. In Aristotle's *Politics*, it is precisely "habit" (*ēthos*) that mediates between nature (*physis*) and reason (*logos*) as the three things "which make men good and excellent" (1332a40); on this topic, see MacIntyre, *After Virtue*, 181–243 and *Whose Justice?* 103–145; C. Taylor, *Secular Age*, 112–136; Dupré, *Passage*, 15–29; and Herdt, *Putting on Virtue*, 23–45. For a different and somewhat perplexing account, see Skinner, who rather improbably reads Hobbes "essentially as a theorist of the virtues, whose civil science centres on the claim that the avoidance of the vices and the maintenance of the social virtues are indispensable to the preservation of peace" (*Reason and Rhetoric*, 11).

represent a sort of impersonal tyranny in relation to which however the personal will represents perfect freedom. The machinery is relentless ... What I am 'objectively' is not under my control; logic and observers decide that. What I am 'subjectively' is a foot-loose, solitary, substanceless will. Personality dwindles to a point of pure will."[9]

What has usurped the place of personhood, virtues, and inner (right) reason is an increasingly monochrome conception of the will as sheer *kinesis,* a mindless force to be conceived solely in terms of efficient causality. Already in *De Cive* (1642; Chapter X, i), Hobbes thus identifies *potentia* with *causa.* As Hannah Arendt notes, long before it came "to be substituted for Reason as man's highest faculty" the will already tended to lack any meaningful relationship to the fullness of past time: "the Will's ability to have present the not-yet is the very opposite of remembrance. Remembrance has a natural affinity to thought" whereas "the will always wills to *do* something."[10] With the peculiar satisfaction of a confirmed pessimist, Hobbes thus remarks that "I *can* do if I *will;* but to say I can *will* if I *will,* I take to be an absurd speech."[11] Paradoxically, even as Hobbes posits the will as the unconditional and indisputable source of the self's *inner reality,* he can do so only at the expense of rendering that source terminally opaque and incommensurable with all propositional or discursive knowledge. Long before Schopenhauer's extraordinary decision to reinterpret the Kantian noumenon as, in fact, the ontological *datum* of the will, Hobbes thus posits the will as an absolute source—unaccountable, inexplicable, and hence beyond the reach of any intellectual, reflexive, or dialectical attempts at sublating it into a rational progression.

For Hobbes, no such remedial strategy can ever succeed, quite simply because he has already determined that ostensibly more complex and self-aware intellectual processes are, in fact, nothing more than "calculative" epiphenoma of the will. Impervious to Platonic "recollection" (*anamnēsis*), Aristotelian "judgment" (*prohairesis*), or Thomistic "deliberation" (*consilium*), mind in Hobbes lacks any distinctive phenomenology. It merely presents as a depthless, self-seeking force begging containment by a stronger counterforce; its relation to other minds is defined strictly in legal terms, that is, as party to political and economic covenants of some kind or other: "The *value* or WORTH of a man is, as of all other things, his price, that is to say, so

---

9. *Sovereignty*, 15–16; of significant influence on Murdoch is Simone Weil, whose discussion of "Human Personality" also stresses (and rejects) the voluntaristic understanding of that term in the modern era: "So far from its being his person, what is sacred in a human being is the impersonal in him ... Impersonality is only reached by the practice of a form of attention which is rare in itself and impossible except in solitude" (*Simone Weil Reader*, 317–318).

10. *Life of the Mind*, 2:20, 37.

11. Quoted in Schneewind, *Invention*, 89; though without attribution, Hobbes's position is restated almost verbatim in Schopenhauer's 1839 *Prize Essay*, 14–17; for a provocative reading of Hobbes within a modern genealogy of skepticism, see Thorne, *Dialectic*, 183–208.

much as would be given for the use of his power; and therefore is not absolute." Likewise, "Dominion, and victory, is honourable, because acquired by power; ... Riches are honourable, for they are power. Poverty, dishonourable" (*Lev.*, 10:16, 39–40). One cannot but be struck by Hobbes's brazenly confident and apodictic style, an idiom "entirely freed from the doubts and hesitancies of the process of thought" and unflinching in its reliance on a basic syllogistic method.[12] This prevailing rhetorical model also shows Hobbes's voluntarist psychology to be anchored in the same framework of corpuscular mechanism that Gassendi had begun to establish for physics. Already, Johannes Kepler (whom Hobbes admired) "had proposed the substitution of the word *vis* [force] for the word *anima* in physics," and Hobbes's "staunchly reductionist reading of mechanism led him close to a materialist theory of the mind."[13] As extrapolated from his arguments in *De Corpore* (1655), mind is conceivable solely as an embodied causal agent, and "the interaction between bodies is restricted to physical contact between their surfaces." Indeed, the notions of *impetus* and *conatus* that Hobbes there develops are deployed "in a completely reductive way." Where Descartes had credited bodies with "a tendency to motion," Hobbes refuses to credit matter with any intrinsic dynamism whatsoever: "he cannot allow a *conatus* without a motion, even an imperceptible—because infinitesimally small—one." In so draining natural substances of even the smallest trace amounts of agency, "Hobbes has reduced the power to produce motion to the motion itself."[14]

Bearing marked affinities to Hobbes's mechanistic idea of matter as full space and as defined by its *impetus* and *conatus*, the human individual in the *Leviathan* likewise constitutes an ipso facto mindless force—self-referential, self-interested, and forever opaque to other "minds." As Quentin Skinner has argued in great detail, Hobbes's deep-seated distrust of man's capacity for intellectual self-governance goes hand in hand with his opposition to humanism's "dialogical and anti-demonstrative approach to moral reasoning" and the "assumption that there are two sides to any question."[15] Though subsequently qualified, Hobbes's early disavowal of humanism's

---

12. Oakeshott, *Hobbes on Civil Association*, 15.

13. Ibid., 21; Gaukroger, *Emergence*, 283. For Oakeshott, "Hobbes's philosophy is, in all its parts, preeminently a philosophy of *power* precisely because philosophy is reasoning, reasoning the elucidation of mechanism, and mechanism essentially the combination, transfer, and resolution of forces. The end of philosophy itself is power—*scientia propter potentiam*" (*Hobbes on Civil Association*, 19). In his "Introduction to *Leviathan*" (1935) Oakeshott convincingly situates Hobbes in the tradition of nominalism, especially the "Scotist belief that the natural world is the creation *ex nihilo* of an omnipotent God, and that therefore categorical knowledge of its detail is not deducible but (if it exists) must be the product of observation. Characteristically adhering to the tradition, Hobbes says that the only thing we can know of God is his omnipotence" (ibid., 26–27n).

14. Gaukroger, *Emergence*, 287–289.

15. Skinner, *Reason and Rhetoric*, 299.

rhetorical approach to moral self-governance still resonates in his later denial that the concepts and problems associated with a *scientia civilis* exhibit any historical depth and hermeneutic complexity. His paring down of individual consciousness to an embodied and overwhelmingly reactive will incapable of self-transcendence voids the person of all temporal continuity and historical awareness. Hence his repudiation of that quintessential objective correlative of Renaissance humanism's optimistic and integrative view of history: the book. "Those men that take their instruction from the authority of books, and not from their own meditation [are] as much below the condition of ignorant men, as men endued with true Science are above it" (*Lev.*, 4:13). Whatever rationality we may ascribe to sovereign power in Hobbes's *scientia civilis* no longer derives from the interpretive and rhetorical skill with which it establishes its view of political order, but only from the effectiveness with which it imposes that order on the body politic. Pragmatics has displaced cognition.

The sovereign will is thus characterized by just the kind of autistic constitution that it ascribes to the myriad individual wills whom it seeks to contain; and it is this basic template that explains the prevalence of "war" as Hobbes's preferred metaphor. War becomes the quintessential negative whose latency underwrites the positive rule "that men perform their covenants" and, indirectly, the image of civil society at large.[16] In order to safeguard the power that the political covenant has conferred on him, Hobbes's sovereign must steadfastly resist the temptation to enter into any affective or discursive relation with its subjects. The "sovereign cannot be imagined to love his people" since doing so would inevitably lead to "flattery." Conversely, it is a "great [fault] to speak evil of the sovereign or to argue and dispute his power" since doing so is to incite "contempt" of power itself (*Lev.*, 30:8, 9). Yet even if critical reflection on sovereign power might avoid undermining the latter, its character of "deliberation" ultimately finds its natural end in something on the order of a "decision." For sovereign power can manifest itself only as the termination of competing political scenarios. Hence, the rhetoric of "decision" frames all deliberative thought,

---

16. *Lev.*, 15:1. The irrational thrust of Hobbesian voluntarism would eventually be drawn out by Carl Schmitt who notes how war, though not the aim, purpose, or content of politics, "is the leading presupposition which determines ... human action and thinking" (*Concept of the Political*, 34); Schmitt, of course, was intensely aware of the way in which a modern, secular state might eschew "earlier exaggerations of the state ... [and] its claim to possess the monopoly of the highest unity" (44) by defining itself as a pluralist order of nineteenth-century liberal and secular polities. For Schmitt, that is, war is not itself a rational calculation but, rather, the outer boundary (beyond which lies the realm of madness and slaughter) whose fragility forever threatens and, thus, implicitly conditions the rational deliberations of individuals and societies alike with an unspoken ". . . or else." As Strauss notes, "Schmitt's basic thesis is entirely dependent upon the polemic against liberalism," and indeed Schmitt's chiasmic inversion of Hobbes's thinking is striking: "Whereas Hobbes in an unliberal world accomplishes the founding of liberalism, Schmitt in a liberal world undertakes the critique of liberalism." "Notes on Carl Schmitt," in Schmitt, *Concept of the Political*, 84, 92–93.

contemplation, and dispute as mere symptoms of *indecision*. Whatever "deliberation" has preceded the moment of decision must, according to Hobbes, necessarily find its end in "a last appetite ... that we call the Will." As the following two passages, taken from early and late in the *Leviathan,* make clear, Hobbes's will is not merely the ultimate instantiation of power as efficient *force*. Consumed with its own *finality,* Hobbesian "power" appears incommensurable with all rational contemplation and thinking:

> The definition of the *will* commonly given by the Schools, that it is a *rational appetite,* is not good. For if it were, then there could be no voluntary act against reason. For a *voluntary act* is that which proceedeth from the *will,* and no other ... Instead of a rational appetite, we shall say an appetite resulting from a precedent deliberation ... *Will* therefore *is the last appetite in deliberating.* (*Lev.,* 6:53)
>
> Aristotle and other heathen philosophers define good and evil by the appetite of men; and well enough, as long as we consider them governed every one by his own law: for in the condition of men that have no other law but their own appetites, there can be no general rule of good and evil actions. But in a Commonwealth this measure is false: not the appetite of private men, but the law, which is the will and appetite of the state, is the measure. And yet is this doctrine still practised, and men judge the goodness or wickedness of their own and of other men's actions, and of the actions of the Commonwealth itself, by their own passions; and no man calleth good or evil but that which is so in his own eyes, without any regard at all to the public laws. (*Lev.*, 46:32)

Hobbes's well-known rejection of the doctrine of "free decision" (*liberum arbitrium*) stems from his very understanding of the will itself. Forever impelled by antecedent and inscrutable causes, the will in Hobbes defines human agency as compulsive and a-rational, no matter how sophisticated the individual's conscious (and seemingly rational) calculations may be. Conceived as sheer appetition, the will thus stands to be opposed by an equally opaque and inarticulate notion of sovereign power. Wholly self-certifying, devoid of meaning, and hence immune to rational evaluation and potential falsification, power here has been pared down to the occasionalist notion of the particular "decision" or "last appetite" wherein it manifests itself.[17]

---

17. On Hobbes, see Arendt, *Origins of Totalitarianism*, 186–196, whose reading seems less beholden to Hobbes's position than the more pessimistic accounts of Carl Schmitt, Leo Strauss, and Michael Oakeshott; echoing Oakeshott's argument that Hobbes's premise of a "world composed of *individuae substantiae*" ought to be traced to nominalist thought (*Hobbes on Civil Association*, 64), Gillespie has recently focused on Hobbes's debt to medieval nominalism (*Theological Origins*, 207–254).

It is the sheer irrationality and inarticulacy of the will—and the consequent creation "of political hedonism"—that constitutes Hobbes's most significant legacy. Ultimately, his view of man as "the victim of solipsism, ... an *individual substantia* distinguished by incommunicability" not only shapes all positive law but at times even threatens to vitiate Hobbes's idea of natural law.[18] For what sanctions law is not its intrinsic and universal rationality but its efficacy and enforceability: "There is ... requisite, not only a declaration of law, but also sufficient signs of the author and authority." For "law in general is not counsel, but command" (*Lev.*, 26:16; 26:2). To be sure, Hobbes will qualify these particularly blunt statements elsewhere in the *Leviathan* by invoking God as the source of "natural law" and "reason."[19] Yet this paradox only foreshadows Hobbes's vexing bequest to modern political thought, viz., to have "sharply divided the work of reason seen as deliberation from the operation of the will seen as decision."[20] As a result of this division, state interest is the highest (perhaps the only) conceptual framework from which specific arguments and injunctions can derive force and legitimacy. Hobbes's *Leviathan* may well be the first time that the notion of a highest good is no longer being couched in transcendent, metaphysical terms but solely in the language of *interests*—itself a secular variant or extension of a voluntarist and determinist model of agency whose origins arguably predate Hobbes by at least three centuries.

Still, any account of the sort just offered risks lopsidedness insofar as it leaves unaddressed the relationship of the will to natural law, which (in Hobbes's account of it) appears to embody a certain rationality after all. The appeal to nature as a timeless source of rationality was, in any event, central to the undertaking of his contemporaries, Hugo Grotius and Samuel von Pufendorf. By putting the ancient concept of natural law on a new footing, they sought to provide "a rational terrain d'entente" and a "basis for rational agreement" in a world devastated by prolonged confessional strife.[21] Yet to argue that for Hobbes "laws of nature are always obligatory" and that they can be equated with "Natural Reason ... 'written in every man's own heart'" is hardly persuasive.[22] For if that was indeed the case—and if Hobbes's idea of natural

---

18. Strauss, *Natural Right and History*, 169; Oakeshott, *Hobbes on Civil Association*, 44.

19. "What interested [Hobbes] was not the substance of laws but their function as warrants of peace. Their legality did not lie in their substantive qualification but in their source alone, ... the fact that they expressed the will of the sovereign power" (Koselleck, *Critique and Crisis*, 35); on this issue, see *Lev.*, esp. Chapters 31–33; on Hobbes's religious thought, see Gillespie, *Theological Origins*, 246–253; on the command-logic of law in *Leviathan*, see Oakeshott, *Hobbes on Civil Association*, 44–47.

20. McCabe, *On Aquinas*, 85.

21. C. Taylor, *Secular Age*, 127.

22. Hoekstra, "Hobbes on Law, Nature, and Reason," 112, 118; more plausibly, Schneewind maintains that "Hobbes does not think that each individual is to be an interpreter of the laws of na-

law could be shown to comprise a set of distinct propositions or certitudes grounded in "conscience"—then his entire doctrine of the state's absolute and peremptory authority over its subjects would seem rather uncalled for. Yet Hobbes is fully aware of the ways in which voluntarism renders the self irrational, intractable, and terminally opaque to other minds. As a result, any appeal to "conscience," even where it is proposed as the source for the subject's assent to the will of the sovereign and the law, proves unjustifiable on logical grounds and unacceptable on political grounds: "As the judgment, so also the Conscience may be erroneous." Framing the question in exclusively epistemological terms, Hobbes views conscience as nothing more than self-awareness or "consciousness," and he denies it any metaphysical authority or certitude. If in the past people "gave ... their opinions also that reverenced name of conscience," the latter word for Hobbes reflects either an instance of bad faith or of self-delusion. In the realm of private life, there is only "opinion" (*Lev.*, 7:4), and only "the law is the public conscience" (*Lev.*, 29:7). Consequently, "Hobbes does not so much deny personal revelation as bar it from any public use."[23] So as to avoid reintroducing inner states and certitudes as the arbiters of what is to count as rational, Hobbes takes extra care to disaggregate conscience or "private judgments" from reason. Hence, as Michael Oakeshott remarks, Hobbes "does not normally speak of reason, the divine illumination of the mind that unites man with God; he speaks of reasoning." The point is echoed by Reinhard Koselleck, who notes that Hobbes's "distinction between conscience and action ... allowed the substance of an act to be separated from the act itself—the necessary premise of a formal concept of law."[24]

The operative concept of reason and natural law found in the *Leviathan* never stipulates any normative or absolute contents: "Law of Nature (*lex naturalis*) is a precept or general rule, found out by reason, by which a man is forbidden to do that which is destructive of his life or taketh away the means of preserving the same, and to omit that by which he thinketh it may be best preserved" (*Lev.*, 14:3). Hobbes's arguments in this regard take to its logical conclusion a line of thought influentially developed by Grotius for whom natural law (and the concept of individual rights derived from it) curiously vacillates between an Aristotelian, right reason model and an Occamite voluntarism. The latter position, already prevalent in Grotius and the *only* framework for Hobbes's notion of sovereignty, shines through in Grotius's remark that "even the Law of Nature itself, ... may be justly ascribed to God, because it was

---

ture" and that Hobbes specifically "den[ies] that we can appeal to natural law in order to criticize positive law" (*Invention*, 93).

23. Schneewind, *Invention*, 98; see also Koselleck, who remarks on Hobbes's quest for an "extra-political, supra-partisan position ... Unlike his contemporaries, [Hobbes] did not argue from the inside outwards but the reverse, from the outside in" (*Critique and Crisis*, 26, 29–30).

24. Oakeshott, *Hobbes on Civil Association*, 27; Koselleck, *Critique and Crisis*, 36.

his Pleasure that these Principles should be in us." Grotius's often critical editor, Jean Barbeyrac (whose 1715 French translation of *De Iure Belli ac Pacis* was the basis for the first English version, published in 1738), takes Grotius here to be specifically "talking of *Divine Voluntary Law* ... or of that, which, being in its own Nature indifferent, becomes just or unjust, because God has commanded or forbidden it."[25] In fact, Grotius seems confused, even erratic on this crucial point: whether natural law flows from the intrinsic and eternal order of things, or whether its authority hinges on the exegetical and rhetorical effectiveness with which individuals and groups *ascribe* it to a divine origin. Thus he will at times claim that "the Law of Nature is so unalterable, that God himself cannot change it" (1:155), and elsewhere that "the Mother of Natural Law is human Nature itself" (1:93). As regards the former, seemingly Scholastic claim, Jerome Schneewind points out that Grotius's affirmation "is not decisive unless being good is equivalent to, or entails, being obligatory," and on that question Grotius time and again punts.[26] Conversely, the second, manifestly anthropocentric view of divine reason has been supplanted by a hypostatized, immutable human nature of the kind associated with Grotius's Renaissance teachers. Yet he also realizes that human institutions ("often changed, and different in different Places" [1:107]) may often prohibit what natural law seemingly allows or fail to enforce what it commands, such that "right" often "signifies merely *that which is just,* and that too rather in a negative than a positive Sense. So that *the Right of War* is properly *that which may be done without Injustice*" (1:136). As Grotius's opening discussion proceeds, natural law is frequently positioned equivocally vis-à-vis "the Word *Right* ... which relates directly to the *Person*" (1:138), with the result that the very notion of *lex naturalis* takes on an increasingly anthropomorphic and constructed quality: "Natural Law does not only respect such Things as depend not upon Human Will, but also many Things which are consequent to some act of that Will. Thus, *Property* for Instance" (1:154). Among Grotius's more significant bequests to Hobbes is his distinction between natural law and divine voluntary law. The latter supposes that divine command establishes what is to count as rational, good, and just, and in so doing seems to license an analogous command ethic in the finite political realm. By contrast, the idea of a natural law inherent in the structure of the cosmos seems increasingly distant, opaque, and Aristotelian, a point borne out by the rhetorical comportment of *De Iure Belli ac Pacis.* Bulging with thousands of references to classical Roman and Greek thought (many of them scarcely pertaining to the point in ques-

25. Grotius, *Rights*, 1:90n, 91; henceforth quoted parenthetically.

26. Quoting Barbeyrac, Schneewind notes that "if [as Grotius suggests] rules impose obligation independently of the will of God, then it is not clear why God's will must be invoked at all" (*Invention*, 74–75).

tion), the work's parade of classical learning seems rather symptomatic of its author's dawning awareness and unease regarding the essential modernity of his position.[27]

No such compunctions or anxieties plague Hobbes, of course, who no longer operates within a Renaissance humanist framework but, instead, emulates the impersonal methods of Baconian science and a model of efficient causation pioneered by modern physics. As regards questions of (natural) right and the sources of human obligation, Hobbes proceeds analytically, whereas Grotius still proceeds from authority. Thus he distinguishes sharply between notions of wrong ("injury") and violation of property ("damage"), such that a servant refusing his master's command to give alms to a beggar in the street is said to "injure" his master by not honoring the (legal) covenant to obey the latter's orders, while concurrently causing "damage" to the beggar by depriving him of property. As has been noted, this view creates a potentially insoluble problem for Hobbes inasmuch as his *Leviathan* sees the individual both covenanted to the sovereign and to the *persona ficta* of the state. For to suggest that "Robbery and Violence, are Injuries to the Person of the Common-wealth" (*Lev.,* 15:12) implies that in some instances it is not the sovereign who determines the meaning of such acts.[28] Moreover, and for our purposes most importantly, Hobbes implicitly rejects the notion of a covenant that relates individuals to one another as ethical beings, opting instead to restrict interpersonal obligation only to what has been legally enumerated.

In confining the *meaning* of natural law to the intellectual skills of the individual seeking to legitimate an inherently self-interested ("prudent") course of action, Hobbes's text encourages the conclusion, recently drawn by John Deigh, that natural law is consubstantial neither with "reason" nor with "human nature." It acquires meaning only by virtue of *definitions* that invariably arise from (and are legitimated by) the individual's, and especially the sovereign's, appeal to (or manipulation of) the prevailing political and discursive conditions. Given his view of consciousness as a historically specific product determined by the aggregate force of political agents and institutions, Hobbes's appeal to natural law can become an effective part

27. With evident approval, Grotius quotes Anaxarchus, by way of Plutarch: "that GOD does not *will* a Thing because it is just; but it is just, that is, it lays one under an indispensible obligation, because GOD *wills* it" (1:164).

28. Runciman notes that on Hobbes's "account, the victim of violence cannot remit his injury precisely because he has suffered none, having no covenant with his assailant." For Hobbes to stipulate that "there is nonetheless injury done to the person of the commonwealth) ... now makes no sense, for it suggests that each subject has a covenant with the commonwealth itself, which is then master, and the subject servant. This is impossible, because the person of the commonwealth speaks only through its sovereign representative, whom Hobbes is adamant cannot contract with a subject, and without whom, he is equally adamant, there is no person of the commonwealth with whom to contract" (*Pluralism,* 19).

of his intellectual armature only against the backdrop of a latent "state of exception" or, simply, "warre" said to circumscribe all political reasoning. It is a state bound to resurface if ever covenants should no longer be honored and appeals to natural law be contested outright. Both as regards its rationality and force, in other words, natural law no longer reflects a realist (Aristotelian) framework but, instead, exists only by virtue of *ascription.*[29] It does not mean anything per se but only serves as a mute and ineffable origin, the appeal to which licenses a variety of "principles" such as the fundamental one regarding that covenants must be honored. As John Milbank puts it, "natural law transcribes the sealed-off totality of nature, where eternal justice consists in the most invariable rules. These are not derived (as for Aquinas) from the inner tendencies of the Aristotelian practical reason towards the *telos* of the good, but rather from purely theoretical reflections on the necessity for every creature to ensure its own preservation."[30] Unsurprisingly, then, the voluntarist logic of Hobbesian thought also informs his basic concept of natural reason and natural law, thereby demonstrating yet again just how decisively reason has become estranged from nature. It does not suffice to posit (as Hobbes occasionally attempts) a homology of reason with the law of nature, quite simply because it is the principal business of finite, political reason to distill certain binding principles *from* that law. Hobbes's familiar definition of "Reason … [as] nothing but *reckoning* (that is, adding and subtracting) of the consequences of general names agreed upon for the *marking* and *signifying* of our thoughts" (*Lev.,* 5:2) rather bluntly underscores its pragmatic and denatured status.

This remarkable flattening out of human agency confronts us on almost every page of the *Leviathan,* such as in Hobbes's contention that thought acquires meaning and coherence only insofar as it is "*regulated* by some desire and design … From desire ariseth the thought of some means we have seen produce the like of that which we aim at; and from the thought of that, the thought of means to that mean, and so continually, till we come to some beginning within our own power" (*Lev.,* 3:3). The

29. As John Deigh remarks in his controversial article, Hobbes may have "failed to see that the real theorems of ethics were propositions about which principles are laws of nature and not the laws themselves." Alternatively, Hobbes may simply have been "speaking loosely … because for him proving that a principle is a law of nature is as good as proving the principle itself … Unfortunately, a problem of circularity arises if the only definition from which the principle follows is the definition of a law of nature. For the principle meets this definition only if it is found out by reason, yet it is supposed to qualify as a rule of reason because it follows from this definition" ("Reason and Ethics," 45). For critical responses to Deigh, see Hoekstra ("Hobbes on Law, Nature, and Reason") and Murphy ("Desire and Ethics in Hobbes's *Leviathan*"). Strauss had already offered a convincing reading of Hobbes's attempted "restoration of the moral principles of politics, i.e., of natural law, on the plane of Machiavelli's 'realism,'" which is in effect tantamount to "maintain[ing] the idea of natural law but to divorce it from the idea of man's perfection" (*Natural Right and History,* 180).

30. Milbank, *Theology,* 10.

catalyst of all intellectual processes, including those instances where reason identifies a certain principle as (supposedly) licensed by the law of nature, is the will as *desire*—implacable, opaque, and wholly bereft of any framing vision of the good. There being "no natural harmony between the human and the universe," Hobbes places unprecedented stress on the will's restless, at times frantic quest for constructing a second nature—"a City of Man to be erected on the ruins of the City of God":[31]

> the felicity of this life consisteth not in the repose of a mind satisfied. For there is no such *Finis ultimus* (utmost aim) nor *Summum Bonum* (greatest good) as is spoken of in the books of the old moral philosophers ... Felicity is a continual progress of the desire, from one object to another ... I put for a general inclination of all mankind, a perpetual and restless desire of power after power, that ceaseth only in death. (*Lev.*, 11:1, 2).

Leo Strauss's passing remark that in Hobbes "death takes the place of the telos" impresses on us how utterly the very notion of final causes has been emptied of all meaning.[32] Instead of being understood as the completion of a life and the point where its overall meaning and achievement can be articulated, death in Hobbes is but the cessation of a mechanistic and implacable desire; Georg Simmel had called this "the common idea of death as a life-ending cut, like the fates [*Parzen-Vorstellung des Todes*], with a more organic conceptualization in which death is understood as a shaping moment of the continual course of life from its beginning."[33] All but synonymous with the will, and just as enigmatic and implacable in its manifestations, such a model of desire also explains why Hobbes should regard all political association and covenants as infinitely precarious, no matter how elaborately rationalized they may be. Arguably, no one before Hobbes had stated the usurpation of the intellect by the will quite as bluntly as that. Whatever role the intellect may yet play in Hobbes's conception of the individual has been pared down to the computation of cost-benefit and means-end ratios; and at least one consequence of that contraction now stands to be considered.

Arguably the most conspicuous casualty associated with the rise of a secular and voluntarist model of agency is that of "personhood" or (as Coleridge was to term it later) "Personëity."[34] That is, the idea of the individual as centered on a rich, unique,

---

31. Strauss, *Natural Right and History*, 175.
32. Ibid., 181.
33. "Metaphysics of Death," 75.
34. On the concept of "person," see esp. the comprehensive entry *Person* in *Historisches Wörterbuch der Philosophie*. For a classical Thomistic statement on the category of "person," in contradistinction (though not in opposition) to individual, see Maritain, *Person and the Common Good*;

and dynamic spectrum of intellectual and affective dispositions and states—and their experience as both generative and transformative of the very idea of the self *as a person*—has all but vanished. Both in the *Elements of Law* and the *Leviathan,* this disappearance is reflected by the sharp divide between "the person naturall" and a "civil" and "artificial" person or "person in law." What Koselleck describes as Hobbes's fracturing of man "into private and public halves" also explains why in the *Leviathan* Hobbes consistently appeals to a Roman and emphatically pre-Christian conception of personhood while ignoring its powerful revision and deepening in the writings of the Cappadocian fathers, Tertullian, Augustine, and Boethius.[35] With noticeable satisfaction Hobbes thus remarks on the supposed shift from the Greek "*prosopon,* which signifies the *face,*" to the Latin "*persona* ... [which] signifies the *disguise or outward appearance of a man,* counterfeited on the stage," and from there to "any representer of speech and action." Given the overall thrust of his political philosophy, of course, it does not surprise to find Hobbes so elated at the apparent disappearance of psychological depth and individuation that this etymological shift betokens. In his customary, apodictic style, he thus concludes that "a *person* is the same that an *actor* is, both on the stage and in common conversation" (*Lev.,* 16:3). What renders such a historical genealogy appealing to Hobbes is the seeming shift away from divine plenitude to the strictly outward "mask" of the finite human being. Hobbes's argument here seeks to capitalize on the supposedly diminished aura of *prosōpon* ("face"). In its Old Testament usage, frequently embedded in a propositional, genitive construction, *prosōpon* denotes the highest source of meaning and authenticity: thus, *ἀπὸ προσώπου κυρίου τοῦ θεοῦ* (Gen. 3:8; Lat. *Abscondit se Adam et uxor eius a facie Domini Dei* / "turned away from the face of God"). It bears pointing out that the Old Testament in particular seems to support Hobbes's notion of a shift from the authenticity of the divine "face" to the constructed and impersonal logic of roles and masks. For in the Septuagint in particular God's face is frequently conceived as a presence that finite human beings cannot endure and from which,

Clarke, *Person & Being,* esp. 25–42; Sokolowski, *Phenomenology of the Human Person,* 157–176; and Spaemann, *Persons,* 16–33. With good reason, the critique of modern, voluntarist conceptions of agency—framed as an epistemological problem and approached as a methodological rather than hermeneutic challenge—has often converged on the idea of the "person." The result is a rather varied spectrum of counter-Enlightenment (though by no means irrational) strands of thought often loosely gathered under the heading of "Personalism." For a fuller account of the ethics of "personalism" and the intricate evolution of the idea of person, see below, 504–534. For a thoughtful discussion of the notion of "person" within the broader phenomenon of modernity's depleted moral vocabulary—a thesis richly argued by G. E. Anscombe, Cavell, MacIntyre, Murdoch, C. Taylor, and others—see Diamond, "Losing Your Concepts."

35. Koselleck, *Critique and Crisis,* 37; on Hobbes's varying usages of "person" and "persona," see Tricaud, "Investigation," which also takes up Hobbes's reinterpretation of "person" in the context of Trinitarian thought; see also Runciman, *Pluralism,* 6–33.

consequently, they will turn away: thus, *καὶ ἔφυγον ἐκ προσώπου αὐτοῦ* (1 Sam. 19:8; *et fugerunt a facie eius* / "and they fled from his face").[36]

Having construed the authenticity of the "face" (*prosōpon*) as an overwhelming psychological burden and (under the heading of "conscience") also as a political liability, Hobbes naturally welcomes the impersonality of the Latin *persona*. According to a (now discredited) etymology found in Boethius and still invoked by Aquinas (*ST,* I Q 29 A 3), the word is derived from the compound verb *personare,* "to sound through," which had referred to the amplification of the actor's voice by the funnel-shaped opening in the mask. Capitalizing on the analogous usages of *persona* on the stage and in courts of law—viz., as a "role," or *officium,* to be fulfilled without regard for inward struggles, desires, and dispositions—Hobbes places quasi-Ciceronian stress on the relation between *persona,* mask, and official action. Thus *persona* signifies a politically or legally expedient fiction to be enacted rhetorically in ways that acknowledge and reinforce the objectives of the state. Mask, role, and office thus abstract from—indeed, positively disavow—the complex inner dynamics and teleological orientation of the "person." Suspending the contingency and inscrutability of the "soul" (*psychē/anima*), the Roman, alternately legal, theatrical, or rhetorical understanding of *persona* allows Hobbes to construct a strictly public, quasi-behaviorist model of agency.

Such depersonalization clearly supports Hobbes's underlying objective, viz., to confine all political relations to what can be expressed as a legal covenant and, in so doing, to ensure maximum enforceability for the sovereign's decrees. No doubt, the point here is to cordon off the intractable and potentially divisive appeal by so many selves to "inner" (spiritual) certitudes and the kind of non-negotiable political action variously sanctioned by the Catholic magisterium, an Anglican episcopacy, Calvinist presbyters, and a wide and fluctuating spectrum of antinomian communities. Against such richly textured and obstinately entrenched notions of political and religious agency, Hobbes posits a self that strictly warrants consideration as the "representer"

---

36. Needless to say, the New Testament appears to reverse the Septuagint's insistence on Yahweh's "face" as an unknowable and unendurable plenitude; thus Acts 3:20 typologically links the aura of God's face to the Christ's supremely realized personhood: *ut cum venerint tempora refrigerii a conspectu Domini* [*ἀπὸ προσώπου τοῦ κυρίου*] *et miserit eum qui praedicatus est vobis Iesum Christum* / "That when the times of refreshment shall come from the presence of the Lord, and he shall send him who hath been preached unto you, Jesus Christ." For a telling (if likely unwitting) echo of Hobbes's anti-humanist construction of *persona* as pure artifice, mask, and role, see de Man's "Autobiography as De-Facement." Like Hobbes, de Man categorically reads acts of self-reference as performative self-deceptions, such that "the identity of autobiography is not only representational and cognitive but contractual, grounded not in tropes but in speech acts ... Voice assumes mouth, eye, and finally face, a chain that is manifest in the etymology of the trope's name, *prosopon poien* [sic], to confer a mask or a face" (71, 76). On Hobbes's concept of *persona*, see Skinner, *Reason and Rhetoric*, 337–339.

(sic!) of its own or someone else's interests—that is, a self whose inner dispositions and commitments are simply irrelevant to the state. In both political and epistemological senses, the Leviathan recognizes the self solely as the author of "covenants." By emphasizing the etymological conjunction (to which we shall attend again later) between author (*auctor*), authority (*auctoritas*), "action" (*actus*), and property, Hobbes construes all agency as staking a claim to some kind of virtual dominion or material ownership: "That which in speaking of goods and possessions is called an *owner* (and in Latin *dominus,* in Greek *kurios*), speaking of actions is called author. And as the right of possession is called dominion, so the right of doing any action is called AUTHORITY" (*Lev.,* 16:4).

Ultimately, it is less that Hobbesian willing triumphs over the intellect than that Hobbes's *Leviathan* stages the utter collapse of all "thinking" into a strictly outward, performative "representation" of volition. Resembling an early precursor of artificial intelligence, the Hobbesian intellect is merely calculative and as such secures its efficiency by *suspending* the entire philosophical vocabulary of normativity, evaluation, inner dispositions, contemplation, and ethical reflection. Without denying their *reality* within the life of any given individual, the *Leviathan* simply considers them something between a distraction and an encumbrance for its pragmatic objectives. Wholly incapable of hesitation, doubt, and revision, the Hobbesian performer of covenants appears substantially denuded of everything we associate with the category of a "person." In light of our present-day intellectual legacies—including the aesthetic theories of Jena Romanticism and its eventual descendants in 1970s deconstruction—there is ample reason to acknowledge performativity, stagecraft, rhetoric, and the whole counterfeit world of the "mask" (*persona*) as integral features of personhood. At the same time, it would amount to a self-defeating or, rather, empty proposition to construe *all* appeal to inner states as either delusional or an instance of bad faith and, hence, to dismiss inspired action as a mere fiction and agency as sheer impersonation. Hobbes, however, comes very close to doing just that. For unlike other philosophers writing before him, Hobbes appears in a principled way uncurious about the myriad ways in which "person" comprises countless ambivalences, inner conflicts, and minute shifts and, consequently, poses profound interpretive challenges for other subjects within its social orbit.

Hobbes's rigid voluntarism prevents him from distinguishing between the classificatory term of the "human being" and the singularity of personhood as first set forth in Boethius's crucial definition of person as "an individual substance of rational nature."[37] As Robert Spaemann points out, *substantia* here renders the Greek *ousia,*

37. *Persona est definitio naturae rationabilis individua substantia.* In Boethius, *Contra Eutychen et Nestorium,* 3; for a fuller discussion of Boethius and the idea of person in late antiquity—as

which could just as plausibly have been rendered as *essentia*. For Boethius clearly means to suggest that this "'rational nature' exists as a being-in-itself" and, consequently, that it "cannot be displayed in full by any possible description. No description can replace *naming*. A person is *someone*, not *something*, not a mere instance of a kind of being that is indifferent to it." Put differently, "personhood is a mode of existence, not a qualitative state."[38] Yet Hobbes, once again revealing the physical and mechanistic foundations of his psychological theory, accords reality and relevance for his political philosophy only to objects—that is, to entities that can be captured descriptively and that ultimately admit of an exhaustive definition. World for Hobbes means the sum total of all possible objects as the bearers of their inherent force and as arrayed in more or less antagonistic ways vis-à-vis each other. Gone is the Scholastic distinction between "intentionality" (*nomen intentionalis*) and that antecedent, unique reality of the living rational being (*nomen rei*) for which alone the world could ever become an interpretive, epistemological, and moral challenge.

For strategic purposes, it ought to be stressed here that the *singular* nature of the person is not to be taken as prima facie evidence of its incommensurability with other persons and, hence, of its irrationality. For where persons are concerned, singularity is not an epistemological proposition but an ontological reality. The person is unique *not* because I venture to *ascribe* that trait to each and every person as a quality, in the way that incommensurability and indivisibility are predicated of atoms by Leucippus and Democritus. Rather, "persons are singular in an unparalleled fashion ... Their self-identification cannot occur solipsistically. It necessarily implies the existence of others and the possibility of being available to their knowledge."[39] By contrast, the Hobbesian individual is little more than the coefficient of legal constraints and inward compulsions, of outward force and inward motive. As such, it lacks all transcendence vis-à-vis its inner constitution. It merely *is* what it is. Yet precisely this axiom turns out to be deeply incoherent and involves Hobbes in an ongoing performative contradiction. Not only does Hobbes's implicit denial of any qualitative divide between human and animal life cause us to stray into very dangerous moral space; but his basic reductionist account of selfhood as mechanical, embodied desire licenses another claim, soon to follow, to the effect that when all is said and done our conscious and rational awareness is but an epiphenomenon of subterranean drives or insensible impressions nesting in the recesses of our unconscious. That, at any rate, was the pessimistic implication readily seized by his successors, Mandeville, Hume, Schopenhauer, and Carl Schmitt among them.

---

well as the reception and inflection of that intellectual genealogy in the later Coleridge—see below, 535–543.

38. Spaemann, *Persons*, 28–29.

39. Ibid., 35.

The performative contradiction, meanwhile, is as follows: to develop the reductionist position concerning the will and to dismiss ethical accountability as something impossible *by definition* still means to have advanced a sophisticated rational *claim* and, hence, to have engaged in precisely the kind of meta-critical reasoning *about* consciousness that this claim itself asserts to be impossible. Taken as a comprehensive political *argument,* Hobbes's *Leviathan* effectively aspires to reasoned uptake by individuals whose capacity for authentically rational behavior the book consistently disputes. Among Hobbes's most acute early critics, Anthony Ashley Cooper, Earl of Shaftesbury, was quick to deploy his favorite argumentative mode of "raillery" against Hobbes and to muse how, in a world where "*Force* and *Power* … constitute *Right*" a philosopher might conceive any wish to articulate that very knowledge for the benefit of others: "whence is this Zeal in our behalf? What are *We* to *You*? … why all this Pains, why all this Danger on our account? Why not keep this Secret to Your-self? Of what advantage is it to You, to deliver us from the Cheat? … It is directly against your Interest to undeceive Us" (*SC,* 1:58). More recently, Robert Spaemann has flagged the paradox underlying all reductionism by stressing that to concede consciousness to an agent is never simply an attributive but an existential proposition; for to the extent that a person "*has* consciousness at all, he is aware that he *is more than* consciousness."[40] It will not do to construe all cognition mechanistically, viz., as the surface representation of some inscrutable desire, or as a tranquil cover draped over an unconscious and implacable force or will.

What Hobbes, Hume, Schopenhauer, and other radical epistemological skeptics fail to consider is this fundamental "self-differentiation of a human subject from everything that may be true *about* him."[41] At times, Hobbes comes close to grasping this contradiction, and hence struggles throughout his career to adjust the ratio of syllogistic reasoning and rhetorical argument in his philosophy. In his dedicatory epistle to *De Cive,* he insists that he wishes to persuade "by the firmness of *rationes* and not by any outward display of *oratio* [*neque specie orationis, sed firmitudine rationum*]." Yet to proceed in that manner is to reintroduce the "dictates of right reason" (*dictamina rectae rationis*) as a condition of uptake, even as the argument in question takes for its point of departure the lack or, at the very least, the unreliability of right reason in matters of (political) argument.[42] Hobbes's refusal to think through this tension between the propositional and performative dimensions of his political theory is a consequence of his startlingly impoverished concept of the human person. In part because of his vehement rejection of Aristotelian and Scholastic thought

40. Ibid., 10 (trans. modified). Ger.: *Sobald er überhaupt Bewußtsein hat, weiß er, daß er nicht nur Bewußtsein ist*" (*Personen,* 18).

41. Ibid., 14.

42. From the Latin edition of *De Cive,* quoted in Skinner, *Reason and Rhetoric,* 302.

or, at least, of any dialectical engagement with that tradition, Hobbes is unprepared to consider that the human person might be a self-subsisting *reality*, and hence not just another "category" (*praedicamentum*) or aggregation of generic traits such as could be indifferently ascribed to any variety of beings.

In a passing remark, Coleridge nicely flags the over-determined character of Hobbesian voluntarism when observing that "Hobbes repeatedly speaks of the will as compelled by certain causes where an accurate speaker would say impelled."[43] What has dropped out of view, then, is the idea of an inner life—of rational personhood as a reality antecedent to mere computation or "reckoning"—which is to say, the sheer possibility of consciousness operating in transformational and narrative, rather than reactive and mechanistic, fashion. We thus witness in Hobbes the determined expulsion of what the Aristotelian, Augustinian, and Thomistic traditions had variously understood as "soul" (*anima*) and as "thinking" in its various senses—viz., as "feeling," as "contemplation," and as potentially counterintuitive and transformative of the materials with which it is engaged. Were it not for that possibility of "transformation"—which no calculative model of thinking can ever approach—all thinking would remain forever a mere apperception of data and could logically never carry over into an awareness-of-self, not even a fictitious or impersonated one of the kind sanctioned by Hobbes's own account. Not until the work of the late Coleridge, the writings of John Henry Newman, and the psychology of Franz Brentano do we find this fundamental blindspot in Hobbesian voluntarism fully exposed and analyzed. Meanwhile, the most salient point may well be that Hobbes's powerful influence on most subsequent thinking about political community (rational or otherwise) involves his decommissioning the fullness and complexity of what philosophy before him had always taken to be a central category: that of the "person." What remains—and what (mistakenly, I think) Hobbes thought was not only sufficient but preferable to limit himself to—is a strictly outward and voluntaristic notion of the self as the "actor" or "agent" performing covenants such as he deems advantageous in accordance with "motives" whose curious insistence the agent himself or herself remains forever unable to grasp, let alone articulate.

---

43. *Logic*, 123; elsewhere, Coleridge reiterates the distinction between "absolute" and "conditional" necessity (*OM*, 8), and he specifically notes how the careless "use which Hobbes and his disciples made of the terms 'Compulsion' and 'Obligation,' till his antagonist, Bishop Bramhall, had de-synonymized them and thus enabled his readers to call into distinct consciousness the proper contradistinguishing character already implied in the words '*Should*' and the conditional '*Must*,' i.e. 'I *must* do this, though I *can*, perhaps, far more easily and pleasurably—yet still I *must* if I will not forfeit my sole and proper claim to the name of man" (*OM*, 36–37); Coleridge's often polemical and at times reductive accounts of Hobbes repeatedly center on the latter's failure to maintain the distinction between compulsion and obligation and, thus, to elide the entire question of conscience, guilt, and a responsible will; see *BL*, 1:92–94 and *LHP*, I:213, 256 and II:542–549.

It would be a mistake, however, to assume that Hobbes's notion of the will—blind, inarticulate, and compulsive—was the only one conceivable for his contemporaries. As we already saw, his conception of human agency was the result of momentous shifts within Scholastic theology as, early in the fourteenth century, Franciscan intellectuals begin to challenge key aspects of Dominican (Thomistic) thought. Without that background story, one might conclude (as has frequently happened) that the concept of a voluntarist and self-defined subject—the "punctual self" (Charles Taylor) observable in the political theories of Machiavelli, Hobbes, James Harrington, and Locke—might have leapt into existence *ex nihilo,* which supposition in turn might lead one to conclude (just as mistakenly) that the Aristotelian and Thomistic models of agency do not so much *differ* from these projects as they are simply *unrelated* to them. Such a view fails on two counts. First, it remains premised on a strictly voluntarist source of order, say of the kind that at times emerges from accounts of the divine will in William of Ockham, Nicholas of Autrecourt, Gabriel Biel, and later nominalists. Yet even the neo-Stoic projects of the early Enlightenment that seek to recover from the more dire implications of Hobbes's secular voluntarism—viz., by situating the modern individual in a meliorist narrative of economic self-discipline and rational self-possession—still accept one of the *Leviathan*'s crucial premises. Thus the descendants of Hobbesian naturalism, which beginning with Hume, the late Edmund Burke, and all the way through Schopenhauer, Nietzsche, Oswald Spengler, and Carl Schmitt takes on an increasingly pessimistic cast, yet also those Enlightenment thinkers committed to a progressive notion of civil society (Adam Ferguson, Adam Smith, James Steuart, Kant, Bentham, Mill) equally struggle with *conceptualizing* the will. They can only *posit* it as a non-cognitive and inarticulate force—to be either duly constrained or subjected to a painstaking disciplinary regimen of "improvement" for which writers from Locke and Kant to John Rawls and Jürgen Habermas have mobilized any variety of economic, legal, and affective tropes.

Needless to say, there has been a lively and ongoing debate about the plausibility of grounding modern, liberal-secular community in a social contract *voluntarily* entered into, in a cosmopolitan public sphere to be realized by deontological ethics, or as a deregulated, discursive space whose inhabitants (whatever their inner lives may dictate) will acknowledge one another in what Habermas apostrophizes as an "ideal speech situation." As remains to be seen, one of modern liberalism's most neuralgic points concerns the fact that the *systems* that it conjures up so as to contain and/or remediate the individual's baser instincts do themselves require for their very implementation that the proposed therapy has already taken place and been successful. In *positing* the individual will as categorically beyond the realm of the intelligible—an atavistic, entrenched, and inscrutable force that can be neither demonstrated nor falsified—modern political philosophy in effect *imitates* it. Destined to terminate in a radically instinctual and implacable agency on the order of Schopenhauer's will or Freud's unconscious, modern voluntarism constitutes a *petitio principii* on a large scale. For voluntarism *infers* a wholly irrational and inscrutable source of energy simply because it finds its own strategies of containment and remediation persistently frustrated or to have failed outright. Having denuded the will of all intellectual substance and self-awareness, modernity unsurprisingly wrestles with what now proves at the very least an irrational and, quite possibly, an indemonstrable and ultimately empty notion. There is something circular and perilously imitative about sequestering the human person in an irrational and inarticulate will. For aside from precluding any independent verification or falsification, such a position can only get started as a retroactive inference drawn by modern reason, which now appears consumed with containing the will as its (putative) other.

In rather pointed form, Theodor Adorno and Max Horkheimer thus characterize the Enlightenment's de-mythologizing fervor as "mythic fear turned radical."[44] The main fallacy of modern voluntarism thus involves its assumption that the will is wholly self-originating and self-certifying, and that its ostensibly psychological epigenesis can be captured by some epistemological procedure or method—such as Johann Gottlieb Fichte's *Tathandlung,* Ludwig Wittgenstein's *Sprachspiel,* John Austin's "performative utterance," or any variety of post-Freudian tropes. To sharpen the point: voluntarism as *theory* effectively imitates what it seeks to demonstrate about the self. Rejecting out of hand the apparent fact that all self-formation takes place in a deep and richly textured, social, historical, and affective space—and, hence, arises in response to a *hermeneutic* challenge rather than a methodological impasse—modern voluntarism and its numerous descendants have essentially foreclosed on the possibility of grasping the human being as an ethical reality, that is, as a "person."

Incapable of appraising the self's teleological orientation toward rational, supra-individual, and transcendent ends, medieval voluntarism and its secular descendants wax understandably anxious about how to defend against the spectral vision of a self that is nothing more than a quasi-kinetic and irrational speck of consciousness forever incapable of achieving enduring, rational self-awareness. Moreover, if the peculiar *fort/da* game that modern reason is playing with the will at times takes on paranoid overtones, this is the case because any rational method that identifies a primal and ineffable will as its core premise and antagonist also betrays confusion about the nature of concepts in general. What the theorists of modern autonomy fail to recognize is that to start out a conceptual claim on a (supposedly) non-conceptual basis is, in fact, quite impossible. Hegel exposed that fallacy in his account of

---

44. *Dialectic,* 16.

"sense-certainty" early in the *Phenomenology*. To be sure, I am not imputing to the concept of the will some ideal and trans-historical (as it were Platonic) quality. Rather, following Hans Blumenberg, I also believe that "the continuity of history across the epochal threshold lies not in the permanence of ideal substances but rather in the inheritance of problems" (*LMA*, 48). Simply put: understanding the self means understanding its relations, its constitutive embeddedness and participation in the world. For the self cannot coherently be thought of as a bundle of transient impressions autonomously processed by an apperceptive *punctum*. Rather, it is above all *person*—that is, a being endowed with reason and, hence, intrinsically self-interpreting *and* at least potentially aware that its own place within the world hinges on its practical, symbolic, and discursive engagements, which present it with an ongoing interpretive challenge. To grasp the primacy of the "person" is to recognize that selfhood is prima facie not an epistemological category but a hermeneutic event. Yet to take that view—as Hobbes and his liberal descendants from Locke to Kant were no longer prepared to do—is to recognize that the will, too, is not some primal agency operating independent of consciousness, that it is not a *concept* but an embedded facet of the total person, and that it consequently partakes of a long, convoluted, and sometimes agonistic history of interpretation.

In the immediate aftermath of Hobbes, it was the Cambridge Platonists—particularly Ralph Cudworth (1617–1688)—who pointed out many of the biases and deficiencies of modern voluntarism. Responding to Hobbes's 1654 letter "Of Liberty and Necessity," Cudworth's *Treatise of Freewill* (published only posthumously in 1838) offers a condensed version of arguments that also inform his 1678 *True Intellectual System of the Universe*, a work whose scope and ambition, as well as deeply critical view of modernity, calls to mind the intellectual Urtext of the Counter-Reformation, Cardinal Robert Bellarmine's *Disputationes* (1581–1593). Yet there is nothing dogmatic about Cudworth, whose brand of Christian Platonism would undoubtedly have been perceived as worrisomely heterodox by the Catholic magisterium, and whose unparalleled linguistic and philological range made him a natural choice for Brian Walton's implicitly ecumenical project of a "Poly-Glot Bible" (1654–1657). A graduate of Emmanuel College, Cambridge, master of Clare Hall and, subsequently, of Christ's College, and from 1645 until his death Regius Professor of Hebrew, Cudworth spent much of his life critically engaging the determinist and empiricist turn that philosophy had taken with Hobbes, Gassendi, and Locke. This basic concern is manifest in all of his writings, though Cudworth takes pains to situate his modern antagonists within a long philosophical tradition that spans from the "Democritick" atomists through Lucretius to Ockham's voluntarism and, eventually, to his contemporary Puritans, whose doctrinaire tendencies he had already criticized in a precariously timed sermon given to the House of Commons in 1647. In the *Treatise of Freewill*, Cudworth not only postulates "something ἐφ ἡμιν, *In nostra potestate*,

In our own power" and "self-active" per se, but he also marshals strong evidence that the opposing argument for the total extrinsic determinacy of all processes animate and inanimate would inevitably put an end to intellectual activity of any kind.[45] Precisely this capacity of the will to *elect* a course of action, and thus to assent to and partially transcend the conditions and constraints with which it is prima facie presented defines the very essence of the intellect and, ultimately, of reason itself: "Rational beings, or human souls, can extend themselves further than necessary natures, or can act further than they suffer, ... [and] are not necessarily determined by causes antecedent" (*Treatise of Freewill*, 14). Yet for this view to amount to more than a mere assertion it is crucial to overcome the Hobbesian scenario of a strictly *reactive* intellect by showing how it fails on its own terms. Cudworth does just that by arguing that the Hobbesian position ultimately makes it impossible to sustain the conceptual integrity of the very terms it purports to elucidate. A rigidly deterministic and mechanist (mono-causal) account of the will inexorably corrodes the broader semantic field of mind, consciousness, reason, deliberation, judgment, and a host of peripheral psychological concepts such as had ensured a nuanced and differentiated understanding of the inner life since Plato. If "mind" is to signify at all, Cudworth insists, it must be generative, just as the will as *liberum arbitrium* cannot be imagined as a merely indifferent faculty. For "it is very absurd to make active indifference blindly and fortuitously determining itself, that is, active irrationality and nonsense, to be the hegemonic and ruling principle in every man" (34).

What in true Plotinian terms Cudworth calls "an ever bubbling fountain in the centre of the soul, an elater or spring of motion" (28), reflects a fundamentally evaluative and incipiently normative orientation of the whole person toward a specific *telos*. A person's identity and reality originate neither in some agnostic computational act nor in a contingent volitional spike as Hobbes had argued. For if that were to be the case, the faculty of the intellect would be marooned on a desert island of endless prevarication, just as "that which willeth in every man will perpetually will not only it knows not why, but also it knows not what."[46] Willing, Cudworth here implies, is inseparable from having a *representation* both of what is willed and of that for the sake of which it is being pursued. To be sure, that representation need not necessarily be entertained in fully conscious and transparent form, but it must at least be qualitatively distinct from the act of will itself. For only so is it possible for the subject of the will to have a sense of her or his own agency as continuous over

---

45. *Treatise of Freewill*, 1, 9.

46. Ibid., 23; as Cudworth here recalls, "Aristotle himself determines that βουλή, counsel, cannot be the first moving principle in the soul, because then we must *consider*, to *consider*, to *consider infinitely*" (ibid., 28).

time. Likely thinking of Ockham, Cudworth thus remarks on the problematic tendency of Scholastic nominalism to disaggregate things by proliferating distinctions and, in so doing, fragmenting the reality of the person into ostensibly unrelated and incommensurable acts of intellect and will: "to attribute the act of intellection and perception to the faculty of understanding, and acts of volition to the faculty of will, or to say that it is the understanding that understandeth, and the will that willeth" is tantamount to saying "that the faculty of walking walketh, and the faculty of speaking speaketh." In short, it is to create "two *supposita,* two subsistent things, two agents, and two persons, in the soul ... But all this while it is really the man or the soul that understands, and the man or the soul that wills, as it is the man that walks and the man that speaks" (25).

Numerous worrisome consequences arise from modernity's fragmentation of the person. Above all, we perceive an erosion of key concepts indispensable for humanistic inquiry—all integral to evolving Platonic, Stoic, and Augustinian traditions. Such erosion threatens as soon as it is supposed that change can only ever be caused *ab extra,* that there is only one type (i.e., efficient) of causation, and that a cause can never partake of the order of mind but, instead, determines the latter's movements in a strictly adventitious manner. Cudworth's theological arguments in support of a universe in which there is at all times an element of contingency at work recall almost verbatim Stephen J. Gould's thesis about "replaying life's tape"—which is to say, finding that evolutionary history, if repeated from the very beginning, could never take the exact same course it did.[47] In a dystopic vision that conjures up Nietzsche's "eternal return" or, at a less exalted level, Harold Ramis's 1993 movie *Groundhog Day,* Cudworth thus imagines how "after the conflagration of this earth, [God will] put the whole frame of this world again exactly in the very same posture that it was in at the beginning ... [and that] it should continue or run out such another period of time as this world has lasted before, seven thousand years or more." Then would "everything, every motion and action in it be the very same that had been in the former periodic revolution." Both on theological and physical grounds, however, this scenario of a (post-lapsarian) world spooling out in the exact same way as before toward the appointed day of judgment strikes Cudworth as unacceptable. For it would suggest a God either cruelly indifferent to the possibility of improvement or incapable of effecting it. In fact, Cudworth insists, a world denuded of all "contingent liberty" would ultimately be meaningless (*Treatise of Freewill,* 9, 13).[48] In the case of the widely misconstrued *liberum arbitrium,* Cudworth thus points out that when

---

47. Gould, *Wonderful Life*, 46–53.

48. For a fuller discussion of Cudworth, see Darwall, *British Moralists*, 109–148; Schneewind, *Invention*, 194–214; and E. Cassirer, *Platonic Renaissance*, esp. 42–85.

faced with seemingly "equiponderant" choices, the mind (unlike Jean Buridan's ass facing two equidistant bales of hay and so supposedly doomed to starvation) is perfectly able "to cast in some grains into the scale ... Here, therefore, is a sufficient cause which is not necessary, here is something changing itself, or acting upon itself" (15–16). For Cudworth, at any rate, it is just this spontaneous and creative dimension of human consciousness that alone makes it meaningful to speak of "mind" at all. For in the obverse scenario, which would find "our souls in a constant gaze or study, always spinning out a necessary thread or series of uninterrupted concatenate thoughts ... [we] could never have any presence of mind, no attention to passing occasional occurrences, always thinking of something else, or having our wits running out a wool-gathering, and so be totally inapt for action" (27–28). Though Cudworth is taking his cue from Hobbes, his critique of the deterministic model of mind and the pervasive, deleterious consequences that he shows to be unwittingly produced by its naïve acceptance will be even more applicable in the case of Locke, Mandeville, and Hume—whose conceptions of the will qua "passion" now remain to be considered.

Yet Cudworth's most valuable insight concerns the extent to which a methodologically self-conscious and hyper-rational modernity inadvertently produces deeply irrational outcomes. It does so by failing to notice that its reductionist understanding of all causation as "efficient," its fragmentation of the person into ostensibly unrelated powers or faculties, and its conflation of action with sheer mechanical motion effectively erases much of Western philosophy's conceptual legacy. If philosophy starts out from the premise of a blind, indeterminate, and compulsively reactive will, Cudworth argues,

> then is all consideration and deliberation of the mind, all counsel and advice from others, all exhortation and persuasion, nay the faculty of reason and understanding itself, in a man, altogether useless, and to no purpose at all. Then can there be no habits either of virtue or of vice, that fluttering uncertainty and fortuitous indifference, which is supposed to be essential to this blind will, being utterly uncapable of either. Nor, after all, could this hypothesis salve the phænomena of commendation and blame, reward and punishment, praise and dispraise. (23)

What Cudworth is sketching here is the dystopia of a pervasive loss of conceptual histories and the intellectual orientation they afford. More recently, this very scenario has received two crucially different assessments (by Iris Murdoch, Alasdair MacIntyre, Stanley Cavell, and Charles Taylor, among others), and we need to clarify which of these accounts Cudworth here means to convey. Is it that the philosopher (primarily Hobbes) here being opposed "writes as if he lacked the concept of morality" and, hence, as someone who is de facto "cut off from a mode of application of the moral vocabulary"? And if so, does such a philosophy experience its particular modernity as an instance of conceptual loss and consequent moral inarticulacy and intellectual impoverishment? Put differently, does this portrayal of "a world in which the concept of morality is missing" amount to "a true portrayal of our world, or is it ... a reflection of [a] blindness to what we still have?" Juxtaposed to the debate that Cora Diamond portrays MacIntyre and Cavell to be waging three centuries later, Cudworth's critical perspective proves quite different; for his historical advantage is to recognize and seize the opportunity of articulating at the very moment of its occurrence a seismic shift in the conception of the human will and reason, and to deduce the incipient demise of what Coleridge would call "a responsible will" just as that process is gathering momentum. Unlike Gertrude Anscombe's and Alasdair MacIntyre's dispiriting ex post facto analysis of modern moral philosophy as a deeply incoherent and all but defunct enterprise, Cudworth's claim is not "that we lack the kind of life within which such concepts as we need could be intelligibly applied." Rather he is arguing that the incipient depletion of a nuanced conceptual framework such as enables us to articulate our inner life will, in due course, produce that scenario as its eventual (and by then seemingly inevitable) outcome. To be sure, it is well beyond Cudworth's purview to forecast just when the loss of a rich conceptual framework will have wormed its way into the modern psyche to such an extent that it is no longer experienced as a loss at all—a point when the very notion of a rational, reflectively self-determining, and ethically responsible human being will have undergone a dramatic contraction. In any event, on Diamond's persuasive reading, even an avowed modern Platonist like Iris Murdoch still holds that "the conceptual losses we have indeed suffered have not actually changed us into human beings limited to the interests and experiences and moral possibilities we can express in our depleted vocabulary."[49]

In fact, it is just this palpable asymmetry between a bewildering spectrum of quotidian experience and conceptual resources too atrophied to permit rendering that experience intelligible which by the later eighteenth century furnishes abundant thematic prompts for the modern novel. Acknowledging that asymmetry, we have good reason to suspend, if not reject outright, modernity's endorsement of a self-originating and inarticulate will. Modern secular agency, even (or, perhaps, especially) when advanced in highly sophisticated accounts such as, say, Kant's critical philosophy, exemplifies rather than eschews the failure of Enlightenment liberalism to achieve a timely and comprehensive understanding of its own historicity and hermeneutic continuities. Readers of modern political and moral philosophy—from

49. Diamond, "Losing Your Concepts," 260, 261, 263.

Hobbes forward to Locke, Kant, and the Victorians—would do well to recall Hans-Georg Gadamer's reminder that "*the prejudices of the individual, far more than his judgments, constitute the historical reality of his being.*" Gadamer's caveat that "the fore-meanings that determine my own understanding can go entirely unnoticed" is ignored as much by the overconfidence of Lutheran *sola scriptura* as by modern (secular) liberalism's presumption that social meanings and frameworks may be created instantaneously and from whole cloth. For Gadamer, what "makes us deaf to what speaks to us in tradition" in the first place is a deep-seated and unreflected prejudice against the very idea of "tradition" itself.[50] Lest tradition and what it may have to tell us be rendered unintelligible by its peremptory stigmatization, we should listen in on it again—not as hidebound devotees of the past but as open to its persistent calling, its unfathomable complexity, and its dynamic, steadily evolving contemporaneity.

---

50. *Truth and Method*, 278, 271–272. With varying emphasis and drawing on different methodologies, Charles Taylor, Alasdair MacIntyre, and Hans Blumenberg have all advanced similar arguments; see esp. Taylor, *Sources*, 3–52 and MacIntyre, *Three Rival Versions*, 149–215; for a thorough discussion of Gadamer's conception of "tradition" (*Überlieferung*)—a term that also connotes aspects of "transmission" and "translation"—see Auerochs, "Gadamer über Tradition."

9

THE PATH TOWARD NON-COGNITIVISM

Locke's Desire *and Shaftesbury's* Sentiment

Beginning with the early Enlightenment, particularly in the writings of Locke, Mandeville, and Montesquieu, and culminating in the hybridization of moral and economic theory in Francis Hutcheson, Smith, Ferguson, John Millar, and James Steuart, we can observe a strategic shift in social theory that promises, if not to remedy, then at least to contain the apparent irrationality of the Hobbesian will. As C. B. Macpherson argued some time ago, Hobbes himself had already hinted at such a shift inasmuch as his concept of "possessive individualism" appeared to find its natural complement in a free market system of some kind or other.[1] Yet the major shift arguably occurs after Hobbes, as a new generation of intellectuals replaces his bleak view of human agency with a meliorist conception of individual and communal flourishing mediated above all by economic behavior. It thus cannot surprise that after Hobbes's extreme voluntarism, the concept of the will should have rapidly faded from moral philosophy. Hobbes's troubling bequest to his successors had been a model of human agents determined by blind and unself-conscious compulsion, itself the offshoot of a mechanistic theory of life as sheer tropism triggered by contin-

1. Macpherson, *Political Theory,* 53–70.

gent objects of desire. Such a framework offered no "openings" of any kind, no possibility of self-transcendence, let alone of a gradual ascent toward knowledge and an inner ordering of moral agents by means of focused and deliberate habituation. There is very little in Hobbes to suggest that social processes (education in particular) could ever counteract the rigid inner determinacy of human agents.

The question that came to preoccupy post-Hobbesian thought was how to recover the potential for moral and spiritual flourishing in the wake of Hobbes's uncompromising assault on teleological and Christian-Platonic models of human agency. Beginning in James Harrington's writings and continuing all the way through Francis Hutcheson and David Hume, it is clear that the only way to answer Hobbes was by reclaiming the inner life of the person as an authentic and significant source of moral reflection and responsible action. Yet this prolonged effort at rehabilitating reason as something more than mere calculation—indeed, as substantially *free*—comes with two significant qualifications. First, reason is no longer juxtaposed to the will but, rather to the passions. Second, thinkers from James Harrington and Anthony Ashley Cooper, Earl of Shaftesbury, to Adam Smith and Immanuel Kant replace the Aristotelian-Thomist dialectic that had posited the will in a necessary *relation* with the intellect with an empirical antagonism between selfish passions and rational interests. While this change of basic focus and guiding concepts was to produce significant complications of its own (some of them still bedeviling humanistic inquiry today), it did at least have the advantage of dramatically expanding the intellectual and social constituencies that could reasonably be expected to participate in such debates to begin with. Thus, in contrast to the civic republicanism of James Harrington and Algernon Sidney that directed its post-Hobbesian arguments to a narrow political elite, moral philosophers and essayists of early Georgian England—mindful that an overly interventionist take on *scribere est agere* might lead to much unpleasantness—no longer treat social, economic, and moral theory as a proxy for strident political claims.[2] Rather, the focus of writing shifts from political controversy to more mediated accounts of social, religious, and cultural processes, with quasi-forensic attention being brought to bear on the affective sources of moral and social action. The intent here is descriptive rather than normative, and the new rhetoric—being directed at a broad, mobile, and amorphous readership rather than political elites—presents us with sustained psychological analyses rather than explicit controversy.

2. In lieu of a missing second witness needed to secure Algernon Sidney's conviction for high treason, the prosecution presented (and selectively quoted from) a copy of his *Discourses Concerning Government* (written in 1680, though not published until 1698), which the presiding judge accepted, ruling that "to write is to act." His conviction having been thus secured, Sidney was executed on 7 December 1683.

Against the backdrop of a widespread retreat from all-or-nothing political partisanship, the elemental conflicts of late seventeenth-century politics are supplanted by a new brand of philosophy that abandons the peremptory style of deductive argument from first principles in favor of inductive reasoning that proceeds from "phenomena, or effects, and from them investigate[s] the powers or causes that operate in nature."[3] As Alasdair MacIntyre has shown, it is particularly in Scotland that moral philosophy emerges as a counterweight to the radical (evangelical) branch of Calvinism with its deep-seated suspicion of rational demonstrations of any kind. Having, in Theodor Adorno's memorable phrase, "abandoned the royal road to origins," the essay form in particular comes to exemplify the overall provisional, thesis-like character of early eighteenth-century secular culture. It disavows the previously accepted "distinction between a *prima philosophia* ... and a mere philosophy of culture" and, in so doing, no longer assumes that "a totality is given" at all. The result is a highly self-conscious, mediated rhetoric in which philosophy, theology, psychology, and what was to become sociology constantly intersect, alternately furnishing background information or the central topic under discussion. In Shaftesbury, Mandeville, Swift, Defoe, Addison, Steele, and eventually Hume's highly popular *Essays Moral, Political, and Literary* (1754), among others, an unpredictable enjambment of multiple, previously distinct registers—urbane wit, satire, soliloquy, Stoic sententiousness, Socratic dialogue, Ciceronian exemplarity (all of them liable to be brought up short by grub-street realism)—attest to the greatly expanded rhetorical scope and versatility of writing and its role in shaping a broader culture of literacy.[4] The hortatory and counterfactual idiom that had shaped the competing moral visions of Pascal, Calvin, Bunyan, La Rochefoucauld, St. Ignatius, and Baltasar Gracián, among others, increasingly yields to a proto-sociological analysis of phenomena that today are the province of behaviorism, market research, social psychology, and cultural anthropology.

This reorientation of political thought from constitutional to moral and social theory, and of its idiom from the controversial to the conversational, reflects the pe-

---

3. Colin McLaurin, *An Account of Sir Isaac Newton's Philosophical Discoveries* (1748), quoted in MacIntyre, *Whose Justice?* 250.

4. Adorno, "The Essay as Form," in *Notes to Literature* 1:11–12; on the distinctive type of cultural work performed by the (new) genre of the English essay, see Gigante, *Great Age*, xv–xxxiii. MacIntyre remarks on "how philosophical discussion extended beyond university classes" to private seminars, for tradesmen and youths, to Athenaeum-style institutions such as Edinburgh's Rankenian club (founded in 1718), Glasgow's Literary Society (founded in 1752), and Aberdeen's Philosophical Society (founded in 1758). The result, he notes, was "that rare phenomenon, ... a philosophically educated public, with shared standards of rational justification" (*Whose Justice?* 248). For Hume's particular strategy of coding into his writing a new form of middle-brow literacy, see Christensen, *Practicing Enlightenment*, esp. 21–44.

culiar genius of early Whig policy which, concerning 1688, "preferred to argue that the government was not dissolved, that traditional institutions retained their authority, and that the actions taken and being taken were to be justified by reference to known law."[5] Observing how the financial revolution of the early eighteenth century had "rendered society more Hobbesian than Hobbes himself," J. G. A. Pocock thus sees the auratic power of the state in manifest decline.[6] In its custodial role (later echoed by Otto von Bismarck's famous characterization of the state as a "night-watch"), the meaning of government shifts from the literal (Hobbesian) idea of sovereign dominion to neo-Stoicism's intellectual concern with "managing the passions." Inasmuch as the consolidation of the modern liberal nation-state involves the migration of power from an auratic to a bureaucratic model, the Leviathan is eclipsed by a complex and evolving array of economic practices first mapped by Scottish Enlightenment theorists keen to trace the metamorphosis (at once imperceptible and inexorable) of raw passion into self-disciplined interest. Yet with rationality having been redefined as an exercise in *sublimation,* the capacity of societies to conceive and articulate normative ends (and an overarching framework for practical choices that invariably have to be made) has undergone further erosion. For even as an irrational will is gradually transmuted into the quasi-rationality of self-interest, itself disciplined by the superagency of the modern market, the resulting "element of reflection and calculation" is strictly concerned with means.[7] What is more, the contraction of rationality to a "reckoning" of efficient means already posited by Hobbes gradually atrophies the intellectual scope of individuals and societies to the point that questions concerning ends are themselves marginalized as the eccentric and more or less irrational province of religious culture.

In its broad outlines, this scenario has been the subject of a variety of political, sociological, psychoanalytic, and anthropological accounts tracing the rise of modern disciplinary society (Charles Taylor, Michel Foucault), the migration of the individual from embeddedness in feudal structures to its administration by modern institutions (Max Scheler, Helmuth Plessner, Arnold Gehlen), and the emergence of an introjected subjectivity (Sigmund Freud, Norbert Elias) made possible by the

5. Pocock, *Virtue,* 224. As Pocock goes on to note, this new-found appreciation for historical continuity also explains why beginning around 1698 "the defense of the Whig regime was beginning to find the once Tory feudal interpretation of medieval history usable for its purposes" (ibid., 231); on the shift from controversy to conversation, see Mee, *Conversable Worlds,* esp. Chapter 1, which traces the growing investment in polite and non-committal conversation in the increasingly commercialized world of mid-eighteenth-century England.

6. Pocock, *Virtue,* 112.

7. Hirschman, *Passions,* 32.

"transformation of hetero-constrictions into self-constrictions that characterize the move from the late-classical period to the modern one."[8] As one distinct phenomenon of this broader transformation, moral philosophy arises as a concerted attempt to translate the speculative interests and normative concepts of philosophical theology into a world whose political stability and economic flourishing (after 1700) were felt to require mimetic techniques of observation and description. Over time, the ongoing quest of that fledgling creation called Britain for political legitimation and stability becomes more closely aligned with the diverse projects of moral and aesthetic self-description. The key term here is "harmony"—the word that George Frideric Handel's setting of John Dryden's "St. Cecilia's Ode" (1739) was to explore with such rich melismatic and harmonic effect.[9] This broader trend also explains why much eighteenth-century moral philosophy seems to accept from the outset that reconciling or harmonizing the tension between reason and the passions could only succeed as a subtle but persistent re-description of prevailing empirical practices. Discarding the normative and emphatically counterfactual rhetoric of institutional religion, moral philosophy instead attempts to translate a noumenal and ineffable will into

8. Esposito, *Bios,* 48. See C. Taylor, *Secular Age,* 90–158; Giddens, *Consequences,* 1–29, and, on the transformation of intimacy, 112–150; Elias, *Civilizing Process,* esp. his account of the transformation of external power into self-constraint (2:363–435); on the transformation of experience and "interior stabilization" of human experience by modern institutions, see Gehlen, *Urmensch und Spätkultur,* esp. 40–73 and 238–251. Most of these accounts owe much to the philosophical anthropology first formulated in Herder's *Auch eine Philosophie der Geschichte zur Bildung der Menschheit* and then extended into the specialized, disciplinary vocabulary of modern sociology (Comte, Durkheim, Weber) and Freudian psychoanalysis, esp. in *Civilization and its Discontents* (1929).

9. Especially in the Chorus (no. 3) of "St. Cecilia's Ode" ("From harmony, from heav'nly harmony / This universal frame began"). A cultural history of the term in the first half of the eighteenth century would furnish an abundance of material and conceptual analogues. On the reconceptualization of harmony in music and the beginnings of modernity, see Chua, *Absolute Music,* 13–22. See also Dahlhaus, who notes that the eighteenth-century emphasis on exemplarity and harmonious structures—providentially revealed in equal temperament tuning, the mathematical ratio of intervals, and the balanced symmetries of musical forms, and said to reflect a natural, cosmic order—often works against the very legitimacy of *Musiktheorie:* "In the treatises of this epoch [the eighteenth century] we notice an effort to reconcile what is ordinarily meant by 'theory' with the fixed principles of a normative musical poetics [*Satzlehre*] inherited from the seventeenth century" (*Musiktheorie im 18. und 19. Jahrhundert,* 7; see also his account of "Nature" in musical theory [ibid., 37–42], and Dahlhaus, *Klassische und Romantische Musikästhetik,* 21–49). Still among the best accounts of classical musical form and the "coherence of the musical language," is Rosen, *Classical Style,* esp. 57–98. For analogous explorations of harmony in eighteenth-century painting, one might turn to the frontispieces for Shaftesbury's *Characteristicks,* the studied symmetries of posture and the ambient landscape in Gainsborough's early conversation pieces (e.g., *Mr. and Mrs. Andrews*), yet also the insistent, anti-commercial stress on a different conception of order—arising organically rather than from "improvement"—in Tory landscapes; see Bermingham, *Landscape and Ideology,* 14–54; Everett, *Tory View,* 91–122; Barrell, *Idea of Landscape,* 1–63; and Barrell, *Dark Side of Landscape,* 35–88.

empirically observable passions. The latter, by dint of their commutation into interests and sentiments, furnish evidence for an inner, "moral sense" credited with reliably anchoring the self in moral and social space. Inasmuch as the eighteenth century conceives of morality no longer as knowledge but as feeling, the voluntarist antagonism of will and intellect, while not wholly resolved, has been deflected into the empirical challenge of how to manage the passions. To thus recast the problem as the ancient, Stoic challenge of the self's quest for inner balance, harmony, and the methodical expurgation of false judgments was to reaffirm the inner life (rather than political or religious institutions and frameworks) as the principal focus of moral theory and as the dominant (perhaps the only) source of moral self-legitimation. As we shall see, in Shaftesbury, Hutcheson, and especially in Adam Smith, this meliorist and oblique disciplinary approach to the emotions entails some acute problems of its own; and it was in response to these that the entire project of moral philosophy would eventually be challenged in Ludwig Wittgenstein's critique of "inner" states and in Gertrude Elizabeth Anscombe's view of modern moral philosophy as oblivious of its conceptual inheritance and, hence, as a terminally incoherent enterprise. In particular, Anscombe insists, the "concepts of obligation, and duty—*moral* obligation and *moral* duty, that is to say—and of what is *morally* right and wrong, and of the *moral* sense of 'ought,' ought to be jettisoned if this is psychologically possible; because they are the survivals, or derivatives from survivals, from an earlier conception of ethics which no longer survives, and are only harmful without it."[10]

While the shift in question appears to push the metaphysical concept of the will to the margins of eighteenth-century moral theory, it would be a mistake to conclude that it had ceased to be an issue altogether. Concepts as foundational to human inquiry as those of will or person do not just vanish; rather, their function is either metonymically displaced onto other terms, or they are themselves being absorbed into and redefined by a different intellectual tradition. To some extent, both developments occur in the case of the will after Hobbes. On the one hand, its voluntarist construction before 1700 is now being absorbed into an empiricist and notably hedonistic theory of human action. Hobbes's notion of the will as the "last appetite" had arguably prepared for this shift to a strictly empirical and occasional model of human

10. Anscombe, "Modern Moral Philosophy," 1. Similar arguments have subsequently been advanced by MacIntyre, *AV*, esp. 6–35 and 204–243; for a somewhat different appraisal of Anscombe, see Cora Diamond, who argues that for Anscombe "ethics *can be done without* [the background of Christian moral traditions]" and that "moral evaluation, like 'unjust' or 'courageous,' have no need of the background divine law." Still, she concedes that for Anscombe, as for MacIntyre, the difficulty, perhaps even impossibility, of moral philosophy arises from the fact "that certain concepts require for their content or intelligibility background conditions which are no longer fulfilled" ("Losing Your Concepts," 257).

agency. Yet it is Locke who completes this downward transposition of the will from an active and dynamic metaphysical source to the epistemological zero-degree of literally mindless passions and, in so doing, prepares the ground for Mandeville's scandalous portrait of hedonistic human agents consumed by the eternal present of desire. In his *Essay,* Locke thus takes up the question of "power" that Hobbes had treated so restrictively—viz., as nothing more than a matter of efficient causation and devoid of any human, spiritual, and reflective dimension. Echoing Hobbes's account of the will as the "last appetite," Locke thus asks "whether a Man can *will,* what he *wills*"—a question repeated verbatim in Schopenhauer's 1839 *Prize Essay on the Freedom of the Will.*[11] For Locke, unequivocally, the answer here has to be "no." Yet there's the rub, for what prompts this view is an already implied understanding of the will as nothing more than efficient causation, the mental equivalent of Newtonian force: "We must *remember,* that *Volition,* or *Willing,* is an act of the Mind directing its thought to the production of any Action, and thereby exerting its power to produce it."[12]

Two contradictory movements shape Locke's reasoning here. On the one hand, he insists on the complete absence of any inner, reflective, indeed human dimension from the will, which he construes as nothing but quasi-mechanical efficacy. Yet that does not address the obvious fact that the will constitutes a distinct human attribute, one whose "antecedent probability" (John Henry Newman) has been ratified by some two millennia of philosophical arguments concerned with elucidating its complex phenomenology. For the will does not simply *act* but as an active power is itself a focus *for* human consciousness—viz., as something felt, observed, questioned, *from within.* Its reality is inseparable from its correlative phenomenology. The question thus arises whether the physico-mechanical account of the will as sheer efficient force can ever be anything more than a makeshift metaphor aimed at capturing some ineffable psychological process. To pose that question is to close in on the second, arguably opposed tendency observable in Locke's account of the will. And it is this tendency that ultimately vitiates his implicit understanding of mental and physical processes as wholly convertible and his attempt to furnish a coherent and exhaustive epistemology of the inner life strictly in terms of mechanical causation. Even as this larger objective is pursued with increasing vigor in the later editions of the *Essay,* Locke remains unable to expunge the psychological, affective quality from his ac-

---

11. "The question therefore is not whether a man be a free agent, that is to say, whether he can write or forbear, speak or be silent, according to his will; but whether the will to write and the will to forbear come upon him according to his will, or according to anything else in his power. I acknowledge this liberty, that I can do if I will; but to say I can will if I will, I take to be an absurd speech" (Hobbes, "Of Liberty and Necessity," 16); on Schopenhauer, see the discussion below, 471–477.

12. *Essay,* Bk. II, Ch. 21, §25, §28.

count of the will, for the obvious reason that if he were to do so the very notion of a will would effectively disappear. It would no longer have phenomenological status, would not *appear for the self* and, thus, would once again become an object of metaphysical conjecture, a noumenon of precisely the kind to which Locke's empiricism is so firmly opposed. The dilemma bears tracing a bit further, if for no other reason than that subsequent eighteenth-century moral philosophy can be divided into those (voluntarist, mechanist, and associationist) thinkers who fail to recognize it as a dilemma and others (Platonists, virtue ethicists, Trinitarian personalists) who do. Struggling to minimize the psychological and affective dimension of the will ("it being a very simple Act"), even while retaining some affective notice as prima facie evidence of the will's existence and operation, Locke thus describes "the Will, or Preference" (so his revealing appositive) as "but the being better pleased." With charming circularity, Locke now observes that

> to the Question, what is it [that] determines the Will? The true and proper Answer is, The Mind ... If this Answer satisfies not, 'tis plain the meaning of the Question, *what determines the Will?* Is this, What moves the mind, in every particular instance, to determine its general power of directing, to this or that particular Motion or Rest? And to this I answer, The motive, for continuing in the same State or Action, is only the present satisfaction in it.[13]

As he proceeds to desynonymize the will from desire, Locke appears caught between a strictly occasional conception of the will that Ockham and Hobbes had bequeathed him, and his manifest need to retain some phenomenological evidence of the will *in actu*. The problem with the latter requirement is that it implies a distinction between the mere *fact* of a person willing this or that and some psychological *quality* whereby the same agent knows or at least feels herself or himself to be willing. Were Locke to abandon this latter requirement altogether, the will would quickly atrophy to a mere *inference* drawn from the fact that a person *appears to be acting*. Moreover, the conundrum would not end there, for in the absence of the will as a primary inner reality, we would have no reason to construe certain processes manifestly unfolding before us as bona fide "actions" at all. In the end, mental life will then have been no longer *explained* as a self-sustaining reality; rather, it will have been *dissolved* into a

---

13. *Essay* Bk. II, Ch. 21, §29 The previous quote (". . . but the better pleased") refers to the first edition of the *Essay*, also Bk. II, Ch. 21, §29. For a detailed account of Locke's confusion in this passage, and for his often perplexing adjustments in the second edition of the *Essay*, see Darwall, *British Moralists*, 152–160. Darwall convincingly maintains that Locke's arguments on the will dialectically feed off Cudworth's published and unpublished arguments on free will and moral agency, esp. his *Treatise of Freewill* (published in 1838).

linear, calculable, and seemingly inevitable sequence of force transfers, a process altogether devoid of consciousness, significance, and responsibility.

Unsurprisingly, this fundamental dilemma registers in a number of uneasy verbal qualifiers and rhetorical maneuvers. Consider the following two instances. In the first, Locke stipulates the non-discursive, indeed wholly inarticulate nature of the will: "*Volition* is conversant about nothing, but our own Actions; terminates there, and reaches no farther; and that *Volition* is nothing, but that particular determination of the mind, whereby, *barely by a thought,* the mind endeavours to give rise, continuation, or stop to any Action."[14] The parenthetical qualifier ("barely by a thought") hints at Locke's deeper perplexity as to whether to treat the will strictly as a mechanical force or as a state of mind which, however minimal its content, may yet become itself the focus of conscious "attention." As remains to be seen, Hutcheson's critique of Mandeville and, indirectly, of Locke homes in on precisely this key question: whether the will circumscribes and determines the totality of our inner life or, alternatively, whether the person qua self-conscious agent *knows* herself or himself to be willing something. Ultimately, Locke realizes that some rudimentary mental function needs to be retained independent of whatever happens to be the contingent, present concern of the will. Reluctantly pressing his quest for some inner, phenomenological evidence of the will, he thus wonders just *"what is it that determines the Will in regard to our Actions?"* His telling and consequential answer is that this causal determinant has to be found in "some (and for the most part the most pressing) *uneasiness* a Man is at present under." Yet almost immediately Locke identifies the source of such uneasiness as desire itself: "Desire being nothing but an *uneasiness* in the want of an absent good, ... desire and *uneasiness* is [sic] equal."[15] If the will is the determining, efficient cause of action, uneasiness stands in the same relation to the will, and desire completes this metonymic chain by being the primordial trigger of such unease. While one may fear that Locke is caught in an infinite regress of explanatory concepts, their relation is not altogether equivalent. For in each case, there appears to be a diminished level of awareness accompanying the inner state. To will something necessarily entails the conscious deliberation and ultimate judgment as regards the means requisite for and appropriate to securing the object or end in question. By comparison, this representational and cognitive dimension has all but vanished where a person's consciousness of her or his "uneasiness" is at stake. Here the consciousness of my *feeling* ill at ease and that very state appear only minimally distinct, and in the case of desire the difference has vanished altogether, such that consciousness-of-desire is ultimately desire itself. The alternative, which Locke does

14. *Essay,* Bk. II, Ch. 21, §§29–30; last italics mine.

15. Ibid., §31.

not entertain, would be a conscious, likely fetish-based staging and nurturing of desire, in which case (following Slavoj Žižek) we should speak of "fantasy" rather than desire. Extending Descartes's selective appropriation and re-orientation of ancient Stoicism, Locke here reveals the full extent to which the Platonic view of desire (viz., as an imperfect reaching for the good and the beautiful) had eroded. As Charles Taylor observes, already in Descartes "the metaphysical aura of desire had been discredited as 'total illusion,'" leaving the human agent "utterly unmoved by the aura of desire. In a mechanistic universe, and in a field of functionally understood passions, there is no more ontological room for such an aura. There is nothing it could correspond to. It is just a disturbing, supercharged feeling, which somehow grips us until we can come to our senses."[16]

Locke's argumentation, subsequently known as "psychological hedonism," rests on the premise that all internal, mental action or volition responds to generic cues of either pleasure or pain, which in turn may arise from the senses or be associated with some kind of mental representation. For all his protests against Hobbes, Locke "had in fact taken over Hobbes's belief that human nature is under the governance of two masters, pleasure and pain," and thus had embraced the in essence mechanistic view of mind as inherently *reactive* and incapable of inner causation. Desire thus becomes "a piece of intermediary mechanism, operated on by present pleasures and pains and by the thought of future pleasure and pain, and itself having the power to produce volition or will."[17] Significantly, such a hedonistic model of mind entails the de facto loss of all temporal perspective. To be sure, mind still is cathected onto the future, albeit only as the projected locus of a pleasure it wishes to possess here and now. Conversely, the past is but lapsed time, a distant repository of pleasures and pains devoid of any continued significance. It is this axiomatic confinement of mind qua will or, ultimately, desire to an eternal present that vexed Shaftesbury about his former tutor. As he remarks in a letter to Philip Dormer Stanhope (7 November 1709), if "he [Locke] had known but ever so little of antiquity, or been tolerably learned in the state of philosophy with the ancients, he had not heaped such loads of words upon us."[18] Stephen Darwall's recent claim that Locke, in the second edition of his *Essay*, appears to embrace a model of the will as self-aware and autonomous ("Will *is the faculty of self-conscious self-command*") seemingly in line with Cudworth's neo-Platonist model of mind as self-possessed, perplexes.[19] At the very least, we must

16. C. Taylor, *Secular Age*, 136.

17. Foot, *Moral Dilemmas*, 120, 122.

18. Shaftesbury, *Life*, 416.

19. *British Moralists*, 159. While Darwall is right to detect "several Cudworthian echoes in Locke's thought" (ibid., 163), his attempt at assimilating Locke to his broader narrative about the emergence of "judgment internalism" (9) in modern moral philosophy remains unconvincing in

bear in mind that willing and self-awareness remain, for Locke, at all times circumscribed by the temporal *punctum* of some present desire and "uneasiness" and, thus, cannot give rise to a self-revising and reflective person capable of grasping her or his discrete acts of willing within a broader narrative framework. As long as the question is couched in terms of *present* exigencies and desires, the self's awareness of *having* or, rather, *being in the grip of* such ephemeral states should not be confused with moral self-consciousness.

Manifesting "a very curious blend of hedonism and rationalism," Locke's *Essay* can make its central case only by performing a metonymic slide toward a purely affective, non-representational, and inarticulate source of willing, thereby denuding human personhood and flourishing of all formal inner time consciousness and narrative development.[20] A reductionist account that traces back consciousness and will (Aquinas's "intellectual appetite") to the psychosomatic tropism of "uneasiness" and, ultimately, to desire as the primal energy source of all mental activity implies a total loss of any historical and hermeneutic perspective. Inasmuch as Lockean consciousness is being reborn every minute or second, it is only ever being *informed* but can never recognize itself as having *been formed*. To be sure, Locke grants that sooner or later consciousness will *have* representations, yet on his account it is hard to see how it should ever know itself *to be having them*. Denuded of any abiding content, perspective, or counterfactual scenario to the representations it actually happens to encounter in the here and now—that is, those that sensation and desire have contingently referred to it—the inner life has become all but a fiction. Mind in Locke's *Essay* functions rather like today's microprocessor, a value-neutral relay station for binary notices variously and unaccountably drifting in and out of its dispassionate circuitry. Hume's mind-as-theater metaphor would draw this conclusion openly by conjuring up persons or actors devoid of any perspective on their own reality and development. Yet already in Locke's *Essay*, the skeptic's notion of un-souled human beings, lacking any inner gyroscope, and hence strictly reactive to contingent sensations and desires, looms large. The Lockean self is an epistemological vagrant of sorts, unable to achieve any temporal perspective on her or his flourishing, let alone grasp and articulate normative commitments of any kind. The discretion of such consciousness extends no further than the 0-1 binarism that will have the "self" either embrace or reject bits of information as desirable or undesirable, pleasant or painful. Recoiling from the "Hobbes-Locke thesis of natural indifference," Shaftesbury regards Locke's self as

part because the temporal, narrative dimension of thought—so manifestly absent from Locke's self—is never really identified as missing. Yet it is just this aspect of moral flourishing that lies at the heart of Platonic, Plotinian, and neo-Platonist thought.

20. Foot, *Moral Dilemmas*, 125.

alarmingly bereft of all ideational, aspirational, and temporal dimensions.[21] It may have abstract representations of "the Good" but appears intrinsically agnostic and indifferent to them unless desire suggests that they are within immediate reach, in which case, prompted by a quasi-instinctual "uneasiness," the self may conceivably grope its way toward an object vaguely associated with the good. What is more, the *Essay* affirms as much without even worrying about the costs entailed by this apparent reduction of the inner life to what is immediately and obviously present:

> It may be said, that the absent good may by contemplation be brought home to the mind, and made present. The *Idea* of it indeed may be in the mind, and view'd as present there: but nothing will be in the mind as a present good, able to counter-balance the removal of any *uneasiness,* which we are under, till it raises our desire, and the *uneasiness* of that has the prevalency in determining the *will.* Till then the *Idea* in the mind of whatever good, is there only like other *Ideas,* the object of bare unactive speculation; but operates not on the will, nor sets us to work.[22]

One may naturally wonder why Locke should assume that the mind will only ever espouse those objects and goods ready-to-hand, while maintaining affective neutrality vis-à-vis anything on the order of an idea or good such as might arise from "unactive speculation." The likely answer is that in conceiving of the mental architecture strictly in terms of *efficient* causation, Locke has implicitly deprived consciousness of the capacity to choose or cultivate a good for intrinsic reasons. His naturalist contention that "Good and Evil ... are nothing but Pleasure or Pain, or that which occasions, or procures Pleasure or Pain to us," shows why a strictly epistemological model of mind cannot persuasively account for ethical awareness or commitments.[23] It lacks a sense of obligation, simply because such a "complex" idea should never arise from bona fide "reflection" but only through a (mechanical) combination of simple ideas that Locke views as inherently "mindless" and inaccessible to reflection. Lockean consciousness is not a source of meanings but a product of inherently meaningless (material) causes; hence the sole factor credited with determining the will is the conspicuous and adventitious presence, desirability, and presumed attainability of a distinct object. Questions of content and value are entirely incidental or, at the very least, strictly subordinate by comparison. As the *Essay* puts it,

---

21. C. Taylor, *Sources,* 253. On Shaftesbury's response to Locke, see also Den Uyl, "Shaftesbury," 287–290.

22. Locke, *Essay,* Bk. II, Ch. 21, §37.

23. Ibid., Bk. II, Ch. 28, §7.

> that *good,* the *greater good,* though apprehended and acknowledged to be so, does not determine the *will,* until our desire, raised proportionately to it, makes us *uneasy* in the want of it ... [for] 'tis *uneasiness* alone operates on the *will,* and determines it in its choice ... For the *will* being the power of directing our operative faculties to some action, for some end, cannot at any time be moved towards what is judg'd at that time unattainable: That would be to suppose an intelligent being designedly to act for an end, only to lose its labour; for so it is to act, for what is judg'd not attainable.[24]

As these last lines make clear, Locke's *Essay* signals a significant advance in the transfer of a conception of method first pioneered by Galileo and Francis Bacon from objective to subjective phenomena. Locke's resistance to the possibility that moral meanings, such as a sense of obligation to another human being, might arise from an internal (and causally indeterminate) process of deliberation, judgment, and action rests on the premise that *one* method—conclusive and efficiently mono-causal in nature—must be capable of explaining *all* phenomena, both those objective ones "out there" and the subjective processes of the mind.

More than anything else, it is this commitment to method that drives modernity's strictly epistemological view of knowledge and explanation. Wondering whether moral obligation is something that we arrive at by a process of sustained introspection and interpretive reasoning, Locke ultimately rejects that possibility in Book IV of his *Essay*. Instead, he insists on locating the moral life "*amongst the Sciences capable of Demonstration:* wherein I doubt not, but from self-evident Propositions, by necessary Consequences, as incontestable as those in Mathematicks, the measures of right and wrong might be made out, to any one that will apply himself with the same Indifferency and Attention to the one, as he does to the other of these Sciences."[25] This is an apt illustration of how (as Edmund Husserl was to argue in *The Crisis of the European Sciences*) the triumph of scientific method had created a world of "idealities" whose relation to the world (*Lebenswelt*) of phenomenological experience had become increasingly tenuous. Finding such idealities methodologically intractable, Locke's *Essay* thus declares them altogether irrelevant to scientific knowledge and, ultimately, beyond the scope of rational inquiry. Such a view is indeed a logical entailment of a nominalism to which Locke was unquestioningly committed, so much so that he appears quite oblivious to the fact that nominalism, too, amounted to a complex intellectual tradition in its own right. As his exchanges with some of his

24. *Essay,* Bk. II, Ch. 21, §§35, 36, 40. Schneewind is right to note that the "Lockean will, though an active power different from motives, has no rational ordering principle of its own" (*Invention,* 300).

25. *Essay,* Bk. IV, Ch. 3, §18. On this passage, see Foot, *Moral Dilemmas,* 128f.

contemporary realists (John Norris, Edward Stillingfleet) make apparent, the mere supposition that ideas and (moral) meanings of any kind could ever be conceived other than as abstractions from particular and distinct empirical sensations strikes Locke as so outlandish that he seems literally at a loss for words. Where Stillingfleet posits "that Peter, James, and John ... [share] a common nature, with a particular subsistence proper to each of them," Locke's response amounts to little more than a blank stare in epistolary prose: "I do not understand what subsistence is, if it signify anything different from existence."[26] Two conclusions can be drawn here. First and foremost, we see in Locke's *Essay* the de facto abandonment of the ancient notion of *logos* capable of integrating scientific knowledge *and* human experience. As Husserl puts it, "if man loses this faith, it means nothing less than the loss of faith 'in himself,' in his own true being."[27] The second point follows from the "de facto" phrase; viz., Locke's embrace of voluntarism and nominalism is so complete and, to him, self-evident that he can no longer engage competing traditions of inquiry, most eminently that of realism. A major liability, often unnoticed, of such peremptory commitment to voluntarism and nominalism is that in premising all knowledge on the temporal and logical *punctum* of individual volition and sensation the modern self is rendered strangely inarticulate. It cannot grasp points of view that proceed from a principle other than some singular and isolated external prompt. What is more, a philosopher committed to the axioms of nominalism *sensu strictu* will also be ill-equipped when it comes to grasping the distinction between merely *opposed* and positively *incommensurable* modes of reasoning. For a philosophy that has a strictly occasional understanding of how meanings are formed is necessarily bereft of any meta-conceptual perspective—which is to say, unprepared to perceive how its own intellectual project arises within the long *durée* of a specific tradition. Locke's well-known scorn for the "schoolmen" thus blinded him to empiricism's considerable debt to at least some of them (most prominently Ockham).

Beginning with Shaftesbury and culminating in Smith's *Theory of Moral Sentiments* (1759), the post-Hobbesian rehabilitation of the inner life does indeed appear to come at the expense of a pervasive inarticulacy. In particularly fulsome ways that predicament emerges in the mimetic behavior of Smith's "impartial spectator" and the concurrent displacement of virtue from the realm of action to the social ideal of "praiseworthiness"—a descendant of Aristotle's *megalopsychia,* which had so

26. Quoted in Milton, "John Locke," 133.

27. Husserl, *Crisis,* 13. As he continues, "more and more the history of philosophy, seen from within, takes on the character of a struggle for existence, i.e., a struggle between the philosophy which lives in the straightforward pursuit of its task—the philosophy of naïve faith in reason—and the skepticism which negates or repudiates it in empiricist fashion."

bedeviled the history of virtue ethics for the previous 1,800 years.[28] To begin with, it helps to recall how the civic republicanism of James Harrington had begun, ever so cautiously, to challenge Hobbes's quasi-existentialist and pessimistic view of human psychology. Against Hobbes, Harrington maintains that "the right and wrong of action are determined by the inner condition of the agent, not by law and not by consequences." Hence, "what Hobbes takes as man's natural condition is for Harrington what people become when they are 'corrupt.'" Whereas for Hobbes "desires and impulses do not respond to perceived good," Harrington and, even more emphatically Shaftesbury, insist that feelings "are the sources of good in the world."[29] A far more consequential rejection of Hobbesian voluntarism is found in Shaftesbury's *Inquiry Concerning Virtue or Merit,* written in the early 1690s, first published in 1699, and the centerpiece of his *Characteristicks of Men, Manners, Opinions, Times* (1711)—a work of great influence on the eighteenth century during which it underwent no less than eleven successive editions. Hinting at Hobbes's de facto (if not admitted) atheism, Shaftesbury begins his *Inquiry* by arguing for the embeddedness of every individual within "a SYSTEM of all Things, and a Universal Nature." By definition, every human being is both ordered toward and (at least potentially) capable of participating in that system's "harmony."[30] To suggest otherwise is to hypostatize a human being effectively bereft of all "sense"—a word whose pivotal, albeit ambiguous career in eighteenth-century moral philosophy seems to commence with Shaftesbury.

---

28. The principal source here is Aristotle, *Nicomachean Ethics,* 1123^{b}1–1124^{a}30, with the critical term, *μεγἄλοψῦχία,* which has been variously translated as "magnanimity," "pride," or being "high-souled." On the longstanding debate of how to read Aristotle's apparent characterization of *μεγἄλοψῦχία* as a virtue or at least a complement of the virtues (1124^{a}2), see Herdt, *Putting on Virtue,* 23–44. From within an Augustinian tradition, the generous self-appraisal of moral agents that Aristotle's term implies is, of course, exceedingly close to the vice of pride; and even if such a reading may appear anachronistic, the ambivalence of *μεγἄλοψῦχία* is indeed evident in (not just imported into) Aristotle's argument: "Even if consciousness of one's own moral worth simply supervenes on virtuous activity, it is nevertheless possible to pervert it by treating it as an external good to which the pursuit of virtue can be instrumentalized." Moreover, the claim to autonomy contained in this notion downplays, perhaps even denies, the contingency of one's socialization and character formation. Hence, Aristotelian "magnanimity in fact involves serious self-deception, inasmuch as the magnanimous person fails to remember the goods she has received from others and thus arrives at a false estimate of her own self-sufficient greatness" (ibid., 41–42); on the role of "praiseworthiness" in A. Smith, see Den Uyl and Griswold, "Adam Smith."

29. Harrington, quoted in Schneewind, *Invention,* 293–294, 298.

30. Shaftesbury, *Characteristicks,* 2:11; henceforth cited parenthetically as *SC.* On the significant role of the Cambridge Platonists on subsequent critiques of modern voluntarism and Hobbesian thought, see Schneewind, *Invention,* 194–214 and C. Taylor, *Sources,* 248–259; E. Cassirer, *Platonic Renaissance,* esp. 42–85; and, on Cassirer's ambivalent relationship to Christian Platonism, see Wisner, "Ernst Cassirer." On Shaftesbury's relation to the Cambridge Platonists, see Rivers, *Reason,* 2:127–132; regarding the increasing importance of the Cambridge Platonists (esp. Smith, More, and Cudworth) on Coleridge's later writings, see below, 484–488.

A student of Locke and, just as importantly, an avid reader of the Cambridge Platonists, Shaftesbury first published an anonymous edition of Benjamin Whichcote's *Sermons*. As Shaftesbury insists, to conceive of God as the source of goodness, justice, and order necessarily means according these notions eternal status. They cannot be merely contingent on a (potentially fluctuating) divine will but depend on what he calls a "natural *moral Sense*"—this being the first time that the concept makes its appearance in English philosophy:

> for wher-ever any-thing, in its nature odious and abominable, is by Religion advanc'd, as the suppos'd Will or Pleasure of *a supreme Deity;* if in the eye of the Believer it appears not indeed in any respect less ill or odious on this account; then must *the Deity* of necessity bear the blame, and be consider'd as a Being naturally ill and odious, however courted, and solicited, thro' Mistrust and Fear. But this is what Religion, in the main, forbids us to imagine ... For whoever thinks there is a God, and pretends formally to believe that he is *just* and *good,* must suppose that there is independently such a thing as *Justice* and *Injustice, Truth* and *Falshood, Right* and *Wrong;* according to which he pronounces that *God is just, righteous,* and *true.* If the mere *Will, Decree,* or *Law* of God be said absolutely to constitute *Right* and *Wrong,* then are these latter words of no significancy at all. (*SC,* 2:27, 29)

To suppose that the divine will constitutes the sole source of meanings is to subscribe to a purely occasional and conceptualist idea of reason and to hold that meanings are instituted only in a performative sense, that is, established by *ascription* rather than by *recognition.* Such a universe Shaftesbury regards as intrinsically devoid of value and quite impossible to conceive as an object of worship. In uncharacteristically decisive language, he thus affirms the existence of a basic, innate moral sense (a phrase he hardly ever uses but which, plausibly, Hutcheson and Hume took to be one of Shaftesbury's main intellectual bequests): "Sense of Right and Wrong therefore being as natural to us as *natural Affection* itself, and being a first Principle in our Constitution and Make; there is no speculative Opinion, Persuasion or Belief, which is capable *immediately* or *directly* to exclude or destroy it. That which is of original and pure Nature, nothing beside contrary Habit and Custom (a second Nature) is able to displace" (*SC,* 2:25). Shaftesbury's claim here may seem dogmatic or, at the very least, indemonstrable and of evidently metaphysical provenance. Yet to contest or reject it for that reason as "circular" is to fall back on a Lockean (empiricist) framework whose curtailment of knowledge to methodologically deducible, value-neutral, and plainly demonstrable "data" Shaftesbury here means to oppose.[31] In fact, the impulse

31. Citing the same "remarkable claim" from the *Inquiry,* Yousef sees Shaftesbury's argument "rest on a number of related, often circular, assertions about nature, affection, and the 'good'

to reject the self-authorizing claim that is being made here for "natural Affection" likely arises from the equally problematic—because equally indemonstrable and in origin nominalist—assumption that feeling can by definition only ever amount to something occasional, empirical, and *sui generis*. To the question as to whether "these *mental* Children, the Notions and Principles, of *Fair, Just,* and *Honest,* with the rest of these *Ideas,* are *innate?*" Shaftesbury's alter-ego in *The Moralists,* Theocles, responds by pointing out that what matters is not at what point in time these basic "Notions and Principles" are formed. Rather, the crucial

> question is, whether the Principles spoken of are *from Art,* or *Nature?* If from *Nature* purely; 'tis no matter for the Time: nor wou'd I contend with you, tho you shou'd deny *Life* it-self to be *innate,* as imagining it follow'd rather than preceded the moment of Birth. But this I am certain of; that *Life,* and the *Sensations* which accompany Life, come when they will, are from *mere Nature,* and nothing else. Therefore if you dislike the word *Innate,* let us change it, if you will, for Instinct; and call *Instinct,* that which *Nature* teaches, exclusive of *Art, Culture,* or *Discipline.* (*SC,* 2:229–230)

As it happens, Shaftesbury's contrary affirmation, viz., that the affection in question is "a first Principle in our Constitution and Make," merely transposes into the eighteenth century's emergent "word-cloud" of sense, sensation, sentiment, affection, etc. a key concept of humanistic inquiry that had endured from Plato and Aristotle all the way through Aquinas.

What Locke had disallowed when rejecting "innate ideas" constitutes for Shaftesbury an indispensable, shared ground for understanding the rational and ethical nature of human agency most fully mapped by Aquinas. Inasmuch as this "natural Affection" must be understood as something all human beings have in common, their sociality and rationality are indeed convertible—at least up to a point. Aquinas had analyzed this basic moral sense under the name *synderesis* and, importantly, he classifies it as a "natural habit, not a power" (*synderesis non est potentia, sed habitus*). In defining *synderesis* as a natural habit all but synonymous with natural law, Aquinas emphasizes that this moral sense is but a point of departure and, as such, in need of constant cultivation. It must not be misconstrued as some ready-made, apodictic inner certitude. Notably, in his response to the third objection (which, by way of Augustine, affirms that within "the natural power of judgment there are certain 'rules

---

of individuals," and she specifically seems uncomfortable with his deducing "both the existence of impulses . . . and the specific moral and emotional content of those impulses, from the observation that social life is natural to human beings" ("Feeling for Philosophy," 614). For other strong readings of Shaftesbury, see Darwall, *British Moralists,* 176–206; and Marshall, *Figure of Theater,* 9–70.

and seeds of virtue, both true and unchangeable'") Aquinas emphasizes the complementarity of the practical reason or "judgment" (*prudentia*) and this inner sense. Both here and in his subsequent discussion of conscience (*ST*, I Q 79 A 13), Aquinas thus takes care not to mystify *synderesis* and conscience as a metaphysical "power" of sorts but, instead, to stress its evolving, act-like nature. The very basic moral orientation that is universally infused into every human being ("do good," "avoid sin," etc.) merely constitutes the "seed" (Augustine's term) or dynamic source for the progressive cultivation of rational personhood in dispositions, habits, and the virtues. As such, *synderesis* offers no determinate judgments or rational appraisals of the good. Rather it constitutes a basic phenomenology of what it means for us to be ethical agents—that is, both capable of making moral choices *and* obligated to do so. Thus it manifests itself affectively and with minimal articulacy; in Aquinas's words, "*synderesis* is said to incite to good, and to murmur at evil [*instigare ad bonum, et murmurare de malo*]." It furnishes human life with a basic orientation and with the energy potential required if the narrative of moral self-fashioning is ever to get underway. As he puts it, "those unchangeable notions are the first practical principles, concerning which no one errs [*circa quae non contingit errare*]; and they are attributed to reason as to a power, and to *synderesis* as to a habit. Wherefore we judge naturally both by our reason and by *synderesis*."[32] Even for the skeptic Hume, it is a given that "the mind by an *original* instinct tends to unite itself with the good, and to avoid the evil, tho' they be conceiv'd merely in idea" (*HT*, 280).

Bearing this model of *synderesis* in mind we can pinpoint a crucial shift in the way that theoretical reflection about agency unfolds after Shaftesbury. Prior to Kant, positing a basic, indispensable moral orientation tends to be the norm—with Hobbes, Locke, and Mandeville still something of an exception. What prompts this affirmation of an innate orientation toward the good, however diversely it may be articulated in the empirical world, is that without it theory would quickly lose all model-building force. Its authority would be reduced to a merely reactive (and likely pessimistic) appraisal of what is empirically and, it would seem, irreversibly at hand. No longer would theoretical discourse be able to set forth a coherent and rational framework for human flourishing. Unsurprisingly, then, the shift from a model of

---

32. *ST*, I Q 79 A 12. On Aquinas's notion of *synderesis* and conscience, see McCabe, "Aquinas on Good Sense." Drawing on Aquinas's early *Questio Disputata de Veritate*, McInerny notes how, unlike the "purely cognitive" judgment of conscience, choice "consists in the application of knowledge to affections ... My choices reveal my character, the condition of my appetite, whereas the judgment of conscience reveals my cognitive ability to see that a given act is forbidden" ("Prudence and Conscience," 300). As an act, not a habit, conscience does not err, though "judgment" (*prudentia*) certainly may: "one errs in choice and not in conscience" (Aquinas, quoted in ibid., 301). For a perceptive discussion of how *synderesis* reappears in the Scottish Enlightenment, particularly in the writings of Dugald Stewart, see MacIntyre, *Whose Justice?* 330–334.

vision toward one of critique increasingly drives eighteenth-century theoretical inquiry to challenge *synderesis* and the will as unwarranted metaphysical assumptions and, in the absence of acceptable proof, to reject them outright. It is instructive to ponder this demand for discrete material evidence of any concept, so representative of Enlightenment intellectual culture (and so seemingly rational), with some care and to raise some basic questions. First, can modern critique, which refuses to affirm anything prior to its own operation and outside its own conceptual jurisdiction, ever legitimately *reject* the first principles of a wholly different intellectual project? One naturally thinks here of an ancient, therapeutic model of philosophy rather less concerned with the empirical verification of facts and the formal-syllogistic demonstration of what may be predicated about them. Instead, the focus of premodern thought (Platonic, Stoic, or Augustinian) is unfailingly on knowledge as an integral (though hardly the *only*) feature of human life, understood as a self-revising narrative in which practical and theoretical reason are complementary and of equal dignity. Second, is an Enlightenment model of critique not also obligated to engage the *entire* conception whose premises it questions? For how can modern critique justify its selective questioning of the premises and principles of all other theoretical and moral systems, unless it is also prepared to advance an equally comprehensive, alternative account of human flourishing? Finally, can there even be such a thing as a wholly neutral intellectual stance—a view from nowhere uninformed by some distinct *tradition* of rational inquiry? Beginning with Mandeville's critique of Shaftesbury and Gilbert Burnet's and John Clarke's rationalist objections to the postulate of a moral sense in Francis Hutcheson's 1725 *Inquiry*, what changes is not simply the status and reality of the will and of an innate moral sense but the very nature of theoretical inquiry itself. Echoing Hans-Georg Gadamer, Alasdair MacIntyre insists that there can be no such thing as a critical and rational argument unfolding independent of a specific intellectual tradition: "it is indeed only relative to some competing theory or set of competing theories that any particular theory can be held to be justified or unjustified. There is no such thing as justification as such, just as there is no such thing as justification independent of the context of any tradition." Moreover, MacIntyre notes, "the first principles of such a theory are not justified or unjustified independently of the theory as a whole."[33]

Few eighteenth-century thinkers rival the degree to which Shaftesbury embraces this notion of philosophical argument as embedded within deep intellectual genealogies and unfolding as a sustained and highly self-conscious reflection on the various traditions feeding into it. His indebtedness to neo-Platonic and neo-Stoic thought, fused with a strong investment in classical Roman rhetoric and the Chris-

33. *Whose Justice?* 252.

tian emblematic tradition (vividly engaged in the engravings prefacing the individual parts of *Characteristicks*), is as apparent as his appropriation of these iconic traditions is original.[34] Nonetheless, his philosophical arguments also exhibit a strikingly modern quality, such that the reality of the innate moral sense pivots on the supplement of the aesthetic, itself the indispensable medium for clarifying what *synderesis* contains *in nuce*. In this regard, it is especially noteworthy about Shaftesbury's appeal to the *synderesis*-like "natural Affections" that their authority is no longer complemented by an account of conscience, or by a fully developed model of prudence. Rather than arguing for the cultivation of habits (e.g., conscience) and virtues, particularly the crucial intellectual virtue of good sense or "judgment" (*prudentia*), Shaftesbury's distinctly modern outlook seeks to *mediate* the basic "natural affections" by means of an aesthetic regimen. In other words, the containment and ordering of the will is not entrusted to a process of introspection and self-examination but, instead, is said to hinge on the cultivation of "taste." Hence, even as he follows the Cambridge Platonists in retaining a conception of *logos* that had endured from Plato and Plotinus through Anselm and Aquinas, Shaftesbury "rejects all those mystical conclusions which Henry More, especially, had drawn from this doctrine."[35]

Yet precisely here Shaftesbury's project also reveals some of its nascent tensions. To reject the mystical implications that tend to be an entailment of a Plotinian view of the inner life such as Henry More in particular had argued it is all very well. Yet if Shaftesbury's *Characteristicks* mark "the moment when the internalization of moral theory is established and ... sentiment becomes important as a moral category," the inner coherence of sentiment, its continuity as a "disposition" (as opposed to some fleeting, contentless "state" of passion) has yet to be phenomenologically secured.[36]

---

34. For the engravings, see http://oll.libertyfund.org/index.php?option=com_content&task=view &id=1565&It (accessed 30 November 2010). Even while away in Naples in 1711, Shaftesbury took an active role in shaping (at times also sketching) the frontispieces for *Characteristicks,* as his exchange of letters and some twenty pages of notes (known as the *Riders Diary* and found at the Public Record Office in London; *Shaftesbury Papers,* PRO 30/24/24/13) make clear. He continued to be involved with the three artists (Closterman, Trench, and Gribelin) in the design and execution of these engravings until his death on 15 February 1713. For a detailed account of the production and emblematic significance of these frontispieces, see Paknadel, "Shaftesbury's Illustrations" and, also, Branch, *Rituals,* 123–134.

35. E. Cassirer, *Enlightenment,* 84.

36. C. Taylor, *Sources,* 258. As Den Uyl points out, however, "at issue is not [Taylor's] internalization per se, but rather the type of internalization, which can have a classical form quite different from modern subjectivism" ("Shaftesbury," 302). The question of whether what Shaftesbury calls the "intire Affection" constitutes a state or disposition, and whether the attainment of such affective experiences is itself something motivated or spontaneous will prove crucial to Hutcheson's *Essay on the Nature and Conduct of the Passions and Affections* (1728). Hutcheson's concern with whether there can be such a thing as non-conscious emotions is closely related to the enduring

Volition simply cannot be thought as pure physics, as sheer "passion" (*thūmos*) and quasi-physical "motion" (*kinēsis*). For to so remove it from the domain of reasoned choice and judgment (Aristotle's *prohairesis;* Aquinas's *electio*) is to undermine rationality—understood not as some external attribute or trait contingently ascribed to human beings but as the essence of personhood itself. Shaftesbury thus is fundamentally opposed to the Ockham-Hobbes-Locke school of thought, which is defined by its procedural quest for autonomy and its insistence on the verifiability of all theoretical claims and practical commitments in propositional language. The result of that strain Shaftesbury takes to be a precarious nominalism for which all meanings (moral and otherwise) prove contingent on the authority underwriting them, an authority destined to face the instant and total collapse of such meanings as soon as the divine affirmation that supports them is either withdrawn or somehow placed in doubt. By contrast, the Cambridge Platonists and Shaftesbury (variously drawing on Plato, Plotinus, and Epictetus) understand the *logos* as fusing transcendent order and human articulacy in a single ontology. In his *Treatise Concerning Eternal and Immutable Morality* (published posthumously in 1731), Cudworth reformulates his critique of reductionist and deterministic accounts of human agency from his 1678 *True Intellectual System of the Universe.* Having identified Ockham as a key proponent of that view, he rejects voluntarist attempts at collapsing the intrinsic and immutable rationality of forms into the divine will said to have ordained them; for to do so is to obliterate the distinction between formal and efficient causes: "God himself cannot supply the Place of a formal Cause ... it is impossible any Thing should Be by *Will only,* that is, without a *Nature* or *Entity,* or that the Nature and Essence of any thing should be Arbitrary ... For though the Will and Power of God have an Absolute, Infinite and Unlimited Command upon the Existences of all Created things to make them to be, or not to be at Pleasure; yet when things exist, they are what they are, This or That, Absolutely or Relatively, not by Will or Arbitrary Command, but by the Necessity of their own Nature."[37] Put differently, the idea of reason necessarily implies its unconditional self-identity and continuity and, as such, is properly constitutive of our selfhood rather than some trait predicatively associated with the self as an "accident."

Elsewhere in his *Characteristicks,* Shaftesbury directly engages the voluntarist view of reason as wholly contingent on the affirmations of the will, as well as voluntarism's fragmentation of the will into a series of discontinuous and strictly *sui generis* affective states. To adopt that view, as his interlocutor in *The Moralists: A*

---

distinction between states vs. dispositions or background emotions and episodic emotions; on this issue, see Nussbaum, *Upheavals,* 69–77.

37. Cudworth, *Treatise,* 16; see also 25–27. On the Cambridge Platonists' opposition to voluntarism and its termination in Hobbesian physico-mechanical and empiricist accounts of nature, see E. Cassirer, *Platonic Renaissance,* 42–65.

Philosophical Rhapsody does, is tantamount to collapsing the distinction between form and meaning, between a transcendent order and its contingent instantiation as subjective experience. Responding to the suggestion "that the current Notion of *Good* ..., our real *Good* is PLEASURE," Shaftesbury thus points out that such a position in effect short-circuits our very structures of communication and ultimately ends up with a tautological and solipsistic model of language that Hegel's *Phenomenology* would later subject to a withering critique in its opening analysis of "sense-certainty." To his interlocutor's proposed conflation of the idea of the good with pleasure, Shaftesbury dryly responds with a request:

> If they wou'd inform us "*Which*," said I, "or *What sort*," and ascertain once the very Species and distinct Kind [of pleasure]; such as must constantly remain *the same*, and *equally eligible* at all times; I should perhaps be better satisfy'd. But when *Will* and *Pleasure* are synonymous; when every thing which *pleases us* is call'd PLEASURE, and we never chuse or prefer as *we please*, 'tis trifling to say, "Pleasure is our Good." (*SC*, 2:128)

Inevitably, the radically nominalist and naturalistic approach to moral argumentation devised by Hobbes, Robert Boyle, and Locke expires in an endless series of tautological affirmations wholly bereft of any explicit and intelligible criteria for what is being affirmed ("*We are pleas'd with what delights or pleases us*").[38] Furthermore, the outright conflation of moral meanings with the contentless pleasure (or pain) supposedly occasioned by their experience also highlights the ambivalent role of the word "sense" in eighteenth-century moral epistemology. Any reading of Shaftesbury will thus have to attend to "the rich contradictions between claims of knowledge and assertions of feeling" and to "assumptions about intersubjective knowledge that run counter to the prevailing empiricist epistemologies with which they are contemporary."[39] For the Epicurean, whose "Self-passions" are too strong and who,

---

38. In a 1726 sermon, Joseph Butler remarks that to assert "that no Creature whatever can possibly act but meerly from Self-love" and that, consequently, "every Affection whatever is to be resolved upon into this one Principle" is a deeply incoherent position to take: "this is not the Language of Mankind: Or if it were, we should want Words to express the Difference, between the Principle of an Action, proceeding from cool Consideration that it will be to my own Advantage; and an Action, suppose of Revenge, or of Friendship, by which a Man runs upon certain Ruin, to do Evil or Good to another" (no. 15, "Upon the Love of Our Neighbor," in *Fifteen Sermons*, 205–206).

39. Yousef, "Feeling for Philosophy" (609, 611). In her analysis of the "strangely complementary structure" linking skepticism and sympathy" Yousef over-emphasizes Shaftesbury's "vigorous affirmations of the irresistible self-evidence of moral feeling." One ought to distinguish here between the self-evidence accorded to the feeling as a phenomenon, which cannot be contested without charging that it has been reported in bad faith—itself tantamount to the inference of another type of feeling—and the subsequent cultivation of a specific feeling as an aesthetic simulacrum,

Shaftesbury argues, invariably becomes miserable in the erratic pursuit of pleasure, "sense" is a wholly unstructured, non-teleological, and non-cognitive occurrence. It is a sheer *punctum*, a "sense" pared down to the physiological tropism of mere "sensation," and it leaves the individual "as highly pleas'd as Children are with Baubles" (*SC*, 2:128). In more colloquial terms, Shaftesbury here restates a critique that the Cambridge Platonists, in particular, Cudworth, had advanced against models of human consciousness as a strictly passive and reactive mechanism. These objections warrant recalling, if only to illustrate how modern humanistic inquiry had from the outset constituted itself dialectically, viz., as a series of confrontations with reductionist and mechanist attempts to explain or indeed dissolve the notion of human. Just as Plato's "education" (*paideia*) and Aristotelian inductive reasoning (*epagōgē*) had been shaped in response to the apparent dissolution of reason into a random collision of minimal units of matter (by Protagoras, Democritus, and Leucippus), and just as Erasmus (drawing on Marsilio Ficino and Pico della Mirandola) opposed the fatalist model of agency expounded by Luther, so Shaftesbury is not simply *asserting* a certain concept of moral agency, responsibility, self-awareness, and freedom. Rather, he takes himself to be arguing for the non-generalizable character of human experience by dialectically engaging, and exposing the incoherence of, the intellectual traditions of voluntarism, nominalism, and their fusion in Locke's hedonistic model of the self.

Cudworth's critique of the word "sense" as supposedly circumscribing the entire scope of human consciousness proved crucial here, just as it continues to be of relevance to the latest installment in the debate on the human—that waged between proponents of neuro-scientific reductionism and those keen to defend the distinctive reality and intrinsic freedom of human consciousness.[40] As Cudworth notes, the key trait of human agency is what we would nowadays call its "intentionality"—that is, having a perspective or attitude vis-à-vis those phenomena that contingently present themselves to its attention: "No Sense can judge of itself, or its own Appearances, much less make any Judgment of the Appearances belonging to another Sense, . . . wherefore that which judges of all the Senses and their several Objects, cannot be it self any Sense, but something of a superior Nature."[41] To insist on the "Different Natures of *Sense* and *Intellection*" does not *eo ipso* amount to some gratuitous and seem-

---

which for Shaftesbury is the only way to access its intrinsic significance and social import. In Shaftesbury, then, affective immediacy is a point of departure, not an all-encompassing answer to the specter of Hobbesian and Lockean skepticism.

40. For a concise and effective critique of "neuro-scientism" (i.e., the misguided importation of neuro-scientific concepts and assumptions into humanistic inquiry, including literary study), see Tallis, *Why the Mind is Not a Computer*, 7–36; for an alternative, still science-based defense of the "I" as self-identical, self-aware, and continuous over time, see Tallis, *I Am*, esp. 22–89 and 220–286.

41. Cudworth, *Treatise*, 69–70, 75.

ingly counterfactual claim. For the customary usage of the word "sense" by physico-mechanical models of (human) nature and perception proves incoherent insofar as it does not distinguish between necessary and sufficient conditions of knowledge, which is to say, between a material process and the "event" of intellection whereby the former crystallizes as a focal point of awareness—that is, as an intentional object. Against Cartesian dualism, Cudworth (as also his fellow Platonists, More and Whichcote) thus maintains that the body itself, and all those notices that it appears to receive, is wholly enmeshed with the representational structure of mind:

> for as much as *Sense* is not meer *Local Motion* impress'd from one Body upon another, or a Body's bare Reaction or Resistance to that Motion of another Body, as some have fondly Conceited, but a Cogitation, Recognition or Vital Perception and Consciousness of these Motions or Passions of the Body, therefore there must of necessity be another kind of Passion also in the Soul or Principle of Life, which is vitally united to the Body, to make up Sensation ... Neither is this Passion of the Soul in Sensation a meer naked Passion or Suffering; because it is a Cogitation or Perception which has something of Active Vigour in it.

Building on this model, Shaftesbury's concern in his *Inquiry* is to restore the explicitness of "sense" as a mental phenomenon and, by extension, the temporal continuity of the person as a rational agent intentionally and purposively interacting with persons and saturated phenomena that cumulatively define her or his social world and the moral meanings whose "actualization" (in the sense of Hegel's *Verwirklichung*) alone invests human life with narrative continuity and purpose.

Unlike the neo-Platonists, however, Shaftesbury pursues this objective in distinctively modern, anti-metaphysical ways. Whereas Cudworth ultimately winds up stigmatizing "sense ... [as] a kind of drowsy and Somnolent Perception," Shaftesbury is more invested in the continuity of "affections" than in the cogency of propositions. Key for him is the self-identity of "affection" over time and its aesthetic connectivity with the *qualia* of other minds.[42] While the semantic band-width of "affection" and "sense" in Shaftesbury and some of his successors (Hutcheson and Smith in particular) seems enormous and may well strike readers today as incoherent, a fair-minded approach to the *Characteristicks* must resist the impulse of seeking conceptual and definitional clarity for terms that are after all expressly introduced as *alternatives* to narrowly epistemological and *eo ipso* reductionist accounts of agency. Inasmuch as "natural affections link human beings to one another [and so] direct the moral sense," their vagueness was meant, and colloquially was understood, to cover "an

42. Ibid., 78–79, 90.

entire range of desires, impulses, feelings, emotions, fundamental dispositions, and occasionally even passions."[43]

While the content and semantics of "affection" and "sense" are notably vague, the matter is at least partially remedied by Shaftesbury's uncompromising affirmation of the way that mind at all times knows itself to be *having*, or to be *imbued with*, affective dispositions of a certain kind. Like the Cambridge Platonists before him and Coleridge a century later, he thus insists on self-awareness as an absolute given, a reality ontologically convertible with the very fact of our existence as persons. In his *Miscellaneous Reflections*, Shaftesbury does not so much oppose as declare irrelevant the ongoing skeptical, Pyrrhonist attempt to subject the I, or *cogito*, to forensic scrutiny and, thus, to premise all philosophy on the resolution of this epistemological dilemma of self-reference. To begin with, he points out how "doubt" already implies self-reference: "that there is *something* undoubtedly which *thinks*, our very Doubt itself and scrupulous Thought evinces." Against "the seeming *Logick* of a famous Modern" (i.e., Descartes) and those "nice Self-Examiners, or Searchers after *Truth* and *Certainty*" (*SC*, 3:117–118), Shaftesbury maintains that self-reference is a premise without which philosophy would never have any occasion to begin, and absent which humans would never feel inclined to construe their relation to the world through various, competing accounts of virtue and the good.

Naturally, the continuity of the I may always be potentially vitiated by "false memory," and the truth of self-reference may turn out "no more than *Dream.*" Yet even to experience the Cartesian's defining anxiety about possible (self-)deception is to acknowledge that the I has Being, rather in the Heideggerian sense of *in-der-Welt-Sein*. As a source and focal point of "care" (*Sorge*), the world is indeed given, and the epistemological dilemma (if we consider it to be one) of self-reference only arises because of the I's antecedent embeddedness within, and orientation toward, the world as a source of existential curiosity, desire, and anxiety. Only on the strength of some such ontology could Hume later remark that "all kinds of uncertainty have a strong connexion with fear" (*HT*, 285). Against the Pyrrhonist's apparent vexation—"that *Identity* can be prov'd only by *Consciousness;* but that Consciousness, withal, may be as well false as real, in respect of what is past"—Shaftesbury offers this deceptively simple and disarming argument:

> I take my Being *upon Trust*. Let others philosophize as they are able: I shall admire their strength, when, upon this Topick, they have refuted what able *Metaphysicians* object, and PYRRHONISTS plead in their own behalf. Mean while, there is no Impediment, Hinderance, or Suspension of *Action*, on account of these wonderfully refin'd *Speculations*. Argument and Debate go on still. Conduct is

43. Dupré, *Enlightenment*, 122.

> settled. Rules and Measures are given out, and receiv'd. Nor do we scruple to act as resolutely upon the mere Supposition that *we are,* as if we had effectually prov'd it a thousand times, to the full satisfaction of our *Metaphysical* or *Pyrrhonean* Antagonist. (*SC,* 3:117–119)

The first reality is not the enigmatic structure of the *cogito* but that of life and action, always already unfolding as any self happens upon the "world." Consequently, it is not epistemology but ethics—for Shaftesbury intimately entwined with aesthetics—that should be the principal focus of philosophy. Yet even those "refin'd *Speculations*" that lead Descartes to restrict truth to those certainties with which the *cogito* is left after having subjected appearances to the most rigorous, methodical, and indeed gratuitous doubt support Shaftesbury's claims concerning the primacy of life. Viz., the very quest for an error-proof *demonstration* of the *cogito*'s certainties can only be conceived as a philosophical project by a philosopher who implicitly knows himself to be already enmeshed with the reality of other minds. As we shall see, Shaftesbury thus regards the Stoic genre of the *Meditation* or *Soliloquy,* no less than its Christian successor genre of *Confession,* as intrinsically social and dialogic.

Likewise, Shaftesbury flat-out rejects the Lockean concept of self-consciousness as a merely derivative property contingent on (and delimited by) the empirical state in which it unfolds, such that "*Socrates* asleep, and *Socrates* awake, is not the same Person."[44] By declaring any two discrete mental *states* to be inherently discontinuous, indeed incommensurable, Lockean empiricism had imprisoned itself within "a distracted Universe." Loss of personhood goes hand in hand with loss of world, the inevitable result of both being a pervasive sense of disaffection and anomie: "how little dispos'd must a Person be, to love or admire any thing as *orderly* in the Universe, who thinks the Universe it-self a Pattern of *Disorder* . . . Nothing indeed can be more melancholy, than the thought of living in a distracted Universe" (*SC,* 2:40). Against this dystopic vision of a world peopled by creatures "morose, rancorous and malignant" (*SC,* 2:47), Shaftesbury seeks to reaffirm a finely spun web of inner and outer relations and continuities, a type of "*Moral Arithmetick*" (*SC,* 2:99) that means to affirm the balance, proportion, and harmony of the person's subjective faculties. Virtue is genuinely possible as long as the microcosm of the human individual and the macrocosm of nature and society have been brought into fortuitous and enduring alignment: "to deserve the name of *good* or *virtuous,* a Creature must have all his Inclinations and Affections, his Dispositions of Mind and Temper, sutable, and agreeing with the Good of his *Kind,* or of that *System* in which he is included, and of which he constitutes a part" (*SC,* 2:45).

44. Locke, *Essay,* Bk. II, Ch. 1, §11.

Striking and consequential for the turn taken by moral philosophy throughout the eighteenth century is Shaftesbury's overtly aesthetic concern with the *form* of relations, as well as his notable indifference to the actual content or semantic value of the terms on which his philosophy relies. Indeed, aesthetics for Shaftesbury "occupies the central position of the whole intellectual structure." The quintessentially Platonic view that "everything real partakes of form, that it is no chaotic amorphous mass, but possesses rather an inner proportion and evidences in its nature a certain structure, and in its development and motion a rhythmic order and rule" is, for Shaftesbury, the basic premise, indeed the condition of possibility for all rational agency.[45] Rationality and virtue are indeed a kind of "Moral Arithmetick" inasmuch as "to stand thus well affected, and to have one's Affections *right* and *intire*" (*SC*, 2:45) requires the cultivation of formal harmony and inner beauty. Just as important, however, weighs the fact that in so identifying life with the dynamic and proportionate operation of form, Shaftesbury advances his defense of reason against Hobbes's voluntarism and Gassendi's materialism (both informing Locke's dispersal of the self into a myriad discontinuous, empirical states) to the near exclusion of any identifiable, normative good. In merging reason into the continuity of the "intire Affection"—what Kant's third *Critique* would later elaborate as the "proportionate accord" (*proportionierte Übereinstimmung*) of the faculties that is the basis for all rational judgment—Shaftesbury has effectively secured the coherence of reason only by evacuating it of all content.

The second part of the *Inquiry* thus features a steady stream of references to the "constant relation" between the inner disposition of creatures and "the Interest of *a Species*, or *common Nature*" (*SC*, 2:45), to "the Order or Symmetry of this *inward Part*" (*SC*, 2:48), and to the fundamental integrity, continuity, and self-identity of the inner life. Shaftesbury calls it "*intire Affection*" (*SC*, 2:64) and is quick to link it to the moral "integrity" of consciousness. His further contention that "to have this Intire Affection or Integrity of Mind, is *to live according to Nature*" (*SC*, 2:65) also means to stress the autonomy of the human individual and, in so doing, all but severs the ties between virtue and organized religion. Here again we see how and why the concept of the will all but disappears from the scene so abruptly after Hobbes and why, among eighteenth-century thinkers, only Kant will once again accord it a pivotal role in moral reasoning. For Shaftesbury, the harmonious ordering of the affections unfolds as a strictly *immanent* process, and it is not to be supervised by or answerable to the superego of any institution or transcendent deity. Just as to suppose that "*Religious Affection* or *Devotion* is a sufficient and proper remedy" depends on contingent and notably extraneous factors ("'tis according as the Kind may hap-

45. E. Cassirer, *Enlightenment*, 152.

pily prove"), so any overriding concern with "self-inspection" is liable to be a mere show of virtue ("the vainer any Person is, the more he has his Eye inwardly fix'd upon himself" [*SC*, 2:67–68]). Clearly unsympathetic to clericalism, mechanical habituation, and the external authority of religious institutions, Shaftesbury is a perfect example of what Charles Taylor has recently analyzed as the "buffered, anthropocentric identity" by means of which modern thought seeks to identify and take possession of "a secure inner mental realm" against the inherited order of myth, magic, and enchantment. Moreover, Shaftesbury is an early and highly influential case of what, for Taylor, constitutes "a growing category of people who, while unable to accept orthodox Christianity are seeking some alternative spiritual sources."[46]

Subtly advancing his quest for a position of intellectual and affective autonomy that wishes to eschew Hobbes's bleak finitude, Shaftesbury thus stresses the inner coherence and harmony of the mental life—a virtual reality in which affective and conceptual experiences continually and, in his view, fortuitously merge. Thus he urges that the "*Solutio Continui,* which bodily Surgeons talk of" (*SC*, 2:49) should also be recognized as the premise for any understanding of the inner life. Up to a point, his position in this regard anticipates Coleridge's late, neo-Platonist view of human personhood as constitutively self-aware and continuous, a claim most palpable where the operation of conscience is involved.[47] For Shaftesbury—herein diametrically opposed to Hobbes and Locke—it is axiomatic that "every reasoning or reflecting Creature is, by his Nature, forc'd to endure the *Review* of his own Mind, and Actions; and to have Representations of himself, and his inward Affairs, constantly passing before him, and revolving in his mind" (*SC*, 2:69).[48] What he calls "the *united Structure and Fabrick of the Mind*" and the "necessary *Connexion* and *Balance* of the Affections"

46. *Secular Age*, 301–302. As Dupré notes, "what distinguishes Shaftesbury from most other deists is that, for him, religion is more than a matter of reason" and that, in his largely affective account of religion, he "anticipates Schleiermacher's romantic theory of religion" (*Enlightenment*, 249); likewise, E. Cassirer remarks how for Shaftesbury "concentration on the nature of the Absolute is now replaced by a complete analysis of the formative forces within the ego" (*Enlightenment*, 152). As regards Shaftesbury's account of God, the most detailed remarks are to be found in "The Moralists," esp. *SC*, 2:119–123, 2:159–166, and 2:201–205. Rivers's characterization of Shaftesbury as "the perfect Stoic theist" urging the compatibilism of the "*general* MIND" with the "*particular* MIND" (*SC*, 2:201) seems correct; see also his essay "Deity" in *The Philosophical Regimen* (in *Life*, 13–39).

47. For Shaftesbury, what "is alone properly call'd CONSCIENCE" is not produced by some external, institutional, or metaphysical super-ego: "for to have Awe and Terror of the Deity, does not, of it-self, imply Conscience." The converse is true—viz., that "*religious Conscience* supposes *moral* or *natural Conscience*" (*SC*, 2:69).

48. Ibid.; regarding Shaftesbury's break with his onetime tutor, Locke, see Schneewind who, following Robert Voitle's biography (*The Third Earl of Shaftesbury, 1671–1713*), locates the beginnings of that break during the time that Shaftesbury composed the *Inquiry* in the late 1690s (*Invention*, 296n).

amounts to an ontological framework; it must not be misconstrued as a mere "accident" or as some metaphysical inference ventured about the constitution of the mind. In other words, it is not a "certain" proposition but a case of phenomenological certitude that has Shaftesbury affirm that "we cannot doubt of what passes *within our-selves.* Our Passions and Affections are known to us. *They* are certain, whatever the *Objects* may be, on which they are employ'd" (*SC*, 2:99). It is telling, in this regard, that Shaftesbury's aesthetic intuitionism also leads him to reject Locke for his supposed failure to appreciate the harmony of the cosmos and the inner life: "Had Mr. Locke been a *virtuoso,* he would not have philosophized thus. For harmony is beauty."[49]

The alleged shortcomings of Locke's aesthetic sensibility notwithstanding, Shaftesbury's position remains problematic and elusive in several respects. For one thing, in claiming that "our Passions and Affections are *known* to us" Shaftesbury leaves the epistemological and conceptual status of such knowledge undetermined. The most plausible hypothesis here is that he means to posit a "moral sense"—an inherently latent awareness of the self as *capable of achieving* inner "harmony." What is known so unconditionally here is but the "certitude" of an inner "feeling." Rationality thus seems vindicated, albeit in a strictly intuitionist and prospective manner and at the price of being denuded of all actual content. Once recalibrated as a formal-aesthetic "sense"—that is, as an inner experience that acquires phenomenological distinctness both *in* and *through* our relations with others—rationality can no longer furnish a plausible *foundation* for the commonwealth. Instead, it is reconstituted as but a (contingent) benefit of patterns of sociality such as happen to transpire within an *already* established community. Reason thereby has been transmuted into an eventuality, a utopia that can only fade into view incrementally, and only if the formation and practices of individual moral character are guided by its continual anticipation.[50] The elusive, shape-shifting style of *Characteristicks* compounds Shaftesbury's evidently ambivalent outlook on the metaphysical and theist underpinnings of virtue. As the divided and vexed responses of contemporaries to his work confirm, "the pleasure of reading *Characteristicks,* intended to entice a reluctant, philosophically ignorant audience into the pursuit of virtue, became instead an end in itself, a substitute for moral thought." The key question thus becomes how to appraise Shaftesbury's fusion of traditional concepts (self, virtue, will, etc.) with a notably modern, experimental outlook on philosophy as literature: "is this fusion to be interpreted rhetorically . . . or is it a more fundamental attempt to unite ethics and aesthetics?"[51]

---

49. Letter of 7 November 1709 (*Life,* 416).

50. As Schneewind notes, "Shaftesbury in fact leaves the door open to skepticism by insisting that the capacity to appreciate moral harmony is as much in need of education or training as the capacity for aesthetic judgment" (*Invention,* 305).

51. Rivers, *Reason,* 2:113, 115. Berkeley's vehement objection to Shaftesbury's motto, πάντα ὑπόληψις ("All is Opinion" [*SC,* 2:233]), as the expression of a capricious, self-indulgent subjectiv-

A second, closely related issue concerns the highly speculative argument that aesthetic education and a growing versatility in our relation with *form* will produce the inner harmony that Shaftesbury associates with virtue and morality. Numerous questions are left unanswered by Shaftesbury's account, such as how the beautiful and the merely "agreeable" are related; how the mind advances from the mere contemplation of beauty to the approval of the moral meanings allegedly embodied in it; and how one is to know whether those meanings themselves, or indeed the beauty allegedly mediating them, are more than "an adventitious projection."[52] From Shaftesbury onward via §59 of Kant's third *Critique,* Schiller's *Letters on the Aesthetic Education of Mankind,* and Goethe's *Wilhelm Meister* and writings on the "Metamorphosis of Plants," all the way to its nostalgic obituary in Allan Bloom's *Closing of the American Mind,* this integrative, Platonist narrative of *Bildung* has been advanced countless times. Yet time and again this idea also found itself challenged by a radically disjunctive skepticism that makes the reality of every particular thing or state contingent on our ability to identify the efficient cause that (supposedly) brought it about.

By its very design, however, Shaftesbury's *Characteristicks* neither can nor means to offer such proof. For his vision of moral personhood produced by an aesthetic *Bildung* at once highly mediated and uncertain as regards its anticipated dividends had been shaped precisely by Shaftesbury's opposition to Hobbes's and Locke's deterministic account of mental life. There is indeed good reason, then, to suppose that for Shaftesbury "a *problem* of selfhood need not be understood as a *dilemma* of selfhood" and that in his writings, herein echoing the ancient philosophers he so admired, "the 'self' was more of an achievement than a 'given thing.'"[53] In his later dialogue, *The Moralists: A Philosophical Rhapsody* (1709), Shaftesbury thus strikes a highly self-conscious, oratorical pose as Philocles sketches for Palemon how rational culture and civil order arise aesthetically and progressively widen in scope:

---

ism rather deliberately ignores the phrase's Stoic origins in Marcus Aurelius and, thus, its evident advocacy of "a rigorous mental and moral discipline." Still, Rivers is right to note how "Shaftesbury's excessive self-consciousness about what he is doing in *Soliloquy* deprives his example of practical value" (120), a point also made by Philip Skelton, who in 1744 expresses vexation at how Shaftesbury "so refines the plain and intelligible Science of Morality, that it is impossible for his Reader to find out its foundation, to distinguish, whether it is seated in the rational, or sensitive Part of our Nature" (quoted in Rivers, *Reason,* 2:121).

52. Darwall proceeds to argue that "a number of important Shaftesburean strains come together at just this juncture—his theology, philosophy of nature, and notions of enthusiasm, love, creative inspiration, sympathy, and mind," and that "the resulting doctrine of moral sense ... is far from the empiricist moral sentimentalism of Hutcheson and Hume." In fact, Shaftesbury's is "a *rationalist* theory of moral sense" (*British Moralists,* 187).

53. Den Uyl, "Shaftesbury," 280. While Den Uyl may well be right to identify the *Soliloquy,* rather than the *Inquiry,* as the "central text for Shaftesbury's moral philosophy ... because it alone gives us the method of moral improvement" (283), a fuller account of Shaftesbury's overall oeuvre would quickly exceed the focus and purposes of the present argument.

> Here, in my turn, I began to raise my Voice, and imitate the solemn way you had been teaching me. "*Knowing* as you are," continu'd I, "*well-knowing* and experienc'd in all the Degrees and Orders of Beauty, in all the mysterious Charms of the particular Forms; you rise to what is more general ... Nor is the Enjoyment of such a single Beauty sufficient to satisfy such an aspiring Soul. It seeks how to combine more Beautys, and by what Coalition of these, to form a beautiful Society. It views Communitys, Friendships, Relations, Dutys; and considers by what Harmony of particular Minds the general Harmony is compos'd and *Commonweal* establish'd. Nor satisfy'd even with publick Good in *one* Community of Men, it frames it-self a nobler Object, and with enlarg'd Affection seeks *the Good of Mankind.* It dwells with Pleasure amidst that Reason, and those Orders on which this fair Correspondence and goodly Interest is establish'd. Laws, Constitutions, civil and religious Rites; whatever civilizes or polishes rude Mankind; the Sciences and Arts, Philosophy, Morals, Virtue; the flourishing State of human Affairs, and the Perfection of human Nature; these are its delightful Prospects, and this the Charm of Beauty which attracts it." (*SC,* 2:120)

This remarkable flourish of affirmative rhetoric lies midway between Seneca's *Moral Epistles* and Kant's critical utopia in the third *Critique,* particularly in his account of "aesthetic ideas" as a "representation of the imagination which occasions much thought, without however any definite thought, i.e. any *concept,* being capable of being adequate to it" (*Critique of Judgment,* 157). Indeed, a faint echo of Shaftesbury's idea can still be detected in Stendhal's definition of beauty as "a promise of happiness" (*une promesse de bonheur*).[54] Closer to home, this meliorist view also underwrites Edmund Burke's insistence that the cohesion and durability of any political community is necessarily anchored in its aesthetic appeal: "to make us love our country, our country ought to be lovely." Burke expands this view by arguing that to so conceive of the state as a communal, ritual artifact is to acknowledge God's will. Beauty thus is extolled as an "oblation of the state itself," a ritualized staging of a people's moral *cum* political commitments to "be performed as all public solemn acts are performed, in buildings, in music, in decoration, in speech, in the dignity of persons, ... with mild majesty and sober pomp." In what may well be an inadvertent hint at his widely suspected Catholic sympathies, Burke thus justifies the idea of a polity suffused with aesthetic and affective appeal as "public consolation."[55]

54. Stendhal's aphorism (*La beauté n'est que la promesse du bonheur*) appears in Chapter 17 of *De l'amour* (1822).

55. *Reflections,* 172, 196–197. On the role of aesthetics in Burke's *Reflections,* see Furniss, *Burke's Aesthetic Ideology,* 113–196; Pfau, *Wordsworth's Profession,* 275–302 and, on the defensive (at times paranoid) logic of Burke's aesthetic politics, ibid., 84–91. As is the case in Burke, so "Shaftes-

However inadvertently, Burke's contention that community can only be realized and sustained as a beautiful artifact cannot but point back to a fallen, ectypal world of self-interested, amoral, and disaggregated selves. Such an aesthetic philosophy aims to compensate for modernity's conspicuous inability to imagine a politics that is attuned to a notion of the good. As in Walter Benjamin's account of Baroque melancholy and Theodor Adorno's lifelong attempt to chart the relation of art to a quintessentially "damaged life," Shaftesbury's ironic and Burke's pessimistic appraisal of aesthetics and the beautiful as "consolation" for the terminal loss of the good reveals modern moral philosophy's deep estrangement from a world that should never have turned out the way it did. Like Burke some eighty years later, Shaftesbury also finds himself "promoting [his] central vision in a world not structurally or intellectually hospitable to it."[56] His mournful outlook on a modern (voluntarist) model of agency and the stunning costs of its apparent triumph is signaled in his carefully chosen classical epigraphs and in the sophisticated emblematic frontispieces for his various essays in the *Characteristicks*. Furthermore, the above passage also shows Shaftesbury's quasi-emergentist idea of reason as a highly contingent narrative progression that markedly differs from the determinate and normative view of being as *entelecheia* found in Aristotle's *Physics* and *Metaphysics*. Like Plato, Cicero, and a number of the Stoic thinkers whom he revered (Epictetus above all), Shaftesbury repeatedly insists on the strong "connection between philosophy and character development."[57]

In his enjambment of moral community with aesthetic education—subsequently writ large in the works of Johann Joachim Winckelmann, Johann Gottlieb Herder, Friedrich Schiller, and Johann Wolfgang von Goethe—Shaftesbury's oratory acknowledges the hazards of pursuing a pedagogical project in a world dominated by calculative, instrumental, and abstract notions of rationality. Being equally opposed to "metaphysical speculation, scientific investigation, Hobbesian self-interest, Lockean relativism, the atheistic tendency in freethinking, [and] the mercenary ethics of Anglicanism," Shaftesbury's own thought is profoundly dialectical in its response to intellectual models he deems exhausted, ineffectual, or morally compromised.[58] Writing during a period when the division of moral from political authority seems extreme, his arguments are imbued with a strong anti-institutional

bury's epistemology is psychological, and it depends on a fusion of the vocabularies of ethics and aesthetics, the good and the beautiful, which is ultimately Platonic in origin" (Rivers, *Reason*, 2:141).

56. Den Uyl, "Shaftesbury," 277.

57. Ibid., 285; it cannot surprise that Shaftesbury's aesthetic approach to virtue ethics insists on the primacy of *praxis* over theory and, like Plato and Aristotle, posits that "theoretical insight into moral conduct is in many significant respects the result of, not the precondition for, proper character formation" (289).

58. Rivers, *Reason*, 2:86.

and anti-systematic energy.[59] Crucially, this opposition to "system" prompts Shaftesbury to criticize his own writings, in particular the "*dry* PHILOSOPHY and *rigid Manner*" of his own earlier *Inquiry,* whose author his later "Miscellaneous Reflections" depict "as a plain *Dogmatist,* a *Formalist,* and *Man of Method;* with his Hypotheses tack'd to him and his Opinions so close-sticking, as wou'd force one to call to mind ... some precise and strait-lac'd Professor in a University" (*SC,* 3:117, 84). For Shaftesbury, who had not attended the university and who repeatedly expressed his dismay at how "we have immur'd her [Philosophy] (poor Lady!) in Colleges and Cells," both the analysis and practice of rational self-cultivation cannot possibly be ceded to "mere *Scholasticks*" (*SC,* 2:105). On the contrary, no longer tied to the flourishing of institutions, and quite deliberately exempted from the normative (and allegedly illiberal) authority of ecclesial and secular political discourse, reason is to be achieved by way of an infinitely reflexive and ironic cultivation of the individual as he or she interacts with a rich array of expressive forms.

The anti-ecclesial and anti-institutional pathos of Shaftesbury's *Characteristicks* hints at the beginnings of what by the end of the century emerges as the middle-class program of *Bildung,* pursued with either ironic self-awareness or philistine earnestness. As regards these latter alternatives, Shaftesbury clearly inclines toward an ironic stance. Nearly a century before Friedrich Schlegel was to make irony the linchpin of his Romantic poetics, Shaftesbury's *Soliloquy* (1709) at once advances and undermines the idea of aesthetic pedagogy in an intensely reflexive, opening meditation on whether the study of beauty can ever lead to moral improvement. In a letter of 1712, Shaftesbury identifies four basic literary genres (the demonstrative, epistolary, miscellaneous, and dialogic) as his main resources, leaving out the soliloquy, which he regards "as an aspect of dialogue." At pains to distinguish his own practice "from the Christian tradition of meditation that [he] despises," Shaftesbury views being "*a good Thinker*" all but coterminous with "being a strong *Self-Examiner,* and *thorow-pac'd Dialogist,* in this solitary way" (i.e., of the genre of soliloquy).[60]

Though partial to Epictetus, Shaftesbury may well be echoing Seneca's Epistle 38 ("On Quiet Conversation"), in particular, the author's advice to Lucilius that the entire Stoic regimen of rational self-examination is best served by a process of dialogue, rather than by abstract lessons unilaterally imparted or philosophical reading pursued in seclusion: "The greatest benefit is to be derived from conversation, because it creeps by degrees into the soul [*quia minutatim increpit animo*] ... Words should be scattered like a seed; no matter how small the seed may be, if it has once found

---

59. On the political, cultural, and religious milieu of late seventeenth-century England, see Brewer, *Pleasures,* esp. 3–55.

60. Rivers, *Reason,* 2:106, 108. The 1 September 1712 letter to Micklethwayte is also quoted there.

favourable ground, it unfolds its strength and from an insignificant thing spreads to its greatest growth. Reason grows the same way [*Idem facit ratio*]." A distant echo of Plato's theory of knowledge as a gradual "awakening" (*anamnēsis* [*Meno,* 81^{b}–87^{d}]), the "Stoic idea of learning is an idea of increasing vigilance and wakefulness" whereby reason constitutes itself successively, and inductively, through the methodical evaluation of impressions and our responses to them.[61] In sharp contrast to the top-down model of instruction practiced by the Epicureans, the dialogic model succeeds by inducing its participants to embrace the provisional, evolving, and partial status of their representations, and to form the habit of suspending assent to all impressions. Habits and routines of this type should not be misinterpreted (pace Foucault) as sociopolitical constraints or non-cognitive mechanisms conspiring against some woolly-headed notion of the human. Rather, they are techniques aimed at enabling the individual to participate in the *logos* and, thus, to become a genuine "citizen of the world [*politēs tou kosmou*]."[62] Yet as with the political utopia sketched in Kant's late writings, themselves landmark documents of neo-Stoicism, the method that is to take us there remains forever a work in progress. Indeed, to be true to its spirit, it must also be turned against the dialogic framework whereby Seneca, Cicero, and Epictetus proffer exemplars of their philosophy for their respective disciples.

So there lies the rub: the very genre of dialogue (which in Shaftesbury is barely masquerading as *Soliloquy*) demands an acutely and ongoing hermeneutics of suspicion. Indeed, as Shaftesbury himself points out with great delight and wit, the genre's established conventions—merely by virtue of being recognizably that, conventions—inherently conspire against the task of pushing introspection to the point of genuine discovery. In the end, he realizes, his persistent cultivation of a uniquely ironic and prevaricating rhetoric of introspection may not prove an effective remedy either to the Roman "leprosy of Eloquence." For the genre of the soliloquy is likewise compromised, if not by an author's designs *on* then certainly by his professed indifference *to* the reading public. The writer dispensing advice about how to write is like "*an Empirick* talk[ing] of his own Constitution, how he governs and manages it, what Diet agrees best with it," and so forth. Yet to the reader there is nothing appealing about "the experimental Discussions of his practising Author, who all the while is in reality doing no better, than taking his Physick in publick." Moreover, as an often exhaustively detailed transcript of what "passes in this religious Commerce ... between them and their Soul" would seem to suggest, those indulging "this *self-discoursing Practice*" often pursue their own moral self-discipline with gluttonous and exhibitionist fervor; indeed, they are "a sort of *Pseudo-Asceticks*" (*SC,* 1:101–105). A fuller

61. Nussbaum, *Therapy,* 340.
62. Epictetus, *Discourses,* 2.10.3.

reading of Shaftesbury's *Soliloquy* than can here be pursued would show his intense awareness of the insoluble dialectical entanglement of a hypostatized inner life deemed capable of Platonic ascent toward the good and an inventory of rhetorically over-determined forms and genres such as will invariably compromise that very ideal even as they serve to give expression to it. Yet far from being a logical impasse or metaphysical dilemma, this manifest conflict between medium and message constitutes both an acute challenge *to* and an intriguing thematic proposition *for* Shaftesbury's aesthetic cum moral project. It is only through a continued wrestling with foreign-determined and wayward rhetorical and aesthetic forms that moral character and vision can come into relief and, perhaps, even into positive alignment. Only in Kant's third *Critique* will this central insight of Shaftesbury be fully rehabilitated.

10

FROM NATURALISM TO REDUCTIONISM

Mandeville's Passion *and Hutcheson's* Moral Sense

Before exploring how Shaftesbury's "moral sense" theory is consolidated by Francis Hutcheson and, eventually, critiqued in Adam Smith's *Theory of Moral Sentiments,* some consideration will have to be given to Mandeville's revival of Locke's anti-metaphysical conception of the will, viz., as a strictly empirical and unrelentingly hedonistic "passion." First published in 1714, his *Fable of the Bees* greatly expands the critique of virtue ethics that Mandeville had first presented in his 1705 satirical poem, "The Grumbling Hive." Beginning in 1723, the *Fable*'s second edition secured it and its author considerable notoriety, including its presentation by the Grand Jury of Middlesex as a public nuisance. Like Joseph Addison's and Richard Steele's *Tatler* and *Spectator,* Jonathan Swift's satires, Daniel Defoe's economic writings, Joseph Butler's *Analogy* and, eventually, David Hume's *Essays,* Mandeville's *Fable* attests that during the early decades of the eighteenth century "Britain's intellectual sphere had turned into a competitive market for ideas, in which logic and evidence were becoming more important and 'authority' as such was on the defensive." For the first time, "the word *innovation,* traditionally a term of abuse, had become a word of praise."[1] Further editions, again revised and enlarged, followed in rapid

1. Mokyr, *Enlightened Economy,* 31; Gay, *Enlightenment* (vol. 2), 3.

succession (in 1725, 1728, 1729, 1732, and 1733, the year of Mandeville's death), along with translations into French (1740 and 1760) and German (1761). Even before Mandeville, the Jansenist Pierre Nicole (1625–1695) had already hinted at the possible alignment of virtue and self-interest in an essay "Of Charity and Self-Love" and, in so framing the question, had prepared the ground for a strictly secular conception of virtue. Characteristically, Nicole forgoes the Latinate *virtū* in favor of the French *honnêteté,* a word "evocative of the tradition of courtly civility, of outward courtesy with its preoccupation with form."[2] The shift toward a more empirical tone and, consequently, toward a more skeptical or outright suspicious appraisal of virtue coincides with a deliberate retreat from moral philosophy as a metaphysical enterprise. In its place arise the eighteenth-century's twin preoccupations with the anatomy and classification of contingent and frequently overlapping types of sentiment, and the quasi-behaviorist analysis of autonomous selves as they traverse rapidly changing social spaces in the manner of Democritus's atoms.

Few eighteenth-century works offer a more vivid illustration of what Roy Porter has characterized as the "switch from asking how can I be good to how can I be happy" than Mandeville's *Fable of the Bees.*[3] To which it should be added that the *Fable* effects yet another crucial shift, viz. from a conception of knowledge as intrinsically "responsible" and deeply implicated in some metaphysical framework or other, to one of knowledge as efficient causation and as a quasi-instinctual instrument of "power." The latter view, of course, was hardly new and had already been advanced in those very terms by Hobbes. Yet beginning with the economic manifestos and moral polemics of the early eighteenth century, intellectual argument seems to be asphyxiated by its excessive commitment to deductive reasoning. Increasingly, being committed to one view *eo ipso* means rejecting, rather than dialectically engaging, other perspectives. The quest for economic advancement, which as J. G. A. Pocock has shown was bound to transform the very substance of personality by a proliferation of *speculation, innovation,* and *ingenuity* (all terms undergoing rapid reappraisal after 1700), appears all but incommensurable with languages counseling moral and ethical self-discipline. Still, there is no shortage of religious, political, and rhetorical institutions, modes, and genres urging the importance of virtuous self-governance and extolling a civic-republican commitment to the common weal. In-

2. Herdt, *Putting on Virtue,* 255. As Herdt goes on to note, "there is little sense of the Christian as loved by God and summoned by God into the fellowship of the divine life. Where the capacity for true virtue is understood as contingent on the reception of supernatural grace *and* where ordinary participation in the sacramental means of grace does not lend assurance that this grace has been given, the stage has been set for anxiety and suspicion directed at the apparently graced and apparently virtuous" (256). On Nicole, see ibid., 249–261, and also Schneewind, *Invention,* 275–279.

3. Porter, "Enlightenment in England," 14.

vectives against crudely self-interested, speculative, and rapacious behavior continue to issue from a variety of religious and political communities, ranging from high Anglicanism to the anti-capitalist rhetoric of Puritan, millenarian, or militantly secular radical subcultures, yet also from conservative landed interests alarmed by Britain's imperial ambitions, its involvement in the slave trade, the accelerating dissolution of familial, local, and regional ties, the growing clamor for political "reform," and/or the crudely utilitarian stance toward the "two-legged livestock" of the urban and rural poor. Eventually, Burke, Blake, Priestley, Wordsworth, and the young Coleridge will offer powerful, albeit strikingly diverse arguments against the nominally Anglican apologists of laissez-faire economics (William Paley, Thomas Malthus, Richard Watson, Jeremy Bentham). Yet to the enterprising and growing ranks of the "middling sort," which Isaac Kramnick has aptly characterized as the "bourgeois radicals" (Richard Price, Priestley, Godwin, Mary Wollstonecraft, Joseph Johnson, et al.), the claim for economic opportunity, social mobility, and personal flourishing as bona fide "rights" went hand in hand with a vocal and often inspired critique of established religion as gratuitously authoritarian, doctrinally stale, and prejudicial to those very rights.[4]

In rather prescient ways, Mandeville's strident and uncompromising arguments pointed early on to the apparent incommensurability of economic success and moral self-legitimation, a conflict that was to dominate social, economic, and (to a lesser extent) moral theory throughout the century. On the face of it, the *Fable*'s proto-existentialist critique of virtue ethics offers a powerful revival of Hobbesian psychology. Yet Mandeville significantly reorients that older model toward a rapidly changing discursive landscape in which more than ever social ties and moral norms appear contingent on economic relations, even (or, perhaps, especially) where organized religion might dispute or purport to transcend such a connection.[5] For Mandeville,

---

4. Literature on the rapidly diversifying demographic and ideological profile of eighteenth-century Britain is predictably rich. On the emergence of the middle class, see Wahrman, *Imagining the Middle Class,* esp. 31–73; Thompson, "Eighteenth-Century Society"; and Porter, *English Society in the Eighteenth Century,* esp. 48–97. On the Unitarian apologists for an entrepreneurial, innovative middle-class community, see Kramnick, *Republicanism,* esp. 43–98; on the ideological effects of the commercial and financial revolutions of the eighteenth century, see esp. Pocock, *Virtue,* 103–124 and 230–246; on the radical millenarian fringe, see E. P. Thompson, *Making of the English Working Class,* 26–101, and *Witness,* 10–64; on secular protest movements of the late eighteenth and early nineteenth century, see McCalman, *Radical Underground,* esp. 7–94.

5. Literature on the economic (and financial) revolution unfolding in Britain after 1700 is abundant. Among the more concise, general accounts of note are Porter, *English Society,* 185–250; Colley, *Britons,* 55–100; K. Polanyi, *Great Transformation,* 33–76; for a material history of modern consumerism, Brewer and Plumb, *Birth of a Consumer Society;* on technological innovation and the growing emphasis on "useful knowledge," see Mokyr, *Enlightened Economy,* esp. 17–62; on the affective and discursive nature of the eighteenth-century *homo economicus,* see Rothschild, *Economic*

there is no question that Britain's intellectual superstructure is de facto conditioned by an accelerating and diversifying quest for economic gain ceaselessly pursued by so many possessive individuals. The basic argument of Mandeville's *Fable,* then, is familiar enough and easily summarized. It holds that strict moral self-governance and a quest for virtuous conduct systematically chokes off an individual's congenital desire for economic gain, self-improvement, and a materially constituted vision of happiness. Conversely, "frugality" is being unmasked as Protestantism's low-key version of the Scholastic virtue of *temperantia;* it is but "a mean starving Virtue, that is only fit for small Societies of good peacable Men, who are contented to be so poor they may be easy; but in a large stirring Nation you may have soon enough of it. 'Tis an idle dreaming Virtue that employs no Hands, and therefore very useless in a trading Country" (*MFB,* 1:104–105). Opening a long, programmatic section, Mandeville thus asserts "that no Society can be rais'd into such a rich and mighty Kingdom, or so rais'd subsist in their Wealth and Power for any considerable Time, without the Vices of Man" (*MFB,* 1:229). This is not to claim, however, that all vices are equally conducive to the common good, nor indeed that they should necessarily be allowed to flower unchecked. As regards the widely supposed correlation between virtue and affluence, the *Fable* tends to keep its readers off balance by switching back and forth between scandalous hypotheses (e.g., that "all the worldly Interest of the Nation consists in, depends entirely on the Deceit and vile Stratagems of Women") and pious affirmations ("I have always without Hesitation preferr'd the Road that leads to Virtue" [*MFB,* 1:229, 231]). Such startling reversals, aside from keeping his readers uncertain as to the book's ultimate intent and thus slowly estranging them from their own moral presuppositions, ultimately reflect Mandeville's appraisal of virtue itself as an untrustworthy and equivocal phenomenon.

In a series of ingenious metaphors aimed at illustrating the tension "between Real, and Counterfeited Virtue" (*MFB,* 1:230), Mandeville elaborates the contrast between the strictly ornamental and lifeless nature of virtue and the inside of a living human body, teeming with raw and implacable passions and desires:

> Where would you look for the Excellency of a Statue, but in that Part which you see of it? 'Tis the Polish'd Outside only that has the Skill and Labour of the Sculptor to boast of; what's out of sight is untouch'd. Would you break the Head or cut open the Breast to look for the Brains or the Heart, you'd only shew your Ignorance, and destroy the Workmanship. This has often made me compare the Vir-

---

Sentiments, esp. 7–51 and 116–156. On the first-generation Romantics' equivocal response to modern political economy, one that stresses the continuity of the past and the uses of second nature qua habituation, see Chandler, *Wordsworth's Second Nature,* 156–215; Pfau, *Romantic Moods,* esp. 191–225; and Liu, *Wordsworth,* 311–358.

tues of great men to your large *China* Jars: they make a fine Shew, and are Ornamental even to a Chimney … but look into a thousand of them, and you'll find nothing in them but Dust and Cobwebs. (*MFB,* 1:168)

Could we undress Nature, and pry into her deepest Recesses, we should discover the Seeds of this Passion [sexual desire] before it exerts itself, as plainly as we see the Teeth in an Embryo, before the Gums are form'd … among well-bred People it is counted highly Criminal to mention before Company any thing in plain Words, that is relating to this Mystery of Succession: By which Means the very Name of the Appetite, tho' the most necessary for the Continuance of Mankind, is become odious, and the proper Epithets commonly join'd to Lust are *Filthy* and *Abominable.* (*MFB,* 1:142–143)

Characteristic of the modern anatomist's outlook on virtue as a kind of moral theater or inauthentic surfeit, Mandeville's trope of the body also reveals vestiges of Puritanism's distrust of art, beauty, outward proportion, and allegorical images—precisely the structures of mediation on which Shaftesbury's moral cum aesthetic pedagogy so centrally relies. Notably, though, Shaftesbury himself had pointed to the apparent enigma of "our own Minds" and how difficult it is to "understand what our main *Scope* was … Our Thoughts have generally such an obscure implicit Language, that 'tis the hardest thing in the world to make 'em speak out distinctly." Yet unlike Mandeville's *Fable,* the *Characteristicks* draw entirely different conclusions from this apparent dilemma. Rather than peremptorily discrediting the *existence* and *reality* of the inner life on account of its manifest failure to achieve authentic, transparent, and verifiable status for other minds, Shaftesbury insists that the challenge is to pursue the cultivation of one's aesthetic awareness and expressive gifts: "the right Method is to give 'em [our Thoughts] Voice and Accent" (*SC,* 1:107). Juxtaposing Søren Kierkegaard's vexed affirmations to Ludwig Wittgenstein's cool skepticism, Iris Murdoch captures the salient point quite succinctly: "This running up against the limits of language is *ethics.*" That the good seems to defeat each and every attempt to seize it in propositional form proves nothing, for what matters is the very effort to achieve moral articulacy: "the inclination, the running up against something, *indicates something.*"[6] With Shaftesbury, then, one must learn to ask (and to accept the impossibility of a definitive answer to) this question: "Tell me now, my honest Heart! Am I really *honest,* and of some worth? Or do I only make a fair show, and am *intrinsecally* [sic] no better than *a Rascal*? … Am I not then, at the bottom, the same as he? The same: an arrant Villain; tho perhaps more a Coward, and not so perfect in my kind. If

6. Murdoch, *Metaphysics,* 29.

Interest therefore points me out this Road; whither would *Humanity* and *Compassion* lead me?" (*SC,* 1:109).

Shaftesbury's *Soliloquy* here highlights a gap in modern thinking about moral agency that had steadily been widening. He seems keenly aware that, by his time, a modern, hyper-Augustinian tradition had begun to interpret the tentative and necessarily promissory nature of Platonic *ēros* and Christian soteriology as warrants for an unbridled epistemological skepticism. Consequently, it is no longer the *quality* of the person's inner life but its very *reality* that is under siege. If in the Augustinian-Thomist tradition moral agency had depended on a richly layered consciousness of one's own moral being whose strands could only be disentangled in the course of a person's life—understood as a narrative project of sorts—the tenuous epistemology of the inner (moral) awareness is now being construed (by Hobbes, Mandeville, and Hume) as prima facie evidence of its nonexistence. Unwilling to entrust to time and progress what cannot be conceptually secured in the present, Mandeville thus seizes on the equivocal nature of moral personhood and human self-expression so as to prescind the reality of the inner life altogether. Presaging Hume's opposition between the painter and the anatomist of virtue in the eloquent closing paragraph to his *Treatise of Human Nature,* and even glancing ahead to similarly dystopic arguments advanced in Part III of Nietzsche's *Genealogy,* Mandeville's *Fable* offers a sharply critical account of virtue as mere show and pretence on the part of weak individuals seeking to conceal their self-interest and, in so doing, ensnaring more robust psychological types in a normative morality at once self-effacing and paralyzing.[7] As he suggests, following the gradual establishment of a moral superstructure of virtuous self-denial, particularly resourceful individuals, realizing "that they might reap the Fruits of the Labour and Self-denial of others, and at the same time indulge their own Appetites with less disturbance, ... agreed with the rest, to call every thing, which, without Regard to the Publick, Man should commit to gratify any of his Appetites VICE; ... and to give the Name VIRTUE to every Performance, by which Man contrary to the impulse of Nature, should endeavour the Benefit of others, or the Conquest of his own Passions out of a Rational Ambition of being good" (*MFB,* 1:48–49).

Some sixty years later, the antagonism of outwardly virtuous and institutionalized (religious) forms of morality and the psycho-physical kernel of self-interest that they seek to repress makes its reappearance in some of Blake's *Songs.* Here the an-

7. "The anatomist ought never to emulate the painter; nor in his accurate dissections and portraitures of the smaller parts of the human body, pretend to give his figures any graceful and engaging attitude or expression. There is even something hideous, or at least minute in the views of things, which he presents; ... An anatomist, however, is admirably fitted to give advice to a painter" (Hume, *Treatise,* 395). On the modern anatomical critique of virtue ethics (in Gracián, Pascal, and Mandeville), see Herdt, *Putting on Virtue,* 221–280.

tithesis is troped as the bad faith and cunning of adults suddenly exposed by the guileless candor of children. The poem "A little Boy Lost" bears quoting in full:

Nought loves another as itself
Nor venerates another so.
Nor is it possible to Thought
A greater than itself to know:

And Father, how can I love you,
Or any of my brothers more?
I love you like the little bird
That picks up crumbs around the door.

The Priest sat by and heard the child.
In trembling zeal he siez'd his hair:
He led him by his little coat:
And all admir'd the Priestly care.

And standing on the altar high,
Lo what a fiend is here! said he:
One who sets reason up for judge
Of our most holy Mystery.

The weeping child could not be heard.
The weeping parents wept in vain:
They strip'd him to his little shirt.
And bound him in an iron chain.

And burn'd him in a holy place,
Where many had been burn'd before:
The weeping parents wept in vain.
Are such things done on Albions shore.[8]

The problem, as Blake's poem highlights, is that both organized religion *and* the rationalist critique of it seem to share the same, notably inexplicit framework or "mystery." It is the need to eschew such oppositional thinking (a symptom of what Blake calls "true friendship") which explains the poem's overwrought melodramatic

8. Blake, *CPP,* 28–29.

indictment of the "Priestly care" discharged in the closing image of an auto-da-fé. The poem's final question shows its central thematic proposition—a scene of grotesque institutional "discipline"—to oscillate between deliberate hyperbole, paranoid fantasy, and factual occurrence. In a world in which conventional morality and rationalist demystification either will not or cannot account for the motives and sources of their respective affirmations, the only alternative is a poetry that carefully stages the persistent interference between medium and message. For Blake, a thoroughgoing critique of the present, if it is to have spiritual legitimacy and truth, demands that aesthetic form transcend the notion of a "finite text, contained within a closed circuit of interpretation as defined by some cage of mutually illustrative (and hence reinforcing) words and images, but rather as virtual texts, constituted by, even suspended in, the indefinite gap between words and images."[9]

By contrast, Mandeville seems hardly troubled by the fact that his own indictment of virtue and any non-contingent account of the good rests on some utopian view from nowhere. Inasmuch as he fails to recognize the ideological assumptions coded into conventional satire, his 1705 depiction of the "grumbling hive" and the later, hypertrophic notes fleshing out that initial vision, Mandeville ought to be seen as a transitional figure between the Augustinian "anatomical" critique of virtue ethics that had preceded him (Blaise Pascal, François de La Rochefoucauld) and the nineteenth-century project of "genealogy." The latter, identified above all with Nietzsche, issued its challenge to what it took to be a widespread conformist "belief in the unity of truth and reason which excluded any possibility of the existence of radically incommensurable standards."[10] It is important to note, however, that Mandeville does not argue (as Nietzsche at times seems to) that morality and virtue had been specifically contrived at some particular moment in history. Rather, the gradual process of "improving upon [man's] Dread of Shame" came to be viewed as something that "might be greatly encreas'd by an artful Education, and be made superiour even to that of Death."[11]

Oddly enough, Mandeville's passions, though ostensibly atavistic and self-seeking, turn out to be intensely social and imitative: "We all look above our selves, and, as fast as we can, strive to imitate those, that some way or other are superior to us." Thus envy is said to be both "riveted in Human Nature" *and* to manifest itself as a life "spurr'd on by Emulation" (*MFB*, 1:129, 137). In so happening upon the mimetic nature of a weak and vain human psychology forever craving approval and

9. Makdisi, *William Blake*, 163.

10. MacIntyre, *Three Rival Versions*, 41. "From the standpoint of the genealogist the encyclopedist is inescapably imprisoned in metaphors unrecognized as metaphors" (ibid., 43); MacIntyre subsequently challenges the self-authorizing logic of Nietzsche's model of genealogy (ibid., 196–215).

11. *An Enquiry into the Origin of Honour and the Usefulness of Christianity in War*, 40.

dreading shame, society gradually "had made a Discovery of a real Tie" between the institutional mechanisms of moral discipline and the creation of a tractable, even predictable self. Yet as Mandeville insists, such processes "are the joint Labour of Many. Human Wisdom is the Child of Time. It was not the Contrivance of one Man, nor could it have been the Business of a few Years, to establish a Notion, by which a rational Creature is kept in Awe for Fear of it Self."[12] In seeking to expose both institutional morality and some inward commitment to virtue as a sophisticated pretense, Mandeville repeatedly invokes Locke's widely accepted model of a mind dominated by physiological sensation. Having received medical training as a specialist in nerve and stomach disorders (what he calls "the hypochondriack and histerick passions"), Mandeville remains largely committed to the "humour" theory of temperaments. Opening the *Fable,* he thus "believe[s] Man (besides Skin, Flesh, Bones, &c. . . .) to be a compound of various Passions, [and] that all of them, as they are provoked and come uppermost, govern him by turns, whether he will or no" (*MFB,* 1:39). As F. B. Kaye notes, "it is no great stride from the belief that the soul (rational principle) is dependent on the body for its existence to the belief that the rational faculty cannot help but be determined by the mechanism through which it has its being" (*MFB,* 1:lxxxiv–lxxxv).

Throughout the *Fable,* the body, and its strict metonymic relationship to the passions, thus constitutes the only uncontestable reality. Avowedly concerned with "meer Man, in the State of Nature and Ignorance of the true Deity" (*MFB,* 1:40), Mandeville readily slides into a hyper-Augustinian stance in which nature—both *through* man's actions and, subsequently, *in* his depraved moral constitution—appears inherently fallen and sinful. Yet inasmuch as the depraved and sinful constitution of the modern self stems from the mind's determinacy by passions and desires—and thus no longer involves judgment and choice—the Augustinian stance ultimately frays and unworks itself. What reappears behind it is the pre-Augustinian, latently Gnostic/Manichean view of "matter" (*physis*) as endlessly betraying "form" (*eidos*). Mandeville thus offers an apt instance of what Hans Blumenberg calls the "converted Gnostic," someone who "had to provide an equivalent for the cosmic principle of evil in the bosom of mankind itself. He found it in inherited sinfulness, as a quantity of corruption that is constant rather than being the result of the summation of individual faulty actions" (*LMA,* 53). Indeed, it is in his *Free Thoughts* (2nd ed., 1729) that Mandeville launches into a spirited, if rather belated defense of Marcion's Gnostic heresy, here coupled with an assault on 1,500 years of Christian

12. *Enquiry,* 41. In good Hobbesian language, the *Fable* likewise asserts that "the only useful Passion ... that Man is possess'd of toward the Peace and Quiet of a Society, is his Fear" and that "nothing civilizes Man equally as his Fear" (*MFB,* 1:206, 219).

theological reasoning on the nature of free will as it relates to divine predestination. Yet the arguments presented merely assert that God's deeper intent in having created a being with the potential for sin remains ineluctable.

For Mandeville—who fails to distinguish between predestination and foreknowledge—the only alternative, hardly more palatable, is to imagine mankind in the grip of a wholly determinative model of divine grace such as will rob "the blessed" of free will and "reduce the faithful to the condition of slaves."[13] With characteristic impatience, Mandeville thus declares it "needless to run thro' the several degrees, and the different systems that have been made of it [i.e., free will] by the *Pelagians, Semi-pelagians, Originists, Molinists, Synergists,* and *Arminians.*" Once again, Mandeville's sharp opposition to neo-Platonism and, more immediately, to Shaftesbury's thought does not take the form of a dialectical reading, that is, a point-by-point refutation of the latter's arguments and premises as incoherent or otherwise unsustainable. Rather, in characteristically modern fashion, Mandeville simply rejects the Christian-Platonist conception of the virtues and the good by establishing axioms of his own, a decision that produces two equally unfortunate results. First, it shows how modernity's self-authorizing rhetoric of *epochē* and its disavowal of any involvement with complex traditions necessarily displaces Platonic and Thomist dialectics. As regards this demise of genuinely responsive and responsible forms of inquiry, the contrast between Shaftesbury's ironic and hyper-reflexive employment of rhetorical genres and Mandeville's pugnacious assertiveness could hardly be greater. Second, and closely related, we find a process of forgetting, an eroding awareness of the deep history of concepts to be all but inescapable in the case of Mandeville. That he should have failed to notice the way in which his hyper-Augustinian, and paradoxically secular *and* moralizing argument revives the Gnostic dilemma (from which Augustine struggled to extricate himself) is rather telling. With some justification, Blumenberg points to a "residue of Platonism" being paradoxically "involved ... in 'secularization': Just as the image not only represents the original but can also conceal it and allow it to be forgotten, so the secularized idea, if left to itself and not reminded of its origin, rather than causing one to remember its derivation can serve instead to make such remembrance superfluous" (*LMA,* 72).

Having replaced the Scholastic paradigm of *disputatio* with the modern method of reasoning from unilaterally asserted principles, Mandeville's entire argument hinges on precisely such "forgetting" as its enabling and enduring condition. Thus his central claim that any quest for (self-)transcendence is but an illusion bound to be exposed as such by the wayward motion of bodies, themselves subject to mindless and unrelenting passions, can only be secured by an unflinchingly empirical account of how cognition itself unfolds: "As all our Knowledge comes *à posteriori,* it is im-

13. Mandeville, *Free Thoughts* (2nd ed.), 107, 102–103.

prudent to reason otherwise than from Facts" (*MFB,* 2:261). Perhaps because this radically empiricist and physico-mechanist position is so reminiscent of Hobbes and Locke (and, though not unchallenged, has held sway from Julien Offray de La Mettrie and Claude Adrien Helvetius all the way to contemporary neuro-science), the assumptions underwriting Mandeville's *Fable* ought to be drawn out with some care. A particularly striking trait of the book involves its emphatic disavowal of any presuppositions whatsoever. Inasmuch as for Mandeville all thought merely "follows," with seeming inevitability, from its underlying material conditions and attendant circumstances, anything more "is a vain/EUTOPIA seated in the Brain" (*MFB,* 1:36). Yet to hold that "Facts" should imply their own interpretation is problematic, if for no other reason than that it fails to distinguish between necessary and sufficient causes. Moreover, it would seem that a mechanistic epistemology of the kind espoused by Mandeville could only ever result in a mimetic restatement, presumably in the medium of language, of those very causes that supposedly occasioned some particular observation. The apparent tautology of such a position ("A man's real pleasure are [sic] what he likes") had already been flagged by Shaftesbury, to whose arguments Mandeville would gradually understand himself to be diametrically opposed.[14] As we shall see, Coleridge was among the most astute and voluble critics of the Hobbes-Mandeville ("corpuscular") school, even as he also repudiated Shaftesbury's apparent dissolution of organized (Trinitarian) religion into a purely emotive and implicitly deist position. As regards the former, Coleridge would never tire of flagging what he took to be the elemental failure of necessitarian and associationist strains in eighteenth-century philosophy—viz., to recognize the essentially counterfactual, interpretive, and evaluative nature of all thought. The origin of thinking itself, Coleridge was to argue, cannot be located in some contingent motive or inherently "mind-less" material cause. What truly impels and shapes human reflection, and what inevitably shows it to be of a categorically different quality than its physiological underpinnings, is an implicit view of *and* commitment to a transcendent good. For human reflection to commence and proceed in meaningful fashion at all, there has to be some counterfactual *telos* to whose pursuit a human being must be intuitively pledged if he or she is to take an interest in whatever contingent perception or problem happens to present itself in the here and now.

---

14. Mandeville's opposition to Shaftesbury only emerges in his later years; first references to the *Characteristicks,* still complimentary, can be found in his *Free Thoughts* (1720). Only in the second (1723) edition of the *Fable* does Mandeville reject Shaftesbury's "fancy ... that as Man is made for Society so he ought to be born with a kind Affection to the whole, of which he is a part, and a Propensity to seek the Welfare of it." He goes on to note "that two Systems cannot be more opposite than his Lordship's and mine" and that "the hunting after this *Pulchrum & Honestum* is not much better than a Wild-Goose-Chace" (*MFB,* 1:323–324, 331).

Yet for Mandeville, there is no such thing as an authentic space for inner reflection. His axiomatic rejection of any such possibility stems from his essentially Newtonian understanding of passion as sheer embodied force devoid of any intrinsic end or meaning.[15] Being the only reality, it must also discharge itself in an unbroken causal sequence whereby an inherently non-cognitive force is transferred or, properly speaking, displaced onto deceptively normative and meaningful notions or ideals. Echoing similar views held by Hobbes and Locke, and anticipating even more radical arguments to the same effect in Hume's *Treatise,* Mandeville's thought is the very antithesis of process philosophy. That is, he rejects as a matter of principle any temporal, reflective, and deliberative qualities within the mind. Defined as virtually indistinguishable from implementation, willing "is properly the last result [Hobbes's "last appetite"] of deliberation ... which immediately precedes the execution of, or at least the endeavour to execute the thing will'd: I say the result, ... for, when a will or volition is made long before the execution of the thing will'd, it is only call'd a resolution."[16] Just as by its very nature the will or some specific "passion" (that being Mandeville's preferred metonym) must discharge itself instantaneously, so it is also incapable of any qualitative advancement or ascent. Contrary to Shaftesbury, Mandeville thus insists that "all the passions are suspect" and that "even those that seem amiable, like pity and mother love, 'center in self-love' and cannot be relied on to produce concern for the common good."[17]

The point emerges with particular force in Mandeville's striking appropriation of Boethius's motif of the wheel of fortune, and it is here that the *Fable* reveals a deep and lasting connection between philosophical pessimism and political conservatism, one that has periodically reappeared ever since (e.g., in Burke, Malthus, Schopenhauer, Oswald Spengler, Carl Schmitt, Francis Fukuyama). In defending social inequality ("that the poor should be kept strictly to Work, and that it was Prudence to relieve their Wants, but Folly to cure them"), Mandeville draws on the Boethian conception of *fortuna* as he depicts life as a strictly cyclical and fundamentally non-cognitive process:

> The various Ups and Downs compose a Wheel that always turning round gives motion to the whole Machine. Philosophers, that dare extend their Thoughts beyond the narrow compass of what is immediately before them, look on the al-

---

15. The distinction, pioneered by Aristotle (*Metaphysics,* 9.6) here is between sheer "movement" and "actuality," that is, between *kinēsis* and *energeia.* The modern concept of the will no longer involves "actuality" (or action) but merely triggers a "movement" that, lacking an intrinsic end, is necessarily "incomplete" (*Metaphysics,* 1048^{b}30).

16. *Free Thoughts* (2nd ed.), 96.

17. Schneewind, *Invention,* 328.

> ternate Changes in the Civil Society no otherwise than they do on the risings and fallings of the Lungs; the latter of which are as much a Part of Respiration in the more perfect Animals as the first; so that the fickle Breath of never-stable Fortune is to the Body Politick, the same as floating Air is to a living Creature. (*MFB*, 1:248, 250)

What makes this image so striking is its shrewd enjambment of contingency and necessity. If the overall rotations of the wheel unfold with the clockwork-like necessity of the human body's respiratory system, fortune's "fickle Breath" may yet at any time adversely impact the body politic. Notably, then, Mandeville's wheel of fortune no longer allegorizes human, imperfect understanding but, on the contrary, views cognition as wholly determined by the involuntary ups and downs of the body—which in good Cartesian fashion is here depicted as a "machine."

To some extent, Mandeville's enlistment of an allegorical image dating back to late (Christian) antiquity for a social theory of radical finitude uncovers a weakness also found in the work that had first pioneered the image. In Boethius's *Consolation of Philosophy*, written between his arrest in A.D. 523 and his execution for charges (ultimately proven false) of treason against Emperor Theodoric in 524, the wheel of fortune functions as an allegory of the human being's finite and inadequate perspective on divine reason. To be sure, there is for Boethius no question whatsoever that "the generation of all things, and the whole development of changeable natures ... are given their causes, order and forms from the stability of the divine mind. That mind, firmly placed in the citadel of its own simplicity of nature, established the manifold manner in which all things behave." Principally, then, the image of the wheel of fortune throws into relief how far humans are removed from "the utter purity of divine intelligence, [which] is called providence" (*ipsa divinae intellegentiae puritate conspicitur, providentia nominatur*). Yet precisely this profound gap separating human fate from divine providence renders Boethius's assertion that "the one depends on the other" worrisomely abstract. The *Consolation* does not offer a clear understanding of this correlation. Far from suggesting how finite human existence may yet practically participate in divine reason, Boethius's work posits the disparity between the two realms as something merely to be endured. Thus fate is posited as the trope for a metaphysical enigma, what Boethius calls the "movable interlacing and temporal ordering [*mobilem nexum atque ordinem temporalem*] of those things which the divine simplicity has disposed to be done."[18] In what follows, Boethius thus argues that even those acts that appear conspicuously evil, certainly in consequence if not in intent, merely result from individuals' varying capacity for

18. Boethius, *De Consolatione* (4.6), 359, 361.

apprehending and pursuing the good. Yet even here, the gap between the simplicity of divine reason and a finite, eminently fallible, and distractible human understanding remains seemingly unbridgeable. Philosophy may "console," to be sure, yet on Boethius's account it remains unable to generate a practical, remedial strategy. Even in his last work, Boethius continues to betray the lingering influence of pagan (Stoic and Gnostic) thought.

To understand Mandeville's peculiar appropriation of Boethius's allegory in the *Fable,* it is necessary to recall how medieval Scholasticism had sought to overcome Boethius's disjunctive account of fortune and providence by suggesting a plausible and concrete venue for their reconciliation. Opening his discussion of providence and predestination (*ST,* I Q 22–23), Aquinas observes that to take up these issues is to address "those things which have relation both to the intellect and the will" (*ad ea quae respiciunt simul intellectum et voluntatem*). As one would expect, Aquinas posits that all created things involve "good [*bonum*] not only as regards their substance [i.e., functional design] but also as regards their order toward an end and especially their last end, which, ... is the divine goodness." The alignment of formal with final causes—that is, of the functional, "good" design exhibited by created things and the ultimate end (*finem ultimum*) toward which they are ordered—thus pivots on an underlying and identical conception of the good. For Aquinas, there can be no conflict between an efficacious and an ideational good, for to suggest otherwise is to have lapsed into a strictly occasional notion of the good as contingent on its affirmation by a divine will or transient human preference, respectively. Indeed, the nominalist position inevitably ends up mimicking the very omnipotence and irrationalism that it imputes to God's will. For as an argument it can only proceed by having pre-emptively disrupted the ontological nexus between the good and its eternal self-identity as a substantial form to which, at least in the view extending from Aristotle to Anselm and Aquinas, not even God could ever wish to oppose himself.

While there is no space here to dwell on Aquinas's complex parsing of providence and predestination, what does matter is his response to one particular objection (*ST,* I Q 22 A 1, ad 3) concerning the idea of divine providence.[19] That objection holds that, considering God's ontological simplicity (i.e., non-composite nature), he cannot have "providence" because the latter, by including both intellect and will, constitutes "something composite" (*aliquid compositum*). Aquinas rejects this distinction as misleading, for both faculties are at all times implicated in the "reason of order" (*ratio ordinis*), viz., as the intellectual conception properly called providence and as the will to implement what is thus foreseen, "which is termed government"

19. In his *Free Thoughts on Religion, the Church, and National Happiness* (2nd ed., 1729), Mandeville devotes a long chapter to a discussion of free will and predestination.

(*gubernatio*). Simply put, to apprehend a good qua rational order is also to be committed to its practical realization: "Providence resides in the intellect; but presupposes the act of willing the end. Nobody gives a precept about things done for an end; unless he will that end. Hence *prudence presupposes the moral virtues,* by means of which the appetitive faculty is directed towards good, as the Philosopher [Aristotle] says."[20]

For Aquinas, it is above all the virtues whose cultivation gradually builds up links between a good intellectually apprehended and the practical pursuit of it by means of rational self-governance (*gubernatio*). A fuller account here would have to show how Aquinas's theory of the habits and the virtues (*ST,* Ia IIae QQ 49–67) demonstrates the capacity of human beings possessed of intellect to refashion their own "nature" so as to be able to participate to ever higher degrees in the *summum bonum.* Such a project of a (self-)education into the virtues reveals a very different, indeed far more profound understanding of "autonomy" than the range of finite, self-affirming, and self-fulfilling practices that from Locke to Kant, Nietzsche, and Heidegger have come to define the term. At the very least, the possibility of slowly redefining one's habituations and, in time, achieving a measure of rational self-governance is here understood as a participation *in* rather than as an emancipation *from* divine reason. It also furnishes concrete evidence for what in Boethius's *Consolation* had remained so doubtful: viz., that "when it is said that God left man to himself, this does not mean that man is exempt from divine providence." Finally, and for our purposes most crucially, Aquinas's accounts of the habits and the virtues as a gradated progression shows purity of intention to be a *terminus ad quem* rather than a precondition of moral agency, which it will be for Kant. As Jennifer Herdt points out, "as long as [individuals] wish truly to enter a state of perfection, the failure of their character to measure up to their action is weakness but not itself an additional vice." Inadvertently sowing the seeds of a conflict that was to prove central to the debate between Luther and Erasmus, Aquinas's emphasis (essentially Aristotelian in nature) is on *action,* even imitative action. Simply to perform, regularly and without overt disbelief, a set of actions conducive to one's inner flourishing as well as that of the community of which one is part "will create new habits in this person, transforming her

20. *ST,* I Q 22 A 1 ad 3 (italics mine). Recent analytic philosophy has reassessed the distinction between "goal intentions" and "implementation intentions" and, however unwittingly, echoing Aquinas has characterized intention as a complex intellectual operation that is self-aware ("second-order intentions"), and that should neither be viewed as mindlessly implemented nor subjected to other reductionist accounts. Instead, being inherently open to adjustment in light of changed circumstances (appraised by judgment), intention resembles what is colloquially known as a "resolution." See Holton, *Willing, Wanting, Waiting,* 3–19, as well as his related discussion on "choice" and "judgment" (53–58).

character." That was what Aristotle had in mind when noting that "states arise out of like activities," and that, as regards just and temperate acts, "it is not the man who does these that is just and temperate, but the man who also does them *as* just and temperate men do them."[21]

Yet precisely this possibility of a non-deterministic connection between God and man—and, by extension, of an integrative rather than adversarial relation between intellect and will—Mandeville rejects out of hand. Proceeding in classical voluntarist fashion, his *Fable* instead accords the will qua "passion" absolute primacy and, having done so, anatomizes both intellect and a notably monochrome idea of virtue as mere epiphenomena of the will.[22] Moreover, whereas the Aristotelian and Thomist accounts of virtuous action had always presupposed, indeed were themselves only licensed by, a highest good relative to whose internal simplicity and perfection the finite intellect stands as a teleologically ordered (albeit imperfect) image, Mandeville's account "required not only no recourse to revelation but no recourse even to the concept of Providence."[23] As a result, both action and the entire spectrum of contemplation (*theoria*), introspection, and such reflective states as Keats would later group under the heading of "negative capability" are wholly incommensurable with Mandeville's idea of passion.[24] In fact, Mandeville's passion is its own self-contained and supreme reality and as such is the very antonym of knowledge:

> We meet with Thousands [of Phenomena] every Day to convince us, that Man centers every thing in himself, and neither loves nor hates, but for his own Sake. Every Individual is a little World by itself, and all Creatures, as far as their Understanding and Abilities will let them, endeavour to make that Self happy: This

21. Herdt, *Putting on Virtue,* 82; Aristotle, *Nicomachean Ethics,* 1103^{b}20, 1105^{b}8. On the Luther-Erasmus debate, see Gillespie, *Theological Origins,* 129–169; on Erasmus's assimilation of habituation to outright imitation, see Herdt, *Putting on Virtue,* 101–127.

22. On Aquinas's theory of the virtues, in particular, their reliance on the supplemental role of "action" vis-à-vis grace, see Herdt, *Putting on Virtue,* 72–97; MacIntyre, *Whose Justice?* 183–208; McInerny, *Ethica Thomistica,* 60–89; for a precise analysis of the conception of human action in relation to intention, habit, teleology, Burrell, *Aquinas,* 131–154.

23. Herdt, *Putting on Virtue,* 273.

24. While the contrast between Aristotelian thought and the mechanist theory of mind set forth by Hobbes and Mandeville is rather too obvious to require elaboration, it is yet worthwhile to note how Aristotle's account of "contemplation" and the "excellence of the intellect [as] a thing apart" in Book 10 of his *Nicomachean Ethics* arises specifically out of a parsing of "pleasure" (1174^{a}12–1177^{a}10) as a good devoid of intrinsic meaning. Hence, while admitting the potentially decisive role of "luck" in either enabling or preventing the mind from achieving meaningful and secure fulfillment, Aristotle stresses that "happiness [*eudaimonia*] extends just so far as contemplation does" (1178^{b}25). On the role of luck and contingency in Aristotle's ethics, see Nussbaum, *Fragility,* 318–372.

> in all of them is the continual Labour, and seems to be the whole design of Life. (*MFB*, 2:178)

There has been no shortage of attempts to rehabilitate this position of "doing as one likes" as a prescient and well-intentioned version of modern, liberal, and pluralist society subsequently championed by thinkers from Jeremy Bentham and John Stuart Mill to John Rawls and Jürgen Habermas. Mandeville's editor, F. B. Kaye, thus notes how "to say that welfare, or pleasure, or happiness should be the end of action does not mean the limiting of this welfare, pleasure, or happiness to one particular kind, but may allow the satisfaction of as many kinds as there are people" (*MFB*, 1:lix). Yet if there is a pluralist dimension to Mandeville's moral relativism, it arises merely by default—that is, from the sheer *indifference* and *non-transparency* of individual agents toward one another. Conversely, it is hard to see how pluralism can ever amount to a reasoned position except in a strictly formal sense. For that to happen, a common ground from which a deliberate espousal of pluralism would draw both its intrinsic appeal and legitimation, as well as a common good furnishing the *terminus ad quem* being served by such a position, would already have to be acknowledged. Precisely this idea of a *summum bonum*, however, Mandeville emphatically denies by insisting that the source or, rather, the only substratum and determinant of all social processes and meanings is passion. For Mandeville, passion most vividly reveals the essentially mimetic nature of mind and the narcissistic structure of social interaction: "The highest wish of the Ambitious Man is to have all the world ... of his Opinion. The most insatiable Thirst after Fame that every Heroe was inspired with, was never more than an ungovernable Greediness to engross the Esteem and Admiration of others in future Ages, as well as his own" (*MFB*, 1:54).

Arguing that "it is the Work of Ages to find out the true Use of the Passions" (*MFB*, 2:319), Mandeville positions all ostensibly rational claims in a strictly epiphenomenal relation to this enigmatic force. As he was to put it in his *Origin of Honour*, "all Human Creatures are sway'd and wholly govern'd by their Passions, whatever fine Notions we may flatter our Selves with; even those who act suitably to their Knowledge, and strictly follow the Dictates of their Reason, are not less compell'd so to do by some Passion or other."[25] The irrationalism of this position foreshadows Edmund Burke's misgivings about rationalism in politics, and even glances ahead to Schopenhauer's pessimistic conception of the will. While the conjunction between pessimism and political conservatism has often been remarked upon, what has often escaped notice is the peculiar *fort/da* game that an anthropological pessimism of the kind set forth by Mandeville tends to play with metaphysics. Ostensibly, his account of the

25. *Enquiry*, 31.

passions as materially coded, discontinuous, and devoid of any transcendent perspective on their own occurrence would appear to be the very quintessence of empiricism. Thus to suppose that "no Person can commit or set about an Action, which *at that then present time* seems not to be the best to him" (*MFB,* 2:178; italics mine) is to contract the scope of human intelligence to the merely sentient and by definition unself-conscious fluctuations of present desire. Yet as Schopenhauer's rather more profound philosophical position would admit, to accord passion such absolute sway over all human endeavors means to reinvest a seemingly empirical phenomenon with a noumenal and metaphysical dimension all of its own and, in so doing, to expose the Enlightenment's own mythical underpinnings. For Theodor Adorno and Max Horkheimer, such a radically empiricist method posits that "the guarantee of salvation lies in the rejection of any belief that would replace it: it is knowledge obtained in the denunciation of illusion ... [and] the contesting of every positive without distinction."[26]

Moreover, Mandeville's quintessentially nominalist claim that "every Individual is a little World by itself," a proposition already compromised by its apparent nonfalsifiability in the terms in which it is here advanced, leads to two additional paradoxes. First, it raises the question as to its own, underlying motivation as a philosophical argument. Various critics have read the *Fable* as an attempt to formulate "a bourgeois ethic [where] participating in a flourishing capitalist society takes precedence over securing one's nobility," or as offering "an examination of the social effects of morality."[27] Yet the forensic vigor with which Mandeville pursues his analysis of modern economic and moral practices as springing from the noumenal reality of the passions points to a moralist dimension in his own account—an unexamined and implacable pathos for truth-telling as an open-ended process of demystification. In this regard, his attempt to formulate a rigorously non- or anti-metaphysical ethic has long struck readers as oddly ambiguous. On the one hand, the *Fable* prefigures what would later be billed as "consequentialism" in ethics, viz., by appraising the value of an act in accordance with whether, and how far, it could be shown to be conducive to national prosperity—a concept which, in turn, he unhesitatingly conflates with the idea of "happiness." Characteristic here is the inability of the consequentialist to derive action from the articulation of an a priori, normative good: "though he may speak *against* some action, he cannot prescribe any—for in an *actual* case, the circumstances (beyond the ones imagined) might suggest all sorts of possibilities."[28]

26. *Dialectic of Enlightenment,* 23; on the relation of nominalism to historicism, see Chandler, *England in 1819,* 53–73.

27. Herdt, *Putting on Virtue*, 268; Schneewind, *Invention*, 324.

28. Anscombe, "Modern Moral Philosophy," 13. "If you are a consequentialist, the question 'What is the right thing to do in such-and-such circumstances?' is a stupid one to raise. The casuist

Modern consequentialist ethics amounts to a secular echo of medieval nominalism; for inasmuch as the language of human self-interest has supplanted William of Ockham's inscrutable divine will, which performatively decrees what is to count as rational, it cannot surprise that Mandeville will habitually discredit the inner motives of individuals as so much subterfuge and false consciousness. Quite the moral rigorist (and pessimist), Mandeville thus insists "that it is impossible to judge of a Man's Performance, unless we are throughly [sic] acquainted with the Principle and Motive from which he acts." Anticipating his later, extensive discussion of "Charity and Charity-Schools" (*MFB,* 1:253–322), he even questions an emotion as palpably unselfish as "pity." While "the most gentle and the least mischievous of all our Passions, [it] is yet as much a Frailty of our Nature as Anger, Pride, or Fear." Predominant in the "weakest Minds," pity thus is but "the most amiable, and bears the greatest Resemblance to Virtue." In his quest to unmask all virtue as mere semblance pointing back to those non-cognitive "passions in which the Seeds of most Virtues are contained" (*MFB,* 1:67)—that is, pride and shame—Mandeville reveals his deep affinity with a hyper-Augustinian tradition that runs through modernity from Luther, Calvin, and Pascal through the "secular" moralists (Françoise de La Rochefoucauld, Pierre Nicole, Pierre Bayle) and, in his own time, all but merges with epistemological skepticism.

A second paradox has to do with Mandeville's declared intention of unmasking virtue, benevolence, and indeed the very fact of human sociality as an illusion fostered by the endless dissimulations of solitary individuals seeking to justify their allegedly hedonistic, self-interested pursuits. Having posited a quasi-autistic model of human consciousness, Mandeville finds himself unable to explain his principal contention, viz., "that the Moral Virtues are the Political Offspring which Flattery begot upon Pride" (*MFB,* 1:51) in the first place. On the face of it, his attempt at unmasking virtue, altruism, and self-denial as but an ingeniously conceived and far-reaching ideological superstructure (e.g., virtue ethics) seems a perfectly plausible endeavor.

---

raises such a question only to ask 'Would it be *permissible* to do so-and-so?' or 'Would it be permissible *not* to do so-and-so?' Only if it would *not* be permissible *not* to do so-and-so could he say '*This* would be *the* thing to do" (ibid., 12). Later in her seminal essay, Anscombe sharpens her contrast between a normative and a casuistic approach to ethics: "If 'what is unjust' is determined by consideration of whether it is *right* to do so-and-so in such-and-such circumstances, then the question of whether it is 'right' to commit injustice can't arise, just because 'wrong' has been built into the definition of injustice. But if we have a case where the description 'unjust' applies purely in virtue of the facts, without bringing 'wrong' in, then the question can arise whether one 'ought' perhaps to commit an injustice, whether it might not be 'right' to?" The chief problem, as Anscombe well realizes, arises from the fact that "the man who makes an absolute decision that injustice is 'wrong' has no footing on which to criticize someone who does *not* make that decision as judging falsely" (ibid., 17). For a recent (not altogether persuasive) critique of Anscombe's position, see Wolterstorff, *Justice,* 370–372.

Yet the *Fable* can never quite decide whether the false consciousness in question is the result of efficient (material) *causes* operating on the mind, or of specious *motives* gratuitously embraced by the will. Moreover the very phenomena of moral, virtuous self-representation that Mandeville seeks to discredit only present themselves and continue to flourish because individuals (however self-interested) do in fact take themselves to be inextricably related to and implicated in an idea of the good and, hence, to owe other minds some account of their own pursuits. That the strategies of self-legitimation elected by individuals or entire communities may be disingenuous does not alter the fact that, in so devising them, the selves in question act on the antecedent and non-propositional reality of their relatedness and sociality. As Hutcheson was to point out, a society composed of nothing but self-seeking and morally agnostic individuals would create no incentive for the invention of, or aspiration to, a supra-individual idea of the good: "Had we our selves been *wholly selfish,* and lived in a system of beings *wholly selfish,* without a *Moral Sense,* ... we should have no grounds to have expected any Regard to the Good of each other."[29] Mandeville, that is, cannot account for the *motive* that should prompt individuals to spawn ideas, values, and entire systems of moral self-legitimation, however flawed or inauthentic those may seem.

It is specifically this last paradox that lets us glimpse the intrinsic poverty of Mandeville's account. Even as satire directs its withering attack against perceived hypocrisy, it is unable to identify the sources of its own perspicacity. It can only unfold performatively and repetitively, and in so doing reveal itself to be intrinsically devoid of any temporal, narrative dimension. The "history" of (false) morals it purports to uncover is, evidently, a history that should never have been, indeed in a certain sense never "was"—being, after all, but a fraudulent and self-regarding account of moral progress. Yet the rhetoric of unmasking that drives Mandeville's satiric, and notably repetitive, indictment of traditional virtue ethics arises as it were out of nowhere. It is no bona fide skepticism since even that position (found in La Rochefoucauld) hinges on a minimal set of epistemological commitments and inner certitudes. Instead, Mandeville's stance seems curiously utopian in its own right, for it centers on the indictment of virtue as the subterfuge of passions that he considers to be utterly impervious to rational control and governance. His "commitment to self-knowledge goes hand in hand with a splitting off of the analyst's perspective from that of the agent, a split typical of modern moral thought."[30] The result is something of a paradox, a text marshalling a rich and effective inventory of generic and rhetorical techniques to im-

29. Hutcheson, writing under the pseudonym Philanthropus, in the *London Journal*, 19 June 1725.

30. Herdt, *Putting on Virtue*, 278.

press on its intended audience the supposed reality of man's autistic constitution. Against the intrinsic sociality of human beings that had been urged by Shaftesbury, and that was to be reaffirmed again in Hutcheson's 1728 *Essay on the Nature and Conduct of the Passions and Affections,* Mandeville pits a dystopia of isolated and self-imprisoned human beings whose social practices and norms he seeks to expose as strictly transferential and solipsistic: "does not Man love Company, as he does every thing else, for his own sake?" (*MFB,* 1:341). Yet his objective here is not to deny the phenomenon of sociality per se but, in hyper-Augustinian fashion, to expose the ultimate inauthenticity of all ostensibly social and altruistic practices. The various material and virtual spaces (political, economic, ecclesiastical, and cultural) wherein social life and moral reflection unfold are revealed as pragmatic fictions designed to conceal the fundamentally hedonistic constitution of all human beings and their de facto indifference to the flourishing of others.

As it happens, even those who strenuously repudiated and chastised the *Fable* (Hutcheson and Smith among them) found themselves saddled with a new and unprecedented explanatory burden. For what rendered Mandeville's account of the passions so problematic was ultimately less their non-cognitive, self-centered epistemology, which Hutcheson was able to refute. Rather, the *Fable*'s most vexing bequest had to do with its radical evacuation of final causes, ultimate goods, and normative commitments from human existence. To reintroduce a notion of benevolence, a "moral sense" or some such affective source for the sociality of human beings was all good and well. Yet to take that path, as Hutcheson and Smith variously would, did not alter the fact that moral and rational agency had effectively been redefined as the participant in a polite, sociable, and open-ended conversation whose hoped-for success might license, yet could never compel, happy inferences about the moral character and inner life of those participating in it. Hence the divide between the inner life and the symbolic practices meant to give assurance of its benevolent nature was there to stay. As a result, moral reasoning and practice take on an increasingly transactional character. The notion of the good is no longer understood teleologically but as something whose viability has all but merged with the modern self's ostentatious affirmation of her or his benevolent (inner) dispositions—a practice that naturally comes to rely upon (and be imitative of) contingent, middle-brow aesthetic and rhetorical conventions, manners, and customs. After Shaftesbury and Mandeville, what Iris Murdoch calls the sovereignty of good as *telos* has been absorbed into the theatricality of social "behavior" and a middle-brow regime of "manners." Struggle has yielded to convention, and action—which once had been the consummation of the will qua "intellectual appetite"—now appears to have been supplanted by behavior, a term that remains to be desynonymized from "habit" and "habituation" with which, erroneously, it is often conflated.

Arguably the most influential respondent to Mandeville's controversial arguments, Francis Hutcheson—in his 1725 *Inquiry on Beauty* and the 1728 *Essay on the Nature and Conduct of the Passions and Affections*—seeks to recover a model of free and responsible human agency. Hutcheson and his Scottish fellow moral theorists (Archibald Campbell, William Leechman, John Simon, William Law) were caught between the mechanism of Locke and Mandeville and a Cartesian rationalism (represented by Samuel Clarke) that traced moral concepts back to eternal forms. Yet their principal challenge was political in kind, viz., to legitimate moral philosophy as something to be pursued outside the institutional framework of *Kirk,* presbytery, synod, and general assembly. In this regard, Hutcheson's main contribution to eighteenth-century thought, the idea of a pre-discursive and seemingly universal (though always empirically operative) "moral sense," constituted "a secular counterpart to the appeal to inward feeling so characteristic of the doctrines of Evangelical conversion."[31] Not everyone welcomed this peculiar hybrid—half empirical, half Platonic—and opposition to the suggestion that the moral orientation of the self might at any point depend on factors other than grace and revelation met with considerable opposition from Presbyterian officialdom. Moreover, the memory of Thomas Aikenhead, the last person to be tried and executed in Scotland for heresy (in 1697) certainly had not entirely faded by the time that Hutcheson published his first writings in 1717; and indeed, by 1738, Hutcheson (along with Campbell and Leechman) found himself on trial for heresy, though all three were ultimately acquitted. Notably, one of the two charges against him was that it is possible "to have a knowledge of good and evil, without, and prior to a knowledge of God"—which is to say, the question of whether *synderesis* is part of revelation or an equivalent of natural law.[32] Like Shaftesbury and the Cambridge Platonists, Hutcheson looks askance at the nominalist-empiricist model of philosophy, which rejects moral principles, ideas, or intuitions simply because experience furnishes no direct and transparent evidence for such a claim. At the same time, he also rejects Samuel Clarke's rationalism according to which moral agency hinges on the intellectual clarity and conceptual distinctness of the values and meanings that it seeks to pursue and realize.

Faced with the equally unpalatable frameworks of nominalism and rationalism, Hutcheson "wants to find a *via media* between naïve realism according to which aesthetic and moral qualities exist in objects independently of the involvement of any

31. MacIntyre, *Whose Justice?* 278.

32. Scott, *Frances Hutcheson*, 84. The true point of contention, however, was whether the Presbyterian Church should exercise full authority over all teaching in the realm, a position Hutcheson emphatically challenged by offering assistance even to divinity students on the composition of sermons, scarcely a recognized branch of moral philosophy.

observer, and non-cognitivism according to which moral judgment reduces to the experience or expression of feelings."[33] To this end, and partially drawing on Shaftesbury, Hutcheson threads the needle by distinguishing between two conditions of moral consciousness—viz., the necessary one of the passions and the sufficient condition of an intellectual view or reflection that completes moral agency. As he puts it, "*Desires, Affections, Instincts,* must be previous to all *Exciting Reasons;* and a *Moral Sense* antecedent to all *Justifying Reasons.*"[34] In his prolonged dialectical engagement of both these positions, Hutcheson advances the overall debate in two crucial ways that shall briefly occupy us. Against Samuel Clarke's (in origin Platonic) claim that moral agency presupposes knowledge of the moral law or virtue to be affirmed, Hutcheson's moral sense approach maintains that the cognitive dimension of moral action cannot be the source of our willing. Rational conceptions of morality may well have legitimacy, yet they can never originate a specific course of action: "the *natural Dispositions* of Mankind will operate regularly in those who never reflected upon them, nor form'd just Notions about them. Many are really *virtuous* who cannot explain what *Virtue* is" (*EPA,* 4). Still, anticipating Kant's arguments a half century later, Hutcheson means to oppose Locke's hedonistic account of the will by offering "farther illustration of *disinterested Affections,* in answer to [Locke's] Scheme of deducing them from *Self-Love*" (*EPA,* 6). At the same time, however, the scope of the good toward which Hutcheson's moral sense orients the individual has significantly narrowed in comparison with Shaftesbury. Whereas the latter's *sensus communis* had posited aesthetics not only as the medium of the good but also as prima facie evidence of its attainment (and, at times, seems nearly consubstantial with it), "Hutchesonian moral sense approves of only one basic motive: benevolence, . . . a stand-in for Christian *agapē.*"[35] His intuitionist argument that "our *moral Actions* and *Affections* may be in good order, when our Opinions are quite wrong" (*EPA,* 4) already hints that the affective source of the "moral sense" has only a tenuous relationship to practical reason.

It was this claim that Gilbert Burnet, writing under the pseudonym Philaretus, had first questioned in a public letter to the author of the *Inquiry,* addressed as Britannicus (Hutcheson is not mentioned by name, having published the first edition of

---

33. Campbell, "Francis Hutcheson," 171.

34. Letter to Gilbert Burnet, published in the *London Journal* (1735), quoted in *EPA*, xiv. For a fuller discussion of Hutcheson's philosophy, see MacIntyre, *Whose Justice?* 260–280, who sees Hutcheson as exemplary of Scottish moral philosophy and as providing "an accurate introspective psychology or phenomenology of the passions" (268); and Schneewind, *Invention*, 333–342. Eagleton views Hutcheson as "a civic humanist of a traditional stamp, convinced that the public good is the highest moral end" and as delivering, in all his writings, "a broadside against philosophical egoism" (*Trouble*, 29, 31).

35. Schneewind, *Invention*, 334.

his *Inquiry* anonymously). The exchange is remarkable both for its intellectual probity and for the fact that it was conducted in a weekly, the *London Magazine,* catering to a non-specialist audience. While acknowledging the importance of a "moral sense of Virtue" for ensuring individuals' ability to achieve a coherent moral persona such as will "have a *lasting* and *uniform* Influence on their Actions," Burnet expresses "a secret Uneasiness arising from my Suspicion of its [the moral sense's] not being *right*."[36] To the rationalist Burnet, such fears cannot be allayed as long as the moral sense is conceived in phenomenological, not propositional, terms. It is not enough for it to manifest itself affectively as a certain pleasure, whose value and moral legitimacy Burnet takes to be undecided per se: "The perception of *Pleasure* therefore ... seems to me not to be a certain enough Rule to follow. There must be, I should think, something *antecedent* to justify it, and to render it a *real Good.* It must be a *Reasonable Pleasure,* before it be a *right* one." It will take until Kant (in §9 of the *Critique of Judgment*) for this pivotal question of "whether the feeling of pleasure precedes or follows the judging of the object" to be addressed, and for the antinomy of reason and pleasure to be plausibly resolved. For Burnet, at any rate, it is axiomatic that rationality can be achieved only in propositional language and, consequently, that it cannot simply be anchored in the affective realm of the "moral sense": "the Constitution of all the Rational Agents that we know of is *such* indeed, that *Pleasure* is inseparably annexed to the Pursuit of what is *Reasonable.*" While granting the supplemental value of beauty and pleasure as "strong *Motives to Virtue,*" Burnet insists that their role can only be to deliver an energy charge to a self that has already acquired a conception of the good by purely intellectual means: "I would not have Men depend upon their *Affections* as *Rules* sufficient to conduct them, tho' they are the proper Means to animate them to, and support them in, such a Conduct as *Reason* directs."

An impressive piece of tactful, clear, and concise argument, Hutcheson's response (published under the pseudonym Philanthropus) anticipates several of the claims subsequently advanced in greater detail in his *Essay on the Nature and Conduct of the Passions and Affections* (1728). Reminding his audience of the distinction between theoretical and practical reason, Hutcheson proceeds to make a strikingly modern argument about the apparent multiplicity and potential incommensurability of different rationalities. As he shows, to be rational means to be *already* committed to a specific conception of the highest good, for it is only such an antecedent end that can motivate and orient deliberative and practical activity—which for Hutcheson involves the purposive interaction of will, judgment, and understanding. It is illogical to suppose that the self might *choose* such a hyper-good, as if picking

36. "To Britannicus," *London Magazine,* 10 April 1725, 1–2. Hutcheson's two-part reply appears in the same journal on 12 and 19 June 1725.

from a menu of options. At the same time, reasoning *about* ends and their potential conflict with means variously being considered for their attainment is not only possible but forms an integral aspect of what we call judgment. Implicitly, then, Hutcheson's justification of a moral sense antecedent to practical reason offers a (weak) justification for why there has to be a complementarity of will and intellect, and why it is illusory to entrust the intellect (or reason, as Burnet understands it) with both the discernment of the highest good and our commitment to its practical realization. Undoubtedly, for the Lockean or Mandevillean hedonist "that end is *Reasonable,* which contains a greater Happiness than any other which it could pursue." Moreover, Hutcheson notes, "with such a Being, the Cruelty of the *Means,* or their bad Influence on a *Community,* would never make them pass for *Unreasonable.*" Conversely, individuals who "by the very Frame of their Nature desire the Good of a *Community* ... and [who] have withal a *Moral Sense,* which causes them necessarily to approve such Conduct in themselves or others" will have a radically different appraisal of what constitutes an end and, consequently, what may count as reasonable.

Inasmuch as such beings are simultaneously possessed of self-love, their status as rational agents would involve recognizing and, as much as possible, reconciling self-interest and benevolence in a vision of "*Universal Happiness,* the very pursuit of which is supposed to be the greatest Happiness to the several Agents themselves; for thus both Desires are at once gratified." As this last remark makes clear, the initial impression of an outright symmetry between a self-interested and a benevolent temperament is misleading. For, properly speaking, the hedonistic or possibly sociopathic individual cannot be credited with rationality at all but, at most, with a purely calculative intelligence that, lacking all negative capability, merely seeks to assimilate for its determinate purposes whatever data and objects happen to present themselves to its attention. It would not so much "think" as merely implement a simple and invariant syllogism. Hutcheson's concern ultimately does not lie with either resolving or justifying the enigmatic conceptual status of the moral sense. Rather, he means to foreground the impoverished understanding of reason itself if we were to embrace an ethic that denied the reality of the moral sense merely because of its alleged lack of propositional explicitness. As Shaftesbury before him and Kant a half century later, Hutcheson sees rationality and sociality inseparably entwined, and it is just this affinity that no propositional argument can ever secure. To a wholly dissociated sensibility lacking any intuitive apprehension of the good (what Aquinas calls *synderesis*), the communal, engaged, and deliberative dimension of reason would remain terminally foreclosed:

> If there were any *Natures disjoined from us,* who knew all the *Truths* which can be known, but had no *Moral Sense,* nor any Thing of a superior Kind equivalent to it; such *Natures* might know the Constitution of our Affairs, and what *Publick*

> and *Private Good* did mean; they would grant, that equal Intenseness of Pleasure enjoyed by Twenty, was a greater Sum of Happiness than if enjoyed only by One; but to them it would be *indifferent,* whether One or more enjoyed Happiness ... These *disjoined Natures,* without a *Moral Sense,* would see nothing *Reasonable* in the good Affections of one Man towards another.[37]

This dystopic vision of a world peopled by "disjoined," morally indifferent and socially autistic selves—which has since congealed into the preeminently American ideology of libertarianism—makes for a distressing scenario, to be sure, yet it does not per se secure Hutcheson's claim for a moral sense. Neither does it as yet show how its affective manifestation can be transposed into an enduring commitment to the good on the order of what Shaftesbury had called "our Obligation to VIRTUE" (*SC,* 2:99). Indeed, by conjuring the counterfactual scenario of a morally "disjoined" self as he does, Hutcheson already reveals a latent drift of moral sense theory toward a utilitarian conception. As he notes, "that which produces more Good than Evil in the Whole, is acknowledg'd Good, and what does not, is counted Evil." What may at first blush seem an inadvertent slippage of his intuitionist account into a consequentialist one quickly hardens into formulations that anticipate Bentham's "felicific calculus" almost verbatim: "we are led by our moral Sense of Virtue to judge thus: that in equal Degrees of Happiness, expected to proceed from the Action, the Virtue is in proportion to the Number of Persons to whom the Happiness shall extend ... or that the Virtue is in a compound Ratio of the Quantity of Good, and Number of Enjoyers ... so that Action is best, which procures the greatest Happiness for the greatest Numbers" (*HI,* 125).

How are we to explain this startling shift from an affective, inner certitude regarding virtue to a calculation of "the Consequences of Actions" (*HI,* 125)? Is Hutcheson simply unaware of the tension between these two models, or might he possibly regard them as commensurable after all? Only a few pages earlier, while stressing that "this moral Sense" does not "suppose any innate Ideas, Knowledge, or practical Proposition," Hutcheson had specified that the sense in question simply constitutes a deep-seated receptivity to "amiable or disagreeable Ideas of Actions, when they occur to our Observation, *antecedent to any Opinions of Advantage or Loss*" (*HI,* 100; italics mine); similarly, at the outset of his second *Inquiry,* Hutcheson expressly rejects any proto-utilitarian and consequentialist arguments: "Had we no Sense of Good distinct from the Advantage or Interest arising from the external Senses ... we should have the same Sentiments and Affections toward inanimate Beings, which we have toward rational Agents" (*HI,* 89). If the moral sense is to function as a source of good, the idea of benevolence to be realized by actions taken in accordance with the dictates of

37. Hutcheson, *London Magazine,* 12 June 1725.

that sense cannot, in turn, be calculative, interested, or speculative in nature; hence the rhetorical question: "is not our Love always the Consequent of Bounty, and not the Means of procuring it?" (*HI,* 107). As the editor of the *Inquiry* notes, only by the time of the fourth edition did Hutcheson realize that the "Christian sense of *benevolence* is not an emotion or feeling, but an act of will," and that his notion of love qua benevolence or "Desire of the Good of Others" accords with Aristotle's *orexis bouleutikē,* that is, amounts to "a settled Disposition of the Will, or a constant Determination, or desire to act, ... or a fixed Affection toward a certain Manner of Conduct."[38] What this points to is a gradual realization on Hutcheson's part that the moral sense, like Aquinas's moral virtues, is not intelligible if presented as some mystical, metaphysical power. On the contrary, it constitutes a habit ("settled disposition" being a good translation of Aquinas's *habitus*).

That being the case, the nervously anti-metaphysical tendencies of moral sense philosophy ultimately cannot escape the conception of the will as it is sedimented into the steady state of the benevolent affections that are an actual or potential aspect of human beings—that is, socially related moral agents whose very nature has them interpret and thus transform their life world. Understood as a "habit" of the will, the moral sense is not a value-neutral power enabling the individual to discriminate, seemingly without presupposition, between the good and the bad. Instead, it is what it is only because a sense of good and ill has been slowly and painstakingly cultivated and internalized. Quoting Aristotle (*Metaphysics,* 1022^{b}4), Aquinas thus characterizes habit "as a quality, ... a disposition whereby that which is disposed is disposed well or ill, and this, either in regard to itself or in regard to another: thus health is a habit [*habitus dicitur dispositio secundum quam bene vel male disponitur dispositum, et aut secundum se aut ad aliud, ut sanitas habitus quidam est*]" (*ST,* Ia IIae Q 49 A 1). Moreover, it is precisely by virtue of a habit that "we are directed well or ill in reference to the passions."[39] Logically, a habit cannot be anchored in a passion but instead amounts to a "second nature" such as has gradually been fashioned by rationally ordered acts of will. Unlike a mere transient affect or sentiment, "it belongs to every habit to have relation to act [*convenit omni habitui aliquot modo habere ordinem ad actum*]" (*ST,* Ia IIae Q 49 A 3).

To gauge the strengths and weaknesses of Hutcheson's paradigm, further scrutiny must be brought to bear on the operative nature of this moral sense. To do so is to confront "a decidedly psychological or 'inward' turn" in British philosophy and a pre-Kantian version of autonomy. It rejects the Hobbesian account in which

---

38. *HI,* 107 (*Inquiry,* 4th ed., 1738, p. 195).

39. *ST,* Ia IIae Q 49 A 2; Aquinas is referencing Aristotle, *Nicomachean Ethics,* 1105^{b}25; while Aquinas's discussion of the habits (*ST,* Ia IIae QQ 49–54) is far too rich to be fully explored here, the relation between "habituation" and "behavior" will be taken up below in the context of Adam Smith's *Theory of Moral Sentiments*; see 361–368.

obligation necessarily takes the form of a mind being under the dominion of another's will. Rather, it may arise as a form of deliberate inner, self-imposed causality which (following neo-Platonist thought) Shaftesbury finds embodied in human beings' intuitive attraction to and cultivation of the beautiful.[40] As is the case in Shaftesbury, "sense" in Hutcheson's thought serves multiple and potentially contradictory purposes. Its first and seemingly obvious meaning is that it brings immediately into conscious view some object. As such, "sense" for Hutcheson appears all but identical to Lockean "sensation," viz., a notion or idea derived from the presence of an object to consciousness. Yet this rudimentary conception is implicitly challenged once "the simple Idea of Sensation" is scrutinized with regard to its phenomenology. Shifting the terms ever so slightly, Hutcheson thus observes that, whereas traditional Lockean epistemology tends to conceive "Ideas of Beauty and Harmony" as "Perceptions of the external Senses of Seeing and Hearing, ... I should rather chuse to call our Power of perceiving these Ideas, an Internal Sense" (*HI,* 23). It is only when, and insofar as, the raw data of sense impressions are being phenomenologically registered by this internal sense that the self acquires any reality as an epistemological and, by extension, moral agent.

The word that identifies the moment when data received from the outside are being appraised as formally coherent and commensurable with the mind's internal architecture is, simply, "beauty." As the first *Inquiry* puts it, "this superior Power of Perception is justly called a Sense, because of its Affinity to the other Senses in this, that the Pleasure does not arise from any Knowledge of Principles, Proportions, Causes, or of the Usefulness of the Object; but strikes us at first with the Idea of Beauty" (*HI,* 25). Beauty, the aesthetic *Urphänomen* par excellence, thus prompts Hutcheson to regard fact and value, the apprehension of empirical data and their hermeneutic appraisal, to be wholly entwined. In a seemingly instantaneous and often subconscious manner, empirical "impressions" merge with an evaluative "sense" of how what is apperceptively registered as an "impression" will contribute to (or conflict with) our flourishing as moral agents. Contrary to Locke's ostensibly value-neutral idea of sensation, Hutcheson's "sense" is intrinsically evaluative in that it involves, much in the way the Stoics had argued, the self's "assent" (*sunkatathēsis*) to

40. Darwall, "Motive and Obligation," 133; regarding the shift from a concept of moral obligation anchored in natural law advanced by Grotius, Cumberland, and Pufendorf to the position developed by Hutcheson, Darwall identifies Shaftesbury's surprising elision of the legal aspect of "obligation" as creating the principal explanatory challenge: "how could he have assumed his readers would accept a demonstration of the overriding benefits of a virtuous life as establishing an obligation to virtue?" (ibid., 137). Cumberland's modified account of natural law and obligation, as well as Cudworth's late writings on free will, are significant sources for Shaftesbury's seemingly counterintuitive argument "that the *naturally* good consequences of virtue and ill consequences of vice were sufficient to establish an obligation to virtue without recourse to hellfire or heavenly reward" (ibid., 143).

an impression. However obliquely, Hutcheson's "sense" amounts to a non-propositional kind of moral judgment whereby consciousness appraises an object, less as a neutral fact than as regards the strength with which it feels itself to be related to it. To have a "sense *of*" an impression means not only to be aware of the latter as otherness (*genitivus objectivus*) but also to feel the force with which the object impression affects consciousness itself (*genitivus subjectivus*). Far from responding to the factual presence of an object in strictly neutral, constative fashion, the mind responds to every instant of sensation with an act of approval or disapproval.

There is, then, no fact/value divide in Hutcheson's understanding of sense because even the most rudimentary instance of sensory apprehension of an object reveals a certain fitness or ordering of mind and world toward one another. Anticipating Kant's arguments about the "proportionate accord" of the faculties in the aesthetic reflective judgment, Hutcheson thus sees the warrant for "calling our Power of perceiving the Beauty of Regularity, Order, Harmony, Internal Sense" in this very fitness itself; from which he extrapolates that "Determination to be pleas'd with the Contemplation of those Affections, Actions, or Characters of rational Agents, which we call virtuous" is to be marked "by the name of Moral Sense" (*HI*, 8–9). Importantly, however, "this moral sense has no relation to innate Ideas" (*HI*, 9), nor is it to be thought of as rationally constituted and interested. Indeed, Hutcheson goes so far as to suggest that it need not even be made explicit in order to acquire force for individual consciousness. Inasmuch as Hutcheson regards the self's moral orientation to be logically anterior to its deliberative, interested, and discursive intellect, he is making an early argument for what Martha Nussbaum has called the "intelligence of the emotions" and even for evolutionary psychology. Coming as they do in seemingly infinite shapes and nuances (anger, malice, resentment, envy, jealousy, pride, righteousness, etc.), each emotion or affective response appears to serve some distinct purpose. It may put us on guard or make us aggressive (in situations of danger) or prompt us to exert ourselves to unprecedented degrees and with unfathomed resourcefulness (pride, competitiveness, covetousness, etc.). In each instance, the emotion functions as the catalyst of (and is consummated in) action, be it as a contingent *re*action to a particular social dynamic or as a deliberate and focused regimen of practice aimed at redefining our place in the world over the long term. In either case, it is an emotion, not a motive or interest, which triggers the entire sequence. As Jerome Schneewind notes, the emotional substructure of the self is truly agnostic as regards any moral norms: "the passions are simply causal forces urging us in different directions, no straightforward calculation of amounts of good will suffice for their guidance or for their control."[41]

41. *Invention*, 338.

Yet inasmuch as Hutcheson breaks down all moral affections into two kinds—the altruistic and the egoistic—it is this purely occasional and spontaneous meaning of "sense" that ultimately has to be overcome. Lest we be returned to Mandeville's grumbling hive, there has to be a superego of sorts, a monitoring agency (a.k.a. the "moral sense") capable of attenuating the causal chain between a sense impression and an emotion by evaluating the appropriateness or inappropriateness of our primary affective response. Interposing itself between what might easily deteriorate into a merely hedonistic attachment to our initial emotion and a compulsive action following on its heels, the moral sense mediates between the affective event and its significance within the temporal continuum of an individual life. Seen as a quasi-cognitive agent continually appraising our deceptively neutral sense impressions, Hutcheson's moral sense thus refers an intrinsically aimless set of raw data to the self's ongoing "hermeneutic of happiness."[42] Yet these complex micro-judgments made by the moral sense neither unfold in the social and discursive sphere, nor are they the result of some conceptually driven process of self-reflection. Instead, the moral sense constitutes a basic, evaluative intuition such as ensures that consciousness will not simply be transfixed by the strength with which a given impression affects, moves, or agitates its inner, affective economy.[43] On Hutcheson's account, the moral sense thus allows us to see the value *in* the fact, and it also has us take responsibility for that very appraisal. Put differently, the moral sense serves the ultimate end of a person's moral development by enabling the self to transcend its merely adventitious "sense" of externally occasioned impressions and to move from a causally reactive to an intrinsically significant way of being in the world. The ultimate end of moral agency for Hutcheson thus is a formal (and implicitly secular) *eschaton* of sorts, a state akin to the Stoic objective of *apatheia:* "having the *universal calm Benevolence* ... so as to limit and counteract not only the *selfish Passions,* but even the *particular kind Affections*" (*EPA,* 8).

Hutcheson's paradox (which will recur in Hume) has to do with the fact that, precisely if and when it has been consummately realized as *moral* sense, "sense" no longer furnishes any phenomenological evidence of its actual existence. For its ideal

---

42. On the collapse of the ancient Greek conception of happiness—as articulated by Solon's famous saying: "call no man happy until he is dead" (Herodotus, *Histories,* 1.30–35)—in eighteenth-century narrative and ethics, see Soni, *Mourning Happiness,* esp. his discussion of the modern trial narrative, 234–266.

43. As MacIntyre notes, even as "the perception of those objects causes pleasure and pain, ... the objects themselves are not to be identified with those pleasures and pains ... To be moved by the objects and distinctions disclosed by the moral sense is then to have a way of ordering those other affections which move us to action, and by reflecting upon that ordering we discover that if we allow to self-love and to benevolent generosity that and only that which a regard for moral excellence would allot to them, we shall be pursuing that way of life best designed to ensure our happiness" (*Whose Justice?* 271).

end-point involves the conquest of all partial, interested, or otherwise contingent passions and attachments, a Zen-like expurgation of the self's subjective and hedonistic investments: "Our *moral Sense* shews this to be the highest Perfection of our Nature; what we may see to be the *End* or *Design* of such a Structure, and consequently what is requir'd of us by the Author of our Nature" (*EPA,* 8). Notably, what Hutcheson takes to be an indirect affirmation of the human person's transcendent destiny here constitutes itself as a peculiar nonevent. Instead of a positive revelation or *parousia* of some kind we only have the negation of empirically concrete and phenomenologically distinct affection. Unlike the Christian-Platonist *telos* of a *visio beatifica,* Hutcheson's moral sense furnishes the modern self with a strictly virtual and notably contentless proposition that is neither verifiable nor falsifiable in its own terms. It betokens, in John Keats's memorable phrase, "the feel of not to feel it"—an affectively revealed knowledge said to precede all propositional meaning. Terry Eagleton calls this moral sense "a kind of Heideggerian pre-understanding."[44] Intuitively responding to a crisis of reason that would only be fully outlined by Kant, moral sense theory thus insists on the deeper cognitive value of seemingly nondiscursive modes of awareness that a strict empiricist is bound to reject as outright imaginary: "Our sense of pleasure is antecedent to advantage or interest, and is the foundation of it" (*HI,* 86). The crucial point, here implicitly advanced against Hobbes's and Mandeville's naturalist psychology, is that what truly deserves philosophical attention is not the capacity of objects to trigger (even less determine) our "interest" but the mind's antecedent moral orientation toward a world incessantly soliciting moral discriminations and normative commitments. On Hutcheson's account, then, external objects never simply *produce* "our sense of pleasure," but can only throw into relief its continuous and often unapparent operation.

There is a tendency, observable in all moral sense theorists, to invest this innate faculty not only with the basic moral orientation that Aquinas calls *synderesis,* but to infer (as St. Thomas notably does not) that its presence obviates any further need or desire to reason about our ends. Once again, the price exacted for defending against the Hobbesian-Lockean dystopia of a hedonistic and impulsive self is the attenuation of intellectual curiosity and articulacy. Shaftesbury and Hutcheson obviously reject Hobbes's sociopathic and Locke's hedonistic model of a will relentlessly seeking political or epistemological dominion over other natures. Likewise, they recoil from the further implication that the will per se should prove wholly bereft of any inner nature and lack all attunement to a supra-personal order. Yet they can take this view only by collapsing the will into a nature that they simply choose to interpret as

44. Keats, "In a drear-nighted December" (*Complete Poems,* 288); Keats's friend Richard Woodhouse lamented the poet's stubborn retention of the Cockneyism ("'feel' for feeling. [He] seems fond of it, and will ingraft it 'in aeternum' in our language" [qtd. in ibid.]. Eagleton, *Trouble,* 32.

inherently benevolent, herein faintly echoing the Christian-Platonist view of nature as a divinely ordained and eschatological process, rather than the Aristotelian, teleological view of nature (*physis*) as a self-regulating system.[45] At the same time, Hutcheson also mobilizes a different, seemingly opposed aspect of "sense," viz., as a distinctive inner experience that does not merely occur but of whose presence within ourselves we are also aware. In this apperceptive, proto-Kantian meaning, "sense" does not expire in its very occurrence but, on the contrary, acquires distinctness by virtue of our conscious and explicit ratification of that experience itself. Thus Hutcheson stipulates that "every one calls that Temper, or those Actions *virtuous*, which are approv'd by his *own Sense*" (*EPA*, 8).

Much of Hutcheson's subsequent argument seems to get bogged down in endless distinctions and subdivisions of the affections (Hutcheson himself speaks of his "tedious enumeration" of various kinds of passion [*EPA*, 64]) and on the whole seems a rather muddled affair. As he struggles against the dissolution of moral personhood and action into mindless appetition, Hutcheson probes some of the inherent weaknesses of a radically naturalist account of the will qua self-interested desire. At the very least, he insists, "it requires a good deal of Subtilty to defend this Scheme, so seemingly opposite to *Natural Affection, Friendship, Love of a Country, or Community*, which many find very strong in their Breasts." Yet even if one were to concede the Hobbesian argument, a rather more basic conceptual dilemma remains. For "no Desire of any Event is excited by any view of removing the *uneasy Sensation attending this Desire itself*. Sensations which are *previous* to a *Desire*, or not connected with it, may excite Desire of any Event, apprehended necessary to procure or continue the Sensation if it be pleasant, or to remove it if it be uneasy: But the *uneasy Sensation, accompanying and connected with the Desire itself*, cannot be a Motive to that *Desire* which it presupposes" (*EPA*, 24). Hence the question arises whether the dual use to which the word "sense" is being put by moral sense philosophers (Shaftesbury, Hutcheson, Hume) amounts to purposive equivocation or, alternatively, whether it betrays a prolonged intellectual confusion of sorts. As Gertrude Elizabeth Anscombe (in the even more vexing case of "intention") observes, "where we are tempted to speak of 'different senses' of a word which is clearly not equivocal, we may infer that we are in fact pretty much in the dark about the character of the concept which it represents."[46] While such is indeed the case here, Hutcheson's dual-purpose employment of "sense" also serves a more positive function; for it shows him wrestling with

45. Hutcheson's *Essay on the Nature and Conduct of the Passions and Affections* may well constitute the most extreme statement of a concept of universal benevolence, one whose "mildly ludicrous naivety" Hume had good cause to reject (T. Eagleton, *Trouble*, 37).

46. Anscombe, *Intention*, 1.

questions of willing, intention, and agency for which the prevailing, diametrically opposed languages of rationalism and empiricism prove equally inadequate.[47]

Noting that "Desires are not raised by *Volition*" (*EPA*, 26), Hutcheson sets out to expose the hedonistic account as incoherent and unsustainable on its own terms. As he argues, any attempt to instrumentalize desire as some "Means of private Good" would lead to an infinite regress by making "us *wish* or *desire* to have that advantageous *Desire* or *Affection* ... But if *having the Desire or Affection* be imagined the *Means* of private Good, and not the *Existence of the Event desired,* then from *Self-Love* we should only desire or wish to have the *Desire* of that Event, and should not desire the *Event* itself, since the *Event* is not conceived as the *Means* of Good" (*EPA*, 25). It is hardly self-evident (and certainly not demonstrable) that all desire should be exclusively and necessarily self-focused. What is more, the phenomenology of how desire registers within the self reveals a nascent intellectual dimension; for coded into the inner experience of desire is a prima facie orientation (John Henry Newman will speak of a "view") toward the world as something with which our sense of self and its potential flourishing is always already enmeshed. For Hutcheson, there simply is "no possibility of securing to our selves, in our present State, an *unmixed Happiness* independently of all other Beings" (*EPA*, 82); and whereas "some men have piqued themselves so much upon representing all our Affections as *selfish;* as if each Person were in his whole Frame only a *separate System* from his Fellows" (*EPA*, 54), Hutcheson's *Essay* posits that to feel fundamentally means to "care" much in the same way that Heidegger later speaks of *Sorge* as *Dasein*'s fundamental attunement to the world. That is not to say, of course, that there may not be selfish emotions and desires; there certainly are. Yet it is equally apparent that these desires do not simply unfold like some electrical or neural storm within the human cerebrum but that, like any mental event, they are themselves intentionally grasped and appraised in relation to complex and evolving desires, fears, goals, etc. such as give distinctive and significant reality to the inner life. Even the most inveterate Hobbesian or Lockean philosopher seems to accept the very *datum* of desire or passion as an incontrovertible fact and as the point of departure for his arguments concerning our supposedly irrepressible hedonistic nature. To thus credit desire with actuality, rather

---

47. See MacIntyre, who remarks that Hutcheson's moral sense theory presents "a conception within which incompatibles are united" in two distinct ways, viz., as supplying a "kind of certitude which is grounded in sensory immediacy" and, at the same time, as furnishing not just contingent particulars but bona fide examples of universal moral truths (*Whose Justice?* 277). This tension is also reflected in Hutcheson's intellectual biography, which begins with an eclectic mix of Aristotelian and Calvinist instruction during his early years as a Scottish expatriate in Dublin and, after 1710, at Glasgow University. For a reading of Hutcheson as a quasi-Aristotelian moral realist, see Norton's chapter on Hutcheson in *David Hume*, 55–93.

than second-guessing it as some random fiction or projection, is to link the certitude with which it is experienced to the self-consciousness that such desire is said to inform or condition outright. Yet precisely this kind of "certainty is only attainable by distinct *Attention* to what we are *conscious* happens in our Minds" (*EPA,* 15). Passion never simply eventuates but carries within itself the seeds of intentionality. It is not just *in* but *for* consciousness.

A principal objective of Hutcheson's *Essay* now comes into view, viz., to demonstrate that self-awareness is always already in place as an implicit and necessary part of any account that posits desire as a sufficient condition for human action. What the latter view elides is the difference between the sheer occurrence of desire and a supervening, continuous apperception whereby the event of passion is referred to an agent. This distinction between the necessary and sufficient conditions attendant on (moral) agency finds its corollary in Hutcheson's separation of our "selfish Passions for our own Preservation" from those "*public Passions,* which may engage us into vigorous and laborious Services to *Offspring, Friends, Communities, Countries*" (*EPA,* 46). Extending Shaftesbury's contention that "our Passions and Affections are known to us" and that "*They* are certain" (*SC,* 2:99), Hutcheson (in ways faintly anticipating modern phenomenology) seeks to distinguish between the event of "sensation" and the mental notices whereby that event is reflexively appraised as an object of *attention:* "If, indeed, we confine the Word *Sensation* to the immediate Perceptions of Pleasure and Pain ... then we may denote by the Word *Affection,* those *Pleasures* or *Pains* not thus excited, but resulting from some *Reflection* upon, or *Opinion* of our Possession of any Advantage" (*EPA,* 49) following from it. Indeed, even to speak of sensation requires a point of reference that recognizes the sensation as something present, fading, or past: "every Sensation is accompanied with the *Idea* of *Duration,* and yet *Duration* is not a sensible *Idea,* since it also accompanies *Ideas* of *Internal Consciousness* or Reflection" (*EPA,* 16n). In scrutinizing the intrinsic logic of desire, considered as a primal category of human existence, Hutcheson has established two crucial points: first, that passions and desires are *qualia*—that is, inner experiences of varied, distinctive, and significant character whose very nature presupposes an agent conscious of *having* them and is not simply composed of their ephemeral occurrence. In other words, *qualia* can never be said merely to *befall* consciousness and mechanistically determine it but, in Hegel's parlance, have reality precisely because, and only insofar as, they are *for* a specific consciousness. Second, it is in the nature of all affections to alert the human individual to its constitutive and inescapable sociality. Whether I am embarrassed or pleased by particular cravings and desires I happen to be experiencing, their very appraisal can only unfold because I understand sociality to be a constitutive, albeit not unilaterally determinative, aspect of my being.

11

MINDLESS DESIRES AND CONTENTLESS MINDS

Hume's Enigma of Reason

What Hutcheson is unable to do is to imagine how the self might advance from a strictly apperceptive relation to countless instances of affection to a reasoned and continuous sense of moral agency. To be sure, on Hutcheson's account the self knows itself to be experiencing specific types of affect at specific moments in time, but it no longer appears to know anything else. The gap between the certitude of the moral sense and "disjoined," quotidian existence is too wide, their mediation too tenuous; and with Shaftesbury's aesthetic focus having effectively disappeared, the fragmentation of the person into a series of discrete and seemingly unrelated states looms larger than it ever did. Yet above all, it is the apparent inarticulacy of Hutcheson's moral sense that threatens to wreak havoc with reason. Inasmuch as eighteenth-century moral philosophy, following Locke, credits with reality only those phenomena that can be empirically demonstrated—and, conversely, dismisses as nonexistent anything inaccessible to its empiricist nomenclature of proof—the concepts of person and mind as the putative sources of rational deliberation and moral orientation are put under severe epistemological pressure. This dilemma looms particularly large in David Hume's discussion of the will and the "direct passions" in his *Treatise of Human Nature* (1739–1740). While acknowledging that "tho', properly speaking, it [the will] be not comprehended among the passions," those being his main topic here, Hume also recognizes the need to clarify the relationship between will and passion in more explicit ways than either Shaftesbury or Hutcheson had done. To that end, he opens his chapter by defining the will as a strictly contingent, momentary spike whereby the individual psyche becomes aware of yet another "new" state. Once again we encounter the quintessentially nominalist premise that self-awareness can only arise as a result of a rupture, an instant of conspicuous discontinuity exposing the tattered fabric of human consciousness. In an early letter to Hutcheson, Hume had rejected the notion that human beings possess some *natural*, intrinsically rational purpose, and that they therefore enjoy a basic teleological orientation. "I cannot agree to your Sense of *Natural*. 'Tis founded on final Causes; which is a Consideration, that appears to me pretty uncertain & unphilosophical. For pray, what is the End of Man? Is he created for Happiness or for Virtue? For this Life or for the next? For himself or for his Maker ... these Questions ... are endless, & *quite wide of my Purpose*."[1]

Far from according consciousness any providential design, Hume sees it as but a virtual substratum of discrete experiences and, hence, as liable to being disrupted by so many heterogeneous sensations: "by the *will*, I mean nothing but *the internal impression we feel and are conscious of, when we knowingly give rise to any new motion of our body, or new perception of our mind*" (*HT*, 257). If anything, Hume's *Enquiry Concerning Human Understanding* (1748) states the same point in more uncompromising terms yet: "The motion of our body follows upon the command of our will. Of this we are every moment conscious. But the means, by which this is effected; the energy, by which the will performs so extraordinary an operation; of this we are so far from being immediately conscious, that it must for ever escape our most diligent enquiry." To clinch his point, Hume mounts an argument whereby conscious, deliberative, and reflective action is gradually lost in the fog of physiological, noncognitive processes. Thus a command of the will does not move the leg to which, ostensibly, it is directed but only "certain muscles, and nerves, and animal spirits, and, perhaps, something still more minute and more unknown." In what resembles the nightmarish bureaucracy of Franz Kafka's *Castle*, any edict issued by the mind will produce "immediately another event, unknown to ourselves, and totally different from the one intended ... This event produces another, equally unknown." From which Hume concludes that the power of volition is itself unintelligible, for "were it known, its effect must also be known, since all power is relative to its effect. And *vice versa*, if the effect be not known, the power cannot be known or felt."[2]

1. Hume, *Letters*, 1:33; for a discussion of Hume's changing view of Hutcheson's moral sense arguments, including letters written in 1739 and 1740, see Rivers, *Reason*, 2:242f.

2. *Enquiries*, 65–66.

The obvious descendant of such naturalist and atomist claims is Nietzsche who, in *Beyond Good and Evil,* summarily dismisses the concept and word "will" as nothing more than a "popular prejudice [*Volks-Vorurteil*]" prone to nurture the illusion of a unified, causally efficient, and self-conscious agent. Sounding uncharacteristically sanguine about the "progress" that modernity has supposedly made in this regard, Nietzsche elsewhere remarks that "the old word 'will' only serves to describe a result, a type of individual reaction that necessarily follows from a quantity of partly contradictory, partly harmonious stimuli. [Nowadays] the will does not 'affect' anything, does not 'move' anything any more."[3] Yet for Nietzsche, as previously for Hume, what prompts and licenses this critique is the unexamined assumption that human agency is just another instance of efficient causation, albeit one deluded by supposing itself capable of originating values and, therefore, trapped in the "error of false causation [*Irrtum einer falschen Ursächlichkeit*]."[4] Inevitably, the modern skeptic's critique of the will carries over into a naturalist account of person, consciousness, and responsibility that inevitably ends up repudiating the very possibility of intelligent life itself: "It is, therefore, a *falsification* of facts to say that the subject 'I' is the condition of the predicate 'think.' It thinks [*Es denkt*]: but to say the 'it' is just that famous old 'I'—well that is just an assumption or opinion, to put it mildly, and by no means an 'immediate certainty.' In fact, there is already too much packed into the 'it thinks': even the 'it' contains an *interpretation* of the process, and does not belong to the process itself."[5] Ultimately, Nietzsche concludes (not quite savoring the irony of his "concluding" anything), the very hypothesis of human thought as capable of originating meaning vanishes altogether: "There are no mental causes whatsoever!"[6] One may, perhaps, chuckle at the obvious performative contradiction involved in such a forensic dismantling of the very idea of a conscious, thinking, and willing subject. More serious, however, is the rhetorical sleight of hand with which Nietzsche rejects the formal integrity of "it thinks" while eagerly crediting the mistaken identity of thought with a reality all its own. What, after all, could possibly justify his challenge to the formal integrity and temporal continuity of thought, if in the next sentence he then validates that error itself as the "*interpretation* of a process"? In the end, the vehemence of Nietzsche's irrationalism is but a symptom of a pervasive, anxiety-inducing challenge that modern epistemology had obtruded on human conscious existence itself—viz.,

3. *Anti-Christ,* 12 (no. 14).

4. *Beyond Good and Evil,* 18–19 (no. 19). *Twilight of the Idols,* 177 (no. 3); see also *Gay Science,* 121–122 (no. 127), where Nietzsche interprets the will as one of the "after-effects of the oldest religiosity," a product of pre-modern, "magical" thinking as yet unfamiliar with "the concept of mechanics."

5. *Beyond Good and Evil,* 17 (no. 17).

6. *Twilight of the Idols,* 178 (no. 3).

having to *legitimate* itself in the manifestly inadequate terms of efficient causation, immediate certainty, apodictic evidence, etc. Indeed, to burden existence with the task of self-legitimation is fundamentally misguided, if for no other reason than that *existence* is not just another *proposition* whose correctness we ought to demonstrate by ordinary quantitative or syllogistic methods.

With good reason, Aristotle in his analysis of voluntary bodily movement in *De Anima* (433ᵃ10–ᵇ30) had disambiguated "that which originates movement" into "thought" (*νοῦς*) or some type of "imagination" (*φαντασία*)—that is, "something which itself is unmoved"—and the appetite (*ὄρεξις*), "which at once moves and is moved." Approaching a naturalistic stance not unlike that of Hume and Nietzsche, Aristotle even concedes that "thought is never found producing movement without appetite." Simply put, thought on its own will not move anything. Yet to make appetite the sole (efficient) cause of movement will never do since all appetite-induced movement necessarily presupposes some psychological or representational dimension. Were it otherwise, we would no longer have reason to speak of voluntary movement but merely of a physical reaction triggered by some external and antecedent force. Yet in that case, there neither is anything *to* explain nor *has* anything been explained. In fact, "inasmuch as an animal is capable of appetite it is capable of self-movement; it is not capable of appetite without possessing imagination; and all imagination is either calculative or sensitive."[7] Aristotle's key insight here—subsequently lost sight of by Hume and Nietzsche—is that naturalism (not a bad position to inhabit per se) must resist the temptation of outright reductionism and determinism. For the latter erroneously suppose that to have identified the appetitive as the necessary and seemingly mechanical cause of all movement is also to have isolated the *sufficient* condition for action. As it happens, though, to resolve appetite into the merely physiological inexorably obliterates a number of crucial distinctions and conceptions, and it ultimately jeopardizes the very possibility of philosophical reflection.

To begin with, a reductionist can no longer distinguish between action and process; on his or her view, things merely "happen." Yet to accept such an account of life renders it difficult to see why we should ever seek an "explanation" of animated, "moving" life at all. This takes us to the second paradox, viz., that to view "action" as but the (nonrepresentational) transmission of some efficient, mechanical force confines us to a deterministic ratio of means to ends and, thus, explains, literally, nothing. For in the realm of animated life, explanatory efforts are necessarily solicited by something ambiguous or complex. Explanation here operates within a hermeneutic, not a computational framework. By contrast, the reductionist picture of animated

7. *De Anima,* 433ᵇ30.

movement effectively dissolves the very domain of practical rationality itself. Thus, even as Aristotle concedes that "it is the object of appetite which originates movement," he is quick to add that "to produce movement the object must be more than [a real or apparent good]. It must be [a] good that can be brought into being by action [*πρακτὸν*]; and only what can be otherwise than as it is can thus be brought into being."[8] Not until the later nineteenth century, especially the early phenomenological projects of Herman Lotze and Franz Brentano, does Aristotle's methodological caveat receive its due—viz., that inasmuch as "the instrument which appetite employs to produce movement is bodily, ... the examination of it falls within the province of the functions *common to body and soul.*"[9]

To return to Hume's *Treatise*, its principal objective is not to reveal the tenuous and inscrutable nature of certain effects but, ultimately, to expose the enigmatic nature of consciousness itself. His pessimistic conclusions inevitably follow from the premise (never really argued per se) that "we must ... know both the cause and effect, and the relation between them." Given that to "know," for Hume, means nothing less than for a connection to be immediately and transparently obvious, this epistemological requirement will naturally go unsatisfied since "connection" belongs to the order of conceptual thought rather than that of sense impressions. By its very nature it presupposes what Hume here insists ought to be shown: "Do we pretend to be acquainted with the nature of the human soul and the nature of an idea, or the aptitude of the one to produce the other?"[10] Strikingly prescient of contemporary "neuromythology" (to use Raymond Tallis's phrase), Hume here offers an early version of the reductionist argument that—alternately unfolding within a linguistic or neurological framework—has steadily evolved from Nietzsche to Frege, Wittgenstein, Ryle, Derrida, Lyotard, Daniel Dennett, and David Chalmers. In increasingly strident manner, this tradition posits that consciousness and the entire mentalistic and phenomenological word cloud of will, awareness, intentionality, *noetic* functions, etc. that had gradually formed around the concept of mind is to be rejected outright on account of its alleged indemonstrability in the terms of the modern scientific method, whose exclusive epistemological authority had first been asserted by Galileo, Robert Boyle, and Isaac Newton.

8. Ibid., 433^{a}30; concurring with Warren Quinn's "Rationality and the Human Good," Foot also notes that to accept the Humean picture is not to take a new or different turn in practical philosophy; rather, it is to dissolve the concept of it altogether: "what then would be so important about practical rationality? In effect, [Quinn] is pointing to our taken-for-granted, barely noticed assumption that practical rationality has the status of a kind of master-virtue " (*Natural Goodness,* 62).

9. *De Anima,* 433^{b}20.

10. *Enquiries,* 68.

Yet the naturalist alternative of a strictly computational model of intelligence turns out to be just as speculative, albeit far less supple when it comes to accounting for inner states; and the naturalist's irrepressible optimism as to a future where all remaining technical issues will have been sorted out seems to reproduce the same quasi-religious enthusiasm with which he charges anyone prepared to credit mind with self-awareness, *qualia*, and introspective operations in an *ontological* (rather than contingent, predicative) sense. As Charles Taylor wryly remarks, "it is as though [naturalists] had been vouchsafed some revelation a priori that it *must* all be done by formal calculi."[11] Some such spurious certainty (masking an incapacity to accept any ontological givens of any sort) has Hume wonder how there can be such things as ideas ("a production of something out of nothing") when the "power" or cause supposedly capable of such effects "is not felt, nor known, nor even conceivable by the mind. We only feel the event, namely, the existence of an idea, consequent to the command of the will: but the manner in which this operation is performed, the power by which it is produced, is entirely beyond our comprehension."[12] A reductionism of the sort encountered in Hume's *Treatise* not only involves a strikingly dogmatic and unexamined insistence on mono-causal forms of explanation; it also confuses the nature of practical reason with the procedural skepticism that has come to define the modern epistemological method.

In fact, as Philippa Foot has noted, "where practical reason is concerned no such causal story is needed" since operations of practical reason are "on a par with speculative reasoning. And it is as odd to ask for a causal account of one as of the other." After all, who "would suppose, in the case of speculative reasoning (whether deductive or inductive), that after asserting the premises the reasoner *finds himself* with a belief in the conclusion, as if by a causal mechanism which he might always discover not to have worked."[13] Undeterred by Foot's sensible caveat, John Biro and other contemporary readers remain firmly committed to Hume's skeptical disjunction of common-sense belief from philosophical knowledge—a tendency that had notably set Hume apart from his Scottish contemporaries (James Beattie; Thomas Reid;

---

11. "Overcoming Epistemology," in *Philosophical Arguments*, 6.

12. Hume, *Enquiries*, 65–68; on "neuro-mythology," see Tallis, *Why the Mind is not a Computer*, esp. his discussion of the metaphorical slippages in contemporary neuroscience's reductionist accounts of consciousness (Dennett, Chalmers, Churchland, et al.) whereby descriptions of neuronal activity rely on anthropomorphic and psychological tropes, while simultaneously accounting for operations formerly ascribed to consciousness or "mind" by means of tropes borrowed from the computer and information sciences. For a critique (from a phenomenological tradition) of what might be called contemporary neuro-scientology, see Sokolowski, *Phenomenology of the Human Person*, 204–224.

13. *Moral Dilemmas*, 140.

Henry Home, Lord Kames), who "were centrally concerned to refute scepticism."[14] The famous paragraph in which Hume critiques causality as "deriv'd from nothing but custom" thus finds him completing that very argument by remarking that "*belief is more properly an act of the sensitive, than of the cogitative part of our natures*" (*HT,* 123). Indeed, any instance that finds us drawing some causal inference ultimately reveals the very process of "reasoning" leading up to it to be but "a species of instinct or mechanical power, that acts in us unknown to ourselves."[15] Ultimately, the very notion of a *logos* achieving self-presence within the human mind collapses under the onslaught of a strictly procedural skepticism whose insistence on instantaneous and definitive "justification" appears itself curiously exempt from that very requirement.

Hume's overt querying of belief thus is part and parcel of a much more sweeping discomfort with the very possibility of reason, universals, judgment, and authoritative meanings. It simply will not do, then, to argue that Hume's "recognition of our unreflective, instinctive, and unavoidable acceptance of certain basic beliefs must not be confused with claiming to have a philosophical justification of those beliefs."[16] Needless to say, what renders such a remark circular is the fact that it rests on an oddly restrictive notion of "philosophical justification," which in due course prompts Hume (and his latter-day proselytizers) to define anything not commensurable with that procedure as mere belief and to reject it as supposedly "unreflective" and "instinctive." What drops out of the picture is the entire Platonic-Christian narrative of progressively deeper and more significant introspection, a trajectory premised on the certitude of beliefs that stand to be gradually illuminated by the person's intellectual and spiritual development and hence are not subject to some methodological mandate of instantaneous verification. Notably, then, this alternative conception is not being dialectically engaged, let alone refuted, presumably in fulfillment of Hume's youthful resolution never to enter into a dispute.[17] To regain a perspective not yet controlled by the *Treatise*'s restrictive methodology, one may recall Herbert McCabe's insight that the role of moral orientation—what Aquinas calls *synderesis*—constitutes not a proposition, let alone a conclusion such as purports to discriminate between right and wrong. Rather, it is the condition of possibility for any moral (self-)awareness whatsoever, just as the law of contradiction is not a proposition

---

14. Norton, *David Hume,* 196; see also Rivers (*Reason,* 2:238–264) on the tension between skeptical epistemology and moral virtue in Hume and some contemporaries.

15. *Enquiries,* 108.

16. Biro, "Hume's New Science," 45.

17. "I had a fixed resolution, which I inflexibly maintained, never to reply to any body; and … I have kept myself clear of all literary squabbles" (Hume, *Life,* 4). While Hume did not entirely keep to that resolution, he certainly stands out, even within an Enlightenment culture generally inhospitable to dialectical argumentation, for his rigid avoidance of developing his position out of diacritcally engaging the position of others; see Norton, *David Hume,* 192–196.

among others within the domain of logic but the (ontologically given) premise for all reasoning in the first place. Hume's failure to distinguish between the reach of the understanding and that of reason, between the methods devised by epistemology and the a priori status of reason, as both practical and speculative *logos,* effectively creates many of the dilemmas concerning moral agency that he subsequently declares impossible to solve. Both Locke and Hume "tried to find a type of antecedent which invariably preceded volition" and, in so doing, fashioned arguments "that made it impossible for either of them to take proper account of one of the most important types of human activity, namely acting on a reason."[18]

Having restricted from the outset knowledge to what can be ascertained in terms of cause-effect relations, and having furthermore quarantined that relation itself within an instant or temporal *punctum,* Hume's method effectively short-circuits from the outset what it purports to explain. There can be no verifiable cause-effect relations between mind and world (or mind and body), simply because no "relation" can ever be rendered visible, certainly not *for* a mind said to obtain only as the effect *of* such a causal nexus. Furthermore, Hume all but forecloses on the possibility of human flourishing as a narrative development insofar as he is only prepared to credit as genuine knowledge a representation that stands in a 1:1 mimetic ratio to the phenomenon or "impression" begging explanation to begin with. On this radically nominalist supposition, genuine knowledge can only replicate, never transform, what is immediately given. Its gold standard is "information," the solid specie that Hume juxtaposes to the speculative and volatile currency of "ideas" and the virtues. As Coleridge—arguably among the most astute critics of the Hobbes-Locke-Hume tradition of inquiry—would eventually point out, to accept this model of cognition is to collapse all thinking into mere computation and to conceive mind as a mere information registry ("the faculty . . . which judges of truth and falsehood" [*HT,* 268]).

For his part, Hume does not hesitate to reject all attempts to credit intellectual activity with a "value-added" or counterfactual dimension. While his reductionist account of cognition may ultimately strike us as both irrefutable and pointless, it should also be noted that Hume's entire project in the *Treatise* stands and falls on the strength of a single assumption—itself unexamined and, it would seem, seriously misguided: viz., that all mental processes can and must be captured in terms of a cause-effect dyad that has been imported from Newtonian physics. It is this gratuitous premise—sharply critiqued by Reid (notably a clergyman, not a lawyer) and other Scottish common-sense philosophers—which accounts for Hume's habitual recourse to mind/body examples as he seeks to argue the terminal opacity (perhaps

18. Foot, *Moral Dilemmas,* 140.

outright fictitiousness) of both will and person.[19] Thus he stipulates that "the command of mind over itself is limited, as well as its command over the body," and that "our authority over our sentiments and passions is much weaker than that over our ideas." Already earlier in the *Enquiry,* Hume had insisted on the strictly reactive nature of mind, a mere relay-station for information such as happens to pass through it: "Though our thought seems to possess ... unbounded liberty, we shall find, upon nearer examination, that it is really confined within very narrow limits, and that all this creative power of the mind amounts to no more than the faculty of compounding, transposing, augmenting, or diminishing the materials afforded us by the senses and experience."[20]

Having famously qualified causality as something that exists by virtue of ascription rather than being an intrinsic relation between things per se, Hume is naturally reluctant to credit the will with any substantive continuity or self-identity over time. As with any other instance of necessity, the self's continuity exists only at the level of an "inference." Yet inasmuch as such an inference can only be triggered by discrete and ephemeral affective states or passions, the concept of the will quickly dissolves into those discontinuous, kaleidoscopically shifting states that furnish the raw material for its inference: "I shall first prove from experience, that our actions have a constant union with our motives, tempers, and circumstances, before I consider the inferences we draw from it" (*HT,* 258). Having summarily dismissed Samuel Clarke's rationalism as incapable of deriving motivational force from strictly formal-syllogistic operations, Hume is well aware, however, that his critique of causality puts at risk the very idea of rationality as such. As in Hobbes's *Leviathan,* so in Hume's *Treatise* we encounter a strictly situational and performative model of "demonstrative reasoning;" and as "concerns moral motivation and the possibility of demonstrating moral truths on the basis of discoverable moral relations, Hume plainly sides with Hobbes's skeptical position."[21] In virtually all instances, the *Treatise* opposes any metaphysical (realist) conception of reason as a non-contingent order—something in which beings capable of thought and speech may participate, and toward which they may progress, yet which is in no way contingent on such efforts.

---

19. Philosophers like Hume "have taken over what transpires in the movement of bodies, and too rashly applied it to the thought of the soul" (Reid, quoted in Norton, *David Hume,* 198). As its title suggests, Reid's 1788 *Essays on the Active Powers of the Human Mind* rejects the strictly reactive, stimulus-dependent model mind that Hume appears to develop, as well as the non-cognitive moral theory that supplants will and judgment with an amorphous conception of feeling; see Russell, *Riddle,* 239–263.

20. Hume, *Enquiries,* 68–69, 19.

21. Russell, *Riddle,* 241.

Needless to say, an abundant and lively debate over the meaning of Hume's oeuvre, in particular, the Janus-headed *Treatise,* has shuttled back and forth between accentuating the Pyrrhonist skeptic who stole causality from nature and the seemingly more optimistic naturalist who means to trace "the thorough ... subordination of reason to the feelings and instincts."[22] Rather than trying to split the difference between the naturalist and the skeptic Hume, it may be better to think of Hume's naturalism in the *Treatise* as a procedural attempt to recover from the implications of his epistemological skepticism. Anticipating in this regard contemporary neuro-scientific "explanations" of consciousness, the naturalist Hume makes strong deterministic claims for the way that consciousness depends on intrinsically material, non-cognitive causal sequences while postponing the presentation of evidence into the (indefinite) future, simply because the research needed to clinch the naturalist thesis remains as yet incomplete. Such a highly speculative argumentative structure—which in our time also pervades the neuro-scientific naturalism of Dennett, Chalmers, and others—notably resembles the financial pyramid schemes with which Hume's contemporaries were just becoming familiar. That is, while disputing as a matter of principle the validity of any competing (realist or conceptualist) explanation of consciousness in the present, a radical naturalism secures its own conceptual authority not by actual demonstration but by promising wondrous returns at some indefinite point in the future. Yet inasmuch as such dividends cannot be paid out unless the genesis of consciousness has been traced by way of an infinitely subtle and exhaustive mapping of the "functional organization of the brain" (construed as "*the abstract pattern of causal interaction* between various parts of a system"), that future pay-off will in fact never be reached.[23]

While it is well beyond the scope of this book to sift the strident anti-psychologism of Frege, Wittgenstein, Ryle (his disciple, Daniel Dennett), John Searle, Paul Churchland, David Chalmers, and many others, it bears pointing out the origins of that project in Humean naturalism. By steadily dissolving the contiguity of mental pro-

22. Kemp Smith, *Philosophy of David Hume,* 84; major accounts of Hume as epistemological skeptic include Robert Fogelin, *Hume's Scepticism in the Treatise of Human Nature* (1985); Annette Baier, *A Progress of Sentiments: Reflections on Hume's "Treatise"* (1991); following Kemp Smith's reading of Hume as naturalist, and taking a more contextual approach that links Hume with Descartes, Malebranche, Bayle, and others are Barry Stroud, *Hume* (1977), and John Wright, *The Sceptical Realism of David Hume* (1983). Addressing this bipolar interpretive tradition head-on, and concluding that the *Treatise* fundamentally lacks all unity, is Norton, *David Hume,* and, most recently, Russell, *Riddle.* Echoing Jennifer Herdt, *Religion and Faction in Hume's Moral Philosophy* (1997), Russell has recently revived a strain of Hume interpretation that regards the arguments of the *Treatise* as (partially) unified by Hume's "fundamental irreligious intentions *throughout the Treatise*" (*Riddle,* 261); on Hume and religion, see also Rivers, *Reason,* 2:282–329.

23. Chalmers, *Conscious Mind,* 247.

cesses and the coherence of representations not only *in* but *for* consciousness, Hume's *Treatise* paved the way for a "deeply counterintuitive" conception of philosophy that seeks to deny consciousness the reality or substance of its "contents." For the semantic and value-saturated notion of mental representations or specific psychological *qualia,* the ideal of value-neutral and non-perspectival "information" has been substituted—albeit at the expense of neuro-science's habitually taking recourse to various anthropomorphisms while tabulating the effects and capabilities of such information.[24] Part of the sheer implausibility of the project involves a basic logical flaw, viz., that it premises a causal account of consciousness on a detailed analysis of material processes and syntactic *structures* evidently generated by everyday life while at the same time denying these structures any *functional* dimension. In fact, concepts such as structure or system cannot be employed in the absence of some specific function that is being served by them. To suppose otherwise—viz., to argue that all the (allegedly subconscious) infrastructural processes on which neuro-scientific accounts focus so intently are but so much sound and fury signifying nothing is to repurpose much of philosophy to proving a strictly negative proposition—hardly an auspicious undertaking. Not without reason, Plato had insisted that a "sense of wonder is the mark of the philosopher. Philosophy indeed has no other origin."[25] The first and abiding task of philosophy must be to respond to what *is,* that whose manifest existence elicits wonder. Both the reality of that *at* which our wonder is directed, and the reality *of* such "wonder" (*thaumazein*) itself, cannot be disputed. In either case, existence is not to be confused with a proposition.

To return to Hume, then, Book 1 of his *Treatise* expressly links the operation of "reasoning" with a skepticism that relentlessly breaks down antecedent convictions, certitudes, and beliefs: "Let our first belief be never so strong, it must infallibly perish by passing thro' so many new examinations, of which each diminishes somewhat of its force and vigor ... when I proceed still farther, to turn the scrutiny against every successive estimation I make of my faculties, all the rules of logic require a continual diminution and at last a total extinction of belief and evidence" (*HT,* 122). By its very nature, such a method of "demonstrative reasoning" will lead to endless

24. Tallis, *Why the Mind is not a Computer,* 18; Tallis's introduction to this book is both succinct and compelling, particularly in showing how the neuro-philosophical project begs the question on a large scale: "quite apart from its inability to explain 'ground floor' sentience and rudimentary contents of consciousness such as qualia, there is no remotely plausible neurophilosophical account of, for example: a) the intentionality of perceptions and thoughts; b) their (ineliminable) indexicality; c) the unity of consciousness; and d) the architecture of everyday awareness" (ibid., 23). For an incisive critique of naturalism's exclusive focus on "third-person ways of knowing" and its methodological commitment to a "bottom-up or parts-to-whole determinism," see Moreland, *Recalcitrant Imago Dei* (quotes from 8, 17).

25. Plato, *Theaetetus,* 155d.

prevarication; it "never influences any of our actions, but only as it directs our judgment concerning causes and effects; which leads us to the second operation of the *understanding*."[26] The closing, metonymic slippage from "demonstrative reasoning" to the faculty of "understanding"—and the implicit merging of our participation in the *logos* with contingent discursive practice—was to exasperate Coleridge greatly. Yet the demotion of the *logos* to the merely "discursive" and "sciential" operations of the understanding, so vexing to the author of *The Friend* and the *Biographia Literaria,* is the official, indeed eponymous premise of Hume's *Treatise,* viz., the "*Attempt to Introduce the Experimental Method of Reasoning into Moral Subjects*" (*HT,* 1). For Hume, reason only subsists in so many discrete, contingently wrought acts of computation, and any such application of "reasoning" boils down to occasional arguments (enjoying a very short half-life) and a certain pragmatic resourcefulness with which rhetorical and logical schemas are being put to use. In so qualifying causal relationships between discrete entities as occasional and contingent in nature—a matter of observation, inference, and ascription—Hume has changed rationality from the (Platonic and Aristotelian) ontology of timeless and immutable forms to a wholly subjective property of the understanding. To suggest that necessity is but an inference drawn from "the observation of an uniformity in the actions" of particular things and persons is to premise the very fortunes and possibilities of philosophy on the manifestly fickle character of the human individual: "an hour, a moment is sufficient to make him change from one extreme to another, and overturn what cost the greatest pain and labour to establish. Necessity is regular and certain. Human conduct is irregular and uncertain" (*HT,* 259).

Hume's rather perplexing defense against the specter of outright entropy is to harden the analogy between the necessity of natural and intra-psychic processes. Of course, he can do so only by venturing pronouncements about nature as it might be thought to operate independently of human observation. Alternating between affirmations of what is (putatively) self-evident and rhetorical questions insinuating (ultimately indemonstrable) conclusions, Hume's prose seems rather strained as he seeks to extricate himself from his pessimistic view of causally impaired, human knowledge: "*We must certainly allow,* that the cohesion of the parts of matter arises from natural and necessary principles, whatever difficulty we may find in explaining them. And *for a like reason* we must allow, that human society is founded on like

26. *HT,* 266 (italics mine); elsewhere, Hume insists that "demonstrative reason discovers only relations." By contrast, the philosophical realist who hazards pronouncing a particular action "to be virtuous, and such another vicious" oversteps the bounds of reason as Hume perceives it: "for what does reason discover, when it pronounces any action vicious? Does it discover a relation or a matter of fact?" (*HT,* 298–299n); on Hume's frequent slippage from reason to reasoning to understanding, see Norton, *David Hume,* 215f. and 228f.

principles ... *For is it more certain*, that two flat pieces of marble will unite together, than that two young savages of different sexes will copulate?" (*HT*, 258; italics mine). Confronted with the epistemological dead-end of an utterly "irregular and uncertain" human agency, Hume strikes a declarative pose: "*To this I reply*, that in judging of the actions of men we must proceed upon the same maxims, as when we reason concerning external objects" (*HT*, 259; italics mine). It is just this apodictic view of practical reason as a subsidiary or epiphenomenon of scientific understanding that will erode the conceptual depth and psychological complexity in Hume's broader argument. Merely to admire "how aptly *natural* and *moral* evidence cement together" and that "the actions of the will ... arise from necessity, and that he knows not what he means, when he denies it" (*HT*, 261, 260) is to argue from—but, evidently, not prove per se—a view of human nature wholly subject to the exhaustive determinacy of inanimate matter.

As is often the case in philosophy, the force of a particular argument rests on various unacknowledged metaphoric equivalences and transfers. Not only is "regularity" taken to warrant the inference of "necessity," but it is simultaneously construed as the antonym of "liberty": "'Tis commonly allow'd that mad-men have no liberty. But were we to judge by their actions, these have less regularity and constancy than the actions of wise-men, and *consequently* are farther remov'd from necessity" (*HT*, 260; italics mine). Likewise, in a telling paraphrase, Hume remarks that "liberty *or chance* ... is nothing but the want of that determination, and a certain looseness, which we feel in passing or not passing from the idea of one to that of the other" (*HT*, 262; italics mine). In both instances, the assimilation of practical to theoretical rationality, of *phronēsis* to *technē*, drastically foreshortens the scope, complexity, and significance of human agency. Hume's equation of freedom with indeterminacy or outright "indifference"—seemingly licensed by the obverse (and no less specious) supposition that "regularity" constitutes prima facie evidence of necessity—distorts the picture from the outset. After all, "regularity" could itself be something chosen, such as a violinist practicing scales and honing intervals, left-hand shifts, bowing sequences, and so forth so as to habituate the mechanical aspects of playing to the greatest possible degree—and doing so precisely in order to free up mental reserves for interpretive choices and the dynamics of interactive ensemble playing. Behind Hume's theory stands modernity's irrational distrust of the habits and, by extension, the virtues as supposedly coercive, "unfree," and mindless. As we shall see in the next chapter, a critique of this line of argument will have to begin by desynonymizing "habit" from "custom" and Aristotelian and Thomist "action" (*praxis; operatio*) from modern, externally cued "behavior."[27]

---

27. See below, 357–368.

Hume's reigning assumption appears to be that freedom is something people are in the habit of asserting *as a proposition,* rather than drawing on it as an implicit premise for the pursuit of specific and often long-term goals and objectives. Against this view, it ought to be maintained that, far from being concerned with the quasi-fetishistic preservation of their indeterminacy or autonomy, human beings are much more likely to be consumed with the pursuit of specific goals, the cultivation of particular skills, talents, and resources as the means for those goals, and (potentially) with reflecting on the relationship between means and ends in the form of ongoing micro- and macro-judgments or acts of assent. Far from being the object of a practical demonstration that might involve my "choosing" to assist a pedestrian who has slipped on an icy sidewalk, freedom constitutes the background condition for the cultivation of an enduring moral personhood, or "Personëity" (as Coleridge was to call it). Ethics, Charles Taylor notes, "involves more than what we are obligated to do. It also involves what it is good to be ... The sense that such and such is an action we are obligated by justice to perform cannot be separated from *a sense* that being just is a good way to be."[28] Likewise, Iris Murdoch, on whose oeuvre Taylor is approvingly commenting here, had remarked how moral action and freedom of choice are inseparable from precisely those background notions of habit and virtue that are anathema to Hume's "scientific" approach. Far from being the fruits of some speculative excess or a mind expiring in mechanical drudgery, both virtue and "good habit" reflect "a just mode of vision" directed at "the world as it is" and an ethos of sustained "attention" to its possible transformation in the direction of the good: "A philosophy which leaves duty without a context and exalts the idea of freedom and power as a separate top level value ignores this task and obscures the relation between virtue and reality. We act rightly 'when the time comes' not out of strength of will but out of the quality of our usual attachments and with the kind of energy and discernment which we have available. And to this the whole activity of our consciousness is relevant."[29]

Surely, the individual has occasion to prize freedom only insofar as it enables her or him to pursue a specific *telos* to which she or he is either intuitively or deliberatively committed. Yet it is just this notion of a moral sense orienting human agents a priori to, and independent of, the specific relations and circumstances when some kind of action is called for which Hume rejects as but an unlicensed inference. Being committed from the outset to a single, all-explaining model of (efficient) causation, Hume opposes any affiliation of freedom and will with a broader, inherently evalu-

28. *Dilemmas and Connections,* 9 (italics mine).
29. *Sovereignty,* 89.

ative and normative conception of human flourishing.[30] Built into Humean consciousness is a kind of *attention-deficit-disorder,* albeit in Locke's and Hume's cases not a pathology but a willful, even principled refusal to credit mind with anything like rationality, continuity, and self-identity. Such a conception of mind appears bereft of any ability to develop and sustain a reflective and strategic perspective on the data that enter into it, and to monitor the reactions and desires for which these data are seen (by Hume) to be an efficient and instantly determinative cause. Action not only seems indistinguishable from reaction, but it also appears to expire in its very occurrence, leaving no memory trace behind: "Actions are by their very nature temporary and perishing" (*HT,* 264). Once again, we find how repudiating some key concepts of humanistic inquiry—for example, that of the will qua rational appetite, as well as those of self-consciousness, person, and knowledge—inexorably causes other concepts to deteriorate and to lose all significance. Hume's basic stance shows how the reductionist notion of the self as *cogito* and as "constituted by a purely theoretical intention ... involves a withdrawal from action, and so from positive, practical relations with the Other." Ultimately, such a picture leaves us "committed to explaining knowledge without reference to action."[31] It is the constitutively human category of "action" that here begins to fray and unwork itself. By insisting that action should divulge its operative structure as an instantaneous and necessary relation of cause and effect, Hume implicitly removes time and narrative from our appraisal of action; and yet, "we identify a particular action only by invoking two kinds of context, implicitly if not explicitly," Alasdair MacIntyre argues, viz., by "plac[ing] the agent's intentions ... in causal and temporal order with reference to their role in his or her history; [but] we also place them with reference to their role in the history of the setting or settings to which they belong" (*AV,* 208). It follows that to interpret human action necessarily involves a deferral of closure, and a suspension of the expectation of instantaneous verifiability such as is implicit in the Newtonian model of causation (and deemed, by Hume, to be lacking in the operations of mind and the constitution of personal identity).

That model, however valuable to the scientific quest for ever more accurate and particular "information," is fundamentally inapplicable to interpretive, humanistic forms of inquiry, which by their very nature achieve knowledge over time and as something constituted in narrative rather than syllogistic form. Inasmuch as "narrative history of a certain kind turns out to be the basic and essential genre for the

30. Schneewind is right to note that "the will plays no essential role in Hume's theory of action" and that "there is in fact no formula that accounts for our willing." Unlike Locke, however, Hume does not go so far as to restrict causal efficacy to the most recent or "present uneasiness" of some ephemeral passion or desire (*Invention,* 359–360).

31. Macmurray, *Persons in Relation,* 20–21.

characterization of human actions," it follows for MacIntyre "that the concept of an intelligible action is more fundamental than the concept of an action as such" (*AV*, 208–209). Arguing from a similarly Aristotelian perspective, Hannah Arendt had previously maintained that significant action is inseparable from rational agency: "Without the disclosure of the agent in the act, action loses its specific character and becomes one form of achievement among others ... In these instances action has lost the quality through which it transcends mere productive activity." Taking a view diametrically opposed to Hume's, Arendt moreover stresses that action is precisely what reveals the sociality, togetherness, and relatedness of human beings: "Action ... always establishes relationships and therefore has an inherent tendency to force open all limitations and cut across all boundaries" (*HC*, 180, 190). If, then, we accept Aristotle's basic premise concerning the intrinsic rationality of social practice (however much we may demur at particular practices sanctioned by the fourth-century Athenian polis), we can also grasp the close link between classical political thought's "insistence on the living deed and the spoken word" and the ideal of human flourishing qua action. On this view, speech is never (for either Aristotle or Arendt) some incidental and inconsequential enunciation of private opinion by a putatively autonomous "self" (*idios*). Instead, already by virtue of the fact that all speech (however carelessly executed) draws on the rich social, semantic, and rhetorical practices and norms that render meaning intelligible *for others*, speech belongs *eo ipso* to the realm of political action. It "actualizes" (this being Arendt's translation of Aristotle's *energeia*) the rationality of human agency, which never merely aims at some particular objective but—in a more or less articulate and publicly witnessed performance of a purposive action—substantially realizes the end of public reason (*logos*) itself: "the end (*telos*) is not pursued but lies in the activity itself." Ultimately, human action "lies altogether outside the category of means and ends; the 'work of man' is no end because the means to achieve it—the virtues, or *aretai*—are not qualities which may or may not be actualized, but are themselves 'actualities.'"[32]

It is of course hardly surprising that in collapsing the will into a mindless tropism occasioned by the conversion of external stimuli into implacable motives,

32. Arendt, *HC*, 206–207. As Hegel was to point out in a markedly Aristotelian turn in the "Preface" to his *Phenomenology of Spirit*, the concept of action is ultimately inseparable from that of teleology; yet for Hume, being in this regard a true descendant of Hobbes and Locke, "there is no reference ... either to a goal or *telos* of human nature." The disappearance of final causality from eighteenth-century models of reason is, of course, not unique to Hume. Rather, as his subsequent writings on religion—esp. his *Natural History of Religion* (1757) and the posthumous *Dialogues concerning Natural Religion* (1779)—make clear, his rejection of "the Calvinist and Lutheran view of the relation of moral principles to human motivation" is part of a broad-based attempt "to free our understandings of morality ... from any need to appeal to supernatural origins or maintenance" (Schneewind, *Invention*, 368, 358, 369).

Hume should also have lost the concept of action along the way. Anticipating Priestley's and Schopenhauer's later arguments to the same effect, Hume thus contends that proof of free will cannot take the form of some counterfactual experiment, such as "when by a denial of it we are provok'd to try" and show that, say, our hand or leg "moves easily every way." For in all such instances of some "faint motion" or other, we find that "the desire of showing our liberty is the sole motive of our actions [and that] we can never free ourselves from the bonds of necessity." In fact, Hume concludes, the illusion of freedom is but a result of our insufficient awareness of "every circumstance of our situation and temper, and [of] the most secret springs of our complexion and disposition" (*HT,* 262–263). It is here that we begin to see the consequences of Hume's arguments, earlier in the *Treatise* and repeated later in his first *Enquiry,* according to which the only causes admitted to operate efficiently *within* the mind are those that demonstrably work *on* the mind. In other words, for a cause to properly qualify as such it has to be mindless per se. What prompts Hume to assert that "reason alone can never be a motive to any action of the will" (*HT,* 265) is the Newtonian premise that a cause is but a mechanical force, and that it cannot partake of the same immaterial and ideational order that allegedly characterizes mind itself. To the extent that only material causes are said to admit of observation, the effect of this premise is to render doubtful the very reality of mind.

Hume's approach thus conjures up a scenario in which reason and consciousness are at the mercy of fluctuating, external sensations causally impacting the mind as "impressions," which in turn are said to be mimetically converted into motives, passions, and ephemeral (re)actions.[33] Given the *Treatise*'s extreme nominalism, Hume naturally cannot explain just "what kind of experience desiring is," nor devise some conceptual meta-language that would capture the nature of passion and its relation to mind. Still, it does appear that "all connotation of tempestuousness has gone from the word 'passion' as Hume uses it," presumably because there is no longer a center of self-awareness relative to which desire or passion could be appraised as being more or less intense, tempestuous, or placid.[34] Instead, mind itself appears but an epiphenomenon of fluctuating and intrinsically unself-conscious desires or passions and, for that reason, cannot experience them as distinctive *qualia.* Terence Penelhum's claim that "what makes it the passion it is, rather than some other, is therefore the felt quality it has" may well represent Hume's assumption, but it does not quite amount

33. David F. Norton concurs with Norman Kemp Smith's earlier characterization of passion in Hume as a "general title for the instincts, propensities, feelings, emotions and sentiments, as well as for the passions ordinarily so called" (quoted in Norton, *David Hume,* 56).

34. Foot, *Moral Dilemmas,* 133; on the question of intensity and value in Hume's passions, see Penelhum, "Hume's Moral Psychology," 249f., and Árdal, *Passion and Value,* 95ff.

to a coherent argument in the context of Hume's overall psychology. For to characterize each passion as a "distinct feeling" presupposes a phenomenological center sufficiently self-aware as to know herself or himself to be experiencing *this* rather than *that* passion. Yet it is precisely the possibility of such an anterior self-awareness that Hume means to dispute by construing mind as the *effect,* not the judge, of the discrete *qualia* that so inexplicably traverse it.[35] Thus Hume's famous depiction of the "mind as a kind of theatre, where several perceptions successively make their appearance; pass, re-pass, glide away, and mingle in an infinite variety of postures and situations" (*HT,* 165), aside from baiting the anti-theatrical prejudice still running strong in Calvinist Scotland, vividly conjures up the outright phantasmagorical nature of consciousness. At the same time, the radical perspectivalism that lurks in Hume's theatrical metaphor shows him to be truly the heir apparent of the greatest of Sophists, Protagoras. Looking with much diffidence upon "this suppos'd pre-eminence of reason above passion" (*HT,* 265), Hume moves well beyond the anatomical view of moral agency previously taken by François de La Rochefoucauld, Pierre Bayle, Pierre Nicole, and Bernard Mandeville, among others. For he now conceives mind, reason, judgment, and will as so altogether reactive to ephemeral and apparitional stimuli as to render the former terms at best epiphenomenal and, quite possibly, altogether fictitious.

With their relevance to philosophical inquiry steadily dwindling, an entire inventory of nuanced and differentiated concepts is on the verge of disappearing: "Since reason alone can never produce any action, or give rise to volition, I infer, that the same faculty is as incapable of preventing volition, or of disputing the preference with any passion or emotion" (*HT,* 266). In a curious attempt at turning Stoicism's basic view against itself, Hume furthermore notes that passion is not only "accompany'd with some false judgment" but, no matter how unreasonable the passion at stake, that we ultimately realize that "'tis not the passion, properly speaking, which is unreasonable, but the judgment" (*HT,* 267). By disaggregating passion from judgment (and thus overturning a central tenet of Stoic thought), Hume immunizes passion—his anti-foundationalist source of all natural, social, and moral cognition—against all critical scrutiny. Indeed, there can be no counterfactual, evaluative role for rational thinking per se, no remedial method or regimen of logic to sort out the

35. Penelhum, "Hume's Moral Psychology," 247; as Moreland puts it, "mental states are characterized by their intrinsic, subjective, inner, private, qualitative feel, made present to a subject by first-person introspection. For example, a pain is a certain felt hurtfulness. The true nature of mental states cannot be described by physical language, even if through study of the brain one can discover dependency relations between mental/brain states ... Thoughts, beliefs, desires and sensations have intentionality—they are of or about objects—but physical states aren't about anything. They just are" (*Recalcitrant Imago Dei*, 21).

chaos of ephemeral representations. Completing a sustained crescendo of formulations to the same effect, Hume winds up with the notorious observation that "the principle, which opposes our passion, cannot be the same with reason," and that "reason is, and ought only to be the slave of the passions, and can never pretend to any other office than to serve and obey them" (*HT,* 266). Having already sidelined a number of key concepts previously indispensable for understanding moral agency (e.g., judgment, will, personal identity, intention, etc.), Hume's argument here takes a further, seemingly postmodern turn. For in suggesting that passions are at all times potentially convertible with other kinds of passions and thus all but indistinguishable from one another, he now begins to erase the line that had previously demarcated the dispassionate superego of reason from the passions. As he notes, "every action of the mind, which operates with the same calmness and tranquillity," risks being "confounded with reason" inasmuch as "there are certain calm desires and tendencies, which, tho' they be real passions, produce little emotion in the mind, and are more known by their effects than by the immediate feeling or sensation."

Thus it is that reason mutates into a purely virtual and enigmatic notion—a mere inference such as will be erroneously and (this being a particularly ironic twist) precipitously drawn "by all those, who judge of things from the first view and appearance" (*HT,* 268). Philippa Foot is surely right to find Hume unable "to give a proper account ... of the part that *reason* plays in determining human choice." Indeed, if practical reason frequently "has premises and conclusions," in other cases, "we act *on* a reason or *for* a reason, without having been through any reasoning." On these occasions, "practical reasoning is ... the background" (to be thematized only if one is confronted with a differing view of moral action); and yet, neither Hume nor Locke before him is able to "accommodate actions done for a reason" because to take that view is to premise mind as capable of internal causation, of *generating* meanings over and above receiving impulses or experiencing stimuli.[36] The distinction here at issue would later be cannily reformulated as that between "implicit and explicit reason" in John Henry Newman's *Oxford University Sermons.* Arguing that "Reasoning, then, or the exercise of Reason, is a living spontaneous energy within us, not an art," Newman insists that "all men reason, for to reason is nothing more than to gain truth from former truth, without the intervention of sense; to which brutes are limited." At the same time, it is evident that "all men do not reflect upon their own reasonings, much less reflect truly and accurately, so as to do justice to their own meaning; but only in proportion to their abilities and attainments." Yet unlike Hume, Newman does not view that state of affairs to license the sweeping conclusion that whatever act of mind has not been analyzed and rendered explicit is therefore undeserving of

36. Foot, *Moral Dilemmas,* 135–136.

the name reason. Rather, a distinction is to be drawn between "these two exercises of mind as reasoning and arguing, or as conscious and unconscious reasoning, or as Implicit Reason and Explicit Reason." To so distinguish is to understand that "all men have a reason, but not all men can give a reason."[37]

Yet precisely this very possibility that philosophy itself might rest on something permanently given and implicit is anathema to Hume's understanding of the trade. His startling suggestion that philosophy's constitutive act of making reference to reason may in the end have been nothing more than a case of mistaken identity certainly is consistent with the nominalist framework in which all reality—be it material or intellectual in kind—resolves itself into an infinite number of special, distinct, and incommensurable cases. Hence, if one should wish to prod the mass of self-absorbed, self-interested, and distractible human beings in any particular direction, it "will commonly be better policy to work upon the violent than the calm passions" and to take the individual "by his inclination, [rather] than what is vulgarly call'd his *reason*" (*HT,* 269). Behind this pessimistic argument—which Edmund Burke was to reiterate at the opening of his *Thoughts on the Present Discontents* (1770), and which Schopenhauer, feeling much the same way about "the swinish multitude," was to inflect into a pessimism at once metaphysical and existentialist in nature—we already find lurking the Arnoldian crisis of culture and the specter of anarchy. For Hume's suggestion that social exchange pivots on the vicarious manipulation of collective passions spells doom for the Enlightenment project of a rationally deliberative public sphere, even before that concept had been given its consummate, Kantian expression.

Adding further spice to the argument is the fact that Hume's shape-shifting passions are themselves derivatives of (or inferences drawn from) some "violent and sensible emotion of mind." On this account, what separates reason from passion is at most a matter of degree, not kind: "By *reason* we mean affections of the very same kind with the former [i.e., passions]; but such as operate more calmly, and cause no disorder in the temper: Which tranquillity leads us into a mistake concerning them, and causes us to regard them as conclusions of our intellectual faculties" (*HT,* 280). Racked by ambiguous terminology and tenuous predication, Hume's prose toward the close of §2.3.8 rather resembles Jean Tinguely's famous installation at the Museum of Modern Art in 1960 of a machine that progressively destroys itself. Thus we are told that "violent passions have a more powerful influence on the will" than calm ones, only to find that statement disarmed by hasty assurances (without any attempt to argue the point) that the calm passions may yet be able "to controul them in their most furious movements." At the same time, it seems increasingly doubtful that we should be able to tell apart the one from the other; for "what makes this whole affair more uncertain, is, that a calm passion may easily be chang'd into a violent one" as a

37. *Fifteen Sermons* (no. 13), 258–259.

result of a volatile temper or shifting circumstances. In a breathless sequence so ridden with displacement, condensation, and transference as to make Freud's *Interpretation of Dreams* look positively Augustan, Hume goes so far as to cast doubt on the very identity of any particular passion because of its propensity for "borrowing ... force from any attendant passion, by custom, or by exciting the imagination." While the doctrine of the calm passions may well be "Hume's main card in the game against rationalist psychology," difficulties with such a concept go well beyond the fact that, as such, calm passions are unlikely ever to become phenomenally distinct.[38] Inasmuch as they remain "imperceptible" (*HT,* 181) *for* consciousness, the passions are implicitly deprived of all identity and distinctness vis-à-vis one another. Hume's problem here is that he wishes to retain passions as distinct *qualia* while simultaneously construing mind itself as an effect, or epiphenomenon, of their theatrical, shape-shifting efficacy.

It is in his famous discussion of "personal identity" in Book 1 of the *Treatise* that Hume presses the point most aggressively. Extending Locke's reduction of personhood to conscious states that can be expressly indexed *by* the self ("*Socrates* asleep, and *Socrates* awake, is not the same Person"), Hume regards personal identity as nothing more than a belief or "fiction" proclaiming continuity without being able to proffer the requisite evidence. Early in the *Treatise,* Hume had already stipulated that the very category of "duration is always deriv'd from a succession of changeable objects, and can never be convey'd to the mind by any thing stedfast and unchangeable" (*HT,* 30). Not until Kant was the circular fallacy of that skeptical argument exposed—viz., that to speak of a *succession* of states or objects already presupposes the "pure intuition" of time for its condition of possibility. Yet for Hume, philosophical inquiry involves above all the cultivation of a view from nowhere such as will (supposedly) safeguard him against any logically unwarranted commitments or presuppositions (philosophical, cultural, religious). Thus personal identity for Hume is but one more "belief" begging to be scrutinized as regards its etiology: "from what impression could this idea be deriv'd? ... 'tis a question, which must necessarily be answer'd, if we wou'd have the idea of self pass for clear and intelligible" (*HT,* 164). His main point here is plain enough; it concerns the apparent logical impossibility of successful, demonstrable self-reference. Inasmuch as the self is "nothing but a bundle or collection of different perceptions, which succeed each other with an inconceivable rapidity, and are in perpetual flux and movement," all it can index is an externally occasioned perceptual state in the here and now.

Trapped in an endless stutter of self-affirmation, the so-called self forever begs the pivotal question concerning its reality: "I never can catch *myself* at any time

38. Penelhum, "Hume's Moral Psychology," 249.

without a perception, and never can observe any thing but the perception" (*HT,* 165). Inevitably, Hume's critique here not only challenges the temporal continuity of the self but even the basic philosophical concept of "identity" itself, one without which no rational argument can ever take shape. It is one thing to insist that "there is no impression constant and variable," yet quite another that the kaleidoscopically shifting perceptions and inner states can never be discrete, successive, and related "aspects" *of* an identical being.[39] In its concerted attempt at recovering from the more bizarre sophisms to which Humean reductionism had given rise, modern phenomenology from Brentano onward insists that the identity of an object (which the self de facto becomes when it is made into a focal point of sustained introspection) is already presupposed by the very language of "perception." Pablo Picasso's *Le Guitarist* (1910) and the entire cubist aesthetic would seem rather pointless if we did not understand the disassembled "aspects" to be various manifestations *of a single entity*. For his part, Hume presses on toward a logic of radical disintegration by qualifying his earlier analogy of "the mind [as] a kind of theatre" and insisting that "the comparison of the theatre must not mislead us. They are successive perceptions only, that constitute the mind" (*HT,* 265). In other words, unlike the theater, which crucially presupposes the synthesis of perceptions in a unifying plot and as a unified work, Hume rather points toward the absurd, literally demented *durée* of theatrical representation that we find in Pirandello, Artaud, or Beckett.

Following Hume's choice of terms, modern naturalist epistemology has only reinforced an irrational argument that not only jeopardizes the continuity and identity of the person as an epistemological and, by implication, moral agent but, in so doing, also obliterates various key concepts indispensable to rational argument in either field: judgment, responsibility, memory, will, freedom, and even identity.[40] Even a

---

39. Now and then, Hume exhibits uneasiness with his own reasoning, such as in his half-hearted attempt (at 1.4.6.12–13 in the *Treatise*) "to distinguish pretty exactly betwixt numerical and specific identity" and to concede that "an oak, that grows from a small plant to a large tree, is still the same oak; tho' there be no one particle of matter, or figure of its parts the same. An infant becomes a man, and is sometimes fat, sometimes lean, without any change in his identity" (*HT*, 168).

40. For an extreme version, see Derek Parfit's 1984 *Reasons and Persons* (esp. 199–280), which offers a radically discontinuous notion of the individual who is not even the same upon waking up from an afternoon nap, and in five years time is but a "descendant" of the earlier person. For different critiques of this extreme nominalist account, see Spaemann, who notes that Parfit cannot imagine any of the "more profound relations between human beings" (love, fidelity, friendship) inasmuch as doing so presupposes "the continuity of the 'thou.'" Seeing Parfit as the intellectual heir apparent of Hume, Spaemann also notes that neither thinker is able to think of the I-Thou relationship: "It is characteristic that they exclude the I-Thou relation between persons, that being the true locus where personhood is discovered; they only ever speak of persons in the first or third person" (*Grenzen*, 422; trans. mine). From a slightly different angle, Charles Taylor critiques Parfit's logical reductionism by noting that it does not suffice to construe the self, in Lockean fashion, as "a matter

cautious reader such as John Biro all too readily accepts Hume's characterization of self-identity as something that "we come to attribute ... to ourselves." In fact, we do not "attribute" our identity to ourselves, nor do we "imagine," as Hume puts it, that "we are every moment intimately conscious of what we call our SELF" (*HT,* 164). The self is not the object of a proposition; it is not simply another claim to be predicatively secured in an act of self-reference.[41] To his credit, though, Biro acknowledges that Hume's view of the mind as "tracing the succession of time" cannot but "presuppose *its* identity over time," and that even to form hypotheses regarding causal relations requires a mind endowed with temporal continuity and self-identity. For only on that premise is mind capable of *re*-identifying an object or a suspected cause-effect relationship between objects that it takes itself to have observed on previous occasions: "Does such talk not already imply, by virtue of its grammar alone, that there is [an identical self]? One may be tempted to say that if the skeptic were right, 'he' could not state 'his' view."[42] While Biro's subsequent proposal of the self "as a thing having *imperfect* identity," for which the operation of memory and recollection furnishes prima facie evidence, is sensible as far as it goes, his Humean characterization of such self-awareness as a type of belief lingers on. Yet it is nonsensical to argue that in "remembering" and correlating, say, two different types of causal experience I merely *believe* to be remembering, and thus to suspend the self's temporal continuity in a kind of epistemological limbo. For the phenomenology of "remembering" (an object) does not substantially differ from that of "believing" (in self-identity); and neither can be said to differ in any self-evident qualitative sense from the phenomenology of the type of doubting and scrutinizing that defines Humean skepticism.

Perhaps it was only as a result of Locke's, Mandeville's, and especially Hume's naturalist critique of the person that the long-term costs of modernity's repudiation of the will as the embodied, divided, but continuous and self-aware substratum of personhood could fully come into focus. Some of the more sinister implications stand to be considered momentarily. Certainly, it is striking how uncurious any of the writers are as to whether forgetting (or, as the case may be, rejecting on the basis of an artificially narrow epistemology) the metaphysical dimensions of personhood should entail *any* costs whatsoever. At the very least, it seems apparent that the

---

of self-consciousness" only. While "self-perception is the crucial defining characteristic of the person for Locke," the poverty of such accounts has to do with the assumption that this "punctual" or "neutral" self can and must be perceived in isolation from any evaluative framework ("a space of moral issues"), even though the incentive for self-awareness could only ever arise *from* such frameworks (*Sources*, 49).

41. As so often in Hume scholarship, Biro's own language verges on the imitative fallacy: "As Hume *recognizes*, the fundamental *belief* standing in need of an analysis and a genetic account is the belief one has in one's own identity" (Biro, "Hume's New Science," 57; italics mine).

42. Ibid., 57n.

nominalist axiom of radical discontinuity not only vitiates all sustained exchange (and, consequently, any substantive moral commitments) between discrete individuals. It also jeopardizes the basic category of identity (self-identity being really something of a pleonasm) and, as a result, imperils the reality of reason and the very possibility of philosophy: "this struggle of passion and of reason, as it is call'd, diversifies human life, and makes men so different not only from each other, but also from themselves in different times" (*HT,* 280).[43] Not entirely of his own making, to be sure, Hume's dilemma stems in part from the nominalist-empiricist desire to evacuate philosophy of all concepts affiliated with traditional metaphysics and, hence, to dissolve the will into the passions.

The move is not, however, identified as one of outright substitution but, rather, as what eighteenth-century thought (following Newton) takes to be the only kind of rational explanation there is: viz., to link concepts to one another by means of efficient causation. Yet what initially takes shape as an argument about how the passions "inform" (*causa materialis*) and "condition" the will (*causa formalis*) quickly mutates into an argument to the effect that the passions positively determine the *reality* of the will (*causa efficiens*)—from which it is but a small step to conclude that will, reason, and judgment are merely projections or illegitimate inferences drawn from the reality of the passions themselves. To be sure, readers strongly disagree on whether this is Hume's position. David Norton insists that "Hume does not suppose that because we cannot justify our belief in particular entities . . . it is known that there are no such entities."[44] While that may well be the case, it is true that for Hume anything not susceptible of being described in terms of efficient causation constitutes a "belief" and, as such, must be anathemized. Moreover, Hume simply assumes "belief" to be an invariant and non-falsifiable mental state—incapable of progressive clarification or development, and he holds it always to be in latent (if unsuccessful) competition with causally produced representations. Belief for Hume belongs to the factual order of predication and reference, in part because his introduction of a scientific, "experimental method into moral subjects" cannot even recognize the possible existence of another, holistic, dispositional, and hermeneutic form of awareness.

Failing to distinguish between necessary and sufficient causes for mental processes, Hume here does not so much elucidate as dissolve the phenomena at hand. Staying the course, he thus realizes that there is ultimately no warrant for crediting

43. On Hume's multiple and often conflicting meanings of "reason"—as alternately a faculty or its activity, as abstract reasoning, demonstrative reasoning, or particular demonstration, as probable reasoning or probability itself, as "the psychological transition from one perception to another" or, simply, as a "non-inferential present awareness," see the long footnote in Norton, *David Hume,* 96–98.

44. Ibid., 201n.

the passions with the kind of self-identity just denied to consciousness and the will. Hence Hume hints that the concept of "passion" itself has only very tenuous epistemological standing, being apparently but the fleeting, inner notice of some "violent and sensible emotion" that strikes the mind with quasi-mechanical, mindless force. Hume's laconically worded and impersonal predication (especially in §2.3.9) tells this story rather well: "good ... *produces* Joy. When evil is in the same situation there *arises* Grief or Sorrow. When either good or evil is uncertain, it [who? what?] *gives rise to* Fear or Hope, according to the degrees of uncertainty on the one side or the other. . . . The direct passions frequently *arise from* a natural impulse or instinct, which is perfectly unaccountable." Desire for punishment, "hunger, lust, and a few other bodily appetites," etc. constitute passions that "properly speaking, *produce* good and evil, and proceed not from them" (*HT,* 281; italics mine). What is increasingly absent from this theater of the mind is the spectator who could relate to and reflect upon the passion-play that is being acted out before her or him.

Hume's famous observation—so influential for Adam Smith—that "the passions are so contagious, that they pass with the greatest facility from one person to another, and produce correspondent movements in all human breasts" (*HT,* 386) anchors a type of social knowledge that no longer originates in rational deliberation and intention but in the associative, quasi-kinetic transfer of impulses and passions. By his very choice of terms, Hume has significantly compromised the articulacy and commonality of human cognition. There may be (perhaps) knowledge, but in classically nominalist fashion it will be confined within the "breast" of the individual. Conversely, there is community, but it only eventuates as a process whose logic and rationality elude the individual participants and, notably, prove impervious to critical reflection. Thus "the intercourse of sentiments ... in society and conversation, makes us form some general *inalterable* standard, by which we may approve or disapprove of characters and manners" (*HT,* 385; italics mine). Hume's restrictive view of cognition as but an internal play of representations shows his circular nature of a radical epistemological skepticism. In positing from the outset an inchoate, affectively charged "inner" mind haplessly trying to build up a mimetic inventory of inert and alien external "facts," Hume is committed to a rather crude "picture-thinking" whose incoherence Hegel was to exhibit in the *Phenomenology.* As Robert Sokolowski cautions, "mental representations are a deadly trap philosophically: if you start with them, you never get beyond them. They lock us into subjective isolation."[45] Moreover,

45. Sokolowski, *Phenomenology of the Human Person,* 68; on the preceding pages, Sokolowski develops the central point, viz., that knowledge qua predication is always developed "for another," and that "the apparent 'psychological' or 'privately mental' achievement of thinking is really an internalization of what is first and foremost a public activity. We go, in fact, from the outside to the inside, not the other way round" (ibid., 60–63).

the passions—being the primary actors on the stage of Hume's mind-as-theater—would minimally have to exhibit articulacy, continuity, and self-identity. A structured, purposive, and progressive relation between the passions, which is to say a "plot" in the sense of meaningful and significant "action," would greatly enhance the performance.

The deterioration of Hume's argument here ultimately stems from his view that mind can never exercise any causal influence over itself, and that all cause is by definition mindless, extrinsic, and hence aimless. Inasmuch as for Hume "cause" is by definition non-cognitive, reason effectively disappears into the discrete, discontinuous, and heterogeneous "passions" and fluctuating "states" from which rational thought is said to arise as an effect, and a rather spurious one at that. Seeking to recover control over an argument that, as is the Sophist's wont, pivots on the sheer force and charisma with which it is advanced, Hume comes up with a shrewd choice of metaphor: "If we consider the human mind," he muses, "we shall find, that with regard to the other passions, 'tis not of the nature of a wind-instrument of music, which in running over all the notes immediately loses the sound after the breath ceases; but rather resembles a string instrument, where after each stroke the vibrations still retain some sound, which gradually and insensibly decays ... each stroke will not produce a clear distinct note of passion, but the one passion will always be mixt and confounded with the other" (*HT,* 282). What Hume offers here is a gesture, ever so tentative, toward the temporal continuity of passion and, thus, toward the possibility that passion need not necessarily expire in the very instant of its occurrence but, instead, may yet yield something of an afterlife as it is being internalized like Percy Bysshe Shelley's "memory of music fled."

At this point, we may draw something of a balance sheet as regards the implications and consequences of the conceptual transpositions that take place in eighteenth-century moral philosophy. Doing so will help identify the nature of the challenges that Coleridge confronts when, firmly against the grain of the Scottish Enlightenment, he reacquires and importantly redefines will and personhood in his later writings. First, there is what might be called *the Great Divide:* beginning with Locke and culminating in Hume, the projects of modern epistemology and moral philosophy, respectively, no longer unfold in coordinated fashion, and there is growing doubt even about the possibility of their *eventual* convergence or realignment. The prevailing, skeptical tenor of modern epistemology (implicit in Locke and flamboyantly on display in Hume) effectively denies the self the kind of continuity, self-awareness, and, hence, potential for moral responsibility and accountability that moral sense theorists from Shaftesbury to Hutcheson and realists like Reid and Beattie continue to assert as the very essence of the human being's affective constitution.

A significant consequence of this divide concerns the displacement of religion as a framework defined by its universal view of the human condition with modern religious denominationalism. Part of that shift involves the transformation of the inner life or religious subjectivity into what might be called *elective irrationalism*. Given the aggressive challenges issued to the traditional vocabulary of philosophical and theological ethics—that is, the systematic discrediting of *synderesis*, "natural Affections," benevolent sentiments, and the peremptory suspicion of both intellectual and moral virtues—inwardness is being identified with the cultivation of strictly "private," inarticulate, and likely incommensurable beliefs. Owen Chadwick goes so far as to suggest that "Christian conscience was the force which began to make Europe 'secular.'"[46] In an age whose suspicion of metaphysics is nearly universal, spanning the entire spectrum of intellectual sensibilities from Shaftesbury and Hutcheson to Locke, Mandeville, and Hume (to say nothing of Voltaire and the Encyclopedists), it is only the solitary, private self that may yet entertain metaphysical speculations and commitments. It may credibly do so not only because such pursuits are considered to be an inalienable right of "conscience" but, just as importantly, because it is taken for granted that whatever discoveries may yet be made in that realm can never be effectively, let alone authoritatively articulated *to* and *for* others.

Echoing Hobbes's view of the individual's irrational constitution, yet also modifying the latter's aggressive Erastianism, Locke's *Letter Concerning Toleration* (1689) had influentially stipulated that "the church itself is a thing absolutely separate and distinct from the commonwealth." In fact, the *Letter* views religion as nothing more than a quasi-contractual association of like-minded individuals: "a voluntary society of men, joining themselves together of their own accord." Ostensibly personal and private, faith at the same time reveals a paradoxically mimetic quality that will eventually make its (manifestly secularized) reappearance as Smith's "moral sentiments." Presented as the sole remaining (and notably weak) rationale for religious institutions, faith in Locke's influential account warrants recognition only insofar as it pertains to individual salvation. In contrast to interested worldly institutions, faith (in Locke's view) is legitimated less by its intrinsic affirmations and normative commitments than by its (alleged) *irrelevance* to empirical practice and political life. Thus denuded of any specific role within, let alone authority over, quotidian existence, faith all but supplants the authority of religious institutions and teachings. To hold that "whatsoever is practiced in the worship of God, is only so far justifiable as it is believed by those that practice it to be acceptable to him," comes close to making the

46. *Secularization*, 23; see also C. Taylor, *Secular Age*, esp. on "the malaise of immanence" in eighteenth-century Pietism (304–321); and Gregory on the emergence of a modern, "hyperpluralist" society that has channeled religious beliefs into a language of rights as "deified preferences" (*Unintended Reformation*, 180–234).

very reality of God contingent on the subjective faith (or, more likely, plurality of faiths) that affirms him. Faith has been transmuted from a correlate of efficient grace into a reality solely constituted and legitimated by its subjective assertion or its inward, affective experience. Aside from evacuating religion of all normativity—such as had been embodied by communal ritual, established liturgy, and the sacraments—Locke also insists on the absolute segregation of religion from any concern with the common good: "the care of each man's soul, and the things of heaven, which neither does belong to the commonwealth, nor can be subjected to it, is left entirely to every man's self." Whatever is politically desirable, epistemologically demonstrable, and theologically permissible no longer stands in any coherent relationship. As a result, political arguments cannot draw on metaphysical (innate) norms on the order of Aristotelian "right reason" or Aquinas's *synderesis*. Conversely, religious life has been quarantined from the realm of empirical, social practice and, as a result, is unable to configure its soteriological concerns with rationally, prudentially judged action (including the cultivation of habits and virtues).[47]

To get to the third major shift that takes place in eighteenth-century thought, a brief detour is required. Though hardly uncontested, Locke's aggressive partitioning of political practice from epistemological inquiry, and of both from religious and metaphysical speculation, secures further acceptance for the (in origin Hobbesian) naturalist account of human consciousness. Central to that conception is the desire and expectation of taking possession (*dominium*) of the world by framing it as a sum total of objects either demonstrably captured qua "facts" or as yet awaiting to be thus secured. A further premise of this naturalist model involves the assumption that the progressive conquest of the world—taken as an inventory of discrete objects and experiential data—can be effectively separated from the welter of ineffable subjective feelings, desires, and values, and from the hermeneutic traditions with which knowledge of those inner states had previously been bound up. Here it bears recalling how the premodern, Scholastic nomenclature "does not distinguish, as English does, between 'conscious' and 'conscience,' so that the first meaning of *conscientia* is that awareness of what we are doing ... On the one hand, conscience is a judgment before we act which prompts, directs, guides; on the other hand, conscience assesses what we have already done and prompts remorse or satisfaction."[48] By contrast, the post-Cartesian assimilation of human agency to modern scientific standards of detachment, impartiality, and objectivity results in the emergence of the fact/value distinction. Critics of this distinction tend to regard its apparent triumph in the mid-eighteenth century as the result of an artificial and weirdly coun-

47. Locke, *Letter*, 24, 14, 35, 52.

48. McInerny, "Prudence and Conscience," 299.

terfactual turn within philosophy, and there is indeed good reason to take that view. Still, it would be a mistake simply to repudiate this apparent distinction, for doing so would ignore its rather instructive etiology—that is, how in seemingly natural or "self-evident" fashion it arises from a particular set of questions and specific claims about the reach and limits of key concepts, including that of the will.

The *locus classicus* for the distinction is generally agreed to be Hume's *Treatise* (§3.1.1), and it is indeed here that we must begin. Toward the end of that section, Hume professes acute bewilderment at how "every system of morality" sooner or later will leap from a factual claim about what "is or is not" to some proposition "connected with an *ought,* or an *ought not* ... What seems altogether inconceivable [is] how this new relation can be a deduction from others, which are *entirely different from it*" (*HT,* 302; last italics mine). Right away, a basic logical observation seems in order. For on closer inspection it would appear that Hume's partitioning of the order of fact from that of value does not, properly speaking, constitute a distinction at all. Rather, he asserts these two orders to be incommensurable and, consequently, to bear no authentic relation to one another whatsoever. Contrary to a regular philosophical distinction, then, Hume's disaggregation of fact and value is driven by his axiomatic, indeed unprovable pronouncement that the one is "entirely different" from the other. He does not help his case by suggesting how this distressing tendency to move from observable facts to inner *qualia* has supposedly vitiated "every system of morality."[49] For in so expanding the error that he means to rectify to nearly universal proportions he not only magnifies his task, but he implicitly grants that *in practice* (if not in self-description) moral inquiry has always shown human reasoning to be an intrinsically evaluative process. Hume's vexation here comes across as a case of "methinks the lady doth protest too much" in that his strenuous objection against the miscegenation of fact with value (said to violate "the *usual* copulations of propositions, *is,* and *is not*") ultimately constitutes itself an "ought not" type of argument, rather than dispassionately describing a (supposedly) inescapable, factual scenario. Notably, it is Hegel who would offer the definitive critique of Hume's peremptory disaggregation of fact from value by showing how "awareness itself is judgmentally structured, a taking to be such and such, and so is inherently and unavoidably a kind of commitment and so much more like a normative pledge than a mere matter-of-fact psychological event."[50]

Leaving aside the fact that Hume's argument appears on the verge of a performative contradiction, the more instructive question to press is how his quarantining of

49. Quoting this passage, Murdoch notes how Hume "in effect elides morality with the 'soft' concepts of habit and custom" and how, in this case, we see that "one concept may quietly swallow another and obliterate a whole region of thought" (*Metaphysics,* 155).

50. Pippin, *Hegel's Practical Philosophy,* 19.

fact from value is meant to support his broader claim (§3.1.1) that "Moral distinctions [are] not deriv'd from reason." At the heart of his argument lies the supposition that "reason is wholly inactive" and that it "can never immediately prevent or produce any action by contradicting or approving of it" (*HT,* 295). What prompts Hume's blunt appraisal of reason as "utterly impotent" (*HT,* 294) is the nominalist assumption, already firmly in place in Locke's *Essay,* that a world of interconnected, related, and mutually aware and concerned human agents is never *given* (neither as a correlate of divine grace nor in the existentialist sense of the hermeneutic anteriority of Heidegger's "being-in-the-world") but is only ever *made.* Ideas, universals, forms, or whatever other term we might wish to assign to the realm of the normative can never be proper objects for the intellect, quite simply because human intentionality is from the outset conceived as reactive to singular and adventitious external sensations: "the mind can never exert itself in any action, which we may not comprehend under the term of *perception*" (*HT,* 293), from which it follows that "all beings in the universe, consider'd in themselves, appear entirely loose and independent of each other" (*HT,* 300).

It bears remembering here that Hume's sole objective in this section (and also in his later *Enquiry*) is to demonstrate that notions of virtue and vice bear no relation whatsoever to reason, and hence to leave simply undecided the question as to whether value-concepts can ever be anything other than projections by a human intellect distressingly prone to counterfactual speculation. Just as Locke severs all relation between the noumenal affirmations of faith and the nature (and justification) of empirical, political action, so Hume here quarantines the realm of ideas concerning the "eternal fitnesses and unfitnesses of things" and the "immutable measures of right and wrong" from "the calm and indolent judgments of the understanding" (*HT,* 294). Working by fiat rather than by demonstration, the *Treatise* has "opened up a gap between moral judgments and assertions."[51] In so doing, it has effectively suspended the Platonic view of human beings having an indelible awareness of the good (what Aquinas calls *synderesis*), fragments of which had survived into the modern era variously in the tenets of natural law (e.g., in Hugo Grotius), in Erasmus's humanism, and in the Cambridge Platonists' counter-modern affirmation of some "eternal and immutable morality" (Ralph Cudworth). To be sure, vestiges of such foundationalist thinking can be located even in Hume's equivocal appeals to "common-sense," though the practices subsumed under that heading now appear but so many "preju-

51. Foot, *Natural Goodness,* 8; Foot goes on to point out that "a moral evaluation does not stand over against the statement of a matter of fact, but rather has to do with facts about a particular subject matter, as do evaluations of such things as sight and hearing in animals" (ibid., 24).

dices" which, however indispensable for quotidian life, remain permanently bereft of rational justification.[52]

The extreme, not to say absurd conclusions to which Hume's premises compel him, and from which—perhaps more for the sake of consistency than out of conviction (if that word still signifies)—he does not shy away, emerge most clearly in his examples of high crimes (incest, murder, etc.). In disaggregating reason from passion, and arguing that reason is strictly concerned with discerning "*real* relations of ideas, or ... matters of fact," Hume by definition immunizes all passion against the prospect of its rational appraisal: "in other words, our passions, volitions, and actions do not agree or disagree relative to one another, hence cannot be true or false, and hence cannot be the object of reason."[53] Particularly these late portions of his *Treatise* show him to be, in Gertrude Elizabeth's Anscombe's words, "a mere—brilliant—sophist" whose conclusions, "with which he is in love" and which he reaches "by sophistical methods [nonetheless], constantly open up very deep and important problems." Anscombe may be echoing A. E. Taylor who, finding himself unable to see in Hume's thought "the genuine expression of a whole personality," muses in his 1927 Leslie Stephen Lecture whether "Hume was really a great philosopher or only a 'very clever man.'"[54] A case in point might be Hume's seemingly capricious assertion that even where we are confronted with "wilful murder" it is quite impossible to "find that matter of fact, or real existence, which you call *vice*." All that we are presented with are "certain passions, motives, volitions, and thoughts" and, presumably, the empirical axe with which irascible George killed dim-witted Johnny behind the pub. Yet unlike an idea, which has a distinct referential scope, a passion for Hume is "an original existence ... and contains not any representative quality." It is completely self-contained, is non-falsifiable, and cannot be "opposed by, or contradictory to truth and reason" (*HT*, 266). Once again, fact and value are said to bear no relation (positive or negative) to one another: "the vice entirely escapes you, as long as you consider the object. You never can find it, till you turn your reflection into your own breast, and find a sentiment of disapprobation, which arises in you, towards this action. Here is a matter of fact; but 'tis the object of feeling, not of reason" (*HT*, 301). The example is being introduced so as to illustrate Hume's principal contention that "moral good and evil belong only to actions of the mind, and are deriv'd from our situation with regard to external objects" (*HT*, 299).

To an ethicist, the implications of Hume's argument have to be profoundly disturbing. Not only does Hume's casual acceptance of Locke's axiom that the person

---

52. Penelhum, "Hume's Moral Psychology," 245–246.

53. Norton, *David Hume*, 103.

54. "Modern Moral Philosophy," 3; Taylor, quoted in Penelhum, "Hume's Moral Psychology," 238.

asleep is not really a person ("When my perceptions are remov'd for any time, as by sound sleep ... I am insensible of *myself,* and may truly be said not to exist" [*HT,* 165]) implicitly exonerate Macbeth who during that fateful stormy night had plunged the knife not into a person named Duncan but only into a sleeping body. Somewhat less whimsical weighs the fact that in restricting personhood to those who demonstrably enjoy self-awareness and rationality, the naturalist argument opens the door to euthanasia, assisted suicide, and much else. For in disputing the reality and continuity of the person as an embodied will except insofar as it can be verifiably credited with conscious awareness, both Lockean hedonism and Humean naturalism conjure up a conception of social life profoundly ill at ease with the reality of human suffering, mental illness, and death. Evidence of "a liberal, sanitized Christianity, which doesn't quite know what to do with suffering," is copious, from the bowdlerized Shakespearean tragedies (and Samuel Johnson's similarly tendentious essays on the Bard) to a general discomfort with narratives and images of suffering that made the Old Testament Book of Job and, above all, the motif of Jesus's Crucifixion such unsettling experiences.[55] Mired in an impoverished epistemology of the human, post-Lockean thought can only envision collective flourishing as a compound effect of so many isolated individuals caught up in a feverish pursuit of self-fulfillment. The resulting hedonistic vision and the self-cherishing culture of sentimentalism to which it gives rise cannot but treat suffering as sheer disruption or *negation,* just as it regards interpersonal commitments and obligations as strictly elective and is quick to reject any hint of normative (teleological) thinking as an illicit metaphysical encroachment on individual autonomy.

As we shall see in Part IV, to conceive of the human person (and the possibility of murder) as something altogether fictitious and subject to contingent negotiation and incidental definition is to dismiss all normativity as but a conspicuous sentiment; hence "the vice [of murder] entirely escapes you, as long as you consider the object." Whereas Robert Spaemann, among others, emphasizes that "by 'person' we understand the unique human being as such, rather than some particular state" that she or he happens to be in, Hume's (and following him Derek Parfit's) arguments point in the obverse direction.[56] Thus we find Hume drawing a categorical distinc-

55. C. Taylor, *Secular Age*, 318. On Job, see Lamb, *Rhetoric of Suffering;* and Stump, *Wandering in Darkness*, 177–226. Notably, it was Lessing who, having diagnosed the Enlightenment's inability to come to terms with suffering, predicated his main aesthetic treatise, *Laokoön* (1766), on precisely one of the most iconic representations of pain and suffering.

56. Spaemann, *Grenzen*, 427 (trans. mine); Spaemann has developed his critique in a series of powerful review articles on questions of euthanasia and "assisted suicide"—vigorously rejected in the so-called Kinsau Manifesto of 1990, signed by two hundred physicians, philosophers, theologians, and members of the clergy—for which books by Peter Singer, Peter Sloterdijk, Norbert Hoerster, Georg Meggle, and others had sought to create an opening. Spaemann is careful to distinguish

tion between the inner life characterized by unverifiable and ineffable patterns of ideation and a domain of contingent, factually neutral, and affectively constituted object perceptions. In reading Hume's (and Smith's) moral theories as prime instances of an "imaginary" model of ethical relations, Terry Eagleton also happens upon this startling account of murder. Notably, it is toward the end of his account of Hutcheson and Hume that Eagleton moves beyond Hume's "explanation" of murder as but "a matter of fact" for feeling. As he notes, what "the sentimentalists fail on the whole to grasp is that the only authentic moral law is the law of love—but not at all the species of love which can be couched in terms of sensibility." To declare that "sympathy with persons remote from us [is] much fainter than that with persons near and contiguous" (*HT,* 385) is to "think of love in the personal or affective sense, which is not its most fundamental meaning."[57] What Eagleton does not make explicit, however, is the fact that the entire stance of psychological naturalism and its rejection of self's capacity for reflectively transcending the world as it is given blinds Hume (and Smith) to the more profound sense of *agapē*. Neither can imagine the reality of a non-negotiable "obligation to strangers," one that has nothing to do with sympathy but a great deal with a fundamental commitment to the flourishing of the whole.

All too often, "common-sense" philosophy tends to paralyze the critical faculties of its readers by prompting them to acquiesce in the pivotal naturalist axiom that all thinking arises by default of either immutable external realities or inscrutable inner cues. Yet the deceptive solidity and rational self-assurance of such a stance reflects less a compelling solution to persistent epistemological questions than the sheer refusal even to embark on a quest for genuine answers. In his critique of Humean skepticism, Kant had rightly emphasized the lack of analytic scrutiny brought to bear on empiricism's notion of what constitutes a fact. As Maurice Blondel was to sum up the key issue, "the unity of a synthesis does not consist only of an internal relation of the parts; it is the ideal projection of the whole into a center of perception. The *vinculum* (bond) is of an intelligible and, in truth, a *subjective* nature." Where the naturalist sees himself or herself encountering a vast ocean of mute and raw "facts," and where the skeptic gleefully points to a vast *terra incognita* of "sensations" at once

between excessive life-prolonging measures that modern medical technology places at the disposal of physicians treating terminally ill patients and active, life-ending measures undertaken in the same cases: to conflate the two scenarios "by equating disproportionate life-prolonging medical intervention with a lethal injection amounts to a sophism that ultimately only serves to belittle the latter" (ibid., 416; trans. mine); see especially nos. 34–37 in Spaemann, *Grenzen*, 406–428.

57. *Trouble*, 58–59; Eagleton is right to remark on the "affable bourgeois" Hume's "inability ... to see the ascetic virtues as anything but monkish and life-denying, in a misreading of Christianity which is fashionable to this day ... In this respect, the comfortable eighteenth-century clubman is at one with the life-affirming liberals of our own age" (60).

discontinuous and inscrutable, the work of phenomenology begins in earnest: viz., to determine "how we separate what is objective representation and what is subjective act … [or] how phenomena *are interiorized.*"[58] Yet in Hume's orbit, where naturalist and determinist modes of explanation are deemed the only legitimate method for the production of knowledge, the deck is axiomatically stacked against any type of inquiry that accepts the reality of "consciousness" as an incontrovertible given, albeit one that calls for sustained and scrupulous phenomenological description.

For Hume, selfhood is but another proposition, and its philosophical legitimacy extends only as far as the proposition—viz., a predicative act wherein self-reference is the major premise—can compel our notional assent. And yet, to accept that methodological constraint is to have credited the skeptic's *cogito* with a reality that cannot be subverted by any counterfactual scenario. Indeed, just as a law (say, in the realm of physics) can only gain our assent as a proposition about which we are certain based on our antecedent certitude about the very facts which that law seeks to synthesize, so "our consciousness of self is prior to all questions of trust or assent."[59] Only on that basis can philosophical reflection proceed as a meaningful interpretive and "critical" (i.e., counterfactual) undertaking. As Hegel and, eventually, Edmund Husserl had also noted, skepticism per se amounts to an elaborate self-deception by continually accepting as incontrovertibly *given* the very phenomena whose reality and continuity it subsequently disavows. What it cannot explain, however, is its own intrinsic and consuming motivation, its implacable desire to expand the jurisdiction of the negative. Confusing the absence of objective *certainty* with the (alleged) impossibility of inner *certitude*, the modern skeptic is unable to articulate any good to be realized by his or her ruminative questionings. Yet it is just this inarticulacy that bars the skeptic from ever imagining closure to his or her philosophical enterprise. As Maurice Blondel puts it so laconically, "one does not stop at midnight in an open field." For "to play and to enjoy as if we knew, as if we experienced the vanity of all, while we have not experienced it and while we do not know it, because it is impossible to experience it and know it, is to prejudge every question on the pretext of suppressing every question, and to admit by an arbitrary anticipation that there is neither reality nor truth."[60]

In the end, skepticism is above all a refusal to acknowledge and engage the uniquely human and incontrovertible reality of *action*. Blondel here speaks of action as an all-pervading "system of spontaneous or willed movements, a setting of the organism into motion, a determinate use of one's vital strengths, in view of some plea-

58. Blondel, *Action*, 94n, 96.
59. Newman, *Grammar of Assent*, 67.
60. Blondel, *Action*, 28.

sure or interest, under the influence of a need, an idea or a dream."[61] With Descartes and Hume serving as its foil, Blondel's account bears quoting in full, if for no other reason than that it focuses us—with a clarity not seen since Aquinas—on the inseparable bond that links the human being's physiological and intellectual functions and is uniquely manifested in "action":

> Action pertains at once to all the powers in man alien and hostile to one another: through thought, which illuminates its origin and accomplishment, it is of an intellectual order; through intention and good will, it belongs to the moral world; through execution, to the world of science. At one and the same time, it is an absolute, a noumenon, a phenomenon. If, then, there is an antinomy between the determinism of movements and the freedom of intentions; if moral formalism is without relation to the laws of sensibility and of the understanding; if all union is broken between thought, the senses, and voluntary activity; if the body of acts is separated from the spirit that inspires them and if, in this world that is presented as the theatre of morality, man dispossessed of all metaphysical power, excluded from being, and fragmented feels surrounded by impenetrable realities where the most absurd illogicality can reign, then the will to live is broken along with the daring to think.[62]

Of obvious Aristotelian provenance, Blondel's recovery of action as a central category extends into the twentieth-century project of philosophical anthropology, especially in the philosophical anthropology of Arnold Gehlen. Rejecting the Enlightenment legacy of a hierarchical division (*Stufenschema*) of discrete human (somatic, physiological, intellectual) processes and faculties, Gehlen sees human life defined by the "single continuous structural law of action" (*in einem durchlaufenden Strukturgesetz ... der Handlung*). What sets man apart from all other creatures as "an acting being" (*das handelnde Wesen*) is his palpable lack of an organic fit into any specific environment.[63] Drawing on Friedrich Schiller and Johann Gottfried Herder above all, Gehlen notes how in the absence of any instinctual guidance (*instinktentbunden*) every human being "presents a problem to himself" and, thus, "must also develop an attitude toward himself and make something of himself." As such, he must "develop a perspective [*stellungnehmen*] on the outside world," and what we call

---

61. Ibid., 7, 28, 36.

62. Ibid., 40.

63. Gehlen, *Man*, 16. This thesis shows Gehlen's implicit affinity with Aquinas who, against the naturalist (yet partially also Aristotelian) view "that human beings are made up of three substantial forms: vegetative, sensible, and intellectual" (Kerr, *After Aquinas*, 11) viewed the human being as an agent whose specific functions (*operations*) were fully unified and integrated.

action is the result of the human being's essential "incompleteness" (*Unfertigsein*). In sharp contrast to any mechanical, predetermined, or purely reactive "process," *action* involves an element of transcendence—taking the term here not in a metaphysical but in a strictly counterfactual sense. Contrary to some merely physical (or physiological) "doing," action shows our motor system and our intellectual functions to be fully synchronized. It is cued, literally, by *pro*/vidence—by a capacity for imagining, planning, and realizing a state of affairs not given as such: "Man is ultimately a *prospecting* [*vorsehend*] being. Like Prometheus, he must direct his energies toward what is removed, what is not present in time and space."[64] Yet again, Newman anticipates the salient point almost verbatim: "other beings are complete from their first existence, ... but man begins with nothing realized ..., and he has to make capital for himself by the exercise of those faculties which are his natural inheritance ... It is his gift to be the creator of his own sufficiency; and to be emphatically self-made. This is the law of his being."[65]

Certainly Hume and Smith, yet also their readers, today have all but occluded the main alternative concept of agency grounded in what Coleridge was to analyze as a "responsible will consummated in *action*—as opposed to a "self" haplessly replicating, or reacting to, non-cognitive stimuli. Properly speaking, action can only eventuate within a world perceived, considered, and judged to demand its purposive *transformation* "for the sake of" a good that, rather than some hypothesis theoretically entertained, is subject to what Newman means by "real assent." Action, in other words, involves the apprehension and deliberative participation in what is phenomenally *given* and as such has been judged to be susceptible of a perfection as yet not fully realized. Yet precisely this continuity between the factually given and its transformation by imaginative, intellectual, and material action eludes the narrow con-

64. Kerr, *After Aquinas*, 24. Striking, also for its consonance with Blondel's concept of action quoted above, is the following, programmatic passage in Gehlen's 1940 book: "All human actions are twofold: First, man actively masters the world around him by transforming it to serve his purposes for the simple reason that there are no natural, organically fitted conditions of existence into which man might enter, or because the 'natural,' unadjusted conditions of existence are intolerable for him. Second, to accomplish this, he draws upon a highly complex hierarchy of skills and establishes within himself a developmental order of abilities; this order is based on potential usefulness of the skills and must be constructed singlehandedly by man, sometimes overcoming internal resistance to doing so ... In so 'processing' the ambient world, all things are thereby unwittingly endowed with a high degree of symbolism such that, eventually, the eye alone (an effortless sense) can take them in and quickly assess their potential usefulness and value. In succession, objects are experienced by man and then set aside, a process in which these objects are imperceptibly saturated with a highly sophisticated symbolism. Thus the eye, our most effortless sense, may 'survey' them and finally is able to take in casually the utility- and practical value of things that previously had revealed itself only through arduous experience" (Gehlen, *Man*, 29, 32; trans. corrected).

65. *Grammar of Assent*, 274.

fines of Humean rationality. Consistent with his rigid separation of reason from idea, fact from value, Hume insists on quarantining "actions of the mind" from any contextual and evaluative (hermeneutic) awareness:

> As moral good and evil belong only to the actions of the mind, and are deriv'd from our situation with regard to external objects, the relations, from which these moral distinctions arise, must lie only betwixt internal actions, and external objects, and must not be applicable either to internal actions, compar'd among themselves, or to external objects, when plac'd in opposition to other external objects. For as morality is suppos'd to attend certain relations, if these relations cou'd belong to internal actions consider'd singly, it wou'd follow, that we might be guilty of crimes in ourselves, and independent of our situation, with respect to the universe ... Now it seems difficult to imagine, that any relation can be discover'd betwixt our passions, volitions and actions, compar'd to external objects, which relation might not belong either to these passions and volitions, or to these external objects, compar'd among *themselves.* (*HT,* 299)

A first response to this quizzical argument might be to challenge some of its premises and conclusions. Thus we might ask why it should be impossible for us to "be guilty of crimes in ourselves," considering that Hume is quite happy elsewhere to credit an "internal sense" with the capacity of identifying for us cases of vice and virtue. Evidently, Hume means to deny the possibility of a reflexive act of mind unless its occurrence can be linked to an external sensation as its efficient cause; the only exception to that rule would be "some sentiment" (*HT,* 300), provided it is *not* self-conscious and does not result in articulate, quasi-rational claims about supra-personal forms or values. Still, the more instructive approach to this passage will involve mapping how Hume's arguments here foreclose on the semantic depth and potential of several key concepts of practical reason.

A first casualty is what Aquinas had called *synderesis,* that is, the metaphysical premise of "eternal and immutable" forms concerning the "natural fitness and unfitness of things," a notion that Hume explicitly rejects. To grasp the implications of that rejection, we must take care not to confuse *synderesis* with "conscience," a term whose importance actually increases in proportion as modernity either forgets or disputes the reality of the former. Consistent with the thrust of Hume's argument, the editors of the *Treatise* thus index "conscience" as "see, *moral sense.*" For it is Hume's "internal sense" or "sentiment" which, independent of any notional or transcendent meanings, apprises the self of an action's virtuous or vicious tendency. Given its strictly affective phenomenology, Hume's moral sense reinforces the strictly contingent, private, and ineffable nature that Shaftesbury and Hutcheson had already identified as characteristic of all conscience-like sensations. Yet Hume also views this "internal sense" as wholly unrelated to the Platonist and Christian premise of an innate and universal moral awareness or *synderesis,* such as had still survived in seventeenth- and early eighteenth-century accounts of natural law. To grasp the incoherence of Hume's arguments here, we must recall that rational human activity can never unfold without some elemental presuppositions.

As Herbert McCabe notes, *synderesis* does for practical reason what the axiom of non-contradiction does for theoretical reason. We must not confuse it, as Hume does, with some "abstract rational difference betwixt moral good and evil" (*HT,* 299). Neither are we to construe it as a metaphysical premise impermissibly slipped into Aristotle's practical syllogism. Just as the principle of non-contradiction does not compel any specific mathematical or economic syllogisms but, instead, amounts to their condition of possibility, so *synderesis* does not per se entail conclusions of any particular kind whatsoever. Rather, it is "the principle in virtue of which there is any syllogism at all."[66] It is the unconditional point of departure for moral deliberation without which there could, in fact, not even be any elections of the will. Put differently, *synderesis* (McCabe calls it "a piece of fake Greek that seems to have been invented by Latin-speaking medieval philosophers") names the fact that we are not simply neutrally and arbitrarily located *in* a world but, on the contrary, that we *have* that world by being intentionally oriented toward it as interpretive, evaluative, and responsible human beings. A direct descendent of Hobbes, whose "naturalism ... wants to be a physicalism, and like all physicalism follows the model of physical rationality," Hume again reveals how "the separating-off [*Abscheidung*] of the psychic caused greater and greater difficulties whenever problems of reason made themselves felt." In axiomatic ways, the *Treatise* ascribes "a type of being ... to the soul which is similar in principle to that of nature; and to psychology is ascribed a progression from description to ultimate theoretical 'explanation' similar to that of biophysics." In so modeling humanistic inquiry on a scientific methodology, Hume seems driven to repudiate not only practical reason for its alleged "impotence" but theoretical reason, too.[67] That, at least, was Husserl's conclusion in his late work on the *Crisis of*

66. McCabe, *God Still Matters,* 160. As McInerny notes, the "'good' qua end is never merely apprehended cognitively but *as good*" ("Prudence and Conscience," 302). Hence the good cannot itself amount to a conclusion, let alone to the major or minor premise, of Aristotle's practical syllogism. Rather, its presence in all human beings qua *synderesis* is the condition under which practical reasoning alone becomes possible. The Enlightenment utopia of fashioning purely theoretical descriptions of the world, free of any prejudices and presuppositions, naturally blinds it to the possibility that reasoning might ipso facto constitute an evaluative pursuit and that, merely in virtue of its "truth-preserving logic," all theoretical argument is already normatively invested.

67. Husserl, *Crisis,* 62–63. Recent work on Hume has made a concerted effort to salvage him from the reputation of an urbane sophist and (not entirely serious) skeptic; see the essays by Biro and Penelhum. Hume's reliance on a model of causation borrowed from Newtonian physics is well

European Sciences. As he continues, once humanistic value concepts were to be accounted for in strictly factual, quasi-geometric terms, "a paradoxical *skepticism* developed, one that ... directed itself precisely against the models of rationality, mathematics and physics, and even tried to invalidate their basic concepts, indeed even the sense of their domains (mathematical space, material nature) by calling them psychological fictions. In the case of Hume this skepticism was carried through to the end, to the uprooting of the whole ideal of philosophy."[68]

A related consideration points us to the second conceptual loss wrought by Hume's peremptory separation of fact from value. Thus *synderesis* does not simply index us to some transcendent notion of the good but also, and no less crucially, to our own history as practical, moral agents: "what we aim at, what we have *synderesis* of intellectually and intend as a matter of will, may be the result of a previous deliberation. In each bit of practical reasoning, if we take them separately, it is by *synderesis* that we intellectually grasp what by the will we intend, find attractive (i.e., good); and it is by practical reasoning (preferably disposed by good sense) that we decide what we will do about it."[69] Put differently, *synderesis* is affirmed by our continuous and evolving history as moral agents making judgments. It is just this term, "judgment" (*prudentia*), which becomes another casualty of Hume's account. For in quarantining the evaluative, moral dimension of human existence in a non-propositional and self-consuming "inner sense," he also forecloses on its legitimate and rational application to the external world. Being no longer indexed to any supra-personal notion of the good, conscience (Hume's "inner sense") no longer stands in any relationship to "judgment" but, instead, functions as a strictly occasional source of information to be "consulted like a cookbook or a railway timetable."[70] In sharp contrast to Aquinas, for whom conscience is neither a power nor a virtue but an act (*ST,* I Q 79 A 13), eighteenth-century notions such as moral sense, inner affection, moral sentiment, etc. lack all temporal continuity and, consequently, are incapable of any hermeneutic progression. They simply flare up within a hapless, present-tense mind and, in so doing, effectively consume themselves.

known, and in describing his philosophy of mind as a "science" and "Experimental Method of Reasoning" (*HT*, 1), Hume's naturalism also looks forward to "the so-called cognitive sciences of the late twentieth century" (Biro, "Hume's New Science," 40).

68. Husserl, *Crisis*, 67; Norton (*David Hume*, 109) and other more recent interpreters are more cautious, noting that Hume's skepticism often aims to deny reason only *sole* authorship in the domain of morals and, in support, quoting Hume's seemingly conciliatory remark that "vice and virtue are not discoverable *merely* by reason" but that "some impression or sentiment" is required for us "to mark the difference between them" (*HT*, 302; italics mine).

69. McCabe, *God Still Matters*, 160, 159.

70. Ibid., 153.

It is this peculiar failure of moral sense theory to distinguish between the occurrence of some moral affection and its reflexive appraisal as being indexed to an antecedent framework of meanings or a hermeneutic tradition that ultimately deprives the individual of self-awareness as an agent in time. Hume's dilemma here stems from his categorical rejection, earlier in the *Treatise,* of habit and habituation as sheer prejudice or as a logically insupportable custom. What might have seemed a (barely) plausible critique of epistemological practice there now wreaks massive havoc in his account of practical reason. For in depriving the moral self of any temporal dimension—that is, of the ongoing interplay between *synderesis,* judgment, and memory as an integral hermeneutic constellation—and by insisting that "morality is not an object of reason" (*HT,* 301), Hume effectively denudes the self of all articulacy as a responsible agent embedded in a complex web of social, cultural, and political relations with other such agents. His strident enclosure of moral cognition as a strictly subjective and incommunicable affair emerges in his passing remark that "vice and virtue … may be compar'd to sounds, colours, heat and cold, which, according to modern philosophy, are not qualities in objects, but perceptions in the mind" (*HT,* 301). Such a statement reveals a perennial tendency of skepticism to merge with the very irrationalism that it claims to oppose. Inadvertently echoing the example of color in her critique of Hume's fact/value distinction, Iris Murdoch thus wonders: "does not value *colour* almost all our apprehensions of the world?"[71]

What the modern, naturalist account of the self attempts is not so much to undermine or reject outright the possibility of Aristotelian reason and Platonic substantial forms; that, it would seem, is incidental. Rather, Hume's disaggregation of fact from value, and of moral sentiments from "reason or science, [which] is nothing but the comparing of ideas" (*HT,* 301), highlights the utter separation of reason from the richly and continuously evaluative structures of individual consciousness. Ultimately, this means that reason no longer has a history, no memory or reality whatsoever except as an epiphenomenon of contingent sensation and perception. To a significant extent, Hume's rigid and often bizarre insistence on the is/ought distinction is the result of a prolonged forgetting of the rich conceptual legacy of Aristotelian ethics and Christian philosophical theology. Thus Hume "discovered the situation in which the notion 'obligation' survived, and the notion 'ought' was invested with that peculiar force having which it is said to be used in a 'moral' sense, but in which the belief in divine law had long since been abandoned." As a result of such conceptual amnesia, we are presented with "the survival of a concept outside the framework of thought that made it a really intelligible one."[72] Extending Anscombe's

71. Murdoch, *Metaphysics,* 155.
72. Anscombe, "Modern Moral Philosophy," 6.

shrewd intuition, MacIntyre has scrutinized "the claim that no valid argument can move from entirely factual premises to any moral or evaluative conclusion." Indeed, "the disappearance of any connection between the precepts of morality and the facts of human nature already appears in the writings of the eighteenth-century moral philosophers." It is here, he argues, that we find "the emotivist self . . . [to have] lost its linguistic as well as its practical way in the world." Hume's "no 'ought' conclusion from 'is' premises principle . . . signals both a final break with the classical tradition and the decisive breakdown of the eighteenth-century project of justifying morality in the context of the inherited, but already incoherent, fragments left behind from that tradition" (*AV,* 56, 60, 59).

This said, we can proceed to a final observation, which builds on the various reasons just identified—viz., loss of *synderesis,* of judgment, and of a temporal dimension of personhood, as well as the disappearance of an articulate conception of the good, all of which culminate in Hume's disassociation of fact from value as (supposedly) "entirely different." Extending Hobbes's and Locke's reductionist outlook on, and ultimate forgetting of, a differentiated conceptual framework needed to account for the practical and moral dimensions of human life, mid-eighteenth-century thinkers like Hume and Smith thus set the stage for the emergence of new disciplinary formations (social statistics, social psychology, political economy, and, more distantly, modern behaviorism), all of which are designed to render intelligible in the aggregate what has been "proven" to be terminally opaque and non-cognitive at the level of the individual. Moreover, what emerges in Hume's mind-as-theater paradigm of (moral) philosophy no less than in the variously satiric, picaresque, and humorous depictions of a strictly emulative sociability presented in William Hogarth's *Marriage à-la-mode,* Tobias Smollett's *Humphrey Clinker,* Frances Burney's *Evelina,* or Jane Austen's *Mansfield Park* is that attempts to extract moral meanings from actuarial patterns are not confined to social statistics alone. In fact, it is a central proposition of these and other works of fiction that such patterns suffuse the social practice and self-image of individuals who, in classical mimetic fashion, gravitate toward a behavioral "mean" rather than toward articulate moral meanings.[73]

In a scrupulously crafted description, Jane Austen manages to expose the hollow, not to say fraudulent moral epistemology of late Georgian common-sense culture. The world of the Miss Bertrams in *Mansfield Park* is, for much of the novel, alarmingly devoid of narrative development ("winter came and passed without their being called for"). Accordingly, they defend against the encroaching consciousness of time

73. Perhaps there is also a misunderstanding of Aristotle's controversial doctrine of the "mean" in the *Nicomachean Ethics,* 1106^{a}26–1109^{b}25.

incessantly lapsing without any experience of change or meaning by substituting transactionalism for action; theirs is a hideously invariant routine of attending to

> their toilettes, displaying their accomplishments, and looking about for their future husbands ... The Miss Bertrams were now fully established among the belles of the neighbourhood; and as they joined to beauty and brilliant acquirements a manner naturally easy, and carefully formed to general civility and obligingness, they possessed its favour as well as its admiration. Their vanity was in such good order, that they seemed to be quite free from it.[74]

When indexed to growing prosperity, literacy, cultural refinement, consistency and transparency of the law, equality, liberty, etc., the Enlightenment narrative of social "progress" no longer appears to be the product of individual, deliberate action, let alone won by an individual agent's heroic or virtuous commitments. Rather, it involves "gains" brought about by a wholly impersonal, quasi-instinctual logic that, in the later eighteenth century, comes to be associated with such words as "system," "behavior," "common sense," and "propriety." The value of these tropes to the Enlightenment utopia of the Scottish moralists and political economists stems from their seemingly self-evident, factual nature, their imperviousness to rational explication, and (so it was concluded) their immunity to public disagreement.[75] What Adam Smith had famously called the "silent and insensible" operation of commerce characterizes modern systems broadly speaking, including those of such disparate thinkers as Malthus, Hegel, Marx, and Darwin. Again, the point to be stressed here concerns the historical paradox (of which we, too, are heirs) that modern liberalism's Archimedian point of departure—a solitary, "buffered" self defined by economic interests and ineffable sentiments—appears on the stage at the same moment that new critical models (e.g., mechanism, necessitarianism, utilitarianism, speculative dialectics, historical materialism) dispute the modern individual's capacity for timely and comprehensive self-awareness. This paradoxical trend is especially apparent in the widening gap between subjective "interest" and systemic "rationality." It also accounts for the concurrent flourishing of compensatory, intuitionist, and emotivist models of self-

74. Austen, *Mansfield Park*, 63.

75. On the structural conditions contributing to the emergence of "system" as a key concept in the later eighteenth century, see Siskin's amalgamation of "disciplinarity, professionalism, and Literature," and his discussion of "novelism" as a highly reflexive process after 1750; *Work of Writing*, 1–26 and 155–192; for a compressed version of the same argument about the rise and broad ideological appeal of systems in England in relation to the new conception of the novel, see Siskin, "Novels and Systems." On the broader understanding of "system" in eighteenth-century philosophical and scientific inquiry, see E. Cassirer, *Enlightenment*, 8–16 and 37–45.

possession, such as we find them in Hutcheson, Hume, and Smith, yet also in the mid- and later eighteenth-century cult of sensibility, Rousseauvian *sentiment,* and the various antinomian and Pietist denominationalisms and the culture of *Sturm und Drang* that pervade England, France, and Germany, respectively.

The apparent conflict of views and commitments on the part of individuals and communities is not so much a function of a changed political culture, one officially (if not always believably) committed to classical liberal ideals of toleration, pluralism, and personal freedom and "repetitively reminding us of the good old unregenerate idea of diverse human satisfactions."[76] Rather, the emergence of these allegedly value-neutral and timeless notions is itself a result of the modern self's inability to negotiate norms and goods within a culture that has peremptorily disaggregated practical from speculative reason or, in Platonic terms, has estranged the good from the true. Here lie the origins of modern perspectivalism that becomes a raison d'être and formal template for the nineteenth-century novel's ongoing experiments with free-indirect speech, and that Nietzsche would eventually stylize into the one incontrovertible (if paradoxically relativist) truth of philosophical modernity. What ultimately prevents bourgeois culture from deliberating about its various objectives and interests is its refusal to acknowledge that doing so necessarily presupposes a normative (not preference-based) conception of the good and the just. Moreover, the displacement of a culture of moral inarticulacy, judgment, and dialogical reasoning by the hedonism of private "opinion" and moral sentiments also entails the progressive erosion of a differentiated inventory of concepts that enable moral agents to discriminate between normative (rational) goods and merely contingent interests. Inevitably, the rise of a hedonistic (Lockean) epistemology and the subjectivization or sentimentalization of moral agency brings about a pervasive conceptual amnesia of the kind discussed earlier. While passions, even when transmuted into interests, presuppose an underlying and enduring orientation toward the good, the unrestrained pursuit of hedonistic objectives or personal gain is likely to short-circuit rational deliberation and judgment and, in time, to atrophy our capacity for articulate reasoning about ends. The eighteenth century's supplemental world of so many competing interests and ostensibly "tolerated" beliefs held by so many discrete selves or subcommunities thus does not so much reject the question concerning ends as suspend it indefinitely. The price of an age of continual "progress" is that life increasingly resembles a terminally incomplete narrative, with the eventual result that the conceptual resources needed to appraise the value and cogency of the narrative supposedly in progress remain inactive and, like most things unused, gradually lose their cogency and relevance. In passing, and seemingly as a commonplace, Hume thus opens

76. Murdoch, *Metaphysics,* 47.

his discussion of virtue and vice by remarking how "when we leave our closet, and engage in the common affairs of life" the conclusions arrived at by "intense study" of speculative (moral) problems "seem to vanish, like the phantoms of the night on the appearance of the morning" (*HT,* 293). Not only, then, is the individual that has gone to sleep not a person (and thus bereft of all the protection of its dignity and security otherwise due), but upon reawakening she or he will have to construct her world anew and, for the sake of efficiency if nothing else, will forgo the uncertain returns of prolonged speculative reflection for the instant gratification of contingent perceptions and passions as they happen to present themselves. Conceptual amnesia, and the morally sub-literate agency to which it gives rise, appears to be an inevitable result of the *Treatise*'s eponymous "ATTEMPT TO INTRODUCE THE EXPERIMENTAL METHOD OF REASONING INTO MORAL SUBJECTS."

The loss of an operative and nuanced conceptual framework reflects a process of forgetting that stands in close, if causally indeterminate, proximity to modernity's self-image as a narrative of continuous progress toward a *telos* it neither wishes nor is able to articulate. In the absence of an explicit, communicable, and shared idea of the good, the only recognized measure of progress will be the vividly perceptible discontinuity of the present with the past. Indeed, philosophical modernity, from Luther, Hobbes, and Descartes onward, must enact and reaffirm such discontinuity time and again—thereby turning skepticism's founding, iconoclastic gesture into a ritual not at all unlike those that it purports to repudiate. Absent a teleological framework, which is concurrently being supplanted by an emergentist model of rationality, all significant meaning must be traced back to an instance of palpable presence. Under conditions of modernity that seem only to have become more intense and peremptory since Hume's *Treatise,* "mind can never exert itself in any action, which we may not comprehend under the term of *perception*" (*HT,* 293). By contrast, it is in the very nature of ideas that they have a history and, indeed, can only signify to the extent that they do and that we continue to strive for the continued articulation and joint awareness of that history. Naturally, it is precisely for that reason that Hume looks upon the epistemological status and social authority of ideas with acute suspicion and, ultimately, seeks to write them out of the picture altogether. Both on methodological grounds that have to do with the demonstrability of claims and meanings, yet also because of its axiomatic distrust of all things ancient and inherited, the naturalist model of human agency must anchor all motivation in a psycho-physiological stimulus or "perception" that ties the self to the eternal present. It is in Smith's *Theory of Moral Sentiments* (1759) that the concepts of behavior and spectatorship are mobilized so as to recover a model of practical rationality while leaving Hume's dire epistemological critique of the self substantially unchallenged.

12

VIRTUE WITHOUT AGENCY

Sentiment, Behavior, and Habituation in A. Smith

Throughout the *Theory of Moral Sentiments,* there is a marked reversal of emphasis, away from the drama of volatile and non-cognitive passions and toward reaffirming the continuity of a different type of affect. The course correction here takes the form of retranslating the passions—not back into a metaphysics of the will, to be sure—but into a firmly empirical, at times seemingly actuarial understanding of reason as it manifests itself in established customs, prevailing manners, average forms of behavior, and a mimetic conception of virtue. Viewing his arguments as post-metaphysical, yet also wishing to move beyond the rationalist, emotivist, and skeptical critiques of metaphysics that had dominated since the Restoration, Adam Smith seeks to overcome the antagonism of will and intellect—a dilemma that, unbeknownst to him, modernity had not so much discovered as created. To David Marshall, Smith "seems less concerned about the constitution of the self" and indeed "presupposes a certain instability of the self; it depends upon an eclipsing of identity, a transfer of persons."[1] Marshall's compact formula risks obscuring, however, that such a transferential model of sociality achieved by continued imaginative

1. *Figure of Theater,* 177. For another account of Smith's *Theory of Moral Sentiments,* see Griswold, *Adam Smith;* for a brief and wide ranging genealogy of sentimentalism, see Chandler, "Politics of Sentiment."

substitution constitutes something of a logical paradox. For "how can one become another person without suffering the dramatic change that is self-liquidation?" Furthermore, "if my identity is caught up with yours, and yours with another's, and so in a perpetually spawning web of affiliations, how can I ever know that your approving glance is *your* glance, rather than the effect of an unreadable palimpsest of selves?" After all, any such knowledge hinges on "entering into another experience while retaining enough rational capacity of one's own to assess what one finds there. The cognitive distance which such judgements require cuts against the grain of an imaginary ethics."[2]

Arguably, none of these logical paradoxes can be resolved in the terms in which they are here being stated—that is, in a vocabulary still committed to knowledge as propositional and tethered to a solitary and self-aware epistemological agent. Though far from meaning to present an apologia for Smith's transferential conception of moral agency, the reading here undertaken suggests that it is precisely this mentalist idea of knowledge—viz., as a type of intentionality issuing in a distinct representation—which Smith means to leave behind. In fact, his solution to Hume's epistemological dilemma rests on distilling how the inherently non-cognitive conduct of individuals will yield rational, systemic effects that could never be secured if social meaning were to depend on subjective intention. To this end, Smith comprehensively re-describes passion as "sentiment" and, in turn, sentiment as a social transaction or "behavior." The result of Smith's sweeping account of sociality as the circulation and mimetic appraisal of sentiments is a moral theory that bears more than a passing affinity to modern behaviorism.

As is evident from Hobbes's and Locke's strictly epistemological approach to the self, the modern conception of truth as an objective to be realized by specific epistemological *method* is the most significant legacy of late Scholastic nominalism. In repudiating the idea of knowledge as a result of active contemplation (*theoria*) for which the cosmos had once furnished the ontological source and ethical *telos,* modern inquiry after Bacon and Descartes instead proceeds by isolating singular entities as the only viable locus of meaning. The resulting paradigm of knowledge as "information" thus gives rise to a fact/value divide that Hume eventually sets forth as an axiom of modern rational inquiry, the result being that knowledge as an intellectual commodity has become terminally estranged from the broader ideal of wisdom and human flourishing. After 1750, epistemology and ethics are fundamentally conceived as distinct and, increasingly, as unrelated pursuits, and it is in Adam Smith

2. Eagleton, *Trouble,* 71, 75. Eagleton's central objection that "sympathy cannot be entirely spontaneous, since it needs to weigh the merits of its object" is the premise for Lori Branch's recent work on the paradox of spontaneous self-other relations in Shaftesbury and Wordsworth; see *Rituals,* esp. 91–134 and 175–209.

that this bifurcation is completed as the project of moral philosophy migrates from a theological to a sociological, and from a normative to a descriptive endeavor. Such a shift completes a tendency that had first taken shape in the methodological treatises of Bacon and Descartes, and that was subsequently radicalized by Newton and Locke: viz., the wholesale redefinition of knowledge as a quest for "certainty," which is to say, as a possession bound up with its inter-subjective communicability and accredited with reality only insofar as the self can exercise dominion over it in propositional, syllogistic form. Having already traced some of the effects of this anti-humanist strain in modern thought, we might also allude to later, yet more radical versions of epistemological skepticism, such as Frege's and Wittgenstein's systematic de-psychologizing of cognition, of which behaviorist and neuro-scientific models are the two most prominent reductionist offshoots. Notwithstanding their substantially different objectives, all these successors of modern nominalism share at least one aim: viz., to expunge the idea of human interiority and introspection altogether by driving a wedge between the phenomenological *event* of a thought (deemed inaccessible and thus irrelevant to empirical inquiry) and its *content*—judged real and pertinent only insofar as it can be objectively captured either as an ordinary-language proposition, a statistically observable behavioral pattern, or a measurable neural event.

The following reading of Smith's *Theory of Moral Sentiments* seeks to draw attention to the prehistory of these more recent projects by exploring how some key terms (e.g., passion, sentiment, sympathy) traditionally associated with introspective accounts of moral judgment and human flourishing are for the first time being systematically reinterpreted as social and objectively classifiable phenomena. To be sure, Smith's own project rests on a few distinguished precursors, including the Stoics, Shaftesbury, Hutcheson (Smith's teacher), and Hume. While Shaftesbury's role in the *Theory of Moral Sentiments* is oblique and has gone mostly unrecognized, his principal thesis that moral and spiritual self-cultivation is circumscribed by an intricate social cum aesthetic grammar crucially shapes Smith's argument.[3] One need only recall a passage like the following to pick up on the deep connection between Shaftesbury's and Smith's model of benevolence and sympathy; in it Shaftesbury ponders

> the narrowest of all Conversations, that of SOLILOQUY or *Self-discourse*. But this Correspondence … is wholly impracticable without a previous Commerce with the World: and the larger this Commerce is, the more practicable and improving

3. Marshall is among the few critics to point out how Smith's spectatorial doubling of the self reactivates Shaftesbury's "dramatic method" in the *Soliloquy* and his advice there to "divide yourself, or be two" (*Figure of Theater,* 176).

> the other, he thinks, is likely to prove. The Sources of this improving Art of *Self-correspondence* he derives from the highest Politeness and Elegance of antient *Dialogue,* and *Debate* ... And nothing, according to our Author, can so well revive this *self-corresponding* Practice, as the same Search and Study of the highest Politeness in modern *Conversation.* (*SC,* 3:96)

Likewise preferring a meliorist, conversational tone to Hume's sharp-edged analytical idiom, Smith in 1759 locates the "selfish passions" as occupying "a sort of middle place" between "the social and the unsocial" (*TMS,* 40). Long before Kant's late utopian musings advanced "the most extreme counterposition to the [Hobbesian] principle *auctoritas non veritas facit legem,*" Scottish political economists had already begun to shift from a voluntarist command ethic toward a narrative model that conceives the will or, at least, its empirical heirs—the passions—as susceptible of "improvement."[4] The trajectory in question typically proceeds from the mindless force of mute desire, advancing to an expressive but inadequately socialized "passion," and culminating in the eventual lucidity of a stable set of "interests" or, in Hutcheson's phrase, "secondary desires" (*EPA,* 19) responsive to the systemic cues of modern commerce. Even for the skeptical Hume, the project's logic is quite irresistible. Acknowledging that it simply is not feasible to "infuse into each breast ... a passion for the public good," Hume instead suggests that it is "requisite to govern men by other passions, and animate them with a spirit of avarice and industry."[5]

Albert O. Hirschman has called this the "marvelous metamorphosis of destructive 'passions into virtues'" through the new paradigm of "interest."[6] Anticipating Hirschman's thesis, John Pocock had previously qualified this view in one important respect; for even as credit was being "translated into virtue," this "restoration of virtue was subject to a single sharp limitation." Thus, insofar as "virtue was now the cognition of social, moral and commercial reality, ... imagination ... is replaced in the Whig literature of 1710–1711 by nothing more than opinion" and "rationality is only that of opinion and experience."[7] More than any of his contemporaries, Smith

4. Habermas, *Structural Transformation,* 103; for a recent account of political and economic theory, see Mitchell, *Sympathy,* esp. 76–93.

5. Hume, *Essays,* 262–263.

6. Hirschman, *Passions,* 17; in so mediating between reason and passion, the new concept of socio-economic interestedness "was seen to partake in effect of the better nature of each, as the passion of self-love upgraded and contained by reason, and as reason given direction and force by that passion" (ibid., 43). On the emergence of financial and industrial capitalism, see K. Polanyi, *Great Transformation,* 33–42 and 56–76; Pocock, *Virtue,* 103–124, and *Machiavellian Moment,* 462–505; Colley, *Britons,* 55–100; for basic statistical information on this shift, see Porter, *English Society,* 185–213.

7. *Machiavellian Moment,* 456–457.

bears out this insight by systematically disabling the Augustinian conception of virtue as predicated on sustained introspection and standing counterfactually vis-à-vis a fallen world. In its place, Smith ventures a mimetic conception of the good bound up with the vicarious transfer of prevailing, socially sanctioned sentiments, an approach that "made it possible to dismiss the egoistic motivational theory underpinning Mandeville's paradoxes."[8] While Smith's focus on the passions is a familiar feature of eighteenth-century moral philosophy, his objectives differ markedly from those of Hutcheson and Hume. For underlying his concern with the passions, and their potential commutation into durable sentiments, is no longer Locke's or Hume's epistemological quest for a viable and verifiable successor to the metaphysical notion of the will. Rather, Smith seeks to disengage moral reflection from the stranglehold of Hume's radical skepticism, which had pushed Lockean nominalism to such extremes as to render basic humanistic key concepts (e.g., will, action, consciousness, self, introspection, judgment) all but meaningless. Retreating from this philosophical dead-end, Smith finds Stoicism to be of particular value to his objectives, in part because the Stoics, even as they had acknowledged the intrinsic deficiency of human intelligence (moral and otherwise), had also sought to remedy that predicament dialectically by arguing for the complementarity of finite and imperfect human perspectives. From Smith's neo-Stoic viewpoint, this fortuitous alignment of blind volition and systemic rationality reflects a divinely ordained and providential arrangement: "Nature, accordingly, has endowed [man], not only with a desire of being approved of, but with a desire of being what ought to be approved of; or of being what he himself approves of in other men." Moral and economic self-improvement thus follow the same behavioral template, viz., a quest for "self-approbation" whose pursuit constitutes "the principal object, about which [a wise man] can or ought to be anxious. The love of it, is the love of virtue."[9]

At first glance, this new credit-based and self-interested, entrepreneurial self promises to offer "a countervailing strategy" to the Hobbesian, all-consuming will "on a continuing day-to-day basis" rather than demanding for its containment the Leviathan's ad hoc projection of overwhelming force.[10] In short, Hobbesian voluntarism with its strictly negative conception of power periodically reviving the unruly subject's "Feare of Death" is transformed into a systemic principle. For Smith, it is above all sympathy that counterbalances the specter of a radically particularist and

---

8. Schneewind, *Invention*, 380.

9. *TMS*, 117; this neo-Stoic project is echoed time and again in *The Wealth of Nations*, as in Smith's familiar assurance that "the desire of bettering our condition, a desire which, though *generally calm and dispassionate*, comes with us from the womb and never leaves us till we go into the grave" (*Wealth of Nations*, III.2.28; italics mine).

10. *Machiavellian Moment*, 456–457; Hirschman, *Passions*, 32.

pluralist nation where "every individual is ... attached to his own particular order, ... his own interest, [and] his own vanity." In the innumerable, serendipitous encounters of which quotidian sociality is composed, the practice of sympathy gradually effects an enduring, if imaginary model of community. Smith calls it "that more gentle public spirit" (*TMS*, 230–232). Yet to postulate that "every man is ... by nature, first and principally recommended to his own care" (*TMS*, 82) makes clear that voluntarism remains very much the default model of agency. Indeed, far from remedying Hobbes's and Hume's anti-rationalism, the Whig conception of commercial society proved in its own ways just as "fantastic and nonrational" and threatened "to submerge the world in a flood of fantasy."[11] Parallel to this transformation of the will into potentially rational, predictable, and manageable "interests" run various cultural narratives about the rise of modern "refinement," "manners," and "taste," as well as a host of new institutions dedicated to their advancement.[12] To the Scottish political economists of the mid-century it was apparent that governmental authority and individual will were increasingly mediated by the impersonal rationality of complex networks and "expert languages" (Anthony Giddens), rather than by a sovereign and centralized power. As Adam Smith puts it, "in the great chess-board of human society, every single piece has a principle of motion of its own, altogether different from that which the legislature might chuse to impress upon it" (*TMS*, 234).

Though differing widely in their critical stance vis-à-vis the story they wish to tell, Norbert Elias, John Brewer, J. G. A. Pocock, Linda Colley, and Michel Foucault (to name but a few), all converge in interpreting this "rise of disciplinary society" (as Charles Taylor has most recently labeled it) as a process that increasingly disperses and so obscures the force that had loomed so conspicuous and ominous in Hobbes's *Leviathan* and Mandeville's *Fable.*[13] Hutcheson's *Essay* (1728) had pointed the way here, arguing that it was "foolish" to infer "from the universal Prevalence of these Desires [of Wealth and Power] that human Nature is wholly selfish, or that each one is only studious of his *own Advantages;* since Wealth or Power are as naturally fit to gratify our *Publick Desires,* or to serve *virtuous Purposes,* as the *selfish* ones"

---

11. Pocock, *Machiavellian Moment,* 457. Though far more restrained than Burke's *Reflections,* Part VI of *TMS* (added in 1790) shows Smith's clear preference for the fluidity and contingency that Pocock here describes as a "spirit of system" such as might seek to establish an "ideal plan of government ... all at once, and in spite of all opposition" (*TMS,* 234); for a summary of the two alternative scenarios developed by post-Harringtonian, Augustan political theory, see Pocock, *Machiavellian Moment,* 458–460, as well as Pocock, *Virtue,* 103–124.

12. On the gradual contraction of the word "interest" and, especially, its French counterpart (*intérêt*) to a strictly economic understanding, see Hirschman, *Passions,* 31–48; by the time of Hume's essay "Of Interest" (1754) that shift had been all but completed.

13. On the changing scene of artistic production in the eighteenth century, see Brewer, *Pleasures,* 201–324; Barrell, *Theory of Painting,* esp. 1–162.

(*EPA*, 19). With the opportunities for economic and social mobility largely cordoned off by the mercantilist system of his native France, Montesquieu's theory concerning the separation of powers pursues a cognate objective within the field of constitutional philosophy. Thus he also posits a systemic balancing of interests as the most auspicious strategy for containing a will that, like Locke and Mandeville, he also regards as inherently volatile, blind, and irremediably selfish. This shift from an *auratic* to a *technocratic* and from a *personal* to a *systemic* understanding of power would, of course, continue and in time spawn often brilliant analyses of modern constitutional, political, and economic thought (from Montesquieu to the *Federalist Papers* to Hegel's *Philosophy of Right* to Marx's critique of political economy and even Max Weber's sociological account of institutionalized politics, science, and culture).

Their highly diverse ideological commitments notwithstanding, however, the mid-eighteenth-century works of political, economic, and moral theory converge in this one point: they all regard the will as essentially opaque, unfree, and irrational. Such is the case wherever the individual will is subject to the impersonal and remedial discipline of the law and to institutions of education and cultural literacy broadly speaking—all of which are gradually being aligned with the instrumental rationality said to govern economic behavior. Smith's account of the "impartial spectator" perfectly embodies this meliorist logic by positing moral development as a complex process of transference: just as the other's "sympathy makes them look at [suffering], in some measure, with his eyes, so his sympathy makes him look at it, in some measure with theirs, especially when in their presence and under their observation." A new, notably transactional conception of sociality thus takes shape, one that pivots on reconciled customs and manners of mid-eighteenth-century polite and commercial culture, fueled by a rhetoric of "improvement" that oscillates between the obliquely moral and the emphatically sentimental. Under this new dispensation, the containment of the (Hobbesian) will and the restoration of a certain kind of rationality is conceived as a *procedural* question, a matter of the right *technique* being brought to bear on the passions. In classical Stoic fashion, Smith thus views his own philosophical enterprise as therapeutic in nature and, in particular, as a quest for incentives such as will effectively and efficiently regulate *behavior*.[14] No longer considered are Aristotelian and Thomistic conceptions of the will as susceptible of *internal* clarification and as an integral part of moral cognition. Hobbes's view of the will as a strictly appetitive, volatile, and opaque *pathos* devoid of intellectual potential and impervious to introspective remediation thus is not so much opposed by his Enlightenment heirs as it is turned into a bleak premise for their meliorist view of politics,

14. For a detailed account of Hellenistic philosophy as a non-specialist, therapeutic enterprise, see Nussbaum, *Therapy*, esp. 316–438.

culture, and economics as the most auspicious venues for recovering a socially responsible model of human agency.

In fundamental if not entirely obvious ways, Smith accepts Locke's and Hume's account of a self driven by hedonistic pleasures and all but bereft of personal identity due to the discontinuity and incommensurability of all sensation. The self as an inner agent, while perhaps real, is ultimately deemed inaccessible to philosophical speculation; it is constitutively opaque and, to judge by such evidence of the inner life as still reaches us, it lacks any coherent and sustained sense of its reality *as a person*. Hence, Smith's strategic investment in the notion of sympathy arises from his deep-seated conviction "that others' states of mind are naturally inaccessible to us, concealed as they are by the fleshly encasements of their bodies." Moreover, Smith "was not convinced that sympathy could, on its own, maintain social order" and, where the limits of sympathetic sociality arose, was quite willing to maintain such order "through the fear of death operating in conjunction with a sovereign power."[15] It is not merely a breakdown of inter-subjective discourse, then, but the underlying, radically nominalist view of a self dissolved into heterogeneous impressions, sensations, states, desires, and affections that prompts Smith to concede from the outset that "we have no immediate experience of what other men feel" (*TMS*, 9). The apparent tension between Locke's epistemology of the self and his political theory of rational agents electively and deliberatively entering into social relations by means of propositional, quasi-contractual arrangements had been thrown into disarray by Hume's dispersal of personal identity, his atomistic view of the passions, and his disjunction of fact from value. Smith's *Theory of Moral Sentiments* thus attempts to rethink the very project of moral philosophy following the massive onslaught of post-Hobbesian skepticism on the self's epistemological coherence and moral integrity. Crucially, passion in Smith's *Theory* is no longer viewed as a purely impulsive, hedonistic, and self-consuming mental event. Rather its seemingly irrational thrust is being attenuated, even reversed, in that Smith regards emotion as an inter-subjective phenomenon, something to be reconstituted in the simulacrum of "sentiment"—less an expressive act than a behavioral norm designed and displayed so as to maximize prospects of "approval" by others.

While the Stoic project of overcoming the emotions (as de facto misjudgments) remains a central feature of Smith's argument, it is being deployed here for very different strategic purposes. For the objective is no longer, as it had been for the Stoics, a quest for inner balance and wisdom (*apatheia*) but, rather, the smooth operation of social life as an end in its own right. In anticipating and accommodating itself to projected conditions of reception, Smith's individual will "conceiv[e] some degree of

15. Eagleton, *Trouble*, 43; Mitchell, *Sympathy*, 82, 87.

that coolness about his own fortune, with which he is sensible that they will view it" (*TMS,* 22). In sharp contrast to classical Stoicism, then, self-possession and self-mastery (*autarkeia*) are at most incidental to the decidedly un-Stoic project of an affect-based and strictly "imagined" community. Indeed, the meaning, value, and significance of the inner life will only disclose themselves insofar as passion has been successfully converted into the social currency of *behavior,* a term that will occupy us shortly. If a primal passion still furnishes the raw material for this transformation, it signifies and is credited with reality only to the extent that it has been successfully transposed into "sentiment"—that is, a type of behavior sanctioned by the social grammar that undergirds Smith's sympathetic community. The imagined sphere of discursive sociality at once *recovers* passion from the netherworld of mute and unfocused animal desire and in so doing *redeems* it for purposive human life: "Society and conversation, therefore, are the most powerful remedies for restoring the mind to its tranquillity" (*TMS,* 23). Conversely, an inner life that refuses to accommodate the sociality of sentiments is at best value-neutral, though more likely of a pathological or criminal nature—a vagrant, opaque, and intractable symptom on the order of Nietzsche's "extra-moral" (*aussermoralisch*) sense.

It is just this conversion that shows Smith's use of Stoic philosophy to be rather peculiar and selective, and which also reveals the acute modernity of his *Theory of Moral Sentiments.* For the Stoics, who follow Aristotle rather than Plato, lack of self-command or irresoluteness of will (*ἀκρασία*) is not a result of hedonistic and compulsive appetition but of a flawed judgment, which is to say, of precipitous or ill-conceived "assent" to some desire or other. Inasmuch as Stoicism views the mind as substantially one and not to be partitioned in the spirit of modern faculty psychology, the emotions cannot be understood as physiologically conditioned surges intruding on the otherwise distinct and separate superagency of reason.[16] Seneca remarks that "if any one supposes that pallor, falling tears, prurient itching or deep-drawn sigh, a sudden brightening of the eyes, and the like, are an evidence of passion and a manifestation of the mind, he is mistaken and fails to understand that these are disturbances of the body." It is not that the mind is victimized by quasi-physical passions (*pathē*) but, rather, that it "suffers" them to acquire excessive or distorted

16. "There is," Nussbaum argues, "in Greek thought about the emotions, from Plato and Aristotle straight on through Epicurus, an agreement that the emotions are not simply blind surges of affect, stirrings or sensations that are unidentified, and distinguished from one another, by their felt quality alone." Indeed, "it was not an item of unargued dogma for the Stoics that the soul has just one part . . . was a conclusion, and a conclusion of arguments in moral psychology" (*Therapy,* 369, 373). Neither Zeno nor Chrysippus recognized (as some late Stoics would argue) a separate, emotional part of the soul; see Nussbaum's detailed account, 366–386), as well as my discussion above of "judgment" in Aristotle and the Stoics, 88–107.

epistemic and moral force. An emotion thus "does not consist in being moved by the impressions that are presented to the mind, but in the surrendering to these." It is an act of precipitous and misguided "assent," which is to say, an intellectual operation that has eluded the kind of scrupulous supervision that the Stoics mean to instill in their disciples. Speaking of the paradigmatic emotion of anger, Seneca thus notes that it "must not only be aroused, but it must rush forth, for it is an active impulse; but an active impulse never comes without the consent of the will [*numquam autem impetus sine adsensu mentis est*]."[17] While a fuller exploration of this complex and still contested issue of the Stoics' concept of emotion is not feasible here, a number of points can be extracted that will reveal the very different thrust of Smith's seemingly neo-Stoic argument.

First, the Stoics understand all acts of mind—including the passions—as propositional in nature. Second, to the extent that a passion has any hold on the mind it does so solely in consequence of the intellect's assent to the (oblique) proposition with which it is presented. Passion, then, is not a distinct antagonist of reason but evidence of the latter's as yet incomplete cultivation. To the extent that passion holds sway over someone's mind, it points to one's failure to apprehend and appraise the propositional nature of "impressions" with the requisite care. Third, it is just this type of failure that gives rise to a disorder of judgment and, ultimately, to ill-conceived action, a syndrome for which the Stoics adopt the Aristotelian concept of *akrasia*. Here again, it is important not to misidentify *akrasia* as some isolated failure to act on a good clearly perceived; neither is it some deficient act of mind or "opinion" randomly insinuating itself into the otherwise rational and responsible narrative of a life. Far from some contingent psychological mishap obtruded from without, akratic "weakness of will" is integral to human psychology. For Aristotle, *akrasia* thus involves a specific kind of desire (*orexis*), one that is not determinative like the craving for food or sex but, rather, something unpredictable and self-fuelling. Unsurprisingly, it is "anger" (*ira*) that the Stoics, and Seneca in particular, identify as the very embodiment of *akrasia*—viz., a pointedly *a*-social passion that can be either prospective (e.g., revenge) or retrospective (e.g., resentment). In the first case, there is no determinate causal link between the akratic disposition and a specific course of action, for as we well realize, Hamlet may or may not act to avenge his father's murder. Conversely, in the case of looking back in anger it is even more obvious that *akrasia* does not compel a specific action since the past cannot be changed.[18] Crucially, the

17. Seneca, *De Ira* ("On Anger"), 2.3. Similarly, Epictetus remarks that "it is not things themselves that disturb men, but their judgments about things" (Long and Sedley, *Hellenistic Philosophers*, 418).

18. On Stoic accounts of judgment in relation to a disordered, weak will and to desire, see above, 99–107. Amélie Rorty links *akrasia* to "background dispositional patterns of perceptual,

Stoic objective of heading off misjudgment—that is, the precipitous assent to an emotionally colored and distorted perception—was to be realized by methodical and sustained introspection.

Not so in the *Theory of Moral Sentiments;* for while Smith appears to echo the Stoic quest for isolating and disabling the sources of a disordered inner life, his solution is precisely *not* one of introspection but, rather, to declare all passion akratic until and unless it has been socialized as a benevolent sentiment. The result is a marked shift in emphasis, which in turn yields a distinctively modern conception of moral agency. To begin with, it is apparent that Smith no longer views the passions as propositional in nature. Both in its primitive state and once converted into a socially recognized "sentiment," emotion in Smith's *Theory* proves to be altogether extra-rational and non-propositional. The *telos* of Smith's account is thus neither knowledge nor the self's ability to form logical and critical judgments on the quasi-propositional character of the emotions. Rather, Smith seeks to effect a mimetic alignment of the self's emotively charged impressions with what are hypostatized to be the cognate affective experiences of other individuals. Simply put, the modern objective is emulation, not cognition. From this a second point follows; for contrary to views that the Enlightenment came to hold, classical Stoicism's methodical quest for inward composure and rational self-governance had always served the final objective of building up the self's capacity for active and selfless citizenship. Its structured regimen of self-mastery was aimed at furnishing certain and unbiased representations (not sentiments) such as would prove conducive to political "action."[19] Indeed, it is just this framework of a *vita contemplativa* integrated with, rather than opposed to, the *vita activa* that was also shared by Augustine and Aquinas (the latter taking his cue directly from Aristotle rather than the Stoics).[20] By contrast, Smith's

---

cognitive, affective, and behavioral habits" and argues that *akrasia* need not be "self-centered," is not anchored "in a specific desire," and "need not involve any manifest conscious belief" ("Social and Political Sources," 649, 650); see also her discussion of the concept in the context of Aristotle's ethics ("Akrasia and Pleasure"). The once common rendering of *akrasia* as "weakness of will" has recently been called into question; see Holton, *Willing, Wanting, Waiting*, 80–96; and Rorty, "Social and Political Sources."

19. For a powerful (and witty) statement about philosophy as a remedy against suffering—rather than some sterile and inconsequential parsing of propositions—see Seneca, *Moral Epistles*, no. 48. Similarly, Musonius insists on the practical nature of philosophy as a regimen of mental health undertaken for the sake of action: "Whatever arguments [sophists] undertake, I say that these should be undertaken for the sake of deeds. Just as a medical argument is no use unless it brings human bodies to health, so too, if someone grasps or teaches an argument as a philosopher, that argument is no use, unless it conduces to the excellence of the human soul" (qtd. in Nussbaum, *Therapy*, 324).

20. On Augustine's "rehabilitation of the affections" from their apparent proscription in Stoic thought, see Wetzel, *Augustine and the Limits of Virtue*, 98–111; Verbeke, "Augustin et le stoïcisme";

critique of Stoicism as expiring in the languor of "sublime contemplation" and after-the-fact "consolation" is not only misleading but also obscures his own theory's far more equivocal outlook on action. In Smith's rather one-sided portrayal, Stoicism limits the self to a cultivation of "apathy" and to a concerted attempt at "eradicat[ing] all our private, partial, and selfish affections, by suffering us to feel ... not even the sympathetic and reduced passions of the impartial spectator." The result, we are told, is to have removed the self from "every thing which Nature has prescribed to us as the proper business and occupation of our lives" (*TMS,* 292–293).

Notably, Smith's critique of Stoicism in his survey of moral philosophy (added as Part VII to the 1790 edition of his *Theory of Moral Sentiments*) also departs from the neo-Stoic thrust of his book's earlier sections, written some three decades earlier. There Smith had firmly aligned himself with the Stoic ideal of affective self-governance: "to restrain our selfish, and to indulge our benevolent affections, constitutes the perfection of human nature; and can alone produce among mankind that harmony of sentiments and passions in which consists their whole grace and propriety" (*TMS,* 25). Echoing Joseph Addison's *Cato* and anticipating similar claims in Gotthold Ephraim Lessing's *Laokoon,* Smith expresses admiration for "the man who has lost his leg by a cannon shot, and who, the moment after, speaks and acts with his usual coolness and tranquillity." What is put on theatrical display here is the ostensibly anti-theatrical Stoic virtue of "self-command." Yet there is something decidedly equivocal about Smith's claim that "in proportion to the degree of self-command which is necessary in order to conquer our natural sensibility, the pleasure and pride of the conquest are so much the greater" (*TMS,* 147). For it is just this ostensibly Stoic conquest of the passions that, in Smith's account, has to be put on display inasmuch as its success remains in epistemological limbo until and unless it has been confirmed by outside spectators.[21] Elsewhere, Smith thus extols

> that reserved, that silent and majestic sorrow, which discovers itself only in the swelling of the eyes, in the quivering of the lips and cheeks, and in the distant, but affecting coldness of the whole behaviour. It imposes the like silence upon us. We regard it with respectful attention, and watch with anxious concern over

---

and Colish, *Stoic Tradition*, 142–238. On Stoic and Augustinian elements in Descartes's philosophy, see Hanby, *Augustine and Modernity*, 137–144.

21. Though alert to the theatricality of Smith's overall argument, David Marshall views "Smith's endorsement of Stoic ideas ... as the result of an antitheatrical sensibility" (*Figure of Theater*, 184); my argument, by contrast, is that Smith's Stoicism is neither anti-theatrical nor properly Stoic in nature; rather, we are presented with a simulation of Stoic *apatheia* and a theatrically affected anti-theatricalism.

> our whole behaviour, lest by any impropriety we should disturb that concerted tranquillity, which it requires so great an effort to support.[22]

This is a fine instance of what we might call *faux* apathy, a modern, theatrical simulation of the Stoic ideal whose original purposes, however, substantially elude Smith. What stands out is not its professed extirpation of passion but, rather, the manner in which its seeming conquest is put on display. The conspicuous theatricality with which the raw and unfiltered psychic data of passion are shown to have been commuted into the artifact of a "concerted tranquillity" is lauded as a worthy achievement precisely because it enables the spectator to admire the "effort" expended in that very transformation. What accounts for the pivotal role of both behavior and sympathy here is that, being socially intimated with such "propriety," they call for no action whatsoever on the part of the spectator. Following Rousseau, Smith consistently holds "that sympathy be something that one feels rather than something that one does," and as Nancy Armstrong has argued it is this very premise "that our most compelling feelings might have a source external to ourselves [that] becomes especially apparent in writing associated with 'sensibility.'"[23]

As in Smith's later economic arguments in the *Wealth of Nations*—a book now understood to have grown out of his moral philosophy—sympathy gives rise to a distinctly modern conception of sociality as an imagined, indeed simulated lateral bond between anonymous, hermetic, and substantially unrelated individuals.[24] The compact that defines the modern economic and social order is strictly virtual; sympathy here functions as the moral equivalent of the speculative commodity of modern stock. Its value pivots on the shrewd management of how it is socially perceived, that is, on the accommodating and confident "behaviour" (an important word used twice in the above passage) with which it is introduced into social space. Notably, the following rhetorical question (whose closing exclamation mark suggests that it only admits of an affirmative reply) shows Smith entirely blinded to the possibility of deception—collectively felt and reinforced—about the import and value of moral sentiments: "How amiable does he appear to be, whose sympathetic heart seems to re-echo all the sentiments of those with whom he converses, who grieves for their

22. *TMS*, 24; with specific attention to the theatricality of emotion, Smith reiterates this point in Part I: "When we attend to the representation of a tragedy, we struggle against sympathetic sorrow which the entertainment inspires as long as we can, and we give way to it at last only when we can no longer avoid it: we even then endeavour to cover our concern from the company. If we shed any tears, we carefully conceal them" (*TMS*, 46).

23. *How Novels Think*, 14–15.

24. For accounts of this paradigm, see Anderson, *Imagined Communities*, 1–36; Taylor, *Modern Social Imaginaries*, 49–107.

calamities, who resents their injuries, and who rejoices at their good fortune!" (*TMS*, 24). To embrace this stance is to delegate one's moral judgment and orientation to others, to make it dependent on behavioral cues furnished by others, and to collapse the good and the true into the contingent (and possibly opportunistic) affirmation that some particular view is perceived to have been accorded by others. Few models could be farther removed from the Stoic ideal of *autarkeia* than this transferential and manifestly heteronomous cultivation of moral agency.

For a canonical counterexample, one might look at Augustine's suspicious probing of theatrically induced "sympathy" (*misericordia*) in Book 3 of the *Confessions*. Sheer spectatorship, in Augustine's view, does not constitute but merely simulates moral meanings: "What compassion is to be shown at those feigned and scenical passions? For the auditors here are not provoked to help the sufferer [*sed qualis tandem misericordia in rebus fictis et scenicis? Non enim ad subveniendum provocatur auditor*]." Spectatorship by itself is corrosive of action and agency in that it shifts the focus from the practical realization of the good to a narcissistic delight in the very failure of it.[25] If staged suffering elicits sympathy, an increase of it will augment the sympathetic emotion and the spectators' gratification in its experience: "they so much the more love the author of these fictions, by how much the more he can move passion in them [*et auctori earum imaginum amplius favet, cum amplius dolet*]." Indeed, not only do theatrical simulations of suffering beget in the spectator an analogous, virtual attachment to "sympathy" ("Are tears therefore loved, and passions? [*lacrimae ergo amantur et dolores*]"), but they thereby also create an incentive to desist from practical action. The attenuation of moral agency, Augustine contends, stems from the spectator's attachment to the gratifying emotion of sympathy, over and against envisioning himself or herself as capable of achieving the good. To do the latter, in

---

25. Augustine's emotional link between sympathy and narcissism ultimately points back to his underlying preoccupation with *curiositas*. "Curiosity" is always a vice for Augustine, and it is repeatedly linked to voyeurism by way of 1 John 2:16, a passage quoted or alluded to numerous times throughout the *Confessions*: "For all that is in the world is the concupiscence of the flesh and the concupiscence of the eyes and the pride of life, which is not of the Father but is of the world" (*quoniam omne quod est in mundo concupiscentia carnis et concupiscentia oculorum est et superbia vitae quae non est ex Patre sed ex mundo est*). On Book 3, Chapter 2.2, see James O'Donnell's commentary on the *Confessions* at www.stoa.org/hippo/; on curiosity in Augustine, see Griffiths, *Intellectual Appetite*, 1–29. On the particular passage in *Confessions*, Book 3, see also the essay by Breyfogle in Paffenroth and Kennedy, *Augustine's Confessions*, 35–52. Augustine later returns to the question of "compassion" (*misericordia*) as part of his discussion of Stoicism; "and what is compassion but a kind of fellow feeling in our hearts for the misery of another which compels us to help him if we can? This impulse is the servant of right reason when compassion is displayed in such a way as to preserve righteousness" (*Quid est autem misericordia, nisi alienae miseriae quaedam in nostro corde compassio, qua utique, si possumus, subvenire compellimur? Servit autem motus iste rationi, quando ita praebetur misericordia, ut justitia conservetur*). *City of God*, 10.5 (p. 365).

fact, is to conjure the logically absurd scenario of wishing for both the flourishing and the suffering of others at the same time: "For if there be a good will that is ill-willed (which can never be), then only may he, who is truly and sincerely compassionate, wish there might still be some men miserable, that he might still be compassionate [*si enim est malevola benevolentia, quod fieri non potest, potest et ille, qui veraciter sinceriterque miseretur, cupere esse miseros, ut misereatur*]." As embodied by the stage and the hedonistic model of spectatorship to which it gives rise, Augustine's view of sympathy (*misericordia*) is that of a false good, a "joy that enchains" (*vinculum fruendi*), as he puts it with deliberate emphasis on the sexual connotations of *vinculum* (fetters, bondage). For Augustine, emotion is legitimate only inasmuch as it is a source of action, not a self-consuming, narcissistic experience; it enjoins the self "to relieve" (*ad subveniendum*) genuine "suffering" (*miseria*). The anti-theatrical passage that opens Book 3 thus becomes a template for similar scenes in subsequent books, such as Augustine's deeply suspicious hermeneutic of his own grief at the death of a young friend in the book following (*Confessions*, 4.4–7); and it only finds its completion when, mourning the death of his mother in Book 9, Augustine appears at last to have achieved the proper ratio of grief, sympathy, and purposive action.

Fifteen hundred years later, the *Theory of Moral Sentiments* offers a starkly different picture by construing sympathy as a type of virtual action, rather than the inner condition to be complemented by an active quest for providing material relief to those suffering. For Smith, both the spectator and "the sufferer long less for relief from pain than for the relief that is afforded only by sympathy," and the latter necessarily requires some material suffering (or theatrical simulation of it) as its precondition.[26] Discussing sympathy under the heading of the "social passions" (*TMS*, 38–40), Smith is at pains to conceive moral sentiments as the implicit approval of a prevailing social consensus. Having all but lost the hortatory and potentially transformative function as "source" that it held in Augustine, sympathy instead names (and affirms as valid) an existing and manifestly self-certifying social consensus: "Generosity, humanity, kindness, compassion, mutual friendship and esteem, all the social and benevolent affections, when expressed in the countenance or behaviour, even towards those who are not peculiarly connected with ourselves, please the indifferent spectator … We have always, therefore, the strongest disposition to sympathize with the benevolent affections" (*TMS*, 38–39). What facilitates this lateral, albeit inarticulate

26. Marshall, *Figure of Theater*, 173. Hume had previously mused on the "unaccountable pleasure, which the spectators of a well-written tragedy receive from sorrow, terror, anxiety, and other passions on stage" ("Of Tragedy," in *Essays*, 216). A radical Augustinian himself, Blake echoes the point in *Songs of Experience*: "Pity would be no more, / If we did not make somebody Poor: / And Mercy no more could be, / If all were as happy as we" ("The Human Abstract," *CPP*, 27).

comradeship where "these affects, that harmony, this commerce, are felt (*TMS,* 39) is the virtual agency of the impartial spectator, a heuristic fiction that effectively collapses the distinction between self and other, thereby suspending the individual's self-awareness as a responsible being.

A coded answer to Hume's dystopia of a world where empirical fact and normative meaning have become terminally estranged, Smith's moral sentiment goes to the other extreme. By its very nature, he repeatedly argues, sympathy is a psychological fact that implies its own value—something suggested by the equivalent position of "countenance and behaviour" or "sentiments and behaviour" (*TMS,* 162).[27] The moral significance of emotion is thus confined to its inter-subjective realization qua sentiment, a "fellow-feeling" whose meaning and value expire in the narcissistic dramaturgy of sentiments displayed and approved, respectively. David Marshall's well-known account of Smith's theatrical aesthetic seems rather oblivious to the question of just what kind of theater it is that the *Theory of Moral Sentiments* stages. On his account, Smith's "self is theatricalized [sic] in its relation to others and in its self-conscious relation to itself," even as it "must be an actor who can dramatize or represent to himself the spectacle of self-division in which the self personates two different persons who try to play each other's part, change positions, and identify with each other." Quoting Hume on how "the minds of men are mirrors to one another" (*HT,* 236), Terry Eagleton likewise remarks on the narcissistic logic underpinning both Hume's and Smith's social theories: here "the [Lacanian] imaginary ... is a sort of mutual admiration society, in which in a kind of *mise-en-abyme* each act of reflection gives birth to another," thus revealing to us "the cyclical time of the imaginary rather than the linear evolution of the symbolic."[28] In so orchestrating a kind of *égoïsme à deux,* sympathy no longer furnishes the occasion for focused deliberation and judgment as to what action a given situation calls for. Rather, Smith's sociality is theatrical to its very core; here all the world is indeed a stage, albeit with this peculiar qualification that everyone is always on stage, but takes himself or herself to be a mere spectator. At the same time, action and plot have been all but supplanted by an invariant display of epistemological narcissism inasmuch as every individual's performance is mimetically enslaved to the same role and script. We are much closer to the realm of Pirandello and Beckett than to that of Joseph Addison or even Shakespeare. The result is a system based on "substitution as a foundational principle [*eine Art fundamentaler Stellvertretung*]" where the "spectator is above all spectator of himself [*dieser Zuschauer ist vor allem auch Zuschauer seiner selbst*]"; subjective experience and

27. On the derivation of Smith's sympathetic ties from early modern cosmology and natural philosophy, esp. the appeal to a *sympathia naturae vita* in discussions of magnetism, medicine, gravitational physics, and the chemical elements, see Vogl, *Kalkül und Leidenschaft,* 87.

28. Marshall, *Figure of Theater,* 176; Eagleton, *Trouble,* 48.

meaning thus unfold within a hypothetical, indeed, virtual matrix of "as if" relations that render "truth values inseparable from role-playing, illusion, semblance, and stage dynamics."[29]

It is this structural ambiguity of Smith's "moral sentiment" that prompts Vivasvan Soni to speak of a persistent "double meaning of ... sympathy as affective identification and sympathy as pity" and, hence, to read Smith's *Theory* as inaugurating "an epochal shift ... from an ethical to an epistemological and identificatory structure of reading." According to the latter model, for the first time fully realized in Smith's account, "the purpose of a narrative is to allow one to reproduce imaginatively the world and the experiences of the protagonist, instead of producing an ethical relation to the narrative situation of the other." In Smith's post-teleological world, action has been supplanted by the merely transactional, and rational deliberation by a cascade of minute transferences.[30] No longer is human flourishing conceived as a dialectical (and potentially tragic) narrative composed of introspection, (mis-)judgment, and purposive action, however imperfect. Instead, by shifting focus to the socialization of potentially isomorphous selves—types rather than persons—Smith's project pivots on the mimetic acquisition and circulation of virtual sentiments in the guise of approved forms of "behaviour." Action, understood as the teleological fulfillment of rational personhood in Aristotelian and Thomistic thought, has been all but absorbed into this "sentiment of approbation" wherein, as Smith tells us, "there are two things to be taken notice of; first, the sympathetic passion of the spectator; and, secondly, the emotion which arises from his *observing the perfect coincidence* between this sympathetic passion in himself and the original passion in the person principally concerned" (*TMS*, 46; italics mine). More pointedly yet, a later section probes the "difference between the approbation of propriety and that of merit or beneficence," with Smith concluding that "'till I perceive the harmony between his emotions and mine, I cannot be said to approve the sentiments which influence his behaviour" (*TMS*, 78). In both cases, Smith is responding to Hume's astute query whether it is reasonable to suppose that "all kinds of Sympathy are necessarily Agreeable," considering that "the Sympathetic Passion is a reflex image of the principal, [and hence] it

29. Vogl, *Kalkül und Leidenschaft*, 80; likewise, Marshall notes how the transferential operation of sympathy "more than doubles the theatrical positions Smith sees enacted in sympathy by compelling us to become spectators to our spectators and thereby spectators to ourselves" (*Figure of Theater*, 173). Similarly, for Soni, even as sympathy, which "promised to serve as a bridge between self and other, betrays its promise and leaves the self embroiled with its own emotions, which it imagines to have come from the other" (*Mourning Happiness*, 309). Here my account differs in that it views sympathy in Smith as garden-variety transference and projection, simply because the self has no substantive identity prior to or independent of such projection. There is only a virtual self, and its putative inner states are of an equally virtual (or behaviorist) nature, as remains to be shown.

30. Soni, *Mourning Happiness*, 305, 311.

must partake of its Qualities, and be painful where it is so."[31] In the event, Hume's question reflects something of a misunderstanding of Smith's argument, which is no longer premised on some garden-variety intersubjective relation between otherwise autonomous agents but, on the contrary, understands moral agency as something transferentially constituted.

In this regard, the word "sentiment" in Smith's eponymous moral theory is rather misleading in that it seemingly posits an interior and authentic emotional certitude as the epistemological point of departure for its argument. Yet there are multiple indications that this not the case. For one thing, the concept of the will has now definitively vanished from the vocabulary of moral theory, a development only intelligible when seen in the broader context of Smith's overall retreat from an interiorist, *res-cogitans* model of subjectivity. To understand the nature and significance of that shift, it helps to interrogate the work's central concept: what does Smith mean by "sentiment," and what can that word signify in a philosophical context that has rendered the notion of a unified and autonomous self substantially inoperative? In Part VII of the *Theory of Moral Sentiments,* added to the sixth edition (1790), a short but incisive critique of Hutcheson's "moral sense" signals what Smith had taken to be his overall retreat from an internalist account of human agency. As Smith points out, Hutcheson's hypothesis of a moral sense operating wholly unconditioned by external contingency or interest actually breaks down when certain qualities are introduced that, "belong[ing] to the objects of any sense, cannot, without the greatest absurdity, be ascribed to the sense itself" (*TMS,* 323). What Smith objects to is the very supposition that vice and virtue could ever be established by an inner sense operating independent of any contextual awareness or feedback; and his fictitious scenario of a "man shouting with admiration and applause at a barbarous and unmerited execution, which some insolent tyrant had ordered," is meant to illustrate the way that moral judgment is fully enmeshed with some ambient, social dynamic. Characteristically, Smith thus flags the susceptibility of Hutcheson's "moral sense" to possible misjudgments by depicting how the surrounding spectators should "feel nothing but horror and detestation" at such an inappropriate response, and how they are likely to "abominate him even more than the tyrant" who had ordered the execution. As he sees it, absent a reciprocally constituting social awareness, Hutcheson's moral sense lacks all criteria. For Smith, then, there can be no such thing as *synderesis,* for to hypothesize a purely inward source of moral orientation is to hazard a "perversion of sentiment" that may rise to "the very last and most dreadful stage of moral depravity" (*TMS,* 323).

31. Hume to Smith (28 July 1759), in *Letters,* 1:313.

Yet Smith's alternative of moral sentiments bound up with a spectatorial infrastructure that supplants judgment with "wonder and applause" is not without problems of its own. Above all, it is not easy to conceive of "sentiment," as Smith repeatedly insists we must, as socially conditioned and only arising from "mutual regard" (*TMS*, 39). Would not such sentiment be but a more genteel term for a prolonged bout of Freudian "transference" (*Übertragung*) or "projection"? As Vivasvan Soni notes, "affective communion does not occur simply on the evidence of the other's affect; a narrative is required to engender analogous affects in us." Yet if "a narrative understanding of sentiments is already built into [Smith's] theory," the self-certifying nature of affect allows us to take that narrative for granted and allows us to "focus our attention on the emotional state of the other without regard for narrative."[32] To this one might add that a "narrative" that tacitly regulates the specific dynamics of affect, both as it is expressed and received, should perhaps be thought of as an underlying "grammar" rather than narrative. In what is to follow, we shall think of it under the heading of "behavior."

Moreover, how is a spectator to "admire the delicate precision of [someone else's] moral sentiments" if not by autonomously interpreting the symbolic and gestural language ostensibly "expressing" such sentiments? If sentiment is inter-subjectively constituted, would this not preclude any critical or counterintuitive perspective on it? Does the *Theory of Moral Sentiments* recognize the need for a genuine hermeneutic of sentiment? To judge by contemporary responses to Smith's, the answer is "no." Characteristic of early reactions to the *Theory* is a peculiar blend of the enthusiastic and the unreflexive; reporting to the author that his *Theory of Moral Sentiments* "is in the hands of all persons of the best fashion," William Robertson notes "that it meets with great approbation both on account of the matter and stile" and that it is "impossible for any book on so serious a subject to be received in a more gracious manner."[33] A notice published in the *Monthly Review* echoes this appraisal and singles out the author's "agreeable manner of illustrating his argument, by the frequent appeals he makes to fact and experience." While declining to endorse Smith's account of sympathy outright, the reviewer depicts the book's argument as "extremely ingenious and plausible."[34] Finally making good on a promise (to Hume) that he would write to the author of the *Theory of Moral Sentiments,* Edmund Burke is quick to zero in on the favorable ratio between intellectual gain (large) and intellectual effort (small) involved in the perusal of Smith's work: "I do not know that it ever cost me less trouble to admit to so many things to which I had been a stranger before." Like so many

32. Soni, *Mourning Happiness*, 300–301.

33. Letter to Smith (14 June 1759), qtd. in *TMS*, 26.

34. July 1759 (no. xxi); though unsigned, the review is attributed to William Rose; qtd. in *TMS*, 27.

other readers, Burke also dwells on "those easy and happy illustrations from common Life and manners in which your work abounds more than any other that I know by far," and he echoes similar praise by referring to Smith's "lively and elegant" style.

Writing for his *Annual Register,* Burke extends the latter comment in a particularly revealing manner by remarking how Smith's "language is easy and spirited ... it is rather painting than writing."[35] Smith's stylistic accomplishments are well received on the Continent, too, with an early French review approving "*par la beauté et la noblesse des sentiments.*" More surprising might be the same review's affirmation that "religion is respected throughout the work [*que la Religion y est par-tout respectée*]," considering that the same had been affirmed in the *Monthly Review*—viz., that "the strictest regard [is] preserved, throughout, to the principles of religion."[36] That both Anglo-Protestant and Catholic readers should reach such uniform (if notably vague) conclusions on the lingering question of Smith's attitude to religion brings us to the one point observable throughout virtually all of the responses to Smith's book: none of them actually identify, let alone engage, its thesis. Bearing a striking resemblance to Thomas Gray's contemporaneous "Elegy Written in an Country Churchyard," Smith's book draws responses that prove both consistently sympathetic and positively non-cognitive. Working by accretion, the *Theory* sets forth "a system of moral philosophy able to propagate itself by accommodating ever more examples of its central claims. Where Hume had sought, through his *Treatise,* to regulate social systems by means of systematic reflection on sympathy, Smith opted for a form of regulation that denied the appearance of systematicity."[37] The peculiar appeal of the *Theory of Moral Sentiments* stems from its having so successfully harnessed the power of the moral commonplace as both medium and message. Conspicuous for the low interpretive demands it imposes on the reader, Smith's *Theory* truly embodies its implicit conception of virtue, morality, and the good as so many behaviorally instantiated sentiments.

Yet another set of questions to press concerns the traditions of inquiry, ancient and modern, that are being linked by Smith's unprecedented emphasis on the sociality of the emotions. The two conceptual traditions that intersect in his work are classical (and early modern) Stoicism and, from what we may call the "future-past," the twentieth-century project of "behaviorism," whose conceptual roots have often

---

35. Burke, letter and review as quoted in *TMS*, 28–29.

36. *Journal encyclopédique* (October 1760), qtd. in *TMS*, 29.

37. R. Mitchell, *Sympathy*, 89. Similarly, Eagleton notes that in Smith's Lacanian framework, "there is no Other of the Other—no meta-language which would allow us to investigate our intersubjective meanings from a vantage-point beyond them, ... rather as for Adam Smith there is no ground to our world beyond 'the concurring sentiments of mankind'" (*Trouble*, 75–76).

been traced back to Locke's and Hume's epistemology, yet whose conceptual intent is uniquely anticipated in the *Theory of Moral Sentiments.* Though ostensibly consumed with psychological discriminations of all sorts, Smith's argument undertakes a comprehensive de-psychologizing of human agency. Hints to this effect come early, and they typically take the form of shifting focus from the (seemingly inscrutable) inner phenomenology of passion toward an analysis of its social, mediated character: "Sympathy ... does not arise so much from the view of the passion, as from that of the situation which excites it." By choosing emotions such as only ever occur in social settings (flustering, awkwardness, embarrassment, wit, etc.) rather than those liable to crystallize by way of sustained introspection (doubt, despair, hope), Smith draws attention to sociality as positively constitutive of the meaning of "passion"—a term that throughout the *Theory of Moral Sentiments* is all but synonymous with "emotion" yet categorically distinct from "sentiment." Smith's passing observation that "we blush *for* the impudence and rudeness of another" (*TMS,* 12) emphasizes how emotion involves an element of imagination, projection, and transference. Its semantics are conditioned not by an inner certitude but, rather, by an agent's appraisal of a social situation as it is (probabilistically) taken to be perceived by multiple individuals. Moreover, Smith contends that this sublation of individual emotions into socially constituted sentiments is not some adventitious occurrence but is teleologically inscribed into the passions themselves. For intrinsic to every emotion there is a distinctive "motion," a gravitational force or tropism oriented toward the other person, much like Jacques Lacan's account of the "symptom" as something both real and intelligible only insofar as it is directed at an addressee.

Yet for Smith, inasmuch as that other is conjectured to be experiencing the same inner state, she or he is not simply the recipient of some seemingly interior, private feeling. Rather, she mimetically confers reality on that emotion as a socially viable sentiment: "Whatever may be the cause of sympathy, or however it may be excited, nothing pleases us more than to observe in other men a fellow-feeling with all the emotions in our own breast" (*TMS,* 13). If the first part of this sentence discounts the Platonic and Christian conception of emotion as an inner "source," it is not so much that Smith means to reject that view outright but that it no longer holds a central role within his overall philosophical project. Instead, an emotion in his *Theory* acquires distinctness, significance, and prima facie reality only if and inasmuch as it is reflected back by another self. What Smith calls "fellow-feeling" furnishes, if not the material ground (*ratio essendi*), then certainly the ground whereby an emotion becomes positively intelligible and significant (*ratio cognoscendi*). To the extent that "our approbation is ultimately founded upon a sympathy or correspondence [of sentiments]" (*TMS,* 17), an emotion can be credited with epistemological standing and moral significance only insofar as it reflects and reaffirms an underlying social consensus.

Central to the *Theory of Moral Sentiments,* and widely recognized as the book's main contribution to modern thought, is Smith's spectatorial model of moral cognition. Arguing that every sentiment encrypts an instance of approbation or disapprobation, Smith suggests that any judgment ventured about the merits or demerits of one's own (proposed) action is achieved by imagining whether the motive underlying it would meet with someone else's approval. Merely to put the question in that way is to have already migrated from an introspective to a transferential account of moral value. Sentiments in Smith do not identify but constitute their objects; social "facts" do not precede, let alone exist independent of, but are only realized *by* the interpretive and evaluative process of social exchange. Having famously enjoined his readers that "we must become the impartial spectators of our own character and conduct," Smith notes how any instance of subjective "approbation necessarily confirms our own self-approbation." The praise of others "necessarily strengthens our own sense of our praiseworthiness," an interesting update on the problem of Aristotelian *megalopsychia,* which had vexed virtue ethics for more than two millennia. For Smith, praiseworthiness is not inwardly felt, let alone unilaterally asserted by a moral agent about herself or himself. Rather, it is an inference compelled by how others have assessed one's own conduct: "So far is the love of praise-worthiness being derived altogether from that of praise; that the love of praise seems, at least in a great measure, to be derived from that of praise-worthiness" (*TMS,* 114).

Smith's claim that a spectatorial type of moral judgment performatively *creates* the very values to which individuals take themselves to be responding tends to come in two preferred metaphors, the theatrical and the optical. In one of his programmatic accounts of the impartial spectator (109–113), Smith thus insists that this virtual agent "is the only looking-glass by which we can, in some measure, with the eyes of other people, scrutinize the propriety of our own conduct" (*TMS,* 112). That all moral judgment thus rests on, and is conditioned by, the a priori sociality of human beings becomes, according to Smith, evident if one entertains the counterfactual scenario. For a completely isolated self such as the noble savage of Rousseau's second *Discourse* "is provided with no mirror which can present . . . to his view" the "beauty or deformity of his mind" any more than that of his body. Yet "bring him into society, and he is immediately provided with the mirror which he wanted before. It is placed in the countenance and *behaviour* of those he lives with, which always mark when they enter into, and when they disapprove of his sentiments; and it is here that he first views the propriety and impropriety of his own passions" (*TMS,* 112). For strategic reasons, we note the peculiar conjunction of a regime of vision ("countenances" looking out, and being looked into, like looking glasses) with a term suddenly rising to prominence in post-metaphysical social theory: "behavior." Striking about both is the absence of any deliberative, discursively reasoning component. Indeed, behavior and vision alike are valued precisely on account of their supposedly seamless, transparent, and effortless operation. At least for Smith—herein markedly differing from the forensic and evolving concept of vision that we find in Wordsworth, Goethe, Darwin, Ruskin, and Gerard Manley Hopkins—neither sight nor behaviors involve any sustained cognitive effort.[38] Furthermore, both the concept of vision and of behavior can only signify on the basis of an already established (and likely complex) matrix of socio-cultural values. To derive moral approval or disapproval from the gaze of the other, like the kind of implicit social orientation denoted by the concept of "behavior," requires that a grammar of social propriety and order has already been internalized and thus has enabled the self to *look for* specific meanings rather than gazing outward without either purpose or comprehension. Yet to premise moral orientation on a specular and performative model of sociality comes at a price. At the very least, it means that moral cognition and judgment have been demoted from inner reflection to a mere reflex gesture, and that the counterfactual or creative potential of thought has yielded to the mimetic affirmation of some already established, albeit inarticulate notion of social propriety.[39]

To develop a clearer sense of Smith's overarching objectives in the *Theory of Moral Sentiments,* and its overall implications for subsequent accounts of practical reason and ethics, one must scrutinize precisely this transposition of passion into sentiment. What prompts Smith's sweeping re-description of the inner life is a desire to restore to the self the epistemological coherence and moral authority that it had lost in Locke's *Essay,* Mandeville's *Fable,* and, especially, in Hume's *Treatise* and second *Enquiry.* As a result, stress is no longer placed on the ideational content of emotion, a point on which Smith appears cannily agnostic. Rather, scrutiny is brought to bear on the dramaturgy of emotion as it is "actualized" (in the Hegelian sense of *Verwirklichung*) in social exchange. The reality of emotion, Smith suggests, proves altogether inseparable from its social phenomenology, that is, from its mode of appearance as "behavior." For passion to be commensurate with (rather than disruptive of) social relations, it must take the form of a recognized symbolic practice (as opposed

38. On Goethe's (early phenomenological) conception of seeing in his botanical writings, see Pfau, "All is Leaf." On changing models of visual perception in the eighteenth century, see Crary, *Techniques*, esp. 1–24 and (on Goethe) 67–96.

39. Vogl succinctly captures numerous tensions, by remarking how Smith's global principle of sympathy is located at a "precarious threshold": "Sympathy encompasses involuntary responses, and yet its specificity originates in an act of consciousness; it construes as a self-relation what, in fact, becomes manifest in a relation of self to other; it produces realities whose mode of origination is indistinguishable from deception [*Täuschungen*]; it separates the world of functional processes from the world that legitimates these processes, and yet it postulates their interdependency; it locates sociality as an empirical object *sui generis*, and yet it delineates the grounds of legitimacy for that social world; it generates a field in which judgments and natural (physiological, psychological) forces are entwined; and [sympathy] purports to furnish a principle equally capable of formulating criteria for affections and actions" (*Kalkül und Leidenschaft*, 90; trans. mine).

to subjective "expression"). Such practice, Smith shows, is likely to take the form of subtly evaluative discriminations whose referent is no longer some separate object "out there" but, rather, a set of circumstances socially shared and acknowledged as pertinent to the present situation. Insofar as it has been successfully transmuted into outward symbolic practice—viz., has been enacted as a specific "sentiment" *for* others—and only then, does passion acquire meaning and social value. Its true significance resides not in the moral value or appraisal that it ostensibly "expresses" about some particular issue or fact. Rather, sentiment by its very nature accredits and reinforces a shared understanding of social relations as they are presently taken to be constituted.

The language of Smith's sentiment ultimately functions as a meta-language in Roman Jakobson's sense of appraising a prevailing "code" rather than denoting a distinct "referent."[40] As Smith puts it: "to approve of the passions of another ... as suitable to their objects, is the same thing as to observe that we entirely sympathize with them" (*TMS,* 16); and again: "we approve of another man's judgment, not as something useful, but as right, as accurate, as agreeable to truth and reality: and it is evident we attribute those qualities to it for no other reason but because we find that it agrees with our own" (*TMS,* 20). Moral sentiments, then, are not stand-alone units of subjective experience. On the contrary, they constitute prima facie acts of "assent" or value judgments about a social situation, and as such they do not signify an inner (mental) action but a social transaction. By its very nature, such a type of "agreement may produce peace but it cannot produce truth."[41] Anticipating Kant's judgment of taste, which "imputes" (*ansinnen*) universal agreement to those for whom it is voiced, Smith here is able to recover from the extreme nominalism of Hobbes, Locke, and Hume whose strictly epistemological focus had led them to characterize passion as a hermetic, ineffable, and altogether transient phenomenon.

While there is an obvious and significant Stoic dimension to Smith's view of moral sentiments as quasi-judgments, his *Theory* no longer furnishes any frame of reference for such evaluation independent of (or anterior to) the social interaction whose success or failure such judgments ratify. In this regard, Smith's appeal to "the great machine of the universe" and the "secret wheels and springs which produce" its countless appearances (*TMS,* 19), or his later affirmation that "in every part of the universe we observe means adjusted with the nicest artifice to the ends which they are intended to produce" (*TMS,* 87), is less a nod to Stoic physics than an echo of the pervasive deism and the commonplace, "just-so" story endlessly recycled by

40. See Jakobson, "Linguistics and Poetics," esp. his distinction between conative, emotive, referential, and meta-lingual functions (*Language in Literature*, 69–73).

41. Strauss, *Natural Right and History*, 11.

eighteenth-century natural theology, a subject on which Smith repeatedly lectured between 1752 and 1764 as professor of moral philosophy at the University of Glasgow. Particularly revealing in this regard are the uninspired, token references that Smith inserted into the sixth edition (1790) of the *Theory of Moral Sentiments*, such as his tribute to "the great Director of the universe" and to "the idea of that divine Being, whose benevolence and wisdom have, from all eternity, contrived and conducted the immense machine of the universe, so as at all times to produce the greatest possible quantity of happiness." There is a strong deist tendency in this notion of a God who makes sure the cosmic train runs on time but who has delegated day-to-day operations of the material world to "his viceregent upon earth," including the task of "superintend[ing] the behaviour of his brethren" (*TMS*, 130). While affirming that "the administration of the great system of the universe ... is the business of God and not of man," Smith's quick shift of focus strongly suggests that his God remains de facto invisible and, lacking credible revelation, has become all but irrelevant to moral thought: "To man is allotted a much humbler department, but one more suitable to the weakness of his power and to the narrowness of his comprehension; the care of his own happiness, and that of his family, his friends, and his country" (*TMS*, 236–237).

The focus on ethics, which for the ancient Stoics was altogether inseparable from their physics, has here been compartmentalized as a separate province governed by finite, fallible, and self-determining human agents. There is no vertical connection between the judgments and appraisals ventured by the latter and the divine *logos* as disclosed in a cosmological system.[42] Indeed, precisely because judgment appears terminally estranged from any transcendent, metaphysical source, its present-day incarnation as "moral sentiment" can only be focused on, and is exclusively licensed by, endlessly shifting social circumstances. It is just this definitive separation of judgment from ontology that attests to the modernity of Smith's entire project. Additionally, it bears recalling that whatever "social reality" happens to be contingently at issue in a case of moral sentiment does not pre-exist the specific judgment in question—say as a Platonic idea or Aristotelian/Thomist "substantial form." On the contrary, such reality is only instantiated by a vast number of affective micro-judgments and acts of transference.

---

42. On the Stoic view of the cosmos as a living being (*zôion*) "ensouled and rational" (Diogenes Laertius), a unified, interconnected, and living One, see White, "Stoic Natural Philosophy" in Inwood, ed., *The Stoics*, 124–152; Sellars (*Stoicism*, 81–106) views Stoic physics as something inaccessible to the categories of either reductive naturalism or theism, a thoroughly material and empiricist "cosmo-biology" that yet eschews nominalism, and a rationalist cosmogony that steers clear of Platonic forms; on the one hand, they espouse "a theory of rigid causal determinism, but on the other hand, *qua* religious pantheists, they hold a doctrine of divine providence" (99–100).

In this regard, Smith's choice of laughter as an example is both revealing and shrewd inasmuch as it vividly displays the non-propositional and quintessentially social nature of moral cognition:

> He who laughs at the same joke, and laughs along with me, cannot well deny the propriety of my laughter ... If I laugh loud and heartily when he only smiles, or, on the contrary, only smile when he laughs loud and heartily; in all these cases, as soon as he comes from considering the object, to observe how I am affected by it, according as there is more or less disproportion between his sentiments and mine, I must incur a greater or less degree of his disapprobation: and upon all occasions his own sentiments are the standards and measures by which he judges of mine. (*TMS*, 16–17)

In ways that Keats would later explore to rich aesthetic effect, Smith identifies the misalignment of emotion as prima facie evidence for the way that all moral values are social in essence. That is, embarrassment of the kind here described shows Smith to understand moral interaction to be no longer about inner certitudes or metaphysical norms and aspirations. Rather, the objective at hand is the success (or failure) of social interaction itself.[43] Irving Goffman thus notes that contrary to a "breach" of some moral norm, which is likely to "give rise to resolute moral indignation," embarrassment amounts to a constitutively social emotion. For even as "the expectations relevant to embarrassment are moral ... we should look to those moral obligations which surround the individual in only one of his capacities, that of someone who carries on social encounters. The individual, of course, is obliged to remain composed, but this tells us that things are going well, not why. And things go well or badly because of what is perceived about the social identities of those present."[44]

It is no accident that Smith and Goffman should both have happened upon the seeming benefits of embarrassment. Long before there was to be a discipline called sociology and a subsidiary specialization known as behaviorism, Smith offers a sophisticated (if deeply problematic) "account of the formation of groups"; and it is this account's structural reliance on the specter of embarrassment that prompts us to wonder "what sort of unity" sympathy could possibly generate.[45] As it turns out, the psychological constellation known as embarrassment operates at all times, and not merely when it is manifestly unfolding—along with all the outward symptoms of

43. On the sociality of emotion in Keats, see Ricks, *Keats and Embarrassment*, 1–49; and Pfau, *Romantic Moods*, 309–339; on embarrassment as a sociological concept, see Goffman, "Embarrassment," which contains some uncanny echoes of Smith's *Theory of Moral Sentiments*.

44. Goffman, "Embarrassment," 268.

45. Mitchell, *Sympathy*, 82–83.

blushing, flustering, sweaty palms, and stammering speech so prominent in John Keats's canny fusion of depth-psychology and class-consciousness. In a remarkable passage, Smith thus contends that both the outward conduct and inner constitution of the self are at all times circumscribed by the specter of embarrassment. Even at its most introspective, the individual is intelligible to itself only as a socially constituted reality. However free from embarrassment at moments of introspection, it is precisely the constant possibility of some performative misadventure in public that furnishes the motivational prompt for how to cultivate one's persona:

> The man of real constancy and firmness, the wise and just man who has been thoroughly bred in the great school of self-command ... has never dared to forget for one moment the judgment which the impartial spectator would pass upon his sentiments and conduct. He has never dared to suffer the man within the breast to be absent one moment from his attention. With the eyes of this great inmate he has always been accustomed to regard whatever relates to himself. This habit has become perfectly familiar to him. He has been in the constant practice, and, indeed, under the constant necessity, of modeling, or of endeavouring to model, not only his outward conduct and behaviour, but, as much as he can, even his inward sentiments and feelings, according to those of this awful and respectable judge. He does not merely affect the sentiments of the impartial spectator. He really adopts them. He almost identifies himself with, he almost becomes himself that impartial spectator, and scarce feels but as that great arbiter of his conduct directs him to feel. (*TMS*, 146–147)

This is about as vivid and palpable an account of the concept of "introjection" (a term first coined by Sándor Ferenczi in 1909) as one could ask for. Unlike Ferenczi's and Freud's subject, however, Smith's impartial spectator is stripped of all magical or fantasy-like elements and, thus, appears much closer to Lacan's model according to which "introjection is always the introjection of the speech of the other."[46] Introjection thus attests to the incomplete and seemingly illegitimate nature of the individual understood in strictly inward, mentalist terms. Smith thus insists how the virtual superego of the impartial spectator is not merely to be accommodated in "outward conduct and behaviour," but that it is to be internalized as the real substance of moral agency. The "great inmate" or "man within the breast" is not merely some occasional

46. Lacan, *Seminar*, I, 83. See also Eagleton, who notes that "for Smith, as for Lacan, our actions are always at some level a message directed to the Other. It is just that in Lacan's view this dialogue can never be reduced to the imaginary reciprocities of a Smith, for whom each of us thrives under the benignant eye of a collective other" (*Trouble*, 75).

complement of the self, nor is he to be construed as some didactic allegory of righteousness or virtue.

At first glance, the above passage appears to present us with a genuinely (neo-) Stoic model of self-cultivation. Still, the emphasis on "self-satisfaction" stands out as discordant, if for no other reason than that it revives the dilemma that Aristotelian *megalopsychia* had bequeathed Christian virtue ethics, which obviously looks upon pride with acute misgiving. Moreover, Smith here tells essentially a story of emancipation *from* a debilitating inner life. Stoic habituation—whose *telos* was wisdom—yields to modern, neo-Stoic self-discipline, whose *telos* is approval; and it is only for the sake of such approval (and the successful socialization it betokens) that the self embraces the surrogate or virtual interiority of the "man within the breast." The person qua unique, spiritual self is expunged, much in the way that seventeenth-century Calvinist and Jesuit thought had appropriated the neo-Stoic language of self-discipline. With the virtual agency of the impartial spectator having supplanted the metaphysical idea of the person, Smith's *Theory of Moral Sentiments* no longer grants epistemological legitimacy to the idea of the person—understood as a unique, unclassifiable, and "incommunicable" being (as Boethius had defined it). Concurrently, the emotions enjoy epistemic standing only insofar as they can be converted into socially sanctioned sentiments. At the same time, the inner life has reality and legitimacy solely as a socially constituted and accredited performance, a process for which Smith does not (indeed, cannot) identify any *telos* or *terminus ad quem*. Echoing Smith's position, Goffman's closing discussion of the "social function of embarrassment" thus stresses the extent to which this particular affect is properly *generative of the self,* rather than being the inward property of an already constituted individual: "By showing embarrassment when he can be neither of two people, the individual leaves open the possibility that in the future he may effectively be either. His role in the current interaction may be sacrificed, and even the encounter itself, but he demonstrates that while he cannot present a sustainable and coherent self on this occasion, he is at least disturbed by the fact and may prove worthy at another time."[47] Both Smith's impartial spectator as a behavioral template to be introjected, and Goffman's embarrassment as the exemplary (because intrinsically "social") emotion betrays an eagerness to abandon mentalistic and psychologizing conceptions of moral meaning and agency. Crucial to understanding Smith's objectives is that "sentiments" are not, properly speaking, *inner* states but arise from our unwitting negotiation of a complex grammar of social values and behavioral patterns or averages.

One of the abiding consequences of this de-psychologizing, quasi-behaviorist reorientation of moral theory is the contraction of any temporal perspective of

---

47. Goffman, "Embarrassment," 270–271.

agency to the currently prevailing social dynamics. Inasmuch as the teleological nature of "action" has been supplanted by concern with the success of present "interaction," a sentiment-based "appraisal of social facts no longer distinguishes between motive, implementation, and consequences of discrete actions." Instead, the social field appears "determined by complex interdependencies wherein events and affective states reciprocally induce and generate one another."[48] The atrophying of volition, intentionality, and action is palpable when, in his discussion of the impartial spectator, Smith emphasizes the non-participatory nature of this model: "we must imagine ourselves not the actors, but the spectators of our own character and conduct" (*TMS,* 111). To be sure, just a few pages earlier Smith had stressed that sentiment without action is but "indolent benevolence," and that "the man who has performed no single action of importance, but whose whole conversation and deportment express the justest, the noblest, and most generous sentiments, can be entitled to demand no very high reward" (*TMS,* 106). Yet inasmuch as Smith regards the value of action to depend on the motives with which it is being undertaken, it is not the action but the social currency of the motive qua sentiment that furnishes the basis for the agent's moral appraisal by others. Indeed, Smith's concerted attempt to move beyond internalist accounts of moral meaning at times causes him to speak rather diffidently about the intrinsic value of ideas, however large or small. In a revealing aside, Smith thus notes how difference of opinion is easily tolerated where no present business is pending: "I can much more easily overlook the want of this correspondence of sentiments with regard to such indifferent objects as concern neither me nor my companion." Yet Smith here is not thinking of weather patterns in eastern Mongolia or the imbalance of trade between ancient Athens and Sicily but, in fact, of just about *any* topic whatsoever, including "that picture, or that poem, or even that system of philosophy, which I admire. They ought all of them to be matters of great indifference to us both." Indeed, it appears that the intrinsic value and pragmatic relevance of moral sentiments requires the de facto expurgation of all ideational content from them. Moral sentiments merely simulate propositional utterances, while actually furnishing a merely formal assurance of sympathy or "fellow-feeling for the misfortunes I have met with" (*TMS,* 21).

Early on in Smith's *Theory,* this virtual logic of "sentiment" is stated quite explicitly. To be sure, everyone "passionately desires a more complete sympathy" than what can reasonably be expected of others. Yet precisely because this narcissistic desire

48. Vogl, *Kalkül und Leidenschaft,* 89; Soni also notes the attenuation of action in Smith's *Theory,* even as "the unhappiness of the other is an impulsion to moral action." For Smith limits sympathy, and the action it enjoins, strictly to what can be seen: "what cannot be seen does not enter the field of responsibilities," with the result that "Smith, *despite himself,* ends up minimizing the value of action" (*Mourning Happiness,* 315).

("to see the emotions of their hearts, in every respect, beat time to his own") cannot be satisfied in unadulterated form, every individual feels induced to lower his passion to that pitch, in which the spectators are capable of going along with him: "He must *flatten,* if I may be allowed to say so, the sharpness of his moral tone, in order to reduce it to harmony and concord with the emotions of those who are about him." As Smith well realizes, this transposition does not simply involve a reduction in the force of sentiments but amounts to a qualitative transformation. Consequently, the socially successful and, in that rather unique sense, "moral" agent thus is imbued with "the secret consciousness that the change of situations, from which the sympathetic sentiment arises, is but *imaginary*" and thus "not only lowers it in degree, but, in some measure, *varies it in kind,* and gives it a quite different modification."[49] The qualitative shift here is from emotion as a unique, inner, and unverifiable occurrence of questionable rational standing to sentiment as a form of "behavior." By definition, "behavior" is an abstract concept. It signifies a "class" or "type" of outwardly manifest, observable, and recurrent practices and, by implication, it appraises human action only within a matrix of *average practices* or what, in identifying his overarching concern with the "*propriety* of every passion," Smith calls "*emulation* ... of the excellence of others" (*TMS,* 114) and "a certain *mediocrity*" (*TMS,* 27; italics mine). Similar formulations recur throughout the *Theory of Moral Sentiments,* such as in Smith's acknowledgment that, happily, "the reflected passion ... is much weaker than the original one" (*TMS,* 22). Moral agency thus is said to pivot on a kind of actuarial balancing of emotional states, an aspiration still couched in neo-Stoic language as the virtue of "temperance" (*TMS,* 28) or, in statistical terms, as aiming at "*the most ordinary degree* of kindness or beneficence" (*TMS,* 80; italics mine). Just as Locke, Mandeville, and Hume had dissolved the metaphysical concept of the will into a radically nominalist model of ephemeral and heterogeneous passions, so the concept of passion in turn is now systematically being converted from a phenomenon open to introspective appraisal into a reflex-like, behavioral pattern. To sharpen the point, a

---

49. *TMS,* 22; italics mine. Smith's argument clearly furnished the conceptual template for Wordsworth's "Preface" to *Lyrical Ballads* (1800), in particular, the transmutation of "the spontaneous overflow of feelings" into second-order passions: "while he describes and imitates passions, his situation is altogether slavish and mechanical, compared with the freedom and power of real and substantial action and suffering. So that it will be the wish of the Poet to bring his feelings near to those of the persons whose feelings he describes, nay, for short spaces of time perhaps, to let himself slip into an entire delusion, and even confound and identify his own feelings with theirs; modifying only the language which is thus suggested to him by a consideration that he describes for a particular purpose, that of giving pleasure. Here, then, he will apply the principle of selection which has been already insisted upon. He will depend upon this for removing what would otherwise be painful or disgusting in the passion; he will feel that there is no necessity to trick out or to elevate nature" (Wordsworth, *Lyrical Ballads,* 751). For a fuller discussion, see Pfau, *Wordsworth's Profession,* 246–260.

brief and necessarily selective review of behaviorism's central claims and objectives is needed.

Modern behaviorism's principal objective is to render mental phenomena objectively verifiable and actuarially measurable, a project that, in the view of behaviorism's founder, John B. Watson, is likely to succeed only once the whole mentalist vocabulary of consciousness, introspection, emotion, and so forth has been definitively abandoned. It is this proposed conversion of enigmatic *qualia* into quantifiable, social facts that explains the sudden appeal of "behavior" as a new paradigm for the manifestly ailing discipline of experimental psychology just before World War I. As Watson puts it in his 1913 manifesto ("Psychology as the Behaviorist Views It"), "the end is the production of mental states that may be 'inspected' and 'observed.'" In asserting "the independent value of behavior material" and foreswearing traditional introspective concepts ("we will no longer have to work under false pretenses"), Watson intends nothing less than to abandon all mentalist and introspective terms, "to discard all reference to consciousness" and, in quasi-revolutionary fashion, "to throw off the yoke of consciousness."[50] The aspiration here is not merely to extricate social analysis from a half century of inconclusive experimental psychology, which according to Watson "has failed signally" (163) but to expunge a 2,500-year-old tradition (going back to Plato and Augustine) that had traced all moral and social meaning back to a variety of "inner" processes.[51] Vaguely following in the footsteps of Frege, modern behaviorism contests the reality or, at least, the relevancy of mental phenomena and introspective methods. Watson, whose manifesto will have to suffice as synecdoche for the larger movement, thus envisions a psychology that will "never use the terms consciousness, mental states, mind, content, introspectively verifiable imagery, and the like" (166). The basic objection to the entire conceptual inventory associated with inner experience is that the states to which these concepts refer are knowable only insofar as the observer is able to establish a strong correlation between behavioral patterns. At most, then, mind constitutes an "inference" that *may* be drawn on the strength of such patterns, though for Watson, and even more so for his successors, it becomes increasingly unclear why anyone might wish to draw such an inference at all.

In time, a more strident variant known as "analytic" behaviorism abandons any residual mentalist vocabulary by insisting that there is properly no need to speculate on inner states or dispositions. On this view, behavior is no longer evidence *for* a mental process relative to which it might be said to function epiphenomenally (i.e., behavior as the *appearance of* inner states). Instead, analytic behaviorism conceives

---

50. Watson, "Psychology," 163, 160; subsequent references are given parenthetically.

51. For a recent, comprehensive account of that tradition, see C. Taylor's *Sources*; for another, albeit more guarded version of that story, see Schneewind's *Invention*.

behavior as something that "can be described and explained without making ultimate reference to mental events or to internal psychological processes."[52] Here, then, the stabilization of (formerly) psychological, mental states in the form of social conditioning and manifestly associative patterns of social cognition is said to render human behavior intelligible. Yet behaviorism's wholesale abandonment of psychological and spiritual concepts in favor of externally observable and measurable criteria of human behavior is pressed further. True to its origins in a "logical positivism" that had restricted the meaning of scientific statements to experimental conditions such as could be inter-subjectively observed (a.k.a. "verificationism"), modern behaviorism draws on the same spectatorial logic first articulated in Smith's *Theory of Moral Sentiments*. It cannot surprise, then, that behaviorism's disciplinary criterion of strict verifiability reflects an underlying investment in the predictability of its objects of inquiry. Human behavior, if it is to be known at all, must be re-described in such ways as will render its subjects predictably responsive to external mechanisms of control. Watson thus opens his 1913 manifesto by classifying this new field as "a purely objective experimental branch of natural science" whose "theoretical goal is the prediction and control of behavior." What drives behaviorism's proposed dissolution of the enigmatic *qualia* of human consciousness into causally determinate and quantifiable behavioral patterns is a quest for *applicable, instrumental knowledge*. The investigation into predictable and verifiable human responses to stimuli is undertaken solely for the sake of "conditioning." Watson, it seems, might have approvingly chuckled at the old joke, which starts with the question of why of late lawyers have replaced rats as the behaviorist's preferred test subject (Answer: "1. There are more of them; 2. Experimenters run no risk of forming attachments to their subjects; and 3. There are certain things that even rats won't do"). Then again, the joke's underlying premise might have eluded his naturalist viewpoint altogether. For the success of behaviorism as a discipline clearly hinges on pushing the re-description of formerly mental phenomena to the point where the line between animals conditioned by stimuli and humans actuated by motives has been permanently erased. Watson explicitly commits himself to supplanting the hermeneutic concept of "experience" with a strictly material and physiological account of the way that stimuli can be shown to "cause" particular types of behavior.

It is precisely their apparent "mindlessness" that recommends stimuli as the centerpiece of behaviorism's (in origin Hobbesian) attempt to posit a "unitary scheme of animal response, [which] recognizes no dividing line between man and brute."[53] Still, the final challenge that behaviorism has yet to meet involves accounting for

52. "Behaviorism," in *Stanford Encyclopedia of Philosophy* (at www.plato.stanford.edu/entries/behaviorism/), accessed 14 December 2010.

53. Watson, "Psychology," 158.

"more complex forms of behavior, such as imagination, judgment, reasoning, and conception." It is only in a long and truly startling footnote late in the essay that the other shoe drops, with Watson now venturing as "tenable" the

> hypothesis that all of the so-called "higher thought" processes go on in terms of faint reinstatements of the original muscular act (including speech here) and that these are integrated into systems which respond in serial order (associative mechanisms) ... Paucity of "imagery" would be the rule. In other words, wherever there are thought processes there are faint contractions of the systems of musculature involved in the overt exercise of the customary act, and especially in the still finer systems of musculature involved in speech. If this is true, and I do not see how it can be gainsaid, imagery becomes a mental luxury (even if it really exists) without any functional significance whatsoever ... I should throw out imagery altogether and attempt to show that practically all natural thought goes on in terms of sensori-motor processes in the larynx (but not in terms of "imageless thought").[54]

To be sure, as with most other manifestos allowances have to be made for the exuberant, at times utopian fervor with which Watson sketches his arguments here. Moreover, his analeptic conception according to which thought-like effects are produced by "motor habits in the larynx" would appear to receive ample corroboration by what transpires on most Sunday morning political talk shows and most elsewhere in the feeding frenzy of hourly news-cycles today. Nonetheless, Watson's suggestion that "faint contractions" in the musculature of the larynx have, for the past 2,500 years, been misidentified as phenomena of mind, consciousness, imagery, and so forth is likely to disconcert even the most hardened of social constructivists. While there is good reason to view Smith as a precursor of modern behaviorism, it would be gratuitous, not to mention anachronistic, to hold his *Theory of Moral Sentiments* accountable for the conceptual problems eventually experienced by modern behaviorism, a field whose intellectual currency, in any event, has lost nearly all its value during the past few decades.[55]

---

54. Ibid., 174n.

55. Arguably, it is the rise of cognitive-science models—which by now have metastasized across the entire disciplinary spectrum of the humanities, several social sciences, as well as law, economics, and religion—that appears to be the most obvious heir to the behaviorist enterprise. A case in point would be the improbable appeal of the cognitivist model for mental causation, such as Libet's "readiness-potential" or Wegener's related reductionist view of consciousness and will as an "illusion" or an epiphenomenal "feeling of doing"; see Mele, *Effective Intentions*, esp. 31–90; Tallis, *I Am*, 287–324.

The present objective, however, is not another, by now redundant critique of behaviorism. Rather, it is to show that the critique of an interiorist or introspective (Christian-Platonist) concept of agency in the eighteenth century did not always assume the overtly anti-humanist character that Hobbes, Locke, Mandeville, and Hume had given to it. For rather than extending the epistemological critique of that model—which, in the wake of Hume's *Treatise,* seemed hardly possible—the *Theory of Moral Sentiments* opts to reconstitute moral meaning as positively arising from patterns of social transaction such as can be ascertained, quantified, and classified by means of sustained empirical observation. Mind in Smith's account is above all a habit, a virtual relay station for meanings said to originate in social and cultural exchange (speech, manners, gestures, and a vast inventory of codified expressions). Hence, in pressing the question of how Watson could possibly assert that "practically all natural thought goes on in terms of sensori-motor processes in the larynx," one is ultimately confronted with a distinctly modern (associationist) revival of the Aristotelian and Scholastic concept of habit. To connect habit and habituation with the behaviorist vocabulary of "stimulus-response" and "conditioning"—while not intended to rehabilitate Watson's extreme theory—is to recognize a strong link between modern behaviorism and the concept of socially induced sentiment that rises to sudden prominence in Smith's *Theory*. According to Watson, habit comprises the myriad associative motor processes that justify behaviorism's basic hypotheses and research program: for "improvement in habit comes unconsciously. The first we know of it is when it is achieved—when it becomes an object." Without any hesitation, Watson extends this (substantially flawed) observation into an argument to the effect that "the so-called 'higher thought' processes" are no different: "I believe that 'consciousness' has just as little to do with *improvement* in thought processes ... [I]mprovements, short cuts, changes, etc., in these habits are brought about in the same way that such changes are produced in other motor habits. This view carries with it the implication that there are no reflective processes (centrally initiated processes)."[56]

For present purposes the question becomes how "habit" and "behavior" are linked in modern social theory and how that configuration might differ from a premodern account of habit. A first step is to attend to the specific word(s) associated with the two concepts of habit and behavior, viz., the Latin *habitus* and the early modern English *hauyoure,* respectively. In his detailed account of *habitus,* Aquinas opens by observing that "this word *habitus* is derived from *habere* [to have]," and true to form he launches into a complex distinction of meanings, most of which need not concern us.[57] What does matter, however, is his decision to follow Aristotle's cue

56. Watson, "Psychology," 174n.

57. Though substantially correct, the etymology is incomplete. The Latin *habitus* is derived from the Greek *hexis* (itself a derivative of the verb-form *ekhein* = to have). It bears noting that

and define habit as a so-called post-predicament, that is, as a concept made possible by some of the ten categories (*praedicamenta*), such as quantity, quality, relation, priority, posteriority, etc. Which is to say (still nothing original here) that habit is not a substantial form, is not nature or essence, but something "had" only insofar as it has been made. Yet implicitly, this otherwise simple claim also tells us that, being something "constructed," habit derives its intrinsic rationality from the ultimate end whose attainment by the human subject it is meant to facilitate. Having established that habit is something constructed over time rather than ontologically given (in which case nothing could be done about good or bad habits), it follows for Aquinas that the concept can only be understood if integrated into a narrative of human flourishing. To be sure, there is nothing wrong with calling habit second nature provided we understand this to mean that it is, precisely, *not* nature. For present purposes, the most relevant meaning of habit is that of a relation between the "haver and that which is had," which is to say "a quality." Having quoted from Simplicius's commentary on Aristotle's *Metaphysics* (*ST,* Ia IIae Q 49 A 2), Aquinas emphasizes that habits, inasmuch as they bear on theology, arise from strictly "adventitious" dispositions. Like any other quality that has to be cultivated by methodical and sustained practice, we only "have" these qualities in the sense of a possessive relation, which is to say that "they can be lost."[58]

Now, in Aristotelian and Thomistic thought anything "made" or created is rational and intelligible only insofar as it can be shown to have been fashioned in accordance with an overarching teleological framework. Minimally, this means that habit must not be confused with nature, instinct, or anything on the order of a cause that is at once materially heterogeneous and efficient or determinative. Recalling Augustine (no. 36 from the *Book of 83 Questions*), Aquinas thus subtly demurs when it comes to the bishop of Hippo's claim that an animal tamed by means of rewards or incentives has effectively "commuted its nature into something habitual [*quod in earum consuetudinem verterit*]." But the habit is incomplete, as to the use of the will, "for they have not that power of using or of refraining, which seems to belong to the notion of habit: and therefore, properly speaking, there can be no habits in them" [*quia non habent dominium utendi vel non utendi, quod videtur ad rationem habitus pertinere*]." Habit thus is a form of mental causation and in a very specific and limited

another Latin term for "habit," *consuetudo*, operates prominently in Augustine—esp. in *De Vera Religione,* the *Confessions*, and *Opus Imperfectum contra Julianum*. Yet unlike the neutral *habitus*, Augustine's *consuetudo* tends to denote a "bad habit." For a detailed account, see Prendiville ("Idea of Habit"), who also points out significant parallels between Plotinus's theory of purification (*Enneads*, 3.6.5) and Augustine's project of overcoming the soul's habitual fixation on embodied, sensory states (*consuetudo corporum*).

58. *Quaedam autem sunt adventitiae, que ab extrinseco efficiuntur, et possunt amitti.* The point is more fully developed in *ST*, Ia IIae Q 53 ("On the Corruption of Habits").

sense can even be viewed as a precursor of the modern idea of autonomy. Put differently, habit cannot be explained as a result of outward conditioning; it is not, in Aquinas's phrase, "a disposition of the object to the power" but, rather, an act of self-fashioning or "disposition of the power to the object" (*ST,* Ia IIae Q 50 A 4). For Aquinas, certainly, habit and the extended process of habituation that gives rise to it have for their logical premise the essential indeterminacy of the human person—what Nietzsche would much later call that "still undetermined animal" (*das noch nicht festgestellte Tier*).[59] Yet for Aquinas, such indeterminacy constitutes no defect but, on the contrary, is the condition for the human person as a rational, free, and self-creating being. As such, it pertains to intellect and will alike, for both are powers capable of varied development, including the possibility of not properly developing at all or even regressing: "Every power which may be variously directed to act, needs a habit whereby it is well disposed to its act." Crucially, that is, the relation of powers to habits is a result of the fact that powers are not simply "faculties" in the modern, Kantian sense but potentialities in whose nature it is to develop, both in the sense of an ongoing intrinsic cultivation and perfection and with a view to an ultimate end: "Since it is necessary, for the end of human life, that the appetitive power be inclined to something fixed, to which it is not inclined by the nature of the power, which has a relation to many and various things, therefore it is necessary that, in the will and in the other appetitive powers, there be certain qualities to incline them, and these are called habits" (*ST,* Ia II ae Q 50 A 5).

Aquinas now points out that habit, simply because it is formed and sustained by repeated acts, may be misconstrued as something passively and extraneously received: "if the acts be multiplied a certain quality is formed in the power which is passive and moved, which quality is called a habit" (*ST,* Ia IIae Q 51 A 2). Yet this is a misleading view to take, for all repetition *implies* and *produces* difference, rather than amounting to an invariant and mindless mechanism corrosive of rational agency (as most Enlightenment thinkers tend to suppose). For an example, one may think of a violinist practicing over and over her scales or some intractable passage in a score. She will initially do so by monitoring all her movements with great alacrity, deactivating those uncalled for (redundant movements of the bowing arm, for example) and fine-tuning those that are crucial for the end (of playing the piece well). Cumulative practice will gradually diminish the need for such intentional awareness of movement as each instance of repetition imperceptibly stabilizes and perfects interval spacing, position shifts, vibrato control, bow markings, and countless other

59. Nietzsche, *Beyond Good and Evil,* no. 62; the phrase constitutes something of a paradox since by Nietzsche's own account (in this and various other books) it is precisely the attribute of animals to *be* determined, viz., by the guidance of instincts that have long evolved in response to determinate environmental factors.

details susceptible of being assimilated to "muscle memory." The result is a freeing up of mental awareness for higher-level interpretive considerations (dynamics, phrasing, ensemble work, etc.). Habit becomes a substitute for an instinct of which human beings in particular seem to be largely deprived and, as such, rebuilds a strong, seemingly spontaneous connection between will and nature.

Consistent with such a scenario, Aquinas thus remarks that, unlike the stable properties of material substances, habituated actions and passions are susceptible of increase. Put differently, the law of habit is nothing like the law of inertia, for "other qualities which are further removed from substance, and are connected with passions and actions, are susceptible of more or less, in respect of their participation by the subject." Against the second objection (referencing Aristotle's *Physics,* 246ᵃ13) that habit is a perfection term and, thus, does not admit of any "more or less," Aquinas notes that in the sphere of human action perfection is not constrained by an already existing, determinative nature but, on the contrary, continually helps to define that nature more fully: "habit is indeed a perfection, but not a perfection which is terminal in respect to its subject; for instance a term giving the subject its specific being."[60] The following article (Q 52 A 2) reinforces the basic conception of habit and perfection as jointly delineating the teleological character of human development. Within the context of human action and habituation, repetition and perfection cannot be understood as accumulative but, rather, as transformative: "such increase of habits and other forms is not caused by an addition of form to form; but by the subject participating more or less perfectly, in one and the same form [*sed fit per hoc quod subiectum magis vel minus perfect participat unam et eandem formam*]." What stands out about the classical conception of habit presented here, and bearing close affinity to Aristotle's account in the *Nicomachean Ethics,* is the active, self-determining, and teleological character of habit. In its very essence, then, habit is transformational and dynamic, for its very cultivation presupposes an overarching end and, in turn, shapes and develops a conscious intelligence dedicated to its attainment.

If modernity has not looked kindly on habit, this is the case not because of its avowed overcoming of the alleged intellectual shortcomings of habit and habituation but, on the contrary, because if has proven itself unable to grasp their active, evolving, and purposive structure. Descartes and Kant in particular variously identify habit with inauthenticity, mindless routine, a numbing of sense perception, and, especially, purely mechanical and empty repetition.[61] Though just as critical of habit's

---

60. *Quod habitus quidem perfectio est, non tamen talis perfectio que sit terminus sui subiecti* (*ST*, Ia IIae Q 52 A 1).

61. Descartes speaks of "the habit of holding on to old opinions," the "habit of making ill-considered judgments," and he envisions his own philosophical regimen as a way of counteracting the influence of "my habitual opinions ... until the weight of preconceived opinion is counter-balanced and the distorting influence of habit no longer prevents my judgment from perceiving

alleged epistemological limitations, Hume only differs from Descartes and Kant in that he sees habit, or "custom," as providing the self with a kind of shelter from the epistemological paralysis where his own skepticism had left it. To understand Smith's overall affirmative (if deeply problematic) view of habit as "behavior," it helps to complement Aquinas's account of habit with Félix Ravaisson's early phenomenological discussion of habit presented in his remarkable 1838 thesis *De l'habitude* (*Of Habit*), a book that was to exercise considerable influence on Bergson, Heidegger, Sartre, Emmanuel Levinas, and Paul Ricoeur. Like Aquinas, Ravaisson sharply distinguishes between the law of inertia and self-sameness that defines physical substance and the law of habit that cannot be applied to the "empire of immediacy and homogeneity that is the Inorganic realm." Inasmuch as habit is to be understood as the internalization of change as (second) nature, "nothing ... is capable of habit that is not capable of change." Habit is a process, a law of growth and development, and hence a manifestation of reason. Opposing the modern critique of habit as mindless, mechanistic, and invariantly repetitive, Ravaisson emphasizes that habituation is not even conceivable without "a centre at which reactions arrive and from which actions depart," which is to say, consciousness: "in consciousness ... the same being at once acts and sees the act; or, better, the act and the apprehension of the act are fused together."[62]

For Ravaisson, then, the phenomenon of habit is only intelligible within an overarching "economic" model of consciousness that measures how and when different quantities of energy are being expended. Presaging similar arguments in Freud's meta-psychological writings of 1911–1915, Ravaisson thus sees consciousness not only as intentionality but also as a concomitant awareness of the energy required to work through the object or phenomenon with which it happens to be engaged: "its force is its own measure, [and] it is also measured in that it measures itself out sparingly, making its present energy proportional to the resistance it has to overcome."

---

things correctly." Notably, though, his neo-Stoic project of intellectual self-discipline is itself envisioned as a type of habituation: "by attentive and repeated meditation I am nevertheless able to ... get into the habit of avoiding error" (*Meditations*, 23, 56, 15, 43). Kant's outlook on habit is more nuanced, in part because he seeks to move beyond Hume's construal of "habit" and "custom" as pervasive, and epistemologically illegitimate practices justifying his own skepticism; see esp. Kant's discussion of Hume, *Critique of Pure Reason* (B 788–797). Still, Kant's own "critical" rather than "skeptical" perspective on reason begins with a critique of habit. Having restated the central tenets of his transcendental method in the *Prolegomena for a Future Metaphysics*, Kant admonishes "the reader who retains his long habit [*lange Gewohnheit*] of taking experience for a merely empirical composition of perceptions ... to heed well this difference between experience and the mere aggregate of perceptions" (*Prolegomena*, §26, 115). Arguably, Kant's most revealing discussion of habit occurs in his late *Religion within the Boundaries of Mere Reason*, in a "General Remark" appended to Part I (*Religion*, 89–97).

62. Ravaisson, *Of Habit*, 29, 25, 37, 39.

Crucial for understanding habit is to notice how consciousness at all times monitors its own expenditure of energy and, like any economic system, seeks to minimize that expenditure in proportion to the task at hand: "Effort is … not only the primary condition, but also the archetype and essence, of consciousness." Now, if habituation helps to draw down that energy expenditure, it does so without compromising the individual's ability to fulfill a given task; indeed, it facilitates execution of that task while also freeing up consciousness for other purposes. Repetition diminishes both the effort of consciousness and, thus, the consciousness of effort: "As effort fades away in movement and as action becomes freer and swifter, the action itself becomes more of a tendency, an inclination that no longer awaits the commandments of the will but rather anticipates them, and which escapes entirely and irremediably both will and consciousness … In this way, then, continuity or repetition brings about a sort of obscure activity that increasingly anticipates both the impression of external objects in sensibility, and the will in activity." As volition yields to inclination, we can see how fully psychological and physiological processes are enmeshed. No other phenomenon reveals action more clearly to be a fusion of intentionality and execution than habit. Here execution is not some "after-thought" or seemingly mindless implementation of an antecedent conscious process. Rather, habit reveals how motor processes are themselves thoroughly imbued with intelligence. As Ravaisson puts it, "it is not in the will but in the passive element of the movement itself that a secret activity gradually develops. To be precise, it is not action that gives birth to or strengthens the continuity or repetition of locomotion; it is a more obscure and unreflective tendency, which goes further down into the organism."[63]

Precisely this last insight, however, appears to have been lost in the modern era, certainly by the time Descartes disparages habit as the very antithesis of philosophical method. That is, even as habit evacuates the principle of action from conscious volition, it does not constitute a case of mindless and aimless repetition. Conversely, and no less important, Ravaisson (as Aquinas long before him) rejects the modern premise of the body as mere unintelligent, inert "stuff" against whose imposition (or imposture) the Cartesian *cogito* means to assert and defend its autonomy. The modern conception of mind/body is obviously divisive, whereas the stance adopted by the Aristotelian-Thomist tradition and by exponents of modern vitalism and phenomenology (Goethe, Coleridge, John Henry Newman, Félix Ravaisson, Franz Brentano, Maurice Merleau-Ponty, Jean-Luc Marion) is fundamentally integrative. For Ravaisson, to understand habit as "inclination" is to acknowledge both its intelligential nature and its teleological orientation. Hence the body's "sensibility increasingly *demands* this sensation that the will abandons." Even as the habituated movement

63. Ibid., 43, 51–53.

"leaves the sphere of will and reflection, it does not leave that of intelligence. It does not become the mechanical effect of an external impulse, but rather the effect of an inclination that follows from the will." In fact, habituation closes the gap between will and *telos* in what Ravaisson shows to be a metonymic, narrative progression. It intensifies the degree to which consciousness participates in its object or, rather, in the very idea of which the object is the phenomenon: "In the progress of habit, inclination, as it takes over from the will, comes closer and closer to the actuality that it aims to realize; it increasingly adopts its form. The duration of movement gradually transforms the potentiality, the virtuality, into a tendency, and gradually the tendency is transformed into action. The interval that the understanding represents between the movement and its goal gradually diminishes; the distinction is effaced; the end whose idea gave rise to the inclination comes closer to it, touches it, becomes fused with it." Ravaisson's remarkable forensic account of habit culminates in the following extended passage, which is worth quoting in full:

> In reflection and will, the end of movement is an idea, an ideal to be accomplished: something that should be, that can be and which is not yet. It is a possibility to be realized. But as the end becomes fused with the movement, and the movement with the tendency, possibility, the ideal, is realized in it. The *idea* becomes *being*, the very being of the movement and of the tendency that it determines. Habit becomes more and more a *substantial idea*. The obscure intelligence that through habit comes to replace reflection, this immediate intelligence where subject and object are confounded, is a *real* intuition, in which the real and the ideal, being and thought are fused together.[64]

Ravaisson's main point here is to showcase that habit is a particular manifestation of intelligence, and that it consummates the teleological nature of human thought. For the classical tradition—which here would comprise both Platonic and Aristotelian, as well as Augustinian and Thomist, models—it is the essence of all thought to aspire to the most complete participation in an idea and not, as the Enlightenment typically insists, to an open-ended, critical, and prevaricating distance vis-à-vis the object of thought.

It is just this axiomatic disparagement of anything on the order of immediacy, intuition, and unreserved commitment to an idea that prevents modernity from grasping any stance (such as habit) that aims at a "fusion" of being with idea. Habit works by minimizing distance and suspending detachment, concepts that define the intellectual stance of modernity. Another way of drawing the picture might be to un-

64. Ibid., 53–55.

derstand habit as an inconspicuous, almost quotidian form of mystical experience. A writer who feels words, phrases, sentences effortlessly arising and capturing the object of his attention; musicians executing a scrupulously rehearsed piece of chamber music with seeming ease; a practiced long-distance runner conscious of her near-effortless control of pace, posture, breathing, and so forth: they all experience the benefits of habituation and the way that volition and execution have merged in a single, intelligent, and purposive psycho-physiological continuum. Operating outside the modern subject's antagonistic and distrustful outlook on embodied phenomena, habit points a way beyond the subject-object divide. Yet it does so not by investing either the self or some object "out there" with improbable or fantastic powers; rather, it effects a *unio mystica* by moving beyond the gratuitous and constricting premise of the self as a being legitimated solely by its principled distrust of inner and outer phenomena alike. Indeed, there is no other road toward that mystical union than the long, slow, and admittedly uncertain one through habituation. What renders habit a significant philosophical concept is that it opens up a domain of meaning that *cannot be described in terms of autonomy or heteronomy*, of the merely "subjective" or "objective" at all.

Notably, it is just this intelligent, teleological, and metaphysical dimension of habit that vanishes in modern social theory, a shift that brings us to the second term already prominent in Smith's *Theory* and, obviously, in Watson's eponymous disciplinary innovation: "behavior." As it turns out, that concept shares the same etymological root with "habit," being obliquely associated with *habere* and, more obviously, with its Old French derivative, *aveir, avoir*.[65] Appearing a total of eighty-three times in the sixth edition of Smith's *Theory of Moral Sentiments*, "behaviour" already surfaces in the very first chapter ("the furious behaviour of an angry man"); and throughout the book the term denotes not an inner disposition but an outwardly manifested pattern of conduct sufficiently stable to be assigned a proper name (e.g., "we call his behaviour mean-spiritedness" [*TMS*, 35]). Though regularly used as a synonym for "habit," "behavior" in Smith's *Theory* does not denote a dynamic, teleological progression within the self but, instead, refers to the seemingly ubiquitous phenomenon of social adaptation. As such, the term supports Smith's overarching objective of re-describing and thus capturing the individual as an aggregate of statistical probabilities. The logic

65. The roots of the English "behaviour" partially converge with the etymology of "habit" (Lat. *habitus*, from *habere* = to have). Thus the *OED* constructs the provenance of behavior as follows: "by form-analogy with HAVOUR, *hauyoure*, common 15th-16th c. form of the word which was originally *Aver* s.b. (q.v.), *aveyr*, also in 15th c. *avoir*; really in OF. *aveir, avoir*, in sense of 'having, possession,' but naturally affiliated in English to the verb *have*, and spelt *haver, havour, haviour*, etc. Hence, by analogy, *have*: havour, -iour: *behave*: behavour, -iour." Smith is not mentioned, though the subsequent examples of historical usage tend to be clustered around eighteenth-century writers.

of "behavior" in Smith's account is plainly inductive in that it serves to distill concepts from myriad cases whose likeness and meaning, however, are simply assumed rather than interpretively demonstrated. As a result, one might say that for each instance of "behavior" isolated by the colloquial "we" of Smith's meandering account there has to exist within the observer a perceptual *habitus* no less rigid and reflex-like than how the book portrays the cultural conditioning of its empirical subjects. Such is the methodological price exacted by a theory that has reduced moral cognition to mimetic practice, rational action to compliant behavior. A strange hybrid of budding sociologist and virtual superego, Smith's authorial voice (ventriloquized as the "impartial spectator") seeks to extract faintly normative meanings from an abundance of supposedly equivalent and self-interpreting social phenomena.

A few cases will help illustrate the prominent role assigned to moral commonplaces throughout Smith's argument, as well as that peculiarly modern, actuarial deformation of habit as impersonal and invariantly repetitive "behavior." Of the social upstart—no doubt a frequent character in the dynamic entrepreneurial world of mid-Georgian England—Smith notes that, whatever his merits, he is "generally disagreeable." Hence, "if he has any judgment, … he redoubles his attention to his old friends and endeavours more than ever to be humble, assiduous, and complaisant. And this is the behaviour which in his situation we most approve of; because we expect, it seems, that he should have more sympathy with our envy and aversion to his happiness, than we have with his happiness" (*TMS,* 41). By contrast, the coxcomb pretending to nobility and "affect[ing] to be eminent by the superior propriety of his ordinary behaviour" (*TMS,* 54–55) meets with contempt. His failure is to emulate a social and symbolic grammar outside his native domain and, in so doing, to render the theatrical project of social emulation excessively apparent. What Lord Byron would later so cannily identify as Keats's "shabby genteel" also vexes Smith, even as the latter is far more sympathetic than Byron to middle-class "virtues" such as industry, frugality, and mainstream piety. For Smith, there still is such a thing as genuine virtue, though he insists that its authenticity is not achieved by the categorical disavowal of theatrical display but, rather, by its measured and class-specific cultivation. What Smith understands by virtue involves the subtle emulation of aesthetic, rhetorical, and social practices that cumulatively define a class-specific behavioral template. In the *Theory of Moral Sentiments,* the transposition of habit into behavior thus unfolds in strictly mechanical ways. Insofar as the theater of public behavior seeks to merge self and other in the virtual superego of the impartial spectator, habituation involves *repetition without difference.* Its sole objective is the individual's compliance with or internalization of a process of social conditioning that no longer has any intelligible *telos* outside its own operation. Supplanting a premodern, teleological progression that had fused will and object in habituated action, "behavior" in Smith's *Theory* aims at an isomorphism of self and other that strips either individual of the ability to claim credit and take responsibility for the meanings thus generated. Smith's "behaviour" discovers nothing, leads nowhere, but merely reaffirms what is hypostatized or projected as a meaning supposed to be already held by the other: "The man who is conscious to himself that he has exactly observed those measures of conduct which experience informs him are generally agreeable, reflects with satisfaction on the propriety of his own behaviour" (*TMS,* 116).

At times, Smith comes close to perceiving some of his project's more troubling implications, in particular, the threatening disappearance of any line demarcating moral *qualia* from their strictly opportunistic imitation. At what point does the emulation of behavior or "conduct" turn into sheer simulation? And how would a strictly mimetic moral agent fare in a thoroughly corrupt or totalitarian community? It is in a chapter on "the corruption of our moral sentiments, which is occasioned by this disposition to admire the rich and the great, and to despise or neglect persons of poor and mean condition" (*TMS,* 61–66) that we find Smith struggling with the unintended consequences of his spectatorial model. For the first time, the almost complete absence of a narrative, teleological, and counterfactual dimension from Smith's sentimental account presents itself as a serious liability. Once again drawing on theatrical or painterly tropes, Smith offers "two different models, two different pictures ... the one more gaudy and glittering in its colouring; the other more correct and more exquisitely beautiful in its outline." Yet as it turns out, only "the most studious and careful observer, ... a select, though, I am afraid, but a small party" will appreciate the inconspicuous virtues of "temperance and propriety" (*TMS,* 62). The problem, in other words, is that a strictly mimetic conception of moral agency is prone to be drawn to the most glaring and conspicuous patterns of conduct or (what may amount to the same thing) to a type of behavior associated with the most palpable material rewards. The specular mechanism of the "man within the breast" (*TMS,* 130) is not some culturally or morally neutral view from nowhere but has already inscribed within it those moral judgments and valuations to which it is said to give rise. For Smith, the only road out of this dilemma is the one that he was to chart with requisite detail in his *Wealth of Nations* nearly two decades later. Yet the outlines of it are already hinted at in the *Theory of Moral Sentiments,* specifically in Smith's hopeful conjecture that "in the middling and inferior stations of life, the road to virtue and that to fortune, to such fortune, at least, as men in such stations can reasonably expect to acquire, are, happily in most cases, very nearly the same. In all the middling and inferior professions, real and solid professional abilities, joined to prudent, just, firm, and temperate conduct, can very seldom fail of success" (*TMS,* 63).

It is no accident that the word "profession" should show up twice within the same sentence, for its self-regulating view of propriety serves the same purpose in the realm of economics that "behaviour" does in the domain of morals. Clearly, Smith's behaviorism is meant to describe and further consolidate the fluctuating and emergent demographic phenomenon of the middling classes, which lack a viable strategy for self-description and self-legitimation of the kind enjoyed by the "man of rank and distinction." Central to his entire moral argument is a conception of society as a harmonized aggregate of essentially atomistic and selfish individuals: "Every man, therefore, is much more deeply interested in whatever immediately concerns himself, than in what concerns any other man" (*TMS*, 83). With revealing terminological imprecision, Smith thus shifts back and forth between the "impartial" and the "indifferent spectator" (*TMS*, 83, 85). Composed of so many indifferent or "punctual" selves (as Charles Taylor has more recently dubbed them), social space itself has become an extrapolated, virtual domain, rather than a historically grown and socially elaborated, empirical framework.[66] Its evanescent, abstract qualities also explain why the quintessentially modern category of the "social"—so markedly different from the sharp division between the "private" (*oikos*) and the "public" (*politeia*) in Aristotle—can no longer compel individuals to support it as a positive goal but merely insists on the minimal condition of their compliance with the law. Unaware of how a radical, atomistic notion of the individual "vitiates arguments for connections," Smith maintains that, even though "society may subsist among different men, ... no man in it should owe it any obligation."[67]

By positing individual liberty as ontologically prior and substantially unrelated to any shared norms, values, or virtues, Smith conceives of society in strictly contractual, institutional, and pragmatic terms. Its coherence rests not on a normative (Aristotelian) notion of justice and the good but is sustained only by the punitive *post facto* intervention of the law. As Smith concedes, without the specter of legal retribution, modern society cannot exist at all: "If it is removed, the great, the immense fabric of human society ... must in a moment crumble into atoms" (*TMS*, 86). While uneasily retaining ancient Stoic prescriptions as a vague defense against the troubling prospect of a wholly disembedded subjectivity, Smith's classical liberalism ultimately can only conceive the self in actuarial and virtual terms. Anticipating similar concerns in Goethe's *Wilhelm Meister's Apprenticeship* (1796), Smith is intent on bridging the divide that separates the economically self-created "burgher" (*Bürger*) from the public persona (*öffentliche Person*) of the gentry and aristocracy.[68] For the

66. On the shift from historically defined "place" to abstract, bureaucratic and legal "space," see Giddens, *Consequences*, 1–20. On formal-aesthetic echoes of that shift, see my discussion of early Romanticism's transmutation of the ballad form into pastoral elegy in Wordsworth's "Michael," in *Romantic Moods*, 191–225.

67. *TMS*, 86; Ferguson, *Solitude*, 9. What Ferguson identifies as a peculiar feature of Burke's *Enquiry* also haunts Smith's *Theory*; in both cases, the drift of their arguments "makes the impossibility of sustaining claims for a unified and unitary self seem, paradoxically, to emerge precisely out of their basis in individual experience" (ibid., 9).

68. See Pfau, "*Bildungsspiele*."

latter is constitutively identified with the public sphere and its appraisal of conduct as a grammar of meticulously rehearsed and internalized gestures and rhetorical modes. Always on stage, as it were, the public person natively inhabits and projects social authority ("if his behaviour is not altogether absurd" [*TMS,* 51–54]). By contrast, members of "the middling and inferior professions" lack as yet a conceptual matrix and practical regimen by means of which they may understand, shape, and legitimate their persona in "approved" and behaviorally objective ways.

A development already palpable in post-Hobbesian thought thus culminates in the *Theory of Moral Sentiments.* Viz., the displacement of a metaphysical concept of the will by the empiricist (nominalist) notion of contingent and discontinuous desires entails the loss of several other concepts that had previously been just as central to the project of humanistic inquiry. The Aristotelian/Thomist concept of "action" (*praxis, operatio*) loses its rational dimension and, as a result, is absorbed into a Lockean calculus of countless disaggregated "reactions" or, in the case of Smith, into a mimetic logic of "transactions." Concurrently, the concept of deliberation (*phronēsis, prudentia*) and judgment (*prohairesis, electio*) gives way to a proto-behaviorist model wherein the self constitutes itself transferentially—viz., by emulation rather than reflection. By the time that we get to Smith's *Theory of Moral Sentiments,* the (neo-) Platonic idea of mind as truly active and shaping its world has become almost entirely foreign. Perceiving action, judgment, and willing as casualties of a prolonged campaign of epistemological skepticism that begins with Hobbes, mutates in Locke's and Mandeville's hedonistic theory, and reaches its apex in Hume's *Treatise,* Smith acquiesces to work with what is left: a dramatically weakened conception of the human being, substantially devoid of internal development or progression except insofar as it merges with an actuarial and mimetic construct of successful socialization or "behaviour." The striking loss of the founding category of "action" is rendered symptomatic in Smith's *Theory* by his apparent (and decidedly un-Stoic) failure to grasp emotion as an integral part of a unified intellect, and in turn to understand the human intellect itself as teleologically ordered toward active participation in the *ratio* of the world. Further compounding this erosion of a differentiated conceptual apparatus for humanistic inquiry, Smith's *Theory* premises its central category of "behavior" on what turns out to be a misapprehension of the idea of "habit"—viz., as the strictly passive, mechanistic, and non-rational response to external stimuli or some introjected equivalent. Such a view rests on a logical fallacy (endemic to early Enlightenment thought) whereby it is supposed that mind is not only premised on the body as its "material" condition of possibility, but that it is *therefore* also exclusively and exhaustively determined *by* the body.

Though not nearly as dogmatic as Locke, Mandeville, and Hume about reducing all final causation to material and, ultimately, to efficient causation, Smith nonetheless lacks any phenomenological curiosity (as do Watson and his behaviorist heirs) about the operation of mind. Instead, his *Theory of Moral Sentiments* signals moral philosophy's gradual retreat from internalist models of human agency. It also affirms the disappearance of a fully articulated idea of human flourishing by confining human awareness to a continuous present filled with an endless stream of incidental and unreflected desires. Supplanting normative frameworks with contingent interestedness, reasoned deliberation about ends with isolated "decisions" about means, modern liberalism envisions and (if only in an impoverished sense) achieves moral neutrality as a possible, indeed indispensable way of being in the world. What, in a notably Burkean turn of phrase, Hans-Georg Gadamer some time ago had flagged as the "Enlightenment's prejudice against prejudice" thus involves a deep-seated resistance to contemplation, sustained deliberation, and any type of "strong evaluation" (as Charles Taylor calls it). Put in positive terms, beginning with Smith's mimetic and quasi-behaviorist concept of agency, it appears not only safe but decidedly preferable for the modern self to live out its existence unfettered by supra-personal and normative commitments. A significant implication of this "naturalist reduction" (Taylor) involves the pervasive loss of awareness regarding the deep history and hermeneutic complexity of moral concepts. While this implication remains oblique in Smith's *Theory,* it is on full display a generation later, such as in Thomas Paine's characteristically blunt assertion that "every generation is, and must be, competent to all the purposes which its occasions require. It is the living, and not the dead, that are to be accommodated."[69]

Paine highlights a major implication of modernity's reduction of the Aristotelian category of "action" (*praxis*) to a strictly methodological challenge: viz., how to "implement" or "comply" with abstract propositions whose authority and cogency it takes to be substantially unrelated to their material realization. Thus modern liberalism no longer understands "practice" or "doing" as a distinctive realm of contingent experience. Rather, action serves the strictly *instrumental* purpose of securing a non-contingent state captured by Nietzsche's acerbic image of the herd grazing in "security, safety, contentment," reassured that "suffering has been abolished."[70] In pursuit of its speculative utopias, liberalism dissolves action into behavior, that is, into the coordinated, bureaucratic, or habituated *implementation of a concept.* Social and material processes thus come to be seen as but the outward manifestation of a logical ratio of means and ends. Such *Zweckrationalität* (as Max Weber was to christen it) may present itself as the dialectical play of "interests" gradually civilizing the passions in the writings of the Scottish economists or as the abstract "felicific calculus" of Jeremy Bentham's utilitarianism and the harsh social policies seemingly licensed

69. Gadamer, *Truth and Method*, 273; C. Taylor, *Sources*, 25; Paine, *Rights of Man*, 42.

70. *Beyond Good and Evil*, 41.

by it.[71] Yet Smith's *Theory of Moral Sentiments* also reveals the extent to which liberalism's vision of putatively rational and individual agents defined by a single, pervasive, and hence fundamentally predictable template of "behavior" ultimately conspires to dissolve the boundaries of agency and individuality alike. For when transmuted into "labor" and "behavior," respectively, the Aristotelian categories of "work" and "action" are aggregated and tabulated by the metrics of modern statistics, social psychology, and probabilistic reasoning—conceptual and methodological innovations that helped realize the Enlightenment project of stabilizing social life in a set of "average" values, while sequestering contingency as a mere fluctuation of data.[72] Yet as it is captured through disciplinary lenses of its own devising, modern individuality happens upon a model of rational and competitive social processes essentially indifferent to questions of subjective intention and objective responsibility. Modernity's quintessentially linear and progressive (rather than cyclical) conception of time paradoxically pivots on the non-knowledge of individual agency as to the deeper systemic and ideological processes in which it is caught up. Hence the emancipatory claims of liberalism's neutral or "punctual" self make for an odd contrast with the concurrent emergence of systemic (dialectical) models of rationality according to which all historical process is necessarily opaque to the "natural consciousness" (Hegel) so vicariously advancing it.

71. On Weber, see Milbank, *Theology*, 84–100; for Milbank, history can only be captured in the metastases of means/end rationality: "fully objective history (sociology) is *primarily* about economic rationality, formal bureaucracy, and Machiavellian politics. What lies outside these categories cannot be read as a certain distinctive pattern of symbolic action, but only negatively registered." As such, Weber's notion of *Zweckrationalität* is "teasingly ambiguous." "At times ... [it] is a mere matter of methodological convenience, at other times it is the dark business of our Western fate" (ibid., 87).

72. On Aristotle's implicit idea of freedom as "the presupposition of the exercise of the virtues and the achievement of the good," see MacIntyre, *AV*, 146–164 (quote from 159); on Aristotle's concept of "action," see especially Arendt, *HC*, 175–247. Echoing Arendt, Milbank notes how "praxis, in the old Aristotelian sense, referred to a dimension of action which was categorically 'ethical' because it could not be separated from a person's essential being or character (*ethos*); it meant a doing which was also a being. It also implied action directed towards a particular end (*telos*), but an end immanent within the very means used to achieve it, the practice of 'virtue'" (*Theology*, 161).

13

AFTER SENTIMENTALISM

Liberalism and the Discontents of Modern Autonomy

Two major problems now begin to emerge, both of acute concern for the Romantics and, uniquely so, for the later Coleridge. First, it is apparent that, far from being an ontology and "source" of meaning, reason by the late eighteenth century is separating from the interiorist framework that, since St. Augustine, had revolved around a rich pallet of human intentionality that includes notions of will, deliberation, judgment, choice, and so on. Instead, by the late eighteenth century reason tends to be conceived as a type of sublimated socialization (in Adam Smith) or as the adventitious, self-organizing logic of a *System* (in Hegel). In the wake of Smith's *Theory of Moral Sentiments,* Anglo-Scottish liberalism thus tends to conceive of reason *descriptively*—viz., by offering various (quasi-sociological) accounts of our complex, impersonal, and economically driven behavior. No longer does the concept of reason exercise a *normative* function by inducing individuals to dialectically engage and jointly articulate the imperfect rationality of prevailing socioeconomic and cultural practices. Instead, liberalism considers assent to its meliorist narrative and procedural objectives to be purely voluntary and contingent, even as that narrative promises to remedy the self's ephemeral passions and asocial desires. Second, even as the irrational passions that (beginning with Hobbes) had displaced the metaphysical idea of the will are said to be gradually transmuted into rational self-interest by the superego of the modern marketplace, the resulting "element of reflection and calculation" can only ever be concerned with means.[1] Classical liberalism tends to merge the idea of the *logos* with the operation of a strictly interested "understanding" intent on securing those means most apt to help it secure its limited objectives.

As a result, the intellectual and spiritual curiosity of individuals and societies appears to atrophy to the point that questions concerning ends are themselves marginalized as eccentric and more or less irrational. This sets the pattern for nineteenth-century religious culture, characterized by a notably subjective and emotivist (anti-doctrinal) turn in religious practice and theology alike, and by the consequent rise of modern denominationalism that leaves religion little more than a reflex of various socioeconomic *mentalités* or, more feebly yet, as a cult(ure) of mere "opinion" or "private judgment" (as John Henry Newman was to diagnose it in 1841).[2] By the end of the eighteenth century, the shift here summarized leads F. W. J. Schelling, and many of his Romantic contemporaries to express dismay at the chasm between an understanding profoundly transformed by the methodological and scientific advances of the last century, and a model of reason proportionately impoverished. As Coleridge puts it so eloquently, "whatever is achievable by the UNDERSTANDING for the purposes of worldly interest, private or public, has in the present age been pursued with an activity and a success beyond all former experience ... But likewise it is, and long has been, my conviction, that in no age ... have the Truths, Interests, and Studies that especially belong to THE REASON, contemplative or practical, sunk into such utter neglect, not to say contempt, as during the last century" (*AR*, 8). In this chapter, then, we will give some consideration to how these shifts manifest themselves in the understanding of rights, action, and freedom—three conceptions central to the modern subject's self-image and yet, paradoxically, rendered incoherent by the ways in which the liberal-secular nation-state tends to deploy them.

As we saw, the passions in Hume's account appear bereft of any discernible content and as such are destined to expire almost instantaneously because they can only ever occur *in*, but never (as distinct *qualia*) register *for*, consciousness. In response, Smith's *Theory of Moral Sentiments* suggests that the passions may yet be sublated into more focused and communicable interests, provided one is prepared to endorse the behaviorist axioms and conclusions of Smith's moral and economic psychology. Even then, however, a lacuna opens up that goes largely unnoticed by Albert

---

1. Hirschman, *Passions*, 32.

2. On the fragmentation of Christianity into denominations, see Martin, *On Secularization*, esp. the essays on Pentecostalism and on the "Plurality of Faiths," 141–170; Berger, *Sacred Canopy*, 105–125; on anticlericalism in conjunction with secularization and denominational fragmentation, see Gregory, *Unintended Reformation*, 180–234; Chadwick, *Secularization*, esp. 107–139; C. Taylor, *Secular Age*, 352–459; on the anthropological dimension of this shift, see Gauchet, *Disenchantment*, 162–207; on secular communities in British Romanticism, see Jager, *Book of God*.

Hirschman and other intellectual historians who have so persuasively traced modern liberalism to the great economic and financial transformations of the early eighteenth century. Seen as so much combustible material in need of being channeled toward identifiable and socially beneficial ends, the modern individual's discontinuous and opaque passions are regarded as (at best) a kind of motivational fuel for a narrative of "improvement" that, though shrewdly calculative about means, is alarmingly inarticulate about ends. It is only under the purview of probabilistic and statistical systems of explanation that the self can be credited with rationality since on its own it is but a random empirical singularity unwittingly actuated by literally mindless efficient causes or motives. The incommunicable person of the Augustinian and Thomist tradition has morphed into a free-floating particular begging to be sublated into a philosophical, sociological, or statistical calculus.[3] The concurrent expulsion, not only of reflexivity from the modern individual but of wisdom as a legitimate topic from philosophy itself, thus shines through in utilitarian thought (where one naturally expects to find it), yet also in philosophies as disparate as the nationalist and protectionist writings of Johann Gottlieb Fichte, Auguste Comte's *Positive Philosophy,* and the early economic manuscripts of Marx. The same axiom even informs thinkers overtly critical of utilitarian thought, though in their own ways just as committed to the Enlightenment's naturalist epistemology and to *a technocratic rehabilitation of the will* such as it is attempted in the educational and moral philosophy of Rousseau, Mary Wollstonecraft, Andrew Bell, Jeremy Bentham, and even the late Kant. All share a fundamental commitment to disciplining the will by means of a procedure that, much later, Jacques Lacan was to call "introjection."

A diagnostician of rare powers, Alexis de Tocqueville is among the first to recognize the insidious ways in which action is being supplanted by behavior, articulacy about ends by sheer cleverness about means, and independence of thought by petit-bourgeois moral conformism. It is none other than the sovereign power of modern democracy that

> extends its arms over the entire society; it covers the surface of society with a network of small, complicated, minute, and uniform rules, which the most original minds and the most vigorous souls cannot break through to go beyond the crowd; it does not break wills, but it softens them, bends them and directs them; it rarely forces action, but it constantly opposes your acting; it does not destroy, [but] it prevents birth; it does not tyrannize, it hinders, it represses, it

3. On the "mathematization of nature," see Husserl, *Crisis,* §9 (pp. 23–59); the process rests on what M. Polanyi calls the "axiomatization of mathematics" and its continuing strides toward ever-increasing possibilities of non-tautological generalization; see Polanyi, *Personal Knowledge,* 184–193; and also Maritain, *Degrees of Knowledge,* 149–164.

enervates, it extinguishes, it stupefies, and finally it reduces each nation to being nothing more than a flock of timid and industrious animals.[4]

What so troubles de Tocqueville about the "administrative despotism" of the modern liberal nation-state is how its procedural logic encourages the cultivation of a myopic, adaptive (not to say opportunistic), and necessarily short-sighted intelligence at the expense of premodern reason. Naturally, one finds it "difficult to imagine how men who have entirely given up the habit of directing themselves, could succeed in choosing well those who should lead them."[5] Though importantly balanced by a more positive assessment of American-style democracy elsewhere in his magnum opus, Tocqueville's observations here highlight a dilemma that we have had occasion to examine up close. Ultimately sharing the same neo-Stoic framework, the emerging social utopias of the late Enlightenment (Smith, Kant, Comte) all effectively concede the Hobbesian (originally nominalist) premise of the will as the terminally inarticulate and non-transcendable "ground" of all human agency. No longer, then, is "ground" to be understood as Platonic *ratio* but, rather, as unreflected, efficient causation. It is but a hypothesis on the order of Kant's "as-if" (e.g., in his account of teleological reason) or, more mystically, as the primordially "groundless" (*Ungrund*) in Schelling's account of freedom.[6] With the exception of the Cambridge Platonists and Shaftesbury, and possibly Kant, who late in the century seeks to offer a more nuanced appraisal of the voluntarist legacy, the Enlightenment's attempted rehabilitation of the will by means of sociological and behaviorist "engineering" does not seek (indeed cannot even wish) to contest voluntarism's complex legacy. Rather, if only in somewhat jaundiced manner, the Enlightenment identifies a (supposedly non-cognitive) model of the will as the point of departure for its remedial interventions.

The internal complexity of the premodern will—as variously conceived by Aristotelian, Stoic, neo-Platonist, Augustinian, and Thomist traditions of inquiry—appears increasingly alien and illegible to thinkers too deeply implicated in some self-certifying narrative of modernity to do anything but *mimic* the modern will's

---

4. Tocqueville, *Democracy,* 4:1252.

5. Ibid., 4:1260.

6. Following Jakob Boehme, Schelling's 1809 essay *On Human Freedom* develops a mystical (in part Gnostic) account concerning the origin of evil and reason, respectively: "Following the eternal act of self-revelation, the world as we now behold it, is all rule, order and form; but the unruly [*das Regellose*] lies ever in the depths as though it might again break through ... Out of this which is unreasonable, reason in the true sense is born. Without this preceding gloom, creation would have no reality; darkness is its necessary heritage" (Schelling, *Philosophical Inquiries,* 34; Schelling's neologism of "the groundless" (*Ungrund*) appears on p. 87.

performative self-origination.[7] The result is a condition of pervasive conceptual amnesia that continues to haunt modern liberalism to this day. While liberalism has long proven a diffuse conception, its resistance to conceptual explicitness and lack of intellectual coherence is less an accident than a logical consequence of its avowed emergentist logic and pluralist ethic. Still, the present concern is less with offering yet another definition of liberalism than with illustrating how its objectives appear to have been shaped by a persistent struggle with Ockham's and Hobbes's legacy of nominalism and voluntarism. While that legacy has variously been ignored or minimized by staunch adherents of modern liberalism, its underlying conception of human nature as self-interested, hedonistic, and unreflective has been subjected to withering critiques by two schools of thought making their appearance almost at the same time: conservatism and pessimism. This is not the place for a detailed account of the ways in which modern liberalism's naturalist epistemology is unwittingly replicated by some of its most vociferous critics, such as in the pessimistic anthropology of Burke, Schopenhauer, Joseph de Maistre, Hugues-Felicité Robert de Lamennais, Hippolyte Taine, and Thomas Carlyle, among others. We can only attempt to sketch the growing tension, after 1800, between a meliorist, a utopian, and a pessimist conception of agency.

As we have seen, the central dilemma confronted by the liberal critics of Hobbesian voluntarism was how—*without contesting his basic model of human agency*—one might yet imagine a type of civil association premised on rational consensus rather than on legally sanctioned and state-administered force. Might there yet be a way to cultivate individual and compulsively self-seeking agents *in spite of their very nature,* that is, not simply to *contain* but, in effect, *transform* the irrational, compulsive, and inarticulate subject of modern voluntarism? In the event, the Enlightenment's principal response to that challenge—viz., Anglo-Scottish political economy, the twin cultures of sentimentalism and "improvement," as well as the neo-Stoic, millenarian, utilitarian, and cosmopolitan utopias of Rousseau, Priestley, Blake, Godwin, Wollstonecraft, Bentham, and Kant—fail to meet that challenge for a variety of reasons. At least in part, these conceptions had to fail simply because of their inability or outright refusal to credit human agency with a capacity for sustained and evolving introspection. That is, a skeptic like Hume, associationists such as Hartley and Godwin, and social constructionists such as Bentham, Saint-Simon, or Comte fundamentally share the same argumentative stance of no longer engaging (nor indeed betraying any wish to engage) any alternate model of rationality. The one that most obviously would have suggested itself—the teleological model first formulated by Aristotle and,

---

7. On the centrality of performative models of self-origination and speech in the Romantic era, see Esterhammer, *Romantic Performative.*

by way of an intricate genealogy, migrating into thirteenth-century Thomism—would have shown that it was by no means inevitable, indeed ultimately not even justifiable, to conceive the will as the antagonist of reason. For to take that view meant de-potentiating personhood to a mere heuristic *punctum*, an interchangeable, generic "self" devoid of all essence and reduced to an indifferent variable, a mere bearer of claim rights exclusively actuated by inscrutable compulsions and desires.

Smith's paean to a "moderated sensibility" and a mimetically derived "sense of propriety" and "self-command" (*TMS*, 143, 146) is echoed a generation later in Thomas Paine's *Rights of Man*. Yet, in a significant radicalization of classical liberalism, Paine now short-circuits the entire neo-Stoic regimen of mimetic self-examination that defines Smith's "impartial spectator." Instead, Paine simply asserts the equivalence, and hence the de facto indifference, of man as an ontological fact: "Every history of the creation, and every traditionary account, ... agree in establishing one point, *the unity of man;* by which I mean, that men are all of *one degree,* and consequently that all men are born equal, and with equal natural right, in the same manner as if posterity had been continued by *creation* instead of *generation.*"[8] It is just this approach to political argument as a set of strictly performative and apodictic claims seemingly produced *ex nihilo* that prompts Hegel to remark how "the reality which stands in the greatest antithesis to universal freedom ... is the freedom and individuality of actual self-consciousness itself." For Hegel, such an actuarial notion of society as a composite of interchangeable selves and average modes of behavior inevitably culminated in the Terror of the French Revolution. There, Hegel notes, the "sole work and deed of universal freedom is ... death," viz., when the universal will as "self-conscious reality heightened to the level of *pure* thought or of *abstract* matter, changes round into its negative nature and shows itself to be equally that which *puts an end to the thinking of oneself,* or to self-consciousness" (*PS*, 359–361). Extending Hegel's critique, Charles Taylor has offered a succinct formulation of the logical paradox to which the radical egalitarian vision that Paine inherited from Rousseau gives rise. Thus, even as recognition means "that *everyone* should be recognized for his or her unique identity," the acknowledgment of such specificity tends, simply by dint of its proclaimed universality, toward an acute homogenizing of human

8. *Rights of Man,* 66; Godwin echoes Paine's contention in a short chapter on physical and moral "Equality of Mankind" (*Enquiry,* 181–184), though as we shall see he diametrically opposes Paine, Thelwall, and most other English "Jacobins" on the question of rights. Just five years later, Joseph de Maistre vehemently rejects Paine's claim that men are "all of one degree": "there is no such thing as man [*l'homme*] in the world. In my lifetime I have seen Frenchmen, Italians, Russians, etc.; thanks to Montesquieu, I even know that *one can be Persian.* But as for *man,* I declare that I have never in my life met him" (*Considerations,* 53). On Paine, see Kramnick, *Republicanism,* 133–160; on de Maistre, see Pfau, "Rational Theology."

societies: "With the politics of equal dignity, what is established is meant to be universally the same, an identical basket of rights and immunities; with the politics of difference, what we are asked to recognize is the unique identity of this individual or group." The result is a cascade of micro-distinctions, an early version of Sigmund Freud's narcissism of minor differences, which uncannily revives Montesquieu's definition of honor to which Rousseau, Godwin, Paine, and virtually all their contemporaries are so strenuously opposed. Thus, just as *L'esprit des lois* envisions a highly stratified social order based on honor ("*la nature de l'honneur est de demander des preferences et des distinctions*"), Paine's egalitarian dogma ("men are all of *one degree*") ends up in a place not all that different. For if we "give universal acknowledgment only to what is universally present—everyone has an identity—through recognizing what is peculiar to each," then precisely this "universal demand powers an acknowledgment of specificity."[9]

Still, there is something beguiling about Paine's pugnacious assertion of inalienable "natural rights ... which appertain to man in right of his existence" and which are said to differ categorically from "civil rights," such as "appertain to man in right of his being a member of society." The latter, Paine notes, originate in "some natural right pre-existing in the individual" and only accrue to man "after entering into society."[10] Within the unfolding narrative of classical liberalism, a radicalization has clearly taken place in the generation between Smith and Paine; and even if "Smith's and Paine's is the basic liberal vision ... [of] spontaneous and self-regulating mechanisms, peopled by rational, self-seeking individuals," Paine's libertarian project is far more accepting of Locke's hedonistic conception of human agency than is Smith's. To begin with, in Paine's writings the notion of a sympathetic (if only virtual and quasi-behaviorist) community has effectively been abandoned, with the nation now being said to be "composed of distinct, unconnected individuals, following various trades, employments and pursuits; continually meeting, crossing, uniting, opposing and separating from each other, as accident, interest, and circumstances shall direct."[11] Second, and in apparent consequence of this shift toward radical individualism, the language of "rights" has emerged as the centerpiece of a modern liberal polity. For Paine, rights are inherent rather than contingently claimed, though it is obvious to him, too, that this seemingly absolute "fact" had only been discovered at a particular (and quite recent) point in historical time. Paradoxically, this historically specific

9. "The nature of *honor* is to demand preferences and distinctions" (Montesquieu, *Spirit of the Laws,* I.3.7). Taylor, "Politics of Recognition," 38–39. As Taylor goes on to argue, the "distinction" affirmed by an egalitarian politics of recognition is less the manifest reality of the person than his or her presumptive "potential, rather than anything a person may have made of it" (ibid., 41).

10. *Rights of Man,* 68.

11. Kramnick, *Republicanism,* 147; Paine quoted in ibid., 154.

emergence of "rights" within modern political thought takes the form of a non-contingent and seemingly incontestable claim *simultaneously advanced and legitimated* by the modern individual. That is, Paine alternately posits the individual as an absolute or (in his parlance) "originary" value *and* as the sole beneficiary of its own, performative claims concerning the "rights of man."

Here, then, performative self-enactment has supplanted the Platonic model of dialectical reasoning as the basic form in which human agency is to be practically realized within the world. Implausibly, action is conceived as a mode of performative self-realization and instantaneous self-fulfillment. Underlying this fantasy of total self-possession is the assumption that "frameworks are things we invent, not answers to questions that inescapably pre-exist for us, independent of our answer or inability to answer."[12] The tension between a dialectically articulated, humanist ethic on the one hand (elements of which survive in the "bourgeois radicals" of 1790s Britain and in Kant's political thought), and a proto-behaviorist and necessarily inarticulate model of sociality on the other hand, has informed political debate to this day. Thus Jeffrey Stout has formulated a thoughtful alternative to the prevailing conception of modern, secular, and democratic society—identified above all with John Rawls. For Stout, democracy's "ethical substance ... is more a matter of enduring attitudes, concerns, dispositions, and patterns of conduct than it is a matter of agreement on a conception of justice in Rawls's sense. The notion of state neutrality and the reason-tradition dichotomy should not be seen as its defining marks. Rawlsian liberalism should not be seen as its official mouthpiece." Obliquely echoing Hannah Arendt and Michael Oakeshott, Stout thus revives the Aristotelian notion of *praxis* as a coherent and transmissible framework. He thus opposes modern voluntaristic accounts that limit the rationality of an act to the intention or motive said to have *prompted* it and, in so doing, supplant the Aristotelian (teleological) conception of action with a strictly occasional and discontinuous "doing."

As Stout puts it, "my conception of the civic nation is pragmatic in the sense that it focuses on *activities* held in common as constitutive of the political community. But the activities in question are not to be understood in merely procedural terms. They are activities in which normative commitments are embedded as well as discussed."[13] With a rather more overt nod to Aristotle (esp. *Nicomachean Ethics,* 1105^{b}5), Oakeshott insists that "the practical is not a certain kind of performance; it is *conduct* in respect of its acknowledgement of a practice." At the same time, he takes care to desynonymize his understanding of *praxis* from merely mechanical and

12. C. Taylor, *Sources,* 30.
13. *Democracy and Tradition,* 3–5.

mimetic "behavior" (which, unfortunately, he simply calls "habit"). In fact, he emphasizes, *praxis* properly speaking

> does not reduce conduct to a process or impose upon it the character of a *mere* habit. Customs, principles, rules, etc. have no meaning except in relation to the choices and performances of agents; they are *used* in conduct and they can be used only in virtue of having been learned. Nor do procedures prescribe choices or substantive actions. A rule (and a *fortiori* something less exacting, like a maxim) can never tell a performer what choice he shall make; it announces only conditions to be subscribed to in making choices.[14]

The latter point arguably eludes those who, like Paine, reject the very notion of a historically constituted, normative framework (custom, tradition, habit) as but "the manuscript assumed authority of the dead over the living."[15] Likewise, Paine's characterization of man as categorically equivalent and indifferent ("men are all of one degree") notably flattens out the supposed source and beneficiary of that very claim—viz., the modern individual, in effect rendering it a generic article or heuristic fiction.

Recently, John Milbank has offered a brisk genealogy of what, somewhat polemically, he terms "the hidden alliance between liberalism and political absolutism." For Milbank, modernity's "drift to a new despotism" is the result of "an excessive stress upon the isolated individual," a generic fiction made possible by the "dethronement" of the Judeo-Christian "valuing of 'the person' ... shaped through all her interrelationships and yet as a unique 'character' ... transcendently of more value than any conglomerated whole." Liberalism's narrow, monadic stress on the *suum* and on "human rights" as inalienable property "parodies this legacy, because when irreplaceable 'personality' is reduced to an inviolable but inscrutable abstract interiority of negative will, then the *social manifestation* of the individual person can be no more than that of an always replaceable and disposable atom."[16] Milbank's principal debt is

---

14. *On Human Conduct,* 57–58 (italics mine); for a discussion of Oakeshott's theory of moral practice in relation to Wordsworth and Hegel, see Pfau, "Immediacy and Dissolution."

15. *Rights of Man,* 42.

16. Milbank, "Against Human Rights," 2, 5. Milbank rejects Wolterstorff's thesis that modern individual claim rights originate in early Christianity, which entails suspending the Aristotelian-Thomist tradition of "right reason" as bound up with (and derived from) the ontology of the cosmos itself; and he specifically faults Wolterstorff for misconstruing Villey's thesis, which "was *not* ... that a *ius* was anciently a 'thing' on our modern model of thing as *object* that could only be shared in terms of literal partition. To the contrary, it was a 'thing' in the sense of an objective ideality that could be *participated* in.... *Ius* up till at least the 13th century meant always the 'objectively right,' that which was just, and it was linked to a notion of justice as distribution which meant always measuring the proper situation of persons and things in relation to each other" ("Against Human Rights," 8, 14). Regrettably, Annabel Brett's detailed study (*Liberty, Right and Nature*) of the

to a series of influential articles by Michel Villey (especially his 1946 "L'idée du droit subjectif et les systems juridiques"), which had argued that under Roman law and extending well into the seventeenth century, rights were not to be construed as subjective claims. Rather, as in the case of the *ius utendi fruendi, ius* signifies "an array of legal advantages and disadvantages inherent in the property. It is an objectively existing abstract thing, not a power inhering in the owner" and, as such, can only be claimed or exercised "in an objectively just social order."[17] According to Villey, once the idea of "inherent" or "subjective" claim rights takes hold in seventeenth-century legal theory, justice and the good cease to be founding, normative concepts. In the Aristotelian-Thomist tradition (as per Villey's and Milbank's account) claim rights necessarily presupposed and were licensed by a communal and normative conception of the just and the good. By contrast, modern liberal thought supposes that a just and good community will eventuate at some point in the theoretical future when individual rights claims have been duly recognized and balanced. Yet within the "acephalous organism" of the modern nation-state, such balancing now appears altogether adventitious, depending as it does on the disparate influence wielded by individuals within those political and legal institutions tasked with adjudicating individual claims.[18]

---

split between objective right and subjective rights in Scholastic and early modern thought came to my attention too late to be factored into the present discussion.

17. Reid, "Canonistic Contribution," 53–54. Following Brian Tierney's ground-breaking work on *The Origins of Natural Rights Language,* Reid is critical of Villey's initial thesis, and also of his later attempt to connect the emergence of modern rights theory to Ockham's voluntarist epistemology. Some of Reid's many and rich examples support his point better than others; leaving aside the panoply of Latin legal terms often subsumed under the bland word "right" (e.g.., *libertas, potestas, facultas, immunitas, dominium, justitia, interesse,* and *actio* [64]), a bishop's right to tax churches (the so-called *cathedraticum*) would not appear to be a modern claim right since it appends to his person only *ex officio* rather than in virtue of his status as a human being. While conceding that the initial formulation of canon law in Gratian's *Decretum* (ca. A.D. 1140) never yielded an explicit "treatise on rights," Reid nonetheless affirms the link between twelfth-century canon law and modern rights theory: "one should not mistake the absence of theoretical speculation for the lack of a consistently deployed concept" (57). In fact, one should; for the historical effectiveness of an idea or conceptual tradition—what Gadamer calls its *Wirkungsgeschichte*—hinges on the explicit and dialectical ways in which its core propositions or beliefs are probed by successive generations. Intellectual traditions are not something virtual or conjectural but eminently real; prolonged periods of latency tend to undermine their reality. For Wolterstorff, Villey's account is the archetypal "narrative of decline" that proceeds from "the dominance in ancient and early medieval times of the concept of *the right* and the conception of justice as right order to the emergence and eventual dominance in modern times of the concept of *rights* and the conception of justice as grounded in natural rights" (*Justice,* 45–58; quote from 45).

18. The phrase is from Oliver O'Donovan's *The Desire of Nations* (quoted in Wolterstorff, *Justice,* 52).

As Villey before him, Milbank traces the idea of inherent (licit) rights back to Ockham's late "theorization of the notion [of *potestas licita* as] a right *derived* from *de facto* power." For inasmuch as a theory of rights licensed by (contingent) political power views such rights as accidents or predicates of a particular sovereign, it has thereby lost sight of how the legitimacy of that political power necessarily presupposes a notion of "right order"—of *ius* not as a claim right but as *justice,* and hence as something wholly enmeshed with the ontology of a non-contingent order or *logos.* The main weakness of the claim rights model is principally said to stem from a failure to distinguish between the individual and the person, between the abstract species concept of "human being" and the incommunicable human person as a constitutively relational being whose dignity and flourishing are inseparable from the notion of a just community. By contrast, within the nominalist model that would eventually spawn liberalism's notion of absolute claim rights, "the individual is himself seen as a self-sufficient 'sovereign' entity, abstractable from his social insertion [and thus] conversely not essential to the composition of any social aggregate."[19] Where the eighteenth-century discourse of "obligation" (even when contested, as in Mandeville or Hume) still revolved around the primal reality of *the common,* the language of rights tends to proceed by enclosure. Developed in more or less explicit opposition to the notion of inherited realities and obligations, it tends to divide and isolate individuals.

This is not the place to reconsider this multilayered debate in greater depth since it has but a tangential bearing on our main topic—viz., the deteriorating conception of personhood in the modern era. Suffice it to say that on historical grounds Milbank's polemic has been weakened by the fact that Villey's narrative, on which Milbank mainly relies, has largely been disproven by the meticulous research of Brian Tierney, Richard Tuck, Charles Reid, and Charles Donahue. Thus Villey's failure to recognize that the beginnings of natural rights ought to be located in canon law rather than in the philosophical writings of Aquinas or William of Ockham, compounded by the fact that Thomists, Aristotelians, and nominalists "all were speaking of natural rights" as early as the mid-thirteenth century, and that even in Roman law there are several hundred cases in which a right is attached to an individual (*ius esse alicui*), all invalidate the *historical narrative* of decline from right order to individual rights.[20] Nevertheless, the convoluted historical record concerning the genesis of modern natural rights does not automatically invalidate the opposition itself. If, as

19. Milbank, "Against Human Rights," 13, 16.

20. Wolterstorff, *Justice,* 56. Aside from Ockham, whose position in the rights-controversy remains somewhat unique even now, Wolterstorff (drawing on Tierney) mentions numerous other writers, including Dominicans and Aristotelians, who make repeated use of *ius* as a natural right (Hervaeus Natalis, Marsilius of Padua, Godfrey of Fontaine).

Tierney has shown, Ockham's persistent appeal to a distinction between *ius positivum* and *ius naturale,* between right order and individual, "licit power," marks the moment when the language of rights migrates from the juridical domain into that of theology and, eventually, that of political theory, the significance of that transposition will have to be adjudicated on grounds other than those of historical precedent and continuity. For however we choose to assess that shift, our arguments will invariably rest on how we perceive the link between positive claim rights and the flourishing of social and political communities. A historical development, however painstakingly traced, cannot absolve us from having to arrive at a reasoned judgment as to whether we construe communal flourishing to *depend on* (enforced) respect for the inherent rights of individuals, or whether our conception of what those rights are is itself *derived from* a transcendently sanctioned, normative conception of a just (communal) order.

Mindful of these presuppositions, Nicholas Wolterstorff thus notes that to concede "the existence of natural rights leaves open the question of whether natural rights are inherent in the worth of the bearer or conferred on the bearer by some objective norm or standard." Clearly, a strictly historical account, however thorough, cannot per se answer that question any more than a capacity-based model can adequately ground natural human rights.[21] Here one ought to bear in mind (as Wolterstorff does not) that the "capacity" model has dominated Western rights theory beginning with John Locke. This matters since even a casual historical survey shows that, once Locke's tenuous deist framework had melted away, the notion of natural human rights quite effortlessly came to be vested in a human being's capacity for rational agency. In other words, not only was divine love or grace no longer required as the transcendent *source* of natural human rights, but human rational capability, which previously had served as a type of secondary *evidence* for this divine gift, now supervenes on God and becomes the positive *source* of such rights. The narrative of rights, in other words, cannot be disentangled from the historical evolution of the concept of autonomy. Even so, Wolterstorff is quite right to point out how this capability model of rights is bound to marginalize large swaths of the human population, such as the mentally impaired, Alzheimer's sufferers, or infants. His alternative, developed toward the end of *Justice,* holds that natural human rights ought to be conceived as a unique kind of "bestowed worth" (Platonic rather than instrumental) whose source we find in divine love—understood as "attachment" rather than "attraction."

Again, one's agreement in principle is tempered by misgivings about the seemingly ahistorical way in which Wolterstorff makes his case here. It is one thing to

---

21. *Justice,* 63.

commit to a theist (Augustinian) framework and to argue that "if God loves a human being with the love of attachment, that love bestows great worth on that human being," and quite another to speak of such love and bestowal of worth as an instance of "benevolence."[22] That these conceptions are not interchangeable becomes evident once we recall that the language of benevolence rises to prominence precisely as the theist framework—weakened by centuries of voluntarism, Socinian anthropomorphism, and the implicit deism of modern natural theology—is all but collapsing toward the end of the eighteenth century.[23] What makes Wolterstorff's case for natural rights compelling as an argument in philosophical theology is weakened by his apparent inattention to the historical mutability of his key concepts. As it rises to prominence in the later Enlightenment the concept of "benevolence" has all but lost touch with the Augustinian theological framework with which Wolterstorff continues to associate it. Arguably, what accounts for that estrangement is a secular and anthropomorphic idea of moral agency (exemplified by Kant) sponsored by precisely those "individualistic and atomistic modes of thought" that critics of modern natural rights (Michel Villey, Alasdair MacIntyre, Oliver O'Donovan, John Milbank) had foregrounded and that Wolterstorff rejects from the outset early in his study.[24]

Unfolding in the lineage of Locke, Mandeville, and James Steuart (more than Adam Smith), Paine's arguments concerning the rights of man vividly embody that shift, while also exemplifying how the oblique rhetorical maneuvers of "common-sense" or "plain" English indemnify liberalism's axioms from having to prove their validity in a dialectical contest with Judeo-Christian conceptions of personhood and relationality, which they simply displace. Paine's proto-libertarian affirmation of the isolated, autonomous, and generic "self" as the ground-zero for all political theorizing exposes a hidden antinomy between *liberté* and *egalité* that was to prove increasingly vexing to nineteenth-century writers of such disparate sensibility as Blake, Schopenhauer, de Tocqueville, Flaubert, Mill, John Henry Newman, George Eliot, Dostoevsky, and Nietzsche. Yet it is above all Paine's eponymous concept of "rights" that warrants our attention; for not only does it play a crucial role in the genesis of modern liberalism at the end of the eighteenth century but, even today, it constitutes an essential (if woefully unexamined) axiom in popular and academic political discourse. Some time ago, Alasdair MacIntyre called attention to a logical paradox remarkably similar to the one vitiating Paine's concept of agency. In his critique of Alan Gewirth's *Reason and Morality* (1978), MacIntyre scrutinizes Gewirth's central claim that, "since the agent regards as necessary goods the freedom and well-being that constitute the generic features of his successful action, he logically must also hold

22. Ibid., 350–361; quote from 360.
23. See Pfau, "Rational Theology."
24. Wolterstorff, *Justice*, 361.

that he has rights to these generic features and he implicitly makes a corresponding rights-claim." As MacIntyre points out, trouble brews when we take it as an axiom that to identify certain "necessary goods" for the "exercise of rational agency" also licenses *eo ipso* the claim that we have "a right to these goods."[25] As he proceeds to argue, "one reason why claims about goods necessary for rational agency are so different from claims to the possession of rights is that the latter in fact presuppose, as the former do not, the existence of socially established rules. Such sets of rules only come into existence at particular historical periods under particular social circumstances. They are in no way universal features of the human condition."[26]

Though otherwise critical of MacIntyre's Aristotelian orientation, Milbank concurs and elaborates on the historical conditions that favored the emergence of a theory of rights and in time fused it with the modern notion of the individual as a self-certifying and self-possessed agent. At issue is the seventeenth-century redefinition of the classical Roman concept of *dominium*—then understood as rational self-government and (Stoic) mastery of the passions but, by the mid-seventeenth century, recast as *sovereignty* in the sense of sheer self-possession and the unfettered exercise of instrumental reason that dominates the writings of most English radicals, Thomas Paine and John Thelwall most prominently. Instructive here is Thelwall's position as developed in *The Rights of Nature* (1796), a searing indictment of the "hireling" Burke whose *Letter to a Noble Lord* of the same year Thelwall considers the most disgraceful product yet of Burke's "pensioned indolence." For Thelwall, rights are coeval with divine creation itself, such that in endowing man with certain capacities or "means" the creator had implicitly vouchsafed their exercise as an inalienable right: "God created man also, a part of that universe, with all his wants and faculties; and by creating both the wants and the things wanted, HE dictated the rights by the

25. *AV*, 66–67. On Gewirth's (Kantian) attempt to tether rights to the capacity for rational agency, see Wolterstorff's critique, *Justice*, 335–340; echoing MacIntyre's critique of Gewirth, Wolterstorff rejects as "fake" or "pseudo-rights" many of the rights claims found in the Universal Declaration of Human Rights (adopted 10 December 1948). For in many cases (e.g., the right to being educated) "one does not have a right to the life-good" in question, simply because "there is no one against whom one has that right. In that situation, one is not wronged by not receiving such an education" (ibid., 315–317). Yet for Wolterstorff, it is important to distinguish between "inherent rights and human rights," the first of which is contingent on a human being's specific status whereas the latter is "essential to the human being who has that status." As he contends, much of the criticism (including MacIntyre's arguments in *After Virtue*) directed at modern rights theory has ignored that distinction and misconstrued all rights as "inherent."

26. *AV*, 67; MacIntyre's contention invariably (and perhaps uncomfortably) recalls Burke's and de Maistre's critique of modern rights as a frivolous attempt to exempt one's standpoint as a rational agent from the contingent flow of history. As de Maistre had so stridently put it, "the rights of the people are never written, or at any rate, constitutive acts or fundamental written laws are never more than declaratory statements of anterior rights about which nothing can be said except that they exist because they exist" (*Considerations*, 49).

means." Without any hesitation whatsoever, Thelwall here conceives rights as the exercise of *power,* unconstrained in the way it is discharged and without any obligation to give a reasoned account of why it is being exercised thus at a particular moment in time and under specific circumstances. Right here is but a projection of a voluntarist model of power: "Man is the sovereign; the material universe is the subject; his faculties are the powers by which he enforces his authority; and expediency is his rule of right. He is a despot, to the limit of his power, over the physical universe; and he has a right to be so." That the Jacobin sympathizer Thelwall should so unapologetically be channeling Hobbes, or foreshadowing the libertarian gospel of Ayn Rand, ought not surprise. Nor is the matter mitigated by Thelwall's attempt to institute a categorical limit to his equation of right with "expedient" power. For to argue that "this very right precludes him from despotising over his species: for the argument that applies to one, is of force for all, and to know the natural rights of others, it is only necessary to know our own" presupposes (but in no way argues) the *recognition of the other person's ethical and political reality.* Yet nowhere in *The Rights of Nature* do we find a warrant for why such recognition *ought* to precede the exercise of our innate powers/rights, let alone that it actually does.[27] Under the heading of sovereignty, the self's private passions and interests effectively *are* its reason. As a result, rationality takes on an exclusively presentist cast and is equally opposed to "theological dogma," received intellectual traditions, and "state absolutism."[28] Particularly in the writings of Hobbes and Hugo Grotius self-possession is *asserted*—though no longer argumentatively established within a supra-personal, normative framework—as the most elemental or "natural" right of all. Consequently, *ius* no longer denotes "what is 'right' or just, or a 'claim right' to justice, but active right over property. As the traditional link between person and ownership remains, this means that self-identity, the *suum,* is no longer essentially related to divine rational illumination, or ethics, but is a sheer 'self-occupation' or 'self-possession.'"[29]

27. Thelwall, *Politics of English Jacobinism,* 457–458.

28. E. Cassirer, *Enlightenment,* 238.

29. Milbank, *Theology,* 13–14; similarly, Joan Lockwood O'Donovan remarks how "the modern liberal concept of right belongs to the socially atomistic and disintegrative philosophy of 'possessive individualism' ... In this tradition the rights-bearing subject is conceived first and foremost as the immediate, exclusive proprietor of his/her physical and spiritual being and capacities, and derivatively as proprietor of those external objects necessary to their preservation and development. The rights-possessor is portrayed as sovereign over his/her human and non-human environment, in relation to which his/her orientation is controlling, acquisitive, and competitive. He/she is typically occupied in actions of disposing, using, exchanging, commanding, and demanding. The proprietary subject forms social and political relationships through the formal mechanism of the contract, modeled on an economic-legal transaction undertaken from calculations of self-interest" ("Natural Law," 20).

The consequences of Paine's and Thelwall's arguments are both conspicuous and alarming, particularly if one looks ahead to what is arguably their apotheosis: the uncompromising possessive individualism and principled a-sociality that has been enshrined as the political program of contemporary American libertarianism. Here human claim rights have contracted to the unlimited acquisition and possession of commodities, just as the distinction between goods and the good has been definitively erased. No doubt, Paine would have recoiled at the legally sanctioned defeat of public reason by private money that characterizes political and electoral practice in the United States today. Yet this outcome was all but inevitable once his assertion of claim rights as the individual's supposedly "imprescriptible" moral-cum-political estate had established itself as the baseline of modern liberal-republican thought. The contradiction here is that even as such claim rights are invested with seemingly timeless and unconditional force, the individuals advancing them refuse to specify a coherent, let alone normative vision of justice and goodness for the sake of which such claim rights are being invoked. Normativity has retreated to the subjective realm of negative liberty ("freedom from ..."), thus rendering the modern liberal increasingly agnostic and, in time, inarticulate about ends and goods relative to which such rights might plausibly function as appropriate means. It is telling that what renders Paine's assertion of non-contingent, universal rights so powerful (yet also so vulnerable) is precisely its refusal to acknowledge the specific historical moment in which that very claim is embedded and by virtue of which it becomes possible. Indeed, for a conception of universal (inherent) rights to be successfully enacted—as in the Déclaration des droits de l'Homme et du Citoyen (27 August 1789) or the Bill of Rights (15 December 1791)—it is imperative that the historical conditions enabling their conception and performative declaration be neutralized by their allegedly universal and timeless authority. Already for Grotius, "the concept of the law as such is not founded in the sphere of mere power and will but in the sphere of pure reason ... Natural law is not simply the sum total of that which has been decreed and enacted; it is that which originally arranges things. It is 'ordering order' (*ordo ordinans*), not 'ordered order' (*ordo ordinatus*)."[30]

Firmly committed to this self-authorizing mode of reasoning, Paine's and, nearly two hundred years later, Gewirth's assertion of the rights of man as an unconditional (if also strangely empty) "natural" fact is, however, weakened by the absence of any corresponding notion of the good or *telos* for the sake of which the rights in question have been conceived. As Milbank notes, "because it is rooted in an individualistic account of the will, oblivious to questions of its providential purpose in the hands of God, it has difficulty in understanding any 'collective making,' or genuinely social

30. E. Cassirer, *Enlightenment*, 239–240.

process."[31] Beginning with Paine, the language of inherent and claim rights effectively supplants, rather than complements, concerns with moral obligation such as had clearly dominated eighteenth-century philosophy. Making its appearance in European political thought of the late seventeenth century and dramatically flourishing a century later, modern rights theory reflects liberalism's virtual and performative (rather than concretely empirical and morally normative) outlook on politics; as a result, politics increasingly turns into a theoretical endeavor based on a small number of non-negotiable and (supposedly) non-contingent claims.[32] Burke was arguably on to something when indicting modern liberalism as a recklessly speculative venture whose only security is the (likewise speculative) hypothesis that some good will *eventually* derive from the acknowledgment of its abstract claims, even as no normative and practical notion of the good is ever identified *in the present*. For all its paranoid and extravagant rhetoric, the *Reflections* shrewdly identified and challenged a distinctly modern type of political argumentation that, in deploying the non-falsifiable idea of universal, inherent "rights," gives rise to a new model of agency—viz., an abstract self credited with reality independent of any specific human commitments and obligations and established as the sole proprietor, guarantor, *and* beneficiary of those very rights.

Curiously, Burke's misgivings about the apparent disconnect between modern claim rights and a normative conception of the good and reason (absent which we would be hard-pressed to exercise our rights in a meaningful fashion) are echoed by Godwin who, on this crucial issue, clearly breaks with his closest political allies. In so many words, Godwin flat-out rejects any notion of right as pure, unfettered, subjective discretion. As he insists, we only ever act *for a reason*, and by that token alone find our actions dictated *by* that reason: "There is no sphere in which a human being can be supposed to act where one mode of proceeding will not, in every given instance, be more reasonable than any other mode. That mode the being is bound by every principle of justice to pursue." Quite unexpectedly, then, a version of Aristotelian right reason has reappeared and here licenses Godwin's conclusion that the rational and the just cannot be understood as mere epiphenomena of subjective

31. *Theology*, 14.

32. Notoriously, MacIntyre insists that such claims must be classed as "one with belief in witches and unicorns." For just as with these fabled creatures, "every attempt to give good reasons that there *are* such rights has failed. The eighteenth-century philosophical defenders of natural rights sometimes suggest that the assertions which state that men possess them are self-evident truths; but we know that there are no self-evident truths" (*AV*, 69). For a standard account of modern rights theory, see Tuck, *Modern Rights Theories*, and, critical of MacIntyre, Wolterstorff, *Justice;* on natural rights theory in early Romanticism, see White, *Natural Rights*, esp. 1–40; for a concise survey of the evolution of rights from the thirteenth century onward and the recent project of "redeeming the 'Human' through Human Rights," see Assad, *Formations*, 127–158.

volitions, passions, or interest. Rather, the rational is that at which our judgment, however imperfect, must be aimed: "If then every one of our actions falls within the province of morals, it follows that we have not rights in relation to selecting them ... We are bound to regulate ourselves by the best judgment we can exert." By contrast, to assert a right to do as subjective discretion and ephemeral volition might dictate is to abandon one's capacity (and inherent obligation) to exercise judgment, that is, to make reasoned "choice." Consistent with Aristotle's *prohairesis* and Aquinas's *electio,* Godwin understands judgment and choice as circumscribed "by the immutable voice of reason and justice." The obverse case, in which a modern individual invokes the "liberty to regulate his conduct in any instance, independently of the dictates of morality," would, at best, be an "imperfect ... right, the offspring of ignorance and imbecility."[33] As Godwin sums up his case, "there cannot be a more absurd proposition than that which affirms the right of doing wrong." Rarely does Godwin seem more removed from the anarchist program that he is often credited with inaugurating than when, in classical realist and normative fashion, he tethers all human action to acts of judgment and choice that "derive [their] real validity from a higher and less mutable authority."[34]

While the convergence of Godwin's argument and terminology with that employed by writers whom he would ordinarily consider his intellectual and political opponents may be fleeting within the overall project of *Political Justice,* it is of significance nonetheless. For it highlights for us why a modern conception of natural and inherent rights remains incomplete unless it is correlated with a coherent account of judgment and action. Yet if we accept Godwin's objection to a purely voluntarist and discretionary model of rational agency, according to which a specific "right" seemingly relieves a human agent from giving rational account for her or his elections, it is clear that action cannot be pared down to the sheer implementation of subjective interests but requires a supra-individual conception of rationality—a normative framework such as allows movement from the sheer "right" to pursue action X toward showing how that action, rather than originating in some ineffable subjective preference, was chosen *for* a specific reason that can be made intelligible to others. Reason here is not simply the catalyst but the *telos* of action, which is both prompted by and undertaken for the sake of it. To act, then, is not just to do something *for* a reason but to give fuller shape and reality to reason (in the realist sense of *logos* or *ratio*). Once again, the Romantic era offers an unusually diverse range of voices and insights in this regard, with Coleridge (as remains to be seen) arguably the most probing of his generation. Yet we will begin, briefly, by entertaining

33. Godwin, *Enquiry,* 192–194.
34. Ibid., 197.

the seemingly counterfactual case of Blake, a thinker as far removed from the Aristotelian-Thomist tradition of philosophical realism and, it would appear, deeply committed to a particularly radical version of modern individualism.

Like many of the first-generation Romantics, in his early illuminated books Blake seeks to contest the seeming opposition and demonstrate the potential complementarity of the communal and the singular. Appalled by the syncretistic conception of art preached by Joshua Reynolds, whose mobilization of painting for political purposes suggests that he had been "Hired to depress art" (*CPP,* 635), Blake's own aesthetic "criticizes both the radical and conservative views of writing." Viewing formal mediations such as Reynolds's syncretist aesthetics or indeed the productions of commercial print culture with great suspicion, Blake intends to "heal the split between speech and writing, ... [and] close the gap between the pictorial and the linguistic use of graphic figures."[35] He "resolutely attack[s]" the Whiggish narrative of progress, represented by Adam Ferguson and Edmund Burke, which disaggregates "mental from manual labor by concentrating intellectual power in itself and delegating brute physical labor to its remoter appendages—human machines."[36] Yet Blake is no less estranged from the self-authorizing secularism and unimaginative "plain-language" ideal propagated by the bourgeois radicals of the Joseph Johnson circle. Often at his most incisive when thinking through the material and formal challenges faced by the human individual as it seeks to give expression to its radical singularity and spiritual potential, Blake proves a consummate organicist for whom individuality can only be realized in the most "minutely particular" and exact interpenetration of idea and medium. Individuality is "expression" and nothing else: "Ideas cannot be Given but in their minutely Appropriate Words, nor Can a Design be made without its minutely Appropriate Execution" (*CPP,* 576); everything else, of course, is mere copying. Vigorously opposing any conception of liberty as self-interested, rational choice or as elective *compliance* with generic social forms and practices, Blake's ideal is the "strong Man [who] acts from conscious superiority, and marches on in fearless dependence on the divine decrees" (*CPP,* 545), and whose "Vision" is strongly reminiscent of St. Paul's antinomianism.

Yet such a vision clearly challenges, perhaps even shatters any framework of collective responsibility and intelligibility. Blake's claim that it is solely "in Particulars that Wisdom consists & Happiness too" (*CPP,* 560), and that "to Generalize is to be an Idiot" is principally aimed at the modern notion of the "social" that has terminally blurred the line between the private and the public, interests and goods, the expedient

35. W. J. T. Mitchell, "Visible Language," 61–62.
36. Makdisi, *William Blake,* 124.

and the normative. By contrast, Hannah Arendt notes, ancient Greek culture understood "a life spent in the privacy of "one's own" (*idion*) and outside the world of the common, [as] idiotic by definition" (*HC,* 38). Indeed, Blake's conception of "the Human Imagination" as "Divine Vision & Fruition / In which Man liveth eternally" actually is fundamentally consonant with Arendt's understanding of the Aristotelian antithesis of public and private. For in enjoining us to "[d]istinguish therefore States from Individuals in those States. / States Change: but Individual Identities never change nor cease."[37] Blake is anxious to restore the classical Athenian ideal of *public* man, which is to say, to locate authentic and inalienable sources of meaning and so overcome the spurious sociality of the isolated, hedonistic self that Locke and Mandeville had established as the default model of human agency. Far from constituting the *goal* of its affective or economic development, the radical singularity of the Blakean subject is understood as the *source* of its artistic and spiritual strength as a public agent. Hence Blake's affirmation of the artistic temperament as radically singular and non-conformist, rather than conflicting with the classical ideal of the public as the locus of virtue, excellence, and meaning, directs its "honest indignation" at the inherently corrupt public/private distinction and at the procedural notion of sociality that "embraces and controls all members of a given community equally and with equal strength."[38]

Above all, then, art involves constant acts of strong, qualitative "discrimination" as Blake puts it in "A Vision of the Last Judgment" (*CPP,* 560). Opposed to Whigs and Tories alike, Blake's critique of empire, commerce, and commodity-art ultimately rests on an extreme anti-rationalism bound to disquiet even the most sympathetic liberal imagination today. Yet, like Schopenhauer's and Nietzsche's antirationalist arguments later in the nineteenth century, Blake's aesthetic pivots on the radical contingency of the self as *essential.* It resists the Circe-call of instrumental reason no less than it scoffs at the behaviorism of polite manners, commercialized culture, and Blake's contemporaries' class-specific notions of taste and sensibility. Not surprisingly, Blake's "Song of Liberty" in *The Marriage of Heaven and Hell* ties the achievement of genuine, spiritual freedom to the repudiation of precisely those tropes most strongly associated with British "liberty": *commerce* ("O Jew, leave

37. *Milton,* Pl. 32, in *CPP,* 132.

38. MacIntyre, *AV,* 41. On Blake's political vision, see Erdman, *Blake,* esp. 198–279; Makdisi, *William Blake,* esp. 16–77; on Blake's iconoclastic aesthetic in the context of the late eighteenth-century engraving and the commercialization of art vis-à-vis Blake's aesthetic iconoclasm, see Eaves, *Counter-Arts Conspiracy,* esp. 1–91; on the broader institutional and political contexts of late eighteenth-century art, see Barrell, *Theory of Painting,* esp. 1–67 and 222–257; on the rhetorical and figural strategies of Blake's prophetic idiom, see Tannenbaum, *Biblical Tradition,* 8–85 and 124–184; Balfour, *Rhetoric,* 127–172; McGann, *Social Values,* 32–49 and 152–172; and W. J. T. Mitchell, "Visible Language."

counting gold! Return to thy oil and wine"); *modern law* ("the son of fire ... stamps the stony law to dust"); *empire* ("Empire is no more!" [*CPP*, 44–45]). Blake's "Song of Liberty" grasps freedom as the spiritual contrary of the prevailing, strictly secular definition of "liberty" as the commercial and legal project of polite and commercial society and the modern nation-state (or empire). For Blake only genuine vision *is* freedom—realized under the aegis of strong artisanal *praxis* rather than conformist ("social") *behavior*. Blake's account of freedom rejects the false opposition that modern (Whiggish) reason has set up between oppression and emancipation, exclusion and inclusion, prejudice and rational transparency, servitude and rights. For all those generalizing, political-theory types of argumentation inevitably point back to the generalizing mechanism of the "stony Law" as the cause of spiritual and political oppression, even where they urgently draw on the law as a means for obtaining redress for political injustice. Still, in his political, aesthetic, and spiritual commitments, Blake remains deeply at odds with an Enlightenment culture bent on the dismantling and disavowing of all frameworks and traditions since for him freedom is not some agnostic and indeterminate state but a condition of plenitude that "Demands a firm & determinate conduct on the part of Artists" (*CPP*, 580).

More than anything, Blake is dismayed by the apparent evacuation of "action" and "character" from the kind of moral framework envisioned by Ferguson, Smith, and Steuart, as well as the concurrent loss of expressive intensity and commitment of the entire person that he sees perpetrated by the state-sponsored aesthetics of Reynolds & Co. What his stress on "vision" and "energy" opposes is the absorption of moral agency into something called "system," as well as the notion that value and meaning are to be achieved *procedurally*, that is, by *implementing* or *conforming to* a moral law that is little more than a composite of the kind of doing that comes naturally—that is, routine *behavior* that, as Kant puts it, encompasses "no more than what lies in the common moral order."[39] This is not the place to address the longstanding critique of Kant as moral formalist and rigorist whose "rule about universalizable maxims is useless without stipulations as to what shall count as a relevant description of an action with a view to constructing a maxim about it."[40] Certainly,

39. *Religion and Rational Theology*, 93.

40. Anscombe, "Modern Moral Philosophy," 2. The image of Kant as a strict formalist and rigorist (as G. E. Anscombe, Bernard Williams, and Alasdair MacIntyre, among others, had argued) has recently undergone extensive revision; for attempts to locate a "substantive value theory" in Kant's moral philosophy, and to bring transcendental reflection into alignment with the empirical practice of "character formation" and a quasi-Aristotelian initiation into the virtues, see Wood, *Kant's Ethical Thought* and, on Wood, Pippin, "Kant's Theory of Value." For a substantial, rather than formalist, understanding of moral agency in Kant's ethics, see Munzel, *Kant's Conception*, esp. 187–334; Herman, "Making Room for Character" and "Training to Autonomy." On Kant's relation to virtue ethics, see Herdt, *Putting on Virtue*, 322–340; Engstrom, "Happiness and the Highest Good"; and Sherman, *Making a Necessity of Virtue*, esp. 121–186.

that account has been subject to a sweeping revaluation of late, one that seeks to present a kinder and gentler moralist concerned with the (partial) rehabilitation of virtue, the empirical formation of moral character, and the articulation of a "substantive" rather than formal conception of value. While Kant will emerge as a crucial point of reference in Coleridge's critical survey of modern moral philosophy, our broader concern here is with how Smith's mimetic and proto-behaviorist account of moral sentiments had further eroded the meaning and significance of basic humanistic conceptions such as action, person, deliberation, judgment, responsibility, and self-awareness. Their disappearance from moral thought, already intensely disturbing to Blake, will also preoccupy numerous nineteenth-century novelists (from Jane Austen to Stendhal, Gustave Flaubert, George Eliot, Fyodor Dostoevsky, Theodor Fontane, and Thomas Hardy). Concurrently, the specter of a seriously atrophied moral vision, clarity, and articulacy also is of great concern to a remarkably wide spectrum of intellectuals throughout the century—from Heine to Emerson and Nietzsche, and from Coleridge to John Henry Newman, de Tocqueville, and Matthew Arnold, to name but a few.

To sharpen the point, we may recall Hannah Arendt's neo-Aristotelian account of action. "To act," she notes, prima facie "means to take an initiative, to begin ... to set something into motion (which is the original meaning of the Latin *agere*)." Action is never something derivative but positively originates a state. In ways that substantially elude the transactional and behaviorist model of the Scottish Enlightenment, action originates a new and distinct kind of awareness in the same way that "vision" and "making" are thoroughly fused in Blake's aesthetics. Action is never merely a "doing something" in the sense of being busy; nor is it merely the antonym of indolence or idleness (e.g., watching someone else do something or sleeping). Rather, action establishes the agent's involvement with the world in an original and transformative sense. As Arendt puts it, "action can be judged only by the criterion of greatness because it is in its nature to break through the commonly accepted and reach into the extraordinary, where whatever is true in common and everyday life no longer applies because everything that exists is unique and sui generis" (*HC,* 205). Transcending the world as the mere sum-total of facts or an inventory of the mundanely "given," action instead conceives of the world qua vision—that is, action drives toward the world's transformation in word and/as deed. Action pivots on an act of imagination that approaches "world" as a space of sheer possibility, of "play" and consequently of "risk"—a space whose openness the later Friedrich Hölderlin was to trope so poignantly as *das Offene,* albeit without recoiling from it as did Blaise Pascal.[41]

---

41. "*Le silence éternel de ces espaces infinis m'effraie*" (*Pensées,* no. 201, p. 66). On the relation between imagination and risk in a modern conception of "play" aimed at recovering rationality as

Fearful of the causal indeterminacy and unpredictability of action, and wary of committing to the good to be realized through action, the modern individual is prone to accept the world as something "objective," determinate, and non-negotiable. Lived existence thus is de-potentiated into mere "behavior," which acquiesces in facts and circumstances as objective constraints and thus exhausts itself in fine-tuning the myriad protocols (moral, administrative, economic, social, etc.) that govern the relation between self and world. By contrast, action aims to re-imagine and transcend that very relationship, for which reason Aristotle had already linked it to the notion of "excellence" (*aretē*). Similarly, the sociologist Arnold Gehlen links the possibility of action to a "motivational surplus" (*Antriebsüberschuß*), such that "only a being who ... has a surplus of motivation that extends beyond short-term gratification can turn his world-openness [*Weltoffenheit*] into something productive. The motivations for his actions come from outside himself. From the generative, social, and economic context, he creates more sophisticated tasks, which are then reflected objectively in the various social orders."[42] A distant conceptual echo of Aristotelian "excellence" (*aretē*), such motivational surplus explains why action (*Handlung*) ought not to be confused with a merely intentional doing of some kind; for action is not merely oriented toward the manipulation of some specific object or to the attainment of some contingent objective. Rather, *by its very nature,* action transforms the agent's very sense of "being-in-the-world." As Coleridge, Newman, and the great novelists of the nineteenth century grasp so well, the very idea of a *person* as a responsible and (potentially) flourishing being can only arise from action. For, in Arendt's pithy formulation, action "is not the beginning of something but of somebody" and, as such, involves the "disclosure of 'who' in contradistinction to 'what' somebody is" (*HC,* 177). Blake's gnomic pronouncement that "The most sublime act is to set another before you" (*CPP,* 36) furthermore hints that action by its very nature establishes the person's relation to the other, that its focus belongs to the eternal "openness" (and infinite responsibility) of the I-Thou to another person.[43] Hence, Arendt stresses, we must resist the temptation of construing action in terms of efficient, instrumental causality "as a means to an end" or "as a willful purpose" (*HC,* 179).

Unlike mechanical causation, then, action discloses "the unique and distinct identity of the agent." Indeed, "without a name, a 'who' attached to it, [action] is

an *emergent* property and at salvaging reason from the voluntarist fate of a merely ascriptive value, see Pfau, "Appearance of *Stimmung.*"

42. Gehlen, *Man,* 50 (trans. modified)/Ger. *Der Mensch,* 58.

43. As so often, Blake's proverb is deliberately ambiguous in that it also allows "another" to mean "another [act]" rather than "an Other." In that case, the gnomic enjoinder to set "another [act] before you" might ask us to focus on "act" as something more than intention or self-realization, such that the act realizes the ethical substance of the agent, not the other way round.

meaningless" (*HC,* 177–181). Such a model of action has several important characteristics: (1) it transcends the matrix of our present socialization by reimagining community, justice, and human flourishing as anterior to and of more elemental reality than mere subjectivity; (2) it shows "person" or "character" (*ēthos*) to be anterior to self *and,* importantly, sees the meaning of personhood disclosed only in *narrative,* albeit one of which we are "not an author or producer" (*HC,* 184); and (3) the narrative dynamics of action and the life whose contours it fills in are not only unpredictable but will prove only partially legible to the individual agent. All this does not mean, however, that action is free of internal conflict. Thus, as MacIntyre has shown, the heroic model of action first articulated in Homer produces strong contradictions between what are to count as "goods of excellence" and "goods of efficiency."[44] For the sociologist Arnold Gehlen, whom Arendt here cites in an approving footnote, it is above all the concept of *Handlung* ("action") that lays bare the "unique biological position" (*biologische Sonderstellung*) of human beings; Gehlen's project is to identify "the common root of knowledge and action" that coordinates the entire spectrum of human engagement from the motor system to esoteric speculation. Rejecting any hierarchical schemes that divide human activities into "higher" or "lower," Gehlen defines man as "the acting being" (*das handelnde Wesen*). Yet precisely for that reason, man remains necessarily indeterminate and a constant challenge (and potential liability) to himself: "One might also say he is a being who must establish a stance [*stellungnehmend*]. Actions are the expression of man's need to achieve a reasoned hermeneutic view on the outside world," a constant and burdensome existential requirement inasmuch as human beings lack "organic means and instincts." While Gehlen's thesis as to man's "lack of instinctual guidance" (*Instinktentbundenheit*) was certainly challenged by some (eminently by Konrad Lorenz), the basic argument seems sound—viz., of the human being's "unique position" (*Sonderstellung*) in that it is constitutively obligated "to develop its potential" and meet "the challenge of interpreting its own existence."[45] Action is "the key to understanding human impulses" because of its "hermeneutic indeterminacy" (*weltoffen*), thus confirming the foundational importance of interpretive and imaginative activity: "between elemental needs and their external gratification … [there] is interpolated the entire system of world orientation and action."[46]

---

44. *Whose Justice?* 12–46.

45. *Man,* 24, 27/Ger. *Der Mensch,* 32, 34 (trans. modified). Gehlen later qualifies his view of human beings as largely devoid of "instinctual guidance," arguing instead for "a reduction" (*Instinktreduktion*) such that "the instinctual residues found in human beings exhibit high degrees of plasticity and fungibility [*in hohem Grade plastisch und verschmelzbar*] and, to use Freud's expression, are 'convertible'" (*Urmensch und Spätkultur,* 149; trans. mine).

46. Ibid., 45/Ger. 53. Approaching the same issues from the perspective of analytical philosophy, Michael Thompson reaches strikingly similar conclusions; see *Life and Action,* esp. 25–82;

There is reason to view the demise of "action" as a meaningful category not only as a peculiar entailment of Enlightenment moral philosophy but as the final phase in modernity's progressively more stunted conception of the will. What distinguishes the late phase of this development (whose beginnings we had located in Ockham's voluntarism) is the divergence of two distinctive strands within political philosophy that correlate with the emergence of the nineteenth-century nation-state. The first of these continues the liberal Enlightenment project, albeit in a more technocratic and institutionally based form, and it is represented by G. W. F. Hegel, Auguste Comte, Bruno Bauer, Thomas Macaulay, Herbert Spencer, John Stuart Mill, and Max Weber, among others. The other strand rejects the meliorist or utopian aspirations of the Enlightenment by reviving Hobbes's voluntarism and taking his mechanistic conception of the will as the premise for a sweeping philosophical and cultural pessimism. This resurgence of Hobbesian positions is often accompanied by a neo-pagan revival of ancient Greek concepts of necessity, fate, and tragic action. The central representatives of this intellectual orientation—effectively a roll-call of modern anti-liberalism—are de Maistre, Chateaubriand, Schopenhauer, Wagner, Carlyle, Jakob Burckhardt, Hyppolite Taine, and the young Nietzsche, who has not yet discovered the idea of an "overcoming" of man and the "transvaluation" of history's pseudo-rational values. In time, they are succeeded by the young Thomas Mann, Oswald Spengler, and a host of poets and intellectuals (Friedrich Gundolf, Stefan George, Carl Schmitt, Gottfried Benn) some of whose fictive counterparts we find assembled as the ominous clique of Munich's right-wing intelligentsia in Chapter 34 of Thomas Mann's *Doktor Faustus* (1947). In their grandiose and implacable anti-rationalism—"not a one believed any longer in 'free institutions'" and all insisted that Europe was doomed "if one did not simply toss all that emotional stuff about human rights overboard from the start"—they are distant but unmistakable descendants of Schopenhauer.[47]

The anti-liberal implications of Schopenhauer's project did not fully register until after 1848, at which point a failed revolution and the rise of Otto von Bismarck's neo-Machiavellian politics fueled the dystopic narratives of modernity that pervade the writings of Carlyle, Jacob Burckhardt, Taine, Arnold, the early Nietzsche, Ludwig Klages, Georg Brandes, Oswald Spengler, Carl Schmitt, and indeed the young Thomas Mann himself. Their consistent focus is on the stunted intellectual culture of the modern person—formally rational but of dissociated sensibility; politically

---

likewise, David Burrell explores how the Thomistic notion of *actus* as "intentional activity" unfolds "in conceptual independence from accomplishment" and, for that very reason, "opens *actus* to the range of uses it enjoys" (*Aquinas,* 187); see also Blondel, *Action,* arguably still the most comprehensive and thoughtful inquiry into what he convincingly portrays as a basic phenomenon of human existence.

47. Mann, *Doctor Faustus,* 385.

dependable but lacking inner goals; reliably productive but denuded of introspective tendencies—a self plagued by proto-existentialist indifference. Diagnosing such trends in mid-Victorian culture, Elaine Hadley has observed how the quest "for a less-intense, less saturated political domain" rests on liberal principles "that paradoxically pacify the liberal subject in unplanned ways."[48] Similarly, Colin Gunton has rejected the prototypical narrative that liberal-secular modernity tells about itself: viz., that it has replaced the hegemony of the one (transcendent) God with the lateral and pluralist community of the many. In fact, Gunton notes (drawing on voices as disparate as Wordsworth, Kierkegaard, and Mill), modern "individualism is a non-relational creed, because it teaches that I do not need my neighbour in order to be myself." Such an "eschatology of the impersonal" can only furnish the "flat unity of homogeneity" that would be ruthlessly exploited by a modern totalitarianism whose genesis Hannah Arendt had earlier traced.[49] A different narrative than the one here being pursued might well turn up significant affinities between Schopenhauer's and Marx's account of modernity, specifically as regards the question why during those precarious years between 1848 and 1852 both the German and French bourgeoisie were so ready to betray their avowed moral and political ideals for real (economic) interests, thereby revealing the Enlightenment utopia of a rational, deliberative, and transparent commitment to a liberal polity to be unsustainable and likely chimerical.

As remains to be seen, the limitations of Schopenhauer's pessimism and extreme voluntarism tend to become apparent when juxtaposed to the conception of will and person that Coleridge works out in his later writings. Even so, Schopenhauer's dystopic vision of human agents terminally encased in the windowless noumenon of the will furnishes some nineteenth-century novelists with a compelling thematic provocation, perhaps none more than Stendhal for whom the transition from pessimism to a modern existentialist and nihilist stance seems quite effortless and full of

---

48. *Living Liberalism,* 39; Hadley's thesis helpfully corrects Amanda Anderson's generally positive view of nineteenth-century liberalism's cultivation of "impartiality" and "detachment" (*Powers of Distance,* 3–33).

49. Gunton, *The One, the Three, and the Many,* 32–34; Kierkegaard's horror of the quintessentially modern "phantom" of the "public ... made up of unsubstantial individuals who are never united ... and yet are claimed to be a whole" suggests "that when God is displaced as the focus of the unity of things, the function he performs does not disappear, but is exercised by some other source of unity—some other universal" (ibid., 30–31). Kierkegaard's (and Mill's) alarm at mindless conformism routinely mistaken for an expression of unconstrained selfhood (albeit *en masse*) is anticipated by Coleridge, who in 1816 characterizes Britain as a "busy ant-hill in calm and sunshine. By the happy organization of a well-governed society the contradictory interests of ten millions of such individuals may neutralize each other, and be reconciled in the unity of the national interest." Raising the obvious question: "whence did this happy organization first come?" Coleridge adduces various "misgrowths," including the triumph of a "mechanic philosophy, ... an unenlivened general understanding, ... and a Reading Public ... diet[ing] at the two public *ordinaries* of Literature, the circulating libraries and the periodical press" (*CLS,* 21, 28, 36–38).

sinister hilarity. One might recall Julien Sorel's dogged attempt to inspire jealousy in Mathilde de la Môle by copying out reams of love letters to the wholly unattractive Mme de Fervaques. Following some Russian prince's manual for seduction by mail, Julien overcomes excruciating boredom as he transcribes a first love letter "full of virtuous phrases, and killingly dull, [with] several sentences nine lines long." Faintly bemused by the utter insincerity with which he directs these generic missives to "so celebrated a font of virtue," Julien slips into yet another one of his existentialist reveries: "I will be treated with the utmost scorn, and nothing would amuse me more. At bottom it is the only comedy I could appreciate. Yes, to cover the odious object that I call *myself* with ridicule would divert me. If I believed in myself I would commit some crime or other for the sake of amusement."[50] As so often, there is something unrelenting and cruelly transparent about the consciousness of Stendhal's protagonists. Barely sustained by derivative effusions of love to be tendered to a woman for whom he feels nothing, Julien appears crushed by self-loathing, itself the result of an acute sense of the impossibility of meaningful action. Yet Stendhal also shows how the apparent disappearance of meaningful political and social action atrophies several closely related conceptions, including those of person, conscience, and indeed the very possibility of meanings uncontaminated by (self-)interest. More than anything, Stendhal's proto-nihilism stems from his diagnosis of the three basic institutional frameworks—church, state, and family—as irremediably corrupt and broken, thus presaging critiques of the modern, liberal, and secular nation-state's profound loss of moral orientation and authority that were to take center stage decades later.

In his own time, Stendhal—whom Nietzsche was to credit with having "the most thoughtful eyes and ears ... of this century"—appears a prophet of institutional collapse even as Saint-Simon, Comte, Hegel, and the young Mill are engaged in making the best case yet for the necessity of modern secular institutions.[51] That such arguments were urgently needed was certainly apparent, and it has much to do with modernity's progressive disaggregation of reason from nature, and of the will from reason that Hegel in particular had diagnosed as a dangerous legacy of the Enlightenment. This is not the place to take up Hegel's response—consummately set out in his *Philosophy of Right*—to the question that Kant's project had left essentially unresolved: viz., how to imagine a rational community that is neither simply wedded to the empirical status quo and the reactionary preservation of group-specific interests nor in denial about the substantial miscarriage of the French Revolutionary project.[52]

---

50. Stendhal, *Red and the Black,* 424.

51. Nietzsche, *Gay Science,* no. 95 (p. 92).

52. See Pippin, *Hegel's Practical Philosophy,* esp. Part II (121–179) on the psychological and social dimensions of the will; Hegel's grim diagnosis of the French Revolution prompts him to view the modern state "as an organic whole; it cannot be seen simply as an aggregation of its elements, be

How, in other words, can we ever hope to mediate the "is" of the empirical world with the "ought" of a fully rational and moral community? The late Kant's utopian claim that rational morality "cannot be effected through gradual *reform* but must rather be effected through a *revolution* [*Sinnesänderung*] in the disposition of the human being" such as will give rise to "a new man" certainly shows the scope of his ambition.[53] Yet regardless of whether Kant ever succeeded in specifying the pedagogical and institutional process whereby that vision might be realized, there is mounting evidence that beginning with the 1830 Revolution in Paris, this project had effectively foundered. Even if Kant is no longer read as the moral rigorist who seems to present himself on so many pages of his *Grounding for the Metaphysics of Morals* and the second *Critique*, it was this reading that informed the generation that succeeded him. Among them, Schiller and Hegel were especially vexed to find robust conceptions of action, character, judgment, and self-awareness seemingly vanquished by the abstract formalism of a "pure will wholly cleansed [*völlig gesäubert*] of everything which can only be empirical."[54] It was in Hegel's mature work above all that an answer was formulated under the heading of speculative dialectics, itself deemed uniquely capable of grasping rationality as a process rather than as a set of abstractions imposed *ab extra*. As is well known, doing so meant above all to conceive philosophy as a meta-narrative of reason whose dialectical "movement" (*Bewegung*) is fueled by deficient conceptions and articulations such as cannot but produce their own negation and correction over time.

Yet what distinguishes Hegelian dialectics from the meliorist narrative of Scottish political economy is his far more complex understanding of "mediation" (*Vermittlung*). Unlike the meta-agency of the market—which in Smith and Steuart functions as an inherently protean, albeit severe superego—Hegelian mediation does not merely *contain* contingent and insufficiently conceptualized individualities. It becomes a subject in its own right, and we know it by the name of "institution." A number of shifts in political history, theory, and demographics at the beginning of the nineteenth century—some coincidental, others causally linked—contribute to the emergence of strong national institutions as the most promising material and conceptual new strategy for containing a rigidly individuated notion of the will. Perhaps the most striking example here involves Prussia's sudden transformation into a modern nation-state following the collapse of the old, superannuated Reich in 1806 and

---

these groups or individuals ... If we start with men fractioned in individual atoms, no rational state or indeed common life will be possible" (C. Taylor, *Hegel*, 439).

53. *Religion within the Boundaries of Mere Reason*, as translated in *Religion and Rational Theology*, 113; an alternate rendition of Kant's pivotal term, *Sinnesänderung* (*Werke*, vol. 8: 727) would be "conversion."

54. Kant, *Grounding*, 2.

the sweeping territorial reorganization of German-speaking Europe under Napoleon. Concurrent with its prolonged struggle against Napoleonic occupation—and ultimately the foundation for Napoleon's final defeat in 1815—Prussia's aggressive, centralized modernization of its economic, political, and military structures mirrors the concurrent ascendancy of utilitarianism and its legitimation of an impersonal and systemic conception of state power administered by a new type of post-feudal, professional bureaucrat. All these shifts—well-documented by social historians—ultimately rest on one premise that was to define the nineteenth-century nation-state above all: viz., the idea of *institution* as the one truly indispensable, because impersonal, source of social, cultural, and indeed moral meanings.[55] As Arnold Gehlen's brilliant analysis of the logic of modern institutions suggests, a consistent effect of institutions is their systematic atrophying of basic conceptions of agency by way of "autonomizing and habitualizing entire clusters of motives and complex practices; yet also by virtually or metonymically allowing objectives to be supplanted." Defining of a life circumscribed by institutional frames, we observe "that self-certifying autonomy [*selbstzweckhafte Eigengesetzlichkeit*]" and a relentless "proceduralism [*Betrieb*]" whose "supra-personal coordination [*überpersönliches Gefüge*]" will gradually infiltrate and hollow out the conceptual legacy of classical humanism with its stress on contemplation, deliberation, judgment, choice, and individual responsibility.[56] It is this relocation of the agency most centrally responsible for establishing socially relevant meaning (*Sinnstiftung*) to institutions that shows nineteenth-century liberalism addressing the oblique and adventitious logic of Smith's sympathetic community.

Yet insofar as the enlightened, deliberative, and autonomous individual is, paradoxically, both the theoretical premise and the empirical beneficiary of modern (post-Kantian) liberalism, philosophers and intellectual historians of various persuasions (from Hans-Georg Gadamer to Alasdair MacIntyre, Charles Taylor, John Milbank, Louis Dupré, and Brad Gregory) have questioned liberalism's apparent inarticulacy about both its core presuppositions and its ends. Arguably, these failures are to be expected in a stance whose relation to the past is largely one of negation or emancipation and whose speculative outlook on the future is constrained by its present(ist) agendas. Once again, Tocqueville appears especially alert to the incoherence of a political philosophy prone "to take tradition only as information" and arrogating to itself the power of defining and managing what shall count as objective reality. For one thing, such a project (presumably to be launched anew by each suc-

---

55. For standard accounts, see Sheehan, *German History*, esp. 207–388 and 451–487; Wehler, *Deutsche Gesellschaftsgeschichte*, 174–230 and 297–457; and Nipperdey, *German History*, 223–236 and 560–578.

56. Gehlen, *Urmensch und Spätkultur*, 38 (trans. mine).

cessive generation) would seem terribly inefficient: "I find that dogmatic beliefs are no less indispensable for him to live alone than to act in common with his fellows. If man was forced to prove to himself all the truths that he uses every day, he would never finish doing so . . . There is in this world no philosopher so great that he does not believe a million things on the faith of others, and who does not assume more truths than he establishes."[57] Yet in its peremptory view of all political and social meaning as necessarily secular, "democracy diverts the imagination from everything that is external to man, in order to fix it only on man" and, not coincidentally, also "dries up most of the ancient sources of poetry."

A narrowly procedural and institutional conception of rationality—both achieved by and confined to the exigencies and objectives of the present generation—threatens to atrophy the individual of any capacity for transcendence. Tocqueville here does not necessarily have in mind overtly religious or metaphysical matters (on which he can sound surprisingly equivocal) but, more immediately, human beings' distinctive capacity for suspending the immediacy of problems and desires by framing them in a matrix of goods and norms such as can be expected to benefit future generations. Yet because it rejects all transcendence as both conceptually indemonstrable *and* a threat to the fetish of autonomy, modern liberalism typically punts on questions of intergenerational symmetry and, thus, operates with a dramatically impoverished conception of time: "As soon as [people] have lost the custom of putting their principal hopes in the long run, they are naturally led to wanting to realize their slightest desires without delay, and it seems that, from the moment they lose hope of living eternally, they are disposed to act as if they had only a single day to exist ... The instability of the social state comes to favor the natural instability of human desires. In the middle of these perpetual fluctuations of fate, the present grows; it hides the future that fades away, and men want to think only about the next day."[58] Here, then, the aristocratic Tocqueville and the Lambeth radical Blake stage a meeting of true minds; for both understand action as inseparable from transcendence, just as transcendence cannot be achieved without vision. The alternative, already apparent to Blake as he struggles against the usurpation of art by commerce, is a hidebound and self-interested, petit-bourgeois mentality whose relationship to "eternity" (an idea of obvious centrality to Blake) is neither rational nor irrational but, if anything, minimally articulated as skeptical prevarication or agnostic indifference.

That liberalism might suffer from an inadequate grasp of its metaphysical assumptions and of supposed "ends" should not surprise; for neither can easily be articulated by human agents whose intellectual range and curiosity are overwhelmingly defined by

57. Tocqueville, *Democracy*, 3:699, 713–715.

58. Ibid., 3:835, 966–967.

formal procedures and present exigencies. If anything, these strictures on warrantable assertions are more pronounced in the neo-Kantian systems of the late nineteenth century (Heinrich Rickert, Georg Simmel) wherein "values are described as 'irreal,' and the term 'good' for moral value is avoided, precisely because of the traditional metaphysical implication of convertibility with *ens*."[59] The notion of "freedom" that is eponymous to modern liberalism turns out to be *not* coextensive with rationality. On the contrary, freedom presupposes an agent's assent (itself inaccessible to rational discipline) to the reality and apparent significance of phenomena before they can be scrutinized by means of inferential and propositional reasoning. As the radically contingent ground of reason itself, this uniquely human capacity for what Newman would call "real assent" acts as a crucial constraint on the utopian aspirations of Enlightenment rationalism and liberalism. Where it is correlated with assent, judgment, and choice—viz., as acts undertaken by a responsible will— human freedom cannot be reduced to a function of subjective (and likely ephemeral) preference for this or that value or meaning. Rather, it presupposes an intuitive and unconditional commitment to supra-personal ends to be sifted and internalized with increasing clarity by our discursive understanding and deliberative judgment. Newman's *Grammar of Assent* (1870) takes such commitment to be a case of assent, as opposed to inference. Being "in its nature absolute and unconditional," assent commits us to a "view" and so shows mind and world to be engaged and enmeshed from the outset. Assent thus contrasts with the methodical prevarication of the Cartesian *cogito,* which is limited to inferences and "notional assent" because it regards knowledge to be exclusively propositional in nature.

Defined by a perpetual fear of error, the Cartesian subject does not understand knowing as action but, ever so tentatively, as *re*action to phenomena that have been inexplicably received. The opening stance of distance, detachment, and distrust may in time result in the kind of epistemic certainty afforded by some (putatively warranted) act of inference. Yet whereas inference is necessarily confined to the realm of the notional and can only ever be "an acceptance of a proposition," it presupposes what Newman calls "real assent," which is to say our "acceptance of the premisses" of that very proposition. With its "palpable philosophical paranoia," the modern turn that René Descartes is often said to have inaugurated thus "appears locked in a kind of self-created vacuum, determining by argument or reason a method for making claims about the world, but unable to argue convincingly that what results is anything other than what the method tells us about the world, be the real world as it may." In fact, the triumph of Cartesian method as the sole warrant for propositions about the world made it exceedingly difficult "to get us where we wanted to go, 'back'

59. Milbank, *Theology*, 77.

to the world we suspended in the moment of doubt."[60] Yet the appearance that is to be subject to the Cartesian regimen of doubt must, for that to happen, first appear—not as an object or some*thing* given, but as what Jean-Luc Marion calls the phenomenon's sheer "givenness" (*donation*): "this movement of imposing itself on me, of arriving upon me from before or in front of me." Prior to all talk of subject, object, certainty, doubt, proof, and so forth, we "can therefore legitimately posit that *the phenomenon gives itself*."[61] It is this positing that Newman had theorized under the heading of "real assent" in the late 1860s. At that time, both Newman and Gerard Manley Hopkins (then studying at the Birmingham Oratory) anticipate precisely this phenomenological turn, which seeks to lead the modern self out of the methodological dead-end into which it had been led by Descartes. In their sheer givenness and organized presence (what Hopkins develops under the heading of "inscape" and "instress"), *images* above all command our real assent, though Newman acknowledges that assent here "is no warrant for the existence of the objects which those images represent."[62] Anticipating developments in modern phenomenology, Newman's distinction between notional and real assent thus resolves itself into that between "certitude [, which] is a mental state," and "certainty [, which] is a quality of propositions."[63] Though principally concerned with epistemology, Newman's *Grammar* also points to the weakness of a strictly procedural or methodological understanding of rational agency. In so doing, his argument continues a line of questioning begun by the Romantics who had variously taken exception with the Enlightenment's negative definition of freedom as the ability to choose *un*constrained by emotive bias, political authority, or transcendent norms—be they inherited or privately conceived.

Alternatively, a critique of Enlightenment freedom may take a mystical turn, such as when in his 1809 treatise Schelling insists that "only out of the darkness of unreason (out of feeling, out of longing, the sublime mother of understanding) can clear thoughts arise."[64] With his residually Gnostic conception of freedom as "a power for evil," Schelling is only the first of several major philosophers to argue that the Enlightenment utopias of laissez-faire liberalism, state-sponsored utilitarianism, or Kantian cosmopolitanism rest on decidedly shaky foundations.[65] Echoing a point previously made in Burke's *Reflections* and in Goethe's *Wilhelm Meister*—viz., the asymmetry between freedom and rationality—Schelling's famous characterization of

60. Pippin, *Modernism*, 23, 25.

61. *Being Given*, 63, 68.

62. Newman, *Grammar of Assent*, 135, 76, 80. On the phenomenology of the image, with specific reference to G. M. Hopkins and the work of Jean-Luc Marion, see Pfau, "Rethinking the Image."

63. *Grammar of Assent*, 271.

64. Schelling, *Philosophical Inquiries*, 35 (trans. modified).

65. Ibid., 28.

freedom as the "irreducible remainder that cannot be resolved into reason" prepares the ground for Schopenhauer's *Prize Essay on the Freedom of the Will* (1839) and Nietzsche's vehement indictment of free will in *Beyond Good and Evil* (1886), which warrants quoting in full:[66]

> I feel … an *obligation* to sweep away a stupid old prejudice and misunderstanding about all of us that has hung like a fog around the concept of the "free spirit" for far too long, leaving it completely opaque. In all the countries of Europe, and in America as well, there is now something that abuses this name: a very narrow, restricted, chained-up type of spirit whose inclinations are pretty much the opposite of our own intentions and instincts (not to mention the fact that this restricted type will be a fully shut window and bolted door with respect to these approaching *new* philosophers). In a word (but a bad one): they belong to the *levelers,* these misnamed "free spirits"—as eloquent and prolifically scribbling slaves of the democratic taste and its "modern ideas." They are all people without solitude, without their own solitude, clumsy, solid folks whose courage and honest decency cannot be denied—it's just that they are un-free and ridiculously superficial, particularly given their basic tendency to think that *all* human misery and wrongdoing is caused by traditional social structures: which lands truth happily on its head! What they want to strive for with all their might is the universal, green pasture happiness of the herd, with security, safety, contentment, and an easier life for all. Their two most well-sung songs and doctrines are called: "equal rights" and "sympathy for all that suffers"—and they view suffering itself as something that needs to be *abolished.*[67]

A searing indictment of liberalism's axiomatically progressive view of history, the passage reaffirms Nietzsche's fundamentally tragic conception of life as suffering. Not to suffer, it is intimated, would mean being deprived of memory and, ultimately, of "life" itself—a notion that, more than any other modern thinker, Nietzsche tries to rehabilitate as the center of philosophy. Notably, life and consciousness pivot on solitude (*Einsamkeit*) as the apparent basis of all introspection. Rather more surprising in light of Nietzsche's own theses regarding the "overcoming" (*Überwindung*) of man, his argument here also rejects "modern ideas" inasmuch as they purport to improve the human condition by analyzing and gradually transforming "traditional social

---

66. On Schopenhauer, see Pfau, "The Melancholy Gift," and Safranski, *Schopenhauer,* 307–326. In sharp contrast to Kant's notion of freedom, autonomy, and duty, Schopenhauer's account of human freedom rests on the strictly non-cognitive, indeed noumenal causality of the "will." See also below, 471–477.

67. Nietzsche, *Beyond Good and Evil,* no. 44 (pp. 40–41).

structures." Covetous of "security, safety, contentment, and an easier life," modern social psychology is defined, in Nietzsche's view, by its fundamentally escapist or "aesthetic" mode of cognition. The ultimate opiate thus turns out to be "freedom" itself inasmuch as the self's vaunted emancipation from past values and traditions seeks to attenuate the variously bracing or exhilarating, and ever unpredictable nature of life and action. Not without reason, Freud would eventually come to think of modern consciousness as shaped by the overwhelming desire *not* to be stimulated, know, or remember. His vituperations against Augustine elsewhere notwithstanding, Nietzsche here appears as a moralist keen to recover suffering and memory as integral conditions of personhood.[68] His "understanding of enhanced life, which can fully affirm itself, also in a sense takes us beyond life; and in this it is analogous with other, religious notions of enhanced life."[69] In particular, Nietzsche's amalgamation of life to a strong model of "action" (*Handlung*) and "deed" (*Tat*) bears an unmistakable resemblance to Augustine's conception of the divided will, itself a powerful catalyst of suffering and introspection.

Nietzsche's passage is but a particularly strident instance of nineteenth-century writing revealing the experience of freedom as a metaphysical burden, a point echoed by major studies of the concept since.[70] Writing in the momentous year of 1871, Nietzsche's erstwhile teacher Jacob Burckhardt remarks on how, "for two hundred years, people in England have imagined that every problem could be solved through Freedom, and that one could let opposites correct one another in the free interplay of argument." Yet the result has been "a complete disintegration of the idea of authority," as well as the apparent downward transposition of the "idea of goodness ... into the idea of progress, i.e., undisturbed money-making and modern comforts." Notwithstanding the universal, "merciless optimism," he continues, "our assumption that we live in the age of moral progress is supremely ridiculous," being constantly belied by "our vulgar hatred of everything that is different, of the many-sidedness of life."[71] Burckhardt's despondent summation—only "turpitude is immortal on earth"[72]—is famously echoed by Dostoevsky's Grand Inquisitor. For as Ivan Karamazov's famous parable argues, Christ's exemplary rejection of the temporal trappings of power and

---

68. On Augustine, see esp. Nietzsche, *Anti-Christ*, no. 59 (p. 63), as well as his vituperative characterization of the *Confessions* in a letter to Overbeck (31 March 1885), *Sämtliche Briefe* 7:34.

69. C. Taylor, *Secular Age*, 374.

70. Nancy, *Experience of Freedom*, 44–59; Bieri, *Handwerk*, esp. 29–83.

71. Burckhardt (2 July 1871), in *Letters*, 143; last quote from Burckhardt, *Reflections*, 103. In a late letter (17 March 1888), Burckhardt sharpens the point: "Democracy, to be sure, has no sense for the exception, and when it can't deny it or remove it, hates it from the bottom of its heart. Itself the product of mediocre minds and their envy, Democracy can only use mediocre men as tools, and the ordinary careerist gives it all the guarantees it can desire of common feeling" (*Letters*, 225).

72. *Reflections*, 241.

authority ("earthly bread") "in the name of freedom and heavenly bread" only intensified the susceptibility of man to material and ideological temptation: "Now see what you did next. And all again in the name of freedom! I tell you that man has no more tormenting care than to find someone to whom he can hand over as quickly as possible that gift of freedom with which the miserable creature is born."

Just what it is that this vexing freedom consists of can only be inferred from the ways in which, according to Ivan's parable, humans proceed to divest themselves of it: "give them bread and he will bow down to you, for there is nothing more indisputable than bread. But if at the same time someone else takes over his conscience—oh, then he will throw down your bread and follow him who has seduced his conscience ... For the mystery of man's being is not only in living, but in what one lives for."[73] At a more abstract, philosophical level, Ivan Karamazov's parable merely reiterates that the choice was never between the terrifying reality of humans as a predatory species and the Nietzschean "pasture-happiness" supposedly guaranteed by modern liberty, rule of law, and economic prosperity. For it is the latter dispensation that, as Hegel argued, brings the "terror" (*Schrecken*) of "absolute freedom" into full view and thus confronts modernity's disembedded individual with its own isolated and unfathomable singularity.[74] In rather more forthright terms, Hannah Arendt's closing paragraph to *The Human Condition* draws out the central implication of Dostoevsky, Nietzsche, Burckhardt, Eliot, and a host of other eminent intellectuals; as she so laconically puts it, "it is in fact far easier to act under conditions of tyranny than it is to think" (324).

Arguably, it is one of the main achievements of early Romanticism to have located tyranny not merely in the caprices and excesses of the ancient régime or the younger William Pitt's repressive domestic policies but also in the oblique coherence of social conventions, customs, and manners that comprise the humdrum life of the modern petit-bourgeois individual. In so doing, writers like Blake, Coleridge, and Goethe alight on another "unapproachable freedom from which thought itself proceeds," a freedom that can be claimed neither as the ground nor as the object of instrumental thought.[75] Consider Blake's careful parsing of the false opposition between Bishop Watson's *Apology for the Bible* and Paine's *Age of Reason,* Coleridge's brilliantly ambivalent rumination of "free will" in *Aids to Reflection,* John Keats's "negative capability," or Hölderlin's poetic image of "openness" (*das Offene*): in each case, thinking is suffused with creative imagination—that is, a counterfactual, non-

---

73. *Brothers Karamazov*, 254.

74. Hegel, of course, only highlights the "bad infinity" of a freedom thought merely as the absence of all constraint in order to promote his own conception of freedom as the reflexive determination of substance as the universal; see Nancy, *Experience of Freedom*, 5.

75. Ibid., 17.

linear, non-instrumental, and necessarily provisional mode of being. Traversing uncharted and unpredictable terrain, freedom reveals existence to be always something more than mere facticity, self-possession, certainty, or righteousness. We are drawing close here to Heidegger's conception of truth as the dis/closure and un/veiling (*aletheia*), and of lived existence as characterized by Meister Eckhart's *ek-stasis*. Freedom, on that view, "is that which, in thinking and of thinking, must, *simply in order to think*, tend in spite of everything toward a liberation as well as toward the very reality of the existence that is to be thought of. Without this, thinking would have no meaning. All thought, even when skeptical, negative, dark, and disabused, if it is *thought*, frees the existing of existence."[76]

Long before Freud was to make the point in his late work, nineteenth-century literary and philosophical narratives explore just how the infinitely complex textures of social life and religious culture offer humans refuge from the terrifying enigma of freedom. Among the most astute analyses of a social and cultural matrix subtly denuding the individual of its inscrutable singularity and freedom is George Eliot's novelistic oeuvre. Particularly her late novels offer an abundance of examples of quasi-unconscious patterns of socialization and moral inarticulacy powerfully at work. Surely no reactionary in such matters, Mill perceives England as "the native country of compromise" and seems pleased that "there is in the English mind, both in speculation and in practice, a highly salutary shrinking from all extremes." Yet his qualification, immediately following, that "this shrinking is rather an instinct of caution than a result of insight ... [and hence] too ready to satisfy itself with any medium, merely because it is a medium" betrays anxiety about a new type of hidebound and intellectually stunted "herd mentality" that Nietzsche and even a reluctant liberal like Fontane would soon identify as a troubling consequence of modern liberalism.[77] Just how the petit-bourgeois individual's mimetic view of morality obfuscates its

76. For Nancy, thinking thus is "'only' the putting into question of an affirmation, ... [and] freedom is not the freedom *of* this or that comportment *in* existence: it is the freedom of existence to exist, to be 'decided for being,' that is, to come to itself according to its own transcendence" (ibid., 18, 23).

77. Mill, "Coleridge," in *Mill on Bentham and Coleridge*, 117, 134. On the "herd-mentality" in Nietzsche, see *Gay Science*, nos. 116, 354, 369; *Beyond Good and Evil*, nos. 44, 191, 199–202; *Genealogy of Morals*, Pt. 2, nos. 13, 15, 18; the term also figures prominently in Nietzsche's *Nachlass*, esp. his posthumous writings of the 1880s. More surprising is to find a poignant discussion of the apparent demise of bona fide "action" in the modern "herd" in Fontane; see Ch. 38 in *Der Stechlin*, where Pastor Lorenzen admiringly speaks of a "severe action" [*die schwere Gegentat*] that cuts through all convention: "to intuitively feel the right thing in moments like that, and to do something terrible in the conviction of what is right, decisively and unshrinkingly, something that out of context runs counter to honor, that is something which impresses me tremendously. In my eyes, that's the real thing, true courage ... Batallion courage, the courage of the masses, with all due respect for it, it's nothing but the courage of the herd [*Herdenmut*]" (*The Stechlin*, 290/*Der Stechlin*, 344).

covetous and hedonistic psychology emerges vividly in two memorable scenes from *Daniel Deronda*. Showcasing the powerful influence of Darwin's oeuvre, the novel's famous opening chapter at the gambling table in Leubronn offers us a decidedly unglamorous cross-section of Europeans, "Livonian and Spanish, Graeco-Italian and miscellaneous German, English aristocratic and English plebeian." Transfixed by the circular motion of the roulette wheel, those present betray "a certain uniform negativeness of expression which had the effect of a mask—as if they had all eaten of some root that for the time compelled the brains of each to the same narrow monotony of action."

Another instance of what the narrative sardonically characterizes as "a striking admission of human equality" involves an archery meeting at the Brackenshaw estate. Really a Darwinian mating ritual enacted in genteel disguise, this cultivated soirée held in a "carefully-kept enclosure" casts an unsparing light on the uniform, somnambulist placidity of female life in the Victorian upper middle class: "What could make a better background for the flower-groups of ladies, moving and bowing and turning their necks as it would become the leisurely lilies to do if they took to locomotion? The sounds too were pleasant to hear … musical laughs in all the registers and a harmony of happy friendly speeches, now rising towards mild excitement, now sinking to an agreeable murmur."[78] As Eliot (and, at times, Mill too) saw, the effect of modern liberalism involves not "the destruction of consensus," nor indeed "the substitution of another consensus" of equal or perhaps superior articulacy. Rather, it pivots on informed political and social consent having been supplanted by mere "sentiency," by the mindless tropism of individuals as they replicate behavioral conventions in quasi-instinctual form, rather than scrutinizing and where needed transcending what is objectively given by means of action. It is just this progressive extinction of capable agency that shapes the plot of Gwendolen Harleth; and it is her character (seeking counsel from Deronda but lacking the clarity and resolve to act on it) that illustrates how "the result of accepting Mill's advice to decide everything for ourselves is not decision but indecision."[79]

In this regard, George Eliot's focus on the maimed or irresolute nature of the psychology of mid-Victorian women connects her oeuvre to Flaubert's Emma Bovary, Tolstoy's Anna Karenina, Fontane's Effi Briest, and a host of other eponymous, female *névroses* of the nineteenth-century novel—all of whom appear to have seized on emotion as the last remaining avenue toward self-possession. Yet that strategy of conspicuous sentiment, already equivocal and often treacherous in Rousseau's *Julie* and Goethe's *Werther*, has become a dead-end in Flaubert. Misidentifying

78. Eliot, *Daniel Deronda*, 8–9, 100.

79. Chadwick, *Secularization*, 35.

freedom as "choice," Emma frantically pursues various forms of hyper-stimulation (sex and death being uppermost on her list) merely so as to evade her terminally secure and aimless, provincial existence by various means: "She longed to travel; she longed to go back and live in the convent. She wanted to die. And she wanted to live in Paris" or (surely everyone's favorite) "she conceived the idea of becoming a saint."[80] In passing we recall what may justly be termed a philosophical analysis of Emma's predicament, tendered as it were *avant la lettre* in Schopenhauer's 1839 *Prize Essay on the Freedom of the Will*—a work whose diametrical opposition to Coleridge we will have occasion to consider later on. Separating from the very outset the concept of "liberty" and "rights" (which "only refers to an *ability,* that is, precisely to the absence of *physical* obstacles to the actions of the animal") from "moral freedom," Schopenhauer homes in on the one, all-important question, viz., whether "the *will itself* [is] free." The customary protestation by classical liberalism's autonomous subject (viz., "I can *do what I will*") evidently misses the underlying question, viz., "whether the will itself is free" or whether "you can also *will* what you will." For the self-conscious individual to say "I can will, and when I will an action, the movable limbs of my body will at once and inevitably carry it out the moment I will it" is to define freedom strictly as "*being able to do in accordance with the will.*" Such had been the dominant view of Enlightenment thought, as categorically expressed by Voltaire: *vouloir et agir, c'est précisement la même chose qu'être libre.*[81] Yet precisely here trouble lurks; for even as "self-consciousness asserts the freedom of *doing* under the presupposition of *willing,*" Schopenhauer reminds us that "what we have inquired about is the freedom of *willing.*" Consistent with liberalism's erroneous construal of human freedom as "multiple choice," Emma consistently mistakes wishing for willing. It is indeed possible, Schopenhauer notes, to "*wish* opposite things, but [one] can *will* only one of them, and which one is first revealed even to self-consciousness by *the deed.*"[82] Emma's failure to understand as much reflects modernity's fundamental loss of "action" (or "deed") as a meaningful category.[83] Like the equally severe case of

80. Flaubert, *Madame Bovary*, 56, 201.

81. Voltaire, *Traité de Métaphysique*, 32:57; as E. Cassirer notes (*Enlightenment*, 250–251), Voltaire later expressly reversed himself, conceding that the will itself cannot be thought otherwise than as wholly determinate.

82. Schopenhauer, *Prize Essay*, 4, 6, 14–15.

83. Schopenhauer echoes Aristotle, *Nicomachean Ethics*, 1111^{b}20–30; for Aristotle, "wish" and "choice of will" are different genera altogether since a wish is solely concerned with an end (*telos*), whereas "we deliberate not about ends but about what contributes to ends" (1112^{b}10). Noting the "general devaluation of human agency" in Flaubert's prose, Peter Brooks's reading of *L'éducation sentimentale* locates a strikingly analogous confusion in that novel's protagonist, Frédéric Moreau, who is characterized in the novel as "worn out, full of contradictory desires and no longer even knowing what he wanted; he felt an overwhelming sadness, a wish to die" (*Reading*, 181, 185).

Frédéric Moreau in *L'éducation sentimentale,* Emma's life presents a wholly "negative balance sheet, the end of action in *ennui*."[84]

For Robert Pippin, *Madame Bovary* exemplifies "bourgeois culture's growing dissatisfaction with itself, a sense that modernity's official self-understanding—enlightened, liberal, progressive, humanistic—had been a misunderstanding." Fully aware that the dissociated sensibility of mid-nineteenth-century petit-bourgeois individuals had by then permeated every aspect of their existence, Flaubert realized that a fundamentally new conception of narrative art was called for. At the macro-level, this means that "the story is redeemed, rendered worthy of interest, important, only by its telling, not by the discovery of an internal point or purpose to the suffering and misery of the characters."[85] Concurrently, at the micro-level of narrative technique, the correlate of this psycho-historical decline can be found in the "studied irresponsibility" with which Flaubert deploys the *style indirect libre,* a mode of speech that "refuses to designate who is responsible for a given statement." As speech is shown to have deteriorated into "cliché, belonging to everyone and to no one," action has not only been absorbed into "behavior," but the resulting narrative turns out to be conspicuously "anti-novelistic" by converting its protagonist's terminal obtuseness into a frustrating reading-experience. For what supplants action is the mindless *implementation of a subjective attitude or desire* that is almost instantaneously shown to be empty. Consequently, Flaubert's novels forever "preclude turning fascination into knowledge."[86] With nothing more than the hallucinatory and transient force of desire to sustain the bourgeois subject, it makes little difference whether the attitude in question is Kant's formal feeling of respect for the moral law or the wayward, psychosexual cravings and sensations so ubiquitous in Flaubert's prose. Indeed, few writers capture more effectively how Aristotelian *praxis* and its underlying, public and normative sense of a *telos* have been supplanted by modernity's autistic notion of a self defined via its exclusive *dominium* or "right" to access the fantasized, virtual realm of economic and erotic fulfillment.

The price continually exacted from a liberated and emancipated bourgeoisie in the later nineteenth century is that it will become incrementally more conscious of the essential pettiness of its founding vision; having tied the notion of happiness to trivial socioeconomic aspirations (and no less banal psychosexual fantasies) rather than to the fulfillment of a normative good (*telos*), it finds quotidian life aimless, phantasmagorical, and replete with proto-Freudian neuroses and incipient despair that glance ahead to the existential parables of Franz Kafka and Albert Camus: "But

84. Ibid., 203.
85. Pippin, *Modernism,* 31, 33.
86. Brooks, *Reading,* 194, 187.

to [Emma] nothing happened. It was God's will. The future was a pitch-black tunnel, ending in a locked door."[87] Emma's predicament is not that of a supposedly lively and distinctive imagination gratuitously snuffed out by the oppressive humdrum of small-town provincial life and a stultifying marriage. Rather, as her "choice" to marry Charles already suggests, her petit-bourgeois imagination itself is utterly banal and cliché-ridden, a storeroom cluttered with the banal titillation and commodity-like fantasies infused by her desultory perusal of romance literature; she is said to be "deep in Walter Scott," Parisian weeklies, and jejune fantasies of religious rapture.[88] Hence, as the self-styled "bourgeoisophobus" Flaubert takes pains to illustrate at every level of his narrative art, his protagonists do indeed "dwell in possibility" (as Emily Dickinson had put it). Yet in so doing, they turn out to be effectively paralyzed as agents, not to mention the fact that the possibilities in question have themselves been utterly colonized and denuded by commercial culture.[89]

---

87. The image seems to be echoed by one of Kafka's many architectonic images, developed in his "Kleine Fabel": *"Ach," sagte die Maus, "die Welt wird enger mit jedem Tag. Zuerst war sie so breit, dass ich Angst hatte, ich lief weiter und war glücklich, dass ich endlich rechts und links in der Ferne Mauern sah, aber diese langen Mauern eilen so schnell aufeinander zu, dass ich schon im letzten Zimmer bin, und dort im Winkel steht die Falle, in die ich laufe." "Du musst nur die Laufrichtung ändern," sagte die Katze und fraß sie* (*Zur Frage der Gesetze,* 163). Eng. "'Alas,' said the mouse, 'the whole world is growing smaller every day. At the beginning it was so vast that I was afraid, I kept running and running, and I was glad when I saw walls far away to the right and left, but these long walls have narrowed so quickly that I am in the last chamber already, and there in the corner stands the trap that I must run into.' 'You only need to change your direction,' said the cat, and ate it up."

88. *Madame Bovary*, 59, 210.

89. Letter to Louis Bouilhet (December 1852), quoted in Gay, *Modernism*, 6. Perhaps Adorno's best book, *Minima Moralia*, can plausibly be read as a running commentary on the dystopic world of the great nineteenth-century novel.

Part IV

RETRIEVING THE HUMAN

COLERIDGE ON WILL, PERSON, AND CONSCIENCE

A Person is the subject bearing certain capacities [Subjekt des Könnens]. *It thus makes no sense to speak of "potential persons." Person is never a potentiality but always real. Personality does not itself develop but, instead, is what gives a specific human development its distinctive character. Thus we say "I was born on that date," rather than "on that date a human being was born from which I gradually came to be." We don't employ the word "I" to signify* "an *I*" (ein Ich). *There is no such thing as* "an *I.*" *That is but an invention of Descartes. The personal pronoun "I" signifies the human that I myself am. Beings capable of indexing themselves we call* persons. *Yet we also call them persons when under certain circumstances* [aktuell] *they cannot refer to themselves or make reference of some other kind. The word "capacity"* [können] *carries multiple meanings. It makes sense to say: "I can* [kann] *play the piano."*

If subsequently asked to do so even though no piano is at hand, I will answer: "There is no piano. Hence I cannot play." To which it would be impermissible to reply: "But you said that you could play." The capacity of playing the piano is a reality even when, in absence of a piano, it cannot be realized. And so "person" means not what *a human being may come to be* [werden kann], *but instead refers to the human being* who *may come to be something in particular* [aus dem etwas werden kann].

—Robert Spaemann

14

GOOD OR COMMODITY?

Modern Knowledge and the Loss of Eudaimonia

The strain of late eighteenth- and nineteenth-century literary and philosophical narrative briefly indexed here reveals a metaphysical deficit intrinsic to modern liberalism—a deficit certainly unacknowledged, if not outright repressed, and hence steadily more pressing and crippling for the modern individual. The writings in question show the Enlightenment unable to grasp the challenge posed by freedom to its self-satisfied, rationalist trade in non-negotiable and non-contingent "rights" and its reductive understanding of free will as multiple choice and subjective preference. In scrutinizing these structural problems, nineteenth-century literary and philosophical narrative appears wary of liberalism's founding paradigm of agency; the hypothesis of institutionally embedded and ostensibly self-possessed individuals carrying out the work of reason behaviorally, rather than by way of imaginative, transformative, and risk-fraught action, no longer seems inspiring, let alone credible in the way it had been for Locke, Adam Smith, and Kant. Yet the emergent critique of Enlightenment liberalism comes with presuppositions of its own. Thus Schopenhauer's contemptuous view of the modern rational and self-possessed individual—a critique whose political dimensions Edmund Burke's *Reflections* had

Portions of this and the following chapter have appeared in an earlier version in *MLN* and are here being reproduced with the kind permission of Johns Hopkins University Press.

anticipated with cantankerous lucidity—can only reject modern autonomy by viewing the primacy of character, temperament, and sensibility as continuous with ancient Greek notions of "necessity" (*anankē*) and "fate" (*tuchē*). Against modern liberalism's axiomatic view of a world composed of so many autonomous monads that take themselves to be free to join various social, political, religious, and economic communities and associations, Schopenhauer's dystopic account (presaging similar narratives in Stendhal, Flaubert, Wagner, and Dostoevsky) insists that it is only at the level of *action* that the status of the modern individual is decided. As Schopenhauer goes on to argue, the course of action that the modern self "chooses" is the only one she or he could ever have chosen (mere "wishing" being an entirely different matter); and in now embarking on it the individual can only give rise to the (tragic) self-awareness of an implacable and irretrievable determinacy. If liberalism's public face is to be seen in the literary genre of utopia, its unacknowledged Blakean "contrary" is tragedy; and it is no accident that tragedy should have furnished the generic template for the canonical narratives oeuvre of Stendhal, Flaubert, Dostoevsky, George Eliot, Fontane, Hardy, Zola, or Thomas Mann.

For Blake, Goethe, Schopenhauer, Stendhal, Flaubert, Arnold, Eliot, Dostoevsky, and Nietzsche, among others, freedom is a challenge to our capacities for imaginative vision rather than a formal possession or claim right. Yet it is a challenge that their characters rarely tend to meet successfully—some happy exceptions such as Tolstoy's Konstantin Levin and Pjotr Bezuhov or Eliot's Daniel Deronda notwithstanding. Instead, the vast majority of nineteenth-century narratives exhibit a marked disillusionment with modern liberalism's stubborn contrivance of various descriptive and disciplinary procedures, techniques, and systems (e.g., statistics, behaviorism, utilitarianism, quantitative sociology, historicism) aimed at remedying the deleterious effects of the hedonistic, radically singular self that was the epistemological legacy of Hobbesian voluntarism and Lockean nominalism. A great deal of nineteenth-century narrative thus scrutinizes the autonomous individual's repeated failure to achieve a concept or vision of the whole and, especially, of the human other as a "Thou." Behind this failure lurks a story that, so we are given to understand, should never have unfolded in the first place. The basic blueprint in question—encountered in the fiction of Stendhal, Balzac, Flaubert, Dostoevsky, Eliot, Hardy, Fontane, Mann, and numerous others besides—thus reoccupies, however unwittingly, the ancient motif of a world whose material realization inevitably betrays the idea that had given rise to it. Implicitly, then, even as the dystopic arc of modern narrative traces the local failings of specific characters, it also amounts to a symptom of modernity's fateful cultivation of theoretical reason at the expense of practical reason, of affirming singularity over relation, prioritizing self-assertion over obligation, will over meaning, and generally sacrificing vision and participation in what is given to the idol of knowledge as power (*potestas*) and commodity (*dominium*). Hans Blumenberg puts the matter as follows:

> The fundamental Platonic equivocation, that the world of appearance is indeed the reproduced image of Ideas but cannot attain the perfection of the original, is resolved by Neoplatonism in favor of the second aspect: The world appears as the great failure to equal its ideal model. The metaphysical factor in this failure has been prescribed since Plato; it is the *hylē* [matter]. The difference between idea and substratum, between form and stuff, is increased in the Neoplatonic systems; to the *theologizing* of the Idea corresponds the *demonizing* of matter. What could at one time be conceived as the subjection of necessity to rational persuasion, namely, the formation of the world, is now the confinement of the world soul in the womb—or better: the prison—of matter … All of this is still within the realm of discourse laid out by Plato, even if it does, as it were, exaggerate the metaphysical "distances" in the original ground plan … Gnosticism bears a more radical stamp. Even though it employs the Neoplatonist system, it nevertheless is not a consistent extension of that system but rather a reoccupation [*Umbesetzung*] of its positions. The demiurge has become the principle of badness, the opponent of the transcendent God of salvation who has nothing to do with bringing this world into existence. The world is the labyrinth of the *pneuma* [spirit] gone astray; as cosmos, it is the order opposed to salvation, the system of a fall … The downfall of the world becomes the critical process of final salvation, the dissolution of the demiurge's illegitimate creation. (*LMA*, 128)

For a recent illustration of this predicament, one may turn to Ian McEwan's 1998 novel *Amsterdam* where, early in the narrative, protagonist and celebrity composer Clive Linley finds himself taking refuge from his comfortable upper-class West London flat and its predictable array of "design, cuisine, good wine, and the like." Seeking shelter from professional troubles within the poetically charged ambience of the Lake District, Clive first has to endure the passage by train out of North London, a transition that deepens his Gnostic sense of human civilization as an all-encompassing miscarriage. Languidly taking in "square miles of meager modern houses whose principal purpose was the support of TV aerials and dishes; factories producing worthless junk to be advertised on the televisions and, in dismal lots, lorries queuing to distribute it; and everywhere else, roads and the tyranny of traffic," Clive gradually distills all the inchoate perceptions into a comprehensive, dismal allegory. What makes his desultory musings so poignant is, at least in part, their completely unpremeditated character, which so markedly contrasts with the strict means/end rationality governing and visibly misshaping the bustle of economic life without—at once utterly structured and yet entirely devoid of self-awareness:

> It looked like a raucous dinner party the morning after. No one would have wished it this way, but no one had been asked. Nobody planned it, nobody wanted it, but most people had to live in it. To watch it mile after mile, who

> would have guessed that kindness or the imagination, that Purcell or Britten, Shakespeare or Milton, had ever existed? Occasionally, as the train gathered speed and they swung farther away from London, countryside appeared and with it the beginnings of beauty, or the memory of it, until seconds later it dissolved into a river straightened to a concreted sluice or a sudden agricultural wilderness without hedges or trees, and roads, new roads probing endlessly, shamelessly, as though all that mattered was to be elsewhere. As far as the welfare of every other living form on earth was concerned, the human project was not just a failure, it was a mistake from the very beginning.[1]

Intriguing about the passage is its emphasis on the particularity of quotidian life, the frenetic cycles of getting and spending and the consequent denaturing of purposive organic forms into mere vestiges of natural creation ("a sudden agricultural wilderness without hedges or trees"). A veritable allegory of late modern capitalism run amok, the wasteland of North London impresses on Clive "the human project" itself as an impossible one. And yet, by inadvertently echoing a key axiom of ancient Gnostic speculation, Clive's sense of the material world as a cosmic misadventure ("a mistake from the very beginning") also perplexes. For not only should this world, by all appearances, *never have been* in the first place; but the fact of its *manifest persistence* in spite of it all raises the question as to what value there might be left for speculative thinking in a world that has embraced instrumental rationality without reserve. Though likely unaware that his weary meditations retrace Gnostic speculations almost two thousand years old, McEwan's protagonist, quietly dismayed by what he beholds, nonetheless conveys the undiminished force and urgency of Gnostic speculation. What if beneath our self-conscious, post-historical modernity there were to ferment another set of questions, not only distinct from but in actual conflict with what Nietzsche had labeled modernity's "logical optimism"? What if our single-minded embrace of rationality as the efficient, specialized, and institutional/corporate production of knowledge were to have obscured an entirely other dimension of reason? It is the task and privilege of literature and "criticism" in the strong (i.e., extra-professional) sense to pose and explore questions of exactly the kind so vicariously broached by McEwan's dispirited hero.

In what follows, a first exhibit of Coleridge's thoroughgoing critique of modernity will be *The Rime of the Ancient Mariner*, which intones concerns that, decades later, he was to explore with unprecedented rigor and intensity in his *Aids to Reflection*, the *Opus Maximum*, and the notebooks. Yet before taking up these materials, some broader and perplexing questions have to be addressed first. For any critical

1. McEwan, *Amsterdam*, 68–69.

assessment of Coleridge's perspective on modernity as a metaphysical catastrophe (rather than a set of contingent political problems) is complicated by our own situation today. To begin with, as active members of a profession committed to and/or embedded in various disciplinary and institutional pursuits and practices, we ourselves are potentially symptoms of the very modernity that Coleridge found so perturbing. First and foremost, there is the apparent fragmentation of knowledge into so many discrete institutional and disciplinary sub-specializations and, along with it, the dominance of method for which Bacon, Descartes, and Newton had significantly paved the way. As Giambattista Vico, Blake, Coleridge, Goethe, and Schopenhauer, among others, saw it, the gradual extension of modernity's analytic and methodical conception of "science" to *all* areas of knowledge risks fragmenting the world as a whole, to the point that the resulting, utterly particular insight is all but certain to have eroded the human and spiritual significance of the knowledge so obtained. In Coleridge's own times, the most conspicuous instance of methodical, specialized, and institutionally framed knowledge involved a historicism that pervades the emergent disciplines of higher biblical criticism (Christian Gottlob Heyne, Johann Gottfried Eichhorn, Friedrich Schleiermacher, et al.), historical philology (Herder, Lord Monboddo, Jacob Grimm, Franz Bopp, Wilhelm von Humboldt, et al.), and literary history (Herder, Friedrich Schlegel, Coleridge, Wolfgang Menzel, Heinrich Heine, et al.), and post-Kantian aesthetics (Schelling, Schiller, Hegel). Rapidly establishing itself as the very embodiment of "method," historicism also transforms philosophy itself, such that, after Kant, its paradigm of rationality (a.k.a. logic) begins to edge away from a syllogistic, predicative conception of truth and a methodology largely steeped in demonstration by analogy (*more geometrico*). In its place, the early 1800s witness the emergence of an inherently dynamic, temporalized, or "liquefied" (Hegel's expression) paradigm of truth as its own movement (*Bewegung*). Charting the conversion of "substance into subject," the speculative idealism and dialectical materialism of Schelling, Hegel, and Marx recast knowledge as a necessarily transgenerational, historicist progression.[2]

In aligning a reductionist understanding of method with an increasingly professionalized and specialized model of discipline-specific knowledge, the Enlightenment's scientific and epistemological projects revive (however unwittingly) a nominalism first pioneered by Abelard and William of Ockham. For to make an a priori commitment to historicism as a "method" almost inevitably sets inquiry on

2. On Schelling's crucial role in the formation of Marx's thought, see Frank, *Der Unendliche Mangel an Seyn;* on the relationship between Romantic (aesthetic) philosophy and modern critical theory, see Bowie, *From Romanticism to Critical Theory,* esp. 65–89; and Frank, *Einführung,* esp. lectures 1–10; for recent accounts of the changing conception of philosophy in German idealism, see Beiser, *German Idealism;* Pinkard, *German Philosophy;* E. Cassirer, *Erkenntnisproblem,* vol. 3.

a course toward increasing specialization and professionalization such as will inexorably shrink the community for which one's "findings" can have any relevance at all. I say "findings," rather than "arguments," because implicit in Descartes's insistence on the primacy of "method" is the assumption that what legitimates argument is solely the impersonal process by which it is generated; hence, the success of an argument should owe nothing to the rhetorical charisma of its presentation and everything to the methodology that secures the evidence on which modern scientific insight is said to rest. Implicitly, then, the charismatic and necessarily contingent force of rhetorical "argument" is steadily supplanted by the projection of an intersubjective consensus of expert knowledge of what in the final analysis would have to be, literally, "self-evident." Modernity's gradual journey from Cartesian rationalism to Lockean empiricism to nineteenth-century positivism thus revives the nominalist creed (at once irrefutable and pointless) that reality consists only of individual things. It is a position that reverberates in nineteenth-century historicism's conception of knowledge as the methodical disaggregation of proper names, dates, locales.[3]

For Theodor Adorno and Max Horkheimer, the Enlightenment broadly speaking thus amounts to "a nominalist movement" destined to lead its adherents to the threshold of an extreme particularity: "the *nomen,* the exclusive, precisely tailored concept, the proper name [*dem umfanglosen, punktuellen Begriff, dem Eigennamen*]." However dissimilar in their expressive registers, both the proper name and the nominalist concept employ the same strategy of self-legitimation inasmuch as they seek to render intelligible (and so redeem) the matter of history by recasting it as something as yet insufficiently differentiated. For a radically empiricist method "the guarantee of salvation lies in the rejection of any belief that would replace it: it is knowledge obtained in the denunciation of illusion ... [and] the contesting of every positive without distinction."[4] In premising its disciplinary, institutional, and accumulative paradigm of inquiry on an axiomatic and seemingly paranoid suspicion of the world as something given by a conceivably deceptive creator (Descartes's *dieu trompeur*), modernity—certainly by the beginning of the seventeenth century—has effectively abandoned the ancient (Platonic) view of knowledge as the convergence of *theoria* and *eudaemonia*. Knowledge now is construed as necessarily, indeed com-

3. On nineteenth-century historicism and its inflection by (post-) modern new historicism, see Chandler, *England in 1819,* esp. 3–93; Pfau, "Reading beyond Redemption;" Liu, "New Historicism" and "Local Transcendence"; Elam and Ferguson, eds., *Wordsworthian Enlightenment* (esp. the essays by Alan Liu and Kevis Goodman); on the origins of modern historicism in Romantic hermeneutics, see Gadamer, *Truth and Method,* esp. 194–235; for a different, equally acute critique of historicism as "the ultimate outcome of the crisis of modern natural right," see Strauss, *Natural Right and History,* esp. 9–34 (quote from 34).

4. Adorno and Horkheimer, *Dialectic,* 23; on the relation of nominalism to historicism, see Chandler, *England in 1819,* 53–73.

pulsively counterintuitive; it opposes the sheer givenness of phenomena, just as it rejects the metaphysical (in origin Augustinian) conception of the world as a gift. Insisting on the value-neutral and inherently unfinished nature of the material world, the modern, interventionist stance also disavows the ancient, both Stoic and Christian idea of knowledge as *theoria* and as *vita contemplativa*. Speaking of this "most momentous ... reversal of the hierarchical order between the *vita contemplativa* and the *vita activa*," Hannah Arendt elaborates: "the point was not that truth and knowledge were no longer important, but that they could be won only by 'action' and not by contemplation ... The reasons for trusting *doing* and for distrusting *contemplation* or *observation* became more cogent after the results of the first active inquiries." One must not misconstrue this reversal—achieved above all with the help of instruments and, especially, the paradigm of "mathematical knowledge, where we deal only with self-made entities of the mind"—as simply "raising doing to the rank of contemplation as the highest state of which human beings are capable." For as the "handmaiden of doing," all active thinking and its implicit vision of discrete knowledges moving toward a *mathesis universalis* effectively eclipsed the value of contemplation altogether (*HC*, 289–291).

In its classical, Platonic, and (modified) Aristotelian sense, *theoria* had constituted an essentially individual and contemplative relation to the cosmos, one notably accompanied by a sense of wonder (*thaumazein*) for whose continued interpretation Socratic dialogue and Platonic dialectics were thought to furnish the most promising form. Implicitly, the classical notion of *theoria* also posited that "truth in its totality was at the disposition of the individual" and, as a further consequence, that there had to be a strong "association of eudemonia with theory" (*LMA*, 239). Beginning with the Socratic emphasis on self-knowledge, the therapeutic conception of Hellenistic thought (e.g., the *ataraxia* of the Stoics), as well as St. Augustine's qualified and Tertullian's more emphatic restriction of knowledge to matters concerning salvation, the pursuit of theory is framed by "the general suspicion that the temptation to know the material world risks the loss of one's soul."[5]

By contrast, what Jürgen Mittelstrass calls modernity's "reflected" or self-conscious model of "theoretical curiosity" no longer unfolds as the humanistic care of the self, just as it is no longer circumscribed by a sense of metaphysical humility. Furthermore, it harbors no expectation of "wonder," such as would imply some impending, all-consuming revelation. Instead, modern theory is forever fixated on the

---

5. The same restrictive conception of knowledge is observable in some modern thinkers, too, such as Pascal and "the greatest modern Augustinian, Heidegger." Pippin, *Idealism*, 277, 278; on Pascal, see Auerbach's "On the Political Theory of Pascal," in *Scenes*, 101–132.

"never ending question of what will come next."[6] Hence, as Adorno and Horkheimer note with reference to Galileo, Bacon, and Leibniz, "number became the canon of Enlightenment" broadly speaking, such that "the Galilean mathematization of the world" dissolves the identity of objects into "a world of idealities" ultimately bound to supplant nature's material processes with "a rational, systematically unified method … a process of infinite progression."[7] This accumulative or "encyclopedic" impulse accounts for and defines the institutional, professional, and corporate structure of modern intellectual work where "thought inevitably becomes a commodity" that seems "blindly pragmatized."[8] At his most vituperative, Coleridge conjures the nightmare scenario where "Education … [is] defined as synonimous with Instruction," where "the population [is] mechanized into engines for the manufactory of new rich men," and where knowledge has been reduced to individuals engorged with "Idealess facts, misnamed proofs from history, grounds of experience, &c., substituted for principles and the insight derived from them." All this, he imagines is to be wrought by so-called "State-policy, a Cyclops with one eye, and that in the back of the head" (*CCS*, 62–63, 66). If the blind and overweening pragmatism shaping early Victorian educational policy already warranted such an unflattering portrayal, there is reason to suspect that it has returned in full force at the beginning of the twenty-first century, as utilitarian imperatives are incessantly couched in the equally smooth and vapid administrative jargon of "interactivity," "technology-based learning," "problem solving," "learning outcomes," and, of course, the incessant drumbeat of "globalization" and "interdisciplinarity."

For some time now the paradigm of knowledge as a specialized commodity has meant that the resulting spectrum of information "can no longer be surveyed and

6. Mittelstrass, "Bildung und Wissenschaft," 83–104; on the topos-history of "the book of nature," see Blumenberg, *Lesbarkeit*, 162–179.

7. *Dialectic*, 25.

8. *Dialectic*, xi–xii. Some six decades later, the state of the modern university shows Adorno and Horkheimer's critique of commodified knowledge to have been remarkably prescient. On the incoherent curricula of contemporary higher education and the presumed "separateness and separability [of knowledge] both from the uses to which it is put, and from the personal lives, particular beliefs, social practices, and specific commitments of those who create and transmit it," see Gregory, *Unintended Reformation*, 298–363 (quote from 303); on the hollowing out of intellectual passion and labor by corporate structures, see Tuchman, *Wannabe University*, esp. 48–87; Washburn, *University Inc.*, 29–42 and 137–170; and Bok, *Universities in the Marketplace*, esp. 99–121. A revealing instance of the preoccupation with revenue-generating knowledge and the consequent, rampant (and inevitably short-sighted) instrumentalization of inquiry in higher education, see the Strategic Plan by Duke University (September 2006). For a critique of these trends from the perspective of pure (scientific) research and theology, respectively, see M. Polanyi on the limits of doubt and the inescapability of commitment (*Personal Knowledge*, 269–324) and Griffiths, on the centrality of wonder and participation to human inquiry (*Intellectual Appetite*, 75–91 and 124–138).

taken in all at once." Consequently, it had to be reframed in the form of subfields and new disciplines, what Mittelstrass calls "higher-level agencies [*Übersubjekte*]," the result being that the modern producer of knowledge is confronted by an ever increasing and vexing "disproportion between what has been achieved in the way of theoretical insight into reality and what can be transmitted to the individual for his use in orienting himself in his world."[9] A crucial consequence arising from the triumph of modern "theoretical curiosity" is the superior authority accorded to *specialization.* Not only does specialization imply the displacement of the classical, eudaimonistic understanding of *theoria,* it also favors a conception of knowledge as the aggregation of a potentially infinite number of spatiotemporally distinct facts or events. In its principled rejection of universals and the proposition that knowledge ought to be relevant to its individual practitioners, modern, disciplinary, specialized, and methodical inquiry necessarily favors a particularist ideal of knowledge that was to find its consummate expression in the new historicism of the 1980s. And yet, even as this reflexive historicism rejects the grand narrative aspirations of nineteenth-century *Historismus* in favor of tightly circumscribed micro-analyses, its quest for what Alan Liu has termed "local transcendence" rests on largely unexamined ideological commitments of its own. At issue are a small number of axioms concerning the projected benefits of an accumulative (not to say transactional) mode of scholarly production which (with admittedly polemical intent) may be identified as follows:

1. *The Axiom of Specialization:* that specialized research and the recovery of previously "overlooked" materials and sources shall produce a type of knowledge whose legitimacy and significance are taken to be sufficiently guaranteed by the methodological protocols governing the retrieval of the information in question. For our purposes, the most salient example of this outlook involves the assumption—spawned by Lockean nominalism—that "'morality' names a distinct subject matter, to be studied and understood in its own terms," that it has no reality independent of the contingent history of its acquisition by a given individual and, hence, is contingent on the modern self's fluctuating "psychological development." As a result, morality is "apprehended . . . as a set of premises

9. Blumenberg, *LMA*, 238; more than anything, it is the rise of modern professionalism that has vanquished the classical, eudemonistic conception of knowledge as *theoria* and *thaumazein* ("wonder"). On the emergence of modern professionalized modes of production, see Larson, *Rise of Professionalism*, esp. 2–18 and 104–158; Abbott, *System of Professions*, esp. 86–113; on the professionalization of English in print culture, advanced literary study, and global publishing, see Ohman, *Politics of Knowledge*, and Ruth, *Novel Professions*, 1–32; on the beginnings of professionalized models of authorship and literary production, see Pfau, *Wordsworth's Profession*, 25–27 and 105–113.

rather than a conclusion" and, thus, subject to the kind of historical, encyclopedic survey that Alasdair MacIntyre finds supremely realized in Henry Sidgwick's contribution on "Ethics" to the ninth edition of the *Encyclopedia Britannica*.[10] Grasping the specious reasoning that had turned morality into the separate academic precinct of Richard Whately's and Sidgwick's "ethical science," the young Gerard Manley Hopkins formulates the key question ever so succinctly: "in reality the discussion comes to this—Is virtue knowledge or *a* knowledge?" Behind that question stands Hopkins's Socratic distrust of a knowledge so discrete and specialized in its focus and claims as to prove all but incommunicable.[11]

2. *The Axiom of Contextualism, or the Encyclopedic Imperative:* that the "new" materials so recovered largely imply their own causal and argumentative force simply by being exhaustively aggregated and associated with a specific "field" of inquiry whose substance and coherence are either presupposed outright or simply inferred from the interpretive community of specialists currently husbanding it.

3. *The Axiom of Skepticism:* that knowledge, being supposedly incompatible with strong normative and practical commitments of any kind, is inherently skeptical and impersonal; that it amounts to a form of "critique" in the Kantian, jurisdictional sense of delimiting the scope and authority of specific concepts, ideas, and beliefs; and that, consequently, the legitimacy of knowledge hinges on the inward cultivation and professional display of impartiality, distance, detachment, as well as on the express disavowal of any intuitive engagement with (or participation in) the phenomena under investigation.[12]

4. *The Axiom of (Marketplace) Pluralism, or the Ethos of Principled Indifference:* that the "objects" of knowledge have reality independent of any question regarding the *ends* of knowledge. Thus a professionalized and typically historicist procedure, especially in the humanities, embarks on the primitive accumulation and study of any number of discrete (and peremptorily disaggregated) texts on

10. MacIntyre, *Three Rival Versions*, 175, 177.

11. "Knowledge is that wh. can be taught and perhaps is the only thing wh. can be, if teaching is used in the truest sense of getting a certain connection into the pupil's mind ... On this ground imparting opinion is not to be called teaching, for in one way Plato seems to look on opinion as marked off fr. knowledge very much *by its unconnectedness*." Essay written for T. H. Green at Balliol College (Michaelmas Term, 1866), in Hopkins, *Collected Works*, 4:270–271 (italics mine); the preceding essays in this edition (pp. 256–269), written for R. Williams, develop related arguments.

12. See Thorne, who has persuasively argued that "we can discover in early modern philosophy more than the rigid imposition of tyrannical epistemologies, the single-minded clampdown of a claustrophobic empiricism and a one-eyed reason" and who rejects "the assumption that skepticism is inherently 'radical', but also ... the undifferentiated view of early modern intellectual history that makes this claim possible" (*Dialectect*, 15).

the assumption that critical knowledge will spontaneously arise from the open-market interaction of multiple and (presumptively) equivalent perspectives. Given that "contemporary academic practice [is provided] with only a weakly conceived rationality" whose discrete points of view prove "each unable ... to provide conclusive refutations of its rivals," the modern, professionalized conception of knowledge (including humanistic inquiry) has become axiomatically quarantined from questions of value.[13]

5. *The Axiom of Knowledge as Emancipation, or the Ethos of Retroactive Liberation:* that an institutional, professional, and transactional mode of humanistic and social-science inquiry advances modernity's axiomatic and unimpeachable narrative of progress by liberating, bit by bit, historical meanings from their alleged entrapment in religious or ideological norms and values of the past and, thereby, will restore for us their temporarily "missed," yet always "intended" authentic (secular) core.[14]

6. *The Axiom of Critique as a Guarantor of Historical Progress:* that the transactionalism of modern, institutional knowledge effects a teleological progression toward a hypostatized liberal community envisioned as wholly transparent, inclusive, tolerant, and exhaustively informed. Crucially, though, this utopia can only be articulated in a language of permanent deferral and (in what constitutes a diametrical reversal of Aristotelian thought) is being defined either by the presumed *absence* or by the express *rejection* of any specific norms or contents—rather than by the practical acknowledgment of their supra-personal authority.[15]

The self-imposed restriction of recent models of inquiry to tightly localized and circumscribed chronotopes (biographically conceived time spans, the *punctum* of this or that local "event," dates of publication, etc.), as fostered by (new) historicism for some time now, ultimately rests on the same axiom that had underwritten the political and economic projects of classical liberalism and their subsidiary rhetoric of

---

13. MacIntyre, *Three Rival Versions*, 173; see also 23–24; Gunton also identifies "what can be called a pluralism of indifference," a version of "modern liberalism [that] is selective in its tolerance inasmuch as it generates an intolerance of any position which makes claims for truth" (*The One, the Three, and the Many*, 105–106).

14. For an incisive critique of various assumptions supporting the concept of "secularization," see Blumenberg, *LMA*, 3–121; on historicism's implicit ethos of retroactive liberation, see my "Reading beyond Redemption."

15. In the classical era, the most typical expression of normativity involved a conception of the virtues; for strong accounts of the Aristotelian position, see MacIntyre, *AV*, 146–164; more critical of Aristotle's argument on "magnanimity" (*mēgalopsychia*) is Herdt, *Putting on Virtue*, 23–44; see also her subsequent discussion of the emergent critique of virtue ethics after Aquinas and the contraction of the virtues into a single, abstract, and increasingly empty notion.

emancipation, progress, growth, and political "rights" (*liberté, fraternité,* and *egalité,* etc.). Simply put, the (ultimately counterintuitive) axiom in question holds what the French Revolutionary calendar had so categorically stipulated: that the self-creation and self-legitimation of modernity pivots on societies instituting a radical caesura vis-à-vis the past and so freeing themselves from what Thomas Paine calls the "manuscript assumed authority of the dead."[16] Carl Schmitt identifies this "idea of an arbitrary power over history [as] the real revolutionary idea."[17] Defining of modernity—and precisely for that reason *not* a reliable premise for a *critique* of modernity—is a self-conscious and provocative rhetoric of which Paine's *Rights of Man* is just one, albeit a particularly strident example: viz., the rhetoric of the revolutionary check, the caesura (Grk. ἐποχή). It hardly surprises that the latter term—literally, "a suspension of judgment" originally elaborated by Sextus Empiricus as an integral component of skepticism—was to prove crucial for the unfolding self-description and self-legitimation of the modern era (Ger. *Neuzeit*). For unlike all preceding history, modernity stakes its claim to the status of an *epoch* on a self-certifying, all-pervading skepticism.

"The problem of legitimacy, is bound up with the very concept of an epoch itself," Hans Blumenberg notes; for "the modern age was the first and only age that understood itself as an epoch and, in so doing, simultaneously created the other epochs" (*LMA,* 116). In staking the legitimacy of its own scientific and political theories on the repudiation of a cyclical, epiphanic, and recursive understanding of time, modernity gave rise not only to itself as a (putatively) rational and legitimate process but also instituted a view of historical time as linear progression divisible into sharply defined epochs. The very intelligibility of distinctive historical experiences now appeared to call for specific methodical procedures aimed at disaggregating and freezing them in time. If all meaning was now understood as "historical" in its very essence, this axiom also implied an understanding of history as ordinary, mundane, and quantifiable factuality. For history to *mean* anything, the countless events or experiences of which it is composed could never be allowed to mean "too much." Lest it stray into the forbidden world of the epiphanic and normative, historical meaning had to be structurally embedded, familiar, cross-referenced, aggregated. As early as 1798, Friedrich Schlegel had lampooned historicism's irrational desire to "discover" and prove that all things past are familiar and can be assimilated to our conceptual frameworks by means of an ostensibly value-neutral methodology whereby mean-

16. *Rights of Man,* 42.

17. Schmitt, *Political Romanticism,* 62; on the theoretical assumptions standing behind (and propping up the argumentative force) of specific dates (especially in the new historicism), see White, "Imagination's Date." On the theoretical premises of (Romantic) historicism, see Chandler, *England in 1819,* 47–93; on questions of historical time, see Pocock, *Politics,* 233–272.

ings are never creatively generated but, instead, arise by default or imply themselves. As he notes, "the two main principles of the so-called historical criticism are the Postulate of Vulgarity and the Axiom of the Average. The Postulate of Vulgarity: everything great, good, and beautiful is improbable because it is extraordinary and, at the very least, suspicious. The Axiom of the Average: as we and our surroundings are, so must it have been always and everywhere, because that, after all, is so very natural."[18] As a fundamentally sociological (anti-aesthetic, iconoclastic) decoding of any variety of material and symbolic facts, objects, and practices, historicism's commerce with the real is a case of containing and stabilizing otherness in the modality of objective retrieval.

If one accepts Arnold Gehlen's forceful juxtaposition of primitive and modern forms of life, this very impulse turns out to be of archaic provenance. The historicist method of knowing thus would constitute a late instance of what Gehlen calls "transcendence towards *this* world" (*Transzendieren ins Diesseits*), viz., a stabilizing of the past by means of methodical and disciplinary "presentation" (*Darstellung*) whose investment is prima facie not to transform but to disarm the otherness of what has indisputably been bequeathed to us. Modern (historicist) "method" thus can be read as a distant echo of the crucially stabilizing function that "ritual" holds for primitive cultures. In either case, we are presented with modes of practice (*Verhaltensweisen*) "no longer aimed at improving, ennobling, or enriching the objects of practical engagement, no longer concerned with some kind of transformation; and indeed it is apparent that only such a non-transformative practice can provide a foundation for the idea of an enduring trans-temporal existence."[19] Paradoxically, then, the peculiar "creativity" of primitive ritual and modern method lies in its insistence on continuities, recursive patterns, and the strange security and comfort that makes its reappearance in an institutionally embedded proceduralism that Gehlen analyzes under the heading of *Habitualisierung*.

Yet such neutralization of the foreign and the other (Gehlen's *Aussenwelt*) always carries within its bosom the seeds of a new kind of discontent. For neither ritual nor method allows its practitioners to become conscious and articulate about its underlying function. To a significant extent, that is, the peculiar efficacy of ritualized or methodical practice hinges on the non-knowledge of the individuals or communities whose flourishing it helps secure. We recall Jean-Luc Marion's caveat that method "should not ... secure indubitability in the mode of a possession of objects that are

18. *Philosophical Fragments*, 3, no. 25.

19. "[Für] das rituell darstellende Verhalten ... geht es also nicht mehr um ein Verbessern, Veredeln, Anreichern des Gegenstandes dieses Verhaltens, um irgendeine Veränderung, und es ist einsichtig, dass *allein ein solches nichtveränderndes Handeln die Vorstellung eines dauernden, zeitüberlegenen Daseins zu tragen vermag*" (Gehlen, *Urmensch und Spätkultur*, 16).

certain because produced according to the a priori conditions for knowledge. It should ... not run ahead of the phenomenon, by *fore*-seeing it, *pre*-dicting it, and *pro*-ducing it, in order to await it from the outset at the end of the path (*meta-hodos*) onto which it has just barely set forth."[20] And yet, it is just this impression of a closed circuit that modern disciplinary and institutionally framed methods of inquiry often convey. It is no accident, surely, that historicism and liberalism begin to dominate the modern bourgeois imaginary as putatively inevitable developments during the first half of the nineteenth century. If the historicist method construes our commerce with the past as a one-way street whereby unself-conscious, agonistic past events are sublated into safe and familiar meanings by an oddly disembodied and transactional mode of scholarly production, its suitable counterpart is the "infinite conversation" that Carl Schmitt had identified as the vexing legacy of Romantic liberalism. In either case, we are presented with a ritualized, methodical work ethic, even as the "cause" or "end" for the sake of which such inquiry and debate are so ceaselessly being pursued is never properly articulated. There is no normative end, purpose, let alone a concept of the good but only further transaction or busi/ness. The term that encapsulates modernity's striking asymmetry between means and ends, between complex technologies of inquiry and a peculiar inarticulacy about ends and hesitancy about commitment to (or participation in) the phenomena under investigation is *professionalism*. Indeed, the first order of modern professionalism is to generate visible (published) tokens of one's industriousness such as will affirm one's institutional credentials and professional persona.[21] Professionalism involves the outward, accumulative commitment to an idea of knowledge whose ultimate "end" (in the Aristotelian sense) it cannot specify.

The institutional, trans-individual, and accumulative mode of knowledge production just sketched and so unreflexively implemented within contemporary humanistic inquiry is not, however, without its own prehistory. For what Hans Blumenberg calls "the process of theoretical curiosity" *unwittingly transposes the ancient cosmological attribute of infinity into the proceduralism of modern knowledge production.* Already vexing to early modern thinkers, such as Pascal and Hobbes, "infinity is more a predicate of indefiniteness than of fulfilling dignity, more an expression of disappointment than of presumption." As modern rationalism (beginning with

---

20. *Being Given*, 9.

21. On sociology's inability to conceptualize the larger process of modernity (of which sociology is itself a disciplinary and institutional effect), see Milbank's account of Max Weber: "Weberian sociology betrays and subverts history. It takes as an *a priori* principle of sociological investigation what should be the *subject* of genuine historical enquiry: namely, the emergence of a secular polity, the modern *imagining* of incommensurable value spheres and the possibility of a formal regulation of society" (Milbank, *Theology*, 91).

Nicholas Cusanus) began to transpose the attribute of "infinity" from the *object* of Scholastic inquiry (God) to the *process* of knowledge itself, it risked pervasive discontent on account of "the indefinite character of its course, the lack of distinctive points, intermediate goals, or even final goals."[22] With the concept of number and its infinite expandability (*connexio*) serving as its new "metaphysical archetype," modern knowledge thus had to change its "criterion for the general validity of a proposition: for it to be true, the predicate no longer needed to merge without remainder into its subject; rather, what was now required was a self-evident and universal rule of progress guaranteeing that the difference between subject and predicate be steadily diminishing."[23] At the level of terminology, the most conspicuous new term to express the accumulative, impersonal, and abstract mode of knowledge production is that of "system," which arises to prominence in the later seventeenth century and undergoes further scrutiny and differentiation throughout the eighteenth century.[24] Along with the emergence and eventual dominance of "system" as the new, concerted mode of intellectual production we also note the individual's growing alienation from philosophical thinking (in the classical, eudaimonistic sense), specifically its vexing inability to specify any normative objective (in the sense of an Aristotelian *telos*) that is being served by the production of knowledge.

This crucial (if logically flawed) transposition of infinity from a heretofore divine attribute onto modernity's procedural model of inquiry as the accumulation of facts and propositions at once specialized *and* abstract implies that a given instance of cognition may claim only a transitory, occasional, and instrumental role within the process of knowledge itself. The "end" of the process as such remains forever beyond the purview of any individual because "questions of ends are questions of values, and on values reason is silent."[25] Arising in sync with the doctrines of classical liberalism and eighteenth-century political economy, such a conception of rational inquiry tends to conflate meaning with value and value with utility, such that what

---

22. Blumenberg, *LMA*, 84–85, quotes from Pascal's *Traité de Vide* ("On Emptiness") as an early instance of this shift; for a related discussion, see Hölderlin, *Essays and Letters*, 2–11; still the standard account of the emergent, procedural, and administrative model of justice in the modern nation state is Foucault, *Order*, 312–318; for a superb and concise account of the "emergence of objectivity," see Dupré, *Passage*, 65–90.

23. E. Cassirer, *Erkenntnisproblem*, 2:180–181; translation mine.

24. On the rise of system, see Simpson, *Romanticism, Nationalism*, 126–148; Siskin, "Year of the System"; and Mitchell, "Fane of Tescalipoca." Aside from qualifying Siskin's taxonomy of systems in important ways, Mitchell also draws attention to a sacrificial logic integral to both conceiving and maintaining systems.

25. MacIntyre, *AV*, 26; this passage appears to echo a nearly identical comment by Adorno and Horkheimer: "Reason is the organ of calculation; it is neutral in regard to ends; its element is coordination" (*Dialectic*, 88); for an Augustinian radicalization of MacIntyre's Aristotelian critique of modern rationalism, see Milbank, *Theology*, 327–354.

can be known (or, for that matter, is deemed worth knowing) pivots on its perceived expediency or commodity value. Yet the problem with such a self-perpetuating instrumental logic is that in subordinating all its epistemic claims to an abstract and unreflected notion of utility it perpetrates a new type of alienation and meaninglessness on modern productive existence. For inasmuch as the utilitarian "appeal to the criteria of pleasure will not tell me whether to drink or swim and appeal to those of happiness cannot decide for me between the life of a monk and that of a soldier ... the notion of the greatest happiness of the greatest number is a notion without a clear content at all" (*AV*, 64).

In a subtler language that will occupy us again later, Coleridge's *Aids to Reflection* (1825) and his posthumous *Opus Maximum* anticipate MacIntyre's critique of a strictly instrumental notion of rationality and of moral agency. For to posit "virtue as a species of prudence" whose actions "originate in motives supplied by the present state of existence" mires the utilitarian account in a notion of immediacy that runs counter to its calculating implementation. Attempts to respond to this dilemma by drawing a distinction "between Selfishness, or the unconsidered obedience to an immediate appetite or restlessness, and Self-interest, i.e., the extension and modification of the same selfishness by Fore-thought," come to nothing. For as Coleridge notes, "this argument supposes the plenary causative or determining power in these motives or impulses, so that both the one and the other do not at all differ from physical impact as far as the relation of cause and effect is concerned." Clearly, lest it should indeed deserve its eventual label as the "pig-philosophy," utilitarianism at the very least had to grant that "a motive is neither more nor less than the act of an intelligent being determining itself ... i.e., the power of an intelligent being to determine its own agency." Hence the utilitarian's makeshift discrimination between self-interest and selfishness begs the overall question of what prompts our (moral) choices in life; for "we should still have to ask what determined the mind to permit this determining power to these motives and impulses. Or why did the mind or Will sink from its proper superiority to the physical laws of cause and effect, and place itself in the same class with the bullet or the billiard-ball?"[26] There simply is no alternative, Coleridge insists, but to acknowledge that "the Man makes the *motive,* and not the motive the Man" (*AR,* 74), a point developed in John Henry Newman's *Grammar of Assent,* which repeatedly insists that the foundation for all cognition is to be found, not in the "paper logic" of some syllogistic template, but in the realm of personal

26. *OM*, 24–26; as Coleridge elaborates elsewhere, "a Will conceived separately from Intelligence is a Non-entity, and a mere Phantasm of Abstraction; and that a Will, the state of which does in *no sense* originate in its own act, is an absolute contradiction. It might be an Instinct, an Impulse, a plastic Power, and, if accompanied with consciousness, a Desire; but a Will it *could* not be" (*AR*, 141).

judgment or "Illative Sense" roughly analogous to Aristotelian *phronēsis:* "It is the mind that reasons and that controls its own reasonings, not any technical apparatus of words and propositions."[27]

Utilitarianism and classical political liberalism, as well as their twentieth-century sociological and political extensions (e.g., pragmatism, behaviorism, and analytical moral philosophy, anti-foundationalism) prove so frustrating and ineffectual because they refuse as a matter of principle to identify any normative framework of ends (be it "divine law, natural teleology or hierarchical authority") within which individual practices and utterances are to acquire any significance. As MacIntyre puts it, what Enlightenment pluralism ensured was that "each moral agent now spoke unconstrained; . . . but why should anyone else now listen to him?" (*AV,* 68). MacIntyre's and Arendt's critiques of utilitarianism are substantially anticipated by Hegel, who notes how for utilitarian thought the value or significance of anything depends solely on whether it facilitates the strictly formal and contentless end of "pleasure." Hence the "thing" itself (*die Sache selbst*) is "only a pure moment," a merely transitional reality; it can only ever be "absolute *for an other*" and thus has reality merely as a means to a forever unspecified end. For individuals or communities to organize their private or social concerns in this manner is to commit to a wholly inexplicit and unexamined standard of "utility" as the new criterion of value and meaning. Hence, utilitarianism rests on two equally flawed assumptions: (1) what *kinds* of things should count as useful; and (2) that utility (a means/end rationality) should be the *only* standard or measure of value. In leaving these assumptions essentially unexamined or un-reflected, consciousness can only locate this "notion" of utility *in* (or project it *onto*) *an object* outside itself. For Hegel, utilitarianism thus constitutes indeed "a metaphysics, but not as yet the comprehension of it" (*PS,* 354).[28] Utilitarianism's "bad infinity" (*schlechte Unendlichkeit*) thus offers the most conspicuous illustration of how the gradual contraction of reasoning to instrumental "reckoning" and efficient causation had atrophied the life of the mind in its various interlocking dimensions as thinking, willing, and judging—and, by extension, had thoroughly eroded the integrity of the human person as a responsible agent.

What renders modern utilitarian thought both incoherent and dangerous is its principled inarticulacy about *ends.* The single-minded preoccupation with the

---

27. *Grammar of Assent,* 276.

28. As C. Taylor puts it, if "utilitarianism is . . . the ethic of the Enlightenment . . . in which acts are judged according to their consequences" such a conception only ever assigns *instrumental* significance to any particular thing or idea. It therefore is unable to articulate an end or "final purpose; or, as Hegel puts it, this chain of extrinsic justifications does not return to a self, that is to a subjectivity which would encompass the whole development." We thus have "a bad infinity [*schlechte Unendlichkeit*]" (*Hegel,* 181).

maximization of means and efficiencies also happens to be something that utilitarianism has in common with pluralism, with which it thus forms the peculiar symbiosis of modern liberalism. Both utilitarianism and pluralism emphatically repudiate as outright coercive any framework such as seeks to evaluate and normatively appraise actual or proposed economic, cultural, religious, or personal practices—including the methodical, specialized, and professional pursuit of knowledge. What this amounts to is the de facto abandonment of the Platonic and Aristotelian ideal of the *polis* as a community substantially defined by its sustained and explicit deliberation about the right "ends." Instead, modernity—certainly by the time of Bacon, Boyle, Descartes, and Hobbes—expects the self to adjust to *structures* and *disciplines* (of behavior, representation, labor, or moral justification) without making explicit the "end" relative to which specifically *these* structures and practices are indeed the appropriate and ethically justifiable means. For Hannah Arendt, "the much deplored devaluation of all things, that is, the loss of all intrinsic worth, begins with their transformation into values or commodities, for from this moment on they exist only in relation to some other thing which can be acquired in their stead." Hence, "in a strictly utilitarian world, all ends are bound to be of short duration and to be transformed into means for some further ends. This perplexity ... can be diagnosed theoretically as an innate incapacity to understand the distinction between utility and meaningfulness, which we express linguistically by distinguishing between 'in order to' and 'for the sake of'" (*HC*, 165, 154).

It is during the period of European Romanticism (ca. 1780–1830), and specifically in Coleridge's expansive critique of modern reason's dangerously incoherent assumptions about human agency, that we first encounter the thesis (eventually formulated by Hans Blumenberg) that modernity arose by radically overstating its emancipatory and self-authorizing potential. One aspect of modernity's hubris involves its impoverished, flat-line image of time as a single, undifferentiated, progressive vector of strictly anthropomorphic character. What disappears is not only the possibility that time might involve various epiphanic intensities but also the ancient distinction between the *punctum* of biographical time and the long *durée* of cosmological time, a distinction that Blumenberg develops under the heading of *Lebenszeit* and *Weltzeit*. In questioning the modern project's intellectual coherence and moral integrity, Goethe and especially the later Coleridge proceed from the assumption that the proverbial ancient/modern divide amounts less to a decisive break than a prolonged failure to remember traditions, legacies, and debts that, however unrecognized or repressed, continue to operate with undiminished efficacy. For the later Coleridge in particular, there is growing evidence that many of the conflicts and antagonisms vexing modern European society inadvertently re-enact a philosophical dilemma that had haunted Western civilization since the patristic attempts at consolidating Christianity as a coherent theological system in response to the competing philosophical schools of the Hellenistic period (Stoicism, Epicureanism, skepticism, neo-Platonism, and Gnosticism).

Before we explore that dilemma in detail, a methodological reflection is in order. For even as Goethe's and Coleridge's ambitious speculations about modernity are liable to disconcert today's critic with their grand narrative design, it is precisely on account of their provocation to modern, hyper-specialized, and post-metaphysical inquiry that works like *Faust II* or the *Opus Maximum* warrant careful consideration. At the very least, writers as capacious in their range and depth as Goethe and Coleridge prompt us to move away from a parochial and hermetic niche-criticism whose historicist and materialist assumptions and methodologies have steadily diminished the scope and stakes of intellectual argument in the humanities today. Moreover, the two writers just mentioned—of which the remainder of this study can only pursue the case of Coleridge—also throw into relief the regrettable myopia wrought by the late nineteenth-century quarantining of literary history from philosophy and theology, as well as the equally unfortunate, concurrent segregation of national literatures. Coleridge is truly unique as regards the historical range and analytic depth with which he engages intellectual traditions in philosophy, philology, the life sciences, and theology—fields whose bearing on Romantic culture and literature he saw so clearly, perhaps because "literary studies" had not yet branched off into the hermetic pursuit that it was to become a generation later. Indeed, it is the exceptional scope and intensity of Coleridge's intellectual pursuits that allow him to conceive of modernity as a pervasive and potentially irremediable dilemma. As he came to understand it, the modern intellectual's main task was to reappraise the long Enlightenment from Descartes and Hobbes onward, as well as the revolutionary developments at the close of the eighteenth century: not as the apotheosis or fulfillment of modernity's aspirations and promises but as the phase during which the antagonisms and plain incoherence of the modern Enlightenment project reach critical mass, with seemingly catastrophic results. Certainly by the time he leaves for Malta in 1805, Coleridge has effectively abandoned a more conventional and circumscribed identity as poet in favor of philosophical and, eventually, theological speculations that until the early 1810s are now and then punctuated by flurries of more occasional, journalistic prose (gathered as *Essays on his Times* in the Princeton edition of his *Collected Works*). Yet in those more topical writings, too, Coleridge aims to develop a critical perspective on the self-certifying, liberal-progressive optimism that can be traced back, at the very least, to Descartes's pivotal conceptual bequest to modernity: the idea of *skepticism* as a method of epistemological self-creation, and the instauration of *method* as the successor to judgment (*prohairesis*) as it had been realized in

Platonic dialogue and Aristotelian rational, inductive pedagogy (*epagōgē*).[29] For it was above all *the idea of method* that had facilitated the emergence, first in Britain and then on the Continent, of a new type of speculative, entrepreneurial, and self-transforming type of individuality.

---

29. While Aristotle's conception of *epagōgē* is diametrically opposed to the deductive arguments from first principles that characterize modern (Cartesian) rationalism, there has been debate about whether the Aristotelian concept anticipates modern theories of induction. I concur with Engberg-Pederson, who has argued that, contrary to modern inductive methods, Aristotelian *epagōgē* is not burdened with the task of self-legitimation that defines all modern methodologies: "the question of the validity of a given piece of epagoge never arises for Aristotle. There is epagoge when in a debate you make somebody accept some universal point on the basis of a review of particular cases, whether this point be true or false, and there is epagoge when you make somebody see a mathematical truth. Equally, whether you point to one particular case or to many or to all (where that is possible) is quite unimportant: Aristotle does not distinguish between those modern forms of induction ('perfect,' 'intuitive' and the like), quite simply because his concept of epagoge only contains the minimal content that is common to most modern types of induction, viz. coming to see some universal point as a consequence of attending to particular cases" ("More on Aristotelian Epagoge," 307).

15

THE PERSISTENCE OF GNOSIS

Freedom and "Error" in Philosophical Modernity

Coleridge's imaginative tabulation of the "costs of modernity," already on display in some of his poetry but much more expansively in his prose writings beginning with *The Friend* (1808), marks the beginning of a turn, in both philosophy and poetics, away from instrumental and pragmatic models of rationality and toward the (mostly negative) knowledge of history as one all-pervading miscarriage. It is no accident that this shift should have coincided with a rapprochement of philosophy (theology) and poetics in writers like Coleridge, Blake, Hölderlin, Schelling, and Schopenhauer, among others.[1] To inhabit modernity is to find oneself burdened

1. With Schopenhauer and Nietzsche as crucial transitional figures, this shift culminates in by now canonical critiques of modernity that have appeared over the last eighty years or so. Aside from what may well be the Urtext of philosophical critiques of modernity, Heidegger's *Being and Time* (1928), other relevant texts would include Adorno and Horkheimer's *Dialectic of Enlightenment* (1946), Hannah Arendt's *The Human Condition* (1958), Hans-Georg Gadamer's *Truth and Method* (1960), Michel Foucault's *The Order of Things* (1966), Hans Blumenberg's *Legitimacy of the Modern Age* (1966/1976), Alasdair MacIntyre's *After Virtue* (1981), Charles Taylor's *Sources of the Modern Self* (1989), Anthony Giddens's *Consequences of Modernity* (1990), Louis Dupré's *Passage to Modernity* (1993), and John Milbank's *Theology and Social Theory* (1991/2006); it goes without saying that these books follow very different methodological routes and reach often substantially different, at times diametrically opposed conclusions; some of these differences are highlighted by Blumenberg's response to some of his critics in the second (1976) edition of his magnum opus; for a judicious, if

with the complex and permanently incomplete task of theoretical self-legitimation and, hence, to be entangled in the project of modern "critique" (in the Kantian sense) as an ongoing attempt at delineating the limits of reason in new and largely uncharted modes of poetic, philosophical, and theological writing. However different their execution in most respects, critiques of modernity from Blake and Coleridge onward converge in their challenge to the idea of reason as categorically secular, self-legitimating, self-sufficient, and free of historical presuppositions. Especially for Coleridge, undertaking a critique of modernity means, first and foremost, to retrieve and reconsider the intellectual, aesthetic, and theological traditions on whose displacement (or repression) the modern project is premised. In so doing, Coleridge finds that beginning with the theological, cultural, cosmological, and epistemological transformations wrought in the sixteenth and seventeenth centuries, modernity continues to wrest with antagonisms that had already vexed the patristic writers as they sought to demarcate Christian theology from the competing, post-Aristotelian schools of skepticism, Epicureanism, Stoicism, and Gnosticism of the Hellenistic period. Anticipating in particular Hans Blumenberg's account of modernity as a renewed confrontation with the unresolved legacy of Gnosticism, Coleridge interprets modern mechanism, materialism, and voluntarism as latter-day symptoms of lingering problems in philosophical theology—specifically as regards the notions of the will, the person, and conscience. As he sees it, a thorough critique of Enlightenment rationalism must confront the scope and internal antagonisms of early Christian eschatology that, beginning with Descartes, had been displaced into a question of scientific method and the underlying assumption that rational procedure would eventually sort out lingering questions concerning the sources of self-awareness, moral obligation, and the enigma of salvation.

Yet it turns out that the Cartesian *cogito*'s asserted self-identity and freedom, far from settling these questions, opens up an ontological question that, as Blumenberg has argued, had never been conclusively answered since its initial discovery by the Gnostic philosophers of the second and third centuries A.D. Almost exclusively defined by the heresiological writings of the church fathers opposed to it (Irenaeus of Lyon, Hyppolitus of Rome, Tertullian, Clement of Alexandria, Origen), Gnosticism encompasses a considerable variety of positions that straddle both the conceptual and geographical boundaries between the Hellenistic syncretism of the Eastern Mediterranean and the apostolic model of early Western Christianity. What for Irenaeus resembled a "many-headed hydra" was above all understood by its main proponents (Valentinus, Marcion, Menander) as a religion whose members sought

sharply critical account of Blumenberg, see Pippin, "Blumenberg and the Modernity Problem," in *Idealism*, 265–285.

"knowledge" (*γνῶσις*) by esoteric and often multifarious interpretive means.[2] As the Coptic writings in the so-called Nag Hammadi Library make clear, the Gnostics' eponymous stress on "knowledge" (as opposed to "mere" faith) was central to their broader inquiry into the apparent estrangement of nature and matter from the "spirit" (*πνεῦμα*) posited by early Christianity. Far from a self-conscious heresy, that is, "Gnosis ... understood itself as a correct interpretation of Christianity" and as such rejects the absolutism of a faith that "knows nothing concerning itself and remains attached to what is immediately in the foreground."[3] To be sure, Clement of Alexandria (ca. A.D. 140–211) and Origen (d. A.D. 253) had still sought to reconcile the widening breach between a faith exclusively anchored in Christ and the apostles on the one hand, and the Gnostics' more wide-ranging and esoteric speculations on the other hand. Ultimately, though, the powerful label of heresy, firmly in place by the third century A.D., precluded any doctrinal consensus and, in fact, caused the Gnostics' writings to be largely expunged.

Central to Gnosis is its dualist cosmogony, according to which "the world is the product of a divine tragedy, a disharmony in the realm of God, a baleful destiny in which man is entangled and from which he must be set free."[4] Blumenberg seizes on the Gnostic critique of Platonic, Aristotelian, and Stoic cosmology, a critique that focuses on "where, in the process of the world's formation, rational planning and blind necessity, archetype and matter collide" (*LMA*, 127). In radicalizing the neo-Platonist demonization of matter, one early Gnostic, Marcion of Sinope (excommunicated in Rome in A.D. 144), sought to circumvent the apparent tension between spirit (*pneuma*) and matter (*physis*) by disaggregating God the creator from God the

2. *Adversos Haereses* [*Against Heresies*], 1.30.15. For an account of Gnostic cosmogonies, which both Rudolph and Jonas divide into Iranian (Zoroastrian) and Syrian-Egyptian strands, see Rudolph, *Gnosis,* 70–82; still among the best introductions to the Gnostics are Pagels, *Gnostic Gospels,* esp. 119–141, and Jonas, *Gnostic Religion,* esp. 29–99 and (on Marcion), 137–146; for primary texts, see James M. Robinson's translation of the so-called Nag Hammadi Codex, in 1947/1948, arguably the most significant early Christian manuscript discovery of the twentieth century along with the Dead Sea Scrolls; for the best current survey of critical literature on Gnosticism, see Rudolph, *Gnosis,* 390–404.

3. Rudolph, *Gnosis,* 51–52.

4. Ibid., 66; a particularly concise account is found in the so-called Gospel of Philip: "The world came into being through a transgression. For he who created it wanted to create it imperishable and immortal. He failed and did not attain his hope. For the incorruption of the world did not exist and the incorruption of him who made the world did not exist." From this primal miscarriage follow, in order, "anguish," "error" (*πλᾶνή*), which in turn sets to work on "matter" (*ὑλή*) by fashioning a "creature" (*πλάσμα*), a declension characteristic of the progressive deterioration of the "fullness" (*πλήρομα*) of the primal spirit in the Syrian-Egyptian gnosis; *The Nag Hammadi Codex,* II, 3:75, qtd. in ibid., 83.

redeemer.[5] For Marcion, "a theology that declares its God to be the omnipotent creator of the world and bases its trust in this God on the omnipotence thus exhibited cannot at the same time make the destruction of this world and the salvation of men from the world into the central activity of this God" (*LMA*, 129). John Milbank thus views Gnosticism as "an ontological rather than a (pre)historical fall" defined by the "idea of primal disaster within the divine *pleroma*" that requires "the salvation of God himself from his involvement in temporality."[6] By instituting a sharp divide between creator and redeemer, Marcion effectively made the destruction of the material cosmos a requisite outcome, a position that continued to resonate in Christianity's recurrent attraction to chiliastic and millenarian utopias, as well as in Romantic apocalypticism, such as we find it articulated in Blake's prophetic books (especially *The (first) Book of Urizen* and *Jerusalem*), Thomas Malthus's *Essay*, Lord Byron's "Darkness," and Mary Shelley's *The Last Man*.[7] In his 1809 *Investigations into the Essence of Human Freedom*, Schelling notes how "in maintaining freedom, a power which by its nature is unconditioned is asserted to exist alongside of and outside the divine power." A sharp tension thus opens up between modern rationalism's core axiom—the notion of free, self-conscious human agents—and the Judeo-Christian notion of an omnipotent God.[8] Wordsworth's *Prelude* acknowledges the same paradox when framing the emergence of its author-protagonist as that of "A Captive … coming from a house / Of bondage, from yon City's walls set free" (1805, I:6–7), a telling comment that posits freedom as achievable only by escaping the very world of urban commerce in which most contemporaries would ordinarily have sought to realize it.[9] Moreover, in suffusing this opening claim with numerous biblical allusions (here to Exod. 13:14), Wordsworth hints that freedom can be secured only if one is prepared to reject modern culture's enslavement to mimetic desire and its underlying, hedonistic model of agency: "Getting and spending, we lay waste our powers."[10] The *Prelude* in particular attempts to put distance between the self and its socioeconomic

5. On Marcion, see Blackman, *Marcion;* for primary texts, see Marcion, *The Gospel of the Lord;* see also Pagels, *Gnostic Gospels*, 28–32; Rudolph, *Gnosis*, 313–316; and Filoramo, *History of Gnosticism*.

6. Milbank, *Theology*, 304.

7. For a survey of apocalyptic motifs in Romantic writing, see Paley, *Apocalypse and Millennium*, and Beer, "Romantic Apocalypses"; on Blake's millenarian and apocalyptic imagery, see also Goldsmith, *Unbuilding Jerusalem*, esp. 27–84, and Bloom, *Blake's Apocalypse;* on social apocalypse in Malthus and some contemporaries (Wordsworth, Young, Godwin) see Ferguson, *Solitude*, 114–128; Harrison, "Ecological Apocalypse"; Gallagher, "Body versus the Social Body," and Pfau, *Wordsworth's Profession*, 341–370.

8. Schelling, *Philosophical Inquiries*, 10.

9. Unless otherwise referenced, Wordsworth's *Prelude* is quoted from *The Thirteen-Book Prelude*, ed. Mark Reed.

10. "The World is too much with us," in *Poems, in Two Volumes*, 150.

determinants, thus making modernity spiritually habitable by conceiving the realm of the aesthetic as a virtual sanctuary from its more pernicious influences.

That this crisis should often have gone unnoticed, even (perhaps especially) in contemporary literary and cultural studies, is the result of a twofold failure of historical memory. First, the paradox of human freedom, understood as a challenge to the notion of an omnipotent God, reveals within our conception of that God a fundamental dichotomy that ancient Gnosticism had bequeathed modernity. Hans Blumenberg thus reads the "modern age [as] the second overcoming of Gnosticism," an "old enemy who did not come from without but was ensconced at Christianity's very roots" (*LMA*, 126). Second, our current models of disciplinary and institutional knowledge have implicitly embraced modernity's definition of the *vita activa* as the domain of a wholly instrumentalized notion of "production," which in turn rests on a means-end concept of rationality and posits labor as a commodity of exchange. Yet for Hannah Arendt, the concept of "production" "is entirely absorbed in and exhausted by the end product" and thus misses an essential feature of human practice, viz., that as "action ... it is never exhausted in a single deed but, on the contrary, can grow while its consequences multiply" (*HC*, 233). Unsurprisingly, Wordsworth's troubled exploits in the *Prelude*, the irresistible acts of curiosity of William Godwin's *Caleb Williams*, or the materialist hubris of Victor Frankenstein in Godwin's and Mary Shelley's eponymous narratives, all tend to the same point: that "the human capacity for freedom, by producing the web of human relationships, seems to entangle its producer to such an extent that he appears much more the victim and the sufferer than the author and doer of what he has done. Nowhere ... does man appear to be less free than in those capacities whose very essence is freedom and in that realm which owes its existence to nobody and nothing but man" (*HC*, 233–234). Like the God of the Gnostics, unfettered in his freedom and hence fated to entangle himself in the material web of his creation, human agency finds the post-teleological idea of freedom as pure autonomy and self-actualization to be a self-defeating proposition.

As illustrated by sundry crimes and misdemeanors of Wordsworth's child-protagonist throughout Book I of the *Prelude*, modern freedom originates not in a rational or providentially guided self; rather, it stages the wholly extra-moral and extra-rational drama of sheer volition. Reluctantly, Coleridge's *Aids to Reflection* (1825) concedes that, as Schelling had put it, "only out of the darkness of unreason (out of feeling, out of longing, the sublime mother of understanding) grow clear thoughts."[11] In Coleridge's searching formulation, "it *is* in our power to disclaim our Nature as *Moral* Beings. It is possible (barely possible, I admit) that a man may have remained ignorant or unconscious of the Moral Law within him: and a man need

11. Schelling, *Philosophical Inquiries*, 35.

only persist in disobeying the Law of Conscience to *make* it possible for himself to deny its existence, ... Were it otherwise, the Creed would stand in the same relation to Morality as the Multiplication Table." Forty years later, the young Gerard Manley Hopkins would echo the same point: "Conditioned as man is he yet can unweave the web of life for himself."[12] The argument extends a point Coleridge had previously argued in his fragmentary "Essay on Faith" (ca. 1820) specifically with a view to conscience; the latter he defines as "an Act, in and by which we take upon ourselves an allegiance: & consequently, the obligation of *Fealty*. And this Fëalty or Fidelity implying the power of being unfaithful is the first and fundamental sense of Faith" (*SW & F*, 2:836). The Gnostic God's catastrophic entanglement with finite matter has here been transposed into the constant possibility of human beings to disavow the *Logos* in an act of free, sinful choice. Like any genuine theory, which must provide the grounds for its falsification, Christian philosophical theology always retains this elemental possibility of sin, that is, the opportunity for the individual to recuse himself or herself from any spiritual covenant, social embeddedness, and repudiate all ethical obligations. This constant potential for human action to oppose reason defines freedom, and indeed our view of modernity as centrally predicated on that freedom. As Hans Blumenberg puts it in his discussion of Descartes, "man is not free in that he has grounds for his action but rather in that he can dispense with grounds" (*LMA*, 185).

For many Romantics, it is the concept of the will that signifies the modern individual's potential rejection of any antecedent and normative "ground," "reason," or framework for action in favor of a model of freedom as sheer, inarticulate, inexplicable self-assertion. While time and again illustrating Blumenberg's point, Nietzsche's repeated and ambivalent references to the "solitary predatory species of man" also prove notoriously contradictory.[13] More than Freud, whose later writings make strikingly analogous claims, Nietzsche's prose oscillates between positing these feral propensities as something primal and timeless or, alternatively, claiming that civilization has effectively tamed and indeed extirpated these instincts once and for all. Thus he protests that "you utterly fail to understand beasts and men of prey (like Caesare Borgia), you fail to understand 'nature' if you are still looking for a 'disease' at the heart of these healthiest of tropical monsters and growths, or particularly if you are looking for some innate 'hell' in them—: as almost all of the moralists so far have

12. *AR*, 136; Hopkins, notes taken for R. William's lectures on Plato (Michaelmas term 1866), in *Collected Works*, 247; Levinas, whose proximity to Coleridge's conception of person and relationality we shall take up later, echoes Coleridge's point with uncanny precision: "It is certainly a great glory for the creator to have set up a being capable of atheism, a being which, without having been *causa sui*, has an independent view and word and is at home with itself" (*TI*, 58–59).

13. Nietzsche, *Genealogy of Morals*, 114.

done."[14] Yet elsewhere, Nietzsche protests the atrophying of such "nature" by the unrelenting efforts of moral civilization: "Regrettably, man is no longer sufficiently evil; the opponents of Rousseau who claim that 'man is a predator' are, unfortunately, wrong. The true curse lies not in the depravity of man but in his being re-made as a decadent and moral being [*sondern seine Verzärtlichung und Vermoralisierung ist der Fluch*]."[15]

"In the final and highest instance there is no other Being than Will," Schelling remarks even before Schopenhauer was to develop the implications of this thesis to its fullest extent; this "Will is primordial Being [*Urseyn*], and all predicates apply to it alone—groundlessness, eternity, independence of time, self-affirmation." Inasmuch as freedom qua unadulterated volition subsists between the "possibility of good and evil" the premise of a free agent revives an ancient philosophical dilemma: "either real evil is admitted, in which case it is unavoidable to include evil itself in infinite substance or in the Primal Will [*Urwille*], and thus totally disrupt the conception of an all-perfect Being; or the reality of evil must in some way or other be denied, in which case the real conception of freedom disappears at the same time."[16] In an interesting aside on St. Augustine's *liberum arbitrium,* Blumenberg captures a key concern of nineteenth-century pessimism as it contests Romanticism's and classical liberalism's affirmation of free will: "Freedom confirms the goodness of God and His work in every case because it wills itself; indeed it wills itself independently of its moral quality. But falling back upon the reflexive structure of the will, which wills not only this or that but primarily itself as the condition of its concrete acts of choice, only moves the problem a step further back: *The will that wills itself is only free if it can also not will itself.* Here rationality breaks down; reasons cannot be given for self-annihilation" (*LMA,* 134). The modern (implicitly deist) conception of God as First Cause or, in Schelling's phrase, as the architect of "a mere moral world-order," risks the formalization of God as the *actus purissimus* of absolute auto-genesis.[17] It also revives neo-Platonism's demonization of matter as the demiurge's lapsed progeny, an intransigent externality that has betrayed the "primal ground" or "depth" (*βῦθός*)—also known as the "pre-beginning" or "aeon" of Valentinian Gnosis or the "eternals" of Blake's *Book of Urizen.*[18] For Schelling in particular, what renders the rationalism

14. *Beyond Good and Evil,* 84–85.

15. *Sämtliche Werke,* 12:421 (trans. mine).

16. Schelling, *Philosophical Inquiries,* 26.

17. Ibid., 30.

18. As the Secret Book of John, which forms part of the Nag Hammadi Codex, makes clear, Gnostic speculation does not consider it appropriate to think this primal beginning "as a God, or [to suppose] that he is of a (particular) sort: he is a dominion (*archē*) over which none rules; for there is none before him, nor does he need them (the gods); he does not even need life, for he is eternal" (qtd. in Rudolph, *Gnosis,* 63).

that has dominated "the whole of modern European philosophy since its inception (through Descartes)" ultimately suspect is its principled abhorrence of matter, a premise that not only leads to an impoverished view of nature as merely inanimate "stuff" but also deprives God or *logos* of its proper "ground." Schelling's text lets us glimpse at the outlines of a metaphysical problem—most forcefully articulated by Gnostic philosophy—that had long been buried beneath the sands of historical time, and whose unresolved and persistent energy fuels some of the most searching and ambitious philosophical and aesthetic projects of modernity, in particular, those of Schelling and Coleridge.

Arguably one of the most compelling manifestations of Gnosticism in the Romantic period, Blake's *Book of Urizen* conjures up a creator God ("a shadow of horror") anxious to vanquish the "abominable void / This soul-shudd'ring vacuum" that had preceded creation. Seething "in silent activity: / Unseen in tormenting passions," Blake's Urizen dramatizes the primordial act of creation as the epigenesis of reason out of sheer volition: "an activity unknown and horrible; / A self-contemplating shadow." In Blake's agonistic un-writing of the Book of Genesis, Urizen creates matter, form, structure, and order. Yet in so doing he delimits (Urizen = horizon) and ultimately betrays the cosmological attribute of infinity itself. Hence the finitude of the produced world and the bodies that fill it invariably corrupts the raw at the heart of the creative act itself: "Sund'ring, dark'ning, thund'ring! / Rent away with terrible crash / Eternity roll'd wide apart / ... / Departing: departing: departing: / Leaving ruinous fragments of life" (*CPP,* 70–73). Such creative transposition of matter into form amounts to a "primeval" (Blake's pun on *prime evil*) betrayal of eternity by the terminal and singular reality of embodied form. As a result of Urizen's compulsive form-giving eternity is contracted into a single, determinate, and irreversible history—thus betraying all potential worlds for this actual one. By contrast, redemption pivots on a conspiratorial, even paranoid aesthetic, a Blakean "contrary" that shatters the mechanistic and causal stranglehold of empirical history. Noting how "Poetry and criticism after Milton in our language are attempts to *see,* in frequent contradistinction to the main Protestant tradition of listening to the Word," Harold Bloom muses whether this is not "the mark of Gnosis, that seeing is the peculiar attribute of certain spiritualized intellectuals."[19] Though he probably never read Blake, Carl Schmitt captures Blake's Gnostic vision quite accurately when pointing to Romanticism's inverse valuation of reality and potentiality:

> It is not possibility that is empty, but rather reality ... [In] representing possibility as the higher category, ... the romantics ... preferred the state of eternal

19. Bloom, *Agon,* 69; on Blake and Gnosticism, see Pfau, *Romantic Moods,* 98–111, Peterfreund, "Blake, Priestley, and the Gnostic Moment," and Curran, "Blake and the Gnostic Hyle."

> becoming and possibilities that are never consummated to the confines of a concrete reality. This is because only one of the numerous possibilities is ever realized. In the moment of realization, all of the other infinite possibilities are precluded. A world is destroyed for a narrow-minded reality. The "fullness of the idea" is sacrificed to a wretched specificity. In consequence, every spoken word is already a falsehood.[20]

The Gnostic origins of such a position are quite evident, as is Blake's consequent embrace of "an aesthetic that is neither mimetic, like Greek aesthetic from Plato to Plotinus, nor anti-mimetic, like Hebraism from the Bible to Jacques Derrida." For Blake no less than Schmitt, Gnostic knowledge involves a vision that neither appropriates nor indeed learns from what it beholds: "A Gnostic never learns anything, because learning is a process in *time,*" Bloom notes, and "if we were to ask 'What does failed Gnosticism become?' we would have to answer that Gnosticism never fails ... [because] a vision whose fulfillment, by definition must be always *beyond* the cosmos, cannot in its own terms be said to fail *within* our cosmos."[21]

By contrast, a thinker like Hegel is not prepared to follow through on some strong Gnostic elements in his philosophy. Thus, while characterizing nature as "self-alienated spirit [*der sich entfremdete Geist*]" or as "the negative of the Idea," Hegel's speculative method posits that thinking can and will progressively emancipate itself (and so redeem us) from the sheer otherworldliness of Gnostic dualism.[22] Though unconcerned with the Gnostic legacy, Theodor Adorno and Max Horkheimer clearly follow in Hegel's footsteps when interpreting the Enlightenment as a progressive overcoming of the primordial deficiency that modern thought tends to impute to nature *as origin:* "The world becomes chaos, and synthesis [its] salvation" (*Dialectic,* 5). As Blumenberg observes, already the Gnostics posited that the redeeming God had not only "the right to destroy a cosmos that he did not create" but was in fact obligated to do so. For against the betrayal of the infinite potentiality of the idea by a mundane, singular reality, apocalyptic deliverance implies the restitution of eternity over and against the *inter-regnum* of historical time and thus effects "man's enlightenment regarding his fundamental and impenetrable *deception by the cosmos*" (*LMA,* 129–130; italics mine).

This last remark also flags the Gnostic origins of that quintessential modern (Cartesian) preoccupation with "error" and "deception."[23] Thus, as Hannah Arendt

20. Schmitt, *Political Romanticism,* 66.

21. Bloom, *Agon,* 70, 58, 67.

22. *Enzyklopädie,* 25, 30; trans. mine; on the Gnostic aspects in Hegel's philosophy of nature, see O'Regan, *Heterodox Hegel,* 151–169.

23. Blumenberg proceeds to explore how "Gnosticism's systematic intention forced the Church, in the interest of consolidation, to define itself in terms of dogma ... To retrieve the world

argued nearly a half century ago, Descartes's writings are haunted by two "nightmares": first, the possibility that, in the wake of an all-pervading, self-conscious doubt all of reality will prove but an elaborate dream or an episode of madness.[24] Moreover, once the senses had been experienced not just as occasionally unreliable but as the principal source of error and deception, it appeared that "an evil spirit, a *Dieu trompeur,* willfully and spitefully betrays man [rather] than that God is the ruler of the universe." The main casualty of modern rationalism, then, was "not the capacity for truth or faith ... but the certainty that formerly went with it" (*HC,* 277). Descartes's inference that the cosmos had been created in such a way as to allow for the persistence of error registered as all the more troubling because the world had, in fact, *not* come to an end in the way that Marcion's Gnostic vision had implied; as Blumenberg puts it,

> The fact that the expected *parousia* did not occur must have been full of consequences for the transformation of the original teachings ... The world, which turned out to be more persistent than expected, attracted once again the old questions regarding its origin and its dependability and demanded a decision between trust and mistrust, an arrangement of life with the world rather than against it. It is easy to see that the eventual decision against Gnosticism was due not to the inner superiority of the dogmatic system of the Church but to the intolerability of the consciousness that this world is supposed to be the prison of the evil god and is nevertheless not destroyed by the power of the god who, according to his revelation, is determined to deliver mankind. The original eschatological pathos directed against the *existence* of the world was transformed into a new interest in the *condition* of the world. (*LMA,* 131)

While St. Augustine had sought to resolve the suspension of certainty concerning the *eschaton* or *telos* of the cosmos and its inhabitants by devising a sophisticated theory of human freedom, the compensatory doctrinal efforts of the patristic and Scholastic philosophers who succeeded him ultimately fail, at least on Blumenberg's interpretation, on a variety of grounds that need not detain us here. What does matter, how-

---

as the creation from the negative role assigned to it by the doctrine of its demiurgic origin, and to salvage the dignity of the ancient cosmos for its role in the Christian system, was the central effort all the way from Augustine to the height of Scholasticism" (*LMA,* 130); the present argument's concern is with the lasting consequences of Gnosticism for Romantic narrative, which struggles to reflect our vexed, disciplinary, and institutional commitments to modernity's paradigm of critical knowledge.

24. See Derrida's famous discussion of Descartes's *cogito* in the context of Foucault's history of madness, *Writing and Difference,* 48–63.

ever, is that the paradox of freedom or, as Blumenberg calls it, the "senselessness of self-assertion was the heritage of the Gnosticism which was not overcome but only 'translated'" (*LMA*, 136). For if modernity constitutes itself as a renewed confrontation with Gnosticism, it now does so under the "aggravated circumstances" of an unsuccessful Scholastic solution that had "lost its human relevance precisely on account of the absolutism of ... divine grace, that is, on account of the dependence of the individual's salvation on a faith that he can no longer choose to have." Early modern science and philosophy thus respond to the "disappearance of order" by "no longer perceiv[ing] in given states of affairs the binding character of the ancient and medieval cosmos." Instead, inasmuch as modern thought "holds [these states] to be, in principle, at man's disposal," the burden of legitimation has fundamentally shifted. For it now involves "responsibility for the condition of the world as a challenge relating to the future, not as an original offense in the past" (*LMA*, 137).

Arguably, the most seminal consequence to flow from this crucial reversal in how the self achieves stability and legitimacy vis-à-vis its "world" is the rise of classical liberalism as the dominant political theory of post-1750 European societies. Integral to this development is the emergence of fundamentally new strategies and genres of individual and communal self-description, in particular, the idea of *Bildung* as a template for organic, purposive, and transformational development. We encounter it in Goethe's *Wilhelm Meister* and in his botanical writings, yet also in Beethoven's displacement of classical form by a technique of "developmental variation" (*Entwicklungsvariation,* as Arnold Schoenberg was to label it) that dominates his later oeuvre. Common to all of these innovations is an acutely temporalized and emergentist understanding of form whereby the "subject" of the narrative progressively attains (rational) form as simple, motivic propositions begin to unfold their rich and diverse possibilities in organic and epigenetic (rather than mechanistic and predetermined) fashion.[25] To the extent that modern narrative imbues nature itself with agency and history, the Gnostic specter of an ontological struggle between matter (*hylē*) and idea (*eidos*) has been decisively rejected. An early instance of this solution to the lingering Gnostic dilemma can be found in Kant's "Speculative Beginning of Human History" (1786). Noting how the "path that for the species leads to *progress* from the worse to the better does not do so for the individual," Kant sidesteps the Gnostic dilemma with a curious distinction between a twofold beginning: "The history of *nature,* therefore, begins with good, for it is God's work; the history of *freedom* begins with badness, for it is man's work. For the individual, who in the use of his freedom has regard only for himself, such a change was a loss; for nature, whose end for man

---

25. See Pfau, "Of Ends and Endings," "All is Leaf," and "Bildung," as well as the extensive critical literature on *Bildung* in literature, biology, and musical aesthetics cited there.

concerns the species, it was a victory. Man, therefore, has cause to ascribe to himself the guilt for all the evil that he suffers and for all the bad that he perpetrates, while at the same time, as a member of the whole (of the species), admiring and praising the wisdom and purposefulness of the arrangement."[26] No longer the transcript of inner experiences that comprise the *vita contemplativa,* modern narrative mediates and thus secures the reality of the knowledge that it expresses. As such it positions itself as the indispensable (textual) correlate of Blumenberg's "process of theoretical curiosity" that, beginning with Giordano Bruno, Copernicus, and Descartes had decisively transformed the conception of human intelligence. Modernity's reconceptualization of the cosmos as a self-regulating and open-ended dialectical process rests on a number of complex and richly intercalated conceptual traits. The formal shift from epic to novel, so lucidly analyzed in György Lukács's 1914 *Theory of the Novel* and in Walter Benjamin's 1936 "The Storyteller," arises from modernity's decisive abandonment of all ontological guarantees. Whereas in the classical epic the art of telling draws on the authority of received past knowledge (one's ancestry, past debts and crimes, as well as "wisdom" alternately received or acquired as the hero responds to the claims, counsels, and memories of previous generations), modern narrative no longer derives its legitimacy from an appeal to antecedent realities and memories but from its own discontinuous and performative imaginings of an as yet unrealized future.[27]

Likewise, modern political economy repositions the metaphysical category of *providentia* by positing a fundamental convergence (Adam Smith's "invisible hand") between the vicarious purposiveness of compound human interests and "the wisdom of God." Analogously, modern narrative, though perennially vexed by the question of how to achieve formal closure, is sustained and (hypothetically) legitimated by its open-ended acquisition, authentication, and redistribution of an entirely new commodity: "information." Inasmuch as narrative seeks to ascertain and legitimate the intellectual resourcefulness and state autonomy of its subject—which is to say, protagonist *and* reader—closure in the form of a redemptive ending will necessarily prove counterintuitive, indeed would seem to discredit the modern self's claim to intellectual autonomy. Instead, the category of "error" acquires pivotal significance as the principal catalyst of a human intelligence that in time recognizes itself as the primary beneficiary of its own uneven (dialectical) development. It follows that "the final overcoming of the Gnostic inheritance cannot *restore* the cosmos because the function of the idea of the cosmos is reassurance about the world and in the

26. Kant, *Perpetual Peace,* 54–55.

27. For a strident critique of political economy as a specious version of "economic theodicy" (esp. in Adam Smith and James Steuart), see Milbank, *Theology,* 26–47.

world ... The world cannot be made 'good' in itself once more by a mere change of sign *because it would then cease to be man's irritation and provocation*" (*LMA*, 140; italics mine). At the level of political economy, this position will manifest itself in the anti-interventionist rhetoric of classical liberalism, such as Bernard Mandeville's aggressive rejection of the virtues and James Steuart's and Thomas Malthus's more cautious but no less consequential claim that (Catholic) emphasis on charity fails to preserve an element of need among the working poor and so deprives them of a stimulus of "continuous, organized discipline."[28] Designed forever to uproot error *and* at the same time premised on the continued, productive harnessing of further error, modern narrative conceives social processes as inherently dialectical, even (perhaps especially) where they do not understand themselves to be so. Inasmuch as modern narrative's outlook on experience shuttles back and forth between an insistent questioning of what is given and acts of (provisional) inference, it recognizes error as a productive, indeed indispensable component; for "'good error' ... testifies to an ingenious process of inference which overreaches itself by relying partly on mistaken assumptions. Thus the very rise of inferential power brings with it the conjoint capacity for inferential error."[29]

For post-Cartesian modernity, error is the blood sustaining its circulatory system of meaning, even as such meaning is achieved only through our continual anticipation or projection of future states yet to be realized. As the example of Goethe's *Wilhelm Meister* also makes clear, "error" and the modern individual's strictly "errant" developmental trajectory is symptomatic of modern life's infinite precariousness. For rationality here no longer constitutes an ontological framework on the order of the Platonic *logos* but amounts to a merely emergent and terminally contingent property. Reason now will have to be continually recertified and relegitimated through the self's practices of methodical inquiry and mimetic socialization, both of which appear contingent and reactive in nature. Like the *Bildungsroman* genre by which it is so productively harnessed, "error" and the precarious rationality achieved by the self's dialectical recovery from it furnishes a "specific image of modernity ... because of its ability to *accentuate* modernity's dynamism and instability."[30] The wisdom imparted to Goethe's meandering protagonist by the magisterial figure of the Abbé ("the duty of a teacher is not to preserve man from error but to guide him in error, in fact, to let him drink it in, in full draughts") proves both inherently plausible and shrewdly moderate in what it promises.[31] Indeed, strikingly similar conceptions

28. Milbank, *Theology*, 31; see also Herdt, *Putting on Virtue*, 248–282 and 306–321.

29. M. Polanyi, *Personal Knowledge*, 74.

30. Moretti, *Way of the World*, 5; see the above discussion of "error" and modern narrative in Chapter 5.

31. Goethe, *WMA*, 273, 302 (last trans. modified).

inform a great deal of Romantic narrative, such as the dialectically self-correcting *récit* of *The Prelude,* the halting and perpetually vexed inner progression of Jane Austen's Fanny Price or Anne Elliott, or indeed the endlessly self-revising hermeneutic progression imposed on readers by Blake's illuminated books.

Yet Blake's canny hybridization of visual and textual cues whose strategic interference challenges the reader to jettison any static, dualist model of understanding in favor of a dialectic of "contraries," also exposes a fundamental predicament within the modern, emergentist conception of agency. For to conceive the modern self as both constitutively isolated *and* forever provisional as regards its ability to identify, articulate, and justify a *telos* for its very life (a life that naturally continues to unfold all the same), modern dialectical narrative has dramatically widened the gap between the individual's severe epistemological limitations and the sweeping demand for its moral and ethical self-legitimation. Put differently, the scope and authority of practical rationality have been dangerously narrowed by modernity's quintessentially "errant" condition. The self now appears consumed by an insidious *Angst* of taking on commitments as a *practical* agent in ways that may never be fully justifiable in *theoretical* (i.e., epistemological) terms. At the same time, that very anxiety translates into what Augustine had anathemized as the sin of *curiositas*—that is, a compulsion to subject the ambient world to unrestrained theoretical experimentation. That quest, which as remains to be seen is so vividly dramatized by Coleridge's *Rime of the Ancient Mariner,* stems from a perceived need for total control over a distressingly alien and indifferent material and social world into which the modern self finds itself thrust and which it must navigate without guidance by a normative framework on the order of Aristotle's "practical rationality" (*phronēsis*).[32] Consequently, it is the utopian vision of eventual, complete theoretical mastery over that otherness that both impels that very pursuit and serves to justify the horrendous costs of it in the present. More than his Romantic contemporaries or, for that matter, subsequent critics of modernity, Coleridge is acutely aware of and profoundly disturbed by the hegemony of modernity's exclusively epistemological and instrumental outlook on the world, a stance that no longer accepts any antecedent, normative obligations or constraints.

---

32. On the displacement of *phronēsis* by calculative, impersonal, and ostensibly value-neutral modes of knowing—which coincides with the decline of rhetoric in the eighteenth century and is concurrently reflected in the rise of the fact/value distinction—see Murdoch, *Metaphysics,* 25–57; MacIntyre, *Three Rival Versions,* esp. 3–31; Gadamer, *Truth and Method,* esp. 278–306; Dupré, *Enlightenment,* 18–44; and Thorne, *Dialectic,* 178–221. For an earlier instance of this argument (and a remarkable qualification of his lifelong quest for an objectivist phenomenology), see Husserl, "Crisis of the European People," 149–192.

In his "Introduction" to the *Phenomenology of Spirit,* Hegel surmises that "the fear of falling into error sets up a mistrust of Science, which in the absence of such scruples gets on with the work itself and actually cognizes something; ... should we not be concerned as to whether this fear of error is not just the error itself."[33] By showing the Enlightenment's illusion of self-possession to be most palpable in its critique of tradition, Hegel draws attention to modernity's gratuitous distrust in a cosmos whose rational and teleological constitution had formerly been thought as an ontology both independent of and impervious to human intervention. Once again, the Gnostic underpinnings of modernity come into view inasmuch as "the experience of my own error ... at least excludes the interpretation of the postulate of divine benevolence, which had assumed it to be His will that I should *never* be deceived." The fact of error as quite possibly the only constant underlying all of human experience has far-reaching moral and epistemological implications. As Blumenberg continues, it thus appears that if "God, without the cooperation and consent of the subject, can directly produce the latter's acts of perception and thus bring about error without any lapse on the part of the knowledge seeker, then He could also produce morally reprehensible actions, such as hate for one's neighbor and even for God, directly and without the supposed agent being responsible."[34] As we are about to see, it is precisely this theological conundrum wrought by the ontological fact of free, prerational choice that Coleridge's *Rime* captures in all its fatal material and imaginary entailments.

At the level of epistemology, meanwhile, the same "continued experience of error as a fundamental human condition either leads to the hypothesis of god as a *genius malignus* or to the self-conscious Enlightenment project of a radically new beginning, a supersession of divine creation by human self-invention" (*LMA,* 186; trans. modified). Against the Enlightenment and its Hegelian apotheosis, Schopenhauer in 1818 maintains that "discursive concepts of reason," because of their merely derivative and self-certifying nature, are perennially liable to error:

> Perception by itself is enough; therefore what has sprung purely from it and has remained true to it, like the genuine work of art, can never be false, nor can it be

33. *PS,* 47; see also Hegel's later discussion of error in his chapter on "The Struggle of the Enlightenment with Superstition," *PS,* 333–334.

34. *LMA,* 195–196; Blumenberg continues by noting how it is precisely the concept of freedom that in the realm of moral theory seeks to accomplish what Cartesian doubt had sought within formal epistemology: "the introduction of the concept of freedom into the theory of knowledge is an attempt to apply the paradigm of the transcendent incontestability of morality to theoretical self-assertion. A man may be chosen or condemned in the theological sense, destined for salvation or the opposite—but no 'external' agency can make him responsible for such a destiny."

> refuted through any passing of time, for it gives us not opinion, but the thing itself. With abstract knowledge, with the faculty of reason, doubt and error have appeared in the theoretical, care and remorse in the practical. If in the representation of perception *illusion* does at moments distort reality, then in the representation of the abstract *error* can reign for thousands of years, impose its iron yoke on whole nations. (*WWR,* 1:35)

Identifying error as an indelible trait of the modern thought, Schopenhauer offers a predictably bleak view of a mode of existence permanently bereft of any teleological and normative framework. If "the exigency of self-assertion became the sovereignty of self-foundation" and modern rationalism posits a "freedom that does not *submit to* the conditions under which reason has to prove itself radically but *poses* them for itself" (*LMA,* 184), it took until the early Romantic era for modernity to begin tabulating the immense costs and risks associated with this shift. Thus, a formal symptom of modernity's assertion of an all-encompassing autonomy involves the persistent vacillation of its progress narratives between chiliastic utopias and paranoid dystopic scenarios. The peculiar restlessness and overwrought affect of modern narrative stems from the nascent awareness that, in any given account, *everything is at stake.* There is no such thing any longer as *a* story of limited scope or timeless and self-evident validity. Instead, every narrative anxiously probes the validity of a wholly anthropomorphic world and the hermeneutic strategies by which that world might be rendered intelligible.[35] Such a formal shift is bound to leave the modern individual exhausted with the sheer magnitude and terminal uncertainty of the cognitive effort involved, a predicament that shows how in the modern era biographical and cosmological time (*Lebens-* or *Jetztzeit* and *Weltzeit*) no longer stand in any stable and identifiable relation to one another. As Hannah Arendt puts it, modern subjectivity was able to pursue the "enormous enlargement of human capabilities" only at the expense of processes "whose outcome is unpredictable, so that uncertainty rather than frailty becomes the decisive character of human affairs" (*HC,* 232).

Long before the protagonist of Ian McEwan's *Amsterdam* confronts the Gnostic specter of human history as an all-encompassing miscarriage—a materialist dystopia of unfettered production and mindless consumption—Schopenhauer had forcefully

35. The rise of philosophical hermeneutics in the wake of the higher biblical criticism of Ernesti, Wolf, and Schleiermacher thus explains how, beginning with Romanticism, texts are invested with greater authority than readers by testing and correcting their audience's often deficient hermeneutic habits and frames. On this shift, see Rajan, *Supplement of Reading,* 15–35; on the Romantic ballad as a paradigmatic case of a text confounding the hermeneutic expectations of the reader, see Pfau, *Wordsworth's Profession,* 141–259; on Schleiermacher and Romantic hermeneutics, see Frank, *Subject and the Text,* 1–96, and Frank, *Schleiermacher,* 7–67, Pfau, "Immediacy and the Text," and Ellison, *Delicate Subjects,* 45–99.

probed the psychological implications of this development. Mobilizing a key trope that Gnosticism had bequeathed modernity—the image of man's perilous nautical venture into the unknown—the following passage (its fame eventually enhanced by Nietzsche's decision to quote it in full in the *Birth of Tragedy*) conjures up the existentially "anxious" constitution of the modern individual. However "buffered" against contingency, the modern self remains haunted by a spectral, apocalyptic vision in which modernity's elaborate institutional, professional, and conceptual architecture is recognized as a doomed attempt to stabilize and legitimate existence in a world wholly bereft of inner certitudes or metaphysical guarantees. Skillfully enjambing the infinity and anxiety as the joint epistemological and affective dimensions of modern existence, Schopenhauer (like Freud in *Civilization and its Discontents* more than a century later) refuses to answer the question already confronted by Coleridge's Mariner: viz., whether the terror of an empty infinity or of an apocalyptic *plērōma* is ultimately worse:

> Just as the boatman sits in his small boat, trusting his frail craft in a stormy sea that is boundless in every direction, rising and falling with the howling, mountainous waves, so in the midst of a world full of suffering and misery the individual man calmly sits, supported by and trusting the *principium individuationis,* or the way in which the individual knows things as phenomena. The boundless world, everywhere full of suffering in the infinite past, in the infinite future, is strange to him, is indeed a fiction. His vanishing person, his extensionless present, his momentary gratification, these alone have reality for him; and he does everything to maintain them, so long as his eyes are not opened by a better knowledge. Till then, there lives only in the innermost depths of his consciousness the wholly obscure presentiment that all this is indeed not really so strange to him, but has a connexion with him from which the *principium individuationis* cannot protect him. From this presentiment arises that ineradicable *dread,* common to all human beings (and possibly even to the more intelligent animals), which suddenly seizes them, when by any chance they become puzzled over the *principium individuationis,* in that the principle of sufficient reason in one or other of its forms seems to undergo an exception. For example, when it appears that some change has occurred without a cause, or a deceased person exists again; or when in any other way the past or the future is present, or the distant is near. The fearful terror at anything of this kind is based on the fact that they suddenly become puzzled over the forms of knowledge of the phenomenon which alone hold their own individuality separate from the rest of the world.[36]

36. *WWR,* I:353; for a discussion of this famous passage, see Blumenberg, *Shipwreck,* 59–65.

With his stylistic gifts on full display, Schopenhauer here identifies several of the troubling and potentially insoluble antagonisms that define the modern individual: there is its existential isolation and contraction to the logical *punctum* of sheer will; there is its loss of deep and complex memory as a result of modern life's fixation on the "extensionless present"; there is the deep alienation from a world reduced to an infinite sum of objects or inert "stuff" that renders the ancient cosmos "indeed a fiction." And, finally, there is the steadily growing suspicion that the conceptual armature that is to render this strange environment safe, predictable, and habitable will either collapse in times of great duress or, worse yet, will triumph so completely as to denude existence of all curiosity, charisma, and mystery. The "ineradicable dread" of which Schopenhauer speaks can alternatively be interpreted as one of sudden, apocalyptic collapse or of an infinity dulled by the transactionalism of social conformism and impersonal knowledge. In either case, Schopenhauer suggests, the modern individual's existential *Angst* stems from the way in which the scope and meaning of the world has contracted to the chimerical play of "representation" (*Vorstellung*). "World" has been pared down to an anthropomorphism, the tenuous correlate of an act of (Gnostic) creation whose result is barely endurable because it so palpably betrays its original conception.

Prior to Schopenhauer, it is only Coleridge's *Rime of the Ancient Mariner* that delineates these troubling implications of the modern project with such force and, in so doing, confronts yet again the unresolved legacy of ancient Gnosticism.[37] At once a key statement for its period and a parable about the philosophical predicament of modernity, the poem powerfully captures the ontological indeterminacy of the modern autonomous and voluntarist self. Built around one of the most enduring and

37. For repeated, and not always dismissive references to Gnosticism in Coleridge's oeuvre, see his *Lectures 1795*, 195–202; the remarks in that text are largely based on scholarship by his fellow Unitarian, Joseph Priestley, who offers a more comprehensive account of Gnosticism in his 1786 *An History of Early Opinions Concerning Jesus Christ*, vol. I, 139–180; Coleridge returns to Gnosticism in *TT*, 1:35–36 and 158–159, as well as in his parenthetical invocation of "the most cloudy gnostics" (*OM*, 193); like Priestley, for whom all occasional heresy by the apostles "make[s] no more than one system ... which, in the age after the apostles, was universally called *Gnosticism*" (142), Coleridge also recognized Gnosticism as a theological vision largely defined by its detractors and a posteriori ("those, who were afterwards called Gnostics" [*TT*, 1:35]) and expressed regret over the expurgation of their writings: "I regard the extinction of all the Writings of the Gnostics among the heaviest losses of ecclesiastical Literature" (*CM*, 5:624; see also *CM*, 2:713); at times, Coleridge claims a strong intellectual kinship with gnosis: "I solemnly bear witness and declare that every Idea, Law, or Principle in which I coincide with the Cabbala, or the School of Plotinus, or the Christian Gnostics ... I *recognized* in them, as truths already known by me in my own meditation" (*CM*, 4:258–259). In some late notebook entries, Coleridge emphasizes Gnosticism's role in widening the gap between Hebrew (Old Testament) and apostolic (New Testament) theology—"the Object of the Gnostics being to destroy the connection between Christianity and Judaism" (*CN*, no. 5743; see also nos. 5593 and 6309).

ambivalent tropes in Western metaphysical writing—that of the voyage or, rather, shipwreck—the poem's narrative hinges on a single, wholly inexplicable instance of pure and groundless volition.[38] The Mariner's gratuitous shooting of the albatross furnishes us with a parable for the *hubris* that is modernity, specifically its founding, purely volitional act whereby the solitary individual shatters the cosmos by turning it into an inventory of disaggregated objects to be subjected to (inherently skeptical) analysis and experimentation. To do so is to jeopardize the twin theological axioms of a pre-stabilized cosmos and of mankind's eventual salvation, for both can only ever be guaranteed by the recurrent rhythms and norms of past experience and inherited traditions. Yet in the Mariner's singular act of skepticism/killing, all such frameworks are now sacrificed to a radically new, speculative type of curiosity no longer governed by received norms but anxiously cathected onto uncertain future "outcomes."

Far from gratuitously disrupting an otherwise orderly progression, the Mariner's killing of the albatross stands as a synecdoche for the scientific and commercial exploits that modernity so often captures in the master-trope of seafaring and shipwreck. In surveying that trope's pervasive role in Western writing, Hans Blumenberg notes "the ancient suspicion that underlies the metaphorics of shipwreck: that there is a frivolous, if not blasphemous, moment inherent in all human seafaring, on a par with an offense against the invulnerability of the earth, the law of *terra inviolata,* which seemed to forbid cutting through isthmuses or building artificial harbors"; for Blumenberg, two

> assumptions above all determine the burden of meaning carried by the metaphorics of seafaring and shipwreck: first, the sea as a naturally given boundary of the realm of human activities and, second, its demonization as the sphere of the unreckonable and lawless, in which it is difficult to find one's bearings. In Christian iconography as well, the sea is the place where evil appears, sometimes with the Gnostic touch that stands for all-devouring Matter that takes everything back into itself. It is part of the Johannine Apocalypse's promise that, in the messianic fulfillment, there will no longer be a sea (*he thalassa ouk esti etí*). In the purest form, odysseys are an expression of the arbitrariness of the powers that denied Odysseus a homecoming, senselessly driving him about and finally

38. Referencing Purchas, Shelvocke, Cook, and Barents as intertexts mined by Coleridge, David Simpson proposes to read the "Rime" as, among other things, "a romantic variation upon the genre of voyage narratives" ("How Marxism Reads," 153). For a full consideration of the trope, see Blumenberg, *Shipwreck*; on Coleridge's imaginative use of the nautical trope in his 1817 discussion of Spinoza, see Berkeley, "Providential Wreck."

> leading him to shipwreck, in which the reliability of the cosmos becomes questionable and its opposite valuation in Gnosticism is anticipated.[39]

Earlier readers of Coleridge's *Rime* had often mistakenly assumed that prior to the killing of the albatross the crew's nautical exploits were "innocent" or, in any event, unobjectionable. Yet already in the "Argument" prefaced to the poem's original, 1798 version, Coleridge ominously conjures up the transgressive nature of the nautical venture as such by glossing "How a Ship having passed the Line was driven by Storms to the cold Country towards the South Pole." At the risk of running afoul of narrowly historicist protocol, which would hold us to the Unitarian context in which, allegedly, Coleridge was still working out his ideas in 1797–1798, there is ample reason to read Coleridge's archaic and cryptic locutions (reminiscent of the "parabolic style" that Robert Lowth had explored in his *Lectures on the Sacred Poetry of the Hebrews* (Lat. 1753/Eng. 1787) as the parable of a metaphysical dilemma that was to preoccupy him ever more in his later prose, particularly in *Aids to Reflection* and in his posthumous *Opus Maximum,* texts to which I will return shortly.

Before doing so, however, a methodological and interpretive clarification appears in order. In disputing the innocence of the Mariner's journey prior to his killing of the albatross I do not mean to align my reading with the narrowly contextual readings of the *Rime* as a meditation (in prosodic form) on the slave trade, colonial disease, and the moral turpitude of British consumer culture with its seemingly insatiable demand for rum, sugar, cotton, and mahogany.[40] To be sure, Coleridge himself had regularly participated in that debate, both in the years prior to his writing the *Rime* and during the early 1800s. Thus even a cursory review of his statements on Britain's complicity in the slave trade shows that colonial disease (especially yellow-fever) and the practice of slavery furnish a number of metaphoric and symbolic devices used throughout the *Rime*—"parched" throats, "cold sweat," a crew of fully two hundred men (typical only for slave ships), etc. Yet to assume on the basis of these intertexts that in his *Rime* Coleridge simply had "set fever to poetry" is to succumb

39. *Shipwreck*, 10–11; 8.

40. "Fascinated by the idea that the Romantic imagination can reveal things hidden to the naked eye" and, assisted by the axiom that "the Romantic imagination [is] a purely political construct," Lee's *Slavery and the Romantic Imagination* unsurprisingly detects "signs of slavery in imaginative works" (1), including in Coleridge's *Rime*. For a particularly detailed historicist reading of the *Rime* (an argument to which Lee's own reading owes much), see Keane, *Coleridge's Submerged Politics*, 124–165 and 212–353; Keane is quite circumspect with regard to the potential methodological overreach of historicist critique; see also Ebbatson, "Coleridge's Mariner"; Kitson, "Coleridge, the French Revolution and the 'Ancient Mariner'"; and Bewell, *Romanticism and Colonial Disease*, 97–108.

to one of historicism's more basic (albeit persistent) interpretive fallacies.[41] Viz., it conflates figurative and proper meanings by assuming that the field of reference *from* which an expression (allegedly) "derives" its figurative character was therefore also the *sole intended topic* of the literary (symbolic) expression now at issue. Lee briefly acknowledges the volatility and complexity of Coleridge's tropes when noting how, "by marrying the tropes of fever and slavery" the *Rime* "also explores slippages between the walled-off categories of self and otherness," such as when, "in the heat of the poem's fever, the mariner *is identified* with Englishmen *and* slaves."[42] Once again, though, the use of "tropes" for Lee (herein quite representative of historicism's nominalist tendencies) is limited to the binarism of two stable, competing, and equally particular *references* identified on the questionable assumption that all symbolization is but a referential operation in disguise. As a result, the deep inter-implication of spirituality and rhetoric, which was to preoccupy Coleridge throughout his career and eventually prompts his definition of literature as symbolic action or the "translucence of the Eternal through and in the Temporal " (*CLS*, 30), is here pared down to a quasi-sociological accumulation of so many cross-references and decodings. Unable to grasp what, for Coleridge, is the very essence of creative, symbolic action—viz., its capacity "to enunciate the whole"—Romantic historicism's strictly referential approach to literature begs its question on a grand scale, viz., by continually positing the symbolic as but a covert *re*petition or *re*presentation of an already established and familiar field of reference.

Such an assumption, if granted, threatens to dissolve reading into the mere default value of archival industriousness and essentially associative habits of thought; as a result, complex hermeneutic judgments are supplanted by an explanatory regress to antecedent meanings and contexts of ever-increasing particularity and, it would appear, outright irrelevance to the present. Inasmuch as historicism tends to restrict *meaning to reference* and, in apparent circularity, conceives and claims reference as the retrieval of (putatively hidden) antecedent meaning—its methodological confidence and intellectual poverty are two sides of the same coin. Incapable of grasping meaning as an *emergent* property, historicism instead construes Romanticism's key concepts—originality, novelty, imagination, or what Coleridge calls "creative words" (*BL*, 2:129)—merely as so many elaborate, if unwitting, obfuscations of "the real." One has to be surprised, perhaps even dismayed by the persistent appeal of

41. Lee, *Slavery and the Romantic Imagination*, 49. See Coleridge's "Lecture on the Slave-Trade," in *Lectures 1795*, 231–251; a fascinating dream-vision of a post-slavery world is offered in an article of 2 April 1796, published in *The Watchman*, 163–165; Coleridge's entry for 25 March 1796 (no. IV) restates his principal arguments against slavery and scrutinizes arguments offered in favor of continuing the practice of slavery (*The Watchman*, 132–140); on slavery and disease, see Coleridge's "Observations on Egypt" (6 December 1804), in *Essays on his Times*, 3:200–205.

42. Lee, *Slavery and the Romantic Imagination*, 53; first italics mine.

a historicism that claims to expose—and, in so doing, purports to *emancipate us from*—the past, which it prejudges as a far-flung, delicate web of elisions and formal-aesthetic obfuscations that demand unrelenting and painstaking archival and contextual reconstruction. For there certainly have been powerful critiques of the historicist project, beginning with Hegel and extending through Nietzsche and Heidegger to Hans-Georg Gadamer. A half century ago, Gadamer had pointed out that Romantic (old) historicism, while ostensibly "reversing the Enlightenment's presupposition [i.e., of attainable perfection and conclusive emancipation from the past] actually perpetuates the abstract contrast between myth and reason" that the Enlightenment itself had so assiduously propagated. Ultimately, "the historical consciousness that emerges in romanticism involves a radicalization of the Enlightenment. For nonsensical tradition, which had been the exception, has become the general rule for historical consciousness."[43]

Yet to suppose that "the whole of the past—even, ultimately, all the thinking of one's contemporaries—is understood only 'historically'" is to foster the illusion of an observer methodologically immune to and detached from that past. Precisely this "overcoming of all prejudices, this global demand of the Enlightenment, will itself prove to be a prejudice." To tackle it, we must begin by acknowledging our continuity with the past or, in Gadamer's words, "the finitude which dominates not only our humanity but also our historical consciousness." Far from being indemnified by a methodological protocol and the spurious comforts of hindsight and scientific detachment from the phenomena at hand, the historicist would do well to understand himself or herself as working within a "process of tradition" (*Überlieferungsgeschehen*): "We are always situated within traditions, and this is no objectifying process—i.e., we do not conceive of what tradition says as something other, something alien. It is always part of us, a model or exemplar, a kind of cognizance that our later historical judgment would ... regard as the most ingenuous affinity with tradition." Gadamer's view of tradition as tied to the hermeneutic medium of language, text, and interpretation ("tradition is not simply a permanent precondition; rather, we produce it ourselves inasmuch as we understand, participate in the evolution of tradition") thus reveals what he calls "the ontological structure of understanding."[44] As we shall see in the context of Coleridge's exploration of the concept of person, understanding in the human sciences is inherently preoccupied with value concepts. This means that the meaning and significance of such concepts is not independently verifiable by some alternate methodology granting us objective access to their putative "referent." Rather, it is only by tracing the history of such a concept's evolution as an

43. *Truth and Method*, 275, 277.
44. Ibid., 277, 283–284, 293–294.

"idea" (in John Henry Newman's sense) or as a "tradition" (in Gadamer's sense) that we can have any purchase on its meaning and import *for us*. More recently, John Milbank has echoed Gadamer's profound insight into the ontological character of the hermeneutic circle, arguing that "it is impossible to isolate the pre-given, categorical element (which for sociology is schematized as 'society'—as fact or norm) from the flux of becoming. The point here is not that one never has 'unbiased' access to the social genesis, but rather that there *is* no pre-textual genesis: social genesis itself is an 'enacted' process of reading and writing. Curiously enough, it is much easier to talk about the 'social background' of a text when it stands relatively alone; in the mesh of intertextuality provided by a situation of rich evidence, the supposed purely social object much more evidently disappears."[45]

From a somewhat different vantage-point, Adorno and Horkheimer have also pointed to the hubris—and a quasi-mythical fear masked by it—that drives both the Enlightenment and a romantic historicism that is its unwitting apotheosis. Key to the totalizing claims of modern epistemological method even, or perhaps especially, where it surfaces within the human sciences, is a desire to expose the phenomenon at hand as the unacknowledged repetition of some prior constellation. Thus *Dialectic of Enlightenment* scrutinizes the Enlightenment's postulate of "the identity of everything with everything else" as the moment where modernity's pre-emptive commitment to method as salvation loops back into the mythical indifference that the Enlightenment purports to overcome:

> The principle of immanence, the explanation of every event as repetition, which the Enlightenment upholds against mythic imagination, is the principle of myth itself. That arid wisdom that holds there is nothing new under the sun, because all the pieces in the meaningless game have been played, and all the great thoughts have already been thought, ... merely reproduces the fantastic wisdom that it supposedly rejects: the sanction of fate that in retribution relentlessly remakes what has already been. What was different is equalized. That is the verdict which critically determines the limits of possible experience [as] ... universal mediation, the relation of any one existent to any other. (12)

Yet even if we set aside historicism's often unwitting implication in the conceptual aporias of modernity, Coleridge's own writings strongly militate against conflating symbol with allegory and reading with cross-referencing. First and foremost, as the consternation of its first readers makes clear, Coleridge's *Rime* stands well apart from

45. *Theology*, 105, 115; on the conceptual weakness of historicism, see also my "Reading beyond Redemption."

the established tropes and expressive conventions of antislavery poetry that had emerged as a popular genre of middle-class moral sentimentalism during the 1790s and early 1800s. Within the *Rime*—whose tone differs sharply even from Coleridge's own earlier protest poetry (e.g., his 1794 Sonnets "To Kosciusko," "To the Hon Mr Erskine," "To Burke," etc.; "Domestic Peace," "The Destiny of Nations," "Fears in Solitude" etc.)—the slave trade, yellow-fever, and the moral isolationism (represented by the wedding guest) of late eighteenth-century British consumer culture symbolize the systematic and relentless commodification of natural resources and human life alike. The principal target of the poem's elaborate symbolic and narrative patterns thus is not some particular political or moral topic of debate. Rather, Coleridge targets the broader ideological framework of modernity that had produced these distressing empirical scenarios by sacrificing practical reason to unbridled scientific and economic speculation.

Even in his early "Lecture on the Slave-Trade" (1795), Coleridge does not simply follow the prevailing line of argument in antislavery pamphleteering by indicting the systemic cruelty of the slave trade and exposing the self-serving economic and racial arguments invoked by its apologists. Rather, after opening with a quintessentially metaphysical question: "Whence arise our Miseries? Whence arise our Vices?" Coleridge proceeds to argue that a purely Epicurean vision of unbridled and interminable consumerism ("to find Happiness in the complete gratification of our bodily wants") effectively betrays the ontological purpose with which all human life has been invested by its creator: viz., "to busy itself in the acquisition of intellectual aliment" and "to develope the powers of the Creator."[46] This early passage already anticipates Coleridge's figural reading of slavery as a glaring symptom of modernity's contraction of reason to a calculus of means and ends, a shift that also threatens to enslave those ostensibly free in a life of material consumption, intellectual servitude, and spiritual abjection scarcely less horrifying than the depredations visited on indigenous peoples in Africa, the West Indies, and other parts of the globe subjected to the British colonial enterprise. For Coleridge slavery is not simply, nor even primarily, an "injustice" perpetrated within the contingent world of politics and the law; neither is it simply some contingent violation of a people's or individual's "rights." Rather, slavery is essentially a "sin" because it fundamentally negates the human person as a being whose reality and dignity Coleridge had come to perceive as wholly entwined within the person's "recognition" by others; as he formulates the matter in an essay of 1811:

---

46. *Lectures 1795*, 235; for another instance of Coleridge's anti-Epicurean critique of modern consumerism, see "Comparison of France with Rome" (*Morning Post*, 21 September 1802), in *Essays on his Times*, 1:315.

> The Contra-distinction of PERSON from THING being the Ground and Condition of all Morality, a system like this of Hobbes's, which begins by confounding them, needs no confutation to a moral Being. A Slave is a *Person* perverted into a *Thing*: Slavery, therefore, is not so properly a deviation from Justice, as an absolute subversion of all Morality.[47]

It is in this metaphysical rather than occasionalist, legal-political sense, that from the very outset the Mariner's seafaring is depicted as a transgressive pursuit. What troubles Coleridge is not the occasional, wayward act of injustice but sin as a systemic, institutional *practice* licensed by modern instrumental reason. The slave trade merely throws into conspicuous relief what John Milbank calls the "'neo-pagan' character of ... political economy and its outright celebration of what Christian theology rejected, viz., the *libido dominandi*." Cued by modern capitalism's erratic cycles of boom and bust, fluctuating labor costs, and unpredictable swings in the price of essential provisions, Coleridge's *Rime* foreshadows the dystopic economics of Thomas Malthus and David Ricardo far more than it echoes the meliorism of James Steuart and Adam Smith. The poem's apocalyptic imagery and causally unfathomable narrative call into question an eighteenth-century "economic theodicy ... conjoined with an evangelicalism focused on a narrow, individualist practical reason which excludes the generous theoretical contemplation of God and the world." Moreover, Coleridge's unrivaled probing of the psychology of guilt (hardly confined to the *Rime*) also repudiates a version of Christianity that the Unitarianism and deism of his times (and his own youth) had "thinned down to a simple acceptance of positive revealed data which ensures salvation."[48]

In short, the *Rime*'s true concern lies not with the economic, legal, and political wrong of slavery allegedly sequestered behind a "surfeit" of symbolic allusions and metaphysical concerns. Rather, the true catastrophe of modernity lies in its unconditionally espousing a means/end model of rationality as the sole way of being in the world—thereby morphing Aristotelian or Augustinian notions of the good into a strictly economic and utilitarian calculation of contingent advantages. With some

---

47. "The Catholic Petition" (Letter III), in *Essays on his Times*, 3:235n.; the distinction between person and thing had previously surfaced in *The Friend*: "Every Man is born with the faculty of Reason: and whatever is without it, be the Shape what it may, is not a man or PERSON, but a THING. Hence the sacred Principle, recognized by all Laws human and divine, the Principle indeed, which is the *ground-work* of all Law and Justice, that a Person can never become a Thing, nor be treated as such without wrong" (12 October 1809, *CF*, 2:125).

48. Milbank, *Theology*, 37, 47. On Coleridge's evolution from Unitarianism toward High (Tory) Anglicanism, a process significantly driven by his appropriation of Hooker's writings (especially in *CCS*), see Wright, *Coleridge and the Anglican Church*, 185–205.

justification, Stanley Cavell thus reads the *Rime* as an extended figuration of modernity, an epoch at once sinful and *sui generis* in that it involves "a mental line to be crossed that is interpreted as a geographical or terrestrial border."[49] In contrast to Robert Penn Warren's reading of the *Rime* as encrypted theology, Cavell does not approach the poem as a straightforward allegory of the Fall but, "on the contrary, ... take[s] it to provide an explanation of why it fits the Fall, that is, of what the Fall is itself an allegory of." Thus the Fall does not constitute the poem's proper meaning but only serves as our figurative conduit to it. What is being allegorized is, ultimately, "the threat of skepticism [as] a natural or inevitable presentiment of the human mind ... The beginning of skepticism is the insinuation of absence, of a line, or limitation, hence the creation of want, or desire."[50] Cavell here draws on Kant and a well-known passage from the *Biographia Literaria* where Coleridge posits that "the first principle" of a philosophical system is "to render the mind intuitive of the *spiritual* in man (i.e., of that which lies on the other side of our natural consciousness) ... in truth a land of darkness, a perfect *Anti-Goshen* for men to whom the noblest treasures of their own being are reported only through the imperfect translation of lifeless and sightless *notions.*"[51] For Coleridge, the first casualty of modernity is what he calls "the spiritual in man." Its demise is necessarily hastened by modernity's principled embrace of a skeptical (which is to say, inherently *re*active) and methodical paradigm of knowledge that unrelentingly scrutinizes and rejects whatever is "merely" intuitive or ideational and then proceeds to classify it as irrelevant because it is indemonstrable *for others.*

That skepticism is not merely a sudden consequence of the Mariner's capricious act of killing can also be inferred from the cryptic and uneasy geography of the ship's course. To begin with, Coleridge's Mariner and his crew are obviously no ordinary sailors; their world is "a ship with no rank or hierarchy at a time when ships were *all* rank and hierarchy."[52] Furthermore, their palpably anti-realist journey has no stated goal or purpose but—like its compulsive, indeed coercive retelling to the hapless wedding guest—the narrative is but an accretion of isolated episodes strung up like so many beads with the help of the ever-same conjunctive phrases: "*And now* there came both mist and snow, / *And* it grew wondrous cold: / *And* ice, mast-high,

49. Cavell, *Quest,* 46.

50. Ibid., 46–50.

51. *BL,* 1:243; Cavell readily concedes that Kant's writings were not to take full hold of Coleridge's thought until some years later: "I am not saying that when he wrote his poem he meant it to exemplify Kant's *Critique of Pure Reason,* merely that it does so" (*Quest,* 47); for an old historicist reading at the opposite end of the critical spectrum, see Ulmer ("Necessary Evils"), who focuses above all on Coleridge's engagement with necessitarianism and determinism.

52. Simpson, "How Marxism Reads," 156.

came floating by" (ll. 51–53); "*At length* did cross an Albatross" (l. 63); "*And* a good south wind sprung up behind" (l. 71); "*And* the good south wind still blew behind" (l. 87), and so forth. As is evident from its strictly sequential presentation, the poem's organizing trope of seafaring embodies a quintessentially modern paradigm of experience. The quest is for the unconditionally "new," and hence as yet devoid of authoritative (let alone reflected) criteria that would enable the Mariner to articulate the *meaning* of said experience except by compulsively reiterating its distinctive phenomenology. What renders the poem's *récit* so arresting is not the presence of an overarching and identifiable meaning but the sheer charisma or "feel" of an experience that continues to dominate the Mariner's (and his listener's) consciousness with traumatic force. The Enlightenment vision of transparent and incontrovertible meaning has yielded to a vicious cycle of compulsive repetition in the present, punctuated by inarticulate and unfocused hopes for some future *parousia*.

For David Simpson, "the Mariner's return to his 'own countrée' is not an act of reintegration into an intact local community, but a further exacerbation of his isolation and his inability to live in his actual place and time." Likewise, Blumenberg regards "shipwreck [as] something like the 'legitimate' result of seafaring, and a happily reached harbor or serene calm on the sea is only the deceptive face of something that is deeply problematic."[53] In his landmark study, Hans Jonas points to Gnosticism's central notion of the "alien," an "attribute of the 'Life' that is by its nature alien to this world . . . The alien is that which stems from elsewhere and does not belong here. To those who do belong here it is thus the strange, the unfamiliar and incomprehensible; but their world on its part is just as incomprehensible to the alien that comes to dwell here." Coleridge's Mariner reflects Hans Jonas's criterion of "spirit" (*pneuma*) gone astray in an incommensurable world quite precisely. Like the ontologically misplaced subject of Gnosticism, the Mariner "suffers the lot of the stranger who is lonely, unprotected, uncomprehended, and uncomprehending in a situation full of danger. Anguish and homesickness are a part of the stranger's lot," compounded in the *Rime* as in Gnostic thought by a "twofold . . . cosmic terror, the spatial and the temporal."[54] In sharp contrast to the methodological harnessing of "error" as a crucial, positive element in the self-regulating (dialectical) progression of modern reason, the Gnostics saw "error" (*πλἄνή*) very much as Coleridge also sees it, viz., as a wandering, a roaming, a going astray from which the embodied mind can never recover within the material world.

This impasse above all accounts for the *Rime*'s starkly anti-mimetic idiom, as well as for the persistent disequilibrium between tropes of knowledge and desire, respectively, both of which abound throughout the poem. As Harold Bloom notes, "in

53. Ibid., 157; Blumenberg, *Shipwreck*, 10.

54. Jonas, *Gnostic Religion*, 49, 53.

poetry, a 'place' is *where* something is *known*, but a figure or trope is *when* something is willed or desired."[55] The condition underlying the Mariner's opaque nautical explorations, meanwhile, not only recalls the "transcendental homelessness" that Georg Lukács was later to identify as the epistemological signature of modern narrative; it is also informed by an originally Cartesian skepticism from which, as Hegel was to argue, one can never return but which, faced with the impossibility of return, the modern individual must see through to its logical, desperate end: "The skepticism that ends up with the bare abstraction of nothingness or emptiness cannot get any further from there, but must wait to see whether something new comes along and what it is, in order to throw it too into the same empty abyss." It is in the nature of consciousness to "go beyond limits, and since these limits are its own, it is something that goes beyond itself ... Thus consciousness suffers violence at its own hands ... It can find no peace. If it wishes to remain in a state of unthinking inertia, then thought troubles its thoughtlessness, and its own unrest disturbs its inertia." Cartesian skepticism, then, remains a contingent and incomplete practice, a "conceit which understands how to belittle every truth, in order to turn back into itself and gloat over its own understanding, which knows how to dissolve every thought and always find the same barren Ego instead of any content" (*PS*, 51–52).

Why, David Simpson muses, does "Coleridge (in the 1798 poem) make us work out where the ship is going by deduction from the Mariner's report of where the sun rises and sets? Why does he blur the rather simple geography?"[56] Giving rise to itself by its own defiant or skeptical act—a primordial violation of nature—modern consciousness can never again appeal to external realities but must generate and refine a strictly *discursive* map of an exclusively *mental* world devoid of any relation (positive or negative) to so-called nature.[57] The vintage figure here is surely Kant who, more than anyone else, had succeeded at converting space and time into internal properties (*reine Anschauungen*) and, in marked antithesis to Newton and Descartes, had sought to show "that our knowledge of the world clarifies our relation to the world

---

55. Bloom, *Agon*, 69.

56. Simpson continues to muse on the troubled cartography of the poem: "The poem refutes the rationalist cartography that made Cook's voyages possible, as it also complicates any ambition we might have to devise a rationalist theory of the emotions, or of crime and punishment (of the sort that Bentham was setting out to deduce)" ("How Marxism Reads," 155, 157). I regard the Mariner's *disorientation* as emblematic and revealing to the reader how, in the de-centered cosmos of Galileo, Bacon, Descartes, and Hobbes, orientation is no longer achieved intuitively but only inferentially.

57. Perhaps the most rigorous and concise exposition of the phenomenology of *logos* and the sign is Heidegger's *Being and Time*; there Heidegger emphasizes the "detour-character" (*Umwegigkeit*) of *logos*. What stabilizes the sign is not prima facie its referential purchase on external reality but its mediating structure, which renders "visible" something that does not present itself as such. *Being and Time*, 25; for a concise recent account of problems with representationalism, see McDowell, *Mind and World*, 25–65.

rather than defining the nature of the world."[58] With the breakdown of philosophical realism (i.e., models of mind-object correspondence), the process of verification and the notion of truth itself have been decisively altered. Completing the Kantian reflection, Hegel thus notes in the "Introduction" to the *Phenomenology* how "the criterion of testing is altered when that for which it was to have been the criterion fails to pass the test; and the testing is not only a testing of what we know, but also a testing of the criterion of what knowing is" (*PS*, 54–55). The reason why it helps to dwell on Hegel's proposed remedy to this dilemma—viz., that "consciousness provides its own criterion [of knowledge] from within itself, so that the investigation becomes a comparison of consciousness with itself" (*PS*, 53)—is that it essentially re-enacts the Cartesian skepticism at a higher level and so proposes to solve the original dilemma of skepticism (viz., the loss of cosmological stability) by repeating the original transgression. Hence Hegel characterizes the overall narrative project of his *Phenomenology* as "self-perfecting skepticism" (*dieser sich vollbringende Skeptizismus* [*PS*, 50/*PG*, 67]).

It is precisely this Hegelian route toward redemption by "twofold negation" that Coleridge is *not* prepared to take, primarily because of its ethical indifference to the cosmos. For as soon as we shift from speculative thinking to material action, the most obvious equivalent for disputing the reality of otherness would be to analyze, anatomize, and, ultimately, kill. At the very least, in drawing out that analogy we understand how modernity's self-authorizing agent, having created himself by a primal act of skepticism, must henceforth inhabit an ethical limbo that this very act prima facie created. Within the narrative purview of Coleridge's poem and, more emphatically yet, in his later prose writings, skepticism thus proves nothing less than sin. Hence, too, the long tradition of interpreting Coleridge's *Rime* by taking recourse to some causal logic (most famously in Robert Penn Warren's reading) or by adverting to the apparent lack of any causal logic (William Empson, Edward Bostetter) was bound to miss the most salient point.[59] What separates the wedding guest's anxious question ("Why look'st thou so?") and the Mariner's bland response ("With my crossbow / I shot the ALBATROSS!" [ll. 81–82]) is a mere dash that pointedly forecloses on any causal explanation. For the Mariner's act is one of "motive-less malignity," a

58. Dupré, *Enlightenment*, 42.

59. Noting how "the criticism of 'The Rime of the Ancient Mariner' reflects a craving for causes," Frances Ferguson notes that "in Coleridge's work generally, intention and effect are absolutely discontinuous" ("Coleridge and the Deluded Reader," 113, 120). Her subsequent remark that the *Rime* offers "agonizing explorations of the difficulties of recognizing the full implications of an action before it is committed" (122) at least hints that the ontological status of the Mariner's act does not really allow us to speak of intentions in the first place.

radical instance of skepticism that categorically denies reality to another being.[60] The killing of the albatross launches the ship of modernity on its journey into what the likewise seafaring young Wordsworth recalls as "unknown modes of being."

I work the metaphor as hard as I do here to underscore that from here on tropes and linguistic markers are the only remaining substratum wherein one may hope to recover a community or, conversely, mourn its permanent loss. As Daniel Watkins notes, "to follow the Mariner's journey is to witness the breakup of a strong community and the emergence of the isolated individual in history. When the Mariner's ship begins its adventures it leaves a stable and conventional society behind, exemplified most clearly by the Christian values that critics have always recognized; this is followed by the disintegration of community on the ship (seen explicitly in the growing inability of the mariners to speak, that is, in the drying up of meaningful social exchange)."[61] To generations of readers vexed by the sharp asymmetry between the poem and its gloss, to say nothing of the reams of commentary that have accrued around the *Rime* over the past two hundred years, language itself thus appears as a troublingly amorphous sea of differential and often inchoate signs.[62] Unsurprisingly, the only salvation (such as it is) for the Mariner's existential dilemma involves the expressive mobilization of his "strange power of speech." Leaving aside the thorny issue of how the poem's eventual 1817 gloss burdens consciousness with additional exegetical labor and the sheer endless task of parsing the semantic shades of superficially univocal terms (a task that Coleridge's notebooks elaborate under the heading of "desynonymization"), we can already see in the 1798 version that the Mariner's compulsive retelling of his ineffable experiences constitutes a desperate and ultimately doomed attempt at sorting out his own narrative's myriad implications.[63] Cut adrift from all communal and object relations, the Mariner's emblematic impersonation of the modern condition is above all defined and circumscribed by his "strange power of speech." As Hegel was to elaborate in his *Phenomenology,* such a post-lapsarian state of affairs consigns the modern individual to the Sisyphean labor of constantly

60. Cavell, *Quest*, 56.

61. Watkins, "History as Demon," 31.

62. On the imperiled, at times impoverished status of language as expressive act and as meaning, see Modiano, "Words and 'Languageless' Meanings," and McGann, *Beauty of Inflections*, 135–172; in a later essay ("Sameness or Difference?"), Modiano offers a perceptive critique of McGann's historicist account.

63. Properly critical of the gloss and, especially, gloss-based interpretations, Ferguson thus characterizes the *Rime* as "a mini-epic of progress that moves largely by retrogradation" ("Coleridge and the Deluded Reader," 124). On the gloss in relationship to the new, "higher" biblical criticism, see McGann, *Beauty*, 135–172; on Coleridge's idea of desynonymization, see his remark that "all languages perfect themselves by a gradual process of desynonymizing words originally equivalent, as Propriety, Property—I, Me—Mister, Master—&c/." *CN*, no. 4397; the concept is also developed by Coleridge in *BL*, 1:82–84.

having to secure "uptake" or "acknowledgment" for strictly virtual, textually mediated *notions* that are no longer referentially anchored in any objective reality or nature.[64] As a result, "the world has become more mysterious and more threatening, an environment that puts under pressure the homiletic or proverbial rules of operation ('He prayeth best, who loveth best') to which one turns for guidance."[65] We shall yet see to what extent the threatening, anxiety-inducing world of the *Rime* is a consequence of the modern voluntarist conception of the self as pure will.[66]

---

64. For a discussion of Hegel's conception of acknowledgment, as elaborated in the master-slave section of his *Phenomenology*, see Pinkard, *Hegel's Phenomenology*, 46–55; Pinkard's reading elaborates Charles Taylor's earlier account in *Hegel*, 148–157; more recently Taylor has addressed more fully the centrality of "recognition" and "acknowledgment" within modernity's various models of selfhood; see Taylor, *Ethics of Authenticity*, 43–53.

65. Simpson, "How Marxism Reads," 153.

66. See below, 491–503.

16

BEYOND VOLUNTARISM AND DEONTOLOGY

Coleridge's Notion of the Responsible Will

At this point, we can begin to delve into some of Coleridge's late prose in order to draw out a number of related conceptual shifts for which the Mariner's defining act of skepticism furnishes an early and vivid dramatization. Central to this discussion are the concepts of will, person, and conscience—all of them profoundly inter-related in Coleridge's late writings. Beginning around 1804, Coleridge posits as unconditional and anterior to everything else the reality of the will, and by 1825 he apodictically defines it as "pre-eminently the *spiritual* Constituent in our Being" (*AR*, 75).[1] As claims go, such a position is deceptively forthright. For behind it lurks the corollary, so frequently ignored in the wake of modern secular theory, that to assent to the proposition that there *is* a will means by definition to recognize the *spiritual*

1. *AR*, 75. Early in his *Opus Maximum*, Coleridge notes how the ontology of the will licenses most other concepts relevant to the human sciences: "The one assumption, the one postulate, in which all the rest may assume a scientific form, and which granted we may ~~give~~ coercively deduce even those which we might allowably have assumed, is the Existence of the *Will*, which a moment's reflexion will convince us is the same as *Moral Responsibility*" (*OM*, 11). On Coleridge's concept of the will, see Perkins, *Coleridge's Philosophy*, 189–204; Hedley, *Coleridge*, 160–168; Muir, *Coleridge as Philosopher*, 145–161; Evans, "Reading 'Will' in Coleridge's *Opus Maximum*"; and Davidson, "Duty and Power."

foundations of human agency as free and capable of choice: "If there be aught *Spiritual* in Man, the Will must be such. *If* there be a Will, there must be a Spirituality in Man" (*AR,* 135). In so having established this "one great and inclusive postulate and moral axiom—the actual being of a *responsible Will* ... it is at the same time admitted that a something is meant by the Will distinct from all other conceptions" (*OM,* 17). Like any intellectual science, Coleridge insists, moral inquiry also begins with a postulate, "a fact ... taken for granted," which then sets in motion a complex chain of logical and, it is to be hoped, internally consistent operations of thought. That such a postulate should sometimes be called a fact—"an unfortunate word in consequence of its etymology" (*OM,* 6–7)—and at other times a postulate merely reflects the varying degree of phenomenological certitude with which the individual person apprehends the will's operative presence. Coleridge's qualification that the will "may be known, but cannot be understood," does not, however, imply that the postulate of a "responsible will" amounts to a gratuitous and irrational position but merely acknowledges the fundamental constraints under which human reason finds itself in this kind of inquiry. For unlike the *actus purissimus* of God, for whom reality and concept always exist in complete alignment and utter plenitude, human inquiry can advance by gradually and dialectically articulating founding conceptions such as will, responsibility, judgment, etc.

Following Coleridge's preferred method, then, one might begin by desynonymizing the "postulate" of the will from a mere hypothesis. For in the case of a hypothesis, the framework of inquiry within which the truth of that hypothesis remains to be tested already exists. By contrast, the assertion of "a responsible will is not only the postulate of all religion but the necessary datum incapable from its very nature of any direct proof—the datum, we say, and ground of all the reasonings and conclusions, which in the particular religion are assumed as already granted" (*OM,* 32). At issue is not a demonstration of a particular claim, which (as Coleridge well understands) could prove valid only if we are able to entertain a counterfactual scenario and, by implication, the falsifiability of our guiding concepts. Yet in the "moral science" attempted in the *Opus Maximum,* "the conclusion" to be drawn from the opening postulate of a (responsible) will "does not rest on an understood 'if' prefixed ... [T]he truths are not hypothetically true, but ... the necessity arises out of and is commensurate with human nature itself, the sole condition being the retention of humanity" (*OM,* 11). Metaphysical speculation thus proves *eo ipso* to be the kind of "guide to morals" that in the twentieth century it would be for writers like Iris Murdoch, Gertrude Elizabeth Anscombe, and Charles Taylor. To deny the reality of the will—understood not as a hypothesis but as the ontological ground for rational, responsible, and teleologically constituted human action—is to disavow the very notion of the human. To be sure, "a human being may be dishumanized," Coleridge concedes, but even then "it cannot be but by his own act" (*OM,* 11). The logical impasse or performative contradiction, which we had already observed in Hobbes's account of will and person, informs Coleridge's firm conviction "that the truths of Reason are the truths of our distinct humanity." His overarching objective thus is "to substantiate moral thought with the same system of logic as that which substantiates truths of pure intellect."[2]

As Murray Evans notes, "Will so defined must be real and not merely conceivable."[3] Far from a hypothesis, then, the will constitutes the absolute condition of possibility for all moral reasoning. The central question—"have I a responsible will?"—cannot be answered by those who reject the very reality of moral agency and, along with it, the reality and efficacy of ideas. Coleridge's reasoning here is particularly close to Scholasticism inasmuch as the very possibility of rational intersubjective discourse presupposes at least one shared axiom—a joint commitment, most likely to the unconditional (if also indemonstrable) reality of God. Absent this shared position—an act of faith, both in the reality of what it affirms and in the possibility of dialectically reasoning toward a fuller understanding of its object—rational discourse, human community, and indeed the reality of human personhood would be instantly vitiated. Insisting on the profound alignment of reason and faith in an act of real assent—which in time was to dominate John Henry Newman's writings—Coleridge had already pointed out that faith is never an inference or deduction ("a Notion drawn from a Notion"); indeed, faith properly speaking does not even belong to the realm of the notional. Rather, it is an act of assent to an incontrovertible inner certitude, an act that envelops the totality of the human person: "It must *concern* me, as a moral and responsible Being." The certitude in question is a phenomenological *datum*, not a proposition about which one may feel "certain." Hence, the true locus of revelation is indeed subjective, not transcendent, for which reason Coleridge dismisses "the very phrase '*Revealed* Religion' as a pleonasm, [for] a religion not re-

2. Davidson, "Duty and Power," 122.

3. "Reading 'Will' in Coleridge's *Opus Maximum*," 81; Davidson sees a certain "reversal of terms" in Coleridge's postulate of the will, such that in the first instance "the postulated existence of the will leads to moral responsibility, and in the second 'the inclusive postulate' of a '*responsible Will*' leads to the conception of will in general" ("Duty and Power," 124). The reason, I would argue, for this seemingly contradictory approach has to do with the different audiences that Coleridge is hoping to engage. Those who operate within the framework of the Christian *logos* will readily endorse the postulate of the will and grasp the moral ontology behind it ("a Will not personal is no idea at all but an impossible conception" [*OM*, 164]); by contrast, the skeptical reader, perhaps an adherent of the "corpuscular" school that so exercises the later Coleridge, may yet be dialectically led to understand and accept the absolute anteriority of the will. Doing so would likely begin with a phenomenological account of "conscience" and then, in Plotinian fashion, seek to capture the metaphysical presuppositions of our experience of "responsibility." The latter constitutes a "Speculative Disquisition" and, as such, "must begin with postulates, that derive their legitimacy, substance, and sanction from the *Conscience*" (*OM*, 107).

vealed is ... no religion at all" (*AR*, 184). Later on, in the *Opus Maximum*, he puts the matter in the form of an analogy:

> <As little can> the truth or falsehood of Christianity even commence its pleadings before a judge who had refused to acknowledge the existence of a responsible Will in man. For the judge himself the conclusion pre-exists in his premise, and to the other party nothing remains but to ~~disclaim the right authority~~ remove his cause to some other court where at least a trial was possible, by virtue of some principle admitted ~~in common~~ on ~~all~~ both sides. (*OM*, 54)

Considering the centrality of the will in the genesis and legitimation of philosophical and scientific modernity (in Ockham, Hobbes, and Locke, among others), Coleridge's decision to make that term the starting point for his exploration of the human is at once shrewd and courageous. Shrewd, because in so identifying the will as an ontology—"it cannot be an object of conception" (*OM*, 18)—Coleridge seemingly endorses the modern, Kantian view of human agents as strictly self-legislating and self-legitimating . Courageous, because Coleridge, in fact, conceives the will in distinctly Augustinian and Scholastic terms as uniquely spiritual and indexed to the normative (divine) authority of the Platonic *logos*. For the will to have any reality, it has to be "responsible" and, thus, imbued with self-awareness. As "an abiding intentionality of the soul," the will for Coleridge presupposes relationality, that is, the person's enduring conscious, ethical orientation toward the other.[4] It cannot be *re*active, which is what is implied by modern philosophy's habitual conflation of the will with the merely appetitive. It cannot be sheer compulsion because the human will, properly so called, always initiates or "originates" a state. Put differently, an act of will is a form of intentionality; as such, it is consummated in some action that reflects an underlying, steady-state, richly layered, though not necessarily explicit interpretive take on one's life-world or what Newman will eventually call a "view." By contrast, to suppose that the will is but an occasional flicker of consciousness (not self-consciousness), a random compulsion or mental spasm on the order of Locke's "uneasiness" is less to have tendered a mechanist cum materialist account of mind than to have begged its reality on a grand scale. It is to denature the life of the mind into a random cascade of disaggregated, isolated, and irrational occurrences and, by extension, to blur the boundaries between a bona fide act of will and an ephemeral impulse or "wish."

Before returning to Coleridge, we should briefly consider the special case of Schopenhauer, for it is here that we find how a strictly deterministic naturalist and

4. Hedley, *Coleridge*, 161.

anti-humanist understanding of the will, rather than expiring in a vulgar materialism of the kind previously ventured by Claude Adrien Helvetius and Julien Offray de La Mettrie, may actually end up reifying matter itself as a kind of noumenal, metaphysical superagency. In his *Prize Essay on the Freedom of the Will* (1839), Schopenhauer elaborates such a view by refining arguments first advanced in his 1813 doctoral thesis on causality and, subsequently, in Book 2 of *The World as Will and Representation* (1818). Concerning himself specifically with "moral freedom" (in contradistinction to physical or intellectual freedom), he categorically denies freedom of will on the grounds that the very term "will" constitutes an instance of efficient causation. Part of the simplicity or, as Ludwig Wittgenstein felt, "crudeness" of Schopenhauer's argument has to do with his axiom (imported from Joseph Priestley) that "every event is *necessary* in regard to its cause," even as the operative presence of those causes ("everything else with which it coincides in time and space") is itself "contingent" (*zufällig*). Modern liberalism's credo—viz., "I can do what I will"—thus fails on Schopenhauer's account because it has not properly grasped the iron-clad determinism intrinsic to the very idea of a "will." Hence his sarcastic question: "Can you also *will* what you will?" For Schopenhauer, anyone not properly perplexed by that question and inclined to respond with an emphatic "yes" inhabits a logically incoherent position. For to suppose that the will is subject to the interventions of a human consciousness that (certainly in Schopenhauer's view) always depends on motives furnished to it *by the will* is a classic instance of the *proton pseudos*. Indeed, it is to imagine an "absolutely contingent" event devoid of any causality whatsoever. Oddly, Schopenhauer's rejection as irrational of "a free will ... that was determined by nothing at all"—a hypothesis that in his view merely illustrates how "clear thinking is at an end"—rests on a hyper-deterministic view that deprives human agency of all capacity for rational thought, deliberation, judgment, conscience, and bona fide action.[5]

Unsurprisingly, Schopenhauer's next step is to argue the absolute dependence of self-consciousness on the will: "self-consciousness is very greatly, properly speaking even exclusively concerned with the *will.*" Its only legitimate objects, he suggests, are the immediate movements of the will, "more or less weak or strong, stirrings at one moment violent and stormy, at another mild and faint." Precisely because "self-consciousness contains only the willing, not the determining grounds for the willing," its habitual assertion of freedom of willing begs the question altogether. For it

5. *Prize Essay*, 6–8. Within English thought, Locke comes closest to anticipating Schopenhauer's position when arguing that "Man in respect of that act of *willing* [which results in action], is under a necessity, and so cannot be free ... This then is evident, That in all proposals of present Action *a Man is not at liberty to will, or not to will, because he cannot forbear willing*" (*Essay*, Bk. II, Ch. XXI, §§23–24).

can only ever claim "the freedom of *doing* under the presupposition of *willing;* but what we have inquired about is the freedom of *willing.*" Ultimately, Schopenhauer maintains, this imagined and vaunted "freedom" must be understood as the result of a category mistake, viz., a confusion of willing with wishing.[6] Schopenhauer's target here is clearly the modern notion of a solitary (or "punctual"), self-conscious, and autonomous individual supposedly capable of interacting with other such individuals in transparent and purposive fashion and deliberating on its own goals and projects without external constraints. What renders that conception incoherent is that it assumes an agency *unique and distinctive* on the one hand, and yet simultaneously considered *equivalent to and indistinguishable from* all other such agents. For Schopenhauer, the fundamental flaw of the modern liberal model of agency lies with its failure to grasp its own ontological presuppositions. It does not remember that "every *existentia* presupposes an *essentia;* that is to say, everything that is must also be *something,* must have a definite essence. It cannot *exist* and yet at the same time be *nothing,* thus something like the *ens metaphysicum,* i.e., something that *is* and only *is,* without determinations and qualities, and consequently without the definite mode of action that flows from these." Surprisingly unaware of Schopenhauer's argument, Jean-Luc Nancy restates the point almost verbatim and, in so doing, reaffirms the nominalist ethos behind both the *Prize Essay* and his own account: "Freedom … is the fact of existence as the essence of itself; … existence as its own essence—the singularity of being."[7]

For his part, Schopenhauer continues by arguing that "all this is just as true of the human being and his will as of all other beings in nature. He too has an *essentia* in addition to an *existentia,* i.e., he has fundamental essential qualities that constitute his very character and require only occasioning from without to come forth. Consequently, to expect a human being on the same occasion to act at one time in one way and at another in an entirely different way would be like expecting the same tree which bears cherries this summer to bear pears the next."[8] Yet like Hobbes and Priestley, whose works Schopenhauer greatly admired and whose voluntaristic accounts of the will as an unfathomable and implacable *daimon* of sorts he means to extend, his argument suffers from the performative contradiction that afflicts most deterministic models of human agency. Thus his *Prize Essay* begins to reveal a growing disequilibrium between its own rhetorical discretion and intellectual cogency *as a*

6. Ibid., 10, 15, 14. "Imagine that, in a given case, opposite acts of will are possible, and [the uninitiated] boast of their self-consciousness which, they imagine, asserts this. Thus they confuse wishing with willing; they can *wish* opposite things, but can *will* only one of them; and which one it is is first revealed even to self-consciousness by *the deed*" (15).

7. *Experience of Freedom,* 11.

8. *Prize Essay,* 51.

philosophical argument and the central claim advanced in it—viz., to deny the human individual all discretion, self-awareness, or capacity for self-transformation. Echoing Priestley's and Godwin's nominalism, Schopenhauer views human consciousness as utterly *consumed* by its representations and, hence, as incapable of achieving any distance vis-à-vis its contents. Just as Godwin a generation earlier had insisted that "the mind is always full," Schopenhauer conceives mental life as strictly one-dimensional and devoid of any meta-conscious or reflexive potential: "only *one* image at a time can be present in [man's] imagination."[9]

At the same time, it is precisely Schopenhauer's widely acknowledged rhetorical verve and creativity, once again on full display in his *Prize Essay,* which steadily undermines his premises and ultimately derails his argument. A great stylist who recoiled from the turgid prose of his philosophical contemporaries, Schopenhauer certainly presents his case in terms so vivid, inventive, and compelling as to bring to the fore precisely those two terms that his own account is unable to accommodate: "language" and "imagination." Two exemplary instances of metaphor will have to suffice, and both are introduced so as to clinch Schopenhauer's axiomatic view of human consciousness as wholly bereft of any capacity for self-transcendence or self-transformation. It is an odd view to hold for a writer who never tired of extolling the superiority of classical philosophy, and of Plato in particular. The first example is found as Schopenhauer once again seeks to drive home the point about the utter immutability of our "inborn character." Relative to the noumenal essence of its own will, the conscious individual is a mere epiphenomenon—a case of *tuchē,* something fated. Incapable of authentic and timely self-awareness, and hence devoid of any inner progression or transformation, the individual's character "remains the same throughout his whole life. Under the changeable mask of his years, his circumstances, and even his cognitions and views, we find the real identical human being, *like a crab in its shell,* quite unchangeable and always the same."[10] One wonders what to make of the apparent, performative contradiction here, which simultaneously seeks to assimilate the human to the animal, yet which can clinch its point only by means of a figurative rhetoric that so obviously sets it apart from forms of animal life. For to engage life and human consciousness *as a philosophical problem* is to stand necessarily at some remove from it and to have achieved a certain measure of transcendence. It is above all in language and rhetoric that this transcendence manifests itself: viz., as our capacity for viewing objects counterintuitively, from various perspectives, and thus

---

9. Godwin, "On the Mechanism of the Human Mind," in *Enquiry,* 367; Schopenhauer, *Prize Essay,* 37.

10. *Prize Essay,* 46, 44 (italics mine).

to conceive, shape, and qualify an object's putative hold on us as an "original impression" of some kind.

The second example surfaces as Schopenhauer presses his view of mental life as an elaborate, albeit invariant projection by the will. Once again, he asserts the strictly epiphenomenal status of logic and thought, both of which merely "furnish for th[e] external world concepts, the world of thoughts, thus in turn the sciences, their achievements, and so on. Therefore great brightness and clarity are there for it to see *outside;* but *inside* it is dark, like a well-blackened telescope. No principle *a priori* illuminates the night of its own interior; these lighthouses shine only outward." What was to fascinate and inspire the young Nietzsche is this audacious denial of the human being's capacity for self-transcendence and timely self-awareness, as well as the uncompromising rejection of civilization and progress as nothing more than a complex delusional formation. Yet as Nietzsche was also quick to perceive, the intricacy and sophistication of human autonomy—even or especially when understood as a pervasive *illusion*—actually proves a great deal more interesting than the noumenal black hole of Schopenhauer's will per se. For his part, Schopenhauer admits a certain progressive refinement and quasi-evolutionary development of humans' cognitive powers: as "the beings of nature become more complex, ... their susceptibility is enhanced and refined from the merely mechanical to the chemical, electrical, irritable, sensible, intellectual, and finally rational." Yet just as the intellectual powers increase, so the "nature of the *operating causes* must also keep pace with this enhanced susceptibility."[11] What for Wagner and Nietzsche was to be the alluring and opulent flower of *decadence* is effectively brushed aside by the *Prize Essay* as nothing more than an elaborate sideshow of civilization. All that this increase of our intellectual abilities does is to widen the distance between "motive" and "action" and to extend the "guiding wire" of a causality that remains, even now, absolutely implacable and efficient—albeit also increasingly oblique.[12]

Two terminological slippages account for Schopenhauer's existentialist, not to say fatalistic conclusions. One involves his de facto conflation of "cause" and "motive," while the other has to do with his understanding of "thought" as strictly reactive—a mental process both cued and delimited by whatever motives happen to present themselves to our attention: "just as ... is the case with causes in the narrowest sense and with stimuli, so too is it equally the case with *motives;* for in essence, motivation is not different from causality, but is only a form of it, namely, causality that passes through the medium of cognition." On that premise, thinking lacks any

11. *Prize Essay,* 19, 39. Schopenhauer had developed this proto-evolutionary argument more fully in his 1836 essay *On the Will in Nature.*

12. *Prize Essay,* 31.

generative, imaginative, or counterfactual dimension. Inasmuch as thought appears bereft of all discretion, judgment, and counterfactual perspective on whatever issue it happens to engage, it no longer belongs to the realm of action but, rather, to the domain of *mechanē*. Moreover, it has lost all hermeneutic depth and historical scope. For Schopenhauer, thought is neither *interpretive* nor does it understand itself to be situated in, and responding to, complex traditions of inquiry. Instead, in what amounts to an extreme case of nominalism, the scope of thinking has contracted to the *punctum* of a present decision between various competing motives. Being subject to non-transcendable causes/motives, consciousness thus becomes virtually synonymous with the experience of suffering: a "*conflict of motives* ... is very often painful."[13] Schopenhauer's rigid computational portrait of a *cogito* besieged by a phalanx of motives invests the latter with an *immediate* and *unconditional* presence in the mind. Motives are ineffably present *in* but never distinct representations *for* consciousness. Embodiments of a pre-linguistic, *essential* "force" (*Kraft*), motives are said to operate prior to and independent of all interpretive and rhetorical practice. It bears pointing out here that Schopenhauer's account leaves both consciousness *and* motive curiously depleted. Reduced to strictly causal efficacy, Schopenhauer's motive amounts to what Jean-Luc Marion calls "a fallen [*déchu*] phenomenon, [which] because it appears as always already *expired* [*échu*], nothing new can happen to it anymore." In the *Prize Essay*, motives thus operate as wholly generic, deterministic, and alien noumena seemingly bereft of all dynamic quality and distinct phenomenology. Schopenhauer's concept of a motive thus stands in diametrical opposition to what Coleridge will explore under the heading of "conscience," itself a vivid case of what Marion means by a "saturated phenomenon." For as the manifestation of alterity both *within* and *for* the self—and hence properly constitutive of a self as *this* person—conscience is an event at once *experienced* so powerfully as to be recognized as transcendently *given*. Unlike abstract motives, that is, phenomena associated with conscience impinge on the self, such "that their mode of phenomenalization will not only open ... access to their original self but render it incontestable."[14]

If in the *Prize Essay* "thought becomes *motive*,"[15] Coleridge offers us a precisely inverse scenario. In a passage to which we shall return later, he poses the question thus: "what is a Motive? Not a thing, but the thought of a thing. But as all thoughts are not motives ... a motive must be defined as a determining thought. But again, what is a Thought? Is this a thing or an individual? What are its circumscriptions, what the interspaces between it and another? Where does it begin? Where does it

---

13. Ibid., 32.

14. *In Excess*, 36, 31.

15. Ibid., 41, 32, 31; for a critique of Schopenhauer's naturalist view of consciousness, see Murdoch, *Metaphysics*, 57–80.

end? ... A motive is neither more nor less than the act of an intelligent being determining itself" (*OM*, 25–26). For Coleridge, a motive belongs to the order of appearances or phenomena and, as such, is subject to a complex and open-ended hermeneutic process. As it unfolds, that process in turn enables the interpretive agent to develop a reflexive or meta-discursive perspective on the way in which thought frames the phenomenon at hand. Inasmuch as "thought" is itself a form of sustained action (though potentially unapparent to others), it is consummated in expression. As Coleridge had observed early in his career, "language & all symbols give outness to Thoughts / & this the philosophical essence & purpose of Language."[16] Once we factor in this crucial symbolic dimension—so completely elided in Schopenhauer's argument—we understand why for Coleridge a motive constitutes the beginning of thought—and, potentially, of an intellectual and moral ascent—rather than its point of termination. Put differently, a motive can never be invested with noumenal authority but, as a phenomenon, prompts us to develop a reasoned perspective on its noumenal sources. Like all phenomena, it is a trace that appears *for* human consciousness by dint of its "saturated" or "excessive" presence *as* a phenomenon.[17] It is subject to a range of possible *representations* but, crucially, can reveal its intrinsic reality only by appearing *for* consciousness. Even so, Coleridge appears to agree with Schopenhauer on the centrality of "action." Thus, in *Aids to Reflection*, he notes that just "as we know what Life is by *Being*, so we know what Will is by *Acting*." Yet unlike Schopenhauer and, I would argue, far more persuasively than the latter, Coleridge also insists on "the peculiar self-consciousness preceding and accompanying [the *Act*]" of will (*AR*, 269).[18] He makes the point again, this time augmented by a categorical reflection on the incommensurability of ideas and concepts, in *On the Constitution of Church and State:* "Speak to a young Liberal, fresh from Edinburgh or Hackney or the Hospitals, of Free-will, as implied in Free-agency, [and] he will perhaps confess to you with a smile, that he is a Necessitarian,—proceed to assure you that the liberty of the will is an impossible conception, *a contradiction in terms* ... or as it may happen, he may declare the will itself a mere delusion, a non-entity, and ask you if you have read Mr. Lawrence's Lecture." Coleridge's ironic sketch of an optimistic,

16. *CN*, no. 1387; for a contextualist discussion of Coleridge's philosophy of language as "high instinct" and "vehicle" of thought, see McKusick, *Coleridge's Philosophy of Language*, esp. 119–148; for a deconstructive account of Coleridge's linguistic theory at work, see Christensen, *Blessed Machine*, esp. 186–269.

17. For a compelling recent account of "saturated phenomena" and their disclosure as "event," see Marion, *In Excess*, esp. 30–53 and *Visible and the Revealed*, 18–48; in sharp contrast to Schopenhauer's appraisal of the phenomenon as efficient "cause," Marion persuasively revives an Augustinian understanding of the visible and the apparent as a "gift" (*donum*).

18. For a concise juxtaposition of Schopenhauer and Coleridge on the will, see McFarland's "Prolegomena" (*OM*, clxxxv–clxxxviii).

vivacious, and well-educated young professional serenely denying free agency or, possibly, the reality of the will itself is supplemented by a footnote that extracts the main philosophical point here: "See AIDS TO REFLECTION ... where it is shown to be one of the distinguishing characters of *ideas,* and marks at once of the difference between an *idea* (a *truth-power* of the reason) and a conception of the understanding; viz. that the former as expressed in words, is always, and necessarily a *contradiction in terms*" (*CCS,* 17). For Schopenhauer no less than for the young, self-styled necessitarian, the will cannot be an idea, quite simply because it has already been quarantined as non- or, rather, pre-conceptual, a function of sheer instinct and compulsion. Yet such a characterization of the will not only fails to recognize its spiritual, ethical, and ideational dimensions; it also fails on its own terms, Coleridge insists, because it collapses willing with a strictly non-cognitive and quasi-somatic energy. Coleridge calls it "spontaneity" and is quick to desynonymize it from willing: "Spontaneity ... we should not call a Will: we perceive at once its inferiority" (*OM,* 140). In so reaffirming the spiritual and intellectual nature of the will—and doing so in sharp opposition to theological voluntarism and its secular (Hobbes), deontological (Kant), or pessimistic (Schopenhauer) descendants—Coleridge reinvests human agency with an intellectual cum spiritual dimension of which the will had been stripped since the advent of Franciscan, voluntarist theology in the early fourteenth century.

Meanwhile, Coleridge's forceful double-claim concerning the primacy of the will and the ontology of the human qua spiritual reveals yet another premise, just as non-negotiable as the other two. For Coleridge also posits that within the domain of the human only that can be real which is fully individualized, what he calls man's "proper Individuality or ... his Personëity." On ontological grounds, that is, the life of the mind must be grasped as wholly sui generis. Mental life is a truly self-actuating and self-organizing process (not merely an *effect*); it is concerned not only with facts but always also with their complex evaluation (i.e., is *spiritual*); and, lastly, it is both dynamic and unique (i.e., an individual *will* rather than a mere *type* of occurrence). To identify these three interlocking postulates from the outset helps us understand why Coleridge was so strenuously opposed to any philosophy that would construe mind as a secondary, abstractive, or derivative thing—a *class* or *type* of sorts rather than an imperfect echo of the divine. In his view, at any rate, only something truly unique and individuated can be said to reflect "an eternal origin in the Divine Idea" (*SW & F,* 1:429). As he puts it in *Aids to Reflection,* the work that shows Coleridge pursuing these questions most fully, "we are responsible Agents; *Persons,* and not merely living *Things*" (78). In particularly imaginative ways, Coleridge later on sharpens the contrast between life as mere carbon-based, protein-churning biomass and the idea of uniquely self-determined and responsible, human agents. Whereas in any variety of organized bodies, "Polypi for instance ... their motive powers are all from without," a slow evolutionary progression subtly refines both the *form* of (human) life and our conception of it:

> As life ascends, nerves appear; but still only as the conductors of an *external* Influence; next are seen the knots or Ganglions, as so many Foci of *instinctive* Agency, that perfectly imitate the yet wanting *Center*. And the Reservoir of Sensibility and the imitative power that actuates the Organs of Motion (the Muscles) with the net-work of conductors, are all taken inward and appropriated; the Spontaneous rises into the Voluntary, and finally after various steps and a long Ascent, the Material and Animal Means and Conditions are prepared for the manifestation of a Free Will, having its Law within itself and its motive in the Law—and thus bound to originate its own Acts, not only without but even against alien Stimulants.[19]

As Coleridgean speculation goes, this is a particularly hazardous instance. For the strategic objective of his later philosophy—viz., to demonstrate the uniquely self-originating nature of the human will as the ontological ground of all thought and action—is here seemingly *deduced* from a long evolutionary process. The metonymic pattern asserted, while said to result in a uniquely human and fully individuated being *in the end*, effectively casts that very outcome as the final stage of countless gradations and imperceptible transitions. By its very nature, metonymy not only does not support the categorical distinction that Coleridge wishes to draw; it is positively corrosive of it. That Coleridge nonetheless should have opted for this naturalist line of argument, rather than relying on the traditional Christian creation-myth, reflects his underlying view of life and *logos* as a single, integrated, and mutually reinforcing system. The tendency toward greater self-organization, though rudimentary in many of the "lower" species of animal life, is not simply a secondary attribute to be ascribed *to* life but, for Coleridge, the very essence of it. Hence its operation is not to be disputed anywhere, not even in the most primitive protozoa—a view that once again reveals deep affinities between the later Coleridge and Plotinus.[20]

Already in April 1819, while examining John Abernathy's *Physiological Lectures* and still under the impression of Abernathy's Hunterian Oration, which he had attended on 15 February of that year, Coleridge proves acutely sensitive to the risks of (over-)extending a model of life as self-organization into what eventually would come to be known as "emergentism" in the work of John Stuart Mill, C. D. Broad,

---

19. *AR*, 97–98; Hedley is right to stress Coleridge's general indebtedness to Plotinus, Cudworth, and (less obviously, perhaps) Kant as regards his opposition of materialist and naturalist accounts of human life. Yet by juxtaposing the passage "to a Platonic view of ethics where the will is not so much unimpeded activity as concentrated attention" (Hedley, *Coleridge*, 166), Hedley underplays Coleridge's recognition of the somatic foundation of the (Plotinian) "soul," and of continuities between animal and human life.

20. "Form is certainly in some way present to everything. How then is it present? As one life: for life in a living being does not reach only so far, and then is unable to extend over the whole, but it is everywhere" (Plotinus, *Ennead*, 6.5.11).

and Samuel Alexander. Recalling arguments from his unpublished "Theory of Life" (1816), his notebook entry of April 1819 (no. 4517) sums up the increasingly popular scientific view thus: "The common sense of mankind ... never attributes Life without Being. So again Life without sensation (in Plants) Sensation without Consciousness (as in the Zoophytes) Consciousness without Self-consciousness; but not vice versa. Nature leaves nothing *be~~ing~~hind* but still takes up the lower into the higher, still refining and ennobling what it elevates." Yet in the subsequent entry he proceeds to flag the risks of *inferring* from the manifestly self-organizing processes of organic life their de facto origination. Phenomenality and ontology remain, for Coleridge, absolutely distinct and, in keeping with Scholastic thinking, he never supposes that the reality of God has to be inferred, in the manner of natural theology, from the coherent and purposive design of the phenomenal world.[21] Coleridge is particularly vexed by what he takes to be Abernathy's position on the different kinds and dignities of organic life: "he declared all the rest, the superiority of animals, ~~to~~ one to the other, and of men to animals, resulted wholly from their organization" (*CN*, no. 4518). For Coleridge, this is a vintage case of reversing antecedent and consequent, a "*hysteron proteron* which involves all the mystery of ~~the~~ Spiritualism with the preposterousness of ~~the~~ Materialism—in the latter ('Mind is the result of Structure,' *says M*r *Lawrence*) the House begets the Mason." Against "the supposition that the functions are the offspring of the structure, and 'Life the result of organization,' connected with it as effect with cause" (*SW & F*, 1:501–502), Coleridge maintains that to employ the concepts of "organism" and "organization" *eo ipso* means to concede a strictly inner causality and to grasp life as a process in which cause and effect must themselves be understood as categorically isomorphous with life itself: "Life only can assimilate Life."[22]

Rather surprisingly, Coleridge's misgivings have recently been echoed from within the precincts of analytic philosophy. Thus Michael Thompson notes how the "notion of order or organization is a very abstract or generic one, and that, left abstract, it does not make sense to think of a standard of more-and-less in respect of it." Homing in on the specious explanatory value of reductionist models, Thompson

21. Buckley's argument (*Denying and Disclosing God*, 48–69) that natural theology as a *proof* of God's existence originates in Aquinas is substantially wrong; in fact, the argument from design only serves to articulate for the fallible and temporal intelligence of human beings the reality of God, a point crucially elaborated in the first of the so-called five ways of Aquinas's *Summa Theologiae* (see esp. *ST*, Ia Q 2). For a thorough reading of this momentous passage, see te Velde, *Aquinas on God*, esp. 37–64.

22. *CN*, no. 4521, f91v. In the *Enneads* (esp. 4.6), Plotinus had already wrestled with the fact that, while obviously dependent on the body, the "soul" (*psychē*) remains categorically distinct from it. In both diction and conception, Plotinus notably anticipates Kant's and Coleridge's strictures on empiricist and associationist attempts to derive mental phenomena from mechanical, extrinsic causes.

cautions that "there is an obvious sense in which *no such succession of [material] goings-on will add up to a single process.*" The hypothesis of life as a "process" of "self-organization" tends to blur or, rather, collapse discrete events into a (supposedly) single and unified process. Yet a bona fide conception of life must be concerned "not with a special *nexus of events,* but with a special *nexus of thing and event.*" Thompson here conjures up the mildly improbable scenario of a bird suddenly flying out of the baseball stadium, either because a batter mistook it for a fastball (now seemingly turned into a home-run) or because the bird had suddenly adjusted its flight trajectory in search for better food. If asked "what the difference is between the[se] cases," Thompson observes, we find that "an appeal to 'self-movement' is not illuminating. The reflexive is simply one of the means our language gives us for marking the different relation posited between subject and predicate, thing and event. It does not by itself tell us what this relation is. Self-movement, self-organization, etc. do not enable us to grasp the rationale for the distinction between a mechanical occurrence and a process of life" but, in fact, remain "completely empty ... Something must fall between the would-be agent and what it does—something that, as 'cause,' in a pre-given sense, of the latter happening, gives the whole ensemble the special character of rational or intentional agency in the one case, and of animal movement in the other."[23]

Consideration of a later notebook entry from 1823 further illustrates Coleridge's persistent attempts at drawing more precisely the delicate line between strictly animate and positively human life, between raw events and cases of bona fide action. To grasp the difference is of strategic importance for Coleridge since it enables us to distinguish between the reactive and pragmatic intelligence of the *understanding* and an exclusively human knowledge reflecting on its own limits and ends: *reason.* As that entry also shows, the later Coleridge's attempts at recalibrating the ratio between scientific and theological cognition bears markedly Gnostic features—concurrently observable in Hegel's *Philosophy of Nature*—according to which "mind" (Plotinus's *nous*) is to redeem "nature" (*physis*) from its material scaffolding. What, in his marginalia on the Cambridge Platonist Henry More, Coleridge describes as "the great

23. M. Thompson, *Life and Action,* 38, 40, 43, 45–46. The concepts of self-origination (*epigenesis*) and self-organization remain in wide use, however; see, for example, Müller-Sievers, *Self-Generation,* 48–64; and van de Vijver, "Kant and the Intuitions of Self-Organization." Kant's careful distinction between explanation and judgment, that is, between reflective and determinative types of judgment, partially safeguards him against Thompson's critique of self-organization. For Kant's notion of a "natural purpose" (*Naturzweck*) merely serves to elucidate "the enigma of intrinsic purposiveness ... [and] the mysterious reciprocal determination between the parts and the whole" (van de Vijver, "Kant and the Intuitions of Self-Organization," 147); its focus is on describing functional relationships rather than making absolute claims about the sources from which such a "natural purpose" might have sprung.

redemptive Process, the history of LIFE which begins in its detachment from Nature and is to end in its union with God" (*CM,* 3:919) is unfolded with great acuity in a detailed notebook entry from mid-1823. It opens with Coleridge arguing that the apparent isomorphism of certain primates with human beings is frequently misconstrued as evidence for their substantial identity: "The apparent approximations of the Simia to the Man are … the *perfection* of the animal, the fullest evolution of its possibilities under the most favorable circumstances—whereas the Approximations of certain *Tribes* (N.b. not of Races) to the highest order of the Simia are manifest *Depravations* of the Man, aberrations from proper Humanity and in the literal sense of the term, Degeneracy" (*CN,* no. 4984, f86[v]). To be sure, any skilled anatomist will readily identify various "Semblances of humanity" in certain Simia, such as "the four hands, the climbing prehensile powers and habits, with the position and direction of the Eyes rendered necessary by the frequent necessitated (not voluntary) erect posture," and so forth. The category mistake, meanwhile, occurs when substantial physiological resemblances between Simia and Man are construed as a sufficient basis for asserting their substantive *identity,* that is, when the anatomist proclaims "Semblances of humanity … to be resemblances of the animality of Man, that which is common to ~~him~~ Man and Beast, and not of his Humanity, or that which constitutes him *Man* in contra-distinction from the Beast" (f87). For Coleridge, the logical blunder, at once verbal and conceptual, involves collapsing "the analogy which cannot be denied" into the very thing *to which* the first term is said to bear a resemblance. For to do so is to ignore "the impassable Chasm between the highest Orders of Animals and the Man," a chasm that "can not be filled up or bridged over." Shifting attention toward the decisive *tertium quid* that accounts for this chasm—viz., language as *logos* and reason—Coleridge emphasizes that, however highly evolved, animals are only ever "*Types* not Symbols, dim Prophecies, not incipient Fulfilments [whose] purpose for themselves is wholly diverse from the *meaning* which they convey" (f87[v]). He elaborates the point in *Aids to Reflection,* pointing out how animals notably lack will, reason, and hence the capacity to act and the freedom to sin; instead, in animals "the Will is hidden or absorbed in the Law. The Law is their *Nature* … in irrational Agents the Law constitutes the Will. In moral and rational agents the Will constitutes, or ought to constitute, the Law" (300n).

Only at the level of the human does life exhibit complex symbolic patterns of self-description, self-awareness, and the progressive articulacy of the *logos.* That "the Law is a *Law* for you; that it acts *on* the Will not *in* it" does not, in Coleridge's view, justify a naturalist account of human agency. On the contrary, it merely "proves the corruption of your Will" (*AR,* 301). If "Life itself is not a *Thing*—a self-subsistent *Hypostasis*—but an *Act* and *Process*" (*SW & F,* 1:557), a further shift takes place once the "long Ascent" from a purely foreign and materially determined existence to the genuinely "Voluntary" has been completed. For at that point agency is no longer a mere churning of carbon-based matter but will actually have become conscious of its own, process-like nature. A fragment from 1821 finds Coleridge inveighing against an evolutionary theory that would "trac[e] us back to the bestial, as to our Larva, [and] contemplates the Man as the last metamorphosis, the gay *Imago* of some lucky species of Ape or Baboon." Yet even here Coleridge retains as indispensable the Platonic notion of a gradual ascent and perfection of nature:

> Our belief, that Man first appeared with all his faculties perfect & in full growth, by virtue of the supernatural Act of Creation in no wise contravenes or weakens the assertion, that these faculties, maturely considered, presupposed, and in each succeeding Individual born according to nature must be preceded by a process of growth, and consequently a state of involution or latency correspondent to each successive Moment of Development. (*SW & F,* 2:894, 898)

In modern parlance, Coleridge's evolutionary account of the will as free choice involves a progressive "disembedding" whereby the human being develops awareness not simply as a member of its species but also self-awareness as a unique and distinctive "Personëity."

The extraordinary depth of the idea of person, as well as its arduous recovery from near oblivion in Coleridge's late writings, will occupy us shortly. For now, what matters is his emphasis on the human being's unique and distinctive capacity of speech, understood as reflexive, meta-lingual, and significant expression rather than as the instinctive emission of sounds indexing some basic existential concern (e.g., warning-, distress-, or mating-calls). For Coleridge, human language furnishes the most compelling evidence of metaphysical creation: it "must have existed before it declared itself in Sound: for the Sound is to the Speech as the Effect to its Cause, and begins to exist, as Sound, in the living Ear; as significant Sound, in the pre-conformed instinct; as intelligible Sound in the pre-possessed and pre-possessing Intelligence."[24] To be sure, like Plotinus, Coleridge never denies the somatic and physiological continuities between animal and human life; neither does he dispute that human beings might choose to disavow their moral and intellectual capacities, be it as a merely

24. *CN,* no. 4984, f88v. The motif of the *logos* as the manifestation of a "divine energy" within the finite human being is, of course, absolutely central to Coleridge's philosophy; as early as 1805, he ties human salvation to rational articulacy when arguing that "the moment we conceive the divine energy, that moment we co-conceive the Λογος … the redeemed & sanctified become finally themselves Words of the *Word*—even as ~~lou~~ articulate sounds are made by Reason to represent Forms, in the mind, and Forms are a language of the notions—Verba significant phænomena" (*CN,* no. 2445). On the rich eighteenth-century debate concerning the origin of language (in Locke, Condilliac, Maupertius, Rousseau, Monboddo, Herder, et al.), see E. Cassirer, *Symbolic Forms,* vol. I, chap. 1; Hudson, "Theories of Language"; and Borst, *Turmbau von Babel,* esp. vol. III/2:1395–1629.

contingent lapse of the will or as a principled commitment to some version of the "corpuscular philosophy" that extends from the Greek atomists via Epicurus and Lucretius to the voluntarism of Ockham and Hobbes and, eventually, to the skepticism of Hume. Moreover, Coleridge often dwells on the persistent phenomenon of skepticism, be it accidental or principled, as a symptom of the human being's ontological indeterminacy and spiritual volatility: "I believe in an apostasis, absolutely necessary, as a possible event, from the absolute perfection of Love and Goodness, and because WILL is the only ground and antecedent of all Being" (*CL,* 6:897).

Yet precisely this potential for "degeneracy" and sin—the alternately casual or deliberate disavowal and denigration of human life as a moral and ethical reality—dialectically confirms the ontology of the Trinitarian God in the very act of its disavowal. That sin is an act undertaken by a "responsible Will," which in turn is inseparably bound to God's "fullness" (*plērōma*), Coleridge finds confirmed by an intricate phenomenology of conscience to which we shall turn later. Hence, if human "speech (Sermo, Verbum, Logos *sensu infinito*) denotes the essence of the filial Deity, ... the Mediator between God and Man, and the Redeemer of Man" (*CN,* no. 4984, f89), it is also true that the *logos* "in its finite and derivative existence" will manifest that metaphysical relation in all its instability. Expanding his earlier analogy between man and animal, Coleridge thus argues that

> as Man finds his redemption from the Captivity of his own Will in the Divine Humanity, so ... does the whole inferior Creation, which fell not willingly, seek yea, yearn and groan (i.e. significantly, ~~not~~ tho' inarticulately, utter its desire) for the ~~unloosing~~ Redemption in the Human Animal. In man is the solution of their dark Enigma ... we know that even in the elemental Mass the Spirit pleadeth to the Spirit with groans unutterable, that the whole Creation groaneth to be redeemed—and that the Spirit can not groan unheard! (*CN,* no. 4984, f89, f90[v])

Overtones of neo-Platonic mysticism (Plotinus, Jacob Boehme, and F. W. J. Schelling) are unmistakable here. Indeed, the *Rime*'s overall dramaturgy of a self, a community, and indeed the human project as a whole foundering in consequence of a gratuitous and violent *curiositas* (notably a vice in Christian moral philosophy) recalls Plotinus's use of a nautical analogy to explain the elusive relation of body and soul:

> it is also said that the soul is in the body as the steersman is in the ship; this is a good comparison as far as the soul's ability to be separate from the body goes, but would not supply very satisfactorily the manner of its presence, which is what we ourselves are investigating. For the steersman as a voyager would be present incidentally in the ship, but how would he be present as a steersman?

> Nor is he in the whole of the ship, as the soul is in the body. Are we then to say that it is present as the skill is in the tools, in the rudder for instance?[25]

The problem for Plotinus lies with the fact that the relation of body and soul, of a generic substratum of matter inhabited by a uniquely "ensouled" agent, does not seem to admit of metaphoric characterization. For Plotinus, the breakdown of the nautical trope exposes body and soul as fundamentally incommensurable categories.

This persistent conceptual impasse also compromises a line of argument according to which Coleridge "believed he could show Logos to be the supreme philosophical principle through which the reality of life and mind could be communicated [and that] the external world *in* which humanity finds itself, and the world of thought *through* which it finds itself, share the same source."[26] To be sure, in engaging the middle period of Schelling's writings (then increasingly influenced by Friedrich Heinrich Jacobi and exhibiting markedly voluntaristic and irrational tendencies), Coleridge had grown wary of the static and potentially irreconcilable polarity that separates God from his material creation. As Douglas Hedley cautions, "distinctions between light and darkness, love and hate as finite oppositions cannot be legitimately projected into the divine life."[27] Nor, indeed, do such antitheses adequately describe the human being, for "all that is characteristic of his Nature as Man, is seated in the incommunicable part of his Being, of which we know that it is not his Body, nor of it; tho' it may well be, that his Body is *of* it (Light *cannot* be the offspring of Shade; but Shade *may* be the offspring of Light)" (*CN,* no. 3962). Coleridge's lifelong ambivalence concerning neo-Platonic thought reflects his (correct) intuition that in Plotinus and some of his heirs (including Boehme and even Kant), there is at times a tendency toward static, dualist models that risk a resurgence of Gnostic beliefs even as they seek to overcome them. While often enthralled by "the spirit-fettering, spirit-awakening … fantastic laboratory—the Neo-platonic philosophy" (*CN,* no. 4213),

---

25. Plotinus, *Enneads,* 4.3.21.

26. Perkins, *Coleridge's Philosophy,* 22; for a discussion of the Coleridgean *logos* in relation to language, see ibid., 25–47; for a more skeptical account of Coleridge as endlessly divided and vacillating, a kind of philosophical Hamlet acutely conscious of the "undesigningness of my mind" (*CN,* no. 5226), see Seamus Perry, who argues for a Coleridge defined by the "experience and exploration of division, rather than in the unity of any triumphantly pulled-off conclusion" (*Coleridge and the Uses of Division,* 17).

27. Hedley, *Coleridge,* 85; for the same reason, Coleridge also takes care to distinguish "between the notional One of the Ontologists, and the Idea of the Living God." The former "would no longer be the God, in whom we *believe;* but a stoical Fate, or the superessential One of Plotinus, to whom neither Intelligence, nor Self-consciousness, nor Life, nor even *Being* can be attributed" (*AR,* 175, 169); on Coleridge's reading of Schelling, see Orsini, *Coleridge and German Idealism,* 192–237.

Coleridge also understands that neo-Platonism's tendency toward an undifferentiated and impersonal (in tendency pantheist) model of God threatens to foreclose on the possibility of the human person's rational ascent and salvation and thus turns God into an unfathomable (and potentially irrational) noumenon.[28] Whereas "in the works of Plato and Aristotle you see a painful and laborious attempt to follow thought after thought, and to assist the evolution of the human mind ... in the works of Plotinus it is all beginning, no middle, no progress" (*LHP,* 1:321–322). Even that distinction between Plato's esoteric yet profoundly illuminating dialogues and their subsequent mystification was hard-won since the eighteenth century had proven overwhelmingly hostile or, at best, indifferent to Plato. By 1795, as Coleridge first engages the Platonic tradition, "Plato was unfashionable ... the universities ignored him; the encyclopedists dismissed him as a dreamer; [and] the Dissenters blamed him as a source of Trinitarianism."[29] Such pervasive disregard left its imprint even on a thinker of such deep Platonist leanings as Coleridge, for as far as we know he had only read some of the dialogues and, not infrequently, had distilled his Plato from later sources—including the Alexandrian fathers, the Cambridge neo-Platonists, and Wilhelm Gottlieb Tennemann's history of philosophy.

Returning to his account in the 1823 notebook entry, we find Coleridge acknowledging an evolutionary view of organic life. Evolution here does not involve the eventual, seemingly inevitable Darwinian idea of natural selection that would

---

28. On "the excellence of the doctrine of Plato, or of the Plotino-platonic Philosophy," see *CN,* no. 3935; in his marginalia on Marsilio Ficino's *Platonica Theologia* (1525), Coleridge remarks the present, scientific age's "*supercilium* towards Platonism/the primary cause is that Impatience itself which characterizes Europe, & in a growing ratio from the days of Verulam to Condilliac—occasioning & occasioned by, the passion for merely sensuous phænomena. Finger-active, brain-lazy we ~~grin~~ look with the same arch scorn at ancient philosophy" (*CM,* 2:648); in an early letter (18 February 1801), Coleridge already remarks on the gradual loss of understanding that has befallen Plato's doctrine of ideas: "By the usual Process of language Ideas came to signify not only these original moulds of the mind, but likewise all that was cast in these moulds, as in our language the Seal & the Impression it leaves are both called Seals. Latterly, it wholly lost it's [sic] original meaning, and became synonimous sometimes with Images simply (whether Impressions or Ideas) and sometimes with Images in the memory; and by Des Cartes it is used for whatever is immediately perceived by the mind" (*CL,* 2:682–683). Coleridge's exploits in Platonic and neo-Platonist philosophy, which in turn influenced later nineteenth-century thought (see A. Taylor, *Coleridge's Defense,* 11) begin with intensive reading of Thomas Taylor's (much-reviled) translations of the *Cratylus, Phaedo, Parmenides,* and *Timaeus* between 1794 and 1800; another intensive engagement appears to have occurred around 1810, as suggested by notebook entries of that year (*CN,* nos. 3802–3935), followed by Coleridge's *Lectures on the History of Philosophy* of 1818–1819; as late as April 1830, Coleridge notes Platonism's crucial role in his intellectual formation, remarking how he had "read Plato by anticipation" (*TT,* 2:77). On that claim, see Vigus, *Platonic Coleridge,* 23–30; on the later Coleridge's engagement of Plato, see ibid., 93–165, and Perkins, "Coleridge and the 'Other Plato.'"

29. Vigus, *Platonic Coleridge,* 22–23.

soon destroy Anglo-Protestantism's problematic and unwise grounding in natural theology as articulated from John Ray to William Paley. Rather, it concerns the progressive, if always partial and fragmentary manifestation of the divine idea as eternally self-explicating:

> The Organic Forms ~~of~~ animal life, evolved from the Mollusca to the Ape, each in its kind the Product and Exponent of some selfish Appetance, assailing or resisting, pursuing or escaping, … like Fragments prepared for some harmonious Whole that ~~impatient ere the~~ had burst into particular life. . . . all these in the human Organism are reconciled and balanced into Subordination and Symmetry, and understand one another. In Man, the Vowels have been found, the hidden Letters have become Consonants—the Sound there of is heard, and the interpretation is made. So is it even now in the outward Man—but in the inward and in the World of History which is its Phænomenon, Man is to the Idea as the inferior Creatures to their Co-organization in the Human Frame.[30]

While the underlying topos of man as a microcosm of creation, and of nature's ultimate apotheosis in man, is old and well-known, Coleridge develops his theory of polarity so as to avoid the pantheist or emanationist slant that philosophers from Plotinus to Baruch Spinoza had repeatedly given to it.[31] His preoccupation with the will is deeply enmeshed with his eventual speculations regarding the polarity of life and his late theory of personhood and the Trinity; concurrently, Coleridge's view of life is inseparable from a coherent understanding of process, and such an account in turn cannot be tendered without an internally differentiated model of agency such as it is ultimately realized by the elections of the will. "The mere act of growth does not constitute the idea of Life" (*SW & F*, 1:508), since the tropism observable in all kinds of biomass is merely a matter of "degree" and, as such, tells us as yet nothing about the essence or kind of phenomenon *of* which it is a merely quantitative manifestation. In his "Theory of Life," Coleridge thus follows Kant's *Critique of Judgment* (esp. §§62–68) by "defin[ing] life as *the principle of individuation*, or the power which unites a given *all* into a *whole* that is presupposed by all its parts." What mediates parts and whole in this type of teleo-mechanist model of causation is "the *tendency to individuation*" (*SW & F*, 1:510–511), which Coleridge proceeds to chart as a

30. *CN*, no. 4984, f89^{v}; arguably Coleridge's most forceful statement on evolution, is found in the British Museum MS Egerton, no. 2801, quoted in Muir, *Coleridge as Philosopher*, 132–133, a text whose "bravado … perhaps masks his deep anxiety" (McFarland, *OM*, clxx).

31. "Man himself is a syllepsis ["a taking together, a summary" (*OED*)], a compendium of Nature—the Microcosm!" (*SW & F*, 1:551); see Perkins, *Coleridge's Philosophy*, 105–117; McFarland, *Coleridge and the Pantheist Tradition*, 107–190.

long and complex ascending scale leading from mineral formations to the fully and uniquely self-conscious organism of the human being. Drawing on a vocabulary strongly reminiscent of Goethean metamorphosis, Coleridge thus characterizes life as an unbroken sequence of "transitional states" involving "the perpetual reconciliation and perpetual resurgency of the primary contradiction"—a trajectory exhibiting both "the unceasing *polarity of life, as the form of its process, and its tendency to progressive individuation as the law of its direction.*"[32]

"At the apex of the living pyramid ... is Man, ... referred to himself, delivered up to his own charge" (*SW & F,* 1:551). Quoting nearly verbatim from F. W. J. Schelling and, especially, Henrik Steffens, Coleridge nonetheless is able to enlist the insights of German *Naturphilosophie* and Schelling's metaphysics of human freedom for his own distinctive purposes. For the supreme manifestation of the human being's precarious situation is found in the confrontation with her or his *will.* Both life's polarity and its tendency toward individuation thus present a crucial predicament for the self-conscious individual capable of choice. All choice amounts to an unconditional act of will or, as Coleridge puts it, "*the power of originating a state.*" It is "an *ens simplicissimum,* and therefore incapable of explication or explanation." Revealing yet again his deep kinship with Platonic thought, Coleridge insists that all philosophy commences with an act of self-creation, a choice, a conception and thus, ultimately, with an idea: "in the strict and purest sense of the term, all ideas are unique, and by their very unicity are contradistinguished from all images, conceptions, theorems, and notional forms. Thus life is in its idea inconceivable." In remarking on the will's "absolute antecedency in the necessity of thought and without any relation to time," Coleridge concedes that this idea constitutes "the most abstruse of all metaphysical speculations and the one great mystery of the mind" (*OM,* 18–19); and, like anything "unconditional," it proves *eo ipso* "inexplicable and incomprehensible" (*OM,* 32). Within the ectypal world of the human person, all choice is by definition ideational and original, rather than externally caused: "In Man the Will as Will first appears,

32. *SW & F,* 1:534, 537, 533. On teleo-mechanical thought in Blumenbach and Kant, see Lenoir, *Strategy of Life,* esp. 17–53; on Kant's ingenious fusion of teleology (as a hypothetical form of explanation) and mechanism (as the material engine of causation underlying all organic processes), see also Kolb, "Kant, Teleology, and Evolution"; van de Vijver, "Kant and the Intuitions of Self-Organization"; Ginsborg, "Two Kinds of Mechanical Inexplicability"; and Kreines, "Inexplicability of Kant's *Naturzweck.*" On the broader relevance of Kantian teleo-mechanical thinking for the emergent discipline of biology, see Grene and Depew, *Philosophy of Biology,* 92–127; for Kant, the main objective was to overcome the dogmatism of natural theology's creationism: "there cannot be an argument *from* design to a designer unless there is an argument *to* design in the first place" (Grene and Depew, *Philosophy of Biology,* 99). While Coleridge concurs, he seems doubtful that Kant's understanding of the biological organism (human or animal) as both "an *organized* and *self-organizing* being" (*Critique of Judgment,* 220 [§65]) is of sufficient explanatory force.

enough for him that he hath a Will at all; for ~~in~~ this is the condition of his responsibility, his humanity ... Both how and what we should do are both secondary questions that have no meaning except in reference to the former, incomparably more awful one: namely, what we should be?" (*OM*, 144).

Yet what is so chosen or conceived as an idea remains inherently unfulfilled. While Coleridge's desire "to form the human mind anew after the DIVINE IMAGE" (*AR*, 25) is never in doubt, that very aspiration can only be properly understood as (indeed, is essentially motivated by) an ongoing struggle with the gravitational pull of life as animated *matter*—alternately wayward, sluggish, or defective, and forever clinging to its own, self-consuming physicality. As a result, Coleridge's affirmations of faith in the logos come across as rather frantic—a case of "methinks the lady doth protest too much"—such that the central assertion of the divine "I AM" pivots on the constant awareness of embodied life as that which has as yet failed to make itself adequate to this ontology:

> the faith that He is God, the I AM, the God that heareth prayer—the Finite in the form of the Infinite = the Absolute Will, the Good; the Self-affirmant, the Father, the I AM, the Personeity; —the Supreme Mind, Reason, Being, the Pleroma, the Infinite in the form of the Finite, the Unity in the form of the Distinctity; or lastly, in the synthesis of these, in the Life, the Love, the Community, the Perichoresis, or Inter[cir]culation—and that there is one only God! (6 April 1832; *CL*, 6:897)

Such high-pitched affirmations notwithstanding, Coleridge's basic view of the human person as an *imago dei* stands firm. All finite, human agency thus is self-actuating not only "without but even against alien Stimulants," which is to say, grasps and inhabits the world as an "open," as yet "unfinished" sphere whose value is not objectively given but depends on the individual's capacity to saturate it with symbolic meanings. By contrast, anything that belongs to a class or species and thus is the result of an abstraction cannot be credited with reality at all but must be considered derivative of some anterior being to the comprehension of which thinking simply has not yet advanced: "The Soul must not only be distinct & therefore distinguishable *from* other Souls & other Objects, but she must be capable of actually <so> distinguishing herself" (*SW & F*, 1:428). Characteristic of Coleridge's position are two features: first, he holds it to be an unconditional *truth* (which *eo ipso* lies beyond the realm of empirically verifiable *facts*) that the human and the nonhuman realms are ontologically distinct. Second, he identifies the will as the distinguishing characteristic of a specifically *human* type of agency, even as the will's proclaimed, absolute anteriority also renders it inaccessible to the abstract and conceptual tools of rational discourse.

Coleridge's account of human agency in general, and of the will in particular, would seem little more than the kind of mystification often imputed to Coleridge, a kind of metaphysical haze famously subjected to various satirical (if also superficial) jibes by Lord Byron and Thomas Carlyle. Yet to take the trouble to examine Coleridge's account of the will with some care shows it to be a remarkably probing and imaginative enterprise. Moreover, his philosophical theology implicitly rejects the modern liberal, pluralist, and secular nation-state—that enigmatic, administrative space purporting to be home to "individuals" defined by their myriad competitive economic interests, and demographically partitioned into either indifferent or positively antagonistic cultural, religious, linguistic subgroups. Even as they continue to take themselves to be part of a single and supposedly coherent social formation, an imagined national community, these individuals are consternated by their apparent inability to articulate (when pressed to do so) any supra-personal reasons or "ends" for such a community; to sharpen the point, one may recall Michael Oakeshott's distinction between *societas* and *universitas,* derived from Roman law and in time appropriated by the competing models of the modern European nation-state. The first signifies a "partnership" "of agents who, by choice or circumstance, are related to one another so as to compose an identifiable association." They do so "not to act in concert but to acknowledge the authority of certain conditions in acting." In a *societas* loyalty is above all expressed as the members' joint acquiescence in specific conditions of legality, thus rendering this community "a formal relationship in terms of rules, not a substantive relationship in terms of common action." By contrast, a *universitas* is principally conceived for the purpose of a "joint enterprise" of some kind. It is an "aggregate [of persons] associated in respect of some identified common purpose, in the pursuit of some acknowledged, substantive end, or in the promotion of some specified enduring interest."[33] Much of Coleridge's writing, beginning with *The Friend* (1808), is driven by an anxiety that the commitments and ends supposedly underwriting the nation as a type of *universitas* will have been atrophied by a rampant pluralism and, ultimately, indifference that renders human relations strictly adventitious and, consequently, prevents the human person from grasping and articulating her or his essential relatedness to others.

A central feature of Coleridge's theory of the will involves the radical *singularity* of each human being. It is instructive to observe how he balances this apparently nominalist claim against what he frequently calls "Epicurean" philosophy, which

33. Oakeshott, *On Human Conduct,* 201; on Coleridge's political thought, see Muir, *Coleridge as Philosopher,* esp. 162–194; Edwards, *Statesman's Science,* esp. 111–132; on Coleridge's repudiation of Paley and Hobbes from an increasingly Platonist point of view in *The Friend,* see Morrow, *Coleridge's Political Thought,* 73–82; a detailed account of the later Coleridge's religious *cum* political thought, see Gregory, *Coleridge and the Conservative Imagination,* esp. his discussion of the *Lay Sermons,* 119–195.

would seem an inescapable entailment of any nominalist and empiricist account of the mind-world relation. If it is to signify at all, Coleridge's "one great and inclusive postulate and moral axiom"—viz., "the actual being of a *responsible Will,* distinct from all other conceptions"—will have to be contradistinguished from the notion of "instinct." For that term "implies a necessitation, 'Instinctus,' a goading or pricking" which, though "accompanied with sensation and consciousness, still we do not designate it as a will as long as it is contemplated as an effect, the <sufficient> cause of which pre-existed in an antecedent" (*OM,* 17). By contrast, the will is a primal and ineffable force that creates a new reality rather than reacting instinctively or compulsively to the one given. It *transforms* the real or, at least, envisions that the human individual which, very much in the sense of Heidegger's *Geworfenheit,* has been "thrust" into the world is capable of inner transformation. By its very nature, the self stands in an open, interpretive, and constantly evolving relation vis-à-vis the real. The self's contingent freedom—"contingent" in that it can only respond to the world at hand and, moreover, can never outright change or disavow its own, contingently given nature—is actualized by the will's "*power of originating* an act or state" (*AR,* 268). If this demiurgic conception of the will links Coleridge to a metaphysical dilemma first broached by Gnosticism, it does so in two distinct, indeed contrasting ways that throw into relief the metaphysically ambivalent situation of the human person. Moreover, Coleridge's evolving appraisal of the will also explains the inevitability of his shift from a quasi-deist (Unitarian) to a scrupulously articulated Trinitarian position that eventually had him declare outright that "Unitarianism is not Christianity" (*AR,* 211).

First there is the Augustinian conception of the will as fallen and inherently sinful. On that view, the modern individual's self-origination through an unconditional act of will reoccupies the ambivalent psychology of the Platonic demiurge-creator. Commentators on the *Rime* thus have remarked on the sudden deterioration of a putatively benevolent deity into a menacing and demonic force. For Edward Bostetter, "the rulers of the universe ... are revealed as holding the same contempt for human life that the Mariner held for the bird's life."[34] Likewise, Daniel P. Watkins notes how the Mariner, "no longer an integral part of his community," comes to represent "individualism at its most vicious" and how the "killing of the Albatross, an apparently arbitrary act[,] ... sets into motion the transmogrification of Christian power into demonic power."[35] While Coleridge would likely have agreed with that reading, his later writings also suggest that he would have strongly disputed the apparent premise, viz., that the particular nature of the Mariner's act had positively *caused* this

34. "Nightmare World," 69.
35. "History as Demon," 31, 26.

scenario of a "Christian universe gone mad."[36] Even more so, Coleridge would have disputed the further implication that some other act could have spared the Mariner (and his repeatedly co-opted audience) all the trouble. In fact, the poem's central conflict does not amount to a logical puzzle but, instead, restages the ontological dilemma confronting human agents once they have *actively* intervened in the cosmos by assuming the role of its unsuccessful demiurge-creator. As Coleridge puts it in *Aids to Reflection,* "a Sin is an Evil which has its ground or origin in the Agent, and not in the compulsion of Circumstances. Circumstances are compulsory from the absence of a power to resist or control them" (*AR,* 266–267).

To be sure, there is evil that arises from circumstances, but "such evil is not *sin*" inasmuch as true sin "can never be applied to a mere *link* in a chain of effects" (*AR,* 266–267). The words "origin, original, or originant" thus are strict corollaries of the idea of sin. Indeed, Coleridge notes,

> the phrase, Original Sin, is a Pleonasm ... For if it be Sin, it must be original: and a State or Act, that has not its origin in the will, may be calamity, deformity, disease, or mischief; but a *Sin* it cannot be ... Sin is Evil having an *Origin.* But inasmuch as it is *evil,* in God it cannot originate: and yet in some *Spirit* (i.e. in some *supernatural* power) it *must.* For in *Nature* there is no origin. Sin is therefore spiritual Evil: but the spiritual in Man is the Will ... the corruption must have been self-originated.[37]

In so manifesting the ontological condition of the human ("If there be aught *Spiritual* in Man, the Will must be such" [*AR,* 135]), the will points to a radical and profoundly unsettling freedom. "The Will is ultimately self-determined, or it is no longer a *Will* under the law of perfect Freedom, but a *Nature* under the mechanism of cause and effect" (*AR,* 285). Hence the will necessarily eludes analysis and causal representation, for "it is evidently not the result or aggregate of a composition but an *ens simplicissimum,* and therefore incapable of explication or explanation" (*OM,* 18); it eludes self-knowledge, for it "cannot be an object of conception" inasmuch as it has "absolute antecedency in the necessity of thought and [is] without any relation to time" (*OM,* 18–19); and, finally, the will also eludes deontological framing inasmuch as it is

36. Bostetter, "Nightmare World," 75.

37. *AR,* 270–273; see also the passage just preceding: "Now should you ever find yourself in the same or in a similar state, and should attend to *the Goings-on* within you, you will learn what I mean by *originating* an act. At the same time you will see that it belongs *exclusively* to the Will (*arbitrium*); that there is nothing analogous to it in outward experiences; and that I had, therefore, no way of explaining it but by referring you to an *Act* of your own, and to the peculiar self-consciousness preceding and accompanying it. As we know what Life is by *Being,* so we know what Will is by *Acting.* That in *willing* ... we *appear* to ourselves to constitute an actual beginning" (*AR,* 269n).

not cued (positively or negatively) by anything anterior and exterior to its own enactment.

It is only through its self-originating act—and, hence, discovering the consubstantiality of will, freedom, and sin—that human consciousness grasps its precarious spiritual constitution. In a famous instance of this figure of thought, Johann Wolfgang von Goethe's Faust notoriously rewrites the Gospel of St. John by changing the opening verse, "In the beginning was the word" (1:1), to "In the beginning was the deed." It bears pointing out that Goethe's unusual choice of proposition—*im* rather than *am* (*Im Anfang war die* Tat!)—slants the meaning of *Anfang* away from mere "point of departure" and toward the notion of "origin" as something purely volitional and consubstantial with the "deed" (*Tat*) itself.[38] Hence, the "deed" is not so much located at the "beginning" of historical time as it reveals the dynamic and metaphysically equivocal nature of all origination. Like Goethe, Coleridge understands hubris to be the inescapable signature of a modernity that continually asserts (and takes pride in) its self-originating power: "I place my principle in an *act*. In the language of grammarians, I begin with the verb, but the act involves its reality—it is the act of being, a verb substantive" (*OM*, 72–73). Such voluntaristic and necessarily sinful self-origination ultimately brings about the deeply problematic compartmentalization of the mind into consciousness and conscience. For "becoming conscious of a conscience partakes of the nature of an act. It is an act, namely, in which and by which we take upon ourselves an allegiance, and consequently the obligations of fealty" (*OM*, 72).

It is this split within the modern psyche that reveals the persistence of the Gnostic dilemma within European modernity. For Daniel Watkins, "the *Rime* comes across as being schizophrenic to a degree beyond what even the most extreme psychological interpretations have suggested . . . This schizophrenic quality is intensified by the fact that th[e demonic] presence cannot be located within an individual disturbed mind, but is projected as an external reality whose power is increased by its ability to evade all rational analysis and explanation."[39] In the modern clinical vocabulary of schizophrenia we thus trace the persistent dilemma first articulated by Gnosticism and never successfully solved by Augustinian or Scholastic thought. The dilemma in question originates neither in artistic caprice, nor does it contingently

---

38. Goethe, *Faust*, Part I, line 1237.

39. "History as Demon," 30; Watkins reads the *Rime* as expressive of a "demonic" energy that manifests itself repeatedly, such as in the Mariner's apparently heretical blessing of the water snakes, which he does "in the same way he shot the Albatross, unthinkingly and impulsively. What is more, the objects of his blessing, the water snakes, have obvious connections to the biblical serpent, and to bless them is to abandon the system of belief to which he had previously adhered, not to submit to it" (ibid., 28).

arise from within the Mariner's psyche or gratuitously befall him from without. Rather, the schizophrenic "division" (Coleridge appears to echo Schelling's sense of a primordial crisis [Grk. *krinein* = division]) presents itself as an ontological condition as Coleridge explores the metaphysical paradox of free agency within a supposedly pre-stabilized, benevolent cosmos. For it is by virtue of its own self-originating act that the modern self finds itself permanently divided into consciousness of the act proper and a conscience that intensifies once the costs of that act begin to come into focus. The third part of the *Rime* thus captures the emergence of conscience from within a self-consciousness that had been originated by a strictly motiveless act of will. For the first time, the Mariner now confronts his own nautical transgression of "the Line" with penitential intensity: "As if through a dungeon-grate he peered, / With broad and burning face. / Alas! (thought I, and my heart beat loud) / How fast she nears and nears" (ll. 179–182).

In sacrificing the reality of the cosmos to a network and replacing it with an evolving network of causally warranted propositions forever subject to modern protocols of verification or falsification, reason has in Coleridge's view become hostage to a self-originating act of will alternatively discharged in the open-ended skepticism of the Cartesian *cogito* or in the feral appetites of Hobbesian man. In either case the modern will—far from being a value-neutral, let alone positive "scientific" breakthrough—reveals its underlying and irremediable sinful constitution. In his marginalia to Descartes, Coleridge thus identifies the rigorously disjunctive logic of Cartesian reflection as its basic "sin": "This utter disanimation of Body, and its, *not* opposition, but contrariety … to Soul, as the assumed Basis of Thought and Will … is the *peccatum originale* of the Cartesian System."[40] The will denatures, indeed destroys the world that was *given,* converting it from an inherited "dwelling" into alien or virtual matter for experimental speculation, a shift whose dissociative quality is starkly illustrated by the notorious roll of the dice for the souls of the crew in the *Rime.* Being "not a mere mode of our consciousness, but presupposed therein" (*OM,* 73), conscience relates to consciousness as does the redeemer to the demiurge in Gnosis. With characteristic impatience, Coleridge strains to articulate this Gnostic crisis within the modern self, even as "our present language fails in affording a term sufficiently discriminative." For if we are to understand "self-knowledge in this latter, higher sense of the term 'Self,'" one must understand *conscire* not merely as an apperceptive or sentient state (which animals also possess) but as the human being's capacity

---

40. *CM,* 2:170–171; elsewhere, Coleridge singles out "Descartes [as] the first man who made a direct division between man and nature, the first man who made nature utterly lifeless and Godless" (*LHP,* 2:565).

> to know something in relation to myself in and with the act of knowing myself as acted on by that something . . . thus: the third pronoun *"he," "it,"* etc. could never have been contradistinguished from the first, but "I," "me," etc. ~~but~~ by means of the second. There could be no *"He"* without a previous *"Thou."* (*OM,* 73–75)

It is the "Thou" whose ethical reality the Mariner seeks to reaffirm time and again, in apparent compensation for his primal transgression. The spellbound wedding guest thus allows the Mariner to mediate or transpose the consciousness of his volitional and sinful act of self-origination into responsible knowledge (*conscire*), viz., by recognizing that in the absence of community there can be no such thing as conscience, remorse, or love.

As Coleridge's vivid imagery (the "glittering eye") suggests, his Mariner is no conventional, "realist" character but a quintessentially modern *type*—what Georg Simmel calls "the stranger"—a being whose identity is suspended between that of the mere traveler and that of a person "truly at home."[41] Coleridge's Mariner thus appears a philosophical allegory of sorts, a strange enjambment of Cartesian skepticism and Hobbesian voluntarism, both of which license an ethos of unbridled acquisition and dominion and, in so doing, prima facie created the specter of a menacing and irrational universe against which the modern self now struggles to protect itself. For his part, Coleridge's Mariner can only grasp at (or, as the case may be, defend against) the meaning of his self-creating deed through symbolic and compulsive acts of narration. Yet the resulting story cannot name its own cause, cannot settle the underlying account, but instead compels endless acts of textual exegesis (e.g., the Gloss in the 1817 *Rime*) and a wide array of secondary, methodological reflections. Yet these supplemental practices cannot but perpetuate the Mariner's original and all-consuming "anxiety," a holistic mood that defines the modern individual as it grapples with the consequences of its self-creating hubris. Beginning with repeated references to the "bright-eyed" Mariner and culminating in his tormented final appearance ("this frame of mine was wrenched / With a woeful agony" [ll. 578–579]), it is this "anxiety" that constitutes the affective signature of the modern psyche. Daniel Watkins speaks of "the existential angst of the Mariner," an observation whose deeper significance, however, is readily lost if one attempts to explain such *Angst* in merely causal terms—that is, as supposedly arising from an empirical conflict between base and superstructure. Premised on vulgar Marxism's false methodological choice of proceeding "historically rather than psychologically," such reasoning is palpably unaware of its own metaphysical presuppositions: "The narrative [of the *Rime*] is a

41. "The Stranger" and "The Metropolis and Mental Life," in Simmel, *Essays on Interpretation,* 324–339.

symbolic formulation of the contradictions and struggles within history, and that these historical pressures are antecedent to, and indeed are the primary source of meaning behind, all plot-level representations."[42] Such reasoning merely replaces one set of (theological) causal explanations of the type offered by Robert Penn Warren with another (historical) set; to argue that the Mariner "subtly and brilliantly co-opts the vocabulary of Christian value for the sake of undermining and redefining that value" is to ignore Coleridge's profound and expansive theological inquiry into the ontological status of human freedom, volition, and ethics. The *Rime*'s contradictions, then, are not glossed over by "Christian values" but, on the contrary, turn out to have haunted Christianity itself ever since its first consolidation as a coherent theological system during the early patristic era. The standard Marxist or social-science reading of the *Rime*'s religious imagery and symbolism as but an ideological stalking horse for real, material contradictions also fails to note how historical and material (no less than intellectual) antagonisms also beset the Church, such as when it (unsuccessfully) struggled with the revived ideal of apostolic poverty by thirteenth- and fourteenth-century Franciscans. In fact, far from being some Platonic ideological superstructure glossing over real material tensions, as vulgar Marxism continues to suppose, it was the papacy of the high Middle Ages that "embarked on a course of adjudication which more and more forced it to decide issues in terms of formal rights ... [Thus] it was first of all the Church, the *sacerdotium,* rather than the *regnum,* which assumed traits of modern secularity—legal formalization, rational instrumentalization, sovereign rule, economic contractualism."[43] David Simpson rightly cautions that the *Rime* "is not about a conventional 'experience,' and thus cannot be simply hooked into a straightforward realist exegesis whereby we can test out what we think really 'happened' against what is described in the poem."[44]

Existential *Angst* constitutes an all-encompassing phenomenon throughout the poem—taken as both the *récit* of a bizarre story and a recurrent narrative performance forever vexing and paralyzing its listeners. Just before the apparent death of his shipmates, the Mariner recalls the terror of inhabiting a world without community, devoid of any accepted ethical norms, obligations, or communal ties. In so doing, he depicts a world (like that of Shakespeare's *King Lear*) characterized above all by the utter absence and failure of love: "Fear at my heart, as at a cup, / My life-blood seemed to sip! / The stars were dim, and thick the night, / The steersman's face by his lamp gleamed white" (ll. 208–211). Above all, it is the curious deployment of the Eucharist ("as at a cup, / My life-blood seemed to sip") as a simile for an all-

42. Watkins, "History as Demon," 24.
43. Milbank, *Theology,* 17–18.
44. "How Marxism Reads," 154.

encompassing fear that raises questions about the poem's superficially redemptive turn here. Notwithstanding the Mariner's apparent spiritual restoration at the end of section IV (ll. 292–295), such *Angst* will remain in effect throughout the poem and beyond, where it will metastasize to countless instances of future retelling. For whatever spirituality the modern individual may be able to achieve, its creed will categorically differ from the one that had prevailed prior to modernity's sinful inauguration of the free will. In consummate agreement with Archbishop Robert Leighton, *Aids to Reflection* eventually affirms Coleridge's long-held view that the "root" of all consciousness and human exchange is found in the phenomenological certitude of conscience:

> How deeply seated the conscience is in the human Soul is seen in the effect which sudden Calamities produce on guilty men, even when unaided by any determinate notion or fears of punishment after death. The wretched Criminal, as one rudely awakened from a long sleep, bewildered with the new light, and half recollecting, half striving to recollect, a fearful something, he knows not what, but which he will recognize as soon as he hears the name, already interprets the calamities into *judgments,* Executions of a Sentence passed by an *invisible* Judge ... Remorse is the *implicit* Creed of the Guilty. (127–128)

The perpetrator of a volitional and literally groundless skepticism that has unhinged and denatured all cosmological order, Coleridge's modern individual has also thereby brought about its own irreversible psychological instability. Invariably, then, the modern *vita activa* will repeat the primordial transgression whose consequences its frantic epistemological and material productivity means to contain. The transgression at issue here does not infract on some known positive law or injunction; nor indeed does the modern self will some contingent or absolute evil. In an early fragment, entitled "On the Concept of Punishment" (1794), Friedrich Hölderlin had already drawn attention to the non-causal and non-intentional "mysterious origin" of the ancient concept of *Nemesis;* any attempt at a causal explanation of punishment succumbs to circular argumentation; to define punishment simply as "the suffering of legitimate resistance and the consequence of evil acts" is to be driven to the *ex post facto* conclusion that "evil acts, then, are those followed by punishment. And punishment follows where there are evil acts." Yet such reasoning "could never offer a self-sufficient criterion for an evil act." By contrast, "in moral consciousness ... the moral law announces itself negatively and, as something infinite, cannot announce itself differently."[45] Reflecting distinctly neo-Platonic elements in their early reflections,

45. Hölderlin, *Essays and Letters,* 35.

both Hölderlin and Coleridge insist that ethics involves not some intellectual or conceptual scheme autonomously conceived and articulated. The good is not some notion or idea but is bound up with its realization. It belongs to the order of practical reason, and thus depends on how the human being recognizes and acknowledges *in reasoned practice* her or his communion with the finite and unique other. For Coleridge, philosophy begins with, and must at all times keep in view, the Platonic triad of the good, the true, and the beautiful. For there cannot be knowledge without responsibility, just as what Edmund Husserl terms the "truth of correctness" seems arid and pointless unless correlated with the "truth of disclosure." It is this ontology that prompts Coleridge's characterization of "original sin" as a pleonasm, and that reveals the frightful repercussions of the Mariner's capricious assertion of speculative curiosity aimed *not at participating in the cosmos but asserting dominion over a universe of discrete objects*. In the end, sin for Coleridge names a spiritual condition under which action is not undertaken for the sake of a rational *telos* but, simply, out of curiosity to see what might happen next.

The result is a rich spectrum of psychological disequilibria that may take the form of anxiety, paranoia, melancholy, anomie, or a host of neuroses that would in time fuel the pervasive transformations of philosophy, literature, and psychology that tend to be associated with post-Nietzschean modernism. Freud's early patients and early modernism's tormented images and sound-scapes in the works of Edvard Munch, Knut Hamsun, Arnold Schoenberg, Georg Trakl, Franz Kafka, and Robert Musil, among others, offer vivid updates on Hegel's "unhappy consciousness," now presented in all its quotidian, alternately banal, petty, or horrifying particulars. Coleridge's Mariner thus becomes the archetypal figure for a modern, deracinated, and aimless existentialist vision, the nightmare of Charles Taylor's "buffered self" that ever since has been the stuff of dystopic (proto-)modernist narrative from Coleridge through Kafka, not because the vision had failed but, on the contrary, because it had succeeded on its own impoverished terms. Within philosophical modernity, this predicament has received particularly searching expression in Heidegger's exploration of *Angst* as an ontological disposition. For Heidegger, "*that about which one has* Angst *is being-in-the-world as such*" (*BT*, 174/*SZ*, 186) and, again: "*That about which* Angst *is anxious is being-in-the-world itself.* Being anxious discloses, primordially and directly, the world as world ... *Angst* as a mode of attunement first discloses the *world as world*" (*BT*, 175/*SZ*, 187). As a "mood" or negative "attunement" (*Stimmung*) rather than a contingent and remediable instance of "fear," Heideggerian *Angst* captures the total narrative thrust of Coleridge's *Rime* remarkably well. For it is the Mariner's ontologically isolated (and only intermittently socialized) self whose calamitous journey has at last confronted him with the poverty and, indeed, sinfulness of a strictly voluntarist conception of freedom as sheer, unaccountable self-assertion or "experiment." For Heidegger,

[*Angst*] reveals in Da-sein its *being toward* its unique potentiality of being, that is, *being free for* ... choosing and grasping itself. *Angst* brings Da-sein *before its being free for* .... (*propensio in*), the authenticity of its being as possibility which it always already is. In *Angst* one has an "uncanny" feeling. Here the peculiar indefiniteness of that which Da-sein finds itself involved in with *Angst* initially finds expression: the nothing and nowhere. But uncanniness means at the same time not-being-at-home ... Everyday familiarity collapses. Da-sein is individuated, but *as* being-in-the-world.[46]

Angst thus appears as an ontological condition likely to play itself out in any variety of expressive and conceptual settings—such as Romantic melancholy, Marxist alienation, modernist anomie, etc.

By contrast, Theodor Adorno and Max Horkheimer's *Dialectic of Enlightenment* locates *Angst* in the obverse scenario—viz., one characterized by the total interconnectivity and implicit equivalence of all particulars under the methodological guidance of Francis Bacon's *mathesis universalis.* A precise inversion of Heideggerian *estrangement, Angst* here defines a world exhaustively framed within a single conceptual matrix: "Man imagines himself free from fear [*Furcht*] when there is no longer anything unknown. That determines the course of demythologization, of enlightenment, which conflates the animate with the inanimate just as myth conflates the inanimate with the animate. Enlightenment is mythic anxiety [*Angst*] turned radical. The pure immanence of positivism, its ultimate product, is no more than a so to speak universal taboo. Nothing at all may remain outside, because the mere idea of outsideness is the very source of anxiety [*Angst*]."[47] Approaching the issue from a very different (Platonist) view, Iris Murdoch reaches similar conclusions about modernity's dominant intellectual stances of empiricism and existentialism. Both philosophies, in her view, are informed by an acute dread of anything unknown, "a terror of anything which encloses the agent or threatens his supremacy as a center of significance. In this sense both philosophies tend toward solipsism."[48] Solipsism thus appears a psychological corollary of nominalism's disjunctive epistemology and the concurrent isolation of the will from any normative and inherited rational framework that threatens the modern self's proclaimed autonomy.

Readings of the *Rime* such as have been offered by Bostetter or, more recently, by Daniel Watkins as a "portrayal of ... social relations in crisis" are fundamentally correct. Yet to isolate, as Watkins does, the crisis of the *Rime* as reflecting the

---

46. *Being and Time,* 176; trans. modified.
47. *Dialectic,* 16; trans. modified.
48. *Existentialists and Mystics,* 269.

"triumphant individualism of the 1790s" is to remain identified with the historicist (and in tendency nominalist) paradigm of knowledge as the institutional synthesis of so much impersonal, specialized, and dissociated information.[49] Yet as Coleridge, long before Adorno and Horkheimer's critique of the Enlightenment, had contended (and I agree), such a procedure unwittingly applies modernity's mythic quest for the total and preemptive methodological stabilization of (aesthetic) experience—"conflat[ing] the animate with the inanimate"—to a poem that is itself deeply critical of precisely that kind of procedure. As both its agonizing frame-narrative (i.e., the Mariner detaining a wedding guest with a narrative about the collapse of community) and its nautical master-trope makes clear, Coleridge's *Rime* signifies less by some straightforward referential commerce with the real than by performatively tracing what Nancy calls the "inoperative community" to the deleterious impact of individual, skeptical, and instrumentalized rationality. The crisis explored in the *Rime* ultimately harkens back to the collapse of the public/private distinction during the Hellenistic period and the early eighteenth-century idea of the "social," a concept that already announces the defeat of normativity by utility, of intuition by institution, and of virtuous action by demographic behavior.

In constituting itself as a progressive, transformative, and self-legitimating *epochē,* modernity relies on a handful of mutually reinforcing notions (freedom, rights, individuality, productivity, utility, progress) all of which implicitly presuppose a means/end rationality was being aggressively promoted (and legitimated as "natural" and "moral") after 1688. In his critique of modernity's quintessential theoretician of means/end rationality (*Zweckrationalität*), Max Weber, John Milbank provocatively argues that Weber's commitment to *sociology* as the quintessence of modern disciplinarity begs its questions on a grand scale. For in his *Religionssoziologie,* for example, Weber "confines himself to the vague, unhistorical level of 'elective affinity' between religious belief and economic practice, and sees Protestantism's uniqueness as lying in its transference of asceticism to a totally 'this-worldly' sphere of activity." Yet precisely the category of a "'this-worldly' sphere," Milbank argues, is "assumed by Weber *a priori* ... By contrast, the point about theological influence on modern economic practice was not the transference of asceticism to this world, but rather the theological *invention* of 'this world,' of the secular as a realm handed over by God to human instrumental manipulation."[50] The resulting, radically contingent act of self-origination thus lacks both a coherent representation of the state to be achieved (formal cause) and a normative good for the sake of which a specific process is being initiated (final cause). This double inarticulacy concerning both the form and the dignity of the quest that has been embarked on thus burdens modern

49. Watkins, "History as Demon," 31–32.
50. Milbank, *Theology,* 91.

subjectivity with a kind of ontological *Angst*. In Coleridge's parable of the modern, post-Cartesian self, the latter is thus perpetually haunted by the awareness that its self-originating individuality may announce the return of a long repressed heresy, that of the "free," Gnostic demiurge. Ultimately, such *Angst* was brought about by the migration of "infinity" from a divine name (viz., the idea of God's eternal plenitude) into the realm of modern, procedural time. In the wake of this shift, which Ernst Cassirer sees significantly accelerated by Nicholas of Cusa, existential anxiety and empty (formal) infinity emerge as disturbing corollaries of modern subjectivity.[51] The modern self now appears terminally caught up in a dialectic between an experimental (and potentially irresponsible) *vita activa* and the endless task of containing *ex post facto* the unpredictable outcomes of its own industry. The latter takes the form of converting acts into "experiences," and of deflecting responsibility for what Coleridge calls "the elections of the will" onto the rhetoric of testimony to some inexplicable, traumatic occurrence now requiring sustained narrative and exegetical effort. The "contrapuntal structure of daring and inquietude" so characteristic of Coleridge's Mariner thus exemplifies the Sisyphean consequences of modern instrumental reason forever trying to close the wound that it itself had inflicted.[52]

A capacious understanding of Romanticism's place within modernity will elude us as long as our disciplinary, professional, and institutional habits are unreflexively premised only on such notions as the public sphere, possessive individualism, an axiomatically secular (means/end) model of rationality, and a disciplinary and professional concept of "labor" alternately fashioned or critiqued by the modern discourses of political economy and academic Marxism. As Hannah Arendt has shown, both discourses prove equally oblivious of the premodern, albeit enduring distinction between labor and work (*HC*, 136–174). As long as these and similar categories intrinsic to modernity remain unreflected and are simply deployed, our own disciplinary practices will simply replicate modernity's core assumption: viz., that transformational processes are intrinsically and unconditionally good, and that they may be

51. Cassirer notes how Nicholas of Cusa's theory of "learned ignorance" (*docta ignorantia*) incrementally shifts focus from a categorical to an aggregative form of cognition, that is, from reasoning on the basis of certain principles to reasoning as an endless series of minute inferences: "all knowledge thus is converted into a mere 'hypothesis' [*blosse Annahme*] liable to be displaced by other, future hypotheses." As a result, "the character of infinity has migrated from the object of knowledge to the function of inquiry itself [*Der Charakter der Unendlichkeit ist von dem Gegenstand der Erkenntnis auf die Funktion der Erkenntnis übergegangen*]" (*Erkenntnisproblem*, 1:25, 28). In time, this leads to Newton's paradoxical postulate of "absolute time and space, which as Louis Dupré points out, Newton took to be real even though they were not derived from experience." In the *General Scholium* to the *Principia* Newton thus engages in strange, paradoxical speculations to the effect that infinity is neither an ontological predicate of God nor "a reality beside God." Rather, he postulates that "absolute time and space merely constitute an empty infinity *within* which God creates" (*Enlightenment*, 22, 24).

52. Rosen, *Hermeneutics as Politics*, 19.

adequately legitimated by our appeal to and speculative reliance on strictly hypothetical future outcomes. Once committed to this key premise (and along with it to an utopian streak that equally informs Scottish political economy, Godwinian anarchism, Blakean millenarianism, Painite radicalism, as well as Fabianism and the contemporary socialist visions of Robert Owen, Fourier, Saint-Simon, Marx, et al.), modernity will also, however unwittingly, reoccupy an early Christian eschatological model. By dint of its linear and teleological architecture, eschatological thought necessarily invests history with a number of "constants," quite regardless of the particular project of self-description and self-legitimation it helps sponsor. This holds especially true for modernity's grand narratives of "secularization" (Hegel, Comte, Marx, Max Weber, Émile Durkheim), which in the absence of such constants could never read history as a self-regulating, pluralistic, and teleological progression, nor indeed articulate their own disciplinary and institutional role within that process.

In supplanting the Stoic notion of providence (*pronoia*) with eschatology, early Christianity had initially established the conditions for a process of secularization that, much later, would "transpose eschatology into a progressive history" (*LMA*, 32). Two of the central implications (or dilemmas) of early Christian theology, which had arisen in response to a number of "heresies" (Gnosticism, Stoicism, Manicheanism, etc.), prove especially important for Romanticism's critique of, or at least highly ambivalent outlook on, modernity. First, there arose what Hans Blumenberg calls an "eschatological 'state of emergency'" (*LMA*, 45) when the New Testament's "immediate expectation" of the end of world and time (*parousia*) failed to occur, and when that expectation's "untranslatability into any concept of history" had to be confronted. In the absence of the *eschaton*, the sheer durability of the cosmos, its *having* a "history," presented a major challenge to philosophical speculation. What had to be formulated in response was some notion of history as a trans-generational process, one whose metaphysical significance would depend on an active type of speculative curiosity (Descartes's *vigilantia laboriosa*) whose rational explanations would compensate for the conspicuous non-appearance or, at least, indefinite deferral of the *eschaton*. As a result, post-Cartesian thought gradually converts the *eschaton* that had once been the focus of the *vita contemplativa* into an utopia to be ushered in by the rational and methodical consciousness of the modern *vita activa*: "the idea of progress is precisely not a mere watered-down form of judgment or revolution; it is rather the continuous self-justification of the present, by means of the future that it gives itself, before the past, with which it compares itself" (*LMA*, 32).

A second and more troubling implication that arose along with the discovery of "history" as a metaphysical problem confronted speculative thought with the perplexing migration of what, until now, had been the sole attribute of God (which early Christian theology had taken on from Plotinus): the predicate of "infinity." In struggling with a strictly formal and abstract (mathematical) conception of space and time, the modern self (particularly in the work of Galileo, Descartes, Newton, and

Kant) once again encounters "infinity," though now not as plenitude but as sublime and terrifying emptiness. It can hardly surprise that eighteenth-century philosophy and science so often seek to compensate for the unnerving implication of empty, infinite space and time by *inferring* from it, in a curious instance of *post hoc ergo propter hoc* reasoning, "the infinite extent of the divine presence."[53] What such attempts at deducing a divine presence from a material absence (i.e., of limits to space and time) had curiously forgotten was the fact that the alarming "infinity" of historical time and cosmological space had arisen from the *non-appearance* of the redeeming God to begin with. The ascription of infinity to the historical and cosmological worlds of which modernity sought to take progressive control betrays a persistent element of crisis and insecurity in modernity's projects of self-description: "As an attribute of progress," Blumenberg notes, "'infinity' is more a result of embarrassment and the retraction of a hasty conclusion than of usurpation" and "more a predicate of indefiniteness than of fulfilling dignity" (*LMA*, 84–85). Indeed, he continues, "our discontent with progress is discontent not only with its results but also with the indefinite character of its course, the lack of distinctive points, intermediate goals, or even final goals. The recovery of the finitude of history by means of the idea of a final and conclusive revolution that brings the process of history to a standstill is made attractive, as an antithesis to infinite progress, by that very progress itself" (*LMA*, 85–86). Attesting to the persistent and corrosive power of Gnostic speculation, the metaphysical anxiety in question also helps to explain the *ennui*, disorientation, and melancholia of bourgeois individuals that pervades so much nineteenth-century literature. For the attribute of infinity proves logically incompatible with the idea of progress, which demands an intuitive and non-negotiable norm or *telos* (e.g., Aristotelian "virtue" or even Machiavellian "glory") to whose fulfillment an agent or community is pledged and from which the practices of material and social life may derive their legitimation. Yet the attribute of infinity, which since the rise of political economy had in effect created a new psychology exemplified by post-civic man, a creature who "has ceased to be virtuous, not only in the formal sense that he has become the creature of his own hopes and fears" but also in that "he does not even live in the present, except as constituted by his fantasies concerning a future, ... [thus] plac[ing] the performance of covenants forever beyond the new Tantalus's reach and le[aving] him to live by dreaming of it."[54] Coleridge's overall project was to tabulate the costs of a modernity that could only launch itself by dividing the human psyche between skeptical self-assertion and the supplemental creed of "remorse"—thus reviving (however unwittingly) Gnosticism's split between the primal fraud of material creation and our infinitely deferred redemption from the stranglehold of the alien God.

---

53. Kant, *Universal Natural History and Theory of the Heavens,* quoted in *LMA*, 81.
54. Pocock, *Virtue,* 112.

17

EXISTENCE BEFORE SUBSTANCE

The Idea of "Person" in Humanistic Inquiry

Few terms call more urgently for a deep-historical archeology and for patient "desynonymization" (to use Coleridge's term of art) from "subject," self," or "individual" than that of "person." To embark on tracing the term's genesis and progressive clarification is to encounter a vivid example of what John Henry Newman would subsequently conceptualize as the "development" of an idea—a process of progressive reflection and clarification that, taken as a whole, reveals the vitality of an intellectual tradition and through its many hermeneutic turns impresses on us the "antecedent probability" that the original idea had contained not merely *potential* meanings but a positive *truth*.[1] The idea of personhood, or "Personëity," constitutes the fulcrum of the later Coleridge's at times obsessive rumination of the human being's unique constitution. To be human, he insists, is to recognize oneself as an embodied being with a "responsible Will," capable of reflection and providentially alerted to its vertical rapport with the divine *logos* by the unique phenomenology of "conscience." Well before he had immersed himself in the scholarship that would en-

1. See Newman, *Development of Christian Doctrine* (1845), 33–40 (1878 ed.); in preparing for his "assertion of religion," Coleridge remarks how an event's "*anterior probability* . . . is part of its historic evidence and constitutes proof presumptive or evidence à priori" (*OM*, 16; italics mine).

able him to argue the term's centrality with the requisite detail, Coleridge already insists on the indispensability of "person" to moral philosophy, a discipline whose fortunes in the age of William Paley he understood to be acutely imperiled:

> The Contra-distinction of PERSON from THING being the Ground and Condition of all Morality, a system like ... Hobbes's, which begins by confounding them, needs no confutation to a moral Being. A Slave is a *Person* perverted into a *Thing:* Slavery, therefore, is not so properly a deviation from Justice, as an absolute subversion of all Morality.[2]

As so often, it is the failure to maintain a distinction—not only that between person and thing but, just as importantly, that between a contingent "deviation" from and an "absolute subversion of" justice—which serves as Coleridge's point of departure. Conversely, he is just as concerned about the usurpation of person by forms of animated life that do, in fact, not meet the relevant criteria: "Every Man is born with the faculty of Reason: and whatever is without it, be the Shape what it may, is not a Man or PERSON, but a THING" (*CF,* 2:125). If terminological carelessness is evidence of an ethical lapse, Coleridge's own temporary blindness to the way in which this pronouncement strips animals of all ethical standing arguably disconcerts. The apparent restriction of nonhuman life forms to the status of mere things, while obviously troubling in its own right, also hints at one of the more problematic aspects of Coleridge's in many respects impressive theology. For his contention that "trees and animals are *things*" (*CCS,* 15) highlights a lingering anthropomorphism in his thinking whereby what is divine and transcendent remains forever indexed and restricted to the phenomenology of human interiority, and in particular the potentially erratic operations of conscience. Still, as late as 1825, Coleridge reiterates his contention that "Morality commences with, and begins in, the sacred distinction between Thing and Person: on this distinction all law human and divine is grounded: consequently, the Law of Justice" (*AR,* 327). Here again, Coleridge implies that our ethical being, our capacity

2. *Essays on his Times,* 3:235n; among Coleridge's more forthright statements to the same effect is the following passage from *The Friend:* "The sacred principle, recognized by all Laws, human and divine, the principle indeed, which is the *ground-work* of all law and justice, that a person can never become a thing, nor be treated as such without wrong. But the distinction between person and thing consists herein, that the latter may rightfully be used altogether and merely, as a *means;* but the former must always be included in the *end,* and form a part of the final cause" (*CF,* 1:190; see also, *CF,* 1:191). Notably, Coleridge's distinction reappears in the writings of Simone Weil, who notes "that a human being should be a thing is, from the point of view of logic, a contradiction; but when the impossible has become a reality, the contradiction is as a rent in the soul ... One cannot lose more than the slave loses, he loses all inner life" ("The Iliad, Poem of Might," in *Simone Weil Reader,* 158–159).

for "justice" and for transcending the limits of a merely computational understanding toward the *logos* (reason) of which we are imperfect images, pivots on our ability to keep hold of distinctions without either flattening them into outright sameness or exaggerating them to the point that what is distinguished appears wholly unrelated.

As it happens, it is just that challenge of thinking as the practice of enduring and articulate conceptual differentiation that characterizes the historical evolution of the term "person" and, closely related to it, that of Trinitarian theology. The late Coleridge's observation that "in the Trinity there is, 1. Ipsëity, 2. Alterity, 3. Community" (*TT,* 2:65) succinctly identifies the three concepts that between the Council of Nicaea (A.D. 325) and Scholastic theology of the late thirteenth century would gradually emerge as integral to the idea of person—divine *and* human—and that had to be grasped as jointly present and non-contradictory features of fully realized personhood. In recovering the theological and philosophical traditions that had gradually refined the idea of person as relational, participatory, and internally differentiated, Coleridge places himself in stark opposition to modernity's hermetic and instrumental notion of a "punctual self" (Charles Taylor)—viz., an autonomous individual on the order of René Descartes's *cogito* or Johann Gottlieb Fichte's *Tathandlung* whose specificity and dignity are appraised almost exclusively in epistemological terms. Time and again, Coleridge rejects or, rather, exposes as intrinsically flawed the proposition "that 'I' refers to a purely mental *res cogitans,* or to a bare existent without a nature, which must, as it were, first realize itself *ex nihilo* as some thing with some nature." Even more pervasive today than in Coleridge's time, such reductionist accounts fail because they do not acknowledge, let alone engage, the human person's complex a priori *relatedness* to the world and, especially, to other persons. Nor do they recognize that intentionality is never simply consummated by its "objects" but shows the embodied human mind to stand in a relation of alterity or transcendence vis-à-vis its own representations. To be a person means to "put distance between him- or herself as subject and the whole content of his or her consciousness."[3]

While a fuller phenomenological exploration of this issue will have to wait, an initial juxtaposition helps capture the salient point here. Thus, whereas our awareness of chickens in the barnyard or neighborhood traffic patterns refers to facts that have reality independent of our awareness, it is a different matter with our manifest "awareness of psychological states. Our knowledge of these is part of the states themselves."[4] As we shall see, Coleridge's strategic concern with the "notices" of conscience—in contrast to the (supposed) value-neutrality of representations of

3. Spaemann, *Persons,* 68; W. Norris Clarke sees "*relationality* [as] a primordial dimension of every real being, inseparable from its substantiality, just as action is from existence" (*Person & Being,* 13f.).

4. Spaemann, *Persons,* 55.

fact—had adumbrated precisely this isomorphism of reflexive awareness with human life itself. As Robert Spaemann puts it, our

> dawning awareness of hunger *is* hunger. Hunger is not found out about, like some object in the world; it is something that I *have*. The having of hunger is actualized in the dawning awareness of it … *Awareness* of life is the irreducible paradigm for life and experience. Intentionality does not attend to non-intentional experience as an object "out there," indifferent to our awareness of it. Intentionality is simply the most intense mode of experiencing.[5]

For Heidegger who, to be sure, reaches very different conclusions from Spaemann, this framing, holistic awareness, or "attunement" (*Stimmung, Befindlichkeit*) is most consummately experienced as "anxiety" (*Angst*) and "care" (*Sorge*). Either mood shows that our concrete intentionalities are framed by our embeddedness in the world (*in-der-Welt-Sein*) and only so can be properly experienced as *ours*. Consequently, our affective and cognitive intentionalities do not unilaterally constitute or "produce" the subjectivity of our person but, in fact, presuppose and depend on its antecedent reality. For that reason, too, modernity's longstanding preoccupation with establishing stable *criteria* that account for "personal identity"—or, for that matter, skeptical critiques of that very endeavor as pioneered by Descartes and radicalized by Hume, Nietzsche, and a great deal of twentieth-century existentialist and deconstructionist thought—miss the salient point altogether. For such approaches peremptorily de-contextualize and isolate the person as a "denuded self" (Iris Murdoch), a putatively self-evident methodological decision that effectively creates the explanatory burden that such thinkers take themselves to have discovered. For all the lingering "subjectivism" with which *Being and Time* was charged upon its publication in 1927, and from which Heidegger so strenuously sought to distance himself in later years, its argument avoids the dead-end of attempting to grasp person by means of ordinary, criterial definition. Thus, the constitutively engaged and affective nature of "anxiety" or "care" in Heidegger's account adumbrates a person's fundamentally evaluative, intentional stance vis-à-vis the world—with "world" here understood not merely as a sum-total of facts either previously ascertained or presently at hand, but also including the future perfect of as yet unimagined and unconsummated realities.

As Coleridge's urgent strictures against the widespread and ever-increasing conflation of persons with things make clear, to invoke the term "person" at all is to confront a pervasive ethical miscarriage of modernity—and to do so precisely by

5. Ibid., 55–56, and, for a perceptive discussion of Husserl's argument with Brentano about the nature and scope of human intentionality, 48–61.

remembering a term whose complex history and normative authority the modern era had forgotten at its own peril. Just as Coleridge's account of the will in *Aids to Reflection* and *Opus Maximum* amounts to a logically and historically rigorous critique of modern voluntarism and its secular descendants (materialism, mechanism, utilitarianism, liberalism), so his closely related inquiry into the idea of personhood rejects modernity's conceptually weak, a-historical, and reductionist approach to philosophical explanation as but a self-fulfilling prophecy. For in legitimating itself as a unique *epochē,* a break not only with specific traditions but with the very idea of tradition as *the* indispensable framework for responsible knowledge, the modern project at once authorizes itself but, at the same time, sows the seeds of its own progressive intellectual impoverishment. It cannot remember the history of concepts and, thus, cannot orient itself within specific traditions of inquiry whose rehabilitation has more recently been urged again by Hans-Georg Gadamer, Alasdair MacIntyre, Louis Dupré, and Charles Taylor, among others.[6]

For several reasons, then, a (highly compressed) history of person and its terminological ancestry is in order. First, such a review helps illustrate Coleridge's richly informed and subtle navigation of a tradition that in his time had largely been occluded by two centuries of mechanist and materialist inquiry and their prevailing mono-causal and deterministic forms of scientific explanation. Second, even a brief survey of the word's history already intimates the striking richness and complexity of the idea of human personhood, particularly if juxtaposed to Enlightenment ideals such as self-possession, autonomy, or an exclusively rights-based (legalistic and procedural) model of social relations and community. Third and most important, it turns out that there is no alternative to such a historical retrieval unless, of course, one is a priori committed to the modern project and thus only ever prepared to take a passing archival or encyclopedic interest in ideas like free choice, conscience, person, teleology, or judgment. In that case, the mere suggestion that ideas—and, for that matter, art—might have a distinct and surpassing reality will be rejected as both quaint and indemonstrable, and their cognitive value restricted to ancillary, "historical information" about a past that, however alluring, is by definition precluded from having any significant bearing on our own intellectual condition. As should be clear by now, that is certainly not the spirit of the present argument. Rather than approaching intellectual history as the distillation of plausible facts from ideas, beliefs, and problems variously deemed arcane, metaphysical, or otherwise "premodern," the present account posits that to *understand* personhood we must take an active, interpretive, and urgent interest in the idea itself. Though it should always aim to tran-

6. On the concept of tradition, see Gadamer, *Truth and Method,* esp. 267–306; MacIntyre, *Three Rival Versions,* esp. 170–215; and Diamond, "Losing Your Concepts."

scend the realm of ephemeral, idiosyncratic, and private opinion, genuine humanistic inquiry must indeed commit to and embrace the idea of "personal knowledge"—with "personal" here speaking both to our responsibility for the knowledge in question *and* our recognition of its transformative impact on ourselves as ethical beings. Indeed, precisely because the reality of personhood was so widely and acutely felt to have become atrophied by modern inquiry—to the point that its loss effectively structures the plot of many canonical nineteenth-century novels—personhood emerges as an idea whose marginalization, displacement, and threatened oblivion we should view with great concern and distrust. Thus the following, selective retrieval of the idea of person as it had taken shape in uncommonly self-conscious and explicit ways in the tradition of philosophical theology from the second through the fourteenth centuries honors Newman's typically concise and lucid remark that "the present is a text, and the past its interpretation."[7]

On the face of it, a plausible alternative strategy to the historical hermeneutic sketched above might be to undertake a paradigmatic survey of different conceptual and historical strategies of thinking about the concept of person. For an example one might consider a long chapter in Amélie Oksenberg Rorty's *Mind in Action,* which opens with a catalogue of different strategies of conceptualization (27–42) and, as far as it goes, certainly a concise classificatory scheme. Yet such an approach is by its very nature destined to fail, if for no other reason than that, in axiomatically liberal-pluralistic fashion, it treats the various approaches to person as implicitly equivalent. To do so means to have forgone from the very outset the possibility that person might not be a concept at all but, conceivably, an ontological and normative value. While Rorty fleetingly acknowledges such a scenario, she just as quickly absorbs the ontological argument as merely another conceptual variant into her survey of different intellectual "approaches." Tellingly, she vacillates between speaking of person as a "concept" or as an "idea," thereby revealing that for her no categorical difference exists between the ontology of person and any number of (invariably modern) discursive models of the "subject," the "self," or the "individual" as alternately defined by its rights, autonomy, skeptical inwardness, propensity to mystical self-transcendence, performative enactment, etc.

If, however, one were to acknowledge the distinction between Person as a normative (ontological) idea and persons as variously defined and accented "subjects" (or, more rebarbatively, "subject-positions"), it would soon emerge that doing so

7. "Reformation of the Eleventh Century," first published in *British Critic,* 1841, as reprinted in *Essays Critical and Historical,* 2:250. A distant echo of Augustine's *Confessions* (3.7.13), Newman's point is repeated by Ratzinger, who notes that "although this thought [on the idea of Person] has distanced itself far from its origin and developed beyond it, it nevertheless lives, in a hidden way, from its origin" ("Retrieving the Tradition," 439).

means *eo ipso* to commit oneself to a specific (and normative) position or, conversely, to reject the very idea of normativity itself. Either way, it would signify an explicit and consequential choice and so lift one above what Charles Taylor has called the modern "ethic of inarticulacy," which embraces a pluralist and non-evaluative way of being in the world without ever being able to name the sources that should compel us to adopt such a "view from nowhere." Pointed and perhaps even polemical though it may be, the point to be made here is, simply, that to take up the question of person is not simply to single out another "theme" or "topic." In fact, the idea of person is anathema to the pathos of distance, impartiality, and the value-neutral proceduralism that inexorably migrated from seventeenth-century science into adjacent fields, including philosophy. Instead, person belongs to the order of what Ernst Cassirer in some of his posthumous writings was to call *Basisphänomene,* that is, "conditions that we must posit in order to gain some kind of access to 'reality' and in which everything that we call 'reality' is primordially opened up and made accessible ... Each one is a manner or modality of mediation itself [*die Weisen, die Modi der Vermittlung selbst*]."[8]

To her credit, Rorty concedes that attempts at defining person in strictly analytic terms, however supple and adaptive, ultimately "seem unsatisfactory." Yet her proposed remedy—viz., to produce "an account of the many different reasons we have wanted, and perhaps needed, the notion of a person"—hardly seems a viable alternative.[9] For one thing, such an approach assumes from the outset that an intellectual endeavor, such as a coherent account of the idea of person, will ultimately always be outflanked by motives not transparent, certainly not in timely fashion, to those pursuing such an inquiry. Yet to hold that assumption is itself tantamount to denying what we shall find integral to personhood—not as a demonstrable fact, to be sure, but as an ever-present and non-falsifiable possibility: viz., the ontological reality of person as a free, rational, and incommunicable agent. Yet even if the skeptic's prejudgment to the contrary could be resolved, another problem remains. For, to undertake an archeology or meta-history designed to scrutinize the motives and interests shaping various accounts of person is in the end tantamount to writing a *history* of the term. Yet in so doing, one is then bound to come up against an idea whose capacity for eliciting sustained intellectual engagement for nearly 2,500 years, and for undergoing significant development in the course of that period, furnishes prima facie evidence of its truth-value. For neither the admittedly shifting and variable meanings of person, nor the motives said to have underwritten the gradual elaboration of these meanings could themselves ever be quarantined *from* this history. Clearly, though, Rorty, like any good modern-liberal intellectual, exhibits visceral

8. *Nachgelassene Manuskripte und Texte,* 131–132 (trans. mine).
9. *Mind in Action,* 44.

discomfort with all forms of normativity, as well as with the sheer possibility of having to acknowledge the categorical difference between historically contingent meanings and an ontology of truth.

Specifically as regards the idea of the person, to make sense of the term at all we must begin by acknowledging that we ourselves are fundamentally implicated in the larger ethical and religious conceptions of which the term is a seminal manifestation. There simply cannot be any such thing as a disinterested and noncommittal inquiry into it, whether systematic, analytic, and/or historical. Moreover, to decline rethinking the term's history, or simply being oblivious of it, also will not indemnify us against the claim's normativity. Instead, such a stance merely deprives us of bona fide intellectual agency. By contrast, to accept our own implication in the evolving conception of personhood is to recognize, minimally, that "*who* we are is not simply interchangeable with *what* we are," and that, consequently, unlike dog, table, or even human being, "'person' is not a classificatory term ... [It is] distinguished by not being a predicate of any other thing but identifying things that may be the subject of predicates."[10] If only in passing, it ought to be pointed out that if person is not a category or species designation, then its prevailing modern usage as a merely "legal" entity or bearer of "rights" proves deeply and dangerously incoherent; and here it matters little whether the particular "right" at issue happens to be that of unconstrained economic self-interest, freedom from taxation, or a whole garden variety of economic "rights" promoted with evangelical fervor by today's libertarians; or, more equivocally perhaps, the "right" not to be offended by someone else's exercise of the right of free speech or, perhaps, the right not to be discriminated against on account of one's sexual, religious, or ethnic identity. For in each case, the "person" staking a political or ethical claim of some such kind only exists as an abstraction and legal fiction that takes no account of the "incommunicable" (nontransferable) nature of the individual person invoking the specific right in question. With good reason, Amélie Rorty calls this "legal concept of a person ... a retrospective function, defined by the conditions for presumptive agency."[11] For to define person as the bearer of certain rights and, less eagerly, of responsibilities depends for its uptake and acknowledgment *as a claim* on an antecedent idea of person. In other words, simply to introduce some ensemble of "rights" as the principal criterion of personhood is never just a

---

10. Spaemann, *Persons*, 11, 6–7; to be sure, we may ascribe "personality" to a dog, though doing so involves a metaphoric transfer from the realm of the human, which stands out for its unique embodiment of personhood by the face; see Aristotle's remark on how humans and animals differ in that only the former have a "face" (*prosōpon*), *Historia Animalium*, 491^{b}9–11.

11. Rorty, *Mind in Action*, 32; as Rorty goes on to point out, "neither the Kantian regulative principle of respect nor the Christian idea of the immortal soul have any necessary connection with the legal function of the idea of person. Respect for the person doesn't entail any particular legal rights; nor does the assurance of legal personhood assure social or moral respect" (33).

definition but, inevitably, amounts to a normative claim—viz., that this is how personhood *ought* to be understood. And yet, it is precisely this antecedent reality of the individual *as person* that such an approach elides in favor of pleading group-specific and historically mutable interests. Arguably, then, for as long as the concept of "right" is mobilized strictly for the purpose of asserting a specific "interest," public and legal debate will fail to grasp the underlying ethical or (in Coleridge's parlance) "spiritual" dimension of the issue at hand. With good reason, Spaemann thus notes that it is

> a great mistake to think we must suppress observations of human differences if we are to do justice to human dignity. The dignity of the person is not touched by such observations, for the dignity of human beings as persons is not an object of cognition but of *re*cognition [*kein Gegenstand der Erkenntnis, sondern der Anerkennung*] ... If we say that someone is a person, we are saying that he or she is someone, a unique Individual; and this cannot be understood as the chance implication of one predicate, or even of an ensemble of predicates. *What* he or she may be besides does not settle *who* he or she is. The *what* we can observe and comprehend; the *who* is accessible to us only as we recognize something ultimately inaccessible.[12]

"Recognition," a crucial Hegelian term, here obviously demarcates an ethical relation between individuals qua "persons"—which is to say, something ontologically given by employing the very term "person" in the first place, rather than something compelled by positive law or established by the other person's "role" or "office" (both pre-Christian meanings of *persona*). Precisely this anterior, communal dimension of human personhood, its *relational* nature, is what Hegel's master and servant discover anew whenever they attempt to construe their independence or autonomy in positive, legal terms. The foil to Hegel's analysis is, arguably, Rousseau's deeply problematic legacy of construing all inter-subjective relations as instances of de facto "dependency," and interpreting the latter as the straight path toward inauthenticity, *amour propre*, and general moral turpitude. Whereas Herder, Kant, and their contemporaries still operate largely within the Rousseauvian paradigm, Hegel is the first to broach the all-important question: "Why can't there be other-dependence in condi-

12. *Persons*, 39; trans. modified; for Coleridge, who as Pamela Edwards notes, habitually "emphasized the term 'personal' in his consideration of rights and duties"; the idea of 'personal rights' was grounded in a moral vision that the 'personal will' must be able to accomplish its duties to the 'civic' Commonwealth as a prerequisite of any rights. The Coleridgean concept of a 'personal right' tied social entitlement to a more tangible set of 'personal' relationships than a necessarily vaguer 'natural right' or 'right of man' could ... [and it] forged the link between Coleridge's doctrine of the will and his doctrine of rights" (*Statesman's Science*, 114).

tions of equality?" In classical dialectical fashion, his *Phenomenology* demonstrates how a solution to the paradox of radical self-determination can only be produced by showing the Enlightenment's obverse contention (viz., that relationality necessarily equates dependence, and dependence in turn leads to inequality, inauthenticity, etc.) to fail on its own terms. For our autonomy and (self-)mastery pivots on its recognition by the subaltern other, thus leaving the lord unable to account for "the need that sends people after recognition in the first place." As a result, even those individuals who have seemingly achieved lordship over the other remain "more subtly frustrated, because they win recognition from the losers, whose acknowledgment is, by hypothesis, not really valuable, since they are no longer free, self-supporting subjects on the same level with the winners."[13] The structure of Hegel's famous discussion of lordship and bondage thus turns out to be one of unrelenting irony, and the ultimate target of that irony is modernity's utopian model of human agency as hermetic, self-grounding, and self-possessed. Robert Pippin's conclusion that "virtually everything at stake in Hegel's practical philosophy ... comes down finally to his own theory of recognition and its objective realization over time and in modern ethical life" offers valuable insight into Hegel's understanding of the "ethical life" (*Sittlichkeit*).[14]

Yet Hegel's intrinsically relational and social understanding of reason—which "treats *having reasons* as a matter of *participation* in a social practice under certain conditions"—does not address what, beginning with Adam Smith, had established itself as the dominant conception of sociality. Characteristic of that conception is a gradual weakening of the individual's commitment to conceiving, articulating, and realizing meanings and values in a social space that, for Platonic and Christian thought, had always been defined by the reality of other persons and one's progressive initiation into received, complex traditions of rationality. Instead, post-Enlightenment liberalism accepts, and often actively promotes, a laissez-faire ideal of pluralism and freedom that renders individuals and communities increasingly disinclined to give any reasons whatsoever for their practices, values, and commitments. Here, one begins to glimpse the costs of Hegel's having substantially voided the practical reason of Platonic and Christian practices of introspection such as had situated persons not only in a "horizontal comradeship" of the nation as an "imagined community" (to recall Benedict Anderson's account) but, concurrently, orienting them vertically toward a hyper-good. In the *Phaedrus,* Socrates flags this dual orientation of rational human agents, positing that "within each one of us there are two sorts of

13. Taylor, "Politics of Recognition," 45, 50. See also Pinkard, who remarks how "the relation to the other is ... double-edged in that the other both *affirms* and *undermines* the subject's sense of himself" and, in so doing, notably confine the reality of the other to "an *abstract* idea" (*Hegel's Phenomenology,* 54–55).

14. *Hegel's Practical Philosophy,* 29, 24.

ruling or guiding principle that we follow. One is an innate desire for pleasure, the other an acquired judgment that aims at what is best" (237[d]). Plato's soul can only understand itself as a dynamic, narrative principle "passing from a plurality of perceptions to a unity gathered together by reasoning," and in pointing to the deep connection between knowledge and "remembering" (*anamnēsis*) Plato suggests that discrete perceptions can acquire meaning and significance only inasmuch as they are indexed to an antecedent, divine archetype. There can be no lateral knowledge without the knower being vertically oriented toward the transcendent: "Wherefore if a man makes right use of such means of remembrance, and ever approaches to the full vision of the perfect mysteries, he and he alone becomes perfect" (249[c]).

While the Platonic motif of *anamnēsis* is still powerfully at work in Hegel, its narrative and transcendent implications have all but disappeared from the modern liberal community with the founding of which Hegel is often credited.[15] Thus, in a society of proliferating subdivisions, interest-groups, and strictly preference-based notions of value and meaning, the instantaneity of mimetic (and inherently noncognitive) impulses—whereby the self either "identifies with" or professes itself "disaffected from" a particular model of sociality—has effectively vanquished Platonism's emphatically *narrative* conception of knowledge as a form of ascent; and once the relatedness to other persons is deemed merely volitional and elective (as in modern, social-contract theory), the individual no longer takes herself or himself to have any vertical orientation. Now, too, one can see why Rousseau and his heirs should have (mis)construed relationality as sheer heteronomy, to be anathemized as a symptom of narcissistic and inauthentic sentiment. In stark contrast with the modern ideal of autonomy (and the individual's consequent quest for hedonistic self-fulfillment), Plato's *Phaedrus* contends that "soul not only moves itself and soulless things ... [but] can also be moved or affected by the motions of other souls." Motion here is not some mindless tropism, no Lucretian "swerve" confined to the present moment, nor indeed is it something externally caused or the result of some random inner compulsion. Rather, because "our souls wish to 'see,'" both knowledge and love (*ēros*) always involves "several different kinds of motion—vertical, cyclical, and horizontal." To know is, above all, a moment of vision, of "perceiving the image of the eternal beings in a beautiful face ... regard[ed] as an image of a god."[16] What renders

15. The point cannot be developed further here; for a discussion of Hegel's theory of the modern state as a "moral" (*sittliche*) community, see C. Taylor (*Hegel,* 428–461) and Pippin (*Hegel's Practical Philosophy,* 210–238). For a brief treatment of Hegel's debt to Plato, see Taylor (ibid., 512–516).

16. Zuckert, *Plato's Philosophers,* 313–314, 319. "Sight is the keenest mode of perception vouchsafed us through the body; wisdom, indeed, we cannot see thereby—how passionate had been our desire for her, if she had granted us so clear an image of herself to gaze upon" (ὄψις γὰρ ἡμῖν ὀξυτάτη τῶν διὰ τοῦ σώματος ἔρχεται αἰσθήσεων [250[d]]).

our relation to the "face" (*prosōpon*) of the other meaningful (and human beauty so irresistible) is the fact that the good and the true converge in a single intuition. It is the phenomenology of love, rather than the algorithm of epistemological correctness, that teaches us the inseparable bond between lateral cognition (*epistēmē*) and our vertical orientation toward the divine (*ēros*). To participate in the Platonic *logos* means to aspire—notably, through dialogue with others—to a deeper and necessarily supra-individual grasp of the good, the beautiful, and the true. No matter how far he was to push his Trinitarian speculations, Coleridge never lost sight of these Platonic elements in Christian Trinitarian theology.

To return to Hegel's account of sociality once more, Pippin is right to link the "noble nineteenth-century idea that my own freedom depends on the freedom of others" to the Hegelian "notion that the quality of the reasons available to me in understanding and justifying my deeds is not in the deepest sense 'up to me.'" Yet to conclude, as it were by default, that in fact such reason-giving remains "inextricable from the nature of the social practices ... at a time" already presupposes (but does not, in fact, argue the point) that rational agency is *constitutively inclined to undertake this work of reason-giving* in the first place, regardless of the conceptual sources on which such self-legitimation might draw. In fact, the condition of relatedness—and, consequently, a deep-seated obligation to articulate the meanings and ends for the sake of which we act—must already be accepted as the very essence of human personhood, quite independent of its historically specific sociality. As the case of Adam Smith's *Theory of Moral Sentiments* makes clear, absent some transcendent and normative obligation to make oneself intelligible to (and be recognized by) others as a rational, practical agent, it seems entirely possible that moral "practice" might end up merely replicating some established or prevailing type of "behavior," with the result that action is supplanted by mimetic reflexes, and that the cultivation of practical reason is short-circuited by the unthinking emulation of and compliance with prevailing customs, manners, and fashions.

The same predicament of denuding the practical of all vertical, transcendent sources of meaning also complicates Pippin's arguments about "objective reasons" in Hegel's practical philosophy. Though entirely cogent and persuasive as a statement of Hegel's position, Pippin's account again passes over the consequences of a model of *Sittlichkeit* preemptively stripped of any transcendent (and normative) sources and commitments. As Pippin notes, Hegel certainly allows for (intrinsically Aristotelian) forms of argument wherein reasons are presented as commitments buttressed by the agent's quasi-institutional authority: viz., "'because there is a contract,' 'because I am a father,' or 'because I am a citizen,'" etc.[17] In such cases, Hegel rightly believes that he need not "add on 'given that I want to be a good father,' etc. or 'given that I value the

17. Ibid., 27.

role of father as a good in itself'" since doing otherwise "would imply a separation between a subject and his roles that Hegel wants to deny without also collapsing such a subject into such roles, as if thoughtlessly and automatically acting out such roles." And yet, it is just this Platonic *idea* of the good (as in "good father," etc.) which must evidently be presupposed as the *telos* of all our striving, including the striving to legitimate our actions and practices *for others*. For only on that premise does it make sense to appeal to the role of father (and our fulfillment of it) *as a reason* capable of legitimating the specific actions currently being taken or contemplated. Our very quest for (historically specific) forms of self-legitimation as practical, rational agents pivots on a notion of the good transcendent to the practices associated with either Hegel's historically constituted "ethical life" (*Sittlichkeit*) or its discursive legitimation. As Iris Murdoch puts it, "*Good, not will, is transcendent,*" a point bound to be overlooked by thinkers like Hegel who "prefer to talk of reasons rather than of experiences."[18]

In this context it also bears remembering that Western theology's gradual clarification of the very idea of the human person presupposed that the term could not be assimilated to the standard matrix of individual and universal, which to this day undergirds most acts of (discursive) predication and cognition. For *person is not a concept,* certainly not of the ordinary kind, not a quality to be predicatively applied to some set of objects or even a species; neither is it a transcendental (Kantian) category. Rather, in a far more elemental and inescapable sense, its reality is that of a (normative) idea. Hence, in inveighing against slavery as the conversion of a person into a thing, Coleridge intuits that to accept the term "person" at all means *eo ipso* to have acknowledged the antecedent reality of an unconditional good and our inescapable, ethical implication in it. A "value concept" such as that of person, Murdoch insists, cannot be understood "in terms of switching on to some given impersonal network," that is, cannot be captured by some extrinsic, descriptive, or classificatory scheme. In fact, in the realm of the normative "words may mislead us ... since words are often stable where concepts alter." For Murdoch, there is a certain probability that in time a "deepening process" takes place whereby a value concept is more fully grasped than when the word primarily associated with it first entered the scene.[19] In other words, value concepts only disclose their meaning and significance within a historical hermeneutic of the kind outlined by Hans-Georg Gadamer and, somewhat differently, Alasdair MacIntyre—viz., as a sustained and reflexive process of understanding that gradually coalesces into the objectivity of a "tradition." The knowledge here sought is by definition inaccessible to a historicist methodology, for it is not the

18. Murdoch, *Sovereignty,* 68, 82.
19. Ibid., 28.

knowledge of a specific object or referent "out there" and ostensibly separate from us. Rather, "to employ the term 'person' is to acknowledge definite obligations to those we so designate." Being evidently not a descriptive- or species-term, the meaning of the word "person" cannot be realized by telling the story of its putative "referent." Instead, "with terms that have a normative content, ... to make their meaning understood we must again tell a story, but this time it is not the story of the referent, but of the term itself"—and so we begin.[20]

First, of course, there is the etymological aspect, which in the case of "person" was to be greatly complicated by the vexed question of how to translate manifestly related terms (*prosōpon, hypostasis, ousia, substantia,* etc.)—all of which would prove indispensable for grasping the divine *personae* in early Christian theology. Both the Greek *prosōpon* and the Latin *persona* have been traced back to the Etruscan *φersu,* a word written across the face of a masked figure vanquishing an opponent on a fresco found in a tomb north of Rome and dated ca. 550 B.C.[21] The original etymological association of *persona,* persistent throughout the pre-Christian period in Rome, is thus with the mask worn for theatrical performances and featuring a convex opening at the mouth so as to amplify the speaker's voice, which thus could "sound through" (*personare*). By the time of Terence, and culminating in the writings of Cicero, the association of *persona* widens and shifts—moving from a "role" played on stage by a masked actor to a character's functioning in official life, especially in politics and the law: "How many persons/roles (*Quot personae*) are there in a trial?" asks a Carolingian author, likely quoting an ancient Roman source, and answering "four: the accused, the lawyer, the judge, and the plaintiff." Already, a tension opens up between an understanding of *persona* as the sheer performance of a political role (*officium*) and, hence, as something ostensibly *im*personal and, on the other hand, a concomitant awareness that it takes the right temperament and character to meet the requirements of such a role. Yet if, particularly in Roman culture, person is strongly

---

20. Spaemann, *Persons,* 17; what follows draws on a number of seminal accounts of person, with Spaemann's being one of the most compelling. See also Clarke, *Person & Being;* Maritain, *Person and the Common Good;* for attempts to think person non-normatively, see Rorty, *Mind in Action,* 27–98; Johnson, *Persons and Things,* 3–26 and 179–187. For an analytic account, see Chisholm, *Person and Object;* for a major statement starkly different from the argument developed here, see Parfit, *Reasons and Persons,* esp. 199–350; for an exchange of different perspectives, see Cockburn, ed., *Human Beings,* esp. the essays by Cora Diamond and E. J. Lowe.

21. If our grasp of the idea of person hinges on a number of peripheral terms (substance, being, nature, existence, etc.), this project is greatly complicated by problems of translation that arose as Western theology began to switch from Greek to Latin during the third century. For seminal studies of the historical development of the concept "person," see Nédoncelle, "*Prosopon* et person" and the entry on "Person" in *Historisches Wörterbuch der Philosophie* 7:270–337; Rolnick, *Person, Grace, and God,* 10–60; and Spaemann, *Persons,* 16–33.

tied to the performance of social roles and political offices, it just as crucially denotes an actor's underlying and inalienable way of being, such as may enhance or interfere with the performance of a specific role or office. We thus find Cicero characterizing his political opponent, Piso, as incapable of filling the office of consul, "a role of great seriousness and severity" (*tantam personam, tam gravem, tam severam*).[22] The frequent suggestion in contemporary theory that performativity is unrelated to personhood, or indeed that personal identity is but an effect or "construct" of a given performance, ultimately fails on its own terms.[23] For it does not take into account that not every individual may inhabit every role with equal aptitude and success. In fact, the compelling performance of a role *presupposes* personhood as an incommunicable substratum and, by its various degrees of success or failure, intimates the actor's awareness (or lack thereof) of his or her nature and aptitudes.

Two further implications of consequence also begin to fade into view here: first, there is a strong association between *persona* and "action" (*agere, actio, actor*), such that the nature of a specific person is inseparable from his or her distinctive way of acting, albeit not simply as a performer of random roles or duties but as a being whose very nature can only be realized qua action. Second, there is the fact that the meaning of *persona* can only be realized within an ensemble of *other* persons likewise engaged in the active realization of specific roles and duties. This second meaning of person, still current now, first emerges in the formal analyses of Alexandrine grammarians of the third and second centuries B.C. in whose writings the word *πρόσωπον* (*prosōpon*) becomes a central category. Cued by the threefold aspect of personhood on the Greek stage—the person speaking, spoken to, and spoken about, respectively—these grammarians were among the first to institute a tripartite logic of personhood that in due course was to become the focal point of early Trinitarian thought. That is, even in its most abstract and as it were *im*personal sense, the grammatical concept of "person" (*prosōpon*), which in time would be taken over by Roman grammarians as the *persona* of the speaker, can only signify if it is understood to operate *within an ensemble of relations*.[24] In sharp contrast with the heuristic fiction of the modern autonomous self, or *cogito*, personhood is incommensurable with or, rather, antecedent to any type of solipsism: "The idea of a single person exist-

---

22. Cicero, *In Pisonem*, 24. In a programmatic statement in *De Officiis*, Cicero states that "we are invested by Nature with two characters, as it were [*duabus quasi nos a natura indutos esse personis*]; one of these is universal, arising from the fact of our being all alike endowed with reason ... The other character is the one that is assigned to individuals in particular [*altera autem, quae proprie singulis est tribute*]" (Book 1.30).

23. On the role of "personification" or *prosōpoeia* as licensing a constructionist or performative understanding of person, see the above discussion of Hobbes's "artificial persons" in Chapter 8.

24. E.g., Varro, *De Lingua Latina*, 8.20.

ing in the world cannot be thought, for although the identity of any one person is unique, personhood as such arises only in plurality."[25] That relatedness is an ontological feature of human personhood—rather than a secondary, merely empirical transaction or subjective choice—is prima facie borne out by the basic structure of grammar, at least as regards the Indo-European language-system; for the very nature of articulate speech (*logos*) demands an addressee. As we shall see, Coleridge in his time was almost alone in pointing to the essential relatedness of human life and personhood, a point forcefully taken up and variously developed in the twentieth century by Martin Buber, Emmanuel Levinas, and Jacques Lacan.

Formal-grammatical considerations first prompted Tertullian and other church fathers to suppose that God's first-person-plural declaration in Genesis 1:26—"And he said, let us make man to our image and likeness" (Lat. *et ait faciamus hominem ad imaginem et similitudinem nostrum;* Grk. *καὶ εἶπεν ὁ θεός ποιήσωμεν ἄνθρωπον κατ' εἰκόνα ἡμετέραν καὶ καθ' ὁμοίωσιν*)—presupposed at least one addressee, widely surmised to be Christ, or indeed both Christ and the Holy Spirit.[26] Employing a grammatical or "prosopographic" method of exegesis that proceeds by establishing who speaks to whom and about whom (or what), the early church fathers (Tertullian, Irenaeus, Justinus, and Origen in particular) all emphasize the fact that God qua reason (*λόγος*) implies a relation of different persons.[27] It is thus above all the Greek word *prosōpon* that shapes the early Christian understanding of the divine being and its enigmatic triplicity. In the Septuagint, *prosōpon* serves to render the Hebrew word *panim,* which signifies the human face or, as an intensifier, those traits that most clearly establish the identity of a person or thing. Inevitably, much theological controversy extending from the Council of Nicaea (A.D. 325) well into the modern era revolves around the elusive question of how to conceive the relationship between the three persons of the deity. Neo-Platonic subordinationism, which posits the superior dignity of God the Father within the Trinity; Appolinarianism, which grants Christ both body and soul but disputes his human rationality, whose place it claims is taken by the divine *logos;* Praxean "monarchism," which views Son and Spirit merely as different names for God's singular being; monophysitism, which postulates a single

25. Spaemann, *Persons,* 40.

26. Tertullian, *Adversus Praxean,* 12.3 takes the latter view; earlier exegeses pointing strictly to Christ as the addressee in Genesis 1:26 include those offered by Justinus Martyr, Theophilos of Antioch, and Irenaeus of Lyons. Among the best accounts of the complex theological debates of third- and fourth-century Western and Cappadocian theology is Ayres's *Nicea and its Legacy,* which briefly recapitulates Tertullian's early personalist theology (73–75) and discusses the impact of the Nicean resolution on subsequent Trinitarian thought (esp. 278–288), including in Augustine's later writings (esp. 372–381).

27. For a detailed account of "prosopographic" exegesis in relation to the emerging concept of person, see Andresen, "Zur Entstehung und Geschichte," esp. 9–14.

nature for the deity (thereby denying Christ's humanity); Nestorian dophysitism, which insists on the twofold nature of Christ as both human and divine and, from there, reasons to the duality of his person—they all struggle with the fundamental question, most vividly raised by Christ's divinity, of how to imagine the relationship between the being (*ousia, natura*) and personhood (*hypostasis, substantia*) of the Trinity.

This is not the place to reiterate these debates in great detail. Rather, so as to situate Coleridge's long and scrupulous reflection on Trinitarian theology and the concept of person, only a few stages of this exceedingly intricate philosophical and theological debate will be considered. In each case, the objective is to highlight those particular features in the evolving conception of the divine and human person that, in due course, would come to play a crucial role in Coleridge's own recovery of personhood as a cornerstone of his overall critique of materialism and its encroachment on the very idea of life. With the Alexandrian, grammatical provenance of *persona* (*prosōpon*) having faded into oblivion by the time of the church fathers, and given the absence of direct scriptural guidance as regards the relationship of the three divine persons, the challenge of conceptualizing the idea of the divine person(s) was only growing more urgent as time went on. Inevitably, whatever answers might prevail in the long run would play a crucial role in consolidating and legitimizing Catholic Christianity as an institutional and intellectual formation that involved a coherent body of theological propositions above and beyond the ongoing exegetical guardianship of revealed scripture. A momentous conceptual advance was made by the Cappadocian fathers of the fourth century. In the particularly rich Eastern theological language, the all-encompassing divine principle was rendered as *ουσία* (being), whereas the "distinctive aspect of each member of the Trinity" was designated as *ὑπόστασις* (*hypostasis*), thus germinating "our later term *person* ... in the differentiation from nature or essence."[28] Among the most coherent statements is Epistle 38 of St. Basil of Caesarea (ca. A.D. 330–379). Distinguishing between general nouns "predicated of subjects plural and numerically various ... as for instance *man*" and the individual person of Peter or John, Basil not only clarifies how the term *hypostasis* functions (i.e., by designating "that which is spoken of in a special and peculiar manner") but how abstract concepts can only signify predicatively on the basis of this unique quality or "under/standing" (*hypo/stasis*); it is the latter "which by means of the expressed peculiarities gives *standing* and circumscription to the general and uncircumscribed." Crucially, the Cappadocians hold that what rendered each hypostasis distinctive is its *relation* to the other two, even as Basil struggles to articulate the Trinity to his brother; all three divine persons "are, in a certain sense, ineffable and inconceivable, the continuity of nature being never rent asunder by the distinc-

28. Rolnick, *Person, Grace, and God,* 18.

tion of the hypostases, nor the notes of proper distinction confounded in the community of essence." Against neo-Platonic subordinationism, they insist on the equal dignity of each aspect of the divine being, while also opposing any reductionist attempt to construe each hypostasis or "person" as the mere default of its nature.[29]

The Cappadocian position was consolidated by Augustine who, in the wake of the Arian controversy that had only been partially resolved at the Council of Nicaea, focuses especially on the question of Christ's person. In *De Trinitate* (5.6–8), Augustine stresses how the very person of the "Father" presupposes that of the "Son," and vice versa, and how the Spirit is at once that of Father and Son. At the same time, he safeguards the identity and dignity of each person against being epiphenomenally derived from (or deemed contingent on) the relation it bears to the others: "it cannot be what [the Son] is called with reference to the Father that makes the Son equal to the Father. It remains that what makes him the equal must be what he is called with reference to himself."[30] In so insisting on the distinct nature of the person—at once in relation (*relative*) and "substance-wise" (*secundam substantiam*)—Augustine is still responding to the Sabellian (or "modalist") heresy, a position that had already been opposed by Tertullian. Yet in *Adversus Praxean* (A.D. 213), Tertullian had not been concerned with the *equality* of the divine persons but had only sought to argue their "consubstantiality" (a term introduced soon after Tertullian's *Adversus Praxean* was written). A century later, however, the Arian controversy that was to dominate the Council of Nicaea and continued to linger on for centuries thereafter had decisively altered the stakes of Trinitarian thinking. As he builds his case against Arian's denial of Christ's divinity, Augustine argues for an understanding of the Trinity in which the relation of the divine persons does not in any sense *modify* their substance. "Substance-wise," he insists, the identity of the individual person in the Trinity is not altered by the relation, even as the relation is constitutive of the Trinity and must not be thought of as a secondary attribute or trait of originally separate entities. This position, while effective in his debate with several heresies (modalism, Arianism,

29. St. Basil, Letter 38 (www.newadvent.org/fathers/3202038.htm). A more detailed argument would have to take into account the peculiarities of translation, which frequently led an overlapping or outright confusion of the Eastern theological concept of "being" (*ουσία*) with the Western, Latin term *substantia*—literally, that which "underlies" or "sub-tends" something else—which in turn is awkwardly close in meaning to the Greek *hypostasis*. Similar problems bedevil the relationship between *natura* and *essentia;* for a concise account, see Rolnick, *Person, Grace, and God*, 19f. and 26f.

30. The text (*ADT*, 5.7) continues: "But whatever he is called with reference to himself he is called substance-wise. So it follows that he is equal substance-wise" (*Quia vero Filius non ad Filium relative dicitur, sed ad Patrem; non secundum hoc quod ad Patrem dicitur, aequalis est Filius Patri: restat ut secundum id aequalis sit, quod ad se dicitur. Quidquid autem ad se dicitur, secundum substantiam dicitur: restat ergo ut secundum substantiam sit aequalis*).

monarchism) comes at a price, however, for it commits Augustine to a monolithic conception of substance as unalterable: "Anything that changes does not keep its being [*Quod enim mutatur, non servat ipsum esse*], and anything that can change even though it does not, is able not to be what it was; and thus only that which not only does not but also absolutely cannot change deserves without qualification to be said really and truly to be" (*ADT,* 5.3). Compared to the remarkably fluid and inventive conception of life as the ceaseless self-transformation of animate matter set forth by Ovid's *Metamorphoses* some four centuries earlier, Augustine's retention of a rigid and inert concept of *substantia* as one descriptor of person seems rather problematic.

Without doubt the most cogent and forceful construction of Trinitarianism to be found in the first millennium of Christian thought, *De Trinitate* (written between A.D. 400 and 420) seeks to think explicitly what in scripture is only obliquely articulated—viz., that "Father and Son and Holy Spirit in the inseparable equality of one substance present a divine unity; and therefore there are not three gods but one God [*quod Pater et Filius et Spiritus sanctus, unius ejusdemque substantiae inseparabili aequalitate divinam insinuent unitatem; ideoque non sint tres dii, sed unus Deus*]."[31] Augustine's emphasis on the *inseparable* nature of the three persons is meant to guard against a tendency, still observable in Tertullian, to compartmentalize their agency and, thus, to interpret person still in the loosely pagan sense of "role" (*officium*). Yet if "the Father does some things, the Son others and the Holy Spirit yet others; ... the trinity is no longer inseparable" (*jam non inseparabilis est Trinitas* [*ADT.* 1.8]). Augustine's main objective here is to establish a subtle conception of "relation"—such that the relation is not merely an *aggregate* of otherwise discrete and separate identities, nor is misconstrued as positively *producing* the identities of the divine personae; the latter, quasi-structuralist hypothesis is rejected indignantly: "it is ridiculous that substance should be predicated by way of relationship" (*Absurdum est autem ut substantia relative dicatur* [*ADT,* 7.9]). Specifically this remark has occasioned considerable dispute among theologians, for it appears to isolate the divine persons from relation, even as the second half of *De Trinitate* (Books 8–15) seeks to grasp the human person as an image of the Trinity. None less than Joseph Ratzinger here sees Augustine committing "a decisive mistake" inasmuch as "he projected the divine persons into the interior life of the human person and affirmed that intrapsychic processes correspond to these persons." The result, he argues, was a gradual split between increasingly esoteric Trinitarian speculation and the exoteric realities of human personhood. Hamstrung by his retention of a substantialist model of person, and consistently privileging that model—however enlivened by complex accounts of its "interiority"—over the relational model, Augustine (and following him

31. *ADT,* 1.7. Latin quotes follow the edition in Migne's *Patrologia Latina,* vol. 42.

the Scholastic tradition up to and including Aquinas) gave rise to a conception that, in the modern era, would take the form of social-contract theory, viz., by treating relations as strictly secondary, rational-choice type of arrangements. As Ratzinger concludes:

> the anthropological turn in Augustine's doctrine of the Trinity ... was one of the most momentous developments of the Western Church. In fundamental ways it influenced both the concept of the Church and the understanding of the person which was now pushed off into the individualistically narrowed "I and Thou" that finally loses the "you" in this narrowing. It was indeed a result of Augustine's doctrine of the Trinity that the persons of God were closed wholly into God's interior. Toward the outside, God became a simple "I," and the whole dimension of "we" lost its place in theology.[32]

Struggling to navigate between a substantialist model and a relational model, Augustine succeeds only partially in his attempt to think of the Trinity as a unique ontology manifesting itself as an ensemble of functional differences. The growing divide between the Trinitarian ontology and the ectypal realm of human, embodied, and inward personhood also explains Augustine's difficulty with connecting the first and the second half of *De Trinitate*. Even so, the work largely succeeds in articulating the necessity of thinking of God as a relation between three distinct persons in a way that avoids separating or disaggregating the divine personae. One of the principal sources feeding Coleridge's lifelong investment in organicist models of life and meaning is precisely this "unity of the three, incorporeal and unchanging, a nature

32. Ratzinger, "Retrieving the Tradition," 447, 454. From a different (Aristotelian) perspective, A. C. Lloyd and J. Mader had critiqued Augustine's understanding of person as supposedly blurring the categories of substance and relation; for a concise summary and refutation of those critiques, see Burnell, *Augustinian Person,* 67–70. Against Ratzinger's critique of Augustinian interiority, see Mary T. Clark, "Augustine on Person," who values Augustine's "insight into interiority as essential to human personhood" and sees it as "a valuable gift from Plotinus." On this debate, see Rolnick (*Person, Grace, and God,* 31–32); Clark's neo-Platonist appraisal of Augustine's model of person echoes Charles Taylor's influential reading of the *Confessions,* which credits Augustine with "introduce[ing] the inwardness of radical reflexivity." As regards Augustine's "proto-Cartesian move," Taylor remains curiously ambivalent. Thus he sometimes deems this step a "fateful one, because we have certainly made a big thing of the first-person standpoint. The modern epistemological tradition from Descartes, and all that has flowed from it in modern culture, has made this standpoint fundamental—to the point of aberration, one might think." On other occasions, he stresses "the crucial importance of the language of inwardness ... [which] represents a radically new doctrine of moral resources, one where the route to the higher passes within" (*Sources,* 131–132, 139). For critiques of Taylor, stressing the institutional, communal, and participatory nature of the self, see Aers (*Salvation and Sin,* 1–24) and Hanby (*Augustine and Modernity,* 5–10).

consubstantial and co-eternal with itself" (*unitas Trinitatis incorporea et incommutabilis et sibimet consubstantialis et co-aeterna natura* [*ADT,* 1.15]).

Once it is clear that the very concept of "relation" equally precludes the total indifference and the utter disparity of its constitutive members, it also emerges that "every being that is called something by way of relationship is also something besides the relationship" (*quia omnis essentia quae relative dicitur, est etiam aliquid excepto relativo* [*ADT,* 7.2]); or, as W. Norris Clarke notes, "a relat*ed* is not simply identical with its relat*ion,* reducible to it without remainder; it is distinct from it though not separable from it."[33] Simply put, "person" and "relation" are corollaries and, as regards their meaning, mutually conditioning terms. At the same time, *relation* is not to be conceived in any mimetic or imitative sense; the Son "does not do other things *likewise,* like a painter copying pictures he has seen painted by someone else" (*Neque enim alia similiter, sicut pictor alias tabulas pingit, quemadmodum alias ab alio pictas videt* [*ADT,* 2.3]). What was to prove central some eight hundred years later in Richard of St. Victor's account of relationality and personhood, and what would also preoccupy the late Coleridge, is here first attempted: viz., Augustine seeks to envision the Trinity as the archetype of the ideal, organic community wherein the identity of the persons comprising that unity is inseparable from their relations, even as it is neither transferentially projected upon nor mimetically derived from the other persons in that community. The persons or hypostases that comprise this formation are at once distinct, related, and inseparable. As Peter Burnell sums up Augustine's delicate balancing of a substantialist conception and a relational conception of the human being, "both words ["substance" and "person"] denote the determination of a rational nature as an existent being, but whereas 'substance' denotes that determination in reference to the nature itself, 'person' denotes in reference to relation with other beings."[34]

Quoting John 5:19 ("the Son cannot do anything of himself, but what he seeth the Father doing: for what things soever he doth, these the Son also doth in like manner" (Vulg. *non potest Filius a se facere quicquam nisi quod viderit Patrem facientem quaecumque enim ille fecerit haec et Filius similiter facit*), Augustine pushes the limits of intelligibility here by redeploying the same adverb ("likewise"), albeit this time in a positive sense. The challenge posed by the Trinity to finite human beings involves finding ways to grasp or imagine an isomorphism of being and action (*ὁμοίως/similiter*) without positing a merely derivative, mimetic, and/or narcissistic relation between the persons involved. Put differently, we are to conceive of relation as *on-*

33. *Person & Being,* 16; likewise, Rolnick cautions that "personality cannot be reduced to its relationships, even though its life is unthinkable without them" (*Person, Grace, and God,* 209).

34. *Augustinian Person,* 187–188.

tology, and not as a psychological constellation of discrete anthropomorphic agents. The Trinitarian community is not to be imagined as an empirical, secondary construct, and it will elude comprehension altogether as long as it is being construed as but another concept or category, such as substance or accident.[35] There is another, no less crucial reason for why *De Trinitate* takes such pains to delimit the semantic reach of "resemblance" or "similarity" (*ὁμοίως/similiter*) as regards the personae of the Trinity. For unlike the neo-Platonists, who tend to treat the *imago dei* conception as an ontological given, Augustine contends that "the divine beauty in us has been largely (of course not completely) lost, ... and [that] we are so extensively damaged that the similarity [with God] is not quite restored as long as we are in this world."[36] Understanding the magnitude of the task, in both its intellectual and rhetorical dimensions, Augustine repeatedly hints that to articulate the idea of the Trinity is not to indulge in remote intellectual speculation but to map a path toward understanding man as the *imago dei* and, thus, to draw closer to the ultimate mystical experience of a *visio beatifica.* With both Pauline and Platonic overtones, he thus notes how in engaging the Trinity "I forget what lies behind and stretch out to what lies ahead, and press on intently to the palm of the supernal vocation" (*et secundum intentionem sequor ad palmam supernae vocationis*).

Augustine's phrasing here subtly underscores his emphasis on a complex and dynamic model of inwardness; for in transliterating the Greek *κατὰ σκοπὸν* (Philippians 3:14) as *secundum intentionem,* rather than following the Vulgate's *ad destinatum,* Augustine characterizes this quest for understanding as intensely personal. At the same time, with a characteristic rhetorical flourish, the self's subaltern position vis-à-vis the idea of the Trinity is acknowledged in a chiasmic construction that concedes the ineffable nature of the Trinitarian ontology even as it finds profound compensation in the gift of human speech: "I have undertaken, not so much to discuss with authority what I have already learned, as to learn by discussing it with modest piety" (*non tam cognita cum auctoritate disserere, quam ea cum pietate disserendo cognoscere* [*ADT,* 1.8]). Precisely this ubiquitous linguistic constraint—viz., the ultimate inaccessibility of the threefold God to human speech—also accounts for why we habitually speak and think of the three persons not only as *distinct* but (with seeming inevitability) proceed to misconstrue them as *separate:* "in my words Father and Son and Holy Spirit are separated and cannot be said together." Augustine's choice of illustration is telling, for it affirms human personhood as an instance of the same organic, relational model as the Trinity and, hence, as no less difficult to fathom:

35. Though relatively unfamiliar with much of Aristotle's writings, the relevant Aristotelian text for Augustine's argument here, the *Categories,* was well known to him; see *Confessions,* 4.16.28–29.

36. Burnell, *Augustinian Person,* 174; Burnell references *De Perfectionae Iustitiae Hominis,* 11.28 here.

"when I name my memory, understanding, and will, each name refers to a single thing, and yet each of these single names is the product of all three" (*quemadmodum cum memoriam meam et intellectum et voluntatem nomino, singula quidem nomina ad res singulas referuntur, sed tamen ab omnibus tribus singula facta sunt* [*ADT,* 4.30]).

Quite literally, then, engaging the idea of the Trinity confronts us with a notion that, in due course, would emerge as central to the understanding of both divine and human personhood. Viz., as persons we are truly, indeed essentially *incommunicable.* The idea of person, its ontology, hints at something forever inaccessible to human, conceptual mastery—though nonetheless distinctly manifest or revealed, as remains to be seen in the juxtaposition of Coleridge's account of "conscience" and the I-Thou relation with similar models found in Martin Buber, Emmanuel Levinas, and Jean-Luc Marion. Augustine thus remarks how, in engaging the Trinity, we come up against "things which cannot be expressed as they are thought and cannot be thought as they are" (*Quamobrem ut jam etiam de iis quae nec dicuntur ut cogitantur, nec cogitantur ut sunt*).[37] To acknowledge this conceptual impasse does not, however, imply the premature end or outright failure of his inquiry. On the contrary, in a profound sense this acknowledgment of the limits of conceptual language and representation both highlights Augustine's deep commitment to apophatic theology and establishes the ineffability of the Trinitarian God (and, consequently, of the idea of person) as the permanent condition circumscribing all theological and spiritual reasoning and inquiry.

In rendering the three *hypostases* of the Cappadocians as *tres substantiae vel personae,* Augustine reveals a certain hesitation "as to what 'person' (*persona*) could actually signify among Father, Son, and Spirit." His affirmation (*ADT,* 5.3) that "there is at least no doubt that God is substance" (*sine dubitatione substantia*) almost immediately begins to fray and unwork itself: "or perhaps a better word would be being; at any rate what the Greeks call *ousia*" (*vel, si melius hoc appellatur, essentia, quam Graeci οὐσίαν vocant*); and as late as Book 7 we find him still sifting the implications of terms whose historical origins at least in part elude him. Precisely here, in his attempt to establish the exact meaning of "person" within Trinitarian thought, Augustine's argument appears notably unsure of itself. As is his wont, he opens Book 7 exegetically, in this case by parsing possible interpretations of 1 Corinthians 1:24, followed by other passages of scripture. Yet the exegetical approach offers only very limited safeguards against the dangers of outright error and forbidden predication,

37. *ADT,* 5.4; speaking of Augustine's frequent acknowledgment "that the divine reality cannot be put into language," Mary T. Clark also notes that "he never forgot that God is ineffable. He is a cataphatic [affirmative] theologian with respect to the mysterious Trinity only because Arianism and Sabellianism required the use of basic ontological language" ("Augustine on Person," 110).

quite simply because scripture itself remains all but silent on the issue of the Trinity. To the essential question—"So three what, then? If three persons, then what is meant by person is common to all three?"—it offers no answers. Neither does it clarify what is meant by "person" on those rare occasions where the term is being used in scripture: "So this is either their specific or their generic name."[38] Likewise, to the constantly looming, tri-theist heresy ("Why not three Gods?") scripture offers no decisive refutation: "neither do we find scripture talking anywhere about three persons" (*ADT,* 7.7–8). Augustine thus is faced with the vexing task of sifting a theological question of essential importance without being able to rely on exegesis.

Having recognized early on how "the desire to express the inexpressible" (*cum ineffabilia fari cupimus* [*ADT,* 7.2]) tends to lead one into paradoxical and highly questionable speculations, he confines himself to two related arguments. The first is to distinguish sharply between terms denoting a genus and those denoting a species; the second is to correlate that distinction with the two alternative constructions of person in terms of relation and substance, respectively. As regards the first task, Augustine cautiously edges away from the Greek *hypostasis,* the substratum of all accidents and predation (all but transliterated as *substantia,* that which "stands under"): "Perhaps then it is more correct to say three persons than three substances [*Quanquam et illi, si vellent, sicut dicunt tres substantias, tres hypostases, possent dicere tres personas tria prosopa*]. But we must inquire further into this, in case it looks like special pleading for our own usage against that of the Greeks" (*ADT,* 7.11). The objective, however tentatively broached, is to disaggregate the Latin *persona* from the Greek *hypostasis,* so as to allow for a construction of person that recognizes both its substantive identity and its relatedness as distinct but inseparable aspects of full personhood. Proceeding in a cautious, at times hesitant manner, Augustine arrives at the crucial insight that, in the case of person, we are presented with an idea that cannot be conceptualized in standard, predicative fashion at all. Whereas a horse is sensibly conceptualized as a species of the genus animal, "the Father and the Son and the Holy Spirit are not three species of one being." Rather, in the case of person (divine and human), the reality of the presence so designated is not established by an act of predicative cognition but by an act of recognition that is intrinsically evaluative:

> But if you say that the name substance or person does not signify species but something singular and individual, so that the substance or person is not predicated like man, which is common to all men [*Quod si dicunt substantiae vel*

38. At 7.8, Augustine cites 2 Corinthians 2:10: "For, what I have pardoned, if I have pardoned any thing, for your sakes have I done it in the person of Christ" (Vul. *nam et ego quod donavi si quid donavi propter vos in persona Christi.* – Grk. *καὶ γὰρ ἐγὼ ὃ κεχάρισμαι, εἴ τι κεχάρισμαι δι' ὑμᾶς ἐν προσώπῳ Χριστοῦ*).

> *personae nomine non speciem significari, sed aliquid singulare atque individuum; ut substantia vel persona non ita dicatur sicut dicitur homo*]. (*ADT,* 7.11)

In passing, Augustine notes that being and substance (*ousia* and *hypostasis*) are of fairly recent origin, and that "those who spoke Latin before they had these terms … used to talk about nature instead," thus recalling the older notion of individual existence as the manifestation of a distinct "nature."[39] Augustine's closing example of three statues of gold at least partially clinches the main point, viz., that the gold does not relate to the eventual statues as genus to species. Though its applications are limited, the simile at least helps clarify that to assert the Trinitarian model of "three men said to be of one nature" is not to deny their distinctiveness as persons, nor to suggest their separate existence. The point was hard-won for Augustine, whose earlier writings tend to treat body and soul (*anima*) in rather disjunctive, even antagonistic fashion that faintly echoes his youthful Manichean indiscretions. Only in his later writings was he to emphasize "the unity of the human being … as 'a rational soul which has a body'" and to affirm that "the soul which has a body does not make two persons, but one human being."[40]

Provided we acknowledge that in the case of the Trinity "it is quite impossible for any other person at all to emerge out of the same being" (*nullo modo alia quaelibet persona ex eadem essentia potest existere* [*ADT,* 7.11]), Augustine's simile helps illustrate a point of great consequence for later theology and, eventually, for Coleridge's reliance on Trinitarian thought as a key component in his lifelong critique of modernity. Inasmuch as person signifies "something singular and individual" (*aliquid singulare atque individuum*), it proves distinct from a common nature and also lies beyond the reach of ordinary *qualia* and *praedicabilia.* That a person belongs to a particular species and, furthermore, that it shares with other persons an underlying essence or *natura* does not as such capture the meaning of person. Philip Rolnick makes the salient point very well: "As *person* is developed through the centuries, it is consistently differentiated from nature, for a person is an agent who can initiate action in accord with the nature possessed."[41] It is a distinction pointedly reinforced in the momentous question opening Book 9 of Augustine's *Confessions:* "Who am I and what am I" (*quis ego et qualis ego?*). A particularly consequential implication of Cap-

39. *quia et veteres qui latine locuti sunt, antequam haberent ista nomina, quae non diu est ut in usum venerunt, id est essentiam vel substantiam, pro his naturam dicebant* (*ADT,* 7.11). *Natura* will again occupy a central role in Boethius, whose definition of person was to prevail through Aquinas; for a concise account of Aquinas's implicit critique of Augustine's use of the genus/species opposition, see Edmund Hill's commentary (*ADT,* 234n36).

40. *In Johannis Evangelium Tractatus* 19.15, quoted in Teske, "Augustine's Theory of Soul," 116.

41. Rolnick, *Person, Grace, and God,* 28.

padocian and Augustinian theology, then, is to have identified the human person as the unique instance where the identity of a being is not simply the default of its material substratum or, for that matter, of our concept of such a substratum. As the later Coleridge will insist time and again, personhood implies "alterity," that is, an unbridgeable gap between the individual, "incommunicable" agent and his or her nature, species, or substance. Because a person has by definition what Coleridge calls "a responsible Will," it is in a unique, albeit bounded sense *free*.

Ever the rhetorician in search of vivid analogies to present intricate and elusive subjects more effectively, Augustine arguably found the greatest challenge in thinking through the connection between the persons of the Trinity and the meaning of human personhood. Having embarked on an unprecedented exploration of the singularity and evolving nature of the human individual in his *Confessions*, Augustine's pedagogical role as bishop of Hippo frequently called upon him to find new ways to convey how divine and human personhood might be related.[42] It is beyond the scope of this reading, however, to detail how the later books of *De Trinitate* establish a delicate analogy between the transcendent relation of the divine persons and the gradual cultivation of the human soul as *imago dei*. For now, what needs to be kept in mind is the concern (altogether central to Christological and Trinitarian discourse of the fourth and fifth centuries) with the relation between *ousia* and *hypostasis*, terms whose rendition as *substantia* and *persona*, respectively, had done much to confuse the issue. Even so, in the course of these debates, which were only partially resolved at the Council of Chaldecon (A.D. 451), the term perceived to be most urgently in need of further clarification was "nature" (*physis, natura*). To be sure, the formula of Christ as possessing "two natures [*physeis*] and one person [*hypostasis*]," which had won the day at Chaldecon, while leaving the meaning of "nature" rather obscure, had consolidated an aspect of personhood whose centrality has been acknowledged ever since: *person is integrative*. It "unifies" and, in so doing, allows us "to glimpse the ontological and axiological priority of person over nature."[43]

The late Coleridge thus categorically states that person, or "the spiritual" dimension of the human being, differs from nature in *kind* and, hence, is not reducible to an epiphenomenon of somatic, neural, or otherwise materially conditioned processes:

---

42. A well-known instance is that of the fugitive monk Leporius of Gaul, excommunicated for his heretical, radically disjunctive (Nestorian) view of Christ's divinity and humanity. Having fled to Africa, Leporius received instruction from Augustine who, as his disciple's eventual recantation in the *Libellus Emendationis* shows, had led him to realize "that the Incarnation conjoins human nature with the divine *Person*, not with the divine nature" (Rolnick, *Person, Grace, and God*, 23); for a detailed account of Leporius's case, see also Grillmeier, *Christ in Christian Tradition*, 464–467.

43. Rolnick, *Person, Grace, and God*, 34.

"by *spiritual* I do not pretend to determine *what* the Will *is* but what it is *not*—namely, that it is not nature."[44] Whereas a morality anchored in self-interest ("Prudence") "is at least an offspring of the Understanding, " one anchored in non-cognitive sentiments or "Sensibility" is doomed to failure because of its "passive nature" and its utter contingency on "a quality of the nerves, and ... individual bodily temperament."[45] The point can also be made in terms familiar from Platonic dialectics: "nature itself, as soon as we apply reason to its contemplation, forces us back to a something higher than nature as that on which it depends" (*OM,* 140). Nature (*physis*), the ontological substratum of the person, can never be set aside when thinking about personhood, quite simply because there cannot be any hypothetical persons but only the real person whom we encounter and acknowledge as such. As Spaemann puts it, "life *as such* cannot equally well be or not be, for life *is* what it is to be."[46] To be a person, then, means to stand in a unique relationship to—indeed, to have a singular, "incommunicable" perspective on—that underlying nature. A key trait of personhood (not to be confused with the contingent psychological notion of "personality") is thus that of *transcendence,* that is, an awareness that nature and person are neither homologous nor related in linear fashion as cause and effect.

The human individual qua person is not causally conditioned *by* nature, even as it attains its reality only on the basis of it. Unlike non-personal entities such as plants and animals, whose identity is fully circumscribed by their nature, a person "is a being that *relates* itself to its existence [*verhalten sich zu ihrem Sosein*], i.e. to its attributes. Precisely the constellation of attributes is experienced as contingent. But neither is a person *the* Being, expressing itself in finite ways of being. Persons are not the Absolute, since they only have being in the first place through having a kind of being [*Wesen*], a finite set of attributes, a nature ... Personality hovers at a point between being and kind, between absolute and finite. This point of indifference we call freedom."[47] A momentary glance ahead shows the later Coleridge to have grasped this crucial point very clearly, such as when noting that "in irrational Agents the Law

44. *Inquiring Spirit,* 132.

45. "Are not Reason, Discrimination, Law, and deliberate Choice the distinguishing Characters of Humanity? Can aught, then, worthy of a human Being, proceed from a Habit of Soul, which would exclude all these ... Can any thing *manly,* I say, proceed from those, who for Law and Light would substitute shapeless feelings, sentiments, [and] impulses?" (*AR,* 58, 63).

46. *Persons,* 71. Commenting on a discourse by Joseph Hughes in 1831, Coleridge, by then deeply committed to a (neo-Platonic-inspired) Trinitarian position, yet insists on the absolute interdependency of *physis* and *pneuma:* "I find nothing in Reason to authorize me, nothing in Scripture that requires me, to believe an actuality, or full *existency,* of the Soul separate from the Body—even as I am utterly incapable of conceiving a *Body* without a Soul. A carcase is not a Body—any more than a Dendrite is a Plant—tho' it bears the imprint of a Plant ... —But *the Man* by necessity of his finite nature at once potential and Actual exists, as *Soul and Body*" (*CM,* 2:1185).

47. Spaemann, *Persons,* 73; trans. modified.

constitutes the Will. In moral and rational agents the Will constitutes, or ought to constitute, the Law."[48] At times, Coleridge's insistence on this categorical divide between the uniqueness of the human being as possessed of reason and the sentient, even clever nature of other animated life forms is underwritten by a deepening investment in neo-Platonic thought. A long note early in *Aids to Reflection* finds Coleridge enthusiastically remarking on the felicitous Greek original of James 1:25: "But whoso looketh into the perfect law of liberty, and continueth therein, he being not a forgetful hearer, but a doer of the work, this man shall be blessed in his deed" (*ὁ δὲ παρακύψας εἰς νόμον τέλειον τὸν τῆς ἐλευθερίας*):

> The Greek word, parakupsas, signifies the incurvation or being of the body in the act of *looking down into;* as, for instance, in the endeavour to see the reflected image of a star in the water at the bottom of a well. A more happy or forcible word could not have been chosen to express the nature and ultimate object of reflection, and to enforce the necessity of it, in order to discover the living fountain and spring-head of the evidence of the Christian faith in the believer himself, and at the same time to point out the seat and region, where alone it is to be found. *Quantum sumus, scimus.* That which we find within ourselves, which is more than ourselves, and yet the ground of whatever is good and permanent therein, is the substance and life of all other knowledge. (30n)

The key word here, *παρακύψας*, denotes precisely the elemental form of internal perspective, transcendence, or (as Coleridge calls it) "alterity" that defines the person's relation to the reality and nature of its own existence. Anterior to all empirically acquired information, *knowledge* is indelibly woven into the reality of our being, our personhood: "Inasmuch as we have being, we come to know" (*Quantum sumus, scimus*).[49] As the Platonizing concluding sentence ("That which we find within ourselves") makes clear, human consciousness is distinguished by its *active participation in,* rather than passive determinacy *by,* that which it knows. A fuller account of Coleridge's phenomenology of "conscience" (below) shall demonstrate in some detail that this participatory and relational quality of human awareness also reveals a

48. *AR,* 300n; for a fuller discussion of this point, see also *AR,* 247.

49. In a late notebook entry, Coleridge further scrutinizes the interdependency of personhood as a mode of being and the human being's distinctive mode of knowledge—which is not merely informational or computational but *relational:* "Never shall we [be] able to *comprehend* ourselves <better than we now do>, but we may learn what is of far more worth and importance, to *know* ourselves better. *Sis* ut *scias*! [*Be* that you may *know*!]. *Scire* et esse <Sciendi et essendi> eadem est norma / the more we become normal in this respect, the clearer & more distinct is the Image of God in [? Us] of him who eternally ~~becomes~~ affirmeth" (*CN,* 5, no. 6720); for a discussion of this passage, see Hedley, *Coleridge,* 160–161.

profound connection between Coleridge's neo-Platonic philosophical theology and the emergent field of phenomenology, especially in Newman's late *Grammar of Assent* (1870) and Franz Brentano's nearly contemporaneous *Psychologie vom empirischen Standpunkt* (1874). In the above passage, then, Coleridge draws attention to how the pursuit of seemingly "mundane" object-knowledge lets us glimpse something about the very nature of what it means to know, and how that process is wholly enmeshed with our unique mode of being. However contingent and ostensibly foreign-determined, human knowledge always "curves in on" and "looks down into" the noumenal realm *in* which it participates and *to* which (however unwittingly) its discrete epistemological pursuits orient it qua person.[50] One is reminded of Levinas's gnomic observation that "to know or to be conscious is to have time to avoid and forestall the instant of inhumanity."[51]

Late in *Aids to Reflection*, Coleridge offers a quasi-phenomenological account of different levels of human awareness and, in so doing, homes in on the intrinsic "alterity" or transcendence of the human individual qua person vis-à-vis the world. Glossing an aphorism by Jeremy Taylor that ponders how the moral consciousness is awakened by a "perplexing disparity of success and desert," Coleridge notes that it is precisely the institution of slavery—viz., the conversion of persons into things grasped with our hands, viz. "Mancipia = things" (*CM*, 1:35)—which triggers an indelible, indeed primal awareness of sin.[52] In so "forcing the Soul in upon herself," this sinful practice reveals a palpable disequilibrium between material reality and moral conception. Echoing Augustine's much-criticized view of sin as the necessary impetus whereby human beings transformed from a primitive *imago dei* into bona fide searchers for perfection, Coleridge finds slavery (and, metaphorically, his own en-

---

50. There is a marked Platonic, at times even mystical undercurrent in modern phenomenology, even as some versions of phenomenology, Husserl's in particular, appear anxious to contain such implications; the conjunction between phenomenology and Thomistic theology is especially strong in the work of Franz Brentano, Étienne Gilson, and Jacques Maritain—though substantial connections to Judaic thought can also be found in Martin Buber and Emmanuel Levinas. On the former link, see Sokolowski, who remarks that phenomenology "in a way complements ... the Thomistic approach," viz., by approaching philosophy from "within the natural attitude and [then] distinguish[ing] the philosophical from it" (*Introduction to Phenomenology*, 208); on the relation between modern, proto-phenomenological conceptions of form *qua* "appearance" and Thomism, see Pfau, "All is Leaf."

51. Levinas, *Totality and Infinity*, 35.

52. See Coleridge's marginalia on the "report of a Committee of the House of Commons on the Extinction of Slavery" (1833), which has him muse on "how unhappy a state of mind must that of a humane and religious Planter's be." In explaining the use of the word "mancipia," George Whalley notes how Coleridge here and elsewhere echoes almost verbatim Jeremy Taylor's *Polemicall Discourses* (1674), which had insisted on the "sacred distinction between Person and Thing, which is the Light and the Life of all Law, human and divine" (quoted in *CM*, 1:35 n7[2]).

slavement to opium) to precipitate an (ultimately fortuitous) sense of inner estrangement, self-scrutiny, and moral self-awareness.[53] Ever so slowly, he suggests, there ensues

> a steadier and more distinct consciousness of a *Something* in man different *in kind,* and which not merely distinguishes but contra-distinguishes, him from animals—at the same time that it has brought into closer view an enigma of yet harder solution—the fact, I mean, of a *Contradiction* in the Human Being, of which no traces are observable elsewhere, in animated or inanimate Nature! A struggle of jarring impulses; a mysterious diversity between the injunctions of the mind and the elections of the will; and (last but not least) the utter incommensurateness and the unsatisfying qualities of the things around us, that yet are the only objects which our senses discover or our appetites require us to pursue.[54]

As so often in Coleridge, the acuity of his reasoning is bound up with his delicately woven, metonymic characterization of mental life as a gradated progression; "jarring impulses" give rise to a cognitive dissonance that appears veritably constitutive of the inner life: "a mysterious diversity between the injunction of the mind and the elections of the will." All perceptual or object knowledge thus involuntarily pries open a window on self-knowledge, which is to say, on the rationality that lies *beyond mere understanding* and, as such, delineates the true and unique *nature* of the human person as an *imago dei.* As Jacques Maritain puts it, "the resemblance to God is less in the practical than in the speculative intellect."[55] At the same time, Coleridge opposes the naturalist hypothesis that construes the human being strictly in terms of a utilitarian and instrumental "understanding"—that is, as a particularly advanced and

---

53. See Augustine's *Literal Commentary on Genesis,* where he speaks of the human being as "the spiritual creature that . . . is consummated in its relationship with God only by having been brought to the edge of eternal damnation" (quoted in Burnell, *Augustinian Person,* 177).

54. *AR,* 349; the editor, John Beer, references Plotinus, *Enneads,* 3.2.8 as a likely source for Coleridge's argument here. On Coleridge's growing neo-Platonist orientation in the later work, see Hedley, *Coleridge,* esp. 33–65; for a different reading that seeks to de-emphasize Coleridge's Platonism and to disaggregate his theological and philosophical speculations, see Perkins, *Coleridge's Philosophy,* esp. 141–204; still a good introduction to Coleridge's complex intellectual bearings and concerns is Muir, *Coleridge as Philosopher.*

55. *Person and the Common Good,* 25; see also Maritain's later contention "that the deepest layer of the human person's dignity consists in its property of resembling God—not in a general way after the manner of all creatures, but in a *proper* way. It is the *image of God*" (ibid., 42). That point, of course, is altogether central to Coleridge's notion of the symbol; see Halmi, *Genealogy,* esp. 99–132; McKusick, "Symbol"; Perkins, on symbol in relation to *logos,* in *Coleridge's Philosophy,* 25–90; and Brice, *Coleridge and Scepticism,* 94–102.

efficient computational machine whose capacity for highly sophisticated, second-order reflections (e.g., gaming hypothetical scenarios; preparing for impending or even distant disasters, etc.) merely moves farther out the goal-posts of interestedness. For it neither implies nor even acknowledges the ontological fact of the person's self-transcendence—that is, its capacity at all times to shift from a naturalist to a phenomenological stance whereby the immediacy of our intentional commerce with the world is rendered subject to formal scrutiny.[56] Invariably, this "'Necessitarian Scheme' of 'Modern (or Pseudo-) Calvinism' erodes an agent's capacity for differentiated and articulate self-awareness: With such a system not the Wit of Man nor all the Theodices [sic!] ever framed by human ingenuity ... can reconcile the Sense of Responsibility, nor the fact of the difference *in kind* between Regret And Remorse" (*AR*, 159). Against this purely calculative and inherently reactive model, Coleridge maintains that there is no such thing as reflection without transcendence, or vice versa. To be a person means to inhabit a "perspective" that, however absorbing and compelling, can never unilaterally determine, let alone exhaust the rational scope of the person. For at all times, we understand—either intuitively or, less frequently perhaps, explicitly—that perspective to be *ours* and, hence, relate to it as a "responsible Will." The late Coleridge's scrupulous rumination of Trinitarian theology and its bearing on our understanding of human will, conscience, and personhood stems from this (neo-Platonic) insight into the "original unity of being and thought." It is this unity which, even as "it is what we are not, [constitutes] the presupposition of what, as subjects, we are."[57]

---

56. On this basic phenomenological operation of *epochē* or "phenomenological reduction," see Sokolowski, *Introduction to Phenomenology,* 42–51; following Husserl, Sokolowski notes that this procedure—Coleridge's *parakupsas*—is also at times characterized as a "transcendental attitude" (42).

57. Spaemann, *Persons,* 95; similarly, and very much in the spirit of Coleridge, Rolnick notes that "although self-awareness is a sine qua non of the divine gift of personality, several problems emerge in the modernist self-understanding. First, to the degree that the self is abstracted as a starting point, it forgets what it has received, and this omission is already decisive" (*Person, Grace, and God,* 214).

18

EXISTENCE AS REALITY AND ACT

Person, Relationality, and Incommunicability

If one searches back for the moment where *rationality* of this more-than-calculative kind first enters the definition of the human person, an early and seminal text turns out to be a tract by Boethius (A.D. 480–524), directed against the symmetrical fallacies of monophysitism and dophysitism. *Against Eutyches and Nestorius* did much to resolve the perplexities over the relation between person and nature that had lingered after the Council of Chaldecon. Boethius's objective in this treatise is less with articulating a coherent understanding of the Trinity—which he arguably fails to do both here and even in his programmatic *De Trinitate*—than with desynonymizing the concepts of *persona* and *natura*. To that end, he begins by putting much analytic pressure on the concept of *natura* and showing that "nature" signifies a great deal more than some inert and impenetrable mass "out there." Not all nature is merely an "object" and, consequently, does not necessarily belong to the domain of appearance and, potentially, deception. There are, for Boethius, four ways in which the concept tends to operate. First, it may apply to entities (*substantiae*) that "can in some measure be apprehended by the mind," which is to say, as regards substances and accidents.[1] Second, *natura* may designate things capable of acting or being acted upon, which includes "corporeals and the soul of corporeals" (*corporea*

1. Boethius, *Tractates,* quotes from 7–81; henceforth quoted parenthetically as *BTr.*

atque corporeorum anima). The latter definition already edges toward a model of nature as inner determinacy, somewhat on the order of Aristotle's entelechy, an implication fully realized in Boethius's third sense of nature as denoting strictly corporeal substances or "bodies." Here the identity or nature of a body pivots on its integrative or organic, inner determinacy. In such instances, nature signifies a teleologically constituted substance, that is, a body: "Now in accordance with this view, the definition is as follows: 'Nature is the principle of movement properly inherent in and not accidentally attached to bodies'" (*BTr,* 81). This third definition brings into view a dynamic, self-organizing, and teleological dimension which, in turn, opens up a vista on the term that had appeared wholly alien to the two earlier senses of *natura:* reason. Even so, a teleologically constituted body, however successfully it realizes its predetermined developmental trajectory, still does not possess any actual *awareness* of its own constitution and operation.

Such constitutive self-awareness, which by way of translating the *logos* or divine Word Boethius renders as *ratio,* constitutes the fourth and final definition of nature; and it is this sense of *natura* that characterizes Christ's divinity and humanity, while also linking the human person with the divine. According to this definition, nature is the "specific difference that gives form to anything." Important, then, is not to confuse this *specifica differentia* with garden-variety "accidents," such as whether a table is black or white, etc. For "nature" here does not signify some identifiable trait such as could be abstracted *from* the "substance" in question. Whereas a torn or dirty shirt is still a shirt, the sense of "nature" here under discussion cannot be separated from the substance without, in fact, dissolving the latter's very reality. This fourth and pivotal sense of nature conditions the incommunicable reality or identity that we encounter in the living, embodied, and rational "substance" of the human *person.* In this very specialized sense, then, *natura* actually denotes the condition under which alone other substances can become possible objects of concern for a specific human being. Thus "we speak of the different nature of gold and silver [*alia significatio naturae per quam dicimus diversam esse naturam auri atque argenti*], wishing thereby to point [out] the special property of things." Well aware "that nature is a substrate of Person, just as Person cannot be predicated apart from nature" (*BTr,* 83), Boethius embarks on a swift sequence of exclusionary moves designed to narrow down the possible application of "person." It cannot pertain to inanimate things, nor to irrational beasts, though Boethius certainly does not deny the latter understanding or, to use a Coleridgean term, "sentiency" (*viventium aliae sunt sensibiles*). That leaves only God, angels, and human beings as those to whom the concept of person may conceivably apply.

The definitional challenge that now remains, and which Boethius is about to tackle, has been aptly formulated by Philip Rolnick: "Since person can neither be reduced to nature nor be predicated apart from it, the intricate interrelation is critical, as is the distinction. Denying the interrelation leads to absurdity: just try to imagine a person without a nature. Denying the distinctness misses the uniqueness of the unified operation of intellect and will, moving itself not by nature, but in freedom."[2] The principal challenge, then, is one of method—viz., not to confuse mental *distinctions* with real separate existences; or, as Coleridge was to put it in a different context: "what are distinct, yea, different, need not *therefore* be separated."[3] At its heart, Boethius's entire argument hinges on showing that nature is not some mute and alien force either opposing or unilaterally conditioning the human individual. Rather, it is a concept, and a highly differentiated one at that. To suppose otherwise is to imply that person and nature are part of the same continuum, a premise bound to lead to the Nestorian inference that Christ had two natures and, therefore, two persons. Against this position, Boethius advances what is at bottom a strikingly modern argument about the categorical difference between being (*ousia*) and human acts of conceptualization. Thus he notes how "some substances are universal, others are particular. Universal terms are those which are predicated of individuals, as man, animal, stone, stock and other things of this kind which are either genera or species" (*BTr,* 85). Correctly, Boethius here includes the universal "man" or "human being" (*homo*) among such *predicabilia* because (unlike "person") it is indeed a species term or universal. By contrast, "particulars are terms which are never predicated of other things, as Cicero, Plato, . . ." Particularity or singularity thus is incommensurable with predicative knowledge; it is, in Boethius's early Scholastic neologism, "incommunicable" (*incommunicabilis*).[4] That term, which was to play a pivotal role in subsequent thinking about person, first surfaces in Boethius's commentary on Aristotle's *De Interpretatione*. Importantly, *incommunicabilis* not only exempts the substance so designated from the domain of ordinary predicative knowledge, since that would only amount to a negative definition. It also implies "that individual humans, at their deepest conceivable locus of identity, do not just possess but *are* this unique reality."[5]

---

2. *Person, Grace, and God,* 37.

3. *CM,* 1:233; Coleridge's argument about the person's ontological embeddedness in the world, and about "alterity" as an intrinsic feature of intentionality is strongly prescient of modern phenomenology. Like Coleridge, who urges the mind to "distinguish in order that you may understand, remember, and communicate, but still silently subsuming the Unity that you have wherewithal to distinguish" (*CN,* no. 4711), modern phenomenology seeks to safeguard against the tendency to allow the underlying identity or form of a complex perceptual object to be reified. To "make a moment into a piece, an abstractum into a concretum" is to "introduce a separation where we should simply make a distinction" (Sokolowski, *Introduction to Phenomenology,* 26).

4. Rolnick, *Person, Grace, and God,* 41; that person is a distinctive mode of being, rather than a characteristic property attributable to an entity otherwise identifiable, is also one of Spaemann's main contentions; see *Persons,* 16–33, and, below, the discussion of "incommunicabilis" in Boethius and Richard of St. Victor.

5. Rolnick, *Person, Grace, and God,* 41.

Above all, it is the idea of the person that embodies this ontological (rather than derivative or experiential) kind of singularity: "Person cannot in any case be applied to universals, but only to particulars and individuals [*in his omnibus nusquam in universalibus persona dici potest, sed in singularibus tantum atque in individuis*] ... Only the single persons of Cicero, Plato, or other single individuals [*singulorum individuorum personae singulae*] are termed persons" (*BTr,* 85).

Notably, Boethius had previously rejected reason as inadequate for a definition of person, quite simply because "it is a communal property." As Maurice Nédoncelle points out, the young Boethius not only rejects general properties (including reason) as adequate descriptors of person, but he further distances himself from Aristotle's commentators by stressing that "the form of any given individual is not of the kind that originates in a substantial form but, instead, is produced accidentally." Nédoncelle goes on to note how neo-Platonist thought subsequently reasoned from the "accidental" and ephemeral nature of the individual to its sinful and disgraced condition ("*va sanctionner et aggraver en la circonstance la disgrâce où Boèce établit chacun d'entre nous*").[6] In due course, however, Boethius reverses himself and now "deduces the individual no longer from accidents but from substance ... a *quid proprium* of a substantial nature."[7] Having advanced thus far, Boethius then sums up all the criteria for an understanding of nature as an embodied, rational, and unique *mode* of being and, hence, *not* as an extrinsic trait or quality of being. In so doing, he produces a definition whose influence and cogency was not to be surpassed until Richard of St. Victor's account of person in the twelfth century:

> Wherefore if Person belongs to substances alone, and these rational, and if every nature is a substance, existing not in universals but in individuals, we have found the definition of Person, viz.: "The individual substance of a rational nature" [*naturae rationabilis individua substantia*]. Now by this definition we Latins have described what the Greeks call *ὑπόστᾰσις*. For the word person seems to be borrowed from a different source, namely from the masks which in comedies and tragedies used to signify the different subjects of representation. (*BTr,* 85–87)

Being far more familiar with Greek philosophy than Augustine, Boethius is happy to draw on the "richer vocabulary" (*BTr,* 87) of that tradition in order to elaborate further the "nature" that underlies discussions of person and Christ's two *hypostases.* Even so, Boethius only acknowledges the migration of *prosōpon* (which he

---

6. Nédoncelle, "Les Variations," 203, 205. Boethius: *Quod enim unicuique individuo forma est ea non ex substantiali quadam forma species, sed ex accidentibus venit* (quoted in ibid., 203).

7. Ibid., 206–207.

takes to denote the theatrical mask more than the human face) into the Latin *persona,* while apparently unaware of the Roman rhetorical and legal tradition that had interpreted *persona* as a political role, or *officium.* More problematically yet, Boethius's definition does not equally well apply to the persons of the Trinity *and* to human beings, a matter that he notably also fails to resolve in *De Trinitate,* a work that quite strangely makes no mention of the word *persona* at all and has little to say about the relation between the divine persons. Likewise, his understanding of the "relationality" of persons remains inadequate. For here he falls back on the supposition, later found in Martin Buber's *I and Thou* (1923), that relations are not ontologically woven into the very essence of personhood but are merely conspicuous empirical moments in which a certain (ethical) mode of being is realized in especially fulsome ways.

Nonetheless, Boethius's much-quoted definition of person and, even more so, his characterization of person as *incommunicabilis* proved consequential for several reasons. First, Boethius's conception is the fruit of his having successfully sorted through the longstanding confusion of person with substance to which the near-identical meanings of the Greek *hypostasis* and the Latin *substantia* had given rise. And it is here that his intervention, which at first blush appears merely technical and terminological, turns out to be of great consequence for ethics. As Boethius notes, "essences indeed can have a general existence in universals, but they have particular substantial existence in particulars alone [*in solis vero individuis et particularibus substant*]." The risk, then, against which theology and ethics must guard concerns the assumption that any particular is *eo ipso* intelligible only by virtue of its membership in a certain class or species of beings or, more problematical yet, that it ought to be understood as a derivative of some such universal. In fact, Boethius insists, things happen in reverse, "for it is from particulars that all our comprehension of universals is taken." Boethius supports his argument by drawing attention to a distinction, often overlooked, between "subsistence" and "substance." The former is a merely heuristic or, as it were, retroactive concept not to be confused with the ontological status of nature qua "substance"—especially when the substance in question is that of Christ's divine person as *logos:* "Wherefore, since subsistences [*subsistentiae*] are present in universals but acquire substance [*capiant substantiam*] in particulars the [Greeks] rightly gave the name ὑπόστἄσις to subsistences which acquired substance through the medium of particulars" (*BTr,* 87). The Latin *subsistentia,* Boethius notes, translates the Greek term ουσίωσις, whereas the Greek ὑπόστἄσις is "represented by our *substantia* ... For a thing has subsistence when it does not require accidents in order to be, but that thing has substance which supplies to other things, accidents to wit, a substrate enabling them to be; for it 'substands' those things so long as it is subjected to accidents. Thus genera and species have only subsistence, for accidents do not attach to genera and species. But particulars have not only subsistence but substance" (*BTr,* 89) and, only on that premise, enable accidents and changes to be.

Furthermore, the singularity of person not only *differs* from such particulars as can always be subsumed under some universal or other; it positively exceeds the scope and jurisdiction of ordinary predicative knowledge whose categorical determinations necessarily rest on the universal/particular antithesis. To sharpen the point: person as an "incommunicable" being derives its unity and uniqueness not from some act of external ascription or self-definition. Rather, being (onto)logically anterior to all such predicative acts, the unity of the person cannot be analytically separated from its reality or being. Thus "the very name Christ, indeed, denotes by its singular number a unity, ... [since] what is not one cannot exist either; because Being and unity are convertible terms [*esse enim atque unum convertitur*]" (*BTr,* 93–95). For Boethius, what explains the official ecclesial position that in the case of Christ "the Nature is double, but the Person one [*fides catholica pronuntiat geminam substantiam sed unam esse personam*]" (*BTr,* 120–121) is the analytic separation of person and nature, *hypostasis* and *natura* (Grk. *ousia*). Only the latter is capable of plurality, for it alone is something of which we may attain predicative knowledge. By contrast, the unifying logic of the person belongs to the different, indeed axiologically higher plateau of beings whose reality pivots not on cognition but recognition (Hegel's *Anerkennung*) —that is, on a relation that is ethical rather than definitional simply *because relatedness is intrinsic to personhood.*

To be sure, as Coleridge—keenly mindful of his own stunningly tortured, lapsed, and often weak-willed life—had frequent occasion to observe, we may certainly fail to honor that ethical relation, and indeed we all too often do. Yet to the thinker who at one point interprets his own initials, "S.T.C. i.e., Sinful, Tormented Culprit," such failure contains its own irrefutable evidence of the ethical covenant that the human individual has then violated, viz., in what Coleridge regards as the indisputable anteriority of "conscience."[8] For "conscience is not a Result or Modification of Self-

8. *CM,* 3:512. Few individuals of note exhibited a greater tendency toward interpersonal failure and an endemic weakness of temperament than Coleridge, as he himself acknowledged time and again: "A sense of weakness—a haunting sense, that I was an herbaceous Plant, as large as a large Tree, with a Trunk of the same Girth, & Branches as large & shadowing—but with pith within the Trunk, not heart of Wood / — that I had power not strength — an involuntary Imposter — that I had no real Genius, no real Depth / — / This on my honor is as fair a statement of my habitual Haunting, as I could give before the Tribunal of Heaven / How it arose in me, I have but lately discovered / — Still it works within me / but only as a Disease, the cause & meaning of which I know / the whole History of this Feeling would form a curious page in a Nosologia Spiritualis" (to R. Southey, 1 August 1803; *CL,* 2:959); a year later, he again remarks how "There is a something, an essential something wanting in me. I feel it, I know it—tho' what it is, I can but guess" (*CL,* 2:1102); see also his searching account of his will, "the dark and hidden Radical of the bodily Life" (*CN,* no. 5228); and Holmes, *Coleridge,* 288–422.

Consciousness; but its Ground and Antecedent Condition" (*CN,* no. 5167); put differently, "consciousness properly human (*i.e. Self*-consciousness) ... presupposes the Conscience, as its antecedent Condition and Ground" (*AR,* 125). As in all things normative, the axiological priority of conscience over propositional knowledge does not admit of syllogistic demonstration but only of phenomenological clarification: "the experience or inward witnessing of the Conscience, ... if <it be> *at all* must be *unique* and therefore cannot be supported by an Analogon ... [It] therefore may be *monstrated* but cannot be *de*monstrated" (*CN,* no. 4605). In his own language, Coleridge is here retrieving a foundational conception that Aquinas had introduced under the heading of *synderesis,* an awareness *not* of the difference between this or that *particular* right or wrong (which is what conscience pronounces) but of the ways in which the distinction itself is latently and indelibly inscribed within the human person. As we have seen, it was precisely this ontological matrix as a distinguishing mark of *human* personhood that had been lost sight of in post-Hobbesian naturalist and reductionist accounts of human agency. It is thus in the context of a particular choice to be made that conscience draws on an underlying ethical relation *of* which it now makes us conscious. In so doing, conscience also discloses its unique phenomenality *to* the reflective individual and so furnishes evidence that what we thus discover is *real* in an ontological sense rather than the incidental manifestation of some contingent temperament or personality. At his most succinct, Coleridge thus observes how "consciousness of conscience is itself conscience" (*OM,* 21). Hence the very *appearance* of conscience, the "notice" that it so inescapably registers in the person's consciousness, refers us back to something beyond the conscious individual; it shows that individual qua person to be always already related to another and, by extension, to have been the participant in a community. Its belated and surreptitious phenomenology shows conscience to be a variant of Platonic remembering (*anamnēsis*)—that is, not a value-neutral memory of some fact or other opportunely retrieved, but an individual's unanticipated and irrefutable inner awareness of its ontological embeddedness in an absolute good without which it could not have a distinct and meaningful concept of itself as an ethical agent or "responsible will." Conscience unveils the insistence within the self of a transcendent and non-propositional good. Put differently, it marks the moment where the self becomes aware of its rational and incommunicable nature and, thus, of its embeddedness or participation in a normative realm whose reality is most palpable where an "I" encounters a "Thou." In a passage quoted earlier, Coleridge had spoken of that "something" as "that which we find within ourselves, which is more than ourselves, and yet the ground of whatever is good and permanent therein, is the substance and life of all other knowledge" (*AR,* 30).

To return to Boethius's account of the person one last time, it ought to be stressed that *Against Eutyches and Nestorius* is not (and was not meant to be) a systematic theological treatise but, rather, something of a political or diplomatic mission of sorts. Drawing on earlier research, Nédoncelle thus notes how the work "aimed to bring about a rapprochement between Orient and Occident on the basis of a definition worked out by the bishops of Illyria, Scythia, and Thracia."[9] In addition to some of its shortcomings already mentioned above, Boethius also does not offer a conclusive argument on his main topic; thus he only shows why the supposition of Christ's two natures need not *necessarily* entail that he was two persons. Yet to have demonstrated that much is not the same as to have proven why Christ *could only be one* person. The official Church position states "that both natures continue in Christ and that they both remain perfect, neither being transformed into the other, ... [and] that Christ consists both in and of the two natures [*dicit et in utrisque naturis Christum et ex utrisque consistere*]; *in* the two because both continue, *of* the two because the One Person of Christ is formed by the union of the two continuing natures" (*BTr,* 117). Here the integrative character of person serves to clinch the case against Eutyches. Viz., the "Person of Christ is formed by the union of the two continuing natures." To suppose that this is impossible reflects but an overly restrictive conception of "nature" as a sheer external determinacy. The exception would be Boethius's fourth sense of nature as *differentia specifica,* but then that meaning of nature was precisely what Eutychean monophysitism lacked.

Yet nature is a predicate and, hence, belongs to the domain of concepts or "subsistences," whereas person is the subject (*hypostasis*) and thus possesses its singularity as a mode of being rather than by way of ascription. In other words, *natura* is a class, species, or kind, whereas *persona* is a grounding, antecedent reality or "special ineluctable preeminence."[10] Against Eutyches, Boethius thus insists that Christ's two natures are "consistent" or "continuous" (*in utrisque consistere ... utrasque manere*). Being also official Catholic doctrine, this position ensures the possibility of vertical movement between the natures, that is, the kind of spiritual progression that Christ is taken to have supremely exemplified for all human beings. Not to be confused with a merely computational intelligence, or *cogito,* then, Boethius's *persona* is crucially distinguished by its having a *perspective* on the very substratum (*natura, ousia*) from which it arises. One may also characterize this uniquely human capacity for evolving a distinct perspective *on* its own nature—rather than having all its perspectives be merely conditioned *by* nature—as a form of *ekstasis,* or "transcendence." Yet to put it that way is not to impute to the human being some definitive emancipation from its "nature" or some sort of mystical transport. Rather, it is simply to acknowledge that personhood—precisely because it is not causally constrained or determined by its

9. Nédoncelle, "Les Variations," 215.

10. Sokolowski, *Introduction to Phenomenology*, 33.

nature—is inherently capable of internal development. To suppose otherwise—viz., that person and all its cognitive achievements are but the default of its natural, material, or (in contemporary terms) neural and genetic underpinnings—is to relapse into the Eutychean assumption of a strict homology between person (*hypostasis, substantia*) and nature (*ousia, natura*). To each nature (human, divine) there would thus have to belong a distinct person (Christ as divine *logos* vs. Christ as a real, flesh-and-blood human being), and each pairing would by definition prove incommensurable with the other. Within the ongoing Christological and Trinitarian debates of Boethius's time, such a view would have implied that both the natures *and* the persons neither *have,* nor ever *can* enter into, a relation; and that conclusion, in turn, would have almost certainly led to the return of some crudely Platonizing or (more likely) Gnostic conception according to which the only thing predicable of the divine and the human realms is their absolute and terminal incommensurability. The only alternative would likely be a pantheism that fails to understand the inner differentiation of the divine being into three persons by first merging them into a single, indifferent abstraction and, eventually, collapsing its conceptual deity with an equally indifferent concept of nature as the all. Indeed, Boethius's *De Trinitate,* which remarkably pronounces that "Relation ... cannot be predicated at all of God" (*BTr,* 17) shows him at times struggling with that latter model.

What has been said thus far is also meant to underscore that in his exploration of the I-Thou relationship—to which we now turn, and which functions as a vivid instance for his ultimate concern with the phenomenology of conscience—Coleridge views "acknowledgment," "justice," and "dignity" as "perfection-terms" rather than as historically contingent, discursive constructs. His late exploration of inter- and intra-personal relations taps into a specific strand of intellectual history that had stressed the deep affinity of *persona* and *dignitas.* Starting in late antiquity, *persona* is often used interchangeably with *homo,* frequently for purposes of self-reference or emphasis, especially in its ablative form. Less common in non-theological discourse is the use of *persona* for signifying the unique essence of a human being, his or her soul or individuality. Especially widespread in secular rhetoric is the use of *persona* to signify the heightened dignity of the individual so characterized, such as Boniface's reference to a "worldly man of high station" (*laicus magnae personae*) or Alanus ab Insulis's definition of *persona* as "whoever holds a superior station" (*persona dicitur aliquis aliqua dignitate praeditus*).[11] The strong correlation of *persona* with social or political rank and office is negatively confirmed by the common moralizing remark that "God will not give consideration to someone's elevated status": *deus personam hominis non accipit.* Most typically, though, secular discourse tends to employ

11. Quotes from "Person," in *Historisches Wörterbuch der Philosophie,* 7:282.

persona so as to highlight the grounds for a given individual's "recognition" (*Anerkennung*). The dignity of the person became a major focus of Alexander of Hales, though here obviously in a sense that no longer draws on a person's rank, station, or office.[12] Indeed, moral and normative conceptions are only ever subject to recognition, not to demonstration because their meaning presupposes—and, in turn, adjusts and develops—the ontological relatedness of persons. Alexander's concise distinction between individual, subject, and person foreshadows the partitioning of epistemological, natural, and moral reflection in the centuries to come: "'Person' refers to morality and is a moral term [*nomen moris*]; 'individual' refers to rational, and subject to natural matters." A third definition of person thus emerges in which the person's dignity functions as the crucial term: "[Person] may also be defined thus: a hypostasis distinguished by something unique pertaining to the person's dignity" (*hypostasis distinct proprietate ad dignitatem pertinente*).[13]

Having spent much time during his middle-years (ca. 1805–1819) with finding a way out of these fallacies, especially the pantheist one, Coleridge—like Boethius—insists on the distinctness of person and nature as its *ground:* "Does not personality necessarily suppose a ground distinct from the Person, id, per quod sonat A, ab A sive materiâ soni? Conscire, = scio et me et alterum simul vel scio me dum scio alterum—Ergo, Sui Conscientia = scio me quasi alterum. The *Me* in the objective case is clearly distinct from the *Ego*."[14] Yet Coleridge also struggles—far more than Boethius—to develop an idea of person that supports the crucial neo-Platonic insight into the finite, created person's ontological relatedness to the good and to the divine. In a letter of 25 May 1820, Coleridge thus insists in rather unique terms on this connection of divine and human natures through the person of Christ: "The redeemer cannot be merely God—unless we adopt Pantheism, i.e. deny the existence of a God; & yet God he must be, for whatever is less than God, may act on, but cannot act in, the Will of another—Christ must become Man—but he cannot become us, ex-

12. Ibid., 7:288.

13. Ibid., 7:289.

14. *CL*, 4:849; trans.: "a ground distinct from the Person: that through which 'A' sounds; [or] to have sounded from A or matter? Moral awareness [*conscire*] = that I know myself to be both other and simultaneously know myself when I know something other—Hence, by means of its awareness = I know myself quasi as an other." Coleridge was also aware of the etymological relation between "person" and the theatrical mask through whose opening the speaker sounds (*personare*); see *CN*, no. 5297 and his marginalia on Henry More (*CM*, 3:917). Noting that person implies plurality, Spaemann draws the inevitable conclusion: "That is why philosophical monotheism is invariably ambiguous: either it advances to become trinitarianism, or it slips back into pantheism" (*Persons*, 40); in the same vein, Hedley notes that "if God is *all* he cannot be a person in any remoter sense" (*Coleridge*, 70), a conclusion evidently drawn by the later Coleridge who rejects both Spinoza's objective and Schelling's subjective pantheism.

cept as far as we become him."[15] The movement of person, which effectively defines our spiritual being, cannot simply be one of *descent,* of God acquiring reality and presence for us in the *logos.* Revelation also demands a corresponding *ascent,* a spiritual action of whose possibility and desirability the "notices" of conscience furnish compelling, albeit provisional phenomenological evidence. A notebook entry of the same year, 1820, tackles the matter in ways that suggest the influence of Scholastic arguments about personhood:

> What is the definition, not verbal but real, of Personality? What constitutes a Person? The union of a Self-subsistence with a Basis independent of it, so that both, ~~are~~ mutually interpenetrated, ~~and~~re but one Being. —The ~~Ground~~ of God's existence independent of the Ideal Principle, or his Self-subsistence—God therefore in whom the Basis & the Self-subsistent necessarily unite to form one absolute Existence, is in the highest sense a Person, and the very Principle of all Personality. If this be Manicheism, i.e. a doctrine of two Principles—or if it be Pantheism too, as making the *ground* or *Basis* of human Existence the same with the *Ground* or Basis of God's Existence—both which in any offensive or irreligious sense I utterly deny—yet still I affirm that without *such* Manicheism & Pantheism neither can the Personality or Free Will of God be maintained without gross contradictions—not ~~from transce~~ incomprehensible from Transcendence, but comprehensibly Absurd. Hence in the 3 different Systems, the Vulgar, that of perfected Idealism, ~~an~~ such as [Johann Gottlieb] Fichte's, and that of perfected Realism, such as [Baruch] Spinoza's, God is no *living* God, but either an infinite Power without personality or Consciousness, —of which all other things are modifications—infinita cogitatio sine centro—Spinoza, or a mere *Law* of the Universe, Lex generalissima—ordo ordinans. (October 1820, *CN,* no. 4728)

On the face of it, this passage is a fine instance of what John Henry Newman was to call Coleridge's tendency to "indulge a liberty of speculation, which no Christian can tolerate." Yet it also affirms him to be the "very original thinker" whose profound contribution to English philosophical and theological culture Newman clearly

15. *CL,* 5:48; not coincidentally, Coleridge's reflection is preceded in the same letter by yet another statement of his misgivings about any philosophy in which mind is merely construed as an epiphenomenon of matter: "A system of Materialism, in which Organization stands first, whether composed by Nature or God, & Life &c as it's [sic!] results; (even as the Sound is the result of a Bell)—such a system would, doubtless, remove a great part of the terrors which the Soul makes out of itself; but then it removes the Soul too, or rather precludes it. And a supposition of co-existence, without any *Wechselwirkung,* it is not in our power to adopt in good earnest; or if we did, it would answer no purpose. For which of the two, Soul or Body, am I to call 'I'?"

recognized.[16] Coleridge's willingness to accept a (heavily qualified) version of Manicheanism and pantheism is sanctioned by his innermost conviction that spiritual meanings must be grasped as intimately entwined with *life,* indeed as a supreme manifestation of it. The same thinker who as early as 1811 had chastised Richard Baxter for failing to distinguish between ideas and conceptions ("A *Conception* of God ... is an Absurdity" [*CM,* 1:237]) and who insists that "in RELIGION there is no abstraction" (*CLS,* 90) here again stipulates that divine and human personhood must be ontologically related; there is no viable theological alternative to "making the *ground* or *Basis* of human Existence the same with the *Ground* or Basis of God's Existence." For to suppose otherwise is to have merely an abstract *conception* of God, be it as an "infinite" but unintelligible "Power without personality" of the kind found in Spinoza, or to settle for a mere personification of the "*Law* of the Universe." Here as throughout his later oeuvre, Coleridge (like Newman after him) insists on the profound interrelatedness of ideas and life, on what he calls "the union of a Self-subsistence with a Basis independent of it"—and it is the idea of the person that most concretely embodies that union. In his *Opus Maximum,* Coleridge time and again "reminds the reader that the original postulate was that of a responsible Will from which the reality of a Will generally became demonstrable to convince him that ~~if~~ <his> ~~a~~ <~~there be~~> responsible Will ... <is> the <essential, indispensable> ground and condition of his Personality ... that we become persons exclusively in consequence of the Will, that a source of personality must therefore be conceived in the Will, and lastly that a Will not personal is no idea at all but an impossible conception" (*OM,* 164).

Undeniably, these are somewhat remote metaphysical speculations, and it remains to be seen how Coleridge substantiates them by offering a strikingly modern phenomenological account of the human person, defined as a relation of I to Thou or as the principle of "alterity" manifested qua "conscience"—both of which Coleridge interprets as empirical manifestations or echoes of Trinitarian thought. Before doing so, however, it is important to return one more time to the evolving intellectual genealogy of person, this time in the work of Richard St. Victor (d. 1173) whose principal objective, like Coleridge's many centuries later, was to understand person in terms of its real existence. Doing so meant first and foremost to jettison the longstanding association of person with "substance," a concept inapplicable to divine personhood because it signifies a *quid,* whereas *persona* is a singular, individual, and incommunicable entity and thus to be thought only as a *quis.* It constitutes "a particularity that only pertains to a single individual" (*proprietas, quae non convenit nisi uni soli*). To be sure, the fundamental challenges of how to think the idea of person—divine and

16. Newman, *Apologia,* 212.

human—remain the same: "How the unity of substance can be joined with a plurality of persons" (*deinde quomodo possit convenire unitas substantie cum personarum pluralitate*).[17] Yet Richard's approach is strikingly different. Given his overriding concern with new models of contemplative experience, Richard quite flamboyantly departs from the then prevailing tone and terminology of Trinitarian theology: "it is not so much knowledge that lifts me up, but rather the ardor of a burning soul."[18] Both in his confident, ostensibly anti-Scholastic tone and in his preoccupation with charity as the highest divine good and the supreme deiform virtue, Richard shows himself to be steeped in the emergent culture of courtly love and the music of the troubadours; having joined, presumably in the early 1150s, the relatively new abbey of Saint-Victor outside the city-walls of Paris, Richard undoubtedly had regular contact with this new and flourishing culture, not least because it also left its imprint on the works of his contemporaries.[19] Yet there are also strong philosophical reasons for Richard's focus, just as Jacques Maritain would much later remark that "perhaps the

---

17. *De Trinitate*, Bk. 3/i. The Latin text is quoted from the standard edition, *De Trinitate*, ed. Jean Ribaillier, 135. Translations of Book 3 follow those found in Richard of St. Victor, *Twelve Patriarchs* (here 373) and are cited parenthetically, following the pagination of the Latin original. Translations from Book 4 of *De Trinitate* will be my own. Richard's argument unfolds in significant measure as a response to Gilbert of Poitiers (1076–1154), who had drawn on the grammatical distinction between subject and predicate to argue for the categorical difference between the concrete existence of something (*id quod est*) and the form whereby something comes to assume a specific existence (*id quo est*); the former functions like the subject, whereas the latter takes the place of the predicate. Taking issue with what he regards as Augustine's interiorist conception of person, Gilbert only accords *persona* the status of an analogy in Trinitarian thought. Given that the question of the Trinity had been declared a mystery of the faith as early as A.D. 382, Gilbert (like Boethius before him) is understandably reluctant to challenge "the ecclesial wall of forbidden predication" (Rolnick, *Person, Grace, and God*, 42). Whereas in the case of human beings, the singularity of the person arises from their nature, the Trinity greatly complicates matters inasmuch as here "that whereby one person is also sustains the other" (*eodem quo est una est alia*). Gilbert's main point was to show that *persona* is not the same as *essentia*—a claim that incurred him the charge of treating the divine persons as mere "accidents" and thus reviving the Sabellian (or "modalist") heresy of the early third century. Still, his claim that person is distinguished from other being by the fact that it "*has* a nature" (*habet naturam*) rather than being inescapably conditioned *by* its nature, was widely accepted by later thinkers; see "Person" in *Historisches Wörterbuch der Philosophie*, 285. Perhaps over-reacting against pantheist conceptions, Gilbert arguably overstates distinctions between concepts to the point that they threaten to unravel the integrity of the beings so conceptualized. For him, there is no divine Father but merely a divinity that manifests itself in the form of fatherhood, no Son except ... etc.

18. Eng. 374; see also his opening to *De Trinitate*, 3.18: "Let it disturb no one, let no one be indignant if we speak in a human manner [*humano more*] to provide a clearer understanding of divine and supermundane things" (Eng. 391/Lat. 153).

19. In his informative essay on Richard, Ewert Cousins notes several instances of how the culture of courtly love came to suffuse theology in the twelfth century; he lists Bernard of Clairvaux's *Liber de Diligendo Deo*, Aelred of Rievaulx's *Speculum Caritatis* and *De Spirituali Amicitia*, and William of Thierry's *De Natura et Dignitate Amoris*.

most apposite approach to the philosophical discovery of personality is the study of the relation between personality and love."[20] In Book 3 of his *De Trinitate,* Richard focuses much of his thinking about personhood on divine love (*caritas*), and the conclusions to be dialectically drawn from it. Firmly premised on Anselm's ontological proof of God ("a being than which nothing greater can be conceived, and which exists"), Richard's reasoning unfolds as follows. The "supreme and altogether perfect good" must include charity, "for nothing is better than charity, nothing is more perfect than charity." Richard offers neither an elaboration nor any proof for this apodictic claim, simply because he takes himself to be "dealing with a matter of primary value perception—the grasp of a *ratio necessaria* that is a reflection of the *ratio* aeterna of the absolute good."[21] At the same time, Richard's appeal that "each reader search his own consciousness [*Conscientiam suam unusquisque interroget*] for confirmation that there is nothing better than charity" (*De Trinitate,* 3.3) reveals his quasi-phenomenological approach to arguing from the psychological quality of finite, human love toward an understanding of the Trinity in whose image such finite experience is modeled.

Now, charity (*caritas*), which in Richard's *De Trinitate* functions in ways roughly analogous to Platonic *agapē,* cannot only be held privately but must be communicated, and God must wish to do so. For "no one is properly said to have charity on the basis of his own private love of himself [*pro private proprio sui ipsius amore dicitur proprie caritatem habere*]." The very concept of charity thus implies an orientation toward another, a sharing—quite simply because otherwise there would be no charity. Notably, Richard here draws on both "love" (*amor*) and "charity" (*caritas*), with the former term functioning as the concrete manifestation of divine benevolence and the latter as the underlying, spiritual disposition thereby manifested: "it is necessary for love to be directed toward another for it to be charity [*ut amor in alterum tendat, ut caritas esse queat*]." Inasmuch as the meaning of inner states is inseparable from their sociality, their active, free, and deliberate sharing, any wish "to retain those riches greedily" (Eng. 387) would be tantamount to God's neither *knowing* nor *being* charity. It is this divine self-transcendence that is also "at the very core of the human person," thus showing Richard's theology to unfold within the tradition of a "metaphysics of exemplarism and participation that runs in Western thought from Plato, through Plotinus, through Augustine," and that will also dominate the later Coleridge's *logosophia.*[22]

---

20. *Person and the Common Good,* 38.
21. Cousins, "Theology of Interpersonal Relations," 66.
22. Ibid., 67.

The highest good thus demands a "plurality of persons [*pluralitas personarum*]" (*De Trinitate,* 136/374). In time, Richard will show that plurality means not merely two persons but a community. For now, though, he aims to show that charity implies reciprocity: "it is absolutely necessary that there be both one who gives love and one who returns love" (*De Trinitate,* 137/376). Yet the highest charity can only be imparted to a being worthy of it, and thus "it is necessary that a divine person not lack a relationship with an equally worthy person, who is, for this reason, divine" (*De Trinitate,* 137/375). Aside from *reciprocity,* Richard's dialectical argument here also distills *equality* and *recognition* as further implications of personhood; for just "as the particular nature of charity requires a plurality of persons, so the integrity of such love presupposes supreme equality of persons in true plurality [*ejusdem caritatis integritas in vera pluralitate requirit summam personarum equalitatem*]" (*De Trinitate,* 142/380). Moreover, the notion of "proportionate love" (*caritas ordinata*) also implies that one love fully, without reserve or prevarication, for to do anything less is, in fact, not to realize *caritas* at all. Much later, Coleridge will repeatedly draw on the topos of parental love for the infant (the father in "Frost at Midnight" and the mother in the *Opus Maximum*) in an attempt at tracing the phenomenology of the person's inner, spiritual life as it ascends to a higher plateau of reality and greater fullness *through* that particular relation. As for Richard's argument here, it bears pointing out that while his reasoning is certainly concise and lucid, his main objective is not to *prove* these aspects of personhood syllogistically but, rather, to demonstrate that in their absence the highest good (*caritas summa*) could not even be thought to exist. Inasmuch as the *idea* of the *summum bonum* entails a complex model of personhood—viz. involves alterity, reciprocity, equality, and community—the argument that Richard is working out belongs clearly to the domain of what Coleridge will call reason, as opposed to "mere understanding."

Next (*De Trinitate,* 3, Chapter 4), Richard establishes another implication, later amplified in St. Thomas's theology of "action," arguing that by his very nature God can neither be unwilling nor incapable of sharing his love. For to suppose otherwise would mean to deny the divine characteristics of perfection, omnipotence, and benevolence. If, then, God cannot be thought except as a "plurality" of (divine) persons, the latter term belongs to the realm of being and, thus, is categorically distinct from discursive conceptions or constructs, such as consciousness, selfhood, and subjectivity. "Person" is an ontological term, a unique way of being rather than a composite of virtual traits or empirically verifiable qualities. Neither is it conceivable as a merely speculative or hypothetical "notion," that is, as an *ens metaphysicum.* For it is not speculatively *inferred* but, on the contrary, at all times the substratum (or *hypostasis*) of human "action"—in the dual sense of a "doing" (*praxis*) and "realization" (what Hegel will later call *Verwirklichung*) consubstantial with personhood. While Richard initially demonstrates as much in the context of the Trinity, the same point just as emphatically applies to human persons, as is apparent from his chiasmic account of divine and human being: "in the divine nature there is unity of substance; in human nature, unity of person. In the former there is plurality of persons; in the latter, plurality of substances [viz., body and soul]." This *ordo inversus* he takes as confirmation of "how human nature and divine nature seem to be related mutually yet as opposites" (*De Trinitate,* 143/382).

One last implication extracted from the highest good of divine charity and personhood concerns the idea of *communion.* Community has to do with the fact that any proposed restriction of charity to giver and receiver alone would almost certainly lead to a mutually narcissistic relation, an "égoïsme à deux."[23] A "sharing of love cannot exist among any less than three persons" (*De Trinitate,* 147/388), for only in a community can the *meaning,* not just the occurrence, of charity be fully realized. The move from sheer "delight" (*dilectio*) between two persons to "shared delight" (*condilectus*—Richard's neologism) in a community is crucial here. For love to be genuine charity, it must be oriented toward a community of persons as its ultimate end, rather than as a pleasure to be hedonistically consumed by two lovers. While not stated quite as explicitly, the reason for this added stipulation appears to be that the giving and receiving of charitable love is not merely to be an *event* but a *meaning* to be discerned and witnessed by others, which requires that it be shared with them in a spirit of generosity. Hinting as much, Richard quotes Matthew 18:16—a passage occasionally invoked by arguments about the Trinity: "But if he will not hear *thee, then* take with thee one or two more, that in the mouth of two or three witnesses every word may be established" (*ut in ore deorum testium vel trium stet omne verbum*). The implicit connection of person qua relation with the dynamics of witnessing warrants some further scrutiny. Unlike a mere narrative *reportage* of facts, witnessing is not centrally defined by the factual history of events, though it always is premised on such a history as its necessary condition. If we place it under phenomenological scrutiny, we find witnessing to revolve prima facie around a unique dynamic of telling and listening, one that in turn gives rise to a community of persons willing to share (and share *in*) how specific factual events were—and still are—being *experienced.* Central to witnessing, as the extraordinary case of Claude Lanzman's *Shoah* illustrates, is the listener's unconditional acknowledgment of the speaker's reality as person. Enabled by the listener's focused silence, the speech of the witness

23. Quoted in Rolnick, *Person, Grace, and God,* 50; W. Norris Clarke's suggestion that the "older metaphysical tradition of the person ... has left the relational dimension underdeveloped" (*Person & Being,* 5) perplexes, all the more so since his subsequent reading of Aquinas's theology of "action" and "communication" convincingly shows fourteenth-century Dominican theology to have adopted a model of relationality that had been developed in St. Augustine's and Richard of St. Victor's analyses of the Trinity.

constitutes what Martin Buber calls the "founding word of I-Thou" (*das Grundwort Ich-Du*). It can only be "uttered with one's entire being" (*kann nur mit dem ganzen Wesen gesprochen werden*), for it "founds the world of relatedness ... and reciprocity" (*stiftet die Welt der Beziehung ...* [*und*] *Gegenseitigkeit*).[24] Human, articulate speech here brings into light, or clears a space of unique openness and vulnerability—something akin to Heidegger's "clearing" (*Lichtung*) wherein speaker and addressee are revealed to be at once incommunicable and yet primordially related.[25]

For Coleridge, this is the quintessential mystery of human speech—not to communicate information but to transcend itself where one comes face to face with the unspeakable. With unerring instinct, he thus focuses on the closing phrase of Romans 8:26: "Likewise the Spirit also helpeth our infirmities: for we know not what we should pray for as we ought: but the Spirit itself maketh intercession for us with groanings which cannot be uttered" (*similiter autem et Spiritus adiuvat infirmitatem nostram nam quid oremus sicut oportet nescimus sed ipse Spiritus postulat pro nobis gemitibus inenarrabilibus*). As Coleridge notes, to imagine that "the Spirit aid our infirmities" is not to hazard either "Fanaticism or Enthusiasm," provided that "attention be carefully and earnestly drawn to the concluding words of the sentence ... [and] due force and *full* import be given to the term *unutterable* or *incommunicable.*" The Greek ἀλάλητος (rendered as *inenarrabilibus* by St. Jerome) does not, however, signify a sheer impasse but dialectically impresses on the listener the speaker's "incommunicable" reality as a person: "it signifies, that the subject of which it is predicated, is something which I *cannot,* which from the nature of the thing it is impossible that I should, communicate to any human mind ... so as to make it *in itself* the object of his direct and immediate consciousness" (*AR,* 78–79). Much more would have to be said here about Coleridge's conception of finite language in relation to the *logos,* and about the art of reading and interpretation that it calls for. Just as "we are responsible Agents; *Persons,* and not mere living *Things*" (*AR,* 78), so words are not merely mechanical conveyors of residue-free meaning but establish a relation between persons whose potential and depth require that in the language here uttered something ultimately "unnarratable" (ἀλάλητος) be "*inferred ...* from its workings" (*AR,* 79).

Above all, the word asks that we infer the (incommunicable) reality of the speaker. Closely entwined with *caritas,* witnessing is above all an act of communion.

---

24. Buber, *Ich und Du*, 9, 12, 14; translations are my own.

25. The term "clearing" (*Lichtung*) is central to Heidegger's "Origin of the Work of Art," where it notably stands in close proximity to his discussion of "truth" as Αλήθεια—an "uncovering" or "un-concealing" (*Unverborgenheit*). "Vom Ursprung des Kunstwerkes" in *Holzwege*, 38–43. Earlier, Buber had emphasized how in the space of the I-Thou "there may be no prevarication" (*wer sich drangibt, darf von sich nichts vorenthalten*), 17.

Not unlike the rare musical performance that, as Schopenhauer remarks, unexpectedly brings us face to face with the Platonic realm of the noumenon or *nunc stans,* the language of witnessing is not defined by the sheer temporality and historicity of past events. Rather, the listener's rapt and concentrated silence powerfully acknowledges the speaker's absolute reality and presence as person, a point stressed by Buber (as later also by Emmanuel Levinas), who notes how all "actual life is encounter" (*alles wirkliche Leben ist Begegnung*) and how the "present" (*Gegenwart*) only obtains where there is "a being present-to [*Gegenwärtigkeit*], an encounter, [and] a relation." In so telescoping presence, relationality, and reciprocity into an act of elemental articulacy, the witness unveils community as something ontologically given, rather than as a social contract, a tale of state-administered social progress, or some ideological end-of-time utopia as which it is variously being imagined by political theory. Indeed, for one of these latter models to hold any conceivable appeal, we must be able to remember, however fleetingly, the absolute reality of Buber's *Grundwort* of I-Thou and, in so doing, grasp the difference between an ontology of "presence" and the strictly ektypal realm of a "present" that "merely denotes in thought the closure of 'time elapsed'" [*nur den jeweilig im Gedanken gesetzten Schluß der "abgelaufenen Zeit"*].[26] In so transcending the realm of phenomena—*not* by forgetting them but, rather, by capturing their persistence or, rather, *insistence* in the present—witnessing transcends the discrete event in the direction of a community of persons centered around the relational and reciprocal power of the word. Needless to say, what confronts us in Richard of St. Victor's *De Trinitate,* Buber's *I and Thou,* or Coleridge's *Rime of the Ancient Mariner*—arguably one of the most compelling literary instances of witness speech—is a "logocentrism" without reserve. For now, acknowledging it to be so will have to suffice, though we will have opportunity to engage post-structuralism's critique of it later. To sharpen the point or, perhaps, to throw down in more fulsome ways yet the gauntlet for a good bit of twentieth-century theory, the argument just sketched could be summed up as follows. The language of the witness—unreservedly oriented toward *and* met by its addressee—establishes a *communio* whose very possibility hinges on the joint presence of the three theological virtues identified by Aquinas: faith, hope, and charity. It potentially elevates the persons of speaker and listener alike by creating a framework wherein future events and experiences will henceforth be evaluated. For that reason, the community of persons whose ontological reality it reveals for the (necessarily limited) duration of witness and witnessed speech, is inherently normative in its thrust. It dramatizes life in its highest and most aspiring, if also most vulnerable, form—viz., as always

26. *Ich und Du*, 18–19; trans. mine.

unfolding in a "clearing" (Heidegger) or what Friedrich Hölderlin calls "openness" (*das Offene*)—an opening for our participation in the good.

In community, then, the sheer occurrence of charity is "realized" or "actualized" as *the* supreme good. Rather mischievously, and presaging René Girard's eventual arguments about mimetic and triangular desire, John Keats's opening of Part II of *Lamia* (and a great deal of other nineteenth-century narrative besides) seems to derail this conception by suggesting that such sharing is, in fact, furtively or even unconsciously prompted by a wish to stimulate some voyeuristic participation within the third party. However intriguing, this consummately modern construct proves inapplicable. For it peremptorily elides the possibility of disinterested love, giving, and action realized in a condition of *knowledge,* rather than as a displaced manifestation of blind and compulsive *desire.* Yet such a hermeneutics of suspicion, which amounts less to a demonstrable position than a self-certifying and totalizing stance of "bad faith," only holds appeal inasmuch as its negative pathos is already cued by some version of the Platonic *kalon* and *agathon.* Behind this unmasking of a specious good there stands a good of superior authenticity for the sake of which a critique of mimetic desire had to be undertaken. For his part, Girard insists that "mimetic desire" itself cannot be proper and authentic desire, if for no other reason than that it *fails* to recognize itself as the mere mimetic derivative of the anterior template of triangularity. One might thus conclude that in Girard's account desire's de facto inauthenticity arises precisely because of the subject's *failure* to know what, for Richard of St. Victor, the divine persons and any other being capable of charity *do* know.

Again, Coleridge articulates the point with uncommon precision and force in an aside directed against the "doctrines of necessity and materialism," though his point would seem to apply no less to present-day reductionist accounts of human agency by cognitive science. Their main problem is twofold: first, they bring about an instantaneous loss of all moral categories (responsibility, redemption, remorse, etc.) and, in so doing, effectively denude the human realm of all capacity for *articulate meaning,* Will, action, responsibility, etc. simply become "words without meaning," quite simply because there is "no object."[27] Second, there is also "no conceivable agent,"

---

27. *OM,* 177; see also the conclusion to *Aids to Reflection,* where Coleridge remarks on the true materialist's failure to explain just what he means by "mind." Were he to try, "he would find, that as he had described it by negatives, as the opposite of Bodies, *ex. gr.* as a somewhat opposed to solidity, visibility &c. as if you could abstract the capacity of a vessel and conceive of it as a somewhat by itself, and then give to the emptiness the properties of containing, holding, being entered, and so forth" (*AR,* 394–395). Here as in his later critique to materialist and mechanist reductivism, Coleridge is profoundly influenced in conception and terminology by the Cambridge Platonists, in particular, Ralph Cudworth; Hedley notes that Coleridge had found in Cudworth a clear anticipation of his view that "we cannot distinguish between the substance of mind and its attributes such as memory, insight, and will. Mind just *is* constituted by the interrelation and interdependence of its

and here the "dogmatism of the Corpuscular School" (*AR,* 395) entangles itself in a performative contradiction; for its advocates manifestly claim intellectual agency for a position that declares such agency to be null and void, indeed an impossibility. Yet that is not all. Never one to leave a train of thought unfinished, Coleridge is not content with having shown materialism's *conceptual* incoherence but goes on to muse about its motives. Like George Berkeley before him, Coleridge thus notes that the reductionist account of human agency advanced by "the doctrines of necessity and materialism" is not so much defined by its "denial of Will and Spirit, but [by] *the general passion* for it" (*OM,* 178).

More recently, Robert Spaemann has questioned "the claim of reductionism ... that subjective experience is ontologically irrelevant; that is, there is no *sum* to go with the *cogito*. Epiphenomenalism sees subjective experience as standing in a strict one-to-one relation with objectively observable neuronal processes. But the relation is asymmetric: the neuronal processes affect the experience, but the experience does not affect the neuronal processes." Like Coleridge, Spaemann is concerned not only with the argument's obvious *petitio principi* but also with its underlying motives and aims ("There are questions to be pursued about the interest driving the reductionist endeavour"). The ultimate objective driving deterministic reductionist accounts of human agency must be sought in a desire to establish "a tight *nexus* of cause and effect" capable of accounting for "everything there is." Yet this fixation is itself "not a result of empirical demonstration" but only exists as a utopian "postulate" of sorts. What drives the reductionist endeavor is "an interest in the continuing expansion of our mastery of nature and the possibilities of control."[28] Unsurprisingly, the hermeneutic of suspicion that has held sway within the humanities since the late 1960s is now in the process of yielding to "cognitivist" approaches that are being pursued with equally messianic fervor. Even so, the basic assumptions about the structure of the mental life and personal identity of the human being have hardly changed since the time when Coleridge had subjected materialism and mechanism to such powerful critique. Now as then, a first premise is that all processes, mental or otherwise, are to be accounted for strictly in terms of efficient (never *final*) causation; a second premise holds that every "state" identified in such a causal chain must logically be the re-

---

constituents" and, thus understood, furnishes the *imago dei* for a tri-une God whose persons "are not to be thought of as akin to discrete substances with accidental qualities, but a unity constituted by *relations*" (*Coleridge*, 63–64).

28. Spaemann, *Persons*, 52–53; similarly, C. Beha suspects "that scientism's complete inability to account for the central feature of human experience—consciousness—might be a failure more on the part of scientism than on the part of consciousness. If embracing scientism demands giving up the sole means we have for engaging with the world, it's tough to see why one would believe in it, even if it's true—particularly since the view itself eliminates any normative basis for treating truth as inherently better than falsehood" (Beha, "Literary Response to Radical Atheism," 77).

sult of some anterior and isomorphous material cause; a third axiom holds that all causation is not only efficient but sufficient—that a cause is truly deserving of its name only if and when its exhaustive determinative power can be demonstrated. Behind this desire for total closure stands a kind of *Angst* that, as we saw earlier on, constitutes a uniquely modern phenomenon and that, beginning with Coleridge, Schopenhauer, Kierkegaard, and Nietzsche was at last being diagnosed as a unique corollary of modernity's total commitment to *method* as the means by which to take control of the world. In rather Coleridgean terms, Iris Murdoch calls it "a kind of fright which the conscious will feels when it apprehends the strength and direction of the personality which is not under its immediate control."[29]

29. Murdoch, *Sovereignty*, 37.

19

"CONSCIOUSNESS HAS THE APPEARANCE OF ANOTHER"

On Relationality as Love

There are at least three discrete models that Samuel Taylor Coleridge develops by way of articulating the intrinsic relatedness of the human person. A first has to do with the love between human beings and the question of whether the insistence of (potentially unilateral) desire negates or is compatible with personhood. A notebook entry of October 1820 frames the question as follows: "Is *true genuine* Love of necessity RECIPROCAL? that is, can I really be *in Love* with a woman, who~~m~~ I *know,* does not love me?" By "love," Coleridge means "exclusive sexual Attachment ... of a refined & honorably honest Man, which is *exclusively* felt to some *one* Woman; & vice versa." As one would expect, however, Coleridge soon leaves the realm of the conventional and probes into that "where-in ... this Love consist[s]. What is its universal cause, its *indispensable Condition*?" In the first of two answers, Coleridge follows Plato's *Phaedrus* by focusing on the apparent complementarity of the two lovers, "the yearning after that full and perfect Sympathy with the *whole* of our Being which can be found only in a Person of the answering Sex to our own" (*CN,* no. 4730). Arguably, for Plato's "natural union of a team of winged horses and their charioteer" (*Phaedrus,* 246a6–7) heterosexuality does not exactly constitute the strict requirement that Coleridge makes it out to be; even so, the key Platonic conception of love as a "yearning" for the completion of the soul is unaffected by such technicalities and remains a central feature of Coleridge's argument here. Two implications, again Platonic in nature, follow from this scenario: "1st. that no human individual is self-sufficing (*αὐτάρκης*): 2^{ndly}, that the consciousness and impulsive Feeling of this Self-insufficiency ~~increases~~ is more awakened, is stronger & more active in proportion ~~as the~~ to the *natural* Sensibility & *fineness* … of the Individual" ($f7^{v}$). The Platonic notion of an ascent through *ēros* thus not only compels us to infer the fact of our "Self-insufficiency" but, for its ultimate success, also requires our conscious and explicit acknowledgment of that fact: "In a pure & ~~harmonious Being~~ noble mind, the sense of ~~his~~ its Self-insufficingness, the sense that it is of itself *homo dimidiatus,* but *half* of a compleat Being, exists *consciously,* ~~becomes~~ & with reflection."[1] By contrast, in a self actuated merely by an "instinctive Sense of Self-insufficingness" any rational and ethical ascent will be forestalled by a "turbulent Inquietude of mere *Appetite*" that lacks all regard for its object.

Bearing thus far a notable affinity to Hegel's characterization of desire as the first manifestation of self-consciousness (viz., in the modality of a "drive" and thus as yet devoid of its own concept), Coleridge's argument now takes a rather different turn.[2] A desire that has not grasped its own underlying condition signals a twofold lapse or failure of the person, as both a rational and an ethical agent. For if "*to love* ~~means~~ signifies no more than an appetite represented to the eye or Imagination under the form, ~~of~~ which accidentally excites it" (f9), then the conflation of the true notion or *Begriff* (i.e., desire) with its transient and accidental object reveals a failure to "reflect" as a rational agent. As Martin Buber would put it, "love does not attach to the 'I' in such a way that it has the 'Thou' merely for its content or object; rather, it is *between* 'I' and 'Thou.'"[3] Simultaneously, in so confining itself merely to an appetitive taking-hold-of its incidental object of desire, the self also fails to realize that desire cannot be actualized through an act of *possession* but only in the modality of a *relation*—one that would have to involve, and acknowledge as such, the other as a *person* of reciprocal and equal reality. This originally Platonic insight which, as

1. f10; Coleridge's decision to strike out the masculine pronoun, substituting the neuter, reflects his preoccupation in the preceding notebook entry with finding a way to think about "Love as it exists in common both in the Man & in the Woman." To do so, the appropriate term is "Person, instead of 'a man' or 'a woman'—& yet our Language will not permit <us> to say, *It*—/nor the Greek, or Latin, ό or quod as the pron. Rel. to Ἀνθρωπος or Homo—and the same inconvenience is felt when I mean both sexes" (*CN,* no. 4729); for Coleridge's egalitarian views on women in society, see *Inquiring Spirit,* 303–311.

2. "Desire is the form in which self-consciousness appears at the first stage of its development … and [desire] here has as yet no further determination than that of a drive" (. . . *ist die Begierde diejenige Form, in welcher das Selbstbewußtsein auf der ersten Stufe seiner Entwicklung erscheint … [und] hat hier … noch keine weitere Bestimmung als die des Triebes*). Hegel, *Enzyklopädie,* 3:215 (§426; trans. mine).

3. *Ich und Du,* 22; trans. mine.

remains to be seen, makes a striking (albeit inflected) reappearance in Emmanuel Levinas's account of "metaphysical desire," prompts Coleridge to continue as follows:

> ~~rational~~ Living Beings have no other means of Union but Sympathy & Intercommunication ... in the same moment, same kind, same Degree, <and> one and the same Act, to receive & to give, to give what I receive, to receive what I give. ~~Defin~~ Understand Union in this sense—& then I say—that Love is the Desire of ~~my whole~~ all my Being to be united to some other Individual ~~as the~~ (conceived as alone capable of perfecting my being) in its our present finite state, by all the means which Nature ~~dictates~~ Reason, & Duty, permit or dictate. (*CN,* no. 4730, f12^{v})

There will be occasion later on to explore Coleridge's profound conviction that the human person differs from all other animated life by its having reason over and above the computational and calculative faculty of the understanding, as well as his neo-Platonic interpretation of that very fact as evidence of the human being's metaphysical destiny; as he puts it in his *Opus Maximum,* "Reason is the presence of God to the <Human> Will independent of its unity with the divine will" (172). One of his favored ways of drawing the distinction between reason and the understanding is to apply the law of non-contradiction: one individual's understanding of a given issue may certainly differ from that held by any number of other individuals, or indeed from its own, earlier conceptions. Yet reason only signifies to the extent that the beings capable of it are taken to be related in joint acknowledgment of reason as a transcendent, unified, and normative idea. Whereas "a Will that does not contain the power of opposing itself to another Will is no Will at all, ... a Reason that did contain in itself a power of opposing a Reason, or of not being one with it, would be no Reason" (*OM,* 172). Coleridge here touches on the relation of reason to truth, and he specifically rejects the modern perspectivalist and pluralist argument that rationality is itself contingent on, and determined by, inherently non-rational (material) factors such as race, gender, ethnicity, material circumstances, and so forth.

While all of these factors are obviously of enormous influence as regards the kinds of questions particular individuals ask, and the projects they conceive and pursue, attempts to dissolve reason into historically and materially conditioned perspectives fail inasmuch as they cannot account for the human person's fundamental drive to make herself or himself understood, and so achieve reality and recognition as an ethical being, by another person or community of persons. One individual's reason cannot be opposed to someone else's because it "dwells in *us* only as far as we dwell in *it.* It cannot in strict language be called a faculty [because it] ... is incompatible with individuality, or *peculiar* possession" (*OM,* 167). The point builds on *Aids to Reflection,* where Coleridge had specifically noted that in all other contexts of human intelligence "we add the epithet *human* without tautology: and speak of the *human* Understanding ... But there is, in this sense, no *human*, Reason. There neither is nor can be but one Reason, one and the same" (*AR*, 218). For Coleridge, then, both love and reason are inextricably entwined noumena, not phenomena; and it is only as *persons*—not as autonomous "subjects"—that human beings may indirectly participate in that noumenal realm, viz., in the inherently relational modality of action. Neither love nor reason amounts to some subjectively held conception; rather, each manifests the vertical relation that the individual qua person has to the divine. It is no surprise, then, to find Coleridge putting stress on love's non-proprietary, reciprocal, and intuitive acknowledgment of its metaphysical source. With manifestly Platonic overtones, Coleridge notes how the truth of love (as well as of reason) cannot be syllogistically proven but only phenomenologically experienced as a "co-alescence of all our Powers & Receptivities," resulting from "a secret Intuition of a Sympathy" with another being, a union of "the Person with the Person" (*CN*, no. 4730, f14^{v}). Conversely, not even the most sophisticated understanding and discursive intelligence can possibly yield an adequate grasp of person; for "in human law ... <inquiry is not made concerning> the quantity of knowledge and the degree of intelligence, but whether the individual is a *person* or not; and this is determined by the presence of Reason in reference to the Will, and of the Will in its bearings on the Reason" (*OM*, 174). As Coleridge clearly understands, rationality and relationality are corollaries, and to the extent that the individual achieves full personhood it will reflexively come to realize as much.

Before moving on to Coleridge's second instance of person as the *relational* and *rational mode of existence of an incommunicable nature*—viz., in the relation of "I" to "Thou"—we need to return once more to Richard of St. Victor; for it is in Richard's *De Trinitate* that the italicized terms are found to have been jointly and coherently articulated for the first time. Having "move[d] the metaphysical center of the discussion from individual substance to interpersonal love as constitutive of divine being,"[4] Richard in Book 4 of *De Trinitate* scrutinizes person by inquiring into its "mode of being" (*modus essendi*) and its "mode of origination" (*modus obtinendi*). In *De Trinitate*, 4.6, he thus distinguishes between animal (*substantia animata sensibilis*), human being (*homo-animal rationale mortale*) and person. While the latter certainly includes all the criteria already introduced for animal and human being (animation, sensibility, mortality, rationality), it is not adequately captured by them. For by person we "understand a unique and singular substance" (*unam solam substantiam et singularem aliquam*). Like Boethius, Richard distinguishes between the implication of "general" and "special" properties, and "the term 'person' [implies] the property of

4. Rolnick, *Person, Grace, and God*, 47.

individuality, singularity, and incommunicability" (*ad nomen autem persone proprietas individualis, singularis, incommunicabilis* [*De Trinitate,* 4.7]). Yet for Richard, these latter criteria are not traits or "accidents" but, instead, signify a unique and "incommunicable" mode of being. Because substance always names a "what" (*aliquid*) rather than a "who" (*aliquis*), it is inherently inapplicable to an understanding of person. As regards the meaning of person within the context of the Trinity, the temptation to conceive it in terms drawn from our quotidian experience of human persons—who manifestly differ in body and temperament (*in natura humana, quot personae, tot substantie* [*De Trinitate,* 4.8])—would seem to compel assigning each of the three divine persons a separate substance. Though well aware that Boethius had already sought to disaggregate *natura* and *substantia,* Richard nonetheless insists that the latter term should simply be abandoned, if for no other reason than that a definition of person employing it could never be applicable to both the human *and* divine realm: "for it is more correct to call that which is the principle of all substance an essence rather than [again] a substance" (*rectius essentia quam substantia dici potest* [*De Trinitate,* 4.23]).

In lieu of "substance," Richard introduces the concept of "existence" (*De Trinitate,* 4.12–13), which features two aspects, viz., the "mode of being" (*modus existendi*) and the "mode of origination" (*modus obtinendi*). In carefully parsing the "two aspects" (*gemina consideratione*) encapsulated in the term "existence"—viz., a person's per*sistence* or continuity as a singular being (its *modus existendi*) and its origination (*ex/sistere*) as that distinct being—Richard understands person as, literally, a "standing-out-from."[5] In the case of the divine persons, of course, the mode of origination is not only unique but altogether unfathomable and "incommunicable" (*habens divinum esse ex proprietate incommunicabili*). The latter term, first introduced by Boethius, now moves to center stage and becomes effectively coterminous with the meaning of person. In what follows (*De Trinitate,* 4.22–25), Richard skillfully connects all the pertinent criteria in play: person is incommunicable, singular, rational, and its reality is inseparable from its existence. Thus there can be no virtual or hypothetical person, but only the *individua existentia,* for its integrity derives not from some external trait or set of features predicatively ascribed to it but, instead, inheres in its very "mode of being," even as this very fact itself is shared by all persons (*existere per se solum commune est omnibus individuis* [*De Trinitate,* 4.24]). Richard's definition of person "as a rational existence of incommunicable nature" (*rationalis nature incommunicabilis existentia* [*De Trinitate,* 4.23]) thus fulfills one of his main

5. "Thus, under the term 'existence' we may subsume two distinct considerations, namely that which pertains to the mode of existence [*ad rationem essentie*], and another which concerns to the mode of origination [*ad rationem obtinentie*]" (*De Trinitate,* 4.13); see also Rolnick, *Person, Grace, and God,* 53–55.

objectives, viz., to arrive at a definition of person that is equally applicable to God *and* to human beings.[6] Behind that objective stands his assumption—anticipated by Plato and Plotinus, and later echoed by the Cambridge Platonists and, eventually, Coleridge—that the phenomenology of human psychological experience is intrinsically related, indeed metaphysically indexed to the divine realm of the Trinity. Thomas McFarland thus situates Coleridge's phenomenology of the human person as a staging area for his ultimate concern with understanding the Trinitarian God: "the closer Coleridge approached to a rational disposition of the idea of the Trinity ... the more urgent became his investigations into the structure of human personality, which was to be extrapolated into the Trinitarian truth" (*OM*, cxxxvi); and a second instance of that vertical orientation, to be found in Coleridge's *Opus Maximum*, now stands to be considered.

For the most part, of course, the link between Richard of St. Victor and Coleridge has to do with their joint exploration of person as rational, constitutively relational, and actualized interpersonally; such knowledge as Coleridge had of Richard's work was almost certainly arrived at indirectly, through Wilhelm Gottlieb Tennemann's *History of Philosophy*.[7] Still, Richard's central intuition—"that the person is most human—and most divine—when he transcends himself in love for another person"—crucially informs Coleridge's late writings, particularly in the notebooks and the *Opus Maximum*.[8] Specifically, in his discussion of the mother-child relation Coleridge construes the lateral and reciprocal dynamic of parental love as evidence of a vertical, metaphysical connection between finite human beings and a Platonic notion of the good. The same argument also shapes Coleridge's more abstract account of the relation between I and Thou, which in turn can be seen as the template for his rich phenomenological account of "conscience," to be considered later. What Ewert Cousins portrays as Richard's "shift from a spirituality of 'isolationism' or 'rugged individualism' to a community-based spirituality" had also preoccupied Coleridge since his early days as a poet.[9] Beginning with his so-called "Conversation Poems," Coleridge approaches "spirituality" as a wide spectrum of affective and intellectual perplexities that implicitly call for a sustained hermeneutic of the "self"—understood not as

---

6. The term *incommunicabilis* figures prominently in St. Thomas's account of personhood in the *Summa Theologia*, which follows Boethius in understanding "person" not as a predicate (*nomen intentionis*) but as signifying a "reality" (*nomen rei*); see *ST*, Ia, 29, 3; yet Aquinas follows Richard in adopting the concept of "existence" for his own magisterial definition.

7. Coleridge refers to Richard in his *Marginalia* on Lessing, noting that "I far prefer Ricardus di S[t] Victore and the mystical Theologians" to the rationalism of John of Salisbury's *Metalogicon* (*CM*, 3:682); Coleridge responds more critically to a synopsis of Richard's arguments in Tennemann's *Geschichte der Philosophie*, 10 vols. (1798–1817); see *CM*, 5:783–784, 795.

8. Cousins, "Theology of Interpersonal Relations," 56.

9. Ibid., 58.

modernity's autonomous and self-possessed *cogito* but "in the absolute meaning of 'Self' as the perpetual antecedent within us" (*OM,* 31), also known as the will. As we have already seen, in positing the will as the "inexplicable and incomprehensible ... ground of all reasonings and conclusions" (*OM,* 32), Coleridge seeks to avoid voluntarism's conception of the will as a wholly irrational, quasi-mechanistic mental compulsion. As Coleridge notes, virtually all of Christian theology holds that "the absolute Will and the Supreme Reason are One, and it is the identity of these which we ~~att~~ mean, adoring rather than expressing, by the term of God." Precisely that identity is lacking in human agency or in "*a* person" due to both "imperfection and privation." Consequently, the finite, embodied will stands vis-à-vis the good in a relation of continued and palpable estrangement, "palpable" because the loss is intuitively felt as a dissonance of sorts. As a "finite Will," that is, the human person cannot be understood "otherwise than in some relation to a co-present reason, but yet capable of being conceived in a relation of difference and contrariety to it." Conversely, "Reason is the presence of God to the <Human> Will independent of its unity with the divine Will." Regardless of how highly evolved our understanding or computational intelligence, the question of "whether the individual is a *person* or not ... is determined by the presence of the Reason in Reference to the Will, and of the Will in its bearings on the Reason" (*OM,* 169, 172, 174).

To have a will thus means to be conscious, at least nascently so, of that very "difference and contrariety"—that is, of its imperfection, its responsibility, and its being ordered toward a *telos* that it can never reach under its own powers alone. Coleridge uses the Augustinian image of spoiled wine to show that one cannot speak of "corruption without implying the absence of something that should have been" (*OM,* 170). It is just this tension between the actual development that *did* occur and the ideal one occluded by that very fact that also informs Coleridge's Dante-inspired quatrain entitled "Where is Reason?" of 1820–1821:

> Whene'er the Self, that stands twixt God and Thee,
> Defecates to a pure Transparency
> That intercepts no light and adds no stain—
> There Reason is; and then begins *her* reign![10]

Here the "self" is indeed the obstruction, the occlusion or Dantean "false imagination" (*falso imaginar*) that prevents it from seeing. Thus, in what is surely the po-

---

10. Coleridge, *Poetical Works,* 2:994–995; first printed in *On the Constitution of Church & State* (*CCS,* 184–185), the poem is prefaced by an epigraph taken from the opening canto of Dante's *Paradiso* (I, 88–90), lines that Coleridge also reproduces in *CN,* no. 4786.

em's most startling line, "Defecates to a pure Transparency," the self must expel itself as the presumptive focus of attention—of mere "understanding" or "mind of the flesh" (*phronēma sarkos*)—if reason is to be accessed. It cannot abide within the sheer finitude of volition and desire but, instead, must grasp its own will as the true focus of awareness. Coleridge's frequent recourse to Romans 7:24 ("O wretched man that I am! who shall deliver me from the body of this death?" [*Infelix ego homo quis me liberabit de corpore mortis huius*]) in his late notebooks shows him wrestling with the dualist implications of body and soul to the very end. Yet while that antagonism is never resolved, his dark depiction of the body as "a moveable Dungeon with Windows, and Sound-holes" is not taken as evidence (as it had been by Schopenhauer some years earlier) of the soul's determinacy *by* the body. In fact, Coleridge regards the body's material scaffolding as a providential foil whereby the modern, Cartesian fantasy of a pure (unembodied), seemingly deiform human intellect is corrected and the body recognized as "an Interpreter of our communion with lower Natures" (*CN*, no. 5671).[11]

Embodied existence, though inescapable and often a torment (certainly to Coleridge), is less a plain fact than a Wittgensteinian "case" (*Fall*) *for* the human mind. Its phenomenology, not its material ontology, is what occupies, indeed positively enables the development of mind from the merely computational understanding to the responsible will that is the fulcrum of moral personhood. As long as sin "is not asserted to or consented with by the Conscious Will," embodied existence ("the phantom *Self* of his *Nature*—i.e. the Ground, the Hades") proves of vital importance in effecting authentic introspective tendencies: "yet it is most salutary and needful that he should contemplate and regard as Sin" (*CN*, no. 6304). Far from being the simple, efficient cause of Hobbes and Schopenhauer, the embodied, human will amounts to a complex hermeneutic phenomenon; and it is generative of self-awareness because it is a trace of the identity (of will and reason) from which, as something finite and created, the self has become estranged. The self's phenomenology thus shows it to be constantly nudged toward awareness of both its lateral, empirical relation to other persons and, providentially, toward grasping its vertical relation with the divine. Such a movement, however, can only unfold as a conflict registering in the conscience as the presence of "sin" and, belatedly, as "guilt" and "remorse." Coleridge certainly would have concurred wholeheartedly with Kierkegaard's remark that "an ethic which ignores sin is an altogether useless science."[12] Inevitably, then, to accord the human person a will is to invest it with a "responsible

11. For a detailed reading of Coleridge's invocation of Romans 7:24, and on this passage in the notebooks, see Webster, *Body and Soul*, 71–96.

12. Quoted in Murdoch, *Sovereignty*, 46.

Will": "*man is a responsible agent,* and in consequence *hath a Will.* Have I a responsible Will? Concerning this each individual must <himself> be exclusively <both> querist and respondent" (*OM,* 54).

As in St. Augustine's *Confessions,* however, the task of self-scrutiny proves all but inseparable from the relation that the self bears to others and, hence, from the ethical imperative—always felt, though frequently not honored—to acknowledge the reality of the other as person rather than exploit him or her as a means for subjective emotional or sensual gratification.[13] Thus the question, "have I a responsible Will?" makes but explicit a self-other dialectic that, for Coleridge, is not merely an empirical fact but an ontological trait of personal identity. As his *Opus Maximum* transitions into a new chapter (on the "present general education of man in relation to the good"), Coleridge embarks on a long, densely argued passage that aims to challenge outright the quintessential modern fallacy of "plac[ing] the very principle of personal identity in the consciousness" (*OM,* 127–128). Coleridge opens with the familiar Kantian distinction between "the means and conditions of consciousness"—the *Seelenvermögen* or "reflective ~~mechanism~~ processes of the soul" and those external phenomena elucidated by the concrete operations of consciousness. Yet as so often, Coleridge finds prevailing vernacular usage to be the most entrenched source of philosophical misconceptions; or, as he puts it, "it is a short and downhill passage from errors in words to errors in things" (*AR,* 253). In this case, it is the word "form" whose careless deployment complicates the main task at hand, viz., to distinguish "the subjective necessity of apprehending an object with a form from the objective necessity of a form in the object." His proposed remedy is to distinguish between *form* ("that which in all changes of manifestation … remains the same and cannot be thought of but as having been that which is") and *shape,* understood as "the total superficies of the product of this form" (*OM,* 128).

Yet Coleridge's objective here is not to revitalize some version of Plato's or Aristotle's *forma substantialis,* nor indeed to vindicate Berkeleyan idealism (which in passing he singles out for praise). Rather, in desynonymizing form and shape, Coleridge seeks to recover a crucial feature of mental life, one consistently overlooked in modern Cartesian and Newtonian thought: viz., that it is inherently active, and that

13. A justly famous instance would be Augustine's realization, upon the death of his unnamed "friend" in *Confessions,* 4.4.7–4.9.14, that his conception of friendship—as well as the overwrought and unfocused grief following that loss—had been deeply narcissistic all along. Having lost "the source of gladness" (*amiseram gaudium meum*), Augustine immediately questions the spiritual significance of his own tears: "Or is weeping too a bitter thing, whose pleasure lies in the loathing for things we enjoyed previously, and now abhor?" (*an et fletus res amara est, et prae fastidio rerum, quibus prius fruebamur, et tunc, dum ab eis abhorremus, delectat?*). *Confessions,* 4.5.10; see Wetzel's excellent discussion of this episode in "Book Four."

passivity can only ever be ascribed to the external "shape" of phenomena, yet never to the "form" that grants the mind access to such phenomena to begin with. His fundamental commitment, which he shares with Blake, Wordsworth Goethe, and which he bequeaths John Henry Newman, John Ruskin, Gerard Manley Hopkins, and others, is to a quasi-phenomenological conception of mind as transformative vision. For Coleridge there is no question that "an innate affinity for making contact with reality moves our thoughts" and that human sensation is inherently active, engaged, and incipiently transformative of what is phenomenally and materially given. As he puts it elsewhere, "without the potential moulds, *ανευ μορφαις μορφογενεσι* of the Understanding the notices supplied by the Senses would have no *substans,* no substance—could not be *formed* into Experience."[14] Just as in the case of a peach "a mechanical shaping is ... opposed to the intrinsical and causative form that works ab intra" (*OM,* 128), there is an irreducibly and unceasingly active principle of form at work in the human person. "Led by our imagination," Robert Sokolowski notes, "we tend to posit it as yet another shape, albeit a ghostly one. Even the classical names for it, *forma, species, morphē,* and *eidos,* suggest that it is like a shape. But it is an identity of another sort." Hence, "to name a thing is to begin an adventure in manifestation, not to conclude it," an insight that had also proven central to Coleridge's thought at least since the *Biographia.*[15] All of his examples thus affirm things as intrinsically active and dynamic—in contradistinction to modernity's axiomatic view of objects as inert and heterogeneous.

Coleridge's apt choice of example—precariously close, or so it would seem, to the merely passive and vegetative shape of external phenomena—is the infant as it interacts with its mother:

> The infant follows its mother's face as, glowing with love and beaming protection, it is raised heavenward, and with the word "God" it combines in feeling whatever there is of reality in the warm touch, in the supporting grasp, in the glorious countenance. The whole problem of existence is present as a sum total in the mother: the mother exists as a One and indivisible something before the outlines of her different limbs and features have been distinguished by the fixed and yet half-vacant eye; and hence, through each degree of dawning light, the

14. *CN,* no. 4679; as he writes in a letter of 1801, "any system built on the passiveness of the mind must be false, as a system" (*CL,* 2:709).

15. Sokolowski, *Phenomenology of the Human Person,* 110, 112. The juxtaposition between the dynamic and the inert model of things and objects, respectively, had also occupied the late Husserl's discussion of modern dualism in *Crisis,* esp. 60–100.

> whole remains antecedent to the parts, not as composed of them but as their ground and proper meaning.[16]

In what is an inherently conjectural passage—and undoubtedly a veiled reference to Wordsworth's "Bless'd the infant babe" passage from *The Prelude*—Coleridge's remarks here cash in on his earlier distinction between form and shape. The mother's "different limbs and features" can only be "distinguished" by the infant's "half-vacant eye" because they are part of a living *Gestalt* or form, an "antecedent" whole which, as the "ground and proper meaning" of discrete shapes, ensures that the infant's perception is not simply a passive mirroring of dispersed parts but a bona fide act of perception. Receptivity to form and, as the case may be, a dawning, tenuous "imitation" of the mother's movements, sounds, and expressions thus constitutes an *action* that lays the foundation for the infant's consciousness of its own reality. With good reason, then, Coleridge had already desynonymized "recipiency" and "passivity" on the preceding pages: "if to imitate, and if the presence of a something imitable, are proofs of a state purely passive—if the mirror can be fairly said to imitate the form of the objects … we only have to admit that passiveness is the same as recipiency, and that recipiency is in no essential point different from non-recipiency, for the latter is assuredly implied in the reflection of an object" (*OM,* 129).

As this *reductio ad absurdum* shows, in so folding "recipiency" and "imitation" into sheer passivity, the modern conception of mind as strictly mechanistic and reactive fails the most elemental test of logical thinking (i.e., the law of non-contradiction), such that the idea of "recipiency" effectively collapses it into its antonym ("non-recipiency").[17] Having rather unfairly characterized the Newtonian idea of mind as "always passive—a lazy Looker-on on an external World" (*CL,* 2:709), Coleridge

16. *OM,* 131; on this extended passage, see also A. Taylor, *Coleridge's Defense,* 80–85. For a strikingly similar discussion of the child "assent[ing] to his mother's veracity, without perhaps being conscious of his own act," see Newman's *Grammar of Assent;* an instance of what Newman means by "real assent" (in contradistinction to conditional, "notional assent"), the child's recognition of the mother's reality "has a force and life in it which the other assents have not, insomuch as he apprehends the proposition, which is the subject of it, with greater keenness and energy than belongs to his apprehension of the others. Her veracity and authority is to him no abstract truth or item of general knowledge, but is bound up with that image and love of her person which is part of himself, and makes a direct claim on him for his summary assent to her general teachings" (*Grammar of Assent,* 34–35).

17. Earlier in his *Opus Maximum,* Coleridge juxtaposes the subject's receptivity to sense "impressions" with the presentations of "conscience" and notes how "in the facts of conscience we are not only agents[,] but … know ourselves to be such! Nay! We are aware that our very passiveness herein is *an act* of passiveness" (*OM,* 71); as the most conspicuous instance of self-awareness that Coleridge had found adumbrated in the scriptural term *parakupsas,* conscience effectively negates the very notion of mind ever being, or having been, completely passive.

here contends that "recipiency" cannot logically be grasped as the simple imprint of material sensation on a passive receptacle. Rather, the reception of phenomena is accompanied by the consciousness of its very occurrence and, crucially, also feeds back into and positively transforms that very consciousness. Not coincidentally, Coleridge also interprets materialism and mechanism as unwitting descendants of modern Calvinism, which likewise "represents a Will absolutely passive" and, in so doing, "takes away its essence and definition."[18] In either instance, the principal error is to have neglected "the inherent distinction of things and of our notices of things" (*OM,* 118), a distinction that notably anticipates the concept of intentionality later developed in the phenomenology of Franz Brentano and Edmund Husserl. Far from being a crude, mechanical transmission of force, all reception thus involves a distinct quality, a "sentiency" or *relation* that is reciprocal by nature and quietly transformative of consciousness itself. Logically, the dead-end road of Lockean empiricism or Hobbesian and Hartleyan associationism has to be countered by something on the order of modern phenomenology for which by definition all "perception is dynamic, not static."[19] For Coleridge, "the notion of objects as altogether objective begins in the same moment in which the conception is formed that is wholly subjective ... [and] the sum of the objective as object of the sense ... [has] the *esse* contained in the *percipi*" (*OM,* 135). The event of perception simply cannot be disaggregated into minute particulars, nor indeed can it be confined to the actual process of seeing those particulars. Beholding a tree, we experience a "strength ... in the whole" for which we "seek in vain, in the boughs, the sprays, the leaves" (*OM,* 137). To be sure, Goethe would disagree and insist that, in point of fact, the leaf *is* the whole, indeed, is unintelligible *except* as the concrete embodiment—not just a part—of that whole. Yet in the end, Coleridge mounts a similarly Platonic argument in favor of the "form" (*eidos*) having ontological priority over the "shape" (*hylē*), and indeed proving itself to be the latter's condition of possibility. Like Goethe, that is, he places stress on life as something intelligible only qua form (*eidos*) and, as it were, possessed of an integrative dimension "as unity, as plastic, and as invisible" (*OM,* 134). Levinas, whose work offers numerous intellectual affinities with Coleridge's later writings, argues that "the face is a living presence" and, by its very "manifestation ... is already discourse" (*TI,* 66). The Romanticist, invariably, will recall the powerful account in

18. *AR,* 158–159; elsewhere, Coleridge notes that by the will of God the Calvinists, "the Literalizers of half a dozen metaphors ... mean nothing better than the capricious enslaved Wantonnesses of *human* Choice—determinations pre-determined by the appetites or at best by the ignorance of the Agents and the narrow limits of their Agency" (*CN,* no. 5270).

19. Sokolowski, *Introduction to Phenomenology,* 18; regarding Coleridge's rather unconventional assimilation of Hobbes to associationist psychology, see *BL,* 1:95–97 and, for his refutation of associationism as a whole, *BL,* 1:106–115.

Book 2 of *The Prelude* to how Wordsworth, "a Babe, by intercourse of touch / ... held mute dialogues with my Mother's heart" (1850, 2:285–286; C-Stage text).

At first blush, much of the foregoing would seem to be little more than the standard Berkeleyan fare that commentators have had no difficulty identifying as the main intellectual source of Coleridge's argument here. Yet that view is at the very least incomplete and, ultimately, misleading. First and foremost, there is the fact that the "shape" that is received by the perceiving subject is not simply received *as* an isolated particular but, instead, is axiomatically cross-referenced with the hypothesis of the whole *of* which it is a part. Considered within a phenomenological framework whose outlines Coleridge's late writings often anticipate, "consciousness is 'of' something in the sense that it intends the identity of objects, not just the flow of appearances that are presented to it."[20] More than transposing discrete material externals into the realm of subjective "data" or "information," phenomena are themselves a source of *knowledge;* they have agency. What allows Coleridge to draw this crucial inference is the fact that they are not exhausted by their putative referent but are by their very nature an "appearance *for*" someone. As such, phenomena have an actual and *active* presence as catalysts of inner (and likely transformative) experience. Coleridge intimates as much in his passing observation that the proverbial "mind's eye ... is an eye *for* the mind no less than [an] eye *of* the mind" (*OM,* 127; italics mine); or, as Husserl was to put it in 1905, "appearances themselves don't appear; they are experienced" (*Die Erscheinungen selbst erscheinen nicht, sie werden erlebt*).[21] It follows that from the first instance of perception and "recipiency" the consciousness of fluctuating phenomena or "shapes" is thereby also alerted to the substratum of its own Personëity, that antecedent and invariant form wherein, and by virtue of which, phenomena do indeed not merely "appear" but emerge as distinct and continuous focal points of awareness. Lest this insight be absorbed into some strictly interiorist account, however, it should be stressed that "experience" in Coleridge's proto-phenomenological account of it is not to be confused with "inwardness" or some mental possession. Rather, it signifies precisely insofar as *in it and by virtue of it* the self discerns its ontological relatedness to a world of objects and persons.

Coleridge here builds on and extends Kant's basic contention "that empirical knowledge results from a co-operation between receptivity and spontaneity" and that, contrary to the crudely empiricist (Lockean) model, empirical data are susceptible of being apprehended only because of the mind's intrinsic orientation to the world. Up to a point, his position also anticipates John McDowell's recent attempts to

20. Sokolowski, *Introduction to Phenomenology,* 20.

21. Husserl, *Logische Untersuchungen,* 2:350; see also Cutsinger, *Form of Transformed Vision,* esp. 46–72.

navigate between the non-cognitivist (empiricist) "myth of the Given" wherein mind and world are ontologically distinct, and a likewise non-cognitive, "bald naturalism" that treats the "logical space of reasons" as but an epiphenomenon of "the relations … that constitute the logical space of nature."[22] Rejecting the prevailing view of a "dichotomy of logical spaces," McDowell proposes a "minimal empiricism" in which "the idea of experience is the idea of something natural, without thereby removing the idea of experience from the logical space of reason." As he insists, there simply is no warrant for "identify[ing] the dichotomy of logical spaces with a dichotomy between the *natural* and the normative." In fact, as soon as we recall that for human beings "nature includes *second nature,*" which they acquire "by being initiated into conceptual capacities … whose interrelations belong in the *sui generis* logical space of reason," it becomes clear that sense impressions of the Lockean variety involve *eo ipso* micro-judgments or acts of assent whereby raw data are constituted as "cases" (in the sense of Ludwig Wittgenstein's *Fall*). Critically engaging Gareth Evans's work, McDowell zeros in on this "non-conceptual content" by insisting that "the understanding is already inextricably implicated in the deliverances of sensibility themselves" and that the "impressions" that our senses mediate for us "already have conceptual content."[23] However obliquely, McDowell's argument for impression as something constitutively enmeshed with a mental act revives a fundamentally Platonic insight (previously reaffirmed by Coleridge) that to understand the mind-world relation at all we must begin by rethinking the very notion of "receptivity." In particular, we must extricate ourselves from the popular assumption of a mind passively absorbing and naturalistically mirroring some contingent influx of data. For McDowell, recipiency can never be reduced to some unilateral infusion of data into a passive mind but, instead, pivots on countless acts of assent: "How one's experience represents things to be is not under one's control, but it is up to one whether one accepts the appearance or rejects it."[24]

Circumventing both Cartesian dualism and Humean naturalism, McDowell thus insists that "the passive operation of conceptual capacities in sensibility is not intelligible independently of their active exercise in judgment." To take that view—one that Coleridge had worked out almost manically after 1808, and which also subtends the rather more steady presentation of Newman's 1870 *Grammar of Assent*—is to realize that mind, by its very essence and ontology, can only be thought of as "active." As Newman was to put it, "our consciousness of self is prior to all questions

22. McDowell, *Mind and World,* 9, xviii, xv.
23. Ibid., 46.
24. Ibid., 11.

of trust or assent."[25] Indeed, the naturalist or (more recently) neuro-scientific conception of a strictly reactive and stimulus-dependent, "passive mind" ultimately amounts to an outright contradiction of terms. For to have uptake of an appearance, however minimal, is to engage a phenomenon *as* appearance. It means to grasp it as a "case," a state of affairs or *Fall* that by its very nature presupposes a mind having exercised a judgment and, at least potentially, being aware of its own hermeneutic role in the constitution of a specific "life-world." Consequently, "active empirical thinking takes place under a standing obligation to reflect about the credentials of the putatively rational linkages that govern it. Regardless of whether we stand in a relationship of assent or dissent vis-à-vis a given appearance, the latter only acquires an identity within the logical space of mind in virtue of having been engaged *as* appearance. The mere fact that the skeptical stance (i.e., dissenting from or distrusting appearances) is more conspicuous does not license the inference that a contrasting instance of *assent* is a mindless non-event. Indeed, even where appearances are invested with a certain truth value, "there must be a willingness to refashion concepts and conceptions if that is what reflection recommends."[26] Not only, then, is a judgment of assent or dissent focused on intrinsically "saturated phenomena" (to borrow Jean-Luc Marion's apt phrase) that appear *for* us; but any such judgment is also reflexive in that it continually tests and adjusts its own conceptual inventory.

Yet while McDowell, as Coleridge long before him, is treading in Kant's footsteps by arguing for an integrated and collaborative (rather than disjunctive and agonistic) relationship between sensibility and spontaneity, nature and freedom—he ultimately comes down squarely on the side of a modern scientific framework committed to paring down the non-conceptual contents of sense impressions to quasi-mathematical idealities. For him, nature is ultimately "the realm of law" and modern scientific methodology the royal road toward "a new clarity *about nature*."[27] Putting further scrutiny on McDowell's position is instructive here in that it throws into relief

25. *Grammar of Assent*, 67; for McDowell, "the relevant conceptual capacities are drawn on *in* receptivity" such that we understand "what Kant calls 'intuition'—experiential intake—not as a bare getting of an extra-conceptual Given, but as a kind of occurrence or state that already has conceptual content" (*Mind and World*, 9); surprisingly, in moving beyond Sellars's rejection of the "myth of the Given" and toward arguing "that the conceptual contents that are most basic in this sense are already possessed by impressions themselves" (10), McDowell seems to reinhabit (but not acknowledge) a modern phenomenological conception of mind-world relations that has recently been given powerful articulation in the work of Jean-Luc Marion; see *Being Given*, esp. 7–70. For other arguments to the same effect—viz., that "perception is an active and synthesizing operation" and that "attention is in no sense a response to stimulus"—see Hedley (*Living Forms*, 48), who also invokes Collingwood's *Principles of Art* and, above all, Coleridge's distinction between the primary and secondary imagination in the *Biographia* (1:304f.).

26. McDowell, *Mind and World*, 12–13.

27. Ibid., 78.

Coleridge's ultimately very different orientation. For McDowell, what stands to be avoided at all cost is any suggestion that the logical space of reasons is radically separate from, even incommensurable with, the space of sensory contact with nature. Any suggestion "that the structure of the space of reasons is *sui generis*" risks making "our capacity to respond to reasons look like an occult power." Retaining more than a passing allegiance to naturalism, even in its more stridently reductionist form, McDowell rejects any such "rampant Platonism" because it seeks to superimpose "something extra to our being the kind of animals we are." Rather, he insists, the correct view involves an Aristotelian/Kantian model of humans having evolved rational capacities as a kind of "second nature" that remains structurally cognate with (if conceptually superior to) natural processes of any variety. What renders a traditionally Platonist view unfathomable and thus unacceptable to McDowell is that it argues both from and toward a supernatural source—that is, "the idea that our species acquired what makes it special, the capacity to resonate to meaning, in a gift from outside nature."[28] As McDowell's oddly dogmatic rejection of understanding the world as "gift" and the human person as categorically distinct from "our species" makes clear, his commitments clearly outrun his arguments—such that his ultimate objective remains to strengthen the modern project by resolving specific tensions within its conceptual machinery, rather than questioning its basic viability and coherence.

What McDowell does not consider, then, is the possibility (central to my own account) that those contradictions and aporias that a "bald naturalism" or a strict conceptualism was unable to resolve had arisen precisely because of modernity's peremptory rejection of *ideas* (e.g., of the good, person, relation, and a "responsible will"). Many of modernity's conceptual dilemmas and ethical failures, revealed in its increasingly strident disaggregation of fact from value and of theoretical from practical reason, stem from the failure to "realiz[e] that the world is no *factum brutum* but gift."[29] Behind the rejection of the world as a gift and, hence, as revelation of a noncontingent rational order (a cosmos rather than universe), stands a deep-seated fear that makes us look with hesitation, even mistrust, at more mundane gifts: it is the fear of a metaphysical, all-encompassing *obligation* such as can only be discharged by the way we order our lives. Defining of modernity's (anti-metaphysical) embrace of *method* as the exclusive foundation for knowledge is this inability to imagine or acknowledge the reality of the gift. For in its basic ethos, modern method is pledged to

28. Ibid., 83–84, 123. "We need to recapture the Aristotelian idea that a normal mature human being is a rational animal, but without losing the Kantian idea that rationality operates freely in its own sphere" (85); later McDowell refers to his project as a "naturalized version of platonism" (110).

29. Hedley, *Living Forms,* 69; Hedley's engagement of McDowell's work here, though also critical, ultimately reaches somewhat different conclusions from my own.

a wholly self-licensing and acquisitive life; it appropriates and takes possession of the world as the putative fruit and reward of our cognitive labor alone. Once it has been axiomatically posited as sheer otherness, or *Gegenstand,* the allegedly "brute fact" of the material world naturally proves unintelligible to modern methodical inquiry as a gift or as a phenomenon saturated with meaning. The methodical labor of post-Baconian science takes itself solely as the *producer,* never the recipient, of conceptually distinct meanings, and so purports to redeem an allegedly inchoate (Gnostic) *hylē* from being terminally trapped in the a-semantic flux of disjointed appearance.

More overtly treading in the footsteps of Coleridgean Platonism, John Macmurray's 1953–1954 Gifford Lectures had cautioned "that because sense-perception is learned so early in life we are very apt to forget that it has to be learned at all; so that we talk of it as though the power to perceive a world of objects were born in us, and that its 'immediacy' is an original datum of human experience. This is not so. Perceiving by means of the senses is an acquired skill." At first glance, Macmurray's position would seem to be echoed by McDowell, who contends that "impressions can *be* cases of its perceptually appearing ... to a subject that things are thus and so."[30] Yet to put the matter thus is to suppose—though, notably, not to have demonstrated—that our conceptual abilities are merely acquired and cultivated as a prolonged *Bildung* of second nature. Contrary to Macmurray's and, long before, Coleridge's account of the human person, McDowell reverts to an Aristotelian, naturalist view of the infant as but an organism with superior potential: "Human infants are mere animals, distinctive only in their potential, and nothing occult happens to a human being in ordinary upbringing."[31] Yet for Macmurray, as for Coleridge, the animal metaphor is never appropriate to a human person because the latter's cultivation of innate powers is essentially bound up with its relations with another human being, and also because the structure of what Coleridge calls "recipiency" is never passive, as McDowell takes it to be. In fact, the basic phenomenological insight that the appearance *of* a particular shape is intrinsically appraised as *phainomēnon*—viz., the partial manifestation of an identity that does not itself appear as such—shows just how profoundly recipiency and imagination are enmeshed in human consciousness. For a "shape" to appear *for* consciousness, it must have been appraised as the appearance *of* an underlying form or identity: "The unity of the body is constituted, Plotinus observes, by the spatial continuity of divisible chunks of matter. In conscious awareness, however, the soul is capable of perceiving different parts of the body. It is not a part of the soul perceiv-

30. Macmurray, *Persons in Relation,* 53. McDowell, *Mind and World,* xviii, xii, xix–xx.
31. McDowell, *Mind and World,* 123.

ing; the soul is present as a whole in its awareness of bodily parts. This presence of unity in different parts distinguishes the soul as an ontological item from bodies."[32]

This dual operational structure lies at the heart of Coleridge's famous distinction between the primary and secondary imagination—with the former constituting "the living Power and primary Agent of all human Perception, and as a repetition in the finite mind of the eternal act of creation in the infinite I AM." Meanwhile, the secondary imagination disassembles and reconstitutes what active perception has thus presented; "it dissolves, diffuses, dissipates, in order to re-create" (*BL*, 1:304). In its very essence, then, mind is essentially active even in its apparent "recipiency" or responsiveness vis-à-vis an "outside" empirical world, *and* it is also actively self-revising as regards the adequacy of its internal procedures, which are needed for organizing appearances into specific kinds of knowledge. It is, thus, both responsive *to* and responsible *for* the world that it successively composes in this manner. By contrast, mechanist or mono-causal empiricist accounts of a world supposedly obtruding in its alleged radical otherness on some supposedly self-contained and self-identical *cogito* appear weak on both conceptual and ethical grounds. The same is true of radically naturalist and skeptical arguments purporting to overcome the (again, supposed) antinomy of mind and world by dissolving all mental processes into mere effects or epiphenomena of inherently mindless, extrinsic causes. Anticipating modern phenomenological arguments to the same effect (in Franz Brentano, Maurice Merleau-Ponty, Jean-Luc Marion, and others), as well as McDowell's "minimal empiricism," Coleridge time and again emphasizes the integrative, dynamic, and participatory structure of the mind-world relation. As such, the structure of everyday perception—which is inherently a story about how perceptions are *experienced*—unveils within the finite individual those coexistent principles of "1. Ipsëity, 2. Alterity, 3. Community" that Coleridge elsewhere identifies as core implications of the Trinity.

Mounting a more philosophical argument against the modern Cartesian conflation of "consciousness" with "self-identity," Coleridge puts the matter thus: "as in the patient, retrogressive investigation of organic bodily form, so in the mind nothing will appear as the first, as beginning, except only the power of communicating itself in each, relatively to an outward contemplator ... So true is this, indeed, that in the development of the mind, the consciousness itself has the appearance of another" (*OM*, 127). As Levinas was to put it eventually, "the I that thinks and hearkens to

---

32. Hedley, *Living Forms*, 88; drawing on phenomenological rather than Platonist language, Sokolowski makes the same crucial point; see *Introduction to Phenomenology*, 17–41; for a fuller discussion of the reciprocity between the form of the phenomenon as a catalyst for the cultivation of human intelligence in the context of Romantic science and aesthetics, see Pfau, "All is Leaf."

itself thinking or takes fright before its depths ... is to itself an other."[33] One can scarcely think of a writer more alert to the innate terrors of thought—occasioned not by a refractory world of things but by the mind's intrinsic volatility—than Coleridge. Recalling some of his most sensitive lyric works written more than three decades earlier, his *Opus Maximum* once again ponders how both alterity and relationality are liable to register within the young child as a feeling of profound vulnerability and fragility. In a remarkable passage that presages as much modern object-relations theory and childhood psychology (Ronald Fairbairn, Melanie Klein, and D. W. Winnicott in particular) as its opening trope of writing is liable to invite deconstructionist second-guessing, Coleridge evolves the origins of the person or, as he calls it, "Personëity," from the infant's dawning sense that its own reality depends to an often frightening degree on the continued presence of the parent, frightening because that presence is bound to prove discontinuous:

> The same spirit which beholds the parts in the whole, ... finds a bewildering and, as it were, spectral terror, a sense of sinking, resembling that which it had suffered or dreamt of as the mother's knee had suddenly given way from under it ... Even as we sometimes dwell on a word that we had just written till we doubt, first, whether we had spelt it right, and at length it seems to us as if no such word could exist; and, in a kind of momentary trance, strive to make out its meaning out of the component letters, or of the lines of which they are composed, and nothing results! In such a state of mind has many a parent heard the three-years child that has awoke during the dark night in the little crib of the mother's bed entreat in piteous tones, "Touch me, only touch me with your finger." A child of that age, under the same circumstances, I myself heard using these very words in answer to the mother's enquiries, half hushing and half chiding, "I am not here, touch me, Mother, that I may be here!" (*OM,* 131–132)

Vaguely reminiscent of the clocks and furniture melting away on Salvador Dali's canvases, there is something decidedly surreal about Coleridge's opening analogy as a word's semantic identity begins to dissolve and we find ourselves unaccountably estranged from orthographic convention. As so often, we find Coleridge at his most vivid and compelling when scrutinizing moments of psychological instability. Thus, with customary phenomenological and verbal precision, he here traces the inexorable progression of some "momentary trance" spiraling into an ontological crisis or "spectral terror" in which not only our *grasp* of a specific object but its very *existence* seems in doubt. Ingeniously, Coleridge also hints how a crisis experienced by an adult puzzling over the shape, the meaning, and even the very existence of a specific

33. Levinas, *Totality and Infinity,* 36.

word happens to recall, indeed revive the primal trauma when the infant's reality appeared to disintegrate as a result of the mother's withdrawal ("as the mother's knee had suddenly given way from under it"). Rather than probing, in good Lockean manner, the reality of the outside world by touching the mother, the child begs *to be touched so that it may know itself to be real.* In the beginning, then, there is not an autonomous Cartesian self; nor indeed is the young child of three years some embryonic anticipation of it. Rather, there is the reciprocity and acknowledgment of one person by another in a dynamic of ipsëity, alterity, and community that is as profound as it is fragile.

Still, the psychoanalytic drama so vividly sketched remains incomplete because "there was another beside the mother, and the child beholds it and repeats, and ... carries onward the former love to the new object. There is another, which it does not behold, but is above." That is, the child not only seeks the concrete shape of the mother but also intuits behind it "the father and the heavenly father, the form in the shape and the form affirmed for itself ... blended in one, and yet convey[ing] the earliest lesson of distinction and alterity." As in Richard of St. Victor's account of the Trinity, the reality of the person is not merely affirmed—let alone "produced"—by some kind of "intersubjective" relation. Rather, the primary, lateral relation of mother and child points back to the concurrent, vertical relation that both of them bear to "another, which [the child] does not behold," yet whose presence it discerns in "the mother's eye ... turned upward" (*OM,* 132). The discrete elements that define the mother-child relation thus are axially ordered to the divine in the same manner as the diverse shapes of empirical perception are apprehended as parts of a single, integrative being. "Why," Coleridge muses, "have men a Faith in God?" The only answer, he claims, is that man "has a Father and a Mother." If, as he contends, "all begins in instinct," there is yet something unique about "human rational instincts" in that they alone do not merely discharge themselves but are *experienced* as manifestations of a higher order: "Reason itself mutely prophesying of its own future advent" (*OM,* 122).

In phenomenological terms, the appearance of discrete shapes also points back to the non-appearance of that identity—"the form in the shape"—*of which* they are discrete manifestations. Were it otherwise, finite consciousness would never progress beyond disjointed sensations to the knowledge of objects, nor even to an awareness of itself as *having* these sensations. Yet "while we identify cubes, propositions, facts, symphonies, paintings, moral exchanges, and religious things, we also, always, are establishing our identities as the ones to whom these things are given ... as datives of manifestation."[34] For the infant, then, the traumatic but crucially important passage

34. Sokolowski, *Introduction to Phenomenology,* 32; see also Sokolowski's *Phenomenology of the Human Person,* 108–116. As he notes there, "the shape of the thing is ... not the substance of the thing. As a property it points to something more elementary than itself; it points to the thing in its

involves recognizing its ipsëity as something that endures even in the absence of the parent's tactile and expressive affirmations. Having seen that the mother, too, is bound up in relation to another—a father present yet not to be beheld—the young child takes the crucial leap; it

> now learns its own alterity, and ~~as if some~~ sooner or later, as if some sudden crisis had taken place in its nature, it forgets henceforward to speak of itself by imitation, that is, by the name which it had caught from without. It becomes a person; it is and speaks of itself as "I," and from that moment has acquired what, in the following stages, it may quarrel with, what it may loosen and deform, but can never eradicate—a sense of an alterity in itself, which no eye can see, neither his own nor others. (*OM,* 132)

In continuation of a trajectory that had originated in the mother's "dawning presence" to the infant, self-consciousness is achieved as the child begins to understand that relations to others give a sense of fullness, but do not outright constitute, its being; and thus its "conception of life is elevated into that of Personëity" (*OM,* 134). If "Personëity" signifies a stage in which the individual is conscious of himself or herself *as* a distinct person, it does not imply "immediacy" but, on the contrary, is bound up with an ongoing and precarious ascent. Its "transcendence" thus amounts, as Levinas puts it, to a "transascendence" (*TI,* 35). The reality of the person endures, indeed is deepened and properly apprehended only inasmuch as her or his identity is bound up with a dialectic of estrangement and differentiation that Coleridge captures in the term "alterity" and for which infant and early childhood psychology furnish particularly compelling phylogenetic examples.

To highlight the distinctive thrust of Coleridge's conception of the child gradually attaining a grasp of its own Personëity, we may recall some apposite, albeit rather more controlled and optimistic lines from Wordsworth's two-part *Prelude:*

> now a trouble came into my mind
> From obscure causes. I was left alone
> Seeking this visible world, nor knowing why:
> The props of my affection were removed
> And yet, the building stood as if sustained
> By its own spirit.[35]

---

kind, in its essence or nature. To be able to distinguish the shape of the thing from the thing, to see the shape as a property, is an enormous intellectual accomplishment. We could even say that it is the birth of intelligence" (109).

35. Wordsworth, *The Prelude,* 1798–1799, Part 2, lines 321–326; on Wordsworth's transposition of affect into text into a (ostensibly coherent and continuous) self, see my *Wordsworth's Profession,* 302–320.

On the face of it, these early Wordsworthian lines seem to presage the later Coleridge's consideration of the mother-child dyad as a template for his metaphysics of person and the Trinity. Yet thematic resemblances notwithstanding, the connection is problematic at best. Far more obviously—and in ways that would have left Coleridge squirming with philosophical discomfort—Wordsworth's lines betray above all the powerful influence of Rousseau, that "apostle of affliction" (Lord Byron) for whom the insistence of the other and the specter of other-identified existence invariably signify a descent into inauthenticity and social corruption. Indeed, even as the sentimental overtones have been muted by the 1790s, Wordsworth's *Prelude,* Hegel's account of the "unhappy consciousness" or, eventually, Marx's account of "alienation" in *The German Ideology* all attest to Rousseau's enduring influence. For they all look upon alterity as a sign of loss, betrayal, and decline. Wordsworth's assurance that his poem's author-protagonist has at last advanced from "A naked Savage in the thunder shower" (*Prelude* 1799, 1:26)—himself a next-of-kin to Rousseau's "primitive man"—to a "sensitive and creative soul" (*Prelude* 1805, 11:257) unequivocally "sustained / By its own spirit" and cheered on by its own voice never quite rings true.[36] Wordsworth's deeply counterfactual claim to expressive self-creation and self-possession here seems forced and rhetorically overwrought. The text cannot shake off a pervasive and deep-seated existential *Angst* concerning the precarious, not to say improbable, status of modern autonomy—quite simply because this is the kind of claim that will be valid only for as long as it can be performatively sustained by further instances of self-assertion and self-expression. Yet if that is the case, then the Romantic project of an "expressivist" (Charles Taylor) constitution of the self becomes virtually interchangeable with what, a century later, Freud was to analyze as the dynamic of repression and displacement. The urgent affirmations of Wordsworthian "spontaneous overflow" would appear to have morphed while being "recollected in tranquillity," and what Wordsworth himself characterizes as a "species of reaction" shows spontaneity to be a type of defense mechanism that exposes the language of (self-) affirmation as, in fact, symptomatic. Even as the *Prelude*'s subtly wrought cadences simultaneously *instantiate* and *report* on the emergence of an autonomous and sensitive poetic self, they cannot but invite back in with a vengeance the world *from* which that self claims to have become emancipated.[37] While that is a point that Romantic

36. Wordsworth's Rousseauvian image is almost certainly informed by the sensational discovery, in 1797, of the so-called "Wild Boy of Aveyron," a feral child temporarily captured in France, near Saint-Sernin-sur-Rance, and speculated to have lived outside of civilization at least since his fifth year. Following E. M. Itard's book on the case (*An Historical Account of the Discovery and Education of a Savage Man* ..., 1802), Coleridge's notebook entry of early 1803 remarks on "a man who *hypochond.* fancied himself to have been a lonely Savage; and poisoned by civilization / —savage of Aveyron" (*CN,* no. 1348); see also A. Taylor, *Coleridge's Defense,* 35–36.

37. On the aesthetic design of the *Prelude*'s narrative form, its underlying epistemology of the self, and the widely observed "displacement" of the historical, economic, and political forces that

historicism has made with zealous attention to material detail, its full implications are more likely to disclose themselves within a philosophical analysis. Modernity's basic premise of a punctual and autonomous self variously defined as neo-Stoic autarky (Descartes), as a hermetically sealed and mechanical will (Hobbes), or as a self-certifying bearer of rights and agent of contracts (Locke) dreads nothing more than even the most fleeting intimation that the world vis-à-vis which the self seeks to assert its independence should have inhabited the modern individual all along. The sheer suggestion that alterity might prove intrinsic to personhood, an ontological truth rather than something discretionally invited into the self's orbit, also accounts for a growing utopian streak within modern political and economic thought. The main premise here appears to involve the procedural utopias of classical liberalism (from Smith, Ferguson, and Hume forward via Jeremy Bentham, Ricardo, to Mill and Gladstone): viz., that given enough time, bureaucratic and legal finesse and an ingenious array of economic incentives and fantasies of social mobility, our lived existence will eventually merge with the conceptual structures developed for the purpose of its transformation.

Clearly, Coleridge's outlook is starkly opposed to this kind of project. Both in the range and historical depth of those intellectual traditions and conceptual genealogies on which he draws in later work, yet also in his defiantly Scholastic and disputatious exploration of ideas and terms—a far cry indeed from his contemporaries' "skipping, short-winded asthmatic sentences, as easy to be understood as impossible to be remembered" (*CF,* 1:26)—Coleridge charts his own course. His late explorations in Trinitarian theology thus complete a reflection about the "self-insufficingness" of the person that had arisen from a critique of modern, autonomous, and self-conscious agency begun in *The Friend* and continued in the *Biographia* and the *Lay Sermons.* To begin with, Coleridge insists that the very fact of our being "responsible Agents; *Persons,* and not merely living *Things* ... cannot be the object of *my own* direct and immediate Consciousness; but must be *inferred* ... *from* its workings [as] it cannot be perceived *in* them" (*AR,* 78–79). The basic point in contention here had long been a staple of Coleridge's philosophy, viz., that self-consciousness cannot logically be understood as an abstraction from or evolutionary result of some antecedent, rudimentary state. For "if this Something were a negatively simple entity, a Same throughout, there might be a Self, but no *Consciousness* or conscious Self-knowledge—and Knowledge without Consciousness is not *a Knowing* but *a Being.*" As he would later insist, alterity—no less than ipsëity and community—is a corollary of person.

---

conspire in the development of aesthetic autonomy, see Liu, *Wordsworth,* 359–452; de Man, "Autobiography as de-Facement," 67–82; for a critique of Romantic historicism's axiomatically suspicious hermeneutics, see Pfau, *Wordsworth's Profession,* 114–139, and "Reading beyond Redemption."

Yet since the *Critique of Pure Reason,* this very fact had been a source of acute vexation; for Kant and his successors, the persistence of something outside of the self's jurisdiction could not but derail the very notion of an autonomous subject and, ultimately, the very idea of reason (*Vernunft*) exclusively predicated on it. A perceptive reader and commentator of Kant's critical philosophy, Coleridge was well aware that self-consciousness—if explained strictly as the *synthesis* of two "heterogeneous" constituent sources ("pure intuition" and "pure concepts," or categories)—could never be anything more than a *formal* characteristic. That is, to conceive mind as a synthesis of some kind or other is to forgo any possible advance from the postulate of a merely formal unity (a.k.a. "apperception") to an actual consciousness-of-self and the existential reality of personhood. Simply put, the Kantian subject has no soul; for, as Coleridge puts it so concisely, "we must attribute to the Soul, as a self-conscious personal Being, not only a unity that cannot be divided; but this unity must contain distinctnesses that cannot be confounded."[38] This positive and irrefragable unity, not merely in the formal sense but as a vivid center of lived experience, defines what Coleridge calls "Personëity."

The stress on the "distinctnesses" of the "self-conscious personal Being" reveals Coleridge's intellectual debt to nominalist theology, in particular, to Duns Scotus, whose writings he had begun to explore as early as 1801.[39] Yet Coleridge also came to realize that the nominalist project had miscarried in spectacular ways and, indeed, had given rise (however inadvertently) to the materialist, necessitarian or, as he prefers to put it, "Epicurean" strand of modern thought that his own philosophical project seeks to dismantle. Above all, he rejects the tendency of Humean skepticism to disaggregate all forms and notions to such an extent as to negate the dynamic and self-determining nature of the human mind. A notebook entry from October

38. *SW & F*, 1:427–428; in a letter to Thomas Clarkson (13 October 1806), Coleridge distinguishes levels of consciousness, depending on the degree to which each is aware *of*, or indeed capable of reflecting *on*, its own continuity (*CL*, 2:1196–1197). In his "Prolegomena" to *Opus Maximum*, Thomas McFarland notes how "soul seemed, under the progressive arguments of seventeenth- and eighteenth-century rationalism, more and more a metaphorical conception, or at least one that could not be used in cognitive argument," and he also remarks on Descartes's and Locke's analytical fragmentation of "person" as merely a question of "personal identity" (*OM*, cxv); on the "Soul" in Coleridge's thinking, see also Engell, "Coleridge and His Mariner on the Soul." In a notebook entry of 1820, Coleridge jots down a memorandum to himself, viz., "to give a more *plain* as well as a more satisfactory Demonstration than I have hitherto met with … that *all* Consciousness is necessarily conditioned by Self-Consciousness" (*CN*, no. 4717).

39. On Coleridge's reading of Scotus, see his letter to Josiah Wedgewood, 18 February 1801 (*CL*, 2:678–685), which still professes to be unfamiliar with Scotus's works. By July, Coleridge is reading Duns Scotus with a clear, indeed exuberant sense of purpose: "I mean to set the poor old Gemman on his feet again, & in order to wake him out of his present Lethargy, I am burning Locke, Hume, & Hobbes under his Nose—they stink worse than Feather or Assafetida" (*CL*, 2:746).

1809 identifies modernity's peremptory quarantining of knowledge from significance, of fact from value, and, consequently, of knowledge qua verified information from the putative subjectivism and irrationality of belief. In particular, the notebook entry dwells on the assumption that only that can be known which is radically particular, singular, and hence incommensurable with every other entity. A bequest of fourteenth-century nominalism, this tendency toward a strictly disjunctive understanding of knowledge constitutes a development whose myriad implications Coleridge continued to ruminate over the coming decades with acute dismay:

> It is not that the Philosophy of the Fathers or moderate *Realists* is more abstruse or difficult to be believed than that of the Nominalists & *Materialists* (who are indeed *the true Realists*) so far from it that the philosophy of Plato & his systematic followers is only a display of the *possibility* of that which Mankind in general believe to be *real*—such as, that there is some ground in *Nature* or a common essence why Peter & John are two *men*/whereas the Philosophy of the Nominalists is abhorrent from all the common feelings of all mankind—but this it is, that gives the latter its fashion & favor—that … it consists in unbelieving as far as possible—till we come to *words* that convey all their separate meanings at once, no matter how incomprehensible or absurd the *collective* meaning may be—for the collective meaning cannot be inquired after but by an effort of Thought—and to avoid this is the aim of those who embrace this philosophy. (*CN*, §3628)

Most troubling for Coleridge is how modernity's model of institutionalized and methodically accumulative knowledge has effectively abandoned the ancient idea of a fortuitous convergence of *theoria* and *eudaemonia* in a single and absorbing contemplative act.[40] An 1822 fragment on William of Ockham, while complimentary of the latter's contribution, thus shifts attention to the deleterious aftermath of his project: "the Dialectic was soon left behind; and the Logic swam with the tide; and like the pigs cut its <own> Throat as it advanced—and soon sunk & gave way to the three other Factors, the Sense, <the Sensation,> and the Senses <as the union of both>." As

40. Speaking of this "most momentous … reversal of the hierarchical order between the *vita contemplativa* and the *vita activa*," Hannah Arendt elaborates: "the point was not that truth and knowledge were no longer important, but that they could be won only by 'action' and not by contemplation … The reasons for trusting *doing* and for distrusting *contemplation* or *observation* became more cogent after the results of the first active inquiries." One must not misconstrue this reversal—achieved above all with the help of instruments and, especially, the paradigm of "mathematical knowledge, where we deal only with self-made entities of the mind"—as simply "raising doing to the rank of contemplation as the highest state of which human beings are capable." For as the "handmaiden of doing" (*HC*, 289–291) all active thinking and its implicit vision of discrete knowledges moving toward a *mathesis universalis* effectively eclipsed the value of contemplation altogether.

a result, the nominalists' "empassioned Business of detecting Error and *protesting* against the old Usurpaters, Aristotle and the Papacy" soon gave way to "a heartless Scepticism" (*SW & F*, 2:1002). Coleridge's own conception of human agency, and of the will as its unconditional ground, thus unfolds in the most scrupulous opposition to both the Thomistic and the nominalist positions. As regards the latter, Coleridge specifically repudiates Hobbes's voluntarist model of human agents forever "compelled" by seemingly extraneous motives and desires and incapable of transcending the scope and intensity of these factors in either thought or action.[41]

Often at his best when subjecting another writer's position to rigorous logical critique, the late Coleridge of the *Opus Maximum* thus points out how Hobbes's reduction of the self to an embodied will, taken on its own terms, effectively denies the reality of mind itself inasmuch as it collapses thinking into mere computation. To adopt such a perspective on the life of the mind imperils the reality of the individual person in whom that life is said to unfold. It also dissolves the ethical and imaginative dimensions of all thinking—which (if the term means anything) can never be merely calculative but is both generative of new meaning and reflexively aware of that very fact; thinking, in other words, is never merely concerned with the determination of means but with the counterfactual imagination of as yet unrealized possibilities, which in turn can only acquire significance when evaluated within a framework of meaningful and compelling ends. The following, long passage from the *Opus Maximum* thus targets the meliorist psychology of self-interest that the Scottish political economists had advanced, seemingly in opposition to Hobbes though in effect replicating his opaque and inarticulate voluntarism. As in *Aids to Reflection,* the cardinal issue to which Coleridge returns time and again is the categorical inadmissibility of material, non-spiritual factors as supposed causal determinants of human thought. To impute some a priori determinacy to our mental life—be it in the form of external constraints or internal compulsions—inexorably leads one to deny the reality of mind itself. It is a consequence that those targeted by Coleridge's critique clearly do not intend and that ultimately vitiates their own political and moral objectives. Against the post-Hobbesian "scheme which considers virtue as a species of prudence" or self-interest, Coleridge maintains that

> what [the necessitarian] cannot derive from motives of Self-interest he will attribute to impulses of selfishness. Now this argument supposes the plenary causative or determining power in these motives or impulses, so that both the

41. Hedley observes that, "for all his emphasis upon the 'will,' Coleridge is not a voluntarist. He does not affirm the will over against reason" (*Coleridge*, 10); on Coleridge's opposition to the established symbiosis of voluntarism and natural theology (in Boyle, Newton, Paley et al.), see Brice, *Coleridge and Scepticism*, 10–51.

> one and the other do not at all differ from physical impact as far as the relation of cause and effect is concerned. For if it were otherwise, we should still have to ask what determined the mind to permit this determining power to these motives and impulses. Or why did the mind or Will sink from its proper superiority to the physical laws of cause and effect, and place itself in the same class with the bullet or the billiard-ball? It would be most easy to trace this whole mechanical doctrine of causative impulses and determining motives to a mere impersonation of general terms. For what is a Motive? Not a thing, but the thought of a thing. But as all thoughts are not motives … a motive must be defined as a determining thought. But again, what is a Thought? Is this a thing or an individual? What are its circumscriptions, what the interspaces between it and another? Where does it begin? Where does it end? … A motive is neither more nor less than the act of an intelligent being determining itself, and the very watchword of the necessitarian is found to be, in fact, at once an assertion and a definition of frequency, i.e. the power of an intelligent being to determine its own agency. But even this is for us superfluous; it is enough that he who upholds this scheme of universal selfishness or self-interest, not from any corruption but from the original necessity of our nature, implies the denial of a responsible Will.[42]

In his own language, Coleridge here construes the will as an intellectual act in terms familiar from Aquinas's *Summa*. Thinking and willing are at all times distinguished by the way in which they show the mind or person to be at some remove from itself. Human thought is not "locked into" a particular computational matrix; it does not simply "apply" itself to data of supposedly external and immutable character. Rather, it is dynamic and fundamentally "open" toward any variety of possible meanings and valuations. The entire necessitarian language of Newtonian physics, in particular, the causal model of an instantaneous and inexorable transmission of force from A to B, proves in Coleridge's view categorically inapplicable to the mental life of persons.[43] Not only does it lead to an infinite regress ("we should still have to ask what deter-

42. *OM*, 25–26; the critique here offered echoes his earlier, 1816 essay on "Consciousness and Self-Consciousness," where Coleridge had pointed out how "the phrase *'motive'* ~~is~~ has likewise been much abused by the philosophical Necessitarians, as if a motive were a *Thing*, that by impact communicated motion, instead of being a mere generic Term. For what is a motive, but a determining Thought? And what is a Thought but the mind thinking in this or that direction? And what is thinking but the mind acting on itself? A motive therefore = the mind in the act of self-determination" (*SW & F*, 1:399). For an introduction to Coleridge's conception of self-consciousness, see also A. Taylor, *Coleridge's Defense*, 61–86.

43. It is telling that Coleridge here rejects as inapplicable the common analogy of the billiard ball just as emphatically as Schopenhauer insists on its relevance to the mental life. In his 1839 *Prize Essay* Schopenhauer thus remarks how "a human being can no more get up from his chair before a motive pulls or pushes him than a billiard ball can be set in motion before it is struck; but then his getting up is as necessary and inevitable as is the rolling of the ball after it is struck" (39); McFarland

mined the mind to permit this determining power to these motives and impulses"), but it peremptorily denies any awareness-of-self to a mind so conceived and, thus, in effect negates what it purports to explain. For by its very nature, mechanical (or efficient) causality merely *occurs* but cannot know of itself. As Coleridge puts it, "Knowledge without Consciousness is not *a Knowing* but *a Being*" (*SW & F*, 1:427). In sharp contrast with Hobbes's reification of motives as "causes" and, ultimately, mere things, Coleridge insists on their ideational status. Motives are not the other of thinking but a particularly elemental form of it, viz., "a determining thought." By its very nature, then, all thinking amounts to a dynamic and complex sequence of focused imaginings, a "playing out" of various possible realities. As Maurice Blondel was to observe, a "motive is in effect only the repercussion and the synthesis of a thousand mute activities; that is the reason for its natural efficacy. The motive does not appear suddenly, up in the air, so to speak, and as if by spontaneous generation; it is the deputy of a crowd of elementary tendencies that back it and push it . . . Its efficacious charm therefore comes from its expressing and representing precisely that which it moves."[44] Put differently, thinking is prima facie never reactive *to* motives but positively transformative of its (virtual) "object." Far from merely computing what is factually given, it is counterfactual to its very core. As Coleridge also implies, its form is inherently temporal and incipiently narrative ("Where does it begin? Where does it end?") and, thus, stands in sharp contrast to the instantaneity of cause-effect relations in the realm of physics.

With somewhat different emphasis, Coleridge had already formulated his critique of determinism and its "dehumanizing" implications as early as 1803. Here, too, stress is placed on "the individuality of Man"—not "merely man, as every Tyger is simply Tyger"—but as an incommunicable being, a person "more than numerically distinguishable . . . *this* man, with *these* faculties, *these* tendencies," and so on. Yet in what might seem an uncharacteristic concession, Coleridge also notes that how

> each individual *turns out* (Homo Phainomenon) depends, as it seems, on the narrow Circumstances & Inclosure of his Infancy, Childhood, & Youth—& afterwards on the larger Hedge-girdle of the State, in which he is a Citizen born—& . . . the Zone, Climate, Soil, Character of Country and innumerable other factors besides. It thus would appear that the individual is indeed influenced & determined (caused to be what he is, qualis sit = qualified, *bethinged*) by ~~it~~ Universal Nature, its elements & relations.

---

is right to suggest that "the absolute antithesis of Coleridge and Schopenhauer subtends an urgent similarity . . . throughout the entire philosophical spectrum of their thought" (*OM*, clxxxv).

44. Blondel, *Action*, 111–112.

Characteristic of Coleridge's argument is its dialectical, quasi-Scholastic mode of establishing a point by affording the opposing view the strongest possible hearing so as to show that the conclusions to which it gives rise are not the ones ordinarily drawn. The latter, of course, here would be some version of determinism—viz., that the factors shaping the nature of the person constitute an "apparent horizon, & ~~un~~insurmountable," an absolute boundary impossible for the self to transcend. The condition not only frames the human being but, it would seem, drains it of all perspective and of any potential for further development. To quote Coleridge's rather Blakean image, a tough "~~knot~~ Skein of necessities … interwine[s] the slenderest fibres of his Being" and "binds the whole frame with chains of adamant."[45]

Paradoxically, though, it is the very construction of this scenario *as an argument* fashioned by various proponents that shows it to fail on its own terms. That is, rather than articulating a differentiated ontology, the modern materialist and determinist account of life as a seamless continuum of psycho-physiological processes invariably gravitates toward a monistic form of explanation. It construes the myriad manifestations of life—including highly complex instances of human self-consciousness (e.g., anxiety, guilt, desire, hope, despair, love) as discrete and discontinuous "states." Initially, it does so simply by referring these "states" back to specific, quantifiable constellations or spikes taken to stand out from amidst a hypostatized, steady-state chemical and neuro-electric equilibrium. In due course, however, the specific *factum probandum* of a particular state of mind is effectively being *dissolved* into the *factum probans*—say, some measurable chemical "imbalance" or such. This procedure appears incoherent and flawed in several ways. First, as regards method, it is illogical to purport having *explained* something if the explanation tendered consists in denying the phenomenon that had prompted it any reality by discrediting it as a chemically and neurologically occasioned illusion. Second, still at the level of method, a materialist and deterministic "explanation" of human (self-) consciousness is flawed in that it takes *for its point of departure* some concrete *expression* of how the specific psychological state is being *experienced* (e.g., "I am feeling profoundly guilty, anxious, or—worse yet—conflicted"). Extending Hegel's theory of

45. *CN*, no. 4109, f128^{v}; the entry is listed under 1811 but, in the notes, dated for "a period after 1803." Elsewhere, in a critique of Edward Williams's *A Defence of Modern Calvinism* (1812) and *Essay on the Equity of Divine Government* (1809), Coleridge again uses the image of a supposedly "adamantine" chain of logic that fails, not by breaking but by having been attached to the wrong principle; as he argues, Williams lacks "the noble honesty, that majesty of openness, so delightful in Spinoza, which made him scorn all attempts to varnish over fair consequences, or to deny in words what was affirmed in the reasoning … where should I find that iron Chain of Logic, which neither man or angel could break, but which falls of itself by dissolving the rock of Ice, to which it is stapled—and which thou [Spinoza] in common with all thy contemporaries & predecessors didst mistake for a rock of adamant?" (*CL*, 4:548).

recognition and Freud's notion of transference, Jacques Lacan in his Rome discourse on *Speech and Language in Psychoanalysis* (1953) had made a compelling case for how by its very essence human expression and speech unveils "the transindividual reality of the subject." For all speech and symbolic action—simply in virtue of being *action*—attests to a desire for recognition: "the first object of desire is to be recognized by the other." Operations of reference, constative utterances, and the entire, often trivial chatter (Heidegger's *Gerede*) that circumscribes our lived, social existence would be both pointless and unfathomable in the absence of that other. As Lacan puts it, "there is no Word without a reply," and that reply (whether real or merely projected) allows us to understand "the unconscious [as] the discourse of the other." The Freudian symptom thus is not to be construed as some extrinsic material concatenation but, instead, proves isomorphous with the structure of human speech: "the symptom resolves itself entirely in a Language analysis, because the symptom is structured like a Language." Inasmuch as "man speaks ... because the symbol has made him man," consciousness cannot be framed as the unwitting effect of independently operating, efficient causes except insofar as these have been incorporated into the orbit of its hermeneutic and symbolic practice.[46] Invariably, that is, acts of self-description or self-disclosure not only furnish the initial prompt for any account of mind. By dint of their uniquely structured forms of self-representation, they also condition any attempt at isolating the material causes hypothesized to have occasioned the phenomenon under investigation. Reductionism's basic fallacy here is what the ancients called *metabasis allo genos*—an impermissible leaping from one level of reasoning into an entirely different one. For its account of mind misconstrues as neutral, strictly factual, and quantitative "evidence" those aspects and cues furnished only by the subject's richly layered and symptomatic expressive acts. Neutral "facts" and material "evidence" in the analysis of human consciousness are but a willful and one-sided extrapolation of "causes" from human symbolic practice, a realm wherein fact and value are inextricably entwined.

To "understand" someone's claim that she or he feels overwhelmingly anxious, or has religious visions, or fears of persecution or, perhaps, all three at once, is to enter a hermeneutic circle wherein claims of distinctive conscious states are intelligible only if situated within various, often overlapping narrative frames such as correlate with a person's accumulated experiences, habits, goals, fears, desires, etc. One might, for example, wonder whether a claim at emotional distress is expressively shaped with a view to its intended audience, perhaps reflecting the speaker's deep-seated insecurity, now discharged as an unwitting or unconfessed desire to be selected, say, for some neuro-scientific study. In any event, the meaning of a reported experience, as

46. *Speech and Language*, 19, 31, 9, 27f., 39.

well as the distinctive expressive quality that the report itself takes, is saturated by the given individual's history of socialization, education, and a vast array of experiences (both conscious and oblique) that may contingently factor into the particular expressive act now being scrutinized for its somatic underpinnings. For his part, Coleridge proves keenly aware of the methodological and hermeneutic blindness of the "corpuscular philosophy." Indeed, his critique seems rather prescient of the serious conceptual tensions (not always honored in open debate) between variously hermeneutic, phenomenological, and neuro-scientific accounts of human consciousness. As he sees it, any deterministic conception of human life and agency (its own claims notwithstanding) must itself be appraised as a specific hypothesis, a value-driven and value-fraught hermeneutic act for which it is impossible to claim purely factual neutrality. Hence, the very project of a materialist and determinist explanation of the human simply cannot claim to have originated *ex nihilo* and to be free of all hermeneutic "fore-meanings" (Hans-Georg Gadamer). And yet, from Francis Bacon's and Robert Boyle's anti-teleological framing of a natural science to Edmund Husserl's notion of "bracketing," or *epochē,* modernity habitually authorizes itself by some such trope and the extended rhetorical figure of a radical historical caesura, a break *with* the idea of tradition and, hence, by what it takes to be its decisive emancipation *from* and overcoming *of* all historical and hermeneutic contingency.

Offering an early and uniquely incisive critique of these key axioms of modernity, Coleridge certainly understands that to recover a model of rational personhood within the modern era means resisting constructions of the self as a merely contingent realization of material traits underlying the species "human being" at large. To confuse the value-saturated incommunicability of *person* with the generic and abstract species-concept of *human being* contingently realized as so many "individuals" (Schopenhauer dismissively speaks of man as "nature's mass-produced commodity" [*der Mensch, diese Fabrikware der Natur*]) is to have lost one's bearings as a rational and responsible agent.[47] For in failing to grasp this foundational distinction between persons and things, one will implicitly have endorsed the view of the fact/value opposition as something ontologically given rather than a claim contingently advanced at a particular moment in history. Against modernity's self-certifying narrative, which celebrates the transformation of knowledge from an interpretive and morally responsible hermeneutic into an open-ended quest for (ostensibly value-neutral) information, Coleridge insists that the modern project is itself fueled by a deep-seated narrative motivation to transform and control the ways in which its historical conditions of emergence will henceforth be appraised. If one were to acknowledge the modern scientific project, not as a monist truth but as a hermeneutically conditioned

47. Schopenhauer, *Welt als Wille,* 1:268 (§36).

thesis and evolving *argument*—however rigorous and lucid its methodology and practices—any reductionist account of human agency would have to meet two distinct kinds of intellectual and conceptual responsibility. First, it would have to demonstrate the internal coherence and consistency of its premises, procedures, and claims. Yet while these procedural issues have of course been understood as the bread and butter of modern science since Bacon and his contemporaries, another kind of accountability continues to be not so much met as it is peremptorily rejected. For a scientific explanation of human agency to succeed, it will also have to account *for* (not explain *away*) the fact that to experience a particular quality of consciousness—even a sensation or feeling as rudimentary as "hunger" (to recall Robert Spaemann's example)—is not the same as to register a value-neutral shift in one's psychophysiological constitution. For even as hunger in most instances is bound to involve some measurable drop in blood-sugar levels, followed by appropriate neural signals traveling to the relevant circuitry of the brain, the sensation of it is also, and just as immediately, identified as one that I am now *having;* and in so disclosing itself as *my* experience, the sensation is inevitably absorbed into a web of cross-references and nested within continuously unfolding narratives that define me as a person.

Put differently, a specific sensation or thought never takes place as a *punctum*—isolated in space and time from all contiguous matter. Rather, for it to be recognized as *my* hunger means that the sensation of it becomes instantaneously enmeshed with a host of value-representations. If I suffer from an eating disorder, it may trigger feelings of guilt, shame, or even a suspicion that the sensation itself may not be fully trustworthy. Then again, if it happens to be Lent, I may conceivably welcome the feeling of hunger as an opportunity for deepening my religious commitments and spiritual achievement; or hunger may prove acutely unwelcome because I reflexively observe how its nagging presence disrupts my concentration on the sentence I am just now writing; or again it may be something I proceed to scrutinize as I evaluate whether to satisfy it with a quick but uninspired meal at the faculty commons or much better fare at a restaurant that I can only frequent by missing a likely unpleasant scheduled meeting with a student who has been pestering me with emails about a poor grade. Simply put, all conscious states are subject to some form of appraisal, be it reflexive and explicit or more tentative and, perhaps, even subliminal. What the materialist and reductionist account fails to grasp is the categorical divide between the material "event" of consciousness and its infinitely complex, layered, and richly evaluative internalization; Coleridge elsewhere calls it "the natural differences of *things* and *thoughts*" (*BL,* 1:90).

As the notebook entry resumes, Coleridge thus draws attention to the myriad ways in which ostensibly neutral and seemingly implacable constraints *on* conscious human existence will imperceptibly transform into concerns *for* that very consciousness. Any fact is a fact *for* someone and, consequently, is the bearer of value:

> And yet again, the more steadily he contemplates this fact, the more deeply he meditates on these workings, the more clearly it dawns upon him that this conspiration of influences is no mere outward or contingent Thing, that rather this necessity *is* himself, that that without which or divided from which his Being can not be even *thought,* must therefore in all its directions and labyrinthine folds belong to his Being, and ~~enter into~~ evolve out of his essences. Abstract from these—and what remains? A general term, after all the ~~notices~~ conceptions, notices, and experiences represented by it, had been removed—an Ens logicum which instead of a *thought* <or Conception> represents only the act and process of Thinking, or rather the form & condition, under which it is possible to think and conceive at all.[48]

In phenomenological terms, that external "Skein of necessities" is itself an intentional object, less a constraint or "insurmountable" horizon than a "stimulus" for the human intellect to reason upon. Indeed, were it not for those resistances, constraints, and abrasions that a determinist mistakes for absolute and non-transcendable conditions of human life, it is hard to fathom just what it might be that the self qua person could even reflect *on.* Already in the *Biographia,* Coleridge had warned against "the mistaking the *conditions* of a thing for its *causes* and *essence*" (*BL,* 1:123). As it turns out, "condition" is itself a descriptive, interpretive, and ultimately evaluative concept rather than some extrinsic, value-neutral cause. In an argument that curiously foreshadows Nietzsche's critique of objective, "scientific" explanation in *On Truth and Lies in an Extramoral Sense* (1872), Coleridge shrewdly dismantles the deterministic project by showing human beings to have at all times a unique appraisal of their existence. What Heidegger was to call "being-in-the-world" (*in der Welt sein*) points to precisely this ontology of understanding (*Verstehen*) anterior to any particular factual and material constellation that might emerge as its object. Coleridge's notebook entry continues: "The more he reflects, the more he finds it, that the stimulability determines the existence & character of the Stimulus, the Organ the object" (*CN,* no. 4109, f129^{v}). Any totalizing, deterministic conception of human life winds up draining of all meaning that which it purports to explain. Accounts of this type (from Julien Offray de La Mettrie's *L'homme machine* to contemporary cognitivism) logically

48. *CN,* no. 4109, f129; Coleridge here reiterates his earlier discussion of mechanism as in effect negating the very notion of person, soul, will, or some version of *incommunicabilis*: "The soul becomes a mere ens logicum, ... present only to be pinched or *stroked* ... Accordingly, this caput mortuum of the Hartleian process has been rejected by his followers, and the consciousness considered as a *result,* as a *tune,* the common product of the breeze and the harp: tho' this again is the mere remotion of one absurdity to make way for another, equally preposterous. For what is harmony but a mode of relation, the very *esse* of which is *percipi*" (*BL,* 1:117–118).

fail as explanations precisely because in seeking to dissolve human agency into a web of causal determinants they end up denying the very reality of what they claim to elucidate. Such a stance may amount to a hypothesis but surely does not constitute any proof of it. In exposing determinism's habit of dissolving a priori the very reality of the object *of which,* paradoxically, it purports to offer an explanation, Coleridge thus exclaims: "What then remains! O the noblest of all—to *know* that so it is, and in the warm & genial Light of this knowledge to beget each in himself a new man, which comprehends the whole ~~in~~ of which this phænomenal Individual is but a component point, himself comprehended ~~only~~ in God—alone."[49] One cannot but admire the dialectical brilliance and verbal precision with which Coleridge here teases out conclusions at once necessary and yet diametrically opposed to those that the deterministic picture of human agency has usually been taken to license or, indeed, compel.

Coleridge's intimation—later expanded in his scrupulous reflections on the nature of the will—that "Man [makes] the Motive" holds at least two major implications for a fuller understanding of personhood.[50] First, even the most totalizing deterministic account, simply because it is an *account*—that is, an interpretive or hermeneutic stance vis-à-vis the world—necessarily premises the reality of the person as a *self*-conscious being, rather than the merely reactive, sentient, or instinctual "living *Thing*" to which such accounts seek to reduce it. Second, the very etiology of person, if understood (per Richard's definition) as "a rational existence of incommunicable nature" (*rationalis naturae incommunicabilis existentia* [*De Trinitate,* 4.23]), shows it to be constitutively aware of its relational character, its ontological embeddedness in a community of persons and, thus, *having* "world" rather than merely being embedded in a specific kind of "environment." Even in the most hardened reductionist portrayals of human agency as some foreign-determined and self-enclosed biomass (tenaciously clinging to chimeras such as intellectual autonomy, imagination, choice, and responsibility), the gloomy scenario in question only signifies inasmuch as it is elaborated *for another.* Being constitutively interpretive of the world, of others, and of its own incommunicable self, person is ontologically constituted as a "being who must form attitudes [*das stellungnehmende Wesen*]."[51] As Hegel (who along with Nietzsche strongly influenced Arnold Gehlen's philosophical anthropology) would furthermore insist, the "attitudes" or "perspectives" that arise from the human being's hermeneutic constitution belong themselves to the features of its world that solicit further "attitudes" (*Einstellungen*) and responses from others. Our

49. *CN,* no. 4109, f129^{v}; on the self-dismantling logic of materialist and associationist models, see also *BL,* 1:119, 133–136.

50. "If the will originate in motives, in what do the motives originate? ... It is not the motives [that] govern the man, but it is the man that makes the motives" (*OM,* 33).

51. Gehlen, *Der Mensch,* 32; Eng. *Man,* 24.

interpretations are *eo ipso* oriented toward another or, as Coleridge will put it, they presuppose "an equation in which '*I*' is taken as equal to but yet not the same as '*Thou*'" (*OM,* 75). The nature of the human mind as it constructs perspectives on its world thus proves inseparable from its relation to other persons *for* whom it seeks to articulate its interpretations. Yet to do so, the person must unconditionally acknowledge the reality, equality, and dignity of that other. Put differently, the "thesis" of a self-conscious *I* amounts to "an equation of *Thou* with *I* by means of a free act [of the Will] <by> which <we> negate [sic] the sameness in order to establish the equality" (*OM,* 75–76). In working out his post-idealist model of human agency—one in which personality is constitutively (rather than electively) relational—Coleridge may be drawing on F. H. Jacobi. As early as 1785, Jacobi had taken the overtly anti-Cartesian view that "without the *Thou,* the *I* is impossible" and "that the *I* and the *Thou* ... must be present at once in the soul even in the most primordial and simple of perceptions—the two in one flash, in the same indivisible instant." In one of his most incisive works, a critique of *David Hume on Faith* (1787), Jacobi had gone so far as to make the "distinctness" and clarity of the self's representations contingent on a Thou: "the 'I' becomes more distinct in equal measure as the 'Thou' does. There arise concept, word, person [*In demselben Maaße wie das Du deutlicher wird, wird auch das Ich deutlicher. Es entsteht Begriff, Wort, Person*]."[52] With good reason, then, Coleridge's strikingly modern conception of the I-Thou relation functions as a template for the intrapersonal dynamic of "conscience" that dominates much of his late writings. For only by illuminating the ethical, relational, and intellectual structure of "conscience" can that term be salvaged from the self-cherishing and self-licensing notion of religious enthusiasm and hyper-Augustinian evangelicalism clearly on the ascendant in the 1820s; and, more crucially yet, only if the structure of conscience has been properly articulated will Coleridge's ontological characterization of person as a "*responsible* Will" have been completed.

---

52. Jacobi, *Philosophical Writings*, 231, 277, 319; the first two quotes are from Jacobi's highly controversial 1785 letters *Concerning the Doctrine of Spinoza*, addressed to Moses Mendelssohn; on the I/Thou distinction set forth there, see also di Giovanni's fine Introduction to his translation of Jacobi's writings, esp. 63–65 and 92–94.

20

"FAITH IS FIDELITY ... TO THE CONSCIENCE"

Coleridge's Ontology

Like most of those who, since late antiquity, participated in the ongoing clarification of person as an ontological idea, Coleridge emphasizes that the reality of the human being depends on an act of "recognition." Beginning with his sharply worded, though always carefully reasoned arguments against the practice of slavery, Coleridge had understood "recognition" not merely as some abstract metaphysical injunction but, like person itself, as something woven into the very fabric of human existence. It is not something electively introduced into the empirical reality of communities and interpersonal relations but constitutive of that very reality. For even to imagine that one might not acknowledge the other as person but, instead, willfully treat him or her as a "thing" would strike us, no less than Coleridge, as inherently "dehumanizing"—not only of the other's but also of our own self.[1] Such abuses may well happen—indeed they often do—but they can never plausibly be justified as such but only by laborious schemes of circumlocution and re-description. Hence to say that recognition implies the practical acknowledgment of the other as being of equal *dignity* as the "I" means not so much to have advanced a neutral and formally

1. The *OED* credits Coleridge with the first use of "dehumanizing," a word that appears in a notebook entry of 24 March 1808 (*CN*, no. 3281); see A. Taylor, *Coleridge's Defense*, 13.

contestable *claim* but to have grasped how the empirical, inter-subjective realm is saturated with normative values or ideas. Inasmuch as questions of community and relationality are matters of truth, not correctness, normativity signifies not in the manner of a "thou shalt" but, instead, points to the nature of the real itself and thus proves immune to some counterfactual scenario. What "recognition" of the other qua person thus denotes, and what renders it normative or "transcendent"—as opposed to some subjective "moral" choice or preference—is the intercalation of ontology and ethics. Recognition attests to the absolute givenness of community and, consequently, the incontestable reality of the good—to be conceived not as a speculative hypothesis but as an invitation, a possibility, a gift. Negatively put, community is not simply a function of *technē*; it can never be achieved by the conceptual and propositional logic of political argument. Nor should it be reduced to some distant utopia to be realized by the vociferous and adversarial transactionalism of liberalism's so-called public sphere (Jürgen Habermas's *Öffentlichkeit*). For while open and earnest debate over what constitutes a just, equitable, and humane community is a crucial component of our collective flourishing, its underlying prompt is that community is not simply a "construct" (or "contract") but something ontologically given; and human thought can (and ought to) relate to this very givenness with ever increasing articulacy.

For Emmanuel Levinas, community is "produced within the general economy of being only as proceeding from the I to the other, as a *face to face*" (*TI,* 39); its core unit is the reality of the person in relation to a Thou, not an abstract political theorem. Opposing Heidegger's *Fundamentalontologie,* Levinas insists that for us to understand the I as person "it is necessary to begin with the concrete relationship between an I and a world" (*TI,* 37). Though working from within different intellectual and religious genealogies than the ones traced thus far (the exception being a shared engagement with phenomenology), Levinas reaches remarkably similar conclusions to those progressively articulated by Boethius and Richard of St. Victor. First and foremost, the I or self cannot be grasped by a unilateral and self-certifying act of definition. Levinas's objection here is not simply epistemological in nature. Thus he regards the modern ideal of "autonomy" as both epistemologically incoherent and, to the extent that one seeks to pursue it all the same, deeply unethical. An extreme version of that project would be Johann Gottlieb Fichte's deduction of I and non-I in the 1794 *Science of Knowledge,* for it causes alterity to be "reabsorbed into my own identity as a thinker or a possessor." In sharp contrast to such models of selfhood as *proprietas* and *dominium,* Levinas locates the essence of personhood in what he calls "metaphysical desire." Tending "toward *something else entirely,* toward the *absolutely other,*" such desire is aimed at the realization of a potential within the desiring subject, rather than at extinguishing the desire itself through an act of possession: "It is like goodness—the Desired does not fulfill it, but deepens it" (*TI,*

33–34). The Platonic background (frankly acknowledged in the opening pages of *Totality and Infinity*) subtends much of what follows, albeit only up to a point that remains to be demarcated. The relationship between the desiring subject and that which it desires does not pivot on "the disappearance of distance" but on what Levinas calls "generosity." Metaphysical desire is most consummately realized in the reciprocal acknowledgment of the other by the "I" (and, hence, of the I as its own other) in the face-to-face encounter. Crucially, desire of this nature is not pragmatic; it does not seek fulfillment but proceeds "aimlessly ... toward an absolute, unanticipatable reality" (*TI,* 34). The adverbial qualifier ("aimlessly") is crucial here in that it shows the encounter with the other, and indeed with life, to depend on the Zen-like emptying out of self, a suspension of all pragmatic objectives, beginning with modernity's axiomatic objective of total self-possession and its concurrent assertion of rights. In alerting us to the sheer "irreducibility of movement to inward play, to a simple presence of self to self" (*TI,* 35), what Levinas calls "transcendence" or "absolute exteriority" bears remarkable affinities to Coleridge's conception of personhood.

Central to both Coleridge's Platonizing and Levinas's existential approach is the relational and reciprocal embeddedness of persons *in* and *as* a community. Crucially, relation here neither seeks to integrate the other into the self nor to bring about some virtual (imaginary) identification on the order of Adam Smith's sympathetic community. The objective here is not one of appropriation, transference, or some other dialectical or psychoanalytic trope by which to systematize and methodically delimit the I-Thou relation: "The absolutely other is the Other [*L'absolument Autre, c'est Autrui*]. He and I do not form a number. The collectivity in which I say 'you' or 'we' is not a plural of the 'I.' I, you—these are not individuals of a common concept" (*TI,* 39). In ways strikingly analogous to Trinitarian theology, the other saves the I from the perennial threat of solipsistic and totalizing projection. To imagine—whether in Platonic, Judaic, or Christian discourse—an absolute good is, therefore, to imagine that being eternally contains an other, a reality of equal dignity, and that it acknowledges that alterity as such—viz., in the modality of an infinite *relation* rather than some anticipated utopia in which that other is to be re-absorbed into the One.[2]

2. Clearly, it is the question of the *eschaton* that ultimately divides neo-Platonic and Christian thought from its Judaic counterpart. Yet it does so only in the context of the relationship borne by the human to the divine. As for the relational character of the Trinity, no such collapsing of the distinct Persons into one another is either envisioned or even held to be desirable. My argument in what follows is that it is the Trinitarian model that, for Coleridge (as for Richard, Boethius, and Augustine before him) constitutes the ideal of human community. By contrast, Coleridge tends to be ill at ease around end-of-times fantasies of an Evangelicalism just beginning to impact English religious culture, just as he remains wary of the irrational tendencies in Unitarianism, pantheism, and other offshoots of radical Protestantism. Though working overwhelmingly from within a Christian, Trinitarian tradition, the later Coleridge was deeply interested in Judaic thought, and he greatly

Essential to the idea of the divine and the good is thus the notion of infinity; or, as Levinas puts it, "the idea of the perfect is an idea of infinity." By nature, infinity transcends the realm of the discursive and its proprietary outlook on the business of thinking and conceptualization: "perfection exceeds conception, overflows the concept; it designates distance" (*TI,* 41). For Coleridge as for Levinas, this "transcendence" or "transascendence" (*TI,* 35) constitutes the innermost essence of thought itself. It is radically imaginative, counterfactual, and indeed utopian in nature—*not* by opposing the "imperfect" or envisioning its eventual "correction," but as an openness to the unknown, the unpredictable, the unsought-for within the "I" itself. Whereas "negativity is incapable of transcendence," the idea which alone enables mind to participate in the good pivots on the self opening itself to the unconditionally other. Levinas's formulation is strikingly similar to Coleridge's neo-Platonic notion of reason (*logos*) as the *imago dei:* "Transcendence designates a relation with a reality infinitely distant from my own reality, yet without this distance destroying this relation and without this relation destroying the distance" (*TI,* 41).

To help identify the point at which, crucially, Coleridge's and Levinas's models of person and community diverge, some additional moments of significant convergence first need to be specified. Against Heidegger's *Fundamentalontologie,* which "affirms freedom over ethics," thereby "neutralizing the existent in order to comprehend it" (*TI,* 45–46), Levinas develops a fundamentally mystical, anti-modern stance. Though well aware that Heidegger's objective in *Being and Time* had been to undertake a comprehensive critique of modernity's "obliviousness of Being" (*Seinsvergessenheit*), Levinas rejects the abstract and value-neutral understanding of *Sein* that Heidegger seems intent on recovering. Anticipating the sharp critique of the fact/value distinction by writers like Gertrude Elizabeth Anscombe, Iris Murdoch, and Alasdair MacIntyre, Levinas rejects modernity's leading paradigm of knowledge as "thematization and conceptualization." For in taking possession of the object within a categorical framework we invariably account for it in terms of what it is *not* (as per Spinoza's *omnis determinatio est negatio*), thereby conceiving knowledge as something altogether "impersonal." The result, in Levinas's strident phrase, is a "philosophy of injustice" (*TI,* 46). Plato (as indeed Coleridge himself) might have simply

---

valued his friendship with Hyman Hurwitz (1775–1844), a Polish émigré, scholar of Hebrew scriptures, and author of *Vindiciae Hebraicae, Being a Defence of the Hebrew Scriptures as a Vehicle of Revealed Religion;* for information on Coleridge's friendship with Hurwitz, see *CM,* 2:1188–1189, and *CL,* 5:xxxvif. Coleridge translated two of Hurwitz's *Hebrew Dirges* and read through the *Vindiciae Hebraicae* "sentence by sentence." Expressing his delight to John Murray "that a learned, unprejudiced, & yet strictly orthodox Jew may be much nearer in point of faith & religious principles to a learned & strictly orthodox Christian, of the Church of England, than many called Christians" (*CL,* 5:92), Coleridge conducts some of his most searching epistolary discussions of theology with Hurwitz; see esp. his long letter of 4 January 1820 (*CL,* 5:1–9).

called it a philosophy that no longer offers a conceptual or imaginative space for love—which might itself be the most salient characteristic of philosophical modernity. Levinas's alternative project of retrieving "a non-allergic relation with alterity" by maintaining "within anonymous community, the society of the I with the Other—language and goodness," must not be confused with nominalism's quest for the ultimate particular. For the latter's methodological restriction of knowledge to warranted and "certain" propositions about singular entities is itself prompted by the deep-seated anxiety that some aspect of the other—now framed as an "object"—might yet elude our conceptual possession of it.

By contrast, the human person, whose incommunicable and relational nature a millennium of theological reflection had gradually distilled as its essential trait, brings us face to face (no mere figure of speech) with a singularity that can be known only if we relate to it in a non-possessive, non-proprietary mode. For Levinas, the face of the other is the most concrete instance of infinity. Once again, we must be on guard against modernity's almost axiomatic appraisal of infinity as a utopia, the "not-yet" of continual "progress" toward some ultimate, speculative vantage-point where material reality and our conceptualizations of it shall converge. Even a writer as committed to the modern project as George Eliot had questioned precisely this "tendency, created by the splendid conquests of modern generalization, to believe that all social questions are merged in economical science, and that the relations of men to their neighbours may be settled by algebraic equations."[3] The Platonic and Christian (personalist) conception on which, at this point in his argument, Levinas is relying to a surprising degree, does not conceive infinity as mere "deferral" or "projection" but as something in which we may participate to the extent that we reject appropriative models of knowledge. Repeatedly drawing on the *Phaedrus,* Levinas thus notes how to "think the infinite, the transcendent, the Stranger, is … not to think an object" but, rather, to acknowledge that the very "*ideatum* [of infinity] surpasses its idea." Singular, inescapably present, and in its sheer charisma (Grk. *χάρις* = grace, loveliness, gift) soliciting our utmost engagement, the face of the other "concretiz[es] the idea of infinity." It is a "living presence, … expression" and, indeed, "already discourse." Here "meaning is not produced as an ideal [value-neutral, objective, impersonal] essence; it is said and taught by presence" (*TI,* 50, 66).

While overtones of Jacques Lacan (particularly the Rome discourse on "Speech and Language in Psychoanalysis") are unmistakable here, Levinas's aim is fundamentally different. It is not to solve a puzzle or remediate a psychological dilemma or impasse. Nor indeed does he seek to recover through patient exegesis some presupposed *symptomatic* formation. In short, Levinas's "encounter" is not a variation on

3. "The Natural History of German Life," in *Selected Essays,* 112.

modernity's paradigmatic understanding of knowledge as analysis and critique, both fueled by a desire to achieve certainty by means of detachment and prevarication. Instead, and echoing Coleridge's distrust of an interventionist and transactional epistemology pursued by a "finger-active, brain-lazy" understanding (*CM*, 2:648), Levinas seems to be formulating a mysticism of deeply personal and incommunicable dimensions. Its dominant mode, for which Platonic "love" (as *ēros* and *agapē*) provides the initial template, is one of *relation*, which he understands as participation in alterity. To know for Levinas involves neither the attempt nor the presumption to overcome the distance between the self and the face of the other; rather,

> the way in which the other presents himself, exceeding *the idea of the other in me*, we here name face ... The face brings a notion of truth which, in contradistinction to contemporary ontology, is not the disclosure of an impersonal Neuter, but *expression* ... The first content of expression is the expression itself. To approach the Other in conversation is to welcome his expression, in which, at each instant he overflows the idea a thought would carry away from it. It is therefore to *receive* from the Other beyond the capacity of the I, which means exactly to have the idea of infinity. But this also means: to be taught. The relation with the Other is a non-allergic reaction, an ethical relation. (*TI*, 50–51)

In multiple senses, Levinas's argument bears on Coleridge's exploration of the I-Thou dialectic, which in turn elucidates the ontological status of "conscience." First, the above passage reflects his resistance to the axiom that the legitimacy of knowledge pivots on some invariant, abstract, and putatively self-evident method. The mystic in Levinas (no less than in Coleridge) resists the contraction of knowledge to what is methodologically licensed and verifiably apparent. Hence the operative tropes in Levinas's argument all point to a moment of radical *unpredictability* and, consequently, to a quality of excess notably linked to the idea of *ekstasis*—itself a recurrent motif of Christian mysticism from Augustine to Meister Eckhart to William Blake.[4] The true encounter involves not just being face to face with the other; rather, it is realized at the precise moment when he or she "*exceed*[s] the idea of the other in me." In markedly Platonic and, even more so, Augustinian terms, Levinas's other is a gift (*donum*) inasmuch as I "*receive* from the Other beyond the capacity of the I."

Reflecting on the quality that, for Plato, defines the relation of consciousness to ideas, Levinas remarks how "thought, for Plato, is not reducible to an impersonal

4. As Blake puts it on Plate 7 of *The Marriage of Heaven and Hell*, "He whose face gives no light, shall never become a star" (*BPP*, 35); for a (rather breezy) discussion of face and person in modern literature and psychoanalysis, see Johnson, *Persons and Things*, esp. 94–105 and 189–197.

concatenation of true relations, but implies persons and interpersonal relations." He extends this observation into a fiercely anti-rhetorical conception of thought and discourse: "Justice ... is access to the Other outside of rhetoric, which is ruse, emprise, and exploitation. And in this sense justice coincides with the overcoming of rhetoric" (*TI*, 71–72). Such a mystical account, however, seems more Plotinian and neo-Platonist in character than Platonic. Not even Augustine, and certainly not Plato, stigmatizes, let alone proscribes rhetoric in this manner, as even a casual reading of the acutely and self-consciously rhetorical delight in moving beyond rhetoric (e.g., *Phaedrus*) or a more pointed and rhetorically skilled questioning of the Sophists' rhetoric (e.g., *Gorgias, Sophist*) makes abundantly clear.[5] If to receive in this sense means "to be taught," then learning itself must be rethought. No longer does it involve the unilateral acquisition or harvesting of information. Rather, it involves a *transformation*. Crucially, however, we are not to envision here some transformation *of* the self but, rather, the unveiling of something anterior, more significant, more true than anything that could be ascertained in terms controlled by the self. Introducing (however inadvertently) a key trope of high Scholasticism, Levinas calls it "participation"—that is, "a way of referring to the other, ... to have and unfold one's own being without at any point losing contact with the other." Notably, Levinas also speaks of a "conversion of the soul to exteriority, to the absolutely other, to Infinity" (*TI*, 61). Inasmuch as this mystical "encounter" and the community it instantiates appears to belong to the realm of metaphysics, it bears remembering that the encounter can only ever be conceived apophatically and in explicit departure from the conventions of quotidian, referential speech: "to signify is not to give," he insists, yet as Coleridge's example of the infant-mother dialectic seeks to show, the meaning of the other can only be realized as a gift. Opposing strictly transactional models of community, Levinas thus insists that "meaning is not produced as an ideal essence; it is said and taught by presence" (*TI*, 66).

Levinas's overarching concern with "exteriority" bars him from construing the "encounter" or "relation" in terms of inter-subjectivity. In his thought, relation is not a derivative but a founding concept, for it alone constitutes and consummates the existents as ethical beings. By contrast, certainly in Levinas's account, the proposition of a hermetically enclosed, modern subject on the order of Descartes's *cogito*, Locke's "consciousness," Kant's "apperception," or Fichte's *Tathandlung* fatally divides epistemology from ethics. Levinas also holds that an ethical relation cannot be recovered after the fact— say, as some form of social contract or as a sympathetic community anchored in "moral sentiments" supposed to counteract modern society's transformation into a mere enterprise association. For to construe relations solely in legal,

5. On this issue, see Zuckert, *Plato's Philosophers*, 322–332 and 532–545.

affective, deontological, or utilitarian terms is to have already (mis)conceived person as a non-cognitive and somatically conditioned will incapable of self-transcendence and, hence, in need of perpetual supervision and "management" by disciplinary, institutional mechanisms. Such a self, as Plato and, echoing him, Levinas point out, remains forever encased by its "interiority"—itself but an entropic space of unfathomable desires and compulsions and, as Hobbes saw it, leaving the person bereft of the space for (self-)transcendence—viz., the "clearing" (Hölderlin's *das Offene*, Heidegger's *Lichtung*)—where the putative determinants *of* consciousness might be transmuted into possible interpretations, choices, and judgments *by* consciousness. Sociality in the modern sense is thus never thought as anything but a *correlation* of irremediably hermetic, not to say, autistic subjectivities. Yet, as Levinas puts it (in italics), "*correlation does not suffice as a category for transcendence*" (*TI*, 53).

A fascinating convergence of early patristic thought with Jewish mysticism (ancient and modern) opens up here, one that directly bears on the ongoing debate concerning Augustine's alleged understanding of person as "inwardness." In a widely quoted article, Joseph Ratzinger has proposed a conception of personhood as relationality that substantially mirrors Emmanuel Levinas's and Martin Buber's accounts. Recalling the derivation of person from the Greek *prosōpon*, Ratzinger notes how the "dialogical roles introduced by the prophets are not mere literary devices. The 'role' truly exists; it is the *prosōpon*, the face, the person of the Logos who truly speaks here and *joins* in dialogue with the prophet." To inhabit a role is not, as modern theories of performativity frequently imply, evidence of a merely acquired, transient, and *ipso facto* inauthentic role; on the contrary, it points toward a deeper realization of the person as relational, as "pure act-being." In Trinitarian theology, persons thus are emphatically "not substances that stand next to each other, but they are real existing relations, and nothing besides."[6] Even so, the Trinitarian idea of community as relation in its very essence risks presenting us with a vision that is categorically unattainable, which is to say, "transcendent" in the modern sense of something manifestly implausible, if not positively utopian. Ratzinger's overriding concern here is to counter a tendency within formal Trinitarian theology that had "limited these categories [of existence and relation] to Christology," thereby effectively "treat[ing] the whole thing as a theological exception, as it were." Yet Christ, he goes on to note, has to be understood "as the true fulfillment of the idea of the human person, ... not the ontological exception." Then, and only then, does the Trinitarian model of Christ signify *for us*, viz., "as an indication for theology of how person is to

6. Ratzinger, "Retrieving the Tradition," 442, 444. Levinas stresses the infinity that such a model implies by pointing out how, when understood as relationality, "truth ... does not undo 'distance,' does not result in the union of the knower and the known, does not issue in totality" (*TI*, 60).

be understood as such."[7] We recall Coleridge worrying that the exceptionalist views driving conceptual, systematic theology are liable to eclipse the founding ethical motive behind the entire enterprise: "Christ must become Man—but he cannot become us, except as far as we become him" (*CL*, 4:849). Where Levinas rejects the Cartesian *cogito* and the idea of the human being qua "interiority" (a "separation ... from historical time in which totality is constituted" [*TI*, 55]), Ratzinger likewise warns against misconstruing person as "a substance that closes itself in itself." Drawing on Carl Andresen's scrupulous research into the (in part) Judaic origins of the Trinitarian conception of personhood, Ratzinger also acknowledges its sources "in the Judaism of antiquity, in which the idea is already formulated that the emissary, inasmuch as he is an emissary, is not important in himself" and that, consequently, "a word is essentially from someone else and toward someone else; word is existence that is completely path and openness."[8]

That, however, is as far as the similarities go. Levinas's argument now takes a unique turn away from theistic speculation and, in so doing, allows us to mark the divergence between Coleridge's Trinitarian framework and a post-existentialist mysticism set forth in *Totality and Infinity*. For Levinas, the "noumenon" of infinity "is to be distinguished from the *concept of God* possessed by the believers of positive religions ... who accept being immersed in a myth unknown to themselves" (*TI*, 77; italics mine). His arguments in this regard have been more recently taken up and extended by Jean-Luc Nancy who, in *The Inoperative Community*, insists that "the thinking of community as essence is in effect the closure of the political ... because it assigns to community *a common being*, whereas community is a matter of ... existence inasmuch as it is *in* common, but without letting itself be absorbed into a common substance." Like Levinas, that is, Nancy insists on understanding community as the sustained experience and realization of one's "finitude" and, thus, of "*no longer having, in any form, in any empirical or ideal place ... a substantial identity, and*

7. Ratzinger, "Retrieving the Tradition," 449–450. Ratzinger's central claim that "the human person is the event or being of relativity" (452) stands in close proximity to Levinas's claim that "there can be no 'knowledge' of God separated from relations with men. The Other is the very locus of metaphysical truth, and is indispensable for my relation with God" (*TI*, 78).

8. Ratzinger, "Retrieving the Tradition," 446; Ratzinger here references St. Augustine's commentary on John 7:16 ("Jesus answered them, and said: My doctrine is not mine, but his that sent me." Vulgate: *respondit eis Iesus et dixit mea doctrina non est mea sed eius qui misit me* [Tractate 29, Ch. 7]), where Augustine resolves the apparent contradiction by arguing that the "mission" *is* the person and, thus, is never *owned* or (in Christ's case) indifferently *performed* by a person who takes himself to have independent and autonomous reality and standing: "it seems to me that the Lord Jesus Christ said, 'My doctrine is not mine,' meaning the same thing as if He said, 'I am not from myself.'" See also *De Trinitate*, Book 4, Chapter 29, where Augustine expressly states that in the divine missions "sender and sent are one" (*et qui misit et qui missus est unum sunt*).

sharing this (narcissistic) '*lack of identity*.'"[9] Levinas's and Nancy's unyielding resistance to a language that would merge discrete individuals into a transcendent (communal) essence is a shared, quasi-habitual gesture of post-Enlightenment intellectuals. Both thinkers suspect that it is a small step from imagining a real existent community to the violent expurgation of difference in the name of some mythical essence. Such fears are well founded, not only in light of the twentieth century's horrific record of totalitarian ideology but also, if more subtly, as regards strictly fideist, evangelical models of religion that treat belief in such a transcendent essence as some kind of definitive, counterfactual certitude and spiritual property.

Yet Coleridge, who holds that "a *Conception* of God ... is an Absurdity" (*CM,* 1:237), had long looked with great suspicion on the antinomian and irrational tendencies of radical Protestantism. In the *Opus Maximum,* he makes his misgivings especially clear: "no man serves God with a good conscience that serves him against his reason, [and] in no case can true reason and a right faith oppose each other" (*OM,* 57). Like Newman, whose *Oxford University Sermons* were to echo this view so eloquently just a few years later, Coleridge rejects an exclusively noumenal understanding of faith. Demurring neo-Platonism's institution of an "infinite chasm between the Begotten and the Commanded, the eternal Son and the Creature in Time, ... Proclus especially (the Philosophers with increasing extravagance from Plotinus to Proclus) endeavored vainly to fill up with orders & scales of Gods, Ladders resting at the very footstool of the throne." The principal failure of neo-Platonism lies in its inability to extricate itself from the Gnostic legacy and its consequent inability to develop a phenomenology of empirical consciousness and life oriented toward an unconditional, uncreated good. Instead, neo-Platonism's affirmation of a wholly distinct and separate noumenal realm ("the interest in another Life & another world") hinges on the systematic devaluation, even despair of the finite, empirical realm ("when this Life had lost its charms"). Paradoxically, the long-term result is at once an aggressive valorizing *and* emptying out of the noumenon, manifest in the "gradual conjunction of the speculations of Philosophy with the Passions & idols of Superstition ... till it reached its ne plus ultra in Proclus's School, & *went out!*" (*CN,* no. 3824, f112^{v}–113^{v}). After 1805, the critique of fideist, Unitarian, let alone deist models of religion steadily intensifies in Coleridge's writings. It prompts him to desynonymize "enthusiasm" and "fanaticism" (see his Marginalia on Birch, *CM,* 1:495–496), and in his late work prompts him to oppose any form of organized religion that premises its faith on the opposition of faith and reason:

> I would it were as uncommon, as to every well disciplined mind it is fearful, to hear religionists boast of having sacrificed their reason to their faith, and set up

9. Nancy, *Inoperative Community*, xxxviii.

> against a certain pretence and usurpation of the <mere irrational> understanding ... a pretence to sensible raptures, transports of pain or pleasure ... The utter contrast of this habit and of these principles with that individual faith which demands the first fruits of the whole man, of his intellectual powers, therefore ~~command~~ required of us not our sensations but the subjugation always, the exclusion often, and sometimes the entire sacrifice of our sensations and fancies—that full faith in the intelligential.[10]

Coleridge's objections here substantially anticipate the thrust of Levinas's eventual critique, though with the crucial difference that for Coleridge it is yet possible—indeed of pivotal importance—*to think the noumenon through the phenomenon,* understood as something "given" and by no means implying the imposition of a totality. In an 1818 letter to Hyman Hurwitz, he extends this principle to the language of scripture itself, arguing "that the sacred Writers could only have employed the only permanent, infallible, and suo genere most philosophical, Language, that of appearances" (*CL,* 4:871). Coleridge's ontology of conscience is neither prescriptive nor systematic. Rather, it arises from a description of the infinitely complex and layered phenomenology of human, self-conscious existence. For his part, Levinas presses on, arguing that "the idea of infinity ... is the dawn of a humanity without myths" and that insofar as it has been "purged of myths, the monotheist faith ... implies metaphysical atheism. [It is] an ethical behavior, and not theology, not a thematization, be it a knowledge by analogy, of the attributes of God" (*TI,* 77–78). Levinas's rejection of any noumenal terms in philosophy assumes that ontology inevitably drives toward totality, and that it casts before it the shadow of injustice by tending toward the monolithic and the inhuman. Coleridge's own conclusions in this regard are not dissimilar, though his profound and far-flung grasp of philosophical theology also allows him to see that the true source of such misgivings is to be found in Gnosticism's dualist metaphysics. In the modern era, the same dualism reappears in the guise of Socinianism, Unitarianism, and, eventually, as a deism dressed up in the specious garb of William Paley's *Evidences* and *Natural Theology.* As Coleridge so clearly came to understand, metaphysical totalization had migrated from Scholastic voluntarism into various strains of radical Protestantism, Puritan antinomianism, and late seventeenth-century Pietism according to which the integrity of faith

---

10. *OM,* 180–181. In striking anticipation of Newman's critique of "private judgment," Coleridge had already flagged the intellectual and institutional costs of radically fideist models: "where Private Interpretation is every thing and the Church nothing—*there* the Mystery of Original Sin will be either rejected, or evaded, or perverted into the monstrous fiction of Hereditary Sin, Guilt inherited; in the Mystery of Redemption metaphors will be obtruded for the reality; and in the mysterious Appurtenants and Symbols of Redemption (Regeneration, Grace, the Eucharist, and Spiritual Communion) the realities will be evaporated into metaphors" (*AR,* 297–298).

presupposed what Philipp J. Spener termed the believer's "utter abjection" (*vollständige Zerknirschung*). Such a view renders empirical community and divine authority categorically incommensurable, a deplorable outcome for which William of Ockham's divine command ethic had significantly helped pave the way. To inquire into divine personhood and the significance of the Trinity seemed increasingly pointless, simply because the very notion that the *relation* of the three persons might exemplify or prefigure the ethical requirements and potentialities of human community had been rejected out of hand.

Coleridge's deeply considered Trinitarianism after 1805 does not expose him to the otherwise understandable charges of irrationalism and injustice that cause Levinas to reject Christian theology's conception of the noumenal as strictly exceptional and unintelligible. In discussing Coleridge's striking anticipation of Martin Buber's *Ich und Du* (1923), Thomas McFarland notes that, unlike Buber, Coleridge understands the I-Thou relationship as "a deduction from the nature of consciousness, rather than an axiom of experience" (*OM*, cxxxix). This is certainly true, and it further shows the proximity of Coleridge's conception of consciousness to Levinas's assertion of "the primacy of the ethical" and his claim that "it is not the insufficiency of the I that prevents totalization, but the Infinity of the Other" (*TI*, 79–80). Unlike Buber, Coleridge understands the relation of I to Thou not as an experiential datum but as the ontological matrix of human consciousness that he interprets a fortiori as *self*-consciousness. The very notion of self-consciousness can thus be traced back to that of conscience and, when we do so, reveals to us the implicit presence and categorical anteriority of a "Thou" within the "I," it being understood that the "I" must *act* in order to achieve this awareness of its ethical reality and profound relatedness. In pursuing this phenomenology of the "I" as something that acquires reality and meaning only by virtue of its relatedness *to* and participation *in* the reality of the other, Coleridge is able to link the Trinitarian (ostensibly metaphysical) idea of person to the realm of finite, human experience. To adopt this model is

> to know something in its relation to myself in and with the act of knowing myself as acted on by that something, and proceed to prove the dependence of all consciousness on a self-consciousness, thus: the third pronoun *"he," "it,"* etc. could never have been contradistinguished from the first *"I,"* "me," etc. but by means of the second. There could be no *"He"* without a previous *"Thou,"* and I scarcely need add that without a "Thou" there could be no opposite, and of course no distinct or conscious sense of the term *"I"*" as far as the consciousness is concerned, without a *"Thou."* (*OM*, 74–75)

The I-Thou dyad "is the root of all human consciousness, and à fortiori the precondition of all experience; and therefore ... the conscience in its first revelation cannot have been deduced from experience. Q. E. D." This is what Coleridge means by "the necessity and universality of ~~the~~ relations" (*OM,* 76–77). What Coleridge calls the "equation of *Thou* with *I* by means of a free act [of the Will] <by> which <we> negate the sameness in order establish the equality … is the true definition of Conscience" (*OM,* 76). Not only does this argument hold extraordinary implications for thinking about community—to be explored momentarily—but it also tells us that the self and its other are not related to one another simply by contingent "experience" but ontologically. To speak of a self-conscious person is to imply an inner differentiation that already encompasses a *Thou,* thereby anchoring community in the primacy of the ethical rather than in some contingently negotiated framework of political, legal, or human rights.[11]

At the same time, Coleridge acknowledges that experience can often obscure, even obliterate this ontological relatedness or community by foregrounding "other impulses besides the dictates of conscience." To "preserv[e] our loyalty and fealty against these rivals" is for Coleridge the very essence and definition of faith: "Faith is fidelity, but all human fidelity that is consistent with itself is fidelity to the conscience" (*OM,* 78). This is a powerful argument, to be sure, though one with significant hazards of its own. In particular, Coleridge here risks sliding back into a strictly interiorist model of personhood and, in so doing, to conceive of relatedness and community in a distinctly modern sense after all—viz., as an aggregation or "civil association" based on a social contract or other prudential, calculative alignment of interests worked out among individual "subjects" whose motives Ockham's and Hobbes' voluntarism had rendered terminally self-enclosed, opaque, and irrational. Needless to say, this is not a scenario Coleridge means to endorse; still, his understanding of the dynamics of "conscience" and its claims on the person remains as yet unclear. It is one thing to argue that "consciousness properly human (*i.e. Self*-consciousness) … presupposes the Conscience, as its antecedent Condition and Ground" (*AR,* 125). Yet absent a precise phenomenological account of how the relational, participatory reality of the person qua conscience manifests itself little will have been gained. For conscience to be recognized and acknowledged as the source of personhood it cannot merely be *claimed* to be so but must disclose itself in ways at once distinctive and irrefutable. The difference here is that between metaphysics, where claims are *eo ipso* contestable (albeit only by *other* metaphysical claims rather than by outright falsification) and ontology. Inasmuch as the latter stakes out the very nature of rational existence, its affirmations can neither be verified nor opposed because ontology conditions the very possibility of rational argument and discursive practice, as well as

11. For a different argument that views modern rights not only as commensurable with but as originating in the Western Judeo-Christian tradition, see Wolterstorff, *Justice,* esp. 19–131.

those "notices" or phenomena to which a given philosophical position takes itself to be referring. As regards Coleridge's ideas of person and conscience, his argument resembles Anselm's ontological proof of God. It is not that in referring to our conscience we predicatively affirm its existence. Rather, the nonexistence of conscience is not even conceivable without concurrently disputing the reality and distinctiveness of the reasoning agent, the human person as an incommunicable, rational existence possessed of a responsible will.

Kant's objection to Anselm's proof, while formally valid, ultimately misses the point simply because it is not the "existence" but the "necessity" of God that Anselm had meant to affirm.[12] Analogously, Coleridge also holds the supposition of conscience's nonexistence to be impossible. Thus to refer to it is not to make a metaphysical claim, let alone to venture an epistemological hypothesis, but to acknowledge the reality of something that gives or discloses itself prior to labors of propositional and discursive understanding. Conscience cannot be construed as an inward "state" or quality but, in virtue of its distinctive phenomenology, constitutes a real and manifest event. Coleridge thus notes how "the first step that … the becoming conscious of a conscience partakes of [is] of *the nature of an act*" (*OM,* 72; italics mine). As an act, conscience produces its own reality and thus transcends the merely contingent, psychological, and interiorist language of certitudes, convictions, and feelings: "the *Me* in the objective case is clearly distinct from the *Ego*" (*CL,* 4:849). The act of conscience first manifests itself *to* the person, which in turn involves "an act … by which we take upon ourselves an allegiance, and consequently the obligations of fealty" (*OM,* 72). To the extent that this pattern of phenomenologically discrete acts unfolds within us, conscience proves generative of faith and thus mediates or reveals the ontology of the *logos* (reason in the relational modality of the Trinity). Yet if "the reason in man is the representative of the Will of God" (*OM,* 84), conscience—precisely because it relates the finite person to the *logos*—cannot be subject to the human will. Rather, it "subsists in the synthesis of the reason and the individual Will, or the reconcilement of the reason with the Will, by the self-subordination of the Will to the reason."[13] Conscience, in other words, must not be

---

12. On Anselm's ontological proof, its detractors, and the history of its reception, see Murdoch, *Metaphysics,* 391–430; still a strong account of this issue is Tillich, *Systematic Theology,* 1:186–208. Murdoch quotes Simone Weil who, remarking on the distinctive status of this type of proof, observes that "for everything which concerns absolute good and our contact with it, the proof by perfection (wrongly called ontological) is not only valid, but the only proof which is valid. It is instantly implied by the notion of good" (quoted in Murdoch, *Metaphysics,* 401).

13. *OM,* 94; as Crosby points out, conscience—if understood as "the inner sanctuary of the human person"—cannot be yet another intentional object: "How would this inwardness of self-determination be possible if I had to do with myself only as with an object, if I were for myself nothing but another object on which I acted volitionally? … When conscience stirs and the inwardness

misconstrued as some garden-variety intentional experience, such as an act of expressive self-reference or propositional self-explanation. Rather, it is an instance of "self-presence, *in which I encounter myself not objectively but subjectively*."[14]

At first glance, all talk of conscience and its putative transcendence of ordinary intentionality would appear to invite back with a vengeance the post-structuralist critique of logo- and phono-centrism that has so galvanized the philosophically (not to mention theologically) jejune establishment of American literary theory beginning in the late 1960s. Yet as Coleridge and, eventually, John Henry Newman make abundantly clear, the subjective self-presence that transpires both *in* and *as* the phenomenon of conscience is emphatically not an instance of modern autonomy or enlightened self-possession. Rather, to recall Coleridge's memorable formulation, it involves "a struggle of jarring impulses; a mysterious diversity between the injunctions of the mind and the elections of the will; and (last not least) the utter incommensurateness and the unsatisfying qualities of the things around us" (*AR*, 349). The reality of the person that is phenomenologically disclosed in the action of conscience involves the awareness of a conflict—one that is not simply present *to* us but in truly Augustinian fashion is properly constitutive of our being. If conscience involves self-presence, the intentional object here at issue is not a coherent autonomous self but, on the contrary, the growing awareness of that notion's sheer impossibility due to an indelible inner division that cannot be overcome except within a transcendent framework of grace and a Platonic narrative movement toward a fuller comprehension of that fact. With the irrefutable valuations of conscience serving as its phenomenological cue and first object, faith (in Coleridge's account) is emphatically not the antithesis of reason but, on the contrary, the most comprehensive acknowledgment of the ontology of the *logos*. It would thus be a category mistake to try and construe faith as an empirical phenomenon, such as an ephemeral disposition or transient "experience" that occasionally intrudes on an otherwise orderly and neutral consciousness.

Here Coleridge markedly diverges from Kant, whose loosely Platonic account of the "moral law" within us he had otherwise admired. For Coleridge, as for Augustine long before and Newman shortly afterward, any attempt at compartmentalizing the person into an alternately judging, thinking, desiring, or reflecting consciousness is fundamentally misguided.[15] Far more in the spirit of Plato—whose theory of

---

of conscience opens up in us, we subjectively experience our standing in ourselves and our incommunicability" (*Selfhood*, 89–90).

14. Crosby, *Selfhood*, 84.

15. "When I name my memory, understanding, and will, each name refers to a single thing, and yet each of these single names is the product of all three" (*ADT*, 4.30). Taking a more skeptical view of Newman's account of conscience, John Milbank faults the late *Grammar of Assent* for retreating from the proto-phenomenological account of its opening four chapters into "more extrincisist

anamnēsis and *ēros* provides the original template for Coleridge's account of conscience and its orientation toward God—all these qualities of conscience, though distinct, form a single organic continuum. Faith is not an emotion, and neither are emotions something to be proscribed or subjugated in the neo-Stoic fashion that often creeps into Kant's moral theory. Writing to J. H. Green in 1817, Coleridge distances himself from "the German Philosophers" ("much in several of them is unintelligible to me, and more unsatisfactory") while struggling for a nuanced account of Kant:

> But I make a division.—I reject Kant's stoic principle, as false, unnatural, and even immoral, where in his Critik der Practischen Vernun[f]t he treats the affections as indifferent (ἀδιαφορά) in ethics, and would persuade us that a man who disliking, and without any feeling of Love for, Virtue yet acted virtuously, because and only because it was his Duty, is more worthy of our esteem, than the man whose affections were aidant to, and congruous with, his Conscience. For it would imply little less than that things not the Objects of the moral Will or under it's [sic] controul were yet indispensable to it's [sic] due practical direction. In other words, it would subvert his own System.[16]

Like Augustine, Coleridge regards the affective and the cognitive dimensions of human consciousness to be always entwined, and nowhere more so than in the way that we receive the "notices" of conscience as insistent, at times stunning and disorienting *qualia* (guilt, remorse, sorrow, etc.).

It is this continuum or, rather, co-inherence of thought and feeling in a myriad of *qualia*—distinct but organically related—which for Coleridge constitutes an unbroken interpretive and evaluative horizon. As Hans-Georg Gadamer would later note, this hermeneutic "horizon" is itself moving and evolving, a point strongly implicit in the late Coleridge, who conceives "personality in dynamic terms, as a matter of degrees in relation to the unity of will and Reason." In elaborating an "anthropology oriented towards sanctification,"[17] Coleridge thus identifies this evolving totality

---

arguments for which the fact of conscience supports a strongly probabilistic *inference* as to God's existence." On Milbank's reading, Newman's "modern deontological system of conscience is simply one limited and perhaps dubious notional system" and, as such, seems at odds with Newman's "illative sense," a variant of Aristotelian *phronēsis* that "involves a constant reading of the world and *not* simply a listening to an inner voice" ("What is Living and What is Dead," 50–51).

16. *CL*, 4:791–792. Elsewhere, Coleridge notes how "the philosopher of Königsberg and his first disciple and rival, Fichte, have erred and verged towards enthusiasm in their confusion of … the eunöya with the Hedone, the desirable of the intellect with the desirable of the body, and the exclusion of both indifferently from the permanent objects of the rational Will" (*OM*, 47n).

17. Gregory, "That I may be here," 192.

of the person as a hermeneutic, self-interpreting agent oriented toward an ultimate good; he calls it

> faith, (faith, which is used here in the same sense as Kant uses the Will, as the ground of all particular acts of willing) is a *total* act of the soul: it is the *whole* state of the mind, or it is not at all! ... Faith, in all its relations, subsists in the synthesis of the reason and the individual Will, or the reconcilement of the reason with the Will, by the self-subordination of the Will to the reason ... Faith must be an energy, and inasmuch as it relates to the whole moral man, ... [it] must be a ~~T~~total~~;~~ not a ~~P~~partial~~;~~ it must be a *continuous* and *ordinary*, not a *desultory* or *occasional*, energy. (*OM*, 43, 94)

As regards the many permutations of our will (as desire, ambition, etc.), conscience furnishes the consummate instance of the Greek *parakupsas* that Coleridge analyzes elsewhere. It is the act whereby the human person, in addition to its constitutive self-awareness, recognizes its essential embeddedness in an ontology, a framework of values that are revealed, not propositionally asserted. Iris Murdoch, whose Platonist account often exhibits striking similarities with that of the late Coleridge, thus argues for "a moral unconscious" and insists that the "place, where we are at home, which we *seem* to leave and then return to ... has moral colour, moral sensibility. We have a continuous *sense of orientation*. The concept of consciousness, the stream of consciousness, is *animated* by indicating a moral dimension. Our speech is moral speech, a constant use of the innumerable subtle *normative* words whereby (for better or worse) we texture the detail of our moral surround and steer our life of action."[18]

Though Coleridge would likely have demurred at Murdoch's unusual fusion of Platonism and atheism, his view of our "moral orientation" is largely the same. As the incontrovertible echo of God, the *actus purissimus*, conscience belongs to the domain of action and as such involves the entire person. For Coleridge it follows that we can never take a neutral, impersonal, and indifferent view of the *logos* any more than of the conscience *through* which it appears. In fact, the very proposition that we might look upon the *logos* askance or profess an agnostic outlook on it would have

18. *Metaphysics*, 301, 260; Murdoch pursues her critique of Wittgenstein by noting that "we 'interpret' our surroundings all the time, enjoying as it were a multiple grasp of their texture and significance. We are doing it continuously and this includes intense imaginative introspection, evaluation, focusing upon an image, turning thoughts into things" (ibid., 279). While her main claim remains sound, it is less obvious whether it is fair (or, for that matter, necessary) to stage it as a critique of Wittgenstein. On Murdoch's reading of Wittgenstein, see Antonaccio and Schweiker, eds., *Iris Murdoch*, esp. the essays by Diamond and Antonaccio, and Broackes, ed., *Iris Murdoch*, esp. the sharply critical essay by Moran.

struck Coleridge as absurd or, literally, "preposterous" since to look thus indifferently (or even diffidently) on reason would itself presuppose a considered motive, *a reason for* doing so. Instead, the only conceivable and absolutely necessary relation to the *logos* is one of faith and, by implication, love. Relating to the *logos* as "the form of its reception,"[19] faith orients the person toward an unconditional and forever unrealized hyper-good (to borrow Charles Taylor's term). As such, faith (in Coleridge's account) constitutes simultaneously an act of knowledge *and* of practical commitment, of intellect in fullest alignment with will. As Coleridge puts it, "all human fidelity that is consistent with itself is fidelity to the conscience" (*OM,* 78). As he knew all too well, it is of course always possible to ignore the bearing of conscience on one's specific course of action, but any decision to do so will itself be colored and, in time, haunted by its defiance of (or willful indifference to) the *logos.*[20]

Here we need to attend to how normative conscience relates to ostensibly value-neutral (self-)consciousness. To that end, we recall Coleridge's acknowledgment that "the experience or inward witnessing of the Conscience, ... if <it be> *at all* must be *unique* and therefore cannot be supported by an Analogon. [It] therefore may be *monstrated* but cannot be *de*monstrated" (*CN,* no. 4605). Self-consciousness here relates to conscience the way the species-term "human being" relates to the incommunicability of the person. As Coleridge notes, my consciousness of a specific proposition (e.g., Kant's categorical imperative) only holds significance because I know "with the same clearness, that it is a fact of which all men either are or ought to be conscious." Mere sheer awareness of the moral law *as a proposition* does not yet establish my reality as a unique ethical being. Inasmuch as "I possess this consciousness as a man and not as the individual John or James," such consciousness is "distinguished from all other acts of consciousness by its universality" (*OM,* 59). Yet since I also know that a proposition like the categorical imperative presupposes not only the reality of other conscious beings but also their knowledge of that proposition, my consciousness of it here carries over into my self-awareness as a being standing in a unique ethical relation to others.[21] There is no fact/value distinction here, no impersonal perspective on the proposition at hand, no view from nowhere. In fact, ignorance of the Kantian moral law, Coleridge insists, effectively "establishes the non-

19. Gregory, "That I may be here," 196.

20. On the later Coleridge's theory of conscience, see Gregory, "That I may be there"; and Rule, *Coleridge and Newman,* 41–64; Rule rightly notes "the fundamental but ultimately secondary importance of the will" (48) vis-à-vis conscience, a point that emerges quite forcefully in Coleridge's open endorsement of Luther (*OM,* 102).

21. As Coleridge remarks elsewhere, "A male & a female Tyger is neither more or less whether you suppose them only existing in their appropriate wilderness, or whether you suppose a thousand Pairs. But Man is truly altered by the co-existence of other men; his faculties cannot be developed in himself alone, & only by himself" (*CL,* 2:1197).

personality of the *ignorant.*" Hence, and with "good right have mankind designated [this awareness of my ontological relatedness to and participation in the reality of the other] by a particular term and named it the *Conscience.*" Its absence or presence "determines whether any given subject be a thing or a person." That Coleridge should have chosen for his example Kant's categorical imperative is certainly no accident. For in so doing he is able to show that, if the formal propositions of Kant's "moral law" are to secure uptake among their intended audience of rational agents, such a deontological model presupposes a fundamentally Platonic model of the human as communal, relational, and participatory. Against Kant's attempt to produce an account of moral agency that mirrors the latter's putative autonomy by presenting itself free of all metaphysical presuppositions, Coleridge insists that a moral philosophy solely based on rational, self-conscious, and abstract agency is impossible: "Paradoxical as it may sound to describe the conscience as the ground of all proper consciousness—anterior, therefore, to it in the order of thought, i.e., without reference to time—we yet doubt not of establishing the truth … of the underived, unconditional authority of the Conscience" (*OM,* 59–60).

Our quotidian "work of attention," Iris Murdoch notes, "imperceptibly builds up structures of value round about us." She proceeds to offer a counterfactual example so as to illustrate that everyday practice and its encoding in ordinary language affords us orientation, order, and purpose precisely because it is cued by a transcendent good: "What of the command 'Be ye therefore perfect'? Would it not be more sensible to say 'Be ye therefore slightly improved'?"[22] As the intuitively absurd alternative makes clear, normativity by its very nature presupposes a transcendent orientation. Absent an "idea of perfection," it becomes an empty term, and values merely negotiated and affirmed in historically contingent (and likely opportunistic) fashion are not values at all. For in the latter case, Coleridge notes, value remains but a "generalization," which is to say, "a Substitute for Intuition" (*AR,* 275n). The attempt to contain moral reflection within a "prudential" calculus of interests to be negotiated (e.g., in Paley and Bentham), as well as Kant's restriction of moral reasoning to our formal assent to an impersonal "moral law," fail because they cannot specify a "source" or normative good such as would induce us to follow that route. A chasm separates formal-syllogistic claims of the understanding from the ideas of the good and of justice. Belonging to the latter realm, justice cannot be thought in counterfactual or propositional terms. Like Levinas's "infinity," Plato's good (*agathon*) and justice (*dikē*) prove "exceptional in that [in each case the] *ideatum* surpasses its idea." Indeed, for us to conclude that our interpretation of a normative good or *ideatum* conflicts with someone else's we have to be already persuaded that our dispute centers

22. Murdoch, *Sovereignty,* 36, 60–61.

on the same identical notion. As Levinas puts it, "the distance that separates the *ideatum* and idea here constitutes the content of the *ideatum* itself" (*TI*, 49). Likewise, Coleridge insists that "truth is indeed a necessary attribute of goodness, but while we must receive the truth for the truth's sake, we love it only because it is good." Any allegiance to an idea necessarily originates in an act of intuition; we see that it is good "because we need only contemplate it as realized in its effects to perceive that it is necessarily and eminently true" (*OM*, 151).

By contrast, a deontological (Kantian), calculative (Benthamite), or consequentialist moral philosophy, however consistent in its propositional structure, will inevitably fail to secure our allegiance as long as it proceeds by playing off its syllogistic account against the supposed opacity and indefensibility of our moral intuition. Echoing Blaise Pascal ("Principles are felt; propositions proved"),[23] Newman in 1841 remarks how "many a man will live and die upon a dogma: no man will be a martyr for a conclusion. A Conclusion is but an opinion; it is not a thing which *is*, but which *we are 'certain about'* ... Logicians are more set upon concluding rightly, than on right conclusions. They cannot see the end for the process."[24] With that much, it appears, Coleridge would certainly have concurred. Thus, in his *Lectures on the History of Philosophy*, he specifically faults Socrates for assuming that human virtue is seated exclusively in the intellect rather than the will, and that it can be infused didactically rather than by habituation (*LHP*, 1:174–181). Moreover, Coleridge insists that to secure our assent to a specific *view* (as opposed to some conclusion such as may eventually be drawn from it), something more elemental is required in the realm of moral inquiry. Viz., we must acknowledge "the sacred distinction between Thing and Person, [for] on this distinction all Law human and divine is grounded: consequently, the law of Justice" (*AR*, 327). The distinction between a syllogistically effected "agreement" and the "recognition" of the person—in contradistinction to a "thing"—extends far beyond the merely intellectual and discursive. It discloses the categorical divide that separates the strictly discursive, racionative function of the understanding from the ideational and implicitly normative domain of reason.

23. *Pensées*, no. 110, p. 28.

24. "The Tamworth Reading Room," in *Discussions and Arguments*, 294; on the intrinsic aporias of modernity's fact/value antithesis and its problematic application to "modern moral philosophy," see MacIntyre, *AV*, 51–108. Newman's late project of a comprehensive phenomenology of faith and knowledge and the various intermediate forms of "implicit reason" (a notion already advanced in his *Oxford University Sermons*) is substantially consistent with, indeed the most compelling extension of Coleridge's neo-Platonist argument for a "responsible will" and the implicit proof of God by the phenomenon of conscience. Even before completing his *Grammar of Assent* (1870), Newman had made strikingly cognate arguments in an unpublished essay, "Proof of Theism" (reprinted in Boekraad, *Argument from Conscience*, 103–125).

An enduring motif in virtually all of Coleridge's books and manuscripts after 1809, this distinction tends to occur within a more or less overtly Platonizing construction. An especially powerful instance is found early in *Aids to Reflection,* as Coleridge remarks on Robert Leighton's appreciation of God's ineffability vis-à-vis the inherently dissatisfying encounter "with the Objects of our bodily senses" (Aphorism XII[a]). Calling Leighton's remark "ingenious and startling," Coleridge attempts to draw another, "more fruitful, perhaps more solid inference," viz.,

> that there is something in the human mind which makes it know (as soon as it is sufficiently awakened to reflect on its own thoughts and notices), that in all finite Quantity there is an Infinite, in all measures of Time an Eternal; that the latter are the basis, the substance, the true and abiding *reality* of the former; and that as we truly *are,* only as far as God is with us, so neither can we truly *possess* (i.e., enjoy) our Being or any other real Good, but by living in the sense of his holy presence. (*AR,* 92)

Quoted with much approval in Newman's *Grammar of Assent* (1870), Coleridge's passage relies in its opening image on the Platonic notion of *anamnēsis* (*Meno,* 81[b]–86[d]). Moreover, the argument implicitly endorses Plato's doctrine of form as the indispensable ontological framework conditioning our apprehension of finite, empirical reality. Unlike Kant in the third *Critique,* however, Coleridge does not restrict "form" and "idea" (*eidos*) to a merely hypostatized and heuristic relation between the subject's faculties of cognition. Rather, as the trope of "awakening" suggests, self-awareness involves not merely a *functional* synthesis on the order of Kant's "transcendental apperception," a construct that Kant took pains to keep distinct from the (in his view impermissible) inference of a "consciousness-of-self" or anything approaching the concept of person.[25] For Coleridge, by contrast, to have been "*sufficiently* awakened to reflect on its *own* thoughts and notices" (italics mine) carries two

25. Key here is the B-text of Kant's *Critique of Pure Reason,* especially the so-called "transcendental deduction" (§§24–25): "in the synthetic original unity of apperception, I am not conscious of myself as I appear to myself, nor as I am in myself, but am conscious only that I am ... Hence although my own existence is not appearance (still less mere illusion), determination of my existence can occur only in conformity with the form of the inner sense and according to the particular way in which the manifold that I combine is given in inner intuition. Accordingly I have no *cognition* of myself as I am but merely cognition of how I appear to myself. Hence consciousness of oneself is far from being a cognition of oneself ..." (B 158; *Critique of Pure Reason,* 195–196). The specific problem with self-reference and self-awareness, which culminates in the Deduction, has attracted an unusual amount of attention. For major arguments, see Heidegger, *Kant and the Problem of Metaphysics* (Ger. 1928); Ulrich Pothast, *Über einige Fragen der Selbstbeziehung* (1971); Manfred Frank, *Die Unhintergehbarkeit von Subjektivität* (1986); and Pinkard, *German Philosophy,* 26–40. As Manfred Frank has argued elsewhere, Novalis and Hölderlin—responding more to Fichte than to

crucial implications for human consciousness that Kant—consumed with the demonstrability and "certainty" of all epistemological claims and taking in general a dim view of any appeal to inner "certitudes"—simply could not endorse. Thus where Kant ventures the startling assertion that "reason in all its undertakings must subject itself to criticism," and indeed that "reason depends on this *freedom* for its very existence," Coleridge's neo-Platonist account credits the *logos* with an absolute, ontological reality that in no way depends on our cognitive, critical pursuits. Already in 1806, he had written Thomas Clarkson that "Reason is … most eminently the Revelation of an immortal soul, and it's [sic] best Synonime—it is the forma formans, which contains in itself the law of it's [sic] own conceptions." Just like person and existence, so soul is not a proposition to be demonstrated in the logical space of ordinary concepts: "What the Soul *is*, I dare not suppose myself capable of *conceiving* … *Datur*, non intelligitur" (*CL*, 2:1198, 1193).

To be sure, Plato, Aquinas, and Coleridge all maintain that our full realization as rational human agents and persons demands a continued effort at participating *in* the *logos*. Yet because reason and the eternity of forms is never contingent on this at best uneven progression, Coleridge does not endorse Kant's often repeated claim that the only alternative to critique was some version of philosophical dogmatism.[26] On strictly logical grounds alone, such a position is compromised by its manifestly self-certifying nature, for it is precisely our acceptance of the view that dogmatism is something to be *rejected* that licenses and seemingly compels a *critique* of reason in the first place. Kant's proscription of alternative positions as "dogmatic" is not motivated by their point-by-point *refutation* but by their alleged failure to conform to the boundaries and procedures mapped by transcendental philosophy. As Stanley Rosen so succinctly puts it, Kant "*constructs* theoretical entities that serve his purpose. There is no empirical confirmation of Kant's hypothesis, however, since what counts as experience, and also as confirmation, is created by our acceptance of the hypothesis." Historically, then, the very project of a *critique* of reason constitutes indeed a historical caesura, albeit not quite in the way that Kant and his successors preferred to see it. For the crucial "moment of transition [is] *itself produced not simply by historical circumstances but by Kant's will to change those circumstances*."[27] The alternative, never fully acknowledged by Kant, and almost completely foreclosed on in Fichte, would be that the "thoughts and notices" of consciousness might be taken as a metaphysical "gift" of sorts and, as such, might furnish the phenomenological con-

---

Kant—had clearly recognized the impossibility of premising a coherent self on a model of self-reflection; see Frank, *Einführung*, esp. 248–286 and Pfau, *Romantic Moods*, esp. 33–52.

26. *Critique of Pure Reason* A 738/B 766: *Die Vernunft muß sich in allen ihren Unternehmungen der Kritik unterwerfen. . . . Auf dieser Freiheit beruht sogar die Existenz der Vernunft.*

27. Rosen, *Hermeneutics as Politics*, 25, 31.

duit to the Platonic and Christian *logos*. This argument, so powerfully developed in the recent work of Jean-Luc Marion, naturally presupposes that the very notion of reason, which Kant's first *Critique* proposes (or, rather, presupposes) to be in constant need of delimitation might turn out to be inherently evolving and in flux.

Reacting against both Enlightenment rationalism *and* its variously idealist and pessimist counterpoints (Friedrich Heinrich Jacobi's anti-rationalism; Hume's skepticism; Schelling's pantheism, etc.), the late (Christian Platonist) Coleridge here seeks to reclaim an alternative model of reason. On his account, reason is not to be construed as the hegemonic and monolithic absolute that had begun to run amok in Paris sometime after 1792 and the memory of which is still being fought in the critique of logo-centrism and its administrative terrors in the early writings of Michel Foucault, Jacques Derrida, Gilles Deleuze, and others. Stanley Rosen puts it well when observing that "postmodernism has no more rejected Kant than Kant rejected the Enlightenment. We are now [1987] living through the rhetorical frenzy of the latest attempt of the self-contradictory nature of Enlightenment to enforce itself as a solution to its own incoherence." In fact, the later Coleridge's profound exploration of will, conscience, and person develops a genuine alternative—a kind of Blakean contrary that transcends both the earnest but arid formalism of the Kantian subject *and* the rationalist dogmatism and Humean skepticism that Kant's first *Critique* purports to have overcome. As Rosen puts it, "the greatest barrier separating Kant from Plato disappears as soon as we recognize that transcendental doctrine is a myth and also that to recognize it as such is not to abolish the psyche but to return it to itself."[28]

While the beginnings of this argument in Coleridge's oeuvre are subject to some debate, it certainly characterizes his overall project in *Aids to Reflection*, the *Opus Maximum*, and countless notebook entries after 1819. As he came to understand, there was no reason to suppose that the only alternative (if it is one at all) to an allegedly totalizing model of reason had to be a "critique" à la Kant—viz., a re-description of reason as a strictly virtual or "regulative" framework, a modest, strictly procedural utopia stripped of all transcendent meaning. To be sure, Coleridge concurs with Kant that reason must indeed be conceived as a dynamic, open-ended progression. Yet precisely this act of mind wherein reason discloses its progression, both at the level of individual-biographical time and across expanses of historical, trans-generational time, amounts to a real and inherently qualitative state of being, rather than some strictly formal and ostensibly value-neutral correlation of faculties. For Coleridge,

28. Ibid., 49, 55; for Cutsinger, Coleridge's relationship to Kant, though often one of convergence, ultimately remains agonistic, and he insists that "the reason Kant could never have been a Platonist, and therefore a Coleridgean, was that he remained subject despite his sense of mental power, to the patterns and assumptions of a strictly Newtonian universe of mechanistic materialism" (*Transformed Vision*, 51).

moreover, the anteriority, indeed the unconditioned *reality*, of mind as the source of all those "thoughts and notices" on which even the most austere transcendental critique necessarily depends is never in question. Consequently, consciousness does not require its retroactive legitimation in terms acceptable to mere understanding. Where Kant posits apperception as a strictly coordinating *function*, and where Fichte and the young Schelling speak of the "fact of consciousness," Coleridge insists that the phenomenological attention of consciousness to its *own* "thoughts and notices" constitutes a real and significant reality. More than "certainty" about X it furnishes the mind with "certitude" about its own status as a real and responsible agent. The knowledge at issue here is not some sterile *fact* but, however rudimentary, necessarily carries within itself the intimation of a *value* begging to be realized in progressively fuller form.[29]

The above passage from *Aids to Reflection* also implies that self-awareness cannot be construed as a secondary and derivative *form* but, belonging to the domain of *qualia*, is necessarily presupposed in any account of rational human personhood. Once again following Plato, not Kant, Coleridge insists on the absolute primacy of self-consciousness and stipulates that for a human being to be conscious means *eo ipso* to have awareness of oneself as a person—that is, as an incommunicable, rational existence that stands in a richly layered relation to other such beings. In developing this position by way of his powerful account of will and conscience, Coleridge meant above all to oppose (albeit not always in entirely fair-minded and nuanced ways) various materialist and deterministic accounts that sought to construe consciousness as a mere "some*thing*" produced *by* sensation or, at most, adventitiously reflecting *on* sensation. Such a position Coleridge regards as nonsensical, if only because to advance such an argument presupposes *that we are already conscious of having the sensation* in question—which is to say, are self-aware. For only on that premise could we ever wish to assign sensation a pole position in the race for an all-encompassing explanation of life. Given Coleridge's vehement opposition to materialism in all its guises ("any system built on the passiveness of the mind must be false, as a system" [*CL*, 2:709]), the crucial task of philosophy is not to elucidate the relationship of consciousness to the external world but, rather the dynamic and infinitely complex phenomenology of how it relates to the will. Though Newman was an infrequent reader of Coleridge, his one extended comment on the sage of Highgate

29. Homing in on the tension between "certainty" and "certitude" in his *Grammar of Assent*, Newman follows Coleridge by arguing that the very act of affirming a proposition or conception presupposes an inner state of self-awareness: "what to one intellect is a proof is not so to another, and ... the certainty of a proposition does properly consist in the certitude of the mind which contemplates it" (281).

happens to address precisely the above passage from *Aids to Reflection*[30] and, with unerring instinct, distills its abiding import in a series of focused questions:

> What is this an argument for? How few readers will enter into either premiss or conclusion! And of those who understand what it means, will not at least some confess that they understand it by fits and starts, not at all times? Can we ascertain its force by mood and figure? Is there any royal road by which we may indolently be carried along into the acceptance of it? Does not the author rightly number it among his "aids" for our "reflection," not instruments for our compulsion? It is plain that, if the passage is worthy of any thing, we must secure that worth for our own use by the personal action of our own minds ... And our preparation for understanding and making use of it will be the general state of our mental discipline and cultivation, our own experiences, our appreciation of religious ideas, the perspicacity and steadiness of our intellectual vision.[31]

Newman's shrewd focus on the "mental discipline" required to make judicious and effective use of intellectual traditions eschews Coleridge's omnivorous approach to intellectual inquiry. Though Newman, especially in his late *Grammar of Assent*, is just as concerned with notions of practical reason, judgment, responsibility, and will, his altogether different temperament also allows him to perceive the precariousness and partial failure of Coleridge's undertaking. Too much of the traditions in question had irretrievably vanished, and Coleridge's obsessive attempt at reclaiming and interweaving various strands of humanistic and theological thinking often risks obscuring the *terminus ad quem* of his overall enterprise. Too often, that is, the ultimate objective of Coleridge's far-flung "abstruse research" into philosophical theology risks collapsing under the sheer weight of the machinery reassembled for the purpose. The reclamation of intellectual tradition only works if, as Cora Diamond was to note, the very loss of it is still felt, still registers as a palpable deficit in the minds of an envisioned audience. On precisely that point, however, Newman, herein far more the

30. Newman, who in a late letter (17 August 1884) claimed never to have read a word of Coleridge was clearly misremembering. Perhaps as a result of Coleridge's death on 25 July 1834, Newman in early 1835 began to study *On the Constitution of Church and State* and *Aids to Reflection*, as evidenced by his diary entry of 29 March 1835 in which he finds himself "surprised how much I thought mine, is to be found there" (qtd. in *AR*, cxxxvii); on Coleridge's significant impact on the Oxford Movement, see Beer's Introduction to *AR* (esp. cxxxvii–cxxxix), Boekstraad, *Argument from Conscience*, 29–31; Ker, *John Henry Newman*, 173–174; Rule, *Coleridge and Newman*, 25–40; Pattison, though aware of Newman's reading of Coleridge (and late denial of that fact), nonetheless argues that "on close examination, the apparent fellowship of Coleridge and Newman is illusory" (*Great Dissent*, 42).

31. Newman, *Grammar of Assent*, 242.

empiricist and pragmatic tactician than Coleridge, remains doubtful. Though he shares Coleridge's perception of modernity as increasingly bereft of practical reason and all but oblivious of conceptual traditions extending nearly two millennia back, the controversialist Newman proves shrewdly selective when identifying relevant precursors and taking on intellectual debts (e.g., patristic thought, Duns Scotus, British empiricism, Joseph Butler). He also acknowledges that the recovery of any such tradition as a framework enabling the orientation of rational human beings in a social and hence moral space will only succeed if such a frame becomes the object of "real assent." We cannot be argued into accepting a position such as the one Coleridge has so elaborately retrieved from Trinitarian theology and (neo-)Platonism. Rather, "we must secure that worth for our own use by the personal action of our own minds." For if a prodigious mass of learning alone would enable us to reason others into assent, then Coleridge's entire argument for an essentially self-originating will would collapse anyway.

For a variety of reasons, Coleridge's project of reclaiming the idea of practical reason by retrieving and dialectically engaging complex and far-flung intellectual traditions was not taken up by succeeding generations. Overt criticisms and more tacit misgivings about his project originate from various quarters. There are Newman's reservations about the later Coleridge's excessively speculative proclivities and, in particular, about the failure of Coleridge's theology to issue in a coherent and practical account of moral agency. As Newman's implicit counter-position in the *Grammar of Assent* suggests, Coleridge's hyper-Augustinian obsession with sin threatens to overwhelm the inherently dynamic nature of moral vision and agency with retrospective, at times even fatalist ruminations. As Newman sees it in the above passage, *Aids to Reflection,* its often breath-taking insights notwithstanding, lacks a clear objective, something also borne out by the book's peculiar evolution from a florilegium of Archbishop Leighton's theological writings to the "changeling" that eventually appeared in 1825, a work in which the proportion between text and commentary had decisively shifted toward the latter.[32] Furthermore, the later Coleridge's preoccupation with a fundamentally sinful will and his decision to anchor moral personhood almost exclusively in an ontology of conscience creates tensions of its own. For one thing, Coleridge's radically interiorist account of human agency and responsibility does not quite align with his concurrent emphasis on the relationality of human personhood, just as his Platonism at times ultimately remains at odds with his Trinitarianism. Likewise, his preoccupation with the phenomenology of conscience at times

32. On the work's strange textual evolution, see Wright, *Coleridge and the Anglican Church,* 146–158.

also threatens to derail his ongoing arguments for the indispensable role of the *ecclesia* in realizing an authentic moral community. Despite Coleridge's widely recognized significance for the consolidation of the Anglican Broad Church movement, his own reasoning does not extend beyond his enticing (albeit utopian) conception of a future "clerisy" charged with "cultivating and enlarging the knowledge already possessed, and … watching over the interests of physical and moral science, … thus connect[ing] the present with the future" (*CCS*, 43–44). To the end, Coleridge's Augustinianism remains a rather one-sided affair whose emphatically interiorist character shows him, even in his late years, to have remained an exponent of the Romantic movement that he helped found and define during the first half of his career.

For these reasons, Coleridge's project of reviving the ancient notion of practical reason by retrieving and dialectically engaging the complex intellectual traditions that, beginning with Plato, had allowed Western thought to develop a sophisticated conception of the human person as a "dependent rational animal" (to borrow Alasdair MacIntyre's succinct formula) ultimately did not persuade his immediate successors. Nevertheless, the second half of the twentieth century saw a number of thinkers and intellectual historians (Reinhard Koselleck, Hans-Georg Gadamer, Gertrude Elizabeth Anscombe, Alasdair MacIntyre, Charles Taylor, John Milbank, among others) engaging again the logic of intellectual traditions and tabulating the potentially devastating consequences of their wholesale displacement. Time and again, the work of these critics echoes Coleridge's ambitious, if flawed critique of modernity's exclusively anthropomorphic conception of rationality as a "notion" or "system" of exclusively pragmatic value. Yet in the immediate aftermath of Coleridge, what takes shape is a fundamentally different development, a turn not toward complex and deep intellectual genealogies but toward an objective aesthetic, especially in John Ruskin and the pre-Raphaelite movement of the 1840s and 1850s. Here any noumenal or metaphysical commitments, to the extent that they exist at all, come into focus by means of a rigorous phenomenological analysis of the moment of vision as it is structured by the object of its attention. The focus moves from philosophy's indebtedness to received intellectual traditions to a forensic account of present (perceptual) experience. Such a shift intimates the new generation's discomfort with the ways in which the Romantics had sought to reinvest the individual with metaphysical (noumenal) meanings, yet also with Coleridge's Platonizing variant of that project. Still, the change in intellectual orientation that is observable by the early 1840s does not involve a retreat onto the ostensibly safe ground of radical empiricism, materialism, or for that matter the late Enlightenment project of a "critique" of reason. Rather, the strong emphasis on the visual and the image that characterizes Ruskin's aesthetics and that is also observable in its most profound extension, Gerard Manley Hopkins's poetics, reflects attempts to develop a phenomenology of the human person by other means.[33] It will be the matter of another book to show how, feeling at once estranged from the Enlightenment and alarmed by the sheer magnitude and evident incompletion of Coleridge's attempted retrieval and idiosyncratic fusion of distant theological and philosophical genealogies, his heirs proceeded to rethink the human in emphatically objective terms, viz., by embarking on a rehabilitation of the image.

---

33. See Ball, *Science of Aspects*, esp. 4–102 (on Ruskin); on Hopkins's concept of objective vision and his resulting theory of inscape, see Ward, *World as Word*, 158–197, and Sobolev, *Split World*, 27–112; on Hopkins in relation to the visual arts, see Phillips, *Gerard Manley Hopkins*, esp. 41–86 and 245–263. See also Pfau, "Rethinking the Image."

WORKS CITED

Abbott, Andrew. 1988. *The System of Professions: An Essay on the Division of Expert Labor.* Chicago: University of Chicago Press.

Ackrill, J. L. 1980. "Aristotle on Action." In *Essays on Aristotle's Ethics,* edited by A. O. Rorty, 93–101. Berkeley: University of California Press.

Adams, Marilyn McCord. 1986. "The Structure of Ockham's Moral Theory." *Franciscan Studies,* 46:1–35.

———. 2006. "Ockham on Will, Nature, and Morality." In *The Cambridge Companion to Ockham,* edited by Paul V. Spade, 245–272. Cambridge: Cambridge University Press.

Adorno, Theodor. 1991. *Notes to Literature.* 2 vols. Translated by Shierry Weber Nicholsen. New York: Columbia University Press.

———. 1998. *Gesammelte Schriften.* Edited by Rolf Tiedemann. Darmstadt: Wissenschaftliche Buchgesellschaft.

Adorno, Theodor, and Max Horkheimer. 1972. *Dialectic of Enlightenment.* Translated by John Cumming. New York: Continuum.

Aers, David. 2009. *Salvation and Sin: Augustine, Langland, and Fourteenth-Century Theology.* Notre Dame, IN: University of Notre Dame Press.

Allen, Danielle. 2006. "Talking about Revolution: On Political Change in Fourth-Century Athens and Historiographic Method." In *Rethinking Revolution through Ancient Greece,* edited by Simon Goldhill and Robin Osborne, 183–217. Cambridge: Cambridge University Press.

Allen, Richard. 1999. *David Hartley on Human Nature.* Albany: State University of New York Press.

Anderson, Amanda. 2001. *The Powers of Distance: Cosmopolitanism and the Culture of Detachment.* Princeton, NJ: Princeton University Press.

Anderson, Benedict. 2006. *Imagined Communities: Reflections on the Origin and Spread of Nationalism.* New York: Verso.

Andresen, Carl. 1961. "Zur Entstehung und Geschichte des trinitarischen Personbegriffes." *Zeitschrift für neutestamentliche Wissenschaft,* 52:1–38.

Anscombe, Gertrude Elizabeth. 1958. "Modern Moral Philosophy." *Philosophy,* 33:1–19.

———. 1981. "On Promising and its Justice, and Whether it Need be Respected *in Foro Interno.*" In *The Collected Philosophical Papers of G. E. Anscombe,* 3:10–21. Minneapolis: University of Minnesota Press.

———. 2000. *Intention.* Cambridge, MA: Harvard University Press.

Antonaccio, Maria, and William Schweiker. 1996. *Iris Murdoch and the Search for Human Goodness.* Chicago: University of Chicago Press.

Apel, Karl-Otto, ed. 1971. *Hermeneutik und Ideologiekritik.* Frankfurt: Suhrkamp.

Appel, Toby A. 1987. *The Cuvier-Geoffroy Debate: French Biology in the Decades before Darwin.* Oxford: Oxford University Press.

Aquinas, St. Thomas. 2003. *On Evil.* Translated by Richard Regan. Oxford: Oxford University Press.

———. 2008–. *Summa Theologiae.* 5 vols. Scotts Valley, CA: NovAntiqua.

Árdal, Páll Steinthórsson. 1966. *Passion and Value in Hume's Treatise.* Edinburgh: Edinburgh University Press.

Arendt, Hannah. 1958. *The Human Condition.* Chicago: University of Chicago Press.

———. 1971. *The Life of the Mind.* New York: Harcourt.

———. 2004. *The Origins of Totalitarianism.* New York: Schocken.

Aristotle (attributed to). 1953–1957. *Problemata Physica.* Translated by W. S. Hett. Cambridge, MA: Harvard University Press.

———. 1984. *The Complete Works of Aristotle.* Edited by Jonathan Barnes. 2 vols. Princeton, NJ: Princeton University Press.

Armstrong, A. H. 1972. "Neoplatonic Valuations of Nature, Body, and Intellect." *Augustinian Studies,* 3:35–59.

Armstrong, Charles I. 2003. *Romantic Organicism: From Idealist Origins to Ambivalent Afterlife.* London: Palgrave.

Armstrong, Nancy B. 2005. *How Novels Think: The Limits of Individualism, 1719–1900.* New York: Columbia University Press.

Arnold, Matthew. 1993. *Culture and Anarchy & Other Writings.* Edited by Stefan Collini. New York: Cambridge University Press.

Assad, Talal. 2003. *Formations of the Secular.* Stanford, CA: Stanford University Press.

Atherton, Catherine. 1993. *The Stoics on Ambiguity.* Cambridge: Cambridge University Press.

Auerbach, Erich. 1953. *Mimesis: The Representation of Reality in Western Literature.* Princeton, NJ: Princeton University Press.

———. 1984. *Scenes from the Drama of European Literature.* Minneapolis: University of Minnesota Press.

Auerochs, Bernd. 1995. "Gadamer über Tradition." *Zeitschrift für philosophische Forschung,* 49(2):294–311.

Augustine of Hippo, St. 1991. *The Trinity*. Translated and edited by Edmund Hill, OP. Hyde Park, NY: New City.

———. 1993. *On Free Choice of the Will*. Translated by Thomas Williams. Indianapolis: Hackett.

———. 1998. *The City of God against the Pagans*. Edited by R. W. Dyson. Cambridge: Cambridge University Press.

———. 1999. *Answer to the Pelagians, IV: To the Monks of Hadrumetum and Provence*. Translated and edited by Roland J. Teske, SJ. Hyde Park, NY: New City.

———. 2001. *The Confessions*. Translated by Philip Burton. New York: Everyman.

Austen, Jane. 2001. *Mansfield Park*. Edited by June Sturrock. Peterborough, ON: Broadview.

Ayres, Lewis. 2004. *Nicea and its Legacy: An Approach to Fourth-Century Trinitarian Theology*. Oxford: Oxford University Press.

Balfour, Ian. 2002. *The Rhetoric of Romantic Prophecy*. Stanford, CA: Stanford University Press.

Ball, Patricia M. 1971. *The Science of Aspects*. London: Athlone.

Barrell, John. 1972. *The Idea of Landscape and the Sense of Place, 1730–1840*. Cambridge: Cambridge University Press.

———. 1983. *The Dark Side of Landscape: The Rural Poor in English Painting, 1730–1840*. Cambridge: Cambridge University Press.

———. 1986. *The Political Theory of Painting from Reynolds to Hazlitt*. New Haven, CT: Yale University Press.

Becker, Lawrence C. 1999. *A New Stoicism*. Princeton, NJ: Princeton University Press.

Beer, John. 2002. "Romantic Apocalypses." In *Romanticism and Millenarianism*, edited by Tim Fulford, 53–70. New York: Palgrave Macmillan.

Beha, Christopher R. 2012. "Reason for Living: The Good Life Without God." *Harper's Magazine* (July), 73–78.

Beiser, Frederick C. 2002. *German Idealism: The Struggle against Subjectivism, 1781–1801*. Cambridge: Cambridge University Press.

Benjamin, Walter. 1968. *Illuminations*. Translated by Harry Zohn, edited by Hannah Arendt. New York: Schocken.

———. 1999. *The Arcades Project*. Translated by Howard Eiland and Kevin McLaughlin. Cambridge, MA: Harvard University Press/Belknap Press.

Berenson, Bernard. 1957. *Lorenzo Lotto: Gesamtausgabe*. Köln: Phaidon.

Berger, Peter. 1967. *The Sacred Canopy*. New York: Anchor.

Berkeley, Richard. 2006. "The Providential Wreck: Coleridge and Spinoza's Metaphysics." *European Romantic Review*, 17(4):457–475.

Bermes, Christian. 2004. *'Welt' als Thema der Philosophie*. Hamburg: Meiner.

Bermingham, Anne. 1989. *Landscape and Ideology: The English Rustic Tradition, 1740–1860*. Berkeley: University of California Press.

Besançon, Alain. 2000. *The Forbidden Image: An Intellectual History of Iconoclasm*. Translated by Jane Marie Todd. Chicago: University of Chicago Press.

Bewell, Alan. 1999. *Romanticism and Colonial Disease*. Baltimore: Johns Hopkins University Press.

Bieri, Peter. 2007. *Das Handwerk der Freiheit*. Frankfurt: Fischer.

Biro, John. 1993. "Hume's New Science of the Mind." In *The Cambridge Companion to Hume*, edited by David F. Norton, 40–69. Cambridge: Cambridge University Press.

Blackman, E. C. 1948. *Marcion and his Influence*. London: SPCK.

Blake, William. 1982. *The Complete Poetry and Prose*. Edited by David V. Erdman. New York: Doubleday.

Blondel, Maurice. 1984. *Action: Essay on a Critique of Life and a Science of Practice*. Translated by Oliva Blanchette. Notre Dame, IN: Notre Dame University Press.

Bloom, Harold. 1970. *Blake's Apocalypse: A Study in Poetic Argument*. Ithaca, NY: Cornell University Press.

———. 1982. *Agon: Towards a History of Revisionism*. Oxford: Oxford University Press.

Blumenberg, Hans. 1979. *Arbeit am Mythos*. Frankfurt: Suhrkamp.

———. 1983. *Legitimacy of the Modern Age*. Translated by Robert M. Wallace. Cambridge, MA: MIT Press.

———. 1986. *Die Lesbarkeit der Welt*. Frankfurt: Suhrkamp.

———. 1997. *Shipwreck with Spectator: Paradigm of a Metaphor for Existence*. Translated by Steven Rendall. Cambridge, MA: MIT Press.

Boekraad, Adrian J. 1961. *The Argument from Conscience to the Existence of God*. Louvain: Editions Nauwelaerts.

Boethius. 1918. *Tractates, De Consolatione Philosophiae*. Translated by F. H. Stewart and E. K. Rand. New York: Putnam; rept. Cambridge, MA: Harvard University Press, 1973.

Bok, Derek. 2012. *Universities in the Marketplace*. Princeton, NJ: Princeton University Press.

Borst, Arno. 1957–1963. *Der Turmbau von Babel: Geschichte der Meinungen über Ursprung und Vielfalt der Sprachen und Völker.* 4 vols. Stuttgart: Hiersemann.

Bostetter, Edward. 1962. "The Nightmare World of *The Ancient Mariner*." *Studies in Romanticism*, 1:241–254.

Bowie, Andrew. 1997. *From Romanticism to Critical Theory*. New York: Routledge.

Branch, Lori. 2006. *Rituals of Spontaneity: Sentiment and Secularism from Free Prayer to Wordsworth*. Waco, TX: Baylor University Press.

Brandom, Robert. 2004. *Making it Explicit*. Cambridge, MA: Harvard University Press.

Brennan, Tad. 2003. "Stoic Moral Psychology." In *The Cambridge Companion to The Stoics*, 257–294. Cambridge: Cambridge University Press.

Brenner, Dietrich. 1990. *Wilhelm von Humboldts Bildungstheorie*. Weinheim: Juventa.

Brett, Annabel S. 1997. *Liberty, Right and Nature*. Cambridge: Cambridge University Press.

Brewer, John. 2000. *The Pleasures of Imagination: English Culture in the Eighteenth Century*. Chicago: University of Chicago Press.

Brewer, John, and J. H. Plumb. 1982. *The Birth of a Consumer Society*. Bloomington: Indiana University Press.

Brice, Benjamin. 2007. *Coleridge and Scepticism*. Oxford: Oxford University Press.

Broackes, Justin, ed. 2012. *Iris Murdoch: Philosopher*. Oxford: Oxford University Press.

Broad, C. D. 1929. *The Mind and its Place in Nature*. New York: Harcourt, Brace & Co.

Brooks, Peter. 1992. *Reading for the Plot: Design and Intention in Narrative*. Cambridge, MA: Harvard University Press.

Brown, David A. 1997. *Lorenzo Lotto: Rediscovered Master of the Renaissance*. Washington, DC: National Gallery of Art.

Brown, Peter. 2000. *Augustine of Hippo*. Berkeley: University of California Press.

Bruford, W. H. 1975. *The German Tradition of Self-Cultivation*. Cambridge: Cambridge University Press.

Buber, Martin. 1974. *Ich und Du*. Gütersloh: Lambert Schneider.

Buckley, Michael. 1987. *At the Origins of Modern Atheism*. New Haven, CT: Yale University Press.

———. 2004. *Denying and Disclosing God: The Ambiguous Progress of Modern Atheism*. New Haven, CT: Yale University Press.

Burckhardt, Jakob. 1979. *Reflections on History*. Indianapolis: Liberty Fund.

———. 2001. *The Letters of Jakob Burckhardt*. Edited by Alexander Dru. Indianapolis: Liberty Fund.

Burke, Edmund. 1968. *Reflections on the Revolution in France*. Edited by Conor Cruise O'Brien. Harmondsworth: Penguin.

Burnell, Peter. 2005. *The Augustinian Person*. Washington, DC: Catholic University of America Press.

Burrell, David, CSC. 2005. "Analogy, Creation, and Theological Language." In *The Theology of Thomas Aquinas*, edited by Rik van Nieuwenhove and Joseph Wawrykow, 77–98. Notre Dame, IN: University of Notre Dame Press.

———. 2008. *Aquinas: God & Action*. Scranton: University of Scranton Press.

Butler, Joseph. 1726. *Fifteen Sermons, Preached at Rolls Chapel*. London: Knapton.

Campbell, T. D. 1982. "Francis Hutcheson: 'Father' of the Scottish Enlightenment." In *The Origins and Nature of the Scottish Enlightenment*, edited by R. H. Campbell and Andrew S. Skinner, 167–185. Edinburgh: John Donald.

Carr, Thomas K. 1996. *Newman & Gadamer: Towards a Hermeneutics of Religious Knowledge*. Atlanta, GA: Scholars.

Cassiday, Augustine, and Frederick W. Norris, eds. 2007. *The Cambridge History of Christianity*. Vol. 2. Cambridge: Cambridge University Press.

Cassirer, Ernst. 1951. *Philosophy of the Enlightenment*. Translated by Fritz C. A. Koelln and James P. Pettegrove. Princeton, NJ: Princeton University Press.

———. 1953. *The Platonic Renaissance in England*. Translated by James P. Pettegrove. Austin: University of Texas Press.

———. 1955–1966. *The Philosophy of Symbolic Forms*. 3 vols. Translated by Ralph Manheim. New Haven, CT: Yale University Press.

———. 1974. *Das Erkenntnisproblem in der neueren Philosophie*. 4 vols. Darmstadt: Wissenschaftliche Buchgesellschaft.

———. 1995. *Nachgelassene Manuskripte und Texte*. Edited by John Michael Krois and Oswald Schwemmer. Hamburg: Meiner.

Cassirer, Fritz. 1925. *Beethoven und die Gestalt*. Stuttgart: Deutsche Verlags-Anstalt.
Cavell, Stanley. 1989. *In Quest for the Ordinary*. Chicago: University of Chicago Press.
Chadwick, Owen. 1975. *The Secularization of the European Mind in the 19th Century*. Cambridge: Cambridge University Press.
———. 1987. *From Bossuet to Newman*. Cambridge: Cambridge University Press.
Chalmers, David J. 1996. *The Conscious Mind: In Search of a Fundamental Theory*. Oxford: Oxford University Press.
Chamberlain, Charles. 1984. "The Meaning of *Prohairesis* in Aristotle's Ethics." *Transactions of the American Philological Association*, 114:147–157.
Chandler, James. 1984. *Wordsworth's Second Nature*. Chicago: University of Chicago Press.
———. 1998. *England in 1819*. Chicago: University of Chicago Press.
———. 2010. "The Politics of Sentiment: Notes toward a New Account." *Studies in Romanticism*, 49(4):553–575.
Charlton, William. 1980. "Aristotle's Definition of the Soul." *Phronesis*, 25:170–187.
Chisholm, Roderick M. 1976. *Person and Object*. Chicago: Open Court.
Christensen, Jerome. 1981. *Coleridge's Blessed Machine of Language*. Ithaca, NY: Cornell University Press.
———. 1987. *Practicing Enlightenment: Hume and the Formation of a Literary Career*. Madison: University of Wisconsin Press.
Chua, Daniel K. 1999. *Absolute Music and the Construction of Meaning*. Cambridge: Cambridge University Press.
Cicero, Marcus Tullius. *In Pisonem*. Perseus Project: www.perseus.tufts.edu/hopper/text?doc=Perseus%3atext%3a1999.02.0013%3atext%3dPis.
———. 1945. *Tusculan Disputations*. Translated by J. E. King. Loeb Classical Library. Cambridge, MA: Harvard University Press.
———. 2005. *On Duties/De Officiis*. Translated by Walter Miller. Loeb Classical Library. Cambridge, MA: Harvard University Press.
Clark, David W. 1971. "Voluntarism and Rationalism in the Ethics of Ockham." *Franciscan Studies*, 31:72–87.
Clark, Mary T., RSCJ. 1993. "Augustine on Person: Divine and Human." In *Augustine: Presbyter Factus Sum*, edited by Joseph T. Lienhard et al. New York: Peter Lang.
Clarke, W. Norris, SJ. 1993. *Person & Being*. Milwaukee: Marquette University Press.
———. 1995. *Explorations in Metaphysics*. Notre Dame, IN: University of Notre Dame Press.
Cockburn, David, ed. 1991. *Human Beings*. Cambridge: Cambridge University Press.
Coleridge, Samuel Taylor. 1956–1971. *The Collected Letters of Samuel Taylor Coleridge*. Edited by Leslie Griggs. 6 vols. Oxford: Clarendon.
———. 1957–2002. *Notebooks*. Edited by Kathleen Coburn et al. 5 vols. Princeton, NJ: Princeton University Press.
———. 1969. *The Friend*. Edited by Barbara E. Rooke. 2 vols. Princeton, NJ: Princeton University Press.
———. 1971. *Lectures 1795 on Politics and Religion*. Edited by Lewis Patton and Peter Mann. Princeton, NJ: Princeton University Press.
———. 1972. *Lay Sermons*. Edited by R. J. White. Princeton, NJ: Princeton University Press.

———. 1976. *On the Constitution of Church and State*. Edited by John Colmer. Princeton, NJ: Princeton University Press.
———. 1979. *Inquiring Spirit*. Edited by Kathleen Coburn. Toronto: University of Toronto Press.
———. 1980–2001. *Marginalia*. Edited by George Whalley, Heather J. Jackson, et al. 6 vols. Princeton, NJ: Princeton University Press.
———. 1981. *Logic*. Edited by J. R. de J. Jackson. Princeton, NJ: Princeton University Press.
———. 1983. *Biographia Literaria*. Edited by James Engell and Walter Jackson Bate. 2 vols. Princeton, NJ: Princeton University Press.
———. 1990. *Table Talk*. Edited by Carl Woodring. 2 vols. Princeton, NJ: Princeton University Press.
———. 1993. *Aids to Reflection*. Edited by John Beer. Princeton, NJ: Princeton University Press.
———. 1995. *Shorter Works & Fragments*. Edited by H. J. Jackson and J. R. de J. Jackson. 2 vols. Princeton, NJ: Princeton University Press.
———. 2000. *Lectures on the History of Philosophy, 1818–1819*. Edited by J. R. de J. Jackson. 2 vols. Princeton, NJ: Princeton University Press.
———. 2001. *Poetical Works*. 3 vols. Edited by J. C. C. Mays. Princeton, NJ: Princeton University Press
———. 2002. *Essays on His Times*. Edited by David V. Erdman. 3 vols. Princeton, NJ: Princeton University Press.
———. 2002. *Opus Maximum*. Edited by Thomas McFarland and Nick Halmi. Princeton, NJ: Princeton University Press.
Colish, Marcia L. 1985. *The Stoic Tradition from Antiquity to the Middle Ages*. Leiden: Brill.
Colley, Linda. 1992. *Britons: Forging the Nation, 1707–1837*. New Haven, CT: Yale University Press.
Conway, Daniel W. 1994. "Genealogy and Critical Method." In *Nietzsche, Genealogy, Morality*, edited by Richard Schacht. Berkeley: University of California Press.
Cousins, Ewert. 1970. "A Theology of Interpersonal Relations." *Thought*, 45:56–82.
Crary, Jonathan. 1992. *Techniques of the Observer: Vision and Modernity in the Nineteenth Century*. Cambridge, MA: MIT Press.
Crosby, John F. 1996. *The Selfhood of the Human Person*. Washington, DC: Catholic University of America Press.
Cross, Richard. 2001. "'Where Angels Fear to Tread': Duns Scotus and Radical Orthodoxy." *Antonianum*, 76:7–41.
Cudworth, Ralph. 1678. *True Intellectual System of the Universe*. Rept. Hildesheim: G. Olms Verlag.
———. 1979. *A Treatise Concerning Eternal and Immutable Morality* (1731) and *Of Freewill*. (1838). Hildesheim: G. Olms Verlag.
Curran, Stuart. 1972. "Blake and the Gnostic Hyle: A Double Negative." *Blake Studies*, 4:117–133.
Cutsinger, James P. 1987. *The Form of Transformed Vision: Coleridge and the Knowledge of God*. Macon, GA.: Mercer University Press.

Dahlhaus, Carl. 1984. *Die Musiktheorie im 18. und 19. Jahrhundert*. Darmstadt: Wissenschaftliche Buchgesellschaft.

———. 1988. *Klassische und Romantische Musikästhetik*. Laaber: Laaber Verlag.

Dallmayr, Fred R., and Thomas A. McCarthy. 1977. *Understanding and Social Inquiry*. Notre Dame, IN: University of Notre Dame Press.

Darwall, Stephen L. 1989. "Motive and Obligation in the British Moralists." *Social Philosophy & Policy*, 7:133–150.

———. 1995. *The British Moralists and the Internal 'Ought,' 1640–1740*. Cambridge: Cambridge University Press.

Darwin, Charles. 1981. *The Descent of Man, and Selection in Relation to Sex*. Edited by John Tyler Bonner and Robert M. May. Princeton, NJ: Princeton University Press.

———. 2003. *On the Origin of Species*. 1st ed., 1859. Edited by Joseph Carroll. Peterborough, ON: Broadview.

Davidson, Graham. 2006. "Duty and Power: Conflicts of the Will in Coleridge's *Opus Maximum*." In *Coleridge's Assertion of Religion: Essays on the Opus Maximum*, edited by Jeffrey W. Barbeau, 212–244. Leuven: Peeters.

Deigh, John. 1996. "Reason and Ethics in Hobbes's *Leviathan*." *Journal of the History of Philosophy*, 34(1):33–60.

De Maistre, Joseph. 1996. *Considerations on France*. Translated by Richard A. Lebrun. Cambridge: Cambridge University Press.

De Man, Paul. 1982. "The Resistance to Theory." *Yale French Studies*, 63:3–20.

———. 1984. "Autobiography as De-Facement." In *The Rhetoric of Romanticism*, 67–82. New York: Columbia University Press.

Den Uyl, Douglas. 1998. "Shaftesbury and the Modern Problem of Virtue." *Social Philosophy & Policy*, 15:275–316.

Den Uyl, Douglas, and Charles L. Griswold. 1996. "Adam Smith on Friendship and Love." *Review of Metaphysics*, 49(3):609–637.

Derrida, Jacques. 1978. *Writing and Difference*. Translated by Alan Bass. Chicago: University of Chicago Press.

Descartes, René. 1996. *Meditations on First Philosophy*. Edited by John Cottingham. Cambridge: Cambridge University Press.

Diamond, Cora. 1988. "Losing Your Concepts." *Ethics*, 98(2):255–277.

Dihle, Albrecht. 1982. *The Theory of Will in Classical Antiquity*. Berkeley: University of California Press.

Dod, Bernard G. 1982. "Aristoteles Latinus." In *The Cambridge History of Later Medieval Philosophy: From the Rediscovery of Aristotle to the Disintegration of Scholasticism 1100–1600*. Edited by Norman Kretzmann, Anthony Kenny, Jan Pinborg, and Eleonore Stump. Cambridge: Cambridge University Press.

Dostoevsky, Fyodor. 2005. *The Brothers Karamazov*. Translated by Richard Pevear and Larissa Volokhonsky. New York: Everyman.

Dupré, Louis. 1993. *Passage to Modernity*. New Haven, CT: Yale University Press.

———. 2004. *The Enlightenment and the Intellectual Foundations of Modern Culture*. New Haven, CT: Yale University Press.

Eagleton, Terry. 1995. "The Death of Self Criticism." *Times Literary Supplement,* 24 November, 6–7.
———. 1996. *The Illusions of Postmodernism.* Oxford: Blackwell.
———. 2009. *The Trouble with Strangers: A Study of Ethics.* Oxford: Wiley-Blackwell.
Eaves, Morris. 1992. *The Counter-Arts Conspiracy: Art and Industry in the Age of Blake.* Ithaca, NY: Cornell University Press.
Ebbatson, J. R. 1972. "Coleridge's Mariner and the Rights of Man," *Studies in Romanticism,* 11:171–206.
Eberl, Jason T. 2004. "Aquinas on the Nature of Human Beings." *Review of Metaphysics,* 58(2):333–365.
Eck, Caroline van. 1994. *Organicism in Nineteenth-Century Architecture: An Inquiry into its Theoretical and Philosophical Background.* Amsterdam: Architectura & Natura.
Edwards, Pamela. 2004. *The Statesman's Science: History, Nature, and Law in the Political Thought of Samuel Taylor Coleridge.* New York: Columbia University Press.
Eisenach, J. E. 1982. "Hobbes on Church and State and Religion." *History of Political Thought,* 3:215–243.
Elam, Helen Reguiero, and Frances Ferguson, eds. 2005. *The Wordsworthian Enlightenment: Romantic Poetry and the Ecology of Reading.* Baltimore: Johns Hopkins University Press.
Elias, Norbert. 2000. *The Civilizing Process.* Translated by Edmund Jephcott. 2 vols. Oxford: Blackwell.
Eliot, George. 1990. *Selected Essays, Poems, and Other Writings.* Edited by A. S. Byatt and N. Warren. Harmondsworth: Penguin.
———. 1995. *Daniel Deronda.* Edited by Terence Cave. Harmondsworth: Penguin.
Eliot, T. S. 1975. *The Selected Prose of T. S. Eliot.* Edited by Frank Kermode. London: Faber & Faber.
Ellison, Julie. 1990. *Delicate Subjects: Romanticism, Gender, and the Ethics of Understanding.* Ithaca, NY: Cornell University Press.
Engberg-Pederson, T. 1979. "More on Aristotelian Epagoge." *Phronesis,* 24(3):301–319.
Engell, James. 2002. "Coleridge and His Mariner on the Soul: 'As an exile in a far distant land.'" In *The Fountain Light: Studies in Romanticism and Religion,* edited by J. Robert Barth, SJ. New York: Fordham University Press.
Engstrom, Stephen. 1998. "Happiness and the Highest Good in Aristotle and Kant." In *Aristotle, Kant, and the Stoics,* edited by Stephen Engstrom and Jennifer Whiting. Cambridge: Cambridge University Press.
Erdman, David V. 1954. *Blake: Prophet against Empire.* New York: Dover.
Esposito, Roberto. 2008. *Bios: Biopolitics and Philosophy.* Translated and edited by Timothy Campbell. Minneapolis: University of Minnesota Press.
Esterhammer, Angela. 2000. *The Romantic Performative: Language and Action in British and German Romanticism.* Stanford, CA: Stanford University Press.
Evans, Murray J. 2006. "Reading 'Will' in Coleridge's *Opus Maximum:* The Rhetoric of Transition and Repetition." In *Coleridge's Assertion of Religion: Essays on the Opus Maximum,* edited by Jeffrey W. Barbeau, 73–96. Leuven: Peeters.

Everett, Nigel. 1994. *The Tory View of Landscape*. New Haven CT: Yale University Press.

Fairbairn, Andrew M. 1885. "Catholicism and Modern Thought." *Contemporary Review,* 47:652–674.

Ferguson, Frances. 1992. *Solitude and the Sublime: Romanticism and the Aesthetics of Individuation*. New York: Routledge.

———. 1999. "Coleridge and the Deluded Reader." In *The Rime of the Ancient Mariner,* edited by Paul H. Fry, 113–130. New York: Bedford/St. Martin's.

———. 2005. "Organic Form and its Consequences." In *Land, Nation, and Culture, 1740–1840,* edited by Peter de Bolla, David Simpson, and Nigel Leask. London: Palgrave Macmillan.

Filoramo, Giovanni. 1990. *A History of Gnosticism*. Translated by Anthony Alcock. Oxford: Blackwell.

Fink, Eugen. 1960. *Spiel als Weltsymbol*. Stuttgart: Kohlhammer.

Fish, Stanley. 1999. *Professional Correctness: Literary Studies and Political Change*. Cambridge, MA: Harvard University Press.

Flaubert, Gustave. 1966. *Madame Bovary*. Translated by Francis Steegmüller. New York: Everyman.

Fontane, Theodor. 1995. *The Stechlin*. Translated by William L. Zwiebel. Columbia, SC: Camden House.

Foot, Philippa. 2001. *Natural Goodness*. New York: Oxford University Press.

———. 2002. *Moral Dilemmas*. New York: Oxford University Press.

Foucault, Michel. 1994. *The Order of Things*. New York: Vintage.

Frank, Manfred. 1975. *Der Unendliche Mangel an Sein: Schellings Hegelkritik und die Anfänge der Marxschen Dialektik*. Frankfurt: Suhrkamp.

———. 1977. *Schleiermacher: Hermeneutik und Kritik*. Frankfurt: Suhrkamp.

———. 1986. *Die Unhintergehbarkeit von Subjektivität*. Frankfurt: Suhrkamp.

———. 1989. *Einführung in die frühromantische Ästhetik*. Frankfurt: Suhrkamp.

———. 1997. *The Subject and the Text: Essays on Literary Theory and Philosophy*. Translated by Andrew Bowie. Cambridge: Cambridge University Press.

Frede, Michael, and Gisela Striker, eds. 1996. *Rationality in Greek Thought*. Oxford: Clarendon.

Freud, Sigmund. 1963. *General Psychological Theory*. New York: Macmillan.

———. 1969–1975. *Studienausgabe*. Edited by Alexander Mitscherlich. 10 vols. Frankfurt: Fischer.

Funkenstein, Amos. 1986. *Theology and the Scientific Imagination*. Princeton, NJ: Princeton University Press.

Furniss, Tom. 1993. *Edmund Burke's Aesthetic Ideology: Language, Gender, and Political Economy in Revolution*. Cambridge: Cambridge University Press.

Gadamer, Hans-Georg. 1977. *Philosophical Hermeneutics*. Translated by David E. Linge. Berkeley: University of California Press.

———. 2006. *Truth and Method*. Translated by Joel Weinsheimer and Donald G. Marshall. New York: Continuum.

Gallagher, Catherine. 1987. "The Body versus the Social Body in the Works of Thomas Malthus and Henry Mayhew." In *The Making of the Modern Body: Sexuality and Society in the Nineteenth Century,* edited by Catherine Gallagher and Thomas Laqueur, 83–106. Berkeley: University of California Press.
Gauchet, Marcel. 1999. *The Disenchantment of the World.* Princeton, NJ: Princeton University Press.
Gaukroger, Stephen. 2006. *The Emergence of a Scientific Culture: Science and the Shaping of Modernity.* Oxford: Clarendon.
Gay, Peter. 1977. *The Enlightenment: The Science of Freedom.* New York: Norton.
———. 2007. *Modernism: The Lure of Heresy: From Baudelaire to Beckett and Beyond.* New York: Norton.
Gehlen, Arnold. 1978. *Der Mensch: Seine Natur und Stellung in der Welt 1940.* Wiesbaden: Akademische Verlagsgesellschaft Athenaion.
———. 1988. *Man: His Nature and Place in the World.* Translated by Clare McMillan and Karl Pillemer. New York: Columbia University Press.
———. 2004. *Urmensch und Spätkultur.* Frankfurt: Klostermann.
Giddens, Anthony. 1990. *The Consequences of Modernity.* Stanford, CA: Stanford University Press.
Gigante, Denise. 2008. *The Great Age of the English Essay: An Anthology.* New Haven, CT: Yale University Press.
Gillespie, Michael A. 2007. *The Theological Origins of Modernity.* Chicago: University of Chicago Press.
Gilson, Étienne. 2009. *From Aristotle to Darwin and Back Again: A Journey in Final Causality and Species Evolution.* San Francisco: Ignatius.
Ginsborg, Hannah. 2004. "Two Kinds of Mechanical Inexplicability in Kant and Aristotle." *Journal of the History of Philosophy,* 42(1):33–65.
Godwin, William. 1985. *Enquiry Concerning Political Justice.* 1795 edition. Edited by Isaac Kramnick. Harmondsworth: Penguin.
Goethe, Johann Wolfgang von. 1981. *Werke.* Edited by Erich Trunz, Hamburger Ausgabe. 14 vols. Munich: Beck.
———. 1989. *Wilhelm Meister's Apprenticeship.* Translated by Eric A. Blackall. Princeton, NJ: Princeton University Press.
Goffman, Irving. "Embarrassment and Social Organization." *American Journal of Sociology,* 62(3):264–271.
Goldsmith, Steven. 1993. *Unbuilding Jerusalem: Apocalypse and Romantic Representation.* Ithaca, NY: Cornell University Press.
Goodman, Kevis. 2005. "Making Time for History: Wordsworth, the New Historicism, and the Apocalyptic Fallacy." In *The Wordsworthian Enlightenment: Romantic Poetry and the Ecology of Reading,* edited by Helen Reguiero Elam and Frances Ferguson, 158–171. Baltimore: Johns Hopkins University Press.
Gould, Stephen J. 1990. *Wonderful Life: The Burgess Shale and the Nature of History.* New York: Norton.

Gregory, Alan P. R. 2003. *Coleridge and the Conservative Imagination.* Macon, GA: Mercer University Press.

———. 2006. "'That I may be here': Human Persons and Divine Personeity in the *Opus Maximum.*" In *Coleridge's Assertion of Religion: Essays on the Opus Maximum,* edited by Jeffrey W. Barbeau, 187–212. Leuven: Peeters.

Gregory, Brad S. 2012. *The Unintended Reformation: How a Religious Revolution Secularized Society.* Cambridge, MA: Harvard University Press.

Grene, Marjorie, and David Depew. 2004. *The Philosophy of Biology: An Episodic History.* New York: Cambridge University Press.

Griffiths, Paul. 2009. *Intellectual Appetite: A Theological Grammar.* Washington, DC: Catholic University of America Press.

Grillmeier, Alois, SJ. 1995. *Christ in Christian Tradition: From the Council of Chalcedon (451) to Gregory the Great (590–604).* Translated by Pauline Allen and John Cawte. Louisville, KY: Westminster John Knox.

Griswold, Charles L. 1998. *Adam Smith and the Virtues of Enlightenment.* Cambridge: Cambridge University Press.

Grotius, Hugo. 2005. *The Rights of War and Peace.* 3 vols. Edited by Richard Tuck. Indianapolis: Liberty Fund.

Gunton, Colin E. 2005. *The One, the Three, and the Many: God, Creation, and the Culture of Modernity.* Cambridge: Cambridge University Press.

Haakonssen, Knud. 1996. *Natural Law and Moral Philosophy: From Grotius to the Scottish Enlightenment.* New York: Cambridge University Press.

Habermas, Jürgen. 1983. *Philosophical-Political Profiles.* Translated by Frederick G. Lawrence. Cambridge, MA: MIT Press.

———. 1990. *Understanding and Social Inquiry.* Translated by Christian Lenhardt and Shierry Weber Nicholsen. Cambridge, MA: MIT Press.

———. 1991. *Structural Transformation of the Public Sphere.* Translated by Thomas Burger. Cambridge, MA: MIT Press.

Hadley, Elaine. 2010. *Living Liberalism: Practical Citizenship in Mid-Victorian Britain.* Chicago: University of Chicago Press.

Hadot, Pierre. 1993. *Plotinus or the Simplicity of Vision.* Translated by Michael Chase. Chicago: University of Chicago Press.

Halmi, Nicholas. 2007. *The Genealogy of the Romantic Symbol.* Oxford: Oxford University Press.

Hanby, Michael. 2003. *Augustine and Modernity.* New York: Routledge.

Hankinson, R. J. 2003. "Stoic Epistemology." In *The Cambridge Companion to Stoicism,* edited by Brad Inwood, 64–84. Cambridge: Cambridge University Press.

Harrison, Gary. 2002. "Ecological Apocalypse: Privation, Alterity, and Catastrophe in the Work of Arthur Young and Thomas Robert Malthus." In *Romanticism and Millenarianism,* edited by Tim Fulford, 103–120. New York: Palgrave Macmillan.

Hedley, Douglas. 2000. *Coleridge, Philosophy, and Religion.* Cambridge: Cambridge University Press.

———. 2008. *Living Forms of the Imagination.* London. T. & T. Clark.

Hegel, G. W. F. 1952. *Phänomenologie des Geistes*. Edited by Johannes Hoffmeister. Hamburg: Meiner.
———. 1970. *Enzyklopädie der philosophischen Wissenschaften*. Edited by Eva Moldenhauer and Karl Markus Michel. Frankfurt: Suhrkamp.
———. 1977. *Phenomenology of Spirit*. Translated by A. V. Miller. New York: Oxford University Press.
———. 1988. *The Difference between Fichte's and Schelling's System of Philosophy*. Translated by H. S. Harris and Walter Cerf. Albany: State University of New York Press.
———. 1989. *Science of Logic*. Translated by A. V. Miller. Atlantic Heights, NJ: Humanities.
———. 1998. *Aesthetics: Lectures on Fine Art*. Translated by T. M. Knox. Oxford: Clarendon.
———. 1998. *Vorlesungen zur Ästhetik*. Frankfurt: Suhrkamp.
Heidegger, Martin. 1962. *Kant and the Problem of Metaphysics*. Translated by James Churchill. Bloomington: Indiana University Press.
———. 1964. "The Origin of the Work of Art." In *Philosophies of Art and Beauty*, edited by Albert Hofstadter, 650–708. Chicago: University of Chicago Press.
———. 1969. *Identity and Difference*. Translated by Joan Stambaugh. New York: Harper.
———. 1977. "The Age of the World Picture." In *The Question Concerning Technology and Other Essays*. Translated by William Lovitt. New York: Harper & Row.
———. 1978. *Identität und Differenz*. Pfullingen: Neske.
———. 1979. *Sein und Zeit*. Tübingen: Max Niemeyer.
———. 1980. *Holzwege*. Frankfurt: Klostermann.
———. 1988. *Hegel's Phenomenology of Spirit*. Translated by Parvis Emad and Kenneth Maly. Bloomington: Indiana University Press.
———. 1996. *Being and Time*. Translated by Joan Stambaugh. Albany: State University of New York Press.
Heraclitus. 1962. *Heraclitus: The Cosmic Fragments*. Translated by G. S. Kirk. Cambridge: Cambridge University Press.
Herdt, Jennifer A. 1997. *Religion and Faction in Hume's Moral Philosophy*. Cambridge: Cambridge University Press.
———. 2007. *Putting on Virtue: The Legacy of the Splendid Vices*. Chicago: University of Chicago Press.
Herman, Barbara. 1996. "Making Room for Character." In *Aristotle, Kant, and the Stoics*, edited by Stephen Engstrom and Jennifer Whiting. Cambridge: Cambridge University Press.
———. 1998. "Training to Autonomy: Kant and the Question of Moral Education." In *Philosophers on Education*, edited by Amélie O. Rorty. New York: Routledge.
Hirschman, Albert. 1977. *The Passions and the Interests: Political Arguments for Capitalism before its Triumph*. Princeton, NJ: Princeton University Press.
Hobbes, Thomas. 1994. *Leviathan*. Edited by Edwin Curley. Indianapolis: Hackett.
———. 1999. "Of Liberty and Necessity." In *Hobbes and Bramhall on Liberty and Necessity*, edited by Vere Chappell. Cambridge: Cambridge University Press.

Hodge, Jonathan, and Gregory Radick, eds. 2003. *The Cambridge Companion to Darwin.* Cambridge: Cambridge University Press.
Hoekstra, Kinch. 2003. "Hobbes on Law, Nature, and Reason." *Journal of the History of Philosophy,* 41:111–120.
Hofstadter, Albert. 1976. *Philosophies of Art and Beauty.* Chicago: University of Chicago Press.
Hölderlin, Friedrich. 1943–1988. *Sämtliche Werke.* Edited by F. Beissner. 8 vols. Stuttgart: Kohlhammer.
———. 1987. *Essays and Letters on Theory.* Translated and edited by Thomas Pfau. Albany: State University of New York Press.
Holmes, Richard. 1999. *Coleridge: Darker Reflections, 1804–1834.* New York: Pantheon.
Holton, Richard. 2009. *Willing, Wanting, Waiting.* Oxford: Clarendon.
Hopkins, G. M. 2006. *The Collected Works of Gerard Manley Hopkins.* Vol. 4. Edited by Leslie Higgins. Oxford: Oxford University Press.
Hudson, Nicholas. 2005. "Theories of Language." In *The Cambridge History of Literary Criticism,* vol. 4, 335–348. Cambridge: Cambridge University Press.
Huizinga, Johan. 1998; 1st ed. Dutch, 1927. *Homo Ludens: A Study of the Play-Element in Culture.* London: Routledge.
———. 1998. *The Waning of the Middle Ages.* New York: Dover.
Hulliung, Mark. 1994. *The Autocritique of Enlightenment: Rousseau and the Philosophes.* Cambridge, MA: Harvard University Press.
———. 2001. "Rousseau, Voltaire, and the Revenge of Pascal." In *The Cambridge Companion to Rousseau,* edited by Patrick Riley, 57–77. Cambridge: Cambridge University Press.
Humboldt, Wilhelm von. 1980. *Werke in Fünf Bänden.* Edited by Andreas Flitner and Klaus Giel. Darmstadt: Wissenschaftliche Buchgesellschaft.
Hume, David. 1826. *The Life of David Hume, Written by Himself.* London: Hunt and Clarke.
———. 1902. *Enquiries, Concerning Human Understanding and Concerning the Principles of Morals.* Edited by L. A. Selby-Bigge. Oxford: Clarendon.
———. 1932. *The Letters of David Hume.* Edited by John Young Thomson Greig. 2 vols. Oxford: Clarendon.
———. 1985. *Essays Moral, Political, and Literary.* Edited by Eugene F. Miller. Indianapolis: Liberty Fund.
———. 2001. *A Treatise of Human Nature.* Edited by David F. Norton and Mary J. Norton. Oxford: Oxford University Press.
———. 2007. *Dialogues concerning Natural Religion.* Edited by Dorothy Coleman. Cambridge: Cambridge University Press.
Humfrey, Peter. 1997. *Lorenzo Lotto.* New Haven, CT: Yale University Press, 1997.
Husserl, Edmund. 1965. "The Crisis of the European People." In *Phenomenology and the Crisis of Philosophy,* 149–192. New York: Harper & Row.
———. 1969. *Formal and Transcendental Logic.* Translated by Dorion Cairns. The Hague: Martinus Nijhoff.

———. 1970. *The Crisis of the European Sciences and Transcendental Phenomenology*. Edited by David Carr. Evanston, IL: Northwestern University Press.
———. 1980. *Logische Untersuchungen*. 2 vols. 6th ed. Tübingen: Niemeyer.
———. 2006. *The Basic Problems of Phenomenology: From the Lectures, Winter Semester, 1910–1911*. Translated by Ingo Farin and James G. Hart. New York: Springer.
Hutcheson, Francis. 2002. *An Essay on the Nature and Conduct of the Passions and Affections, with Illustrations on the Moral Sense*. Edited by Aaron Garrett. Indianapolis: Liberty Fund.
———. 2008. *Inquiry into the Original of Our Ideas of Beauty and Virtue*. Edited by Wolfgang Leidhold. Indianapolis: Liberty Fund.
Hütter, Reinhard. 2005. "The Directedness of Reasoning and the Metaphysics of Creation." *Reason and the Reasons of Faith*. Edited by Paul Griffiths and Reinhard Hütter. London: T. & T. Clark.
———. 2007. "St. Thomas on Grace and Free Will in the *Initium Fidei*: The Surpassing Augustinian Synthesis." *Nova et Vetera*, 5(3):521–554.
Hyman, Gavin. 2010. *A Short History of Atheism*. London: I. B. Tauris.
Hyppolite, Jean. 1974. *Genesis and Structure of Hegel's Phenomenology of Spirit*. Translated by Samuel Cherniak and John Heckman. Evanston, IL: Northwestern University Press.
Inwood, Brad, ed. 2003. *The Cambridge Companion to the Stoics*. Cambridge: Cambridge University Press.
Jacobi, Friedrich Heinrich. 1994. *The Main Philosophical Writings and the Novel Allwill*. Translated and edited by George di Giovanni. Montreal: McGill-Queen's University Press.
Jaeger, Werner. 1986. *Paideia: The Ideals of Greek Culture*. Translated by Gilbert Highet. Oxford: Oxford University Press.
Jager, Colin. 2007. *The Book of God: Secularization and Design in the Romantic Era*. Philadelphia: University of Pennsylvania Press.
Jakobson, Roman. 1987. "Linguistics and Poetics." In *Language and Literature*. Edited by Krystyna Pomorska. Cambridge, MA: Harvard University Press.
James, William. 1975. *Pragmatism: A New Name for Some Old Ways of Thinking*. Cambridge MA: Harvard University Press.
———. 1992. *Writings, 1878–1899*. Edited by Gerald E. Meyers. New York: Library of America.
Jameson, Fredric. 1993. *Postmodernism, or: The Cultural Logic of Late Capitalism*. Durham, NC: Duke University Press.
———. 2002. *A Singular Modernity: Essay on the Ontology of the Present*. New York: Verso.
Jauss, Hans-Robert. 2005 "Modernity and Literary Tradition." Translated by Christian Thorne. *Critical Inquiry*, 31(2):329–364.
Johnson, Barbara. 2010. *Persons and Things*. Cambridge, MA: Harvard University Press.
Jonas, Hans. 1969. *The Gnostic Religion*. Boston: Beacon.
Kafka, Franz. 1994. *Zur Frage der Gesetze und andere Schriften aus dem Nachlass*. Frankfurt: Fischer.

Kahn, Charles H. 1988. "Discovering the Will: From Aristotle to Augustine." In *The Question of Eclecticism: Studies in Later Greek Philosophy,* edited by J. M. Dillon and A. A. Long. Berkeley: University of California Press.
Kant, Immanuel. 1951. *Critique of Judgment.* Translated by J. H. Bernard. New York: Macmillan.
———. 1965. *Critique of Pure Reason.* Translated by Norman Kemp Smith. New York: St. Martin's.
———. 1981. *Grounding for the Metaphysics of Morals.* Translated by James W. Ellington. Indianapolis: Hackett.
———. 1981. *Kritik der Urteilskraft.* Edited by Wilhelm Weischedel. Frankfurt: Suhrkamp.
———. 1981. *Werkausgabe.* Edited by Wilhelm Weischedel. 12 vols. Frankfurt: Suhrkamp.
———. 1983. *Perpetual Peace and Other Essays.* Translated by Ted Humphrey. Indianapolis: Hackett.
———. 2001. *Religion and Rational Theology.* Translated and edited by Allen W. Wood and George di Giovanni. Cambridge: Cambridge University Press.
———. 2004. *Prolegomena to Any Future Metaphysics.* Edited by Gary Hatfield. Cambridge: Cambridge University Press.
Keane, Patrick J. 1994. *Coleridge's Submerged Politics.* Columbia: University of Missouri Press.
Keats, John. 1970. *The Complete Poems.* Edited by Miriam Allott. London: Longman.
Kemp Smith, Norman. 1941. *The Philosophy of David Hume.* London: Macmillan.
Kenny, Anthony. 1979. *Aristotle's Theory of the Will.* London: Duckworth.
Kent, Bonnie. 1995. *Virtues of the Will: The Transformation of Ethics in the Late Thirteenth Century.* Washington, DC: Catholic University of America Press.
Ker, Ian. 1988. *John Henry Newman.* Oxford: Oxford University Press.
Kerr, Fergus. 2002. *After Aquinas: Versions of Thomism.* London: Blackwell.
Kirk, G. S. 1962. *Heraclitus: The Cosmic Fragments.* Cambridge: Cambridge University Press.
Kirk, G. S., J. E. Raven, and M. Schofield. 1983. *The Presocratic Philosophers.* Cambridge: Cambridge University Press.
Kirwan, Christopher. 1989. *Augustine.* New York: Routledge.
Kitson, Peter J. 1996. "Coleridge, the French Revolution and the 'Ancient Mariner.'" *Coleridge Bulletin,* 7:30–48.
Kleist, Heinrich von. 1988. "Michael Kohlhaas." In *The Marquise of O. and other Stories.* Harmondsworth: Penguin.
———. 1993. "Michael Kohlhaas." In *Sämtliche Werke und Briefe,* edited by Helmut Sembdner. 2 vols. Munich: Hanser.
Kolb, Daniel. 1992. "Kant, Teleology, and Evolution." *Synthese,* 91(2):9–28.
Koselleck, Reinhard. 1988. *Critique and Crisis.* Cambridge, MA: MIT Press.
———. 2004. *Futures Past: On the Semantics of Historical Time.* Translated by Keith Tribe. New York: Columbia University Press.
Kramnick, Isaac. 1990. *Republicanism & Bourgeois Radicalism.* Ithaca, NY: Cornell University Press.

Kreines, James. 2005. "The Inexplicability of Kant's Naturzweck: Kant on Teleology, Explanation and Biology." *Archiv für Geschichte der Philosophie,* 87(3):270–311.

Lacan, Jacques. 1968. *Speech and Language in Psychoanalysis.* Translated by Anthony Wilden. Baltimore: Johns Hopkins University Press.

———. 1988. *The Seminar. Book I. Freud's Papers on Technique, 1953–54.* Translated by John Forrester. New York: Cambridge University Press.

Lakebrink, Bernhard. 1984. *Perfectio Omnium Perfectionum: Studien zur Seinskonzeption bei Thomas von Aquin und Hegel.* Vatican City: Libreria Editrice Vaticana.

Lamb, Jonathan. 1995. *The Rhetoric of Suffering: Reading the Book of Job in the Eighteenth Century.* New York: Oxford University Press.

Larson, Magali. 1977. *The Rise of Professionalism.* Berkeley: University of California Press.

Lash, Nicholas. 1975. *Newman on Development: The Search for an Explanation in History.* Shepherdstown, WV: Patmos.

———. 1978. "Literature and Theory: Did Newman Have a 'Theory' of Development?" In *Newman and Gladstone: Centennial Essays,* edited by James D. Bastable. Dublin: Veritas.

Lee, Debbie. 2002. *Slavery and the Romantic Imagination.* Philadelphia: University of Pennsylvania Press.

Le Goff, Jacques. 1992. *History and Memory.* Translated by Steven Rendall and Elizabeth Clamann. New York: Columbia University Press.

Leighton, Stephen R. 1982. "Aristotle and the Emotions." *Phronesis,* 27(2):144–174.

Lenoir, Tim. 1982. *The Strategy of Life: Teleology and Mechanics in Nineteenth-Century German Biology.* Chicago: University of Chicago Press.

Lennox, James. 1979. "Teleology, Chance, and Aristotle's Theory of Spontaneous Generation." *Journal of the History of Philosophy,* 20(3):219–238.

Levinas, Emmanuel. 1969. *Totality and Infinity.* Translated by Alfonso Lingis. Pittsburgh: Duquesne University Press.

Levine, George. 2006. *Darwin Loves You: Natural Selection and the Re-Enchantment of the World.* Princeton, NJ: Princeton University Press.

Liu, Alan. 1989. "The New Historicism: The Power of Formalism." *English Literary History,* 56:721–771.

———. 1989. *Wordsworth: The Sense of History.* Stanford, CA: Stanford University Press.

———. 1990. "Local Transcendence: Cultural Criticism, Postmodernism, and the Romanticism of Detail." *Representations,* 32:75–113.

———. 2005. "The New Historicism and the Work of Mourning." In *The Wordsworthian Enlightenment: Romantic Poetry and the Ecology of Reading,* edited by Helen Reguiero Elam and Frances Ferguson, 149–157. Baltimore: Johns Hopkins University Press.

Locke, John. 1689. *A Letter Concerning Toleration.* 1st ed. rpt: London: Huddersfield, 1796.

———. 1975. *An Essay Concerning Human Understanding.* Edited by Peter H. Nidditch. Oxford: Clarendon.

Long, A. A., and D. N. Sedley. 1987. *The Hellenistic Philosophers.* Cambridge: Cambridge University Press.

Löwith, Karl. 1960. *Der Weltbegriff der neuzeitlichen Philosophie.* Heidelberg: C. Winter.

Luther, Martin. 1961. *Selections from His Writings*. Edited by John Dillenberger. New York: Doubleday.

MacCulloch, Diarmid. 2010. *Christianity: The First Three Thousand Years*. New York: Viking.

MacIntyre, Alasdair. 1984. *After Virtue: A Study in Moral Theory*. 2nd ed. Notre Dame, IN: University of Notre Dame Press.

———. 1988. *Whose Justice? Which Rationality?* Notre Dame, IN: University of Notre Dame Press.

———. 1990. *Three Rival Versions of Moral Enquiry*. Notre Dame, IN: University of Notre Dame Press.

———. 2009. *God, Philosophy, Universities: A Selective History of the Catholic Philosophical Tradition*. Lanham, MD: Rowman & Littlefield.

Macmurray, John. 1961. *Persons in Relation*. Atlantic Highlands, NJ: Humanities Press International.

Macpherson, C. B. 1964. *The Political Theory of Possessive Individualism*. Oxford: Oxford University Press.

Makdisi, Saree. 2003. *William Blake and the Impossible History of the 1790s*. Chicago: University of Chicago Press.

Mandeville, Bernard. 1729. *Free Thoughts on Religion, the Church, and National Happiness*. 2nd ed. London: John Brotherton.

———. 1732. *An Enquiry into the Origin of Honour and the Usefulness of Christianity in War*. London: Brotherton.

———. 1988. *The Fable of the Bees: or Private Vices, Publick Benefits*. Edited by F. B. Kaye. 2 vols. Indianapolis: Liberty Fund.

Mann, Thomas. 1997. *Doctor Faustus*. Translated by John Woods. New York: Viking.

Marcion of Sinope. 1980. *The Gospel of the Lord: An early version of which was circulated by Marcion of Sinope*. Edited by James Hamlyn Hill. Brooklyn, NY: AMS Press.

Marion, Jean-Luc. 2002. *Being Given: Toward a Phenomenology of Givenness*. Translated by Jeffrey L. Kosky. Stanford, CA: Stanford University Press.

———. 2002. *In Excess: Studies of Saturated Phenomena*. Translated by Robyn Horner and Vincent Berraud. New York: Fordham University Press.

———. 2008. *The Visible and the Revealed*. Translated by Christina M. Gschwandtner. New York: Fordham University Press.

Maritain, Jacques. 1996. *The Person and the Common Good*. Notre Dame, IN: University of Notre Dame Press.

———. 1998. *The Degrees of Knowledge*. Translated by Gerard B. Phelan. Notre Dame, IN: University of Notre Dame Press.

———. 1950. *Three Reformers: Luther, Descartes, Rousseau*. New York: Scribner.

Marquardt, Odo. 1991. "Unburdenings; Theodicy Motives in Modern Philosophy." In *In Defense of the Accidental*. Translated by Robert M. Wallace. New York: Oxford University Press.

Marshall, David. 1986. *The Figure of Theater: Shaftesbury, Defoe, Adam Smith, and George Eliot*. New York: Columbia University Press.

Martin, David. 2005. *On Secularization: Towards a Generalized Theory*. Aldershot: Ashgate.
McCabe, Herbert, OP. 1986. "Aquinas on Good Sense." *New Blackfriars*, 67:419–461.
———. 2002. *God Still Matters*. Edited by Brian Davies. New York: Continuum.
———. 2008. *On Aquinas*. New York: Continuum.
McCalman, Iain. 1988. *Radical Underground: Prophets, Revolutionaries, and Pornographers in London*. Cambridge: Cambridge University Press.
McDowell, John. 1996. *Mind and World*. 2nd ed. Cambridge, MA: Harvard University Press.
McEwan, Ian. 1999. *Amsterdam*. New York: Anchor.
McFarland, Thomas. 1969. *Coleridge and the Pantheist Tradition*. Oxford: Clarendon.
McGann, Jerome. 1988. *Social Values and Poetic Acts*. Cambridge, MA: Harvard University Press.
———. 1989. *The Beauty of Inflections*. Oxford: Clarendon.
McInerny, Ralph. 1974. "Prudence and Conscience." *Ethica Thomistica*, 38:291–305.
———. 1993. "Ethics." In *The Cambridge Companion to Aquinas*, edited by Norman Kretzman and Eleonore Stump. Cambridge: Cambridge University Press.
———. 1997. *Ethica Thomistica: The Moral Philosophy of Thomas Aquinas*. Washington, DC: Catholic University of America Press.
McKusick, James C. 1986. *Coleridge's Philosophy of Language*. New Haven, CT: Yale University Press.
———. 2002. "Symbol." In *The Cambridge Companion to Coleridge*, edited by Lucy Newlyn, 217–230. Cambridge: Cambridge University Press.
Mee, Jon. 2011. *Conversable Worlds: Literature, Contention, and Community 1762 to 1830*. Oxford: Oxford University Press.
Mele, Alfred R. 2009. *Effective Intentions: The Power of Conscious Will*. Oxford: Oxford University Press.
Mendelson, Jack. 1979. "The Habermas-Gadamer Debate." *New German Critique*, 18:44–73.
Milbank, John. 2006. *Theology and Social Theory*. 2nd ed. Oxford: Blackwell.
———. 2009. "What Is Living and What Is Dead in Newman's *Grammar of Assent*." In *The Future of Love: Essays in Political Theology*, 37–59. Eugene, OR: Cascade Books..
———. 2012. "Against Human Rights: Liberty in the Western Tradition." *Oxford Journal of Law and Religion*, 1(1):1–32.
Mill, John Stuart. 1950. *Mill on Bentham and Coleridge*. Edited by F. R. Leavis. London: Chatto & Windus.
———. 1989. *On Liberty*. Edited by Stefan Collini. Cambridge: Cambridge University Press.
Milton, J. R. 1981. "John Locke and the Nominalist Tradition." In *John Locke: Symposium Wolfenbüttel 1979*, edited by Reinhard Brandt, 128–145. New York: de Gruyter.
Mitchell, Robert. 2007. "The Fane of Tescalipoca: S. T. Coleridge on the Sacrificial Economies of Systems in the 1790s." *Studies in Romanticism*, 46(1):105–127.
———. 2007. *Sympathy and the State in the Romantic Era*. New York: Routledge.
———. 2013. *Experimental Life: Vitalism in Romantic Science and Literature*. Baltimore: Johns Hopkins University Press.

Mitchell, W. J. T. 1986. "Visible Language: Blake's Wond'rous Art of Writing." In *Romanticism and Contemporary Theory,* edited by Michael Fisher and Morris Eaves, 46–95. Ithaca, NY: Cornell University Press.

Mittelstrass, Jürgen. 1967. "Bildung und Wissenschaft. Enzyklopädien in historischer und wissens-soziologischer Betrachtung." In *Die wissenschaftliche Redaktion,* edited by Otto Mittelstaedt, 4:83–104. Mannheim: Bibliographisches Institut.

Modiano, Raimona. 1977. "Words and 'Languageless' Meanings: Limits of Expression in *The Rime of the Ancient Mariner.*" *MLQ,* 38:40–61.

———. 1985. *Coleridge and the Concept of Nature.* Tallahassee: Florida State University Press.

———. 1999. "Sameness or Difference? Historicist Readings of 'The Rime of the Ancient Mariner.'" In *The Rime of the Ancient Mariner,* edited by Paul H. Fry. New York: Bedford/St. Martin's.

Mokyr, Joel. 2010. *The Enlightened Economy: An Economic History of Britain 1700–1850.* New Haven, CT: Yale University Press.

Montesquieu, Charles de Secondat, Baron de. 1989. *The Spirit of the Laws.* Translated and edited by Anne M. Cohler et al. Cambridge: Cambridge University Press.

Moore, Andrew. 2007. "Reason." In the *Oxford Handbook of Systematic Theology,* edited by John Webster et al., 394–412. Oxford: Oxford University Press.

Moreland, J. P. 2009. *The Recalcitrant Imago Dei: Human Persons and the Failure of Naturalism.* London: SMC.

Moretti, Franco. 1987. *The Way of the World: The Bildungsroman in European Culture.* New York: Verso.

Morrow, John. 1990. *Coleridge's Political Thought.* London: Palgrave Macmillan.

Muir, John. 1930. *Coleridge as Philosopher.* London: Allen & Unwin.

Müller-Sievers, Helmut. 1997. *Self-Generation: Biology, Philosophy, and Literature Around 1800.* Stanford, CA: Stanford University Press.

Munzel, Felicitas G. 1999. *Kant's Conception of Moral Character.* Chicago: University of Chicago Press.

Muralt, André. 2002. *L'unité de la philosophie politique: de Scot, Ockham et Suarez au Liberalisme Contemporain.* Paris: Vrin.

Murdoch, Iris. 1992. *Metaphysics as a Guide to Morals.* Harmondsworth: Penguin.

———. 1997. *Existentialists and Mystics.* Edited by Peter Conradi. Harmondsworth: Penguin.

———. 2007. *The Sovereignty of Good.* New York: Routledge.

Murphy, Mark C. 2000. "Desire and Ethics in Hobbes's *Leviathan:* A Response to Professor Deigh." *Journal of the History of Philosophy,* 38:259–268.

Nancy, Jean-Luc. 1991. *The Inoperative Community.* Translated by Peter Connor. Minneapolis: University of Minnesota Press.

———. 1993. *The Experience of Freedom.* Stanford, CA: Stanford University Press.

Nédoncelle, Maurice. 1948. "*Prosopon* et person dans l'antiquité classique." *Revue des Sciences Religieuse,* 22:277–299.

———. 1955. "Les Variations de Boèce sur la Personne." *Revue des Sciences Religieuses,* 29:201–238.

Newman, Jay. 1986. *The Mental Philosophy of John Henry Newman*. Waterloo, ON: Wilfrid Laurier Press.

Newman, John Henry. 1871. *Essays Critical and Historical*. 2 vols. London: Basil, Montagu, Pickering.

———. 1979. *An Essay Written in Aid of a Grammar of Assent*. Edited by Nicholas Lash. Notre Dame, IN: University of Notre Dame Press.

———. 1989. *Development of Christian Doctrine*. Edited by Ian Ker. Notre Dame, IN: University of Notre Dame Press.

———. 1996. *The Idea of a University*. Edited by Frank M. Turner. New Haven, CT: Yale University Press.

———. 1997. *An Essay on the Development of Christian Doctrine*. Edited by Mary Katherine Tillman. Notre Dame, IN: University of Notre Dame Press.

———. 1997. *Fifteen Sermons Preached before the University of Oxford*. Edited by Mary Katherine Tillman. Notre Dame, IN: University of Notre Dame Press.

———. 2004. *Discussions and Arguments on Various Subjects*. 1st ed., 1907. Rept. Notre Dame, IN: University of Notre Dame Press.

———. 2006. *Apologia pro Vita Sua*. Edited by Frank Turner. New Haven, CT: Yale University Press.

Nichols, Roger A. 1985. "Thomas Mann and Spengler." *German Quarterly*, 58:361–364.

Nietzsche, Friedrich. 1980. *Sämtliche Werke: Kritische Studienausgabe*. Edited by Giorgio Colli and Mazzino Montinari. 15 vols. Munich: dtv.

———. 1986. *Sämtliche Briefe*. Edited by Giorgio Colli and Mazzino Montinari. 8 vols. Munich: dtv.

———. 1993. *The Birth of Tragedy*. Translated by Shaun Whiteside, edited by Michael Tanner. Harmondsworth: Penguin.

———. 1998. *On the Genealogy of Morals*. Translated by Douglas Smith. Oxford: Oxford University Press.

———. 2001. *The Gay Science*. Translated by Josefine Nauckhoff. New York: Cambridge University Press.

———. 2002. *Beyond Good and Evil*. Translated by Judith Norman, edited by Rolf-Peter Hostmann and Judith Norman. New York: Cambridge University Press.

———. 2005. *The Anti-Christ, Ecce Homo, Twilight of the Idols, and Other Writings*. Translated by Judith Norman. New York: Cambridge University Press.

Nipperdey, Thomas. 1996. *German History from Napoleon to Bismarck: 1800–1866*. Translated by Daniel Nolan. Princeton, NJ: Princeton University Press.

Norton, David F. 1982. *David Hume: Common-Sense Moralist, Sceptical Metaphysician*. Princeton, NJ: Princeton University Press.

Novalis, Friedrich von Hardenberg. 1977. "Fichte-Studies." In *Theory as Practice*. Edited by Jochen Schulte-Sasse. Minneapolis: University of Minnesota Press.

———. 1978. *Werke, Tagebücher und Briefe*. Edited by Hans-Joachim Mähl. 3 vols. Munich: Hanser.

Nussbaum, Martha. 1986. *The Fragility of Goodness*. New York: Cambridge University Press.

———. 1996. *The Therapy of Desire*. Princeton, NJ: Princeton University Press.
———. 2001. *Upheavals of Thought: The Intelligence of the Emotions*. Cambridge: Cambridge University Press.
Nyhart, Lynn K. 1995. *Biology Takes Form: Animal Morphology and the German Universities, 1800–1900*. Chicago: University of Chicago Press.
Oakeshott, Michael. 1975. *On Human Conduct*. Oxford: Clarendon.
———. 2000. *Hobbes on Civil Association*. Indianapolis: Liberty Fund.
Ockham, William of. 1998. *Quodlibetal Questions*. Translated by Alfred E. Freddoso and Francis E. Kelley. New Haven, CT: Yale University Press.
O'Donovan, Joan Lockwood. 1998. "Natural Law and Perfect Community: Contributions of Christian Platonism to Political Theory." *Modern Theology*, 14(1):19–42.
O'Donovan, Oliver. 2005. *The Ways of Judgment*. Grand Rapids, MI: Eerdmans.
Ohman, Richard. 2003. *Politics of Knowledge: The Commercialization of the University, the Professions, and Print Culture*. Middletown, CT: Wesleyan University Press.
O'Regan, Cyril. 1994. *The Heterodox Hegel*. Albany: State University of New York Press.
Orsini, Gian N. G. 1969. *Coleridge and German Idealism*. Carbondale: Southern Illinois University Press.
Paffenroth, Kim, and Robert P. Kennedy. 2003. *A Reader's Companion to Augustine's Confessions*. Louisville, KY: Westminster John Knox.
Pagels, Elaine. 1979. *The Gnostic Gospels*. London: Vintage.
Paine, Thomas 1984. *The Rights of Man*. Edited by Eric Foner. Harmondsworth: Penguin.
Paknadel, Felix. 1974. "Shaftesbury's Illustrations of Characteristics." *Journal of the Warburg and Courtauld Institutes*, 37:290–312.
Paley, Morton D. 1999. *Apocalypse and Millennium in English Romantic Poetry*. Oxford: Oxford University Press.
Parfit, Derek. 1986. *Reasons and Persons*. Oxford: Oxford University Press.
Pascal, Blaise. 1995. *Pensées*. Translated by A. J. Krailsheimer. Harmondsworth: Penguin.
Pattison, Robert. 1991. *The Great Dissent: John Henry Newman and the Liberal Heresy*. New York: Oxford University Press.
Penelhum, Terence. 1993. "Hume's Moral Psychology." In *The Cambridge Companion to Hume*, edited by David F. Norton. Cambridge: Cambridge University Press.
Perkins, Mary Anne. 1994. *Coleridge's Philosophy: The Logos as Unifying Principle*. Oxford: Clarendon.
———. 1997. "Coleridge and the 'Other Plato.'" *European Romantic Review*, 8(1):25–40.
Perry, Seamus. 1999. *Coleridge and the Uses of Division*. Oxford: Clarendon.
Peterfreund, Stuart. 1990. "Blake, Priestley, and the Gnostic Moment." In *Literature and Science: Theory and Practice*, edited by Stuart Peterfreund, 139–166. Boston: Northeastern University Press.
Pfau, Thomas. 1987. *Friedrich Hölderlin: Essays and Letters on Theory*. Albany: State University of New York Press.
———. 1990. "Immediacy and the Text: Friedrich Schleiermacher's Theory of Style and Interpretation." *Journal of the History of Ideas*, 51(1):51–73.
———. 1994. *Idealism and the Endgame of Theory: Three Essays by F. W. J. Schelling*. Albany: State University of New York Press.

———. 1995. "Immediacy and Dissolution: Notes on the Languages of Moral Agency and Critical Discourse." In *Intersections: Nineteenth-Century Philosophy and Contemporary Theory,* edited by Tilottama Rajan and David L. Clark, 222–242. Albany: State University of New York Press.

———. 1997. *Wordsworth's Profession: Form, Class, and the Logic of Early Romantic Cultural Production.* Stanford, CA: Stanford University Press.

———. 1998. "Reading beyond Redemption: Historicism, Irony, and the Lessons of Romanticism." In *Lessons of Romanticism,* edited by Thomas Pfau and Robert F. Gleckner. Durham, NC: Duke University Press.

———. 2005. "From Mediation to Medium: Aesthetic and Anthropological Dimensions of the Image. *Bild* and the Crisis of *Bildung* in German Modernism." *Medium and Message in German Modernism,* a special issue of *Modernist Cultures,* ed. Thomas Pfau, 2(1) www.js-modcult.bham.ac.uk/backissues.asp.

———. 2005. *Romantic Moods: Paranoia, Trauma, and Melancholy, 1790–1840.* Baltimore: Johns Hopkins University Press.

———. 2007. "Of Ends and Endings: Teleological and Variational Models of Romantic Narrative." *European Romantic Review,* 18(2):231–240.

———. 2008. "The Melancholy Gift: Freedom in the Nineteenth Century," in *Romantic Praxis.* www.rc.umd.edu/praxis/philcult/pfau/pfau.html.

———. 2010. "'All is Leaf': Differentiation, Metamorphosis, and the Phenomenology of Life." *Studies in Romanticism,* 49(1):3–41.

———. 2010. "Bildung: Etiology, Function, Structure (with Some Reflections on Beethoven)." In *Die Romantik: ein Gründungsmythos der Europäischen Moderne,* edited by Ulrich Gaier et al., 123–141. Bonner Universitätsverlag.

———. 2010. "*Bildungsspiele:* Vicissitudes of Socialization in *Wilhelm Meister's Apprenticeship.*" *European Romantic Review,* 21(5):567–587.

———. 2010. "The Letter of Judgment: Practical Reason in Aristotle, the Stoics, and Rousseau." *The Eighteenth Century: Theory & Interpretation,* 51(3):289–316.

———. 2011. "The Appearance of *Stimmung:* Play as Virtual Rationality." In *Stimmung: zur Wiederkehr einer ästhetischen Kategorie?* edited by Anna-Katharina Gisbertz, 95–111. Munich: Fink.

———. 2013. "Rational Theology and the Catholic Critique of Modernity, 1780–1830." In *The Oxford Handbook of European Romanticism,* edited by Paul Hamilton. Forthcoming.

———. 2013. "Rethinking the Image (with some reflections on G. M. Hopkins)." *Yearbook of Comparative Literature,* 59. Forthcoming.

Phillips, Catherine. 2007. *Gerard Manley Hopkins and the Victorian Visual World.* Oxford: Oxford University Press.

Pickstock, Catherine. 2003. "Modernity and Scholasticism." *Antonianum,* 78:3–46.

Pinkard, Terry. 1994. *Hegel's Phenomenology: The Sociality of Reason.* Cambridge: Cambridge University Press.

———. 2002. *German Philosophy, 1760–1860.* Cambridge: Cambridge University Press.

Pippin, Robert. 1991. "Idealism and Agency in Kant and Hegel." *Journal of Philosophy,* 88(10):532–541.

———. 1993. "You Can't Get There from Here: Transition Problems in Hegel's *Phenomenology of Spirit.*" In *The Cambridge Companion to Hegel,* edited by Frederick C. Beiser, 52–85. Cambridge: Cambridge University Press.
———. 1997. *Idealism as Modernism: Hegelian Variations.* New York: Cambridge University Press.
———. 1999. *Modernism as a Philosophical Problem: On the Dissatisfactions of European High Culture.* 2nd ed. London: Blackwell.
———. 2000. Kant's Theory of Value: On Allen Wood's *Kant's Ethical Thought.*" *Inquiry,* 43:239–266.
———. 2008. *Hegel's Practical Philosophy: Rational Agency as Ethical Life.* Cambridge: Cambridge University Press.
Plato. 1989. *The Collected Dialogues of Plato.* Edited by Edith Hamilton and Huntington Cairns. Princeton, NJ: Princeton University Press.
Plotinus. 1995. *Enneads.* Translated by A. H. Armstrong. Cambridge, MA: Harvard University Press.
Pocock, J. G. A. 1971. *Politics, Language, and Time.* Chicago: University of Chicago Press.
———. 1975. *The Machiavellian Moment.* Princeton, NJ: Princeton University Press.
———. 1986. *Virtue, Commerce, and History.* New York: Cambridge University Press.
Polanyi, Karl. 1957. *The Great Transformation.* Boston: Beacon.
Polanyi, Michael. 1998. *Personal Knowledge: Towards a Post-Critical Philosophy.* New York: Routledge.
Porter, Roy. 1981. "The Enlightenment in England." In *The Enlightenment in National Context,* edited by Roy Porter and Mikulás Teich. Cambridge: Cambridge University Press.
———. 1982. *English Society in the Eighteenth Century.* Harmondsworth: Penguin.
Pothast, Ulrich. 1971. *Über einige Fragen der Selbstbeziehung.* Frankfurt: Klostermann.
Prendeville, John G. 1972. "The Development of the Idea of Habit in the Thought of St. Augustine." *Traditio,* 28:29–99.
Price, H. H. 1969. *Belief*—the 1960 Gifford Lectures. London: Allen & Unwin.
Priestley, Joseph. 1786. *An History of Early Opinions Concerning Jesus Christ.* 4 vols. Birmingham: n.p.
———. 1977. *The Doctrine of Philosophical Necessity.* Oxford: Clarendon.
Proust, Marcel. 2003. *In Search of Lost Time.* New York: Modern Library.
Rajan, Tilottama. 1990. *The Supplement of Reading: Figures of Understanding in Romantic Theory and Practice.* Ithaca, NY: Cornell University Press.
———. 2002. *Deconstruction and the Remainders of Phenomenology.* Stanford, CA: Stanford University Press.
———. 2010. *Romantic Narrative.* Baltimore: Johns Hopkins University Press.
Ratzinger, Joseph. 1990. "Retrieving the Tradition: Concerning the Notion of Person in Theology." *Communio,* 17:439–454.
Rauch, Angelika. 2000. *The Hieroglyph of Tradition: Freud, Benjamin, Gadamer, Novalis, Kant.* Madison, NJ: Farleigh Dickinson University Press.

Ravaisson, Félix. 2009. *Of Habit/de l'habitude*. Translated by Claire Carlisle. New York: Continuum.
Reid, Charles J. 1991. "The Canonistic Contribution to the Western Rights Tradition: An Historical Inquiry." *Boston College Law Review*, 33:37–92.
Reynolds, Edward. 1658. *A Treatise of the Passions and Faculties of the Soule of Man*. London.
Richard of St. Victor. 1958. *De Trinitate*. Translated by Jean Ribaillier. Paris: Vrin.
———. 1979. *The Twelve Patriarchs, The Mystical Ark, Book Three of the Trinity*. Translated by Grover A. Zinn. New York: Paulist.
Richardson, Lawrence. 2007. *Newman's Approach to Knowledge*. Leominster, Herefordshire: Gracewing.
Ricks, Christopher. 1984. *Keats and Embarrassment*. Oxford: Clarendon.
Rist, J. M. 1977. *Stoic Philosophy*. Cambridge: Cambridge University Press.
Ritter, Joachim, et al., eds. 1971–2007. *Historiches Wörterbuch der Philosophie*. Basel: Schwabe.
Rivers, Isabel. 2000. *Reason, Grace, and Sentiment: A Study of the Language of Religion and Ethics in England, 1660–1780*. 2 vols. Cambridge: Cambridge University Press.
Robinson, James M., trans. 1996. *The Nag Hammadi Library in English*. Leiden: Brill.
Robinson, Marilynne. 2010. *Absence of Mind: The Dispelling of Inwardness from the Modern Myth of the Self*. New Haven: Yale University Press.
Rolnick, Philip A. 2007. *Person, Grace, and God*. Grand Rapids, MI: Eerdmans.
Rorty, Amélie Oksenberg. 1980. "Akrasia and Pleasure." In *Essays on Aristotle's Ethics*, edited by Amélie Oksenberg Rorty, 267–284. Berkeley: University of California Press.
———. 1988. *Mind in Action*. Boston: Beacon.
———. 1996. "Two Faces of Stoicism: Rousseau and Freud." *Journal of the History of Philosophy*, 34(3):335–356.
———. 1997. "Social and Political Sources of Akrasia." *Ethics*, 107(4):644–657.
———, ed. 1980. *Essays on Aristotle's Ethics*. Berkeley: University of California Press.
Rosen, Charles. 1998. *The Classical Style*. Rev. ed. New York: Norton.
Rosen, Stanley. 1987. *Hermeneutics as Politics*. New York: Oxford University Press.
Rothschild, Emma. 2001. *Economic Sentiments: Adam Smith, Condorcet, and the Enlightenment*. Cambridge, MA: Harvard University Press.
Rousseau, Jean-Jacques. 1979. *Émile: or On Education*. Translated and edited by Allan Bloom. New York: Basic Books.
———. 1997. *The Discourses and Other Early Political Writings*. Edited by Victor Gourevitch. Cambridge: Cambridge University Press.
———. 1997. *The Social Contract and Other Later Political Writings*. Edited by Victor Gourevitch. Cambridge: Cambridge University Press.
Rudolph, Kurt. 1987. *Gnosis: The Nature & History of Gnosticism*. San Francisco: HarperOne.
Rule, Philip C., SJ. 2004. *Coleridge and Newman: The Centrality of Conscience*. New York: Fordham University Press.

Runciman, David. 2005. *Pluralism and the Personality of the State*. Cambridge: Cambridge University Press.

Ruskin, John. 1860. *Modern Painters*. Vol. 5. London: Smith, Elder & Co.

Russell, Paul. 2008. *The Riddle of Hume's Treatise: Skepticism, Naturalism, and Irreligion*. New York: Oxford University Press.

Ruth, Jennifer. 2006. *Novel Professions: Interested Disinterest and the Making of the Professional in the Victorian Novel*. Columbus: Ohio State University Press.

Safranski, Rüdiger. 1991. *Schopenhauer and the Wild Years of Philosophy*. Translated by Ewald Osers. Cambridge, MA: Harvard University Press.

Scheibler, Ingrid. 2000. *Gadamer: Between Heidegger and Habermas*. Lanham, MD: Rowman & Littlefield.

Schelling, F. W. J. 1856–1857. *Sämmtliche Werke*. Stuttgart: Cotta.

———. 1992. *Philosophical Inquiries into the Nature of Human Freedom*. Translated by James Gutman. Chicago: Open Court.

———. 2004. *First Outline of a System of the Philosophy of Nature, 1799*. Translated by Keith R. Peterson. Albany: State University of New York Press.

Schiller, Friedrich. 1844. *Sämmtliche Werke*. 10 vols. Stuttgart: Cotta.

———. 2004. *On the Aesthetic Education of Man*. Translated by Reginald Snell. New York: Dover.

Schlegel, Friedrich. 1991. *Philosophical Fragments*. Translated by Peter Firchow. Minneapolis: University of Minnesota Press.

Schmitt, Carl. 1986. *Political Romanticism*. Translated by Guy Oakes. Cambridge, MA: MIT Press.

———. 2004. *Politische Theologie*. Berlin: Duncker & Humblot.

———. 2006. *Political Theology*. Translated by George Schwab. Chicago: University of Chicago Press.

———. 2007. *The Concept of the Political*. Translated by George Schwab. Chicago: University of Chicago Press.

Schneewind, Jerome. 1998. *The Invention of Autonomy*. Cambridge: Cambridge University Press.

Schopenhauer, Arthur. 1969. *The World as Will and Representation*. Translated by E. F. J. Payne. New York: Dover.

———. 1989. *Die Welt als Wille und Vorstellung*. 2 vols. Darmstadt: Wissenschaftliche Buchgesellschaft.

———. 1992. *On the Will in Nature*. Translated by E. F. J. Payne. New York: Berg.

———. 2003. *Prize Essay on the Freedom of the Will*. Translated by E. F. J. Payne. Cambridge: Cambridge University Press.

Schröter, Manfred. 1922. *Der Streit um Spengler: Kritik seiner Kritiker*. Munich: Beck.

Scott, William R. 1990. *Frances Hutcheson: His Life, Teaching and Position in the History of Philosophy*. Cambridge: Cambridge University Press.

Scruton, Roger. 2004. *Death-Devoted Heart: Sex and the Sacred in Wagner's Tristan and Isolde*. New York: Oxford University Press.

Selbmann, Rolf, ed. 1988. *Zur Geschichte des Deutschen Bildungsromans.* Darmstadt: Wissenschaftliche Buchgesellschaft.
Sellars, Jon. 2006. *Stoicism.* Berkeley: University of California Press.
Seneca. 1958. *The Stoic Philosophy of Seneca.* Translated by Moses Hadas. New York: Norton.
Shaftesbury, Anthony Ashley Cooper, third Earl of. 1900. *The Life, Unpublished Letters, and Philosophical Regimen of Anthony, Earl of Shaftesbury.* Edited by Benjamin Rand. London: Swan Sonnenschein & Co.
———. 2001. *Characteristicks of Men, Manners, Opinions, Times.* Edited by Douglas den Uyl. 3 vols. Indianapolis: Liberty Fund.
Shapin, Steven, and Simon Schaffer. 1989. *Leviathan and the Air Pump: Hobbes, Boyle, and the Experimental Life.* Princeton, NJ: Princeton University Press.
Sheehan, James. 1986. *German History: 1770–1866.* Oxford: Oxford University Press.
Sherman, Nancy. 1997. *Making a Necessity of Virtue: Aristotle and Kant on Virtue.* Cambridge: Cambridge University Press.
Simmel, Georg. 1980. *Essays on Interpretation in Social Science.* Translated by Guy Oakes. Manchester: Manchester University Press.
———. 2001. "Zur Metaphysik des Todes." *Gesamtausgabe.* Vol. 12. Frankfurt: Suhrkamp.
———. 2007. "The Metaphysics of Death." Translated by Ulrich Teucher and Thomas M. Kemple. *Theory, Culture & Society,* 7–8:72–77.
Simpson, David. 1993. *Romanticism, Nationalism, and the Revolt against Theory.* Chicago: University of Chicago Press.
———. 1995. *The Academic Postmodern and the Rule of Literature: A Report on Half-Knowledge.* Chicago: University of Chicago Press.
———. 1999. "How Marxism Reads 'The Rime of the Ancient Mariner.'" In *The Rime of the Ancient Mariner,* edited by Paul H. Fry. New York: Bedford/St. Martin's.
Siskin, Clifford. 1998. *The Work of Writing: Literature and Social Change in Britain, 1700–1830.* Baltimore: Johns Hopkins University Press.
———. 1998. "The Year of the System." In *1798: The Year of the* Lyrical Ballads, edited by Richard Cronin. New York: St. Martin's.
———. 2001. "Novels and Systems." *Novel,* 34(2):202–215.
Skinner, Quintin. 1996. *Reason and Rhetoric in the Philosophy of Hobbes.* Cambridge: Cambridge University Press.
Smith, Adam. 1976. *The Wealth of Nations.* Edited by Edwin Cannan. Chicago: University of Chicago Press.
———. 1984. *The Theory of Moral Sentiments.* Edited by D. D. Raphael and A. L. Macfie. Indianapolis: Liberty Fund.
Sobolev, Dennis. 2011. *The Split World of Gerard Manley Hopkins.* Washington, DC: Catholic University of America Press.
Sokolowski, Robert. 2000. *Introduction to Phenomenology.* Cambridge: Cambridge University Press.
———. 2008. *Phenomenology of the Human Person.* Cambridge: Cambridge University Press.

Solomon, Maynard. 2003. *Late Beethoven: Music, Thought, Imagination*. Berkeley: University of California Press.
Soni, Vivasvan. 2007. "Trials and Tragedies: The Literature of Unhappiness. A Model for Reading Narratives of Suffering." *Comparative Literature*, 59(2):119–139.
———. 2010. "Introduction: The Crisis of Judgment." *The Eighteenth Century: Theory & Interpretation*, 51(3):261–288.
———. 2010. *Mourning Happiness: Narrative and the Politics of Modernity*. Ithaca, NY: Cornell University Press.
Sontag, Susan. 1978. *Illness as Metaphor*. New York: Farrar, Straus, and Giroux.
Spaemann, Robert. 1996. *Personen: Versuch über den Unterschied zwischen 'Jemand' und 'Etwas.'* Stuttgart: Klett-Cotta.
———. 2002. *Grenzen: zur ethischen Dimension des Handelns*. Stuttgart: Klett-Cotta.
———. 2006. *Persons: On the Difference between Someone and Something*. Translated by Oliver O'Donovan. Oxford: Oxford University Press.
Spaemann, Robert, and Reinhard Löw. 2005. *Natürliche Ziele: Geschichte und Wiederentdeckung des teleologischen Denkens*. Stuttgart: Klett-Cotta.
Stendhal. 2002. *The Red and the Black*. Translated by Roger Gard. Harmondsworth: Penguin.
Stout, Jeffrey. 2005. *Democracy and Tradition*. Princeton, NJ: Princeton University Press.
Strauss, Leo. 1953. *Natural Right and History*. Chicago: University of Chicago Press.
Stump, Eleonore. 2001. "Augustine on Free Will." In *The Cambridge Companion to Augustine*, edited by Eleonore Stump and Norman Kretzmann, 124–147. Cambridge: Cambridge University Press.
———. 2010. *Wandering in Darkness: Narrative and the Problem of Suffering*. Oxford: Clarendon.
Szondi, Peter. 1974. *Poetik und Geschichtsphilosophie*. 2 vols. Frankfurt: Suhrkamp.
Tallis, Raymond. 2004. *I Am: A Philosophical Inquiry into First-Person Being*. Edinburgh: Edinburgh University Press.
———. 2004. *Why the Mind Is Not a Computer*. Exeter: Imprint Academic.
———. 2010. "The Suicide of the Humanities." Lecture given at the National Humanities Center, Research Triangle Park, NC, 9 November.
Tannenbaum, Leslie. 1984. *Biblical Tradition in Blake's Early Prophecies: The Great Code of Art*. Princeton, NJ: Princeton University Press.
Taylor, Anya. 1986. *Coleridge's Defense of the Human*. Columbus: Ohio State University Press.
Taylor, Charles. 1975. *Hegel*. Cambridge: Cambridge University Press.
———. 1989. *Sources of the Self: The Making of Modern Identity*. Cambridge, MA: Harvard University Press.
———. 1991. *The Ethics of Authenticity*. Cambridge, MA: Harvard University Press.
———. 1994. "The Politics of Recognition." In *Multiculturalism*, edited by Charles Taylor. Princeton, NJ: Princeton University Press.
———. 1995. *Philosophical Arguments*. Cambridge, MA: Harvard University Press.
———. 2004. *Modern Social Imaginaries*. Durham, NC: Duke University Press.
———. 2007. *A Secular Age*. Cambridge, MA: Harvard University Press.

———. 2011. *Dilemmas and Connections*. Cambridge, MA: Harvard University Press.

Te Velde, Rudi. 2006. *Aquinas on God*. Aldershot: Ashgate.

Tertullian. 1948. *Adversus Praxean liber/Tertullian's treatise against Praxeas*. Translated and edited by Ernest Evans. London: SPCK.

Teske, Roland, SJ. 2001. "Augustine's Theory of Soul." In *The Cambridge Companion to Augustine*, edited by Eleonore Stump and Norman Kretzman. Cambridge: Cambridge University Press.

Thelwall, John. 1995. *The Politics of English Jacobinism*. Edited by Gregory Claeys. University Park: Pennsylvania State University Press.

Theunissen, Michael. *Vorentwürfe von Moderne: antike Melancholie und die Acedia des Mittelalters*. Rotterdam: de Gruyter, 1996.

Thijssen, J. M. 1998. *Censure and Heresy at the University of Paris, 1200–1400*. Philadelphia: University of Pennsylvania Press.

Thompson, E. P. 1968. *The Making of the English Working Class*. New York: Vintage.

———. 1978. "Eighteenth-Century Society: Class-Struggle without Class." *Social History*, 3:133–165.

———. 1993. *Witness against the Beast: William Blake and the Moral Law*. Cambridge: Cambridge University Press.

Thompson, Evan. 2007. *Mind in Life: Biology, Phenomenology, and the Sciences of Mind*. Cambridge, MA: Harvard University Press.

Thompson, Michael. 2008. *Life and Action*. Cambridge, MA: Harvard University Press.

Thomson, Keith S. 2005. *Before Darwin: Reconciling God and Nature*. New Haven, CT: Yale University Press.

Thorne, Christian. 2010. *The Dialectic of Counter-Enlightenment*. Cambridge, MA: Harvard University Press.

Tierney, Brian. 2001. *The Idea of Natural Rights: Studies on Natural Rights, Natural Law, and Church Law 1150–1625*. Grand Rapids, MI: Eerdmans.

Tillich, Paul. 1973. *Systematic Theology*. Vol. 1. Chicago: University of Chicago Press.

Tocqueville, Alexis de. 2009. *Democracy in America*. Edited by Eduardo Nolla, translated by James T. Schleifer. 4 vols. Indianapolis: Liberty Fund.

Tolstoy, Leo. 1890. *The Kreutzer Sonata and Other Stories*. Translated by Benjamin Ricketson Tucker. New York: J. S. Ogilvie.

Torrell, Jean-Pierre. 2005. *Aquinas's Summa: Background, Structure, & Reception*. Washington, DC: Catholic University of America Press.

———. 2005. *Saint Thomas Aquinas*. 2 vols. Washington, DC: Catholic University of America Press.

Trebels, Andreas Heinrich. 1967. *Einbildungskraft und Spiel*. Bonn: Bouvier.

Tricaud, François. 1982. "An Investigation Concerning the Usage of the Words 'Person' and 'Persona' in the Political Treatises of Hobbes." In *Thomas Hobbes: His View of Man*, edited by J. G. van der Bend, 89–98. Amsterdam: Rodopi.

Tuchman, Gaye. 2011. *Wannabe University: Inside the Corporate University*. Chicago: University of Chicago Press.

Tuck, Richard. 1979. *Modern Rights Theories: Their Origin and Development*. Cambridge, MA: Harvard University Press.

Ulmer, William A. 2004. "Necessary Evils: Unitarian Theodicy in 'The Rime of the Ancyent Marinere.'" *Studies in Romanticism,* 43(3):327–356.

Van de Vijver, Gertrudis. 2006. "Kant and the Intuitions of Self-Organization." In *Self-Organization and Emergence in Life Sciences,* edited by B. Feltz et al., 143–161. Dordrecht: Springer.

Varro. 1938. *On the Latin Language/De Lingua Latina.* 2 vols. Translated by Roland G. Kent. Loeb Classical Library. Cambridge, MA: Harvard University Press.

Verbeke, Gérard. 1958. "Augustin et le stoïcisme." *Recherches Augustiennes,* 1:67–89.

Vigus, James. 2009. *Platonic Coleridge.* London: Legenda.

Vogl, Joseph. 2008. *Kalkül und Leidenschaft: Poetik des ökonomischen Menschen.* Zurich: Diaphanes.

Voitle, Robert. 1984. *The Third Earl of Shaftesbury, 1671–1713.* Baton Rouge: Louisiana State University Press.

Voltaire. 1784. *Traité de métaphysique.* In *Oeuvres complètes de Voltaire.* Vol. 32. Paris.

Von Molnar, Géza. 1987. *Novalis: Romantic Vision and Ethical Context.* Minneapolis: University of Minnesota Press.

Wahrman, Dror. 1995. *Imagining the Middle Class: The Political Representation of Class in Britain, c. 1780–1840.* Cambridge: Cambridge University Press.

Ward, Bernadette W. 2002. *World as Word: Philosophical Theology in Gerard Manley Hopkins.* Washington, DC: Catholic University of America Press.

Washburn, Jennifer. 2006. *University Inc.: The Corporate Corruption of Higher Education.* New York: Basic Books.

Watkins, Daniel P. 1988. "History as Demon in Coleridge's *The Rime of the Ancient Mariner.*" *Papers on Language & Literature,* 24(1):23–33.

Watson, John B. 1913. "Psychology as the Behaviorist Views It." *Psychological Review,* 20(2):158–177.

Wawrykow, Joseph. 2005. "Grace." In *The Theology of Thomas Aquinas,* edited by Rik van Nieuwenhove and Joseph Wawrykow. Notre Dame, IN: University of Notre Dame Press.

Weber, Max. 1994. *Political Writings.* Edited by Peter Lassman and Ronald Speirs. Cambridge: Cambridge University Press.

———. 2003. *The Protestant Ethic and the Spirit of Capitalism.* Translated by Talcott Parsons. Mineola, NY: Dover.

———. 2004. *The Vocation Lectures: Science as a Vocation; Politics as a Vocation.* Translated by Rodney Livingstone; edited by David Owen and Tracy B. Strong. Indianapolis: Hackett.

Webster, Suzanne E. 2010. *Body and Soul in Coleridge's Notebooks, 1827–1834: 'What is Life?'* Basingstoke: Palgrave Macmillan.

Wehler, Hans-Ulrich. 2007. *Deutsche Gesellschaftsgeschichte,* vol. 1, *Vom Feudalismus des Alten Reiches bis zur Defensiven Modernisierung der Reformära 1700–1815.* Munich: C. H. Beck.

Weil, Simone. 1999. *The Simone Weil Reader.* Edited by George A. Panichas. Wakefield, RI, and London: Moyer Bell.

Wetzel, James. 1992. *Augustine and the Limits of Virtue*. Cambridge: Cambridge University Press.

———. 2002. "Snares of Truth: Augustine on Free Will and Predestination." In *Augustine and his Critics*, edited by Robert Dodaro and George Lawless, 124–141. New York: Routledge.

———. 2006. "Book Four: The Trapping of Woe and Confession of Grief." In *A Reader's Companion to St. Augustine's Confessions*, edited by Kim Paffenroth and Robert P. Kennedy. Louisville, KY: Westminster John Knox.

White, Deborah Elise. 2003. "Imagination's Date: A Postscript to the *Biographia Literaria*." *European Romantic Review*, 14:467–478.

White, R. S. 2005. *Natural Rights and the Birth of Romanticism in the 1790s*. London: Palgrave Macmillan.

Whitehead, Alfred N. 1985. *Process and Reality*. Edited by David R. Griffin and Donald W. Sherburne. New York: Free Press.

Wippel, John F. 1977. "The Condemnations of 1270 and 1277 at Paris." *Journal of Medieval and Renaissance Studies*, 7:169–201.

Wisner, David A. 1997. "Ernst Cassirer: Historian of the Will." *Journal of the History of Ideas*, 58(1):145–161.

Wittgenstein, Ludwig. 2009. *Philosophical Investigations*. Edited by P. M. S. Hacker and Joachim Schulte. London: Wiley-Blackwell.

Wolterstorff, Nicholas. 2008. *Justice: Rights and Wrongs*. Princeton, NJ: Princeton University Press.

Wood, Allen W. 1999. *Kant's Ethical Thought*. Cambridge: Cambridge University Press.

Wordsworth, William. 1977. *The Prelude, 1798–1799*. Edited by Stephen M. Parrish. Ithaca, NY: Cornell University Press.

———. 1983. *Poems, in Two Volumes, and Other Poems, 1800–1807*. Edited by Jared Curtis. Ithaca, NY: Cornell University Press.

———. 1992. *Lyrical Ballads and Other Poems, 1798–1800*. Edited by James Butler and Karen Green. Ithaca, NY: Cornell University Press.

———. 1992. *The Thirteen-Book Prelude*. Edited by Mark Reed. 2 vols. Ithaca, NY: Cornell University Press.

Wright, Luke S. H. 2010. *Coleridge and the Anglican Church*. Notre Dame, IN: University of Notre Dame Press.

Yousef, Nancy. 2011. "Feeling for Philosophy: Shaftesbury and the Limits of Sentimental Certainty." *ELH*, 78(3):609–632.

Žižek, Slavoj. 1993. *Tarrying with the Negative: Kant, Hegel, and the Critique of Ideology*. Durham, NC: Duke University Press.

———. 1997. *The Plague of Fantasies*. New York: Verso.

Zuckert, Catherine H. 2009. *Plato's Philosophers: The Coherence of the Dialogues*. Chicago: University of Chicago Press.

INDEX

THOMAS PFAU

is the Alice Mary Baldwin Professor of English and professor of German at Duke University, with a secondary appointment on the Duke Divinity School faculty. He is the author and editor of a number of books, including *Romantic Moods: Paranoia, Trauma, and Melancholy, 1790–1840.*